Towns, Ecology, and the Land

Towns and villages are sometimes viewed as minor, even quaint, spots, whereas this book boldly reconceptualizes these places as important dynamic environmental "hotspots." Multitudes of towns and villages with nearly half the world's population characterize perhaps half the global land surface. The book's pages feature ecological patterns, processes, and change, as well as human dimensions, both within towns and in strong connections with surrounding agricultural land, forestland, and aridland. Towns, small to large, and villages are examined with spatial and cultural lenses. Ecological dimensions – water, soil, and air systems, together with habitats, plants, wildlife, and biodiversity – are highlighted. A concluding section presents concepts for making better towns and better land. From a pioneer in both landscape ecology and urban ecology, this highly international town ecology book opens an important frontier for researchers, students, professors, and professionals including environmental, town, and conservation planners.

Richard T. T. Forman is Research Professor in Landscape Ecology at Harvard University, where he teaches urban and town ecology. Often considered a "father" of landscape ecology and road ecology, he helped spearhead urban ecology and has received honors, awards, medals, and honorary doctorates from universities across the world. His books include *Landscape Ecology* (1986), the award-winning *Land Mosaics* (1995), *Urban Regions* (2008), and award-finalist *Urban Ecology* (2014). He received an award for excellence in teaching, and has served for many years on boards for town planning and land-protection nonprofit organizations.

Books by the Author

Pine Barrens: Ecosystem and Landscape
Landscape Ecology
Changing Landscapes: An Ecological Perspective
Land Mosaics: Ecology of Landscapes and Regions
Landscape Ecology Principles in Landscape Architecture and Land-Use Planning
Road Ecology: Science and Solutions
Mosaico territorial para la region metropolitana de Barcelona
Urban Regions: Ecology and Planning Beyond the City
Urban Ecology: Science of Cities

Towns, Ecology, and the Land

RICHARD T. T. FORMAN
Harvard University

CAMBRIDGE
UNIVERSITY PRESS

University Printing House, Cambridge CB2 8BS, United Kingdom

One Liberty Plaza, 20th Floor, New York, NY 10006, USA

477 Williamstown Road, Port Melbourne, VIC 3207, Australia

314–321, 3rd Floor, Plot 3, Splendor Forum, Jasola District Centre, New Delhi – 110025, India

79 Anson Road, #06-04/06, Singapore 079906

Cambridge University Press is part of the University of Cambridge.

It furthers the University's mission by disseminating knowledge in the pursuit of education, learning, and research at the highest international levels of excellence.

www.cambridge.org
Information on this title: www.cambridge.org/9781107199132
DOI: 10.1017/9781108183062

First published 2019

Printed in the United Kingdom by TJ International Ltd, Padstow, Cornwall

A catalogue record for this publication is available from the British Library.

ISBN 978-1-107-19913-2 Hardback
ISBN 978-1-316-64860-5 Paperback

Dedicated to Brent Chandlee Forman
and his family
Maura, Trey, Colin

Contents

Color plates can be found between pages 270 and 271.

Foreword

Most ecologists live in towns and cities but study ecology elsewhere. Ecology's traditional focus was on the least altered or most "natural" systems that could be found, reflecting an underlying belief that these systems provide the best opportunity to examine how ecological processes work. Slowly, however, increasing numbers of ecologists realized two things. First, that humans are an important force shaping ecosystems and biological communities, and how humans manage and alter systems is a key factor in influencing how the systems function. Second, that truly pristine, unaltered systems are few and far between – this has been the case historically, but is even more obvious as human influences pervade the planet.

The fields of applied and landscape ecology, among others, have often incorporated this recognition of the importance of humans as actors in ecological systems. The emerging field of urban ecology goes furthest with this by turning the old order on its head – urban ecologists study the towns and cities where they live. And there is no escape from the obvious observation that cities are habitat for humans. Hence urban ecology has to consider the interactions between humans, other species, the patterns of urban development, and the ecosystem and landscape processes within these urban areas.

The field of urban ecology has grown rapidly in the past decade or so, and there are now several excellent books on urban ecology. Hence, I was initially somewhat bemused when Richard Forman contacted me to say that he was nearing completion on a book on town ecology. Wouldn't this just be another text covering pretty much the same ground? However, having known Richard for several decades now, I recognize that he has been at the forefront of thinking in landscape ecology for most of his career and has made many truly landmark contributions across a wide range of topics, including the fundamentals of the field.

True to form, he cited the reason for writing this current book, *Towns, Ecology, and the Land*, as being that this was "a subject that didn't exist." Indeed, taking a quick look at what is currently available, there is an increasing literature on cities but not much at all on smaller forms of human habitation. Towns (roughly 2,000 to 30,000 people), villages, and their interactions with the agricultural and natural land have really not received much scholarly attention. Although home to smaller populations than big cities, towns are far more numerous and are generally more intimately connected with their surrounding landscapes. And, from the material in this book, this all turns out to be much more interesting than I realized.

The book's goals are to gather, sharpen, develop, and integrate scattered relevant concepts and literature for advanced undergraduate students, graduate students, researchers, and professionals, and to provide building blocks to make better towns and better landscapes. Due to the lack of previous syntheses on the subject, Richard decided to visit and take notes on towns of diverse types and different sizes. He ranged widely and visited 55 towns on four continents, in 16 nations, and 14 states in the USA. The coverage in the book is thus truly international.

The approach is deliberately focused on ecology and hence does not delve deeply into issues such as planning, architecture, and the like. Nevertheless, the material is comprehensive and detailed and provides a wealth of insight into the ecology of towns and their interactions with the surrounding landscape. Richard builds on his lifetime of research and thinking in landscape ecology to provide a scholarly, thoughtful, and provoking book that will set people thinking about towns in ways they probably never thought possible. In an increasingly urbanizing world, smaller habitations remain surprisingly important, and this book provides not just a description of how towns are formed and how they work, but also pointers for how towns can be designed and managed better as part of the increasingly humanized future of the planet.

Richard J. Hobbs
Professor and Australian Laureate Fellow
University of Western Australia

Preface

Imagine a planet of water, farmland, and natural land, plus a few percent urban/suburban land. Like stars on a clear night, towns and villages in huge numbers, sometimes with hamlets and hermits, pepper the planet. Every day town residents disperse into the surroundings, growing our food, harvesting our fiber, and changing the land in myriad ways. These towns, each with some 2,000 to 30,000 residents, appear in countless forms – old and new, small and large, growing and shrinking, industrial and coastal, on and on. What do we know ecologically about these population centers, and their effects on the land? Amazingly little. Ecological *terra incognito* surrounding us … beckoning.

The pages ahead introduce the scientific ecology of towns and villages, how they affect surrounding farmland and natural land, and vice versa. Empirical insights and examples are gathered internationally, developed, and integrated with models, providing a foundation for the growth of town ecology. Promising ideas appear for making better towns and better land.

Nearly half the world's population lives in towns and villages. Overwhelmingly, the people surrounded by farmland or natural land reside, shop, go to school, socialize, govern, and work in and around rural or remote communities. We all depend on town residents, not just for food and wood products, but for freshwater, air quality, flood control, biodiversity, recreational opportunity, and glorious landscapes. Towns are distinctive and diverse; few want to be a city, fewer still a village.

Urban scholars and now ecologists address urban areas as scientific frontiers, highlighting solutions for stubborn societal problems. Yet towns (and villages) display countless characteristics distinct from city-centered urban areas. The literature highlights important social, economic, cultural, and aesthetic dimensions of towns, often mentioning key interactions of towns with surroundings. But town syntheses are mainly pre-1970 Earth Day; newer studies in diverse fields may mention ecological issues.

Noticeably scarce are manifold environmental dimensions in the forefront today, to wit: freshwater shortage, environmental sustainability, ecological footprints, greenhouse gases and climate change, sea-level rise, extreme weather events, impervious surface cover, solid waste recycling, constructed stormwater basins/ponds/wetlands, biodiversity, connectivity/corridors for wildlife movement, large natural vegetation patches/areas, protection of prime agricultural soils, groundwater recharge, road ecology, habitat restoration, solar arrays/wind turbines, and ecosystem services. These features all play important roles in the land of towns and villages.

Ecologists traditionally avoid research sites where human effects are conspicuous, though doubtless numerous research graphs contain points, or an outlier, produced by town or village effects. Often, while analyzing a research site, I've considered the nearest town as simply a convenient place to get lunch or fill the car's tank. I do not recall ever meeting an ecologist whose research focused on town or village, or its interactions with farm land and natural land.

Built areas expand and slice up the land, while natural lands shrink and become more remote. Moreover, food-producing land degrades, and seems increasingly inadequate. Of the global land surface, desert and tundra cover about 33 percent, forest 30 percent, pastures 24 percent, cropland 11 percent, and urban areas 2–3 percent. Essentially, all cropland, much pastureland, some forest, and bits of desert surround, and are directly affected by, towns and villages. Therefore, town ecology focuses on about half the global land surface. The cumulative effect of these small places somewhat regularly spread across the land must be enormous.

This book does not highlight architecture, aesthetics, or culture, nor social structure and economic dimensions, nor suburban towns, nor planning (regional, urban, town, town-and-country, countryside, rural), nor landscape architecture, nor the panoply of promising applications of ecology for society. Yet morsels of each rightfully appear, strengthening the presentation.

As an ecological scientist, my recent background is mainly landscape ecology (e.g., *Landscape Ecology*, 1986; *Land Mosaics: Ecology of Landscapes and Regions*, 1995), road ecology (*Road Ecology: Science and Solutions*, 2003), and urban ecology (*Urban Regions: Ecology and Planning Beyond the City*, 2008; *Urban Ecology: Science of Cities*, 2014), plus lots of good interactions with ecologists, geographers, planners, landscape architects, foresters, wildlife biologists, engineers, and experts in other fields. For several years I've taught a Harvard class on urban and town ecology. Apparently not a single book or major article on town ecology has yet appeared, though I accumulated a valuable pile of detailed indirect studies.

So, for this book, I visited and took notes on towns of diverse types, different sizes, and wide distribution: 55 towns (some are villages, some small cities) visited on 6 continents, 16 nations, and 14 US states. These examples, referred to through the text, revealed key features, as well as commonalities and differences. In addition, I learned from research scholars, students, and the insightful people who showed me towns. With these diverse foundations, the text identifies patterns, processes, and changes, often tying town ecology themes into spatial models or principles. Town ecology coalesced, while intriguing research frontiers appeared.

As a pioneering subject, sometimes only a single relevant study exists, and occasionally an urban/suburban example is used. Town population sizes, mostly cited from Wikipedia online, are rough, mainly due to different areas and years considered. In the pages ahead, twenty types of towns are explored, and hundreds of specific places mentioned. Few conclusions apply to every town, and thus many statements technically require: normally, usually, generally, typically, or often/frequently/commonly – with nuanced meanings to be clarified by researchers.

This highly international book is tailored to: (1) ecologists and environmental specialists; (2) geographers; (3) landscape ecologists and conservationists; (4) town, town-and-country, and rural land planners; (5) agriculture specialists; (6) foresters and wildlife biologists; (7) designers; and (8) urban and regional planners. The well-illustrated readable text with ample references will attract advanced undergraduate and graduate students, researchers, and professionals. Town ecology courses will open a frontier of scholarship and solutions for half our global land. Researchers pushing the town ecology frontier at the beginning will be long cited by subsequent scholars. Professionals may help mold better towns.

The rapidly changing global future, with people packed in cities, increasingly depends on the land of towns and villages. Yet on every continent, too many of these communities and lands are shrinking, degrading. It's time to highlight the core values of towns and villages inexorably tied to their surroundings, the natural and agricultural land.

As the old proverb says: "When the camel gets its nose under the tent, the importance of camel inside will quickly grow." Town ecology is launched.

Acknowledgments

I deeply appreciate the following key colleagues and friends who have intellectually contributed to my understanding and writing about towns: Andrew F. Bennett, Michael W. Binford, Mark Brenner, Anthony P. Clevenger, Barry Copp, Sonja Duempelmann, Andrew Finton, Ann Forsyth, David R. Foster, Jerry F. Franklin, Elizabeth F. Harrell, Michael C. Hooper, Margareta Ihse, Jesse M. Keenan, David Kittredge, Wayne Klockner, Pavel Kovar, Gilbert Metcalf, John H. Mitchell, George F. Peterken, Joshua S. Plisinski, Peter G. Rowe, Eva Serra de la Figuera, Daniel Sperling, Hilary M. Swain, Jonathan R. Thompson, Lisa Vernegaard, Jose Vicente de Lucio, and Jianguo (Jingle) Wu.

In addition, I heartily thank each of the following for showing and teaching me about a particular town, thus converting an abstract subject into one also based on real places: Sito Alarcon, Luis Alvarez, Harry Beyer, Michael W. Binford, Grenada's Mr. Boney, Don Boyd, Mark Brenner, Lawrence Buell, Murray F. Buell (deceased), Tomas Chuman, C. Andrew Cole, Scott Collins, Mark Deyrup, Wenche Dramstad, Joan Ferguson, Barbara L. Forman, Brent C. Forman, Caroline L. Forman (deceased), H. Chandlee Forman (deceased), Maura S. Forman, David R. Foster, Larry J. Gorenflo, Nancy Grimm, Piero Guerrini, Elizabeth Haines, Bill Howe, Delia Kaye, Zdenek Lipsky, David Love, Claudio Magrini, Kendall Mainzer, Xavier Mayor F., Mark J. McDonnell, Mike Meranski, Daniel H. Monahan, Marc Montlleo, Philip Moors, Thomas Nash, Joan I. Nassauer, Trinidad "Birdwatcher" Nicolas, James Nighswander, Edith Pakarati, Paul G. Pearson (deceased), Roberta Pickert, Steward T. A. Pickett, Polly C. Reeve, Francisco Rego, Dusan Rommportl, Gordon H. Shaw, Carl Steinitz, Edmund W. Stiles (deceased), Hilary M. Swain, Swakopmund taxi driver, Frederick J. Swanson, Henry Woolsey, and Jianguo (Jingle) Wu.

I greatly appreciate Dominic Lewis, Alan Crowden, and many production-related editors for encouragement, expertise, and noticeably improving my four Cambridge University Press books over 25 years. Cambridge is an impressive publisher. Also, I heartily thank the Zofnass Program for Sustainability of Infrastructure and Cities at Harvard University and Prof. Spiro Pollalis, Founder and Director of the Zofnass Program, as well as Cambridge University Press, for funding support for the book's illustrations. I warmly thank Mark Brenner for useful insights and photographs.

Taco I. Matthews produced the artwork illustrating many of the town ecology concepts, a highlight of this book. I am ever amazed and honored by her thoughtfulness, creativity, care, and results. She continually teaches me about communicating ideas through illustration. Taco's design accomplishments enrich several of my books and articles.

Two key people were the continuous problem-solvers and fountains of insight for *Towns, Ecology, and the Land*. I am deeply grateful to Lawrence Buell and Barbara L. Forman for encouraging me to explore and weave together another "subject that didn't exist."

Part I

Town Patterns, Processes, Change

1 Town, Village, and Land Spatial Patterns

We live so near the flowers in the fields,
So near the howling wind in forest spires,
And the gentle waves, lapping at the shore,
We live so near.

Paraphrased from a Scottish ceilidh song by
Buddy MacDonald, 2018

"So you want to see the land of towns and villages?"

"Sounds good."

"You're at the right place (rural Iowa, Midwestern USA), and Bertha here is the way to go."

My friend and I stand near a giant blubbery-looking object in bright earth colors and bulging from the ground. It shivers. Filling with air, the thing gets huge. A tiny basket hangs from the bottom, and the friendly aeronaut captain nimbly hops in. Quickly he starts the power. Whooeeesh-sh-sh-sh-sh. Gas-fired flames shoot a few feet upward into the mammoth oval.

"Hop in. Those harnesses will hold you. We can control going up and down somewhat, but horizontally, we go with the wind."

Upward we rise. It's absolutely silent up here. Except for the occasional bursts of fire. Whoeeeeeeeeeeessssh. Captain keeps looking sideways at clouds, distant smokestacks, a grassfire, another balloon, a group of circling vultures. Detecting winds of different directions and speed at different levels. We go up a bit to better wind.

I'm surprised by the array of neat farmsteads, all seemingly caught in a web of barbed-wire fences. Endless green pastures – the cows must be happy.

Later we pass lots of striped fields, some with crawling tractors. We look directly down on woods and streams. But towns and villages are the real eye-catcher from up here.

Now back on the ground, we can explore the land of towns and villages with a worldwide literature, rich in useful concepts, models, principles, ideas, and uncertainties. We start with five perspectives: (1) framework; (2) town sizes, types, and forms; (3) anatomy of towns and villages; (4) land surrounding towns; and (5) hybrid natural–Euclid patterns.

Framework

Diverse, big-picture perspectives get us into the subject: (1) a big picture; (2) why important?; (3) a good place to live; (4) problems living here; and (5) features affecting nature in town.

A Big Picture

More and more of the world's people, increasingly squeezed into urban areas, have never lived in a town. Some have never seen a town or village. Others escape briefly from the city. Nonetheless, more of us know less and less about the land of towns and villages upon which we all depend, fundamentally and daily.

Towns appear in countless forms – old and new, small and large, growing and shrinking, industrial and coastal, on and on. What do we know ecologically about these population centers, and their effects on the land? Amazingly little. Ecological *terra incognito* surrounding us – a giant frontier.

A *town* is a compact mainly residential area in agricultural or natural land that contains about 2,000 to 30,000 residents and a local government. Towns are larger than villages and smaller than cities, and usually have a major economic activity other than harvesting surrounding resources. In contrast, a *village* is a group of households with a public building(s) and about 200 to 2,000 residents, and well separated from town or city.

Not just small places, towns and villages act as key hubs of the global land surface, the essential centers of daily activities affecting the land (Figure 1.1). Nearly half the world's population lives in rural or remote towns and villages. Everyone depends on the activities of "townies," not only for food and wood products, but for freshwater, air quality, flood control, biodiversity, recreational opportunity, and glorious landscapes.

Meanwhile cities dominate our headlines. Urban areas mushroom worldwide, with rural to urban movements, informal squatter settlements, outward sprawling, creative innovations, economic engines, social dynamism, magnets for conflict, and mammoth environmental problems. Many ecologists also turn to urban areas as scientific frontier, with good science contributing to solutions for stubborn societal problems.

Towns (and villages) differ in numerous distinctive characteristics from city-centered urban areas, from farmland, and from natural land. An extensive literature highlights the important social, economic, cultural, and aesthetic dimensions of towns (Nolen, 1927; Francaviglia, 1996; Young, 1999; Wilder, 2013; Albrecht, 2014). But syntheses even introducing ecological or environmental dimensions are indeed scarce (Pysek and Pysek, 1990; Sargent *et al.*, 1991; Pysek and Hejny, 1995; Friedman, 2014; Fallows and Fallows, 2018).

Consider a town linked with its surrounding villages, farmland, and natural land. This distinctive *town-centered area* differs markedly from forest, desert, crop field, pasture, and urban areas. Adding town-centric areas together forms the *land of towns and villages*, which covers perhaps half the global land surface, is home to nearly half the world's population, and provides essential resources to everyone else concentrated in urban areas.

Figure 1.1 Town on amphitheater-like slope descending to harbor with boats for fishing, tourists, or shipping goods. St. George's, Grenada, W.I. See Appendix. R. Forman photo. (A black and white version of this figure will appear in some formats. For the color version, please refer to the plate section.)

Urban areas cover a few percent of the land, and natural landscapes remote from towns and villages cover nearly a quarter of the land surface (Muller and Werner, 2010; Hoekstra *et al.*, 2010). Apparently, almost another quarter is pastureland distant from a community. Yet within some regions and mid-sized nations, about 90 percent of the land surrounds towns (Ratajozak, 2013).

This is the area of active human activity, where local people from towns and villages live, grow crops, tend livestock, and harvest wood. Both local and urban residents recreate here.

Town ecology refers to the interaction of organisms, built structures, and the physical environment in population centers of about 200 to 30,000 (villages and towns), and their interactions with the surrounding agricultural or natural land. This concept follows from the basic concept of *ecology* (interactions of organisms and the environment). Also, it builds from the scientific concepts of *urban ecology* (interactions of organisms, built structures, and the physical environment, where people are concentrated), and an *urban region* (area of active interactions between a city and its surroundings) (Forman, 2008, 2014). Analogous definitions apply to the ecology of villages and smaller hamlets.

A central dimension of the land of towns and villages is the interaction between towns/villages and surrounding land. A population center (community) affects the land, and, vice versa, the land affects the community. Town and land are tied together, usually tightly.

Consider a desert town that owes its existence to water (Figure 1.2). An adjoining river provides irrigation water by canal for vegetables, hay, and livestock. Water from a nearby mountainside provides a water supply for residents. Key resources from high school and hospital to industrial center and dump are close by in the town's adjacent zone. The town serves six nearby villages.

Diverse flows and movements connect town and land. Consider a few entering a town: dust, seeds, floodwater, herbivore pests, farm machinery, and rural shoppers. Conversely, flows/movements from town to surroundings include: workers, recreationists, sewage wastewater, industry air pollutants, town species, traffic/trucks/noise, and school buses. Some effects, including stormwater, air pollutants, and sewage wastewater, mostly impact the area close to a town. Yet others, such as groundwater, dust, manufactured products, and vehicle traffic, may continue on beyond even to cities.

Early geographic spatial models provide valuable insights into the general patterns present. *Von Thunen bands* (Cronon, 1991) refer to the concentric zones of influence around a town. Characteristic bands from town edge outward are: intensive gardens/orchards; cropland; livestock pastureland; and wooded land for hunting and wood products. Of course, the concentric bands are commonly modified by topography, such as mountain, river, and coast.

Another spatial model, *central place theory* (Christaller, 1933), adds insight. Where many population centers are present, each is a central place from which socioeconomic effects such as manufactured products, newspaper coverage, and school districts extend outward, creating a so-called hinterland around the center. Where population centers are close together, they compete and surrounding spaces are relatively small and typically devoted to agriculture.

A spatial polycentric hierarchy is superimposed on these broad patterns (even perhaps in suburban landscapes; Harris, 2015). Population centers of different size, i.e., town, village, hamlet, and crossroads, form the hierarchy (Arendt, 2004; Hough, 2004; Friedman, 2014). Smaller population centers tend to be denser near a larger center. Thus, the land has many towns with clusters of small centers nearby each. Few small centers and normally little human activity modify the land far from towns.

Most population centers originated where farmers settled near the confluence of good soil and a waterbody (Figure 1.3). Thus, over time most towns and villages are mainly surrounded by agricultural land, beyond which is natural or semi-natural land such as woodland/forest or desert. The population centers act as local nuclei with effects extending across the land. Although environmental, economic, social, and cultural dimensions differ markedly in different geographic regions (Forman, 2014), these spatial patterns of population centers and land are widespread, even universal.

Why Important?

Urban areas cover a few percent of the land surface, support a bit more than half the human population, and are growing fast in both population and especially area. Society's attention and funding are increasingly focused on the urban area, both to improve

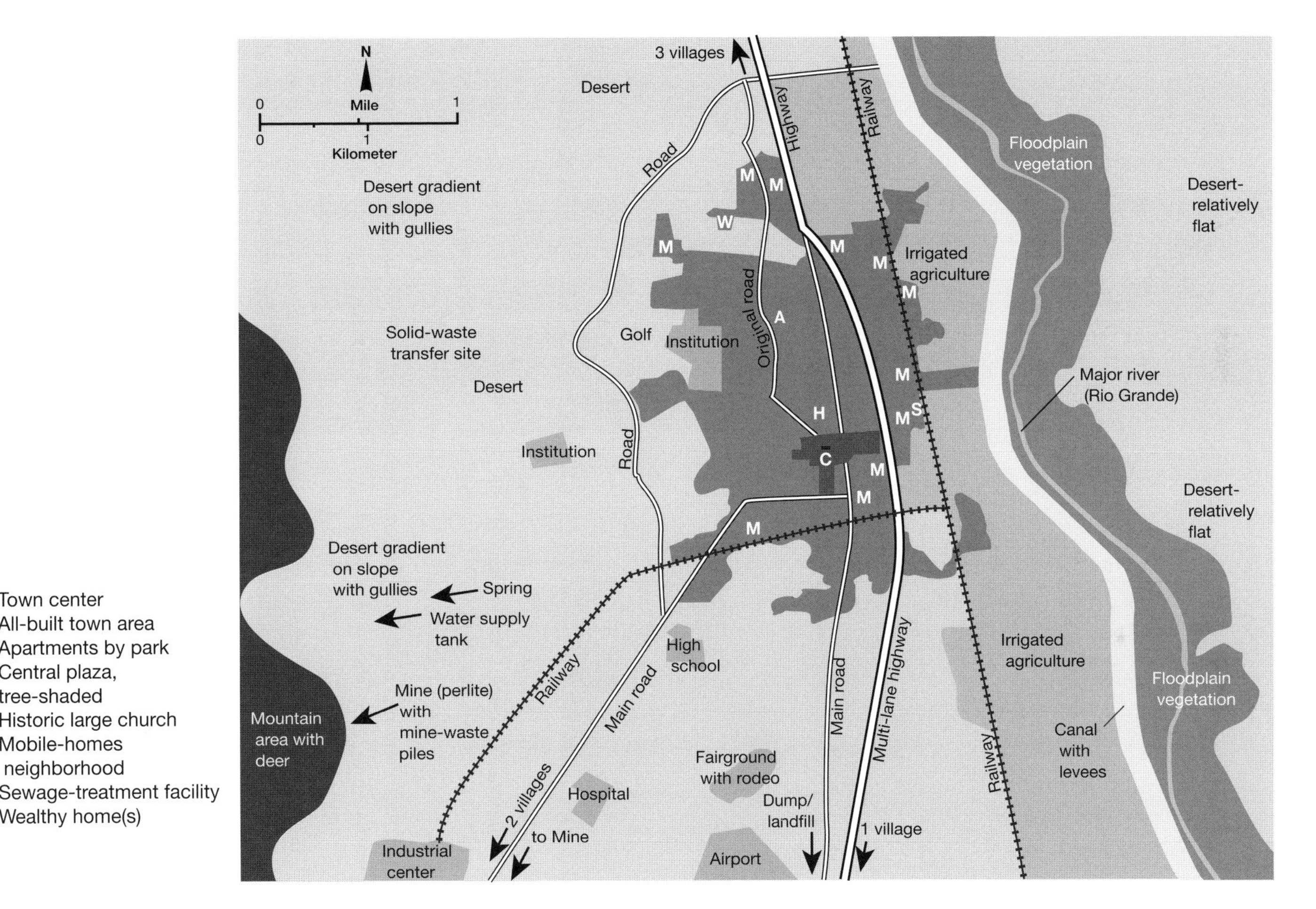

Figure 1.2 Desert town by major river and transportation corridors. Socorro, New Mexico, USA. See text and Appendix. (A black and white version of this figure will appear in some formats. For the color version, please refer to the plate section.)

Figure 1.3 Village with most families returning from city on weekends/holidays to join 89 residents. Abanades, Castilla y La Mancha, Spain. R. Forman photo. (A black and white version of this figure will appear in some formats. For the color version, please refer to the plate section.)

services and increase opportunities. Addressing the large and numerous environmental urban problems is also increasingly on the radar screen.

At the other end of the spectrum are remote natural landscapes which still cover large expanses. Generally of low economic value or too remote, these areas receive little attention and protection. Yet, they are of global importance as reservoirs of clean water, stabilizers of weather patterns and clean cool air, absorbers and storage of greenhouse gases, protection against massive soil erosion and sedimentation, and reservoirs of native wildlife and biodiversity. Furthermore, these large remote natural areas are sources of inspiration for people who treasure them or venture into them for recreation.

The massive middle ground, the *land of towns and villages* where small population centers predominate, is the highlight here. This is home for half the world's people, and provides jobs for almost all of them. Croplands produce most of the food on the table, both for rural centers and urban areas. Pasturelands produce the meat for both targets. Most of society's wood products come from this land. Almost all nature-based recreation for local residents and for urban dwellers exiting a city occurs in this land.

This land of small population centers typically cools the air, facilitates precipitation, absorbs and stores carbon. It is the prime determinant of how much human-caused soil erosion by wind and water occurs. Indeed, the extensive fertile soil loss into rivers and sea basically originates in the agricultural landscapes.

The bulk of the land's biodiversity is present in the land of towns and villages. Doubtless many species of tundra and ice are missing from this area of human activities. Also, a small number of species especially sensitive to human activity are probably missing. But basically, all generalist species are present, some of which benefit from and are attracted to areas of human activity. Also, most specialist species are present at least in low population size in patches within areas of human activity.

In essence, the entire human population depends intensively for food and resources on the land of towns and villages. The environmental dimensions of the area are priceless in their own right, and also intensively affect the entire globe. Understanding the ecological patterns, processes, and changes in this people-and-nature area should be a priority.

Features Affecting Nature in Town

All parts of town are close to the surrounding land, which is commonly agriculture. Typically natural land is farther out, but often readily accessible and used by residents. Small patches and corridors serving as stepping stones in farmland enhance interactions between town and wildlife. Animals with home ranges of more than about 1 km^2 (250 acres) are particularly sensitive to these surrounding patterns.

Towns are located to somewhat fit with the surrounding topography. The town edge is heterogeneous and porous. Relatively few greenspaces, none large, are present in town. Few major green corridors exist. Mature trees are widespread in the older residential area. No extensive impervious surface area is present, so excessive heat buildup and stormwater runoff are much less of a problem than in cities. Towns have limited shallow underground infrastructure, though breaks occur that commonly pollute groundwater. Numerous and diverse microhabitats are present and readily colonized by species from the nearby surroundings.

Towns are suitable habitat for generalist farmland and natural-land species, which readily colonize town areas. Residents plant nonnative and native species, and transportation carries in spontaneous species. But unlike cities, where ships, aircraft, trains, and trucks provide a major stream of incoming nonnatives, and where urban environmental conditions are so inhospitable for species of natural ecosystems, ecologically towns are somewhat closer to agricultural and natural areas. Unlike many cities with extensive descriptions of the nature present, the nature in towns is scientifically little known (Hanski, 1982; Pysek and Pysek, 1990; Pysek and Hejny, 1995).

Town Sizes, Types, and Forms

Arguably towns differ from one another more than do cities. Thus, we examine the town types and forms as follows: (1) differentiating village, town, and small city; (2) types of towns; (3) town shapes and associated processes; and (4) small cities.

Figure 1.4 Main Street with several local shops of village in forested mountains and lakes. Rangeley, Maine, USA. See Appendix. Watercolor by H. Chandlee Forman; R. Forman, owner. (A black and white version of this figure will appear in some formats. For the color version, please refer to the plate section.)

Differentiating Village, Town, and Small City

Villages see moon and stars down to the horizon. Towns watch the Milky Way when sky arches overhead. Cities delight in a star or planet that penetrates the heavy gray.

Hamlets, villages, towns, and small cities are all population centers or communities (Figure 1.4). *Population center* is a spatially separate compact group of inhabitants or residents, while *community* emphasizes the network of social interactions among the inhabitants that produces collective local action.

Population centers or communities from small to large can be differentiated in several ways, including: (1) area; (2) population density; (3) population size; (4) governmental administrative, functional, and locational criteria; (5) grid and mathematical measures; and (6) bioassays. The first two have important ecological implications but are less commonly used in comparative studies. Governmental criteria and names, such as market town in Britain (McGranahan and Marcotullio, 2005) and contrasting sizes of communities, e.g., called cities in Portugal and China, vary widely from nation to nation. Grid measures, for instance with geographical information systems (GIS) (Sorace and Gustin, 2009; Atkinson-Smith, 2014), commonly split a compact community into two or more grid cells. A mathematical formula illustrated by 2, 4, 8, 16, etc. (Forman, 2014) is non-spatial and doesn't mimic biological or social patterns.

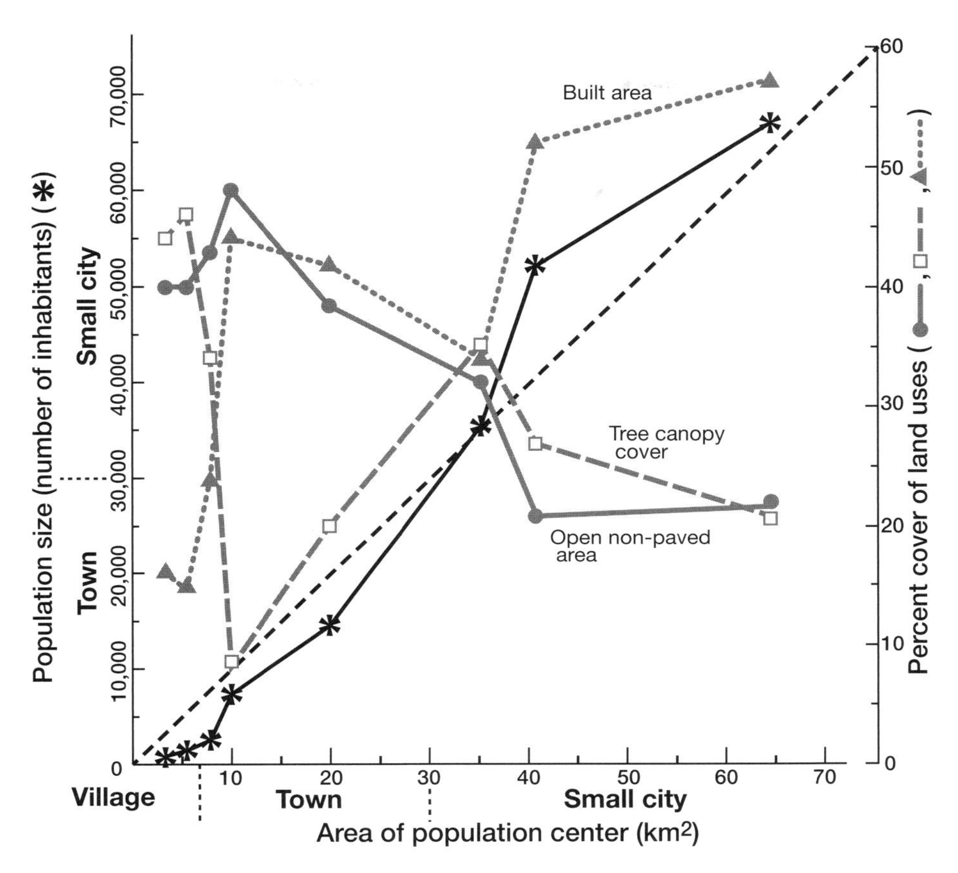

Figure 1.5. Population size and land uses in villages, towns, and small cities. Diagonal line indicates 1,000 residents/km². Area of eight communities in the east-central Pampas Plain of Argentina (small to large: Rivas, Castilla, Rawson, Suipacha, SA de Giles, Chacabuso, Mercedes, Lujan), based on transects containing some agricultural activity and thus are a slight overestimate. Agricultural spaces are omitted in calculating percent cover of land uses (covers). Adapted from Garaffa *et al.* (2009)

In contrast, *population size*, i.e., number of residents, is easily measured and understood, and widely used. Below we will briefly compare the results of population size and the few ecological or bioassay measures. Land uses within a community vary according to population size (Figure 1.5). Yet all communities have a center, e.g., the town center, around which people and activities have evolved. So here we use the number of people aggregated around the center of a compact community as the basic measure.

Villages are smaller than towns, and towns smaller than cities. Cities are often described as small, large, and mega. The dictionary defines *urban* as "of or pertaining to cities," so cities and suburbs are urban, but towns, villages, and hamlets are not. Towns also are commonly differentiated as small, medium-size, and large. Hamlets have a small village-like look, but communities smaller than villages normally have no common building for public meetings (Arendt, 2004). Groups of buildings around a crossroad intersection generally focus around the commercial activity of servicing travelers where roads intersect. An isolated housing development is typically filled with residents commuting to an urban area. Also, individual homes may be dispersed in the land, such as many farmsteads or even where a hermit resides.

Different nations delineate population sizes and ranges for towns differently, sometimes by government decision, sometimes by simple tradition. Sweden, Switzerland, Belgium, Portugal, and Amazonas commonly consider a community of 10,000 to be a small city, whereas in China a small city may reach 100,000 to 500,000 inhabitants.

The following are representative cutoffs differentiating population centers.

Between town and village. Population 1,000: Wales (Atkinson-Smith, 2014). 1,500: England (Sharp, 1946). 2,000: USA (US Census Bureau, 2016); China (Peter Rowe, 2017 personal communication; and Jianguo Wu, 2017 personal communication); Argentina (Garaffa *et al.*, 2009). 2,500: Canada (Friedman, 2014); Europe (Friedman, 2014). 3,000: Poland (Ratajczak, 2013). 4,500: Northern Ireland (Atkinson-Smith, 2014). 5,000: Europe (Vaz *et al.*, 2013a); Poland (Vaz *et al.*, 2013a).

Between town and city. Population 5,000: developing nations (Hardoy *et al.*, 2001; Bell and Jayne, 2006). 10,000: developing nations (Hardoy *et al.*, 2001); Europe (Friedman, 2014). 20,000: developing nations (Hardoy *et al.*, 2001; Satterthwaite, 2006); North Region of Brazil (Browder and Godfrey, 1997); widespread (Muller and Werner, 2010). 25,000: Canada (Friedman, 2014); USA (US Census Bureau, 2016). 50,000: Europe.

This sample from the literature indicates that the median cutoff differentiating town from village is 2,000–2,500 people, and separating town from city is 20,000. Developing nations tend to have lower numbers (Satterthwaite, 2006). For different administrative reasons, China and Britain for years included much larger population centers as towns. Finally, cutoff numbers tend to increase over time as global population rises.

In the Paleolithic era, 15,000 years ago, human aggregations or "villages" had about 6 to 60 families, while in Mesopotamia some 5,000 years ago, hill villages were of about 200–500 people (Mumford, 1961). In the 1940s commonly an English village had 100–1,500 people and a market town 2,000–20,000 (Sharp, 1946), while today some market towns exceed 100,000 people. Four decades ago (Northam, 1979) population sizes were given as follows: hamlet 16–150 people; village 150–1,000; town 1,000–2,500; and small city 2,500–25,000.

Cultural and socioeconomic dimensions strongly imprint a town. Nearby towns tend to share many such characteristics. As cultural influence spreads, town forms spread across the land (Figure 1.6).

Plant diversity and avian communities correlate with population size of communities (Garaffa *et al.*, 2009). Plant species richness correlates with village size (Hanski, 1982), and in communities from 25 to 45,000 people (Pysek and Hejny, 1995). Generalist and specialist predators correlate with population size (Sorace and Gustin, 2009). A bioassay or ecological pattern could be used to help differentiate village, town, and city.

For instance, in northern Argentina, 52 bird species were encountered along gradients from the center to the outside of nine communities: two villages (populations 472, 827);

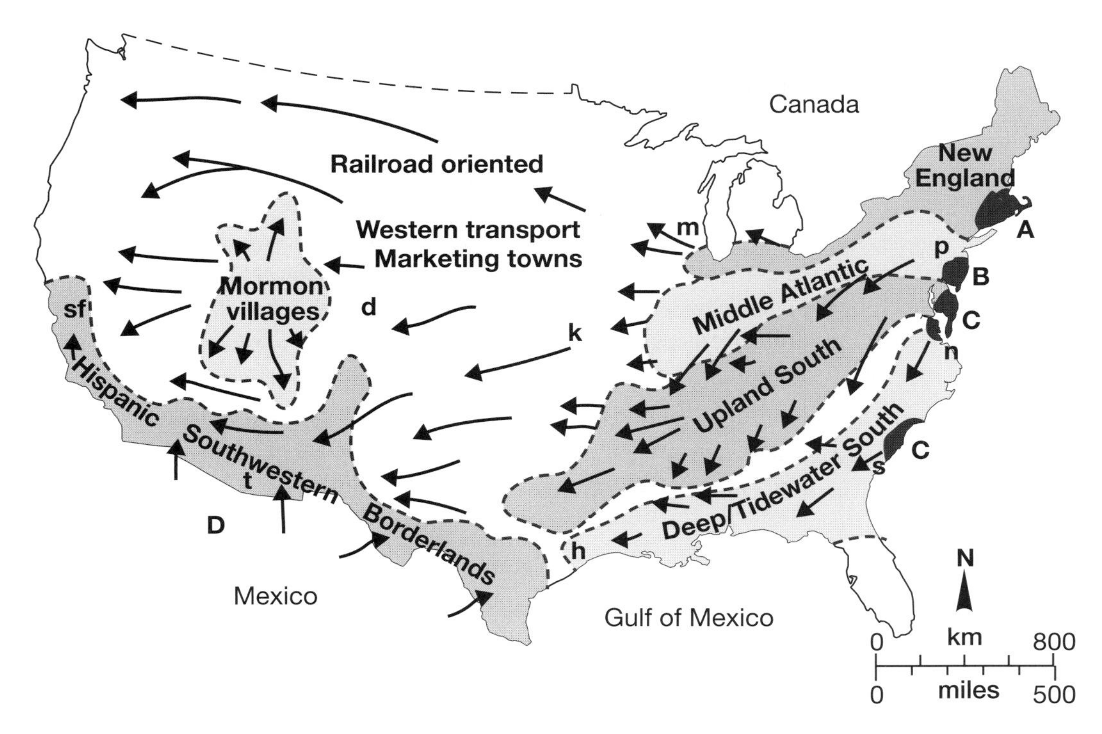

Figure 1.6 Diffusion of townscapes from four original cultural-landscape town/village types. A = New England village; B = Middle Atlantic town; C = Lower Tidewater/Deep South town; D = Southwestern town. Considerable variation exists. Cities marked to aid in locating trends: p = Philadelphia; n = Norfolk; s = Savannah; m = Milwaukee; k = Kansas City; h = Houston; d = Denver; t = Tucson; sf = San Francisco. USA ca. 1770 to 1900. Adapted from Francaviglia (1996).

three towns (2,000–14,000); two small cities ("intermediate sized," 35,000, 52,000); and two larger cities (67,000, 522,000) (Garaffa *et al.*, 2009). For population centers of more than 7,000 (Figure 1.5), avian richness decreased with more urbanized area. In villages and towns, the avian species composition of outer rural areas continued little changed inward into the center. Community species similarity between urban core and peripheral points decreased as the length of the gradient increased, a pattern especially characteristic of cities.

Based on bird species richness, abundance, and community similarity, there may be a *population threshold* between communities of about 7,000 and 35,000 population. Above this population size a significant effect of urbanization on avian community structure was evident, and below it the effect was minimal. A study of Finnish towns/cities also reported a threshold at about 35,000 population, above which bird communities differed markedly from those in smaller population centers (Jokimaki and Kaisanlahti-Jokimaki, 2003). Based on the Argentina and Finland studies, the avian communities in town and city are quite different.

In researching towns, I made observations in more than 50 population centers, ranging from hundreds of people to more than 70,000 (see Appendix). Seemingly obvious villages (e.g., Selborne, Petersham; see Appendix) differed markedly from towns (e.g., Pinedale, Chamusca) which contrasted with small cities (e.g., Rocky Mount, Gavle). Even small towns (e.g., Superior, Plymouth), mid-sized towns (e.g., Tetbury, Tavistock), and large towns (e.g., Tomar, Rock Springs) differed visibly in their attributes.

For communication throughout this book, it is useful to associate population numbers with the concepts of village, town, and city. Two main approaches lead me to the numbers pinpointed: (1) a distillation of the diverse numbers above; and (2) my partially subjective feeling by making observations in, and studying the ecological literature on, numerous population centers worldwide varying in size. The following population sizes seem to apply widely, though of course worldwide variability exists:

Village: 200 to 2,000 people
Town: 2,000 to 30,000 (small town 2,000–5,000, mid-sized town 5,000–20,000, and large town 20,000–30,000)
Small city: 30,000 to >100,000

Types of Towns

In the USA, on average, towns (2,000 to 30,000 population) are 19 km (12 mi) apart, and the median town-to-city (>50,000 pop.) distance is 86 km (54 mi) (US Census Bureau, 2016). In Portugal the median population of cities is only 14,000, which, due to two large cities, is half the average size (Vaz *et al.*, 2013a). In Kerala state in India, most of the more than 30 million residents live in communities of more than 10,000 called villages (Satterthwaite, 2006). In China sometimes towns are mainly considered the nuclei of future cities (Kirby *et al.*, 2006), and a settlement of less than 50,000 may be unable to provide basic services to the population (Peter Rowe, 2017 personal communication). Finally, the urban planner Kevin Lynch (1981) considered the optimum size of cities to be 25,000 to 250,000. However, long ago Plato said that when a city reached 50,000 inhabitants, a new city should be started.

In 38 urban regions worldwide, half of the towns (excluding towns by waterbodies) are mainly surrounded by agricultural land (cropland) (Forman, 2008). Thirty percent are in natural land, and 20 percent are *towns near the boundary* between agricultural and natural land. Towns in agricultural land cause rather little environmental damage, other than removing a bit of farmland. But towns in natural land have significant negative effects on natural systems, including water, soil, wildlife, and biodiversity, including interior species and large home-range species. The many towns near the boundary between agricultural and natural land have diverse resources close by, and seem well located. They displace little valuable land, and outward impacts are mainly on the edges of natural and agricultural areas. This spatial arrangement is consistent with the aggregate-with-outliers model for sustaining natural systems and biodiversity (Forman, 1995).

Many types of towns are simply described by *location* – mountain town (van der Zee and Zonneveld, 2001), hill town, valley town, rural town, desert town (Kosek, 2006), river town, lakeshore town, coastal town (Williams, 2006), urban-fringe town, and

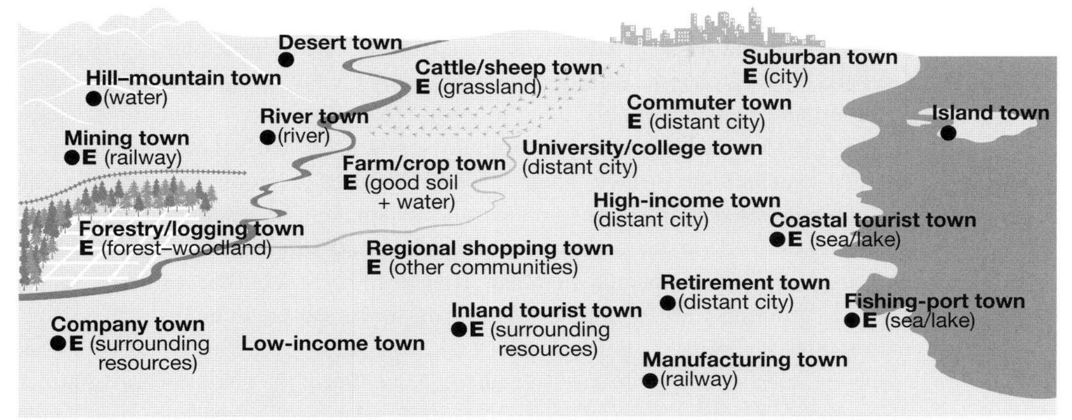

Figure 1.7 Twenty types of towns and key resources upon which they depend. See Appendix.

suburban town (Figure 1.7) (McGranahan and Marcotullio, 2005) (see Appendix). The names provide some insight into the primary economic activities and types of people present, although variability within each type is high.

More informative are town types described by a major *economic activity* or dimension – farm/farm-crop town (Hokanson, 1994), cattle/sheep town, forestry/logging town, mining town (see book cover) (Williams, 2006), manufacturing/industrial town (Williams, 2006), commercial/shopping/market town (Brown, 1986; Powe *et al.*, 2007; Powe and Hart, 2007), university/college town (Winters *et al.*, 2013), commuter town (Powe *et al.*, 2007), fishing town, port town, and tourist/resort town (see Appendix). Some types have two or three predominant economic activities, which periodically change over time. The prime economic activities in turn provide considerable understanding of the kinds of people present, varied social dimensions, and general town form (van Leeuwen, 2010; Ratajczak, 2013; Vaz *et al.*, 2013a).

A third diverse group of specialized town types has town names that suggest *several dimensions* of socioeconomics, activity, and form – retirement town, company town (White, 2012), native people town (Kosek, 2006), wealthy/high-income town, poor/low-income town, historic town, and abandoned/ghost town. Another source uses level of services, tourism, economic structure, commuting, characteristics of residents, and affluence/deprivation to group town types (Powe and Hart, 2007).

Most town plans (Nolen, 1927; Bell and Bell, 1969; Arendt, 2004; Daniels *et al.*, 2007) are a rectilinear grid imposed on curvilinear natural patterns (even though essentially no top-view right angles appear in natural land), and require constant maintenance and repair (Forman, 2014). Planned towns, those planned as a whole at one time, are much discussed in the literature (Easterling, 1993; Duany *et al.*, 2003; Forsyth and Crewe, 2009; Forman, 2014), and thus barely included here. The company town is a special case of a planned town, where a company or government either creates the town or strongly molds it (e.g., Palmerton; see Appendix) (White, 2012).

Overwhelmingly, towns have an organic town layout, a pattern resulting from tradition and trial-and-error (Friedman, 2014). No or little centralized control was involved. Grid and planned patterns using a formal geometry are the basic alternatives.

Suburban towns are highly heterogeneous (e.g., Concord, Issaquah, Cranborne) (Harris, 2015). Starting with the original topography, early land uses, especially farming, changed the surface. Then different mixtures of people populated each suburban town, and development spread in uneven spurts. The amount and arrangement of conservation land, of protected waterbodies and shorelines, and ongoing town and residents' environmental practices lead to today's suburban ecology. Adjacent suburbs commonly differ markedly, creating quite different inter-town ecological interactions.

Desert towns (e.g., Superior, Socorro, Alice Springs) have several distinctive characteristics, including a water source, oasis with trees and shrubs, rare species, small-plot farming, extreme heat, and desiccating winds. A valley or other low location means groundwater for pumping is closer, but also indicates susceptibility for flash floods and flooding. A spring or seep or stream generally has water from mountains near or far.

Today's large cruise ships suggest a floating *boat town* of a few thousand people. An extensive crew creates conditions for the majority to enjoy their holiday. Like a cattle feed lot, the ship tightly packs in the people, pours in the food, collects the waste (for night disposal at sea or sewage treatment on land), has a doctor ready and busy (instead of a veterinarian), and keeps the maintenance/repair crew active. Staff and crew keep a careful watch over the people. It is casino-like inside with countless activities and diversions. The ship's operations, noises, and effluents are muffled. Disasters are normally avoided. In this case, the town moves about (fish patterns on the ship's carpets may tell people which way is front). The ship empties out briefly at ports, and chugs onward to the next shopping center. The population is replaced every few days or weeks, and the place cleaned. Fuel, freshwater, and food are replenished. A "boat town" runs over the sea, repeatedly hooking up with other population centers.

Town Shapes and Associated Processes

Measuring and differentiating the shapes of objects is done in many fields, from the shapes of coral atolls to sand grains, tile patterns, and land uses. Three easily observed variables seem to effectively capture shapes on land (Forman, 1995): *elongation*; *number of major lobes*; and *curvilinearity/straightness of edges* (prevalence or percent of boundary that is straight). Probably all natural land uses fit onto a graph plotting number of major lobes against elongation (length-to-width ratio), and human molded patches mainly have straight boundaries rather than the curves of nature. This approach applies nicely to towns.

A squarish or circular or oval town typically has a length-to-width ratio of <1.5:1 (e.g., Tetbury, Cranborne; see Appendix). Rectangular towns are typically 1.5:1 to 3:1, elongate cases >3:1 (e.g., Scappose, Casablanca), and less common linear towns >6:1 (Williams, 2006). Straight town boundaries are often present, but most town boundaries are finely curvilinear or convoluted, with short straight lines reflecting property boundaries and administrative lines.

Two or three major sections are typical for towns divided by stream or railway, and especially if two or more major corridors are present (e.g., Sooke, Rock Springs). Sections may also result from administrative/jurisdictional divisions over time (e.g., Tomar) (Ferreira and Condessa, 2012). Towns often have about five or six primary roads radiating outward. Prominent strip development (Forman, 2014) typically projects outward as lobes on one to three roads. Towns along a coast, lakeshore, or curving river often appear crescent moon-shaped (Williams, 2006).

Processes Determining Town Shapes

Town shapes reflect key attributes of the site where a town began, especially clean freshwater, good agricultural soil, and land-water location for travel and security (Pacione, 2005; Forman, 2014). Specific features were important in starting a village which grew into a town. Consider: defense; near a spring; by a stream (potentially dammed for a mill); near a ford or bridge; a small knoll in floodable land; a crossroads; near a source of game; around a pond; and/or in a quiet safe spot (Sharp, 1946). Today few of these factors are of central importance in starting a community.

Initially village/town shapes were mainly determined by topography and waterbodies. Waterbodies and flows persist as important variables affecting changes in form. Later pattern changes especially result from administrative lines, transportation ways, and factory areas (Kamra, 1982).

Squarish or rounded towns predominate in plains (Forman, 2014). Elongate ones appear in valleys, and along primary roads and edges of two land-use types. Convoluted towns with lobes and coves thrive on the irregular heterogeneous slopes of hills and mountains.

Also, human construction or activity creates town shapes, e.g., fort, wall, moat (Mumford, 1961; Kamra, 1982), large company/industrial facility, manufacturing center at town edge, commercial center at edge, road grid, radial primary roads extending outward, one or more railway line, crossroads, highway along the edge, adjacent protected forest or natural area, and entrance/exit location of a nearby highway. A dam and millpond, water supply, or irrigation canal also is an effective molder of town forms.

In a town, development follows main roads and tends to avoid flood-prone areas. Shape depends on the angles of radiating primary roads (Sharp, 1946). Town expansion is explored in Chapter 3.

A prominent or unusual site/feature determines the form of many towns. Natural features include waterbody, rock cliff, hot spring, and ancient huge tree, while shrine, castle, religious structure, and town hall illustrate human features.

Town Shapes Determining Processes

The other side of the same linkage is that shapes or forms determine processes, flows, and movements. *Form and function principles* are most useful (Forman, 1995). A *round patch* is best for protecting internal resources, whereas a *convoluted patch* enhances interactions with the surroundings. A *patch with long lobes or strips* has an internal structure for flows/movements. Thus, a starfish-shaped town protects residents around the center, has considerable interactions with the surroundings, and has ample flows in and out along radial lobes or corridors.

Round natural areas best protect internal features, such as biodiversity in a forest or deep-water fish in a pond. Biodiversity and many ecological attributes within a town correlate better with internal habitat conditions, such as one or a few medium-size natural patches, or overall habitat diversity. More sections of towns or convoluted lobes increase habitat diversity.

Also convoluted lobes and coves facilitate more interactions with diverse surroundings close by or farther out. Furthermore, lobes, coves, and narrowness indicate directionality of flows/movements. For instance, traffic noise extends out built lobes, while wildlife enter town in green wedges between lobes. A stream carries wastes from town. Large nearby natural areas serve as species sources for the town (Forman, 2014). Even town strip development out of a road can disrupt broad-scale wildlife movement across the landscape.

Consider some other shape clues that indicate flows/movements. Regularity may lead to low biodiversity and predictable flows. An elongate town may act as both conduit and barrier. Squares and rectangles imply more maintenance/repair, less sustainability. Vegetation over an elongated drainage basin protects water supply. Factory pollution in a town edge produces a plume that degrades downwind air quality, soil, and vegetation. Natural areas close by, or protruding in as a green wedge, recharge groundwater and are a species source. Active nearby farm fields have erosion and wind carrying dust, nitrogen (N) and phosphorus (P) fertilizers, pesticides, and pollen into town. Often buildings along a riverside or coastal strip alter wind and water flows and are prone to severe storms and floods.

In short, the shapes of towns, villages, and other features in the landscape provide a treasure chest of clues for landscape detectives. The diverse movements and flows basically indicate how the landscape works.

Yet shapes change. Start with a small town patch, extend out a road, add more roads to about six, widen some roads with strip development, lengthen the roads, and add some nodes. That process produces an asymmetric six-legged moonwalker or wagon-wheel *hub-and-spokes model*.

Small Cities

Cities typically contain many formal patterns produced by planning, such as circles, ovals, crescents, infills, linear vistas, and ornamented buildings (Nolen, 1927). Such structures commonly appear in small number in large towns of about 20,000 to 30,000 inhabitants.

But small cities of, say, 30,000 to 100,000 (Figure 1.8) (e.g., Valladolid, Gavle; see Appendix), while not the subject of this book, manifest considerable planned formality. Many other characteristics seem to differentiate cities from towns (Bell and Jayne, 2006; Forman, 2014; Friedman, 2014). Consider the following urban attributes: many traffic stoplights; low water table; urban heat island; polluted air; high-income neighborhoods;

Figure 1.8 City street lined with four-level or higher buildings and national commercial enterprises. Pedestrianized Nanjing Road in Shanghai at prime shopping time.
Source: georgeclerk / E+ / Getty Images.

tree-lined boulevard; power-generating facility; regular street-cleaning maintenance, often at night; livestock rarely seen; non-urban wildlife rarely observed; several or many public greenspaces (few in most towns); impervious surface covering most of the central portion of city center blocks; considerable stormwater runoff polluting local waterbodies; considerable area of spontaneous nonnative plants; mainly buildings of three levels or more (above ground surface) on blocks near the central public space (two to three levels in most towns); extensive deep buried infrastructure.

Few people can readily walk entirely across a city. Few people are recognized on a city walk. National/international stores predominate over mom-and-pop shops. Few multi-generation residents are present providing stability to the city.

Anatomy of Towns and Villages

Just as a medical expert knows our anatomy to detect problems and prescribe solutions, we now look inside towns to understand their anatomy (Williams, 2006). Knowing the diverse land uses, objects, and flows/movements, plus variations in these features, is

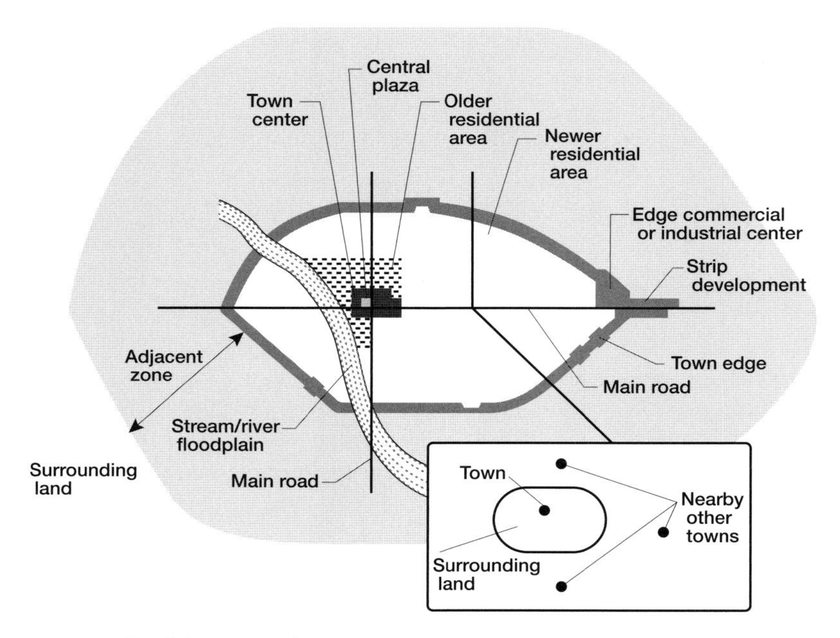

Figure 1.9 Spatial concepts for a town, plus its adjacent zone and outer surrounding land. Inset shows the town's surrounding land relative to other nearby towns.

essential for ecological solutions to problems. Some town features are best understood by walking, others, such as patterns of backyard swimming pools, vehicles in factory car parks, and arrangement of greenspaces, by flying in a balloon or small plane (Fallows and Fallows, 2018).

A Characteristic Town and Its Attributes

A town, as a compact essentially all-built area, around a stream and crossroads provides a good model (Figure 1.9) (Sargent *et al.*, 1991; Ferreira and Condessa, 2012). The town is oblong or rectangular in form, with the long axis north–south or east–west. A main road passes through town, and two to four other primary roads radiate out of town. Roads in town are mostly referred to as streets. Commercial development within the town is particularly prevalent along the main road (Francaviglia, 1996; Bohl, 2002; Arendt, 2004), which often connects to the nearest town or city. Irregularity and asymmetry throughout the town is characteristic, reflecting topographic, transportation, and outside land-use heterogeneity. A stream and roads slicing through town cut it into unequal sections.

The stream is partially straightened and passes under a former mill-dam site in the town center. The stream used to carry off many wastes, a basic town-cleaning process. Buried pipelines pass along the stream corridor and under or alongside primary roads, thus connecting town center with outside town. The main road and primary crossing road are usually the major transportation routes and rather wide, except perhaps in the town center. Short built lobes extending outward along these primary roads as strip (ribbon) development are lined with mixed-use residential, commercial, and sometimes manufacturing structures.

The following components or attributes within towns are widespread and extremely diverse (Brown, 1986). Representative internal attributes are given for each major town area.

Central Plaza

A *central plaza* or square, as a central public space with irrigated plantings, typically one block in size and close to main street, anchors the town (Figure 1.2) (e.g., Clay Center, Palmerton; see Appendix) (Arendt, 2004). The central plaza may appear savanna-like, composed of lawn, paved walkways, scattered trees, and benches. Or these features may surround a prominent government building, such as a town hall or courthouse, or a large religious structure. The central plaza is a prime meeting place for town residents, both frequently and for special events.

Town Center

The *town center* is characterized by a concentration of commercial, residential, civic, and cultural buildings (Figure 1.10) (e.g., Capellades, Astoria, Valladolid) (Bohl, 2002; Urban Land Institute, 2008; Forman, 2014). It encompasses the central plaza and is often elongated in the direction of the main street. In mid-sized towns two-level buildings commonly predominate along several commercial blocks near the central public space (Jackson, 1952). Larger towns usually have mostly three-level central buildings (unless in an earthquake zone). These town-center buildings provide local shops for shoppers, plus upstairs businesses, storage, and some apartments.

Most shops are locally owned and run, with few national or regional retail stores. Local professionals populate the offices. Restaurants feed town residents. Pubs/bars/cantinas/taverns provide drink. But eateries also create food waste, and all commercial activities produce considerable packaging and other solid waste in the town center. Residents from surrounding villages and areas often do weekly shopping here. Typically, few multi-unit housing units are present, and no high rises. An open or enclosed marketplace with stalls for selling food and goods is characteristic (Unwin, 1911; Brown, 1986; Powe *et al.*, 2007). One or more religious structure is typically present.

Unless rebuilt after a major fire, many town-center buildings are old and much renovated. The center of some blocks has considerable plant cover. Some, but often relatively few, blocks are lined with trees providing shade, habitat, and connectivity. The blocks may be single or double tree-lined.

The slightly to quite irregular street grid has the smallest blocks in town. Sidewalks lining streets provide for convenient walking. Car parks are present (Forman, 2014), though adequate parking in the center of a rounded town is a persistent issue. Simple buried stormwater pipe systems complement the dense network of streets, which serve as major conduits carrying water away to a waterbody. The impervious surface cover is fairly large. With a water body in the small area of a town, considerable impervious surface area is directly connected by pipe or ditch to the waterbody.

	Large habitat	Small habitat or structure	Linear habitat	Microhabitat cluster
All three parts of town	—	*Flower gaden Mowed lawn House House plot Shrub border Base of post	*Stream corridor *Railway corridor *Roadside/verge Main road Local street Sidewalk Line of street trees	*Stream corridor *House plot
Town center	Town center Commercial center Industrial site	Central plaza Paved carpark Commercial building Green wall Green roof	—	—
Older residential area	*Older residential area *Neighborhood	*Vegetable garden *Garden pond Green wall	—	*Older residential area *Neighborhood
Newer residential area	*Pond + pondshore *Neighborhood *Low-density residential block *Community garden *Old quarry or gravel pit Cemetery Waterworks Sewage treatment facility Institution Major dump/tip Airport Market place Industrial site	*Vegetable garden *Garden pond *Shrub area	*Town edge *Edge of woods *Farmland edge	*Neighborhood *Low-density residential block *Plant nursery *Community garden *Semi-natural woods *Farmstead in town edge *Farmland edge *Shrub area *Pond + pondshore *Industrial site *Park *Edge of woodland *Old quarry or gravel pit

Figure 1.10 Habitat types and species-rich habitats in different areas of town. * = species-rich habitat. Different spatial scales included.

The countryside is nearby. Farm tractors move through town, dropping seeds and soil, perhaps bringing the aroma of fresh manure. Livestock are commonly seen. Wildlife from the surroundings are often observed. Traffic moves slowly at neighborhood speed.

Several components often present in a town are few in number: multi-unit low-rise buildings; railway; train station; funeral home; marketplace; intersection of main street and a primary crossroads; canal; small pond; fort/castle; waterfront (riverside/lakeshore/coast); boat harbor; factory; town park; unpaved gravel road; prominent bank; central hotel; library; local museum; public rest rooms/loo; police station; ballfield; and school ground (Brown, 1986). Also, some components only exist in relatively few towns: train station; marketplace; canal; fort/castle; and waterfront. Some towns have no traffic stop lights, some a few; most common may be single traffic light in the town center (Jackson, 1952).

Older Residential Area

An *older residential area* mainly or entirely surrounds the town center (e.g., Easton, Capellades, Astoria) (Figure 1.9). Relatively old large single-family homes on small rectangular plots predominate, though some may be subdivided into apartments (Forman, 2014). Stormwater flows down streets, as well as some ditches and pipes, to a local waterbody. Sidewalks line the streets, many with mature street trees, which often have cavities used by birds and mammals. Daily needs can be met by convenient walking. Main street crosses through. A school may remain. A former town-center mill or commercial area may now be housing. Mature trees and shrubs populate the house plots. Thus, the older residential area appears rather well-vegetated, and supports many common bird species.

Newer Residential Area

A *newer residential area* typically covers a majority of a town's surface (Figure 1.9) (e.g., Cranborne, Astoria, Swakopmund). This area contains several neighborhoods of varying size, with local street grids at orthogonal and commonly other angles. Relatively young trees line some streets and are scattered along others. Single-family units on plots often with lawn, flower gardens, and young trees predominate (Girling and Helphand, 1994; Girling and Kellett, 2005; Forman, 2014). Some irrigated vegetable gardens are normally present. Where the newer residential area was former farmland, agricultural weeds are abundant. Some low-rise multi-unit housing is present. A tiny area of high-income homes exists, often on the upwind side or by water. Schools are present, so bus, car, bicycling, and walking routes radiate outward from schools. Ballfields and local parks are often few in number.

Variability in neighborhood size, block size, house plot size, and relative cover of land uses within a plot is high in each case. Some variability results from topographic heterogeneity such as remnant elongated wetlands and flood-prone locations. Further variability follows from past town decisions, plus preferences of individual residents over time. The newer residential area thus is extremely heterogeneous.

Green wedges or corridors (stream, highway, railway) commonly divide the town, and the newer residential area, into sections (e.g., Icod de los Vinos, Sandviken, Rock Springs). A water tower or tank may be conspicuous in this newer residential area.

Edge of Town

The *edge-of-town* is usefully differentiated from the newer residential area because of differences related to the surroundings outside the town (Figure 1.11) (e.g., Kutna Hora, Plymouth). Recall that the town is considered to be the relatively contiguous built area, so scattered buildings beyond the town edge are considered to be close to, but not in, town. The edge of town is mainly lined with residential homes, commonly on large plots, which are heterogeneous, even idiosyncratic, in layout (Figure 1.9). A few dense mobile-home neighborhoods may be present (Figure 1.2). In an agricultural landscape, the town edge may include a few farmsteads formerly surrounded by farmland.

Also, a commercial area, such as a shopping mall, or in a large town a business office center (Serret *et al.*, 2014), commonly occurs in the edge of town. An industrial or manufacturing center is often present in the town edge, especially near a main road

Figure 1.11 View of house backyard with pasture and forest beyond. Harvard Forest near village of Petersham, Massachusetts, USA. See Appendix. R. Forman photo.

(Forman, 2014). One or more institution area is common. Most strip (ribbon) development protruding outward along primary roads is effectively town edge. A number of special features including cemetery, dump, fairground, water tower or tank, sewage treatment facility, plant nursery, and local airport may have originally been constructed beyond the town, and now may be in the town edge (Forman, 2014).

Identifying the town boundary, and understanding ecological patterns along a gradient from town center to surrounding land, can be analyzed with ecological edge/gradient theory (Whittaker, 1975). Rarely is there an even gradient (straight line on a graph) of built-area cover from high to low. Rather, a relatively sharp rate of change somewhere along the gradient identifies the boundary and possible ecological thresholds. A relatively *abrupt edge* could be due to a row of somewhat similar-sized house plots on the outer side of a relatively straight or smoothly curving road. A more *convoluted boundary* exists where the outermost houses are intermixed with other features, such as cemetery, institution, and commercial center. Sometimes a narrow *mosaic-like strip of tiny patches* (Forman, 1995) may describe a portion of the town edge. Here tiny wooded or farmland patches exist in the outermost built strip, and/or tiny built features are present in the adjoining farmland.

Broadening the perspective, in Finland the similarity of bird communities in different habitat types within one town is about the same as that in similar habitats of different towns (Jokimaki and Kaisanlahti-Jokimaki, 2003). Furthermore, bird community similarity between single-family-house areas is higher (54%) than that between different town centers (30%). House sparrow, great tit, and corvid (*Passer domesticus*, *Parus major*, *Pica pica*) predominate in residential areas of the different towns.

Avian species richness varies in different areas of town (Sears and Anderson, 1991). In Cheyenne, Wyoming (USA), bird species richness increased from town-center commercial area to new residential area to medium-age neighborhood to old residential area.

Varied Town Anatomies

An intriguing range of town anatomies (Figure 1.12) can be compared with the preceding representative case (Figure 1.9).

Waterfront Towns. Both the arrangement and flow rate of water are strong determinants of town shape. In a feedback system, the town affects both of the water characteristics.

River, lakeshore, coastal tourist, fishing, and port towns with a boat harbor are basically elongated semicircles, with two long radii and one short radius. A long waterfront means that most buildings are relatively close to water. The main shoreline road often bends away from the waterfront to pass the town close to its edge. The town center is elongated along the waterfront. Short roads connect from the center to other parts of town. The overall design produces considerable traffic congestion. The former old coastal road is lined with dense waterfront development. Sometimes there is a railway along the coast with a station in the town center. These are characteristic of tourist and fishing towns. Riverside towns typically have a bridge. Many older homes and buildings, high-income homes, and multi-unit housing have good water views.

Storms, floods, and damage are common, though often some of the waterfront has been protected for natural conditions. Roads and some pipes drain stormwater and its pollutants to the waterfront. The town fits somewhat with topography, generally with a curvy upslope boundary and parallel drainage channels flowing down to the waterbody. Some industry is commonly by the waterfront and pollutes the water. A coastal town (Seaside, Florida) was planned in a way that extensively degraded coastal habitat (Forman, 2014). To protect nature and provide recreational opportunity along a coastline, priorities from highest to lowest are to (1) limit the number of access points, (2) make the access points wonderful for recreationists, and (3) control the number of recreationists (Forman, 2010).

If there is no protected harbor on a coast or lakeshore, jetties typically extend out to create a harbor against storm waves. Docks (wharfs, quays) and warehouses are habitats with distinctive species thriving. The waterbody is contaminated by upslope surface, ditch, and pipe runoff of wastewater, stormwater, agricultural chemicals, eroded

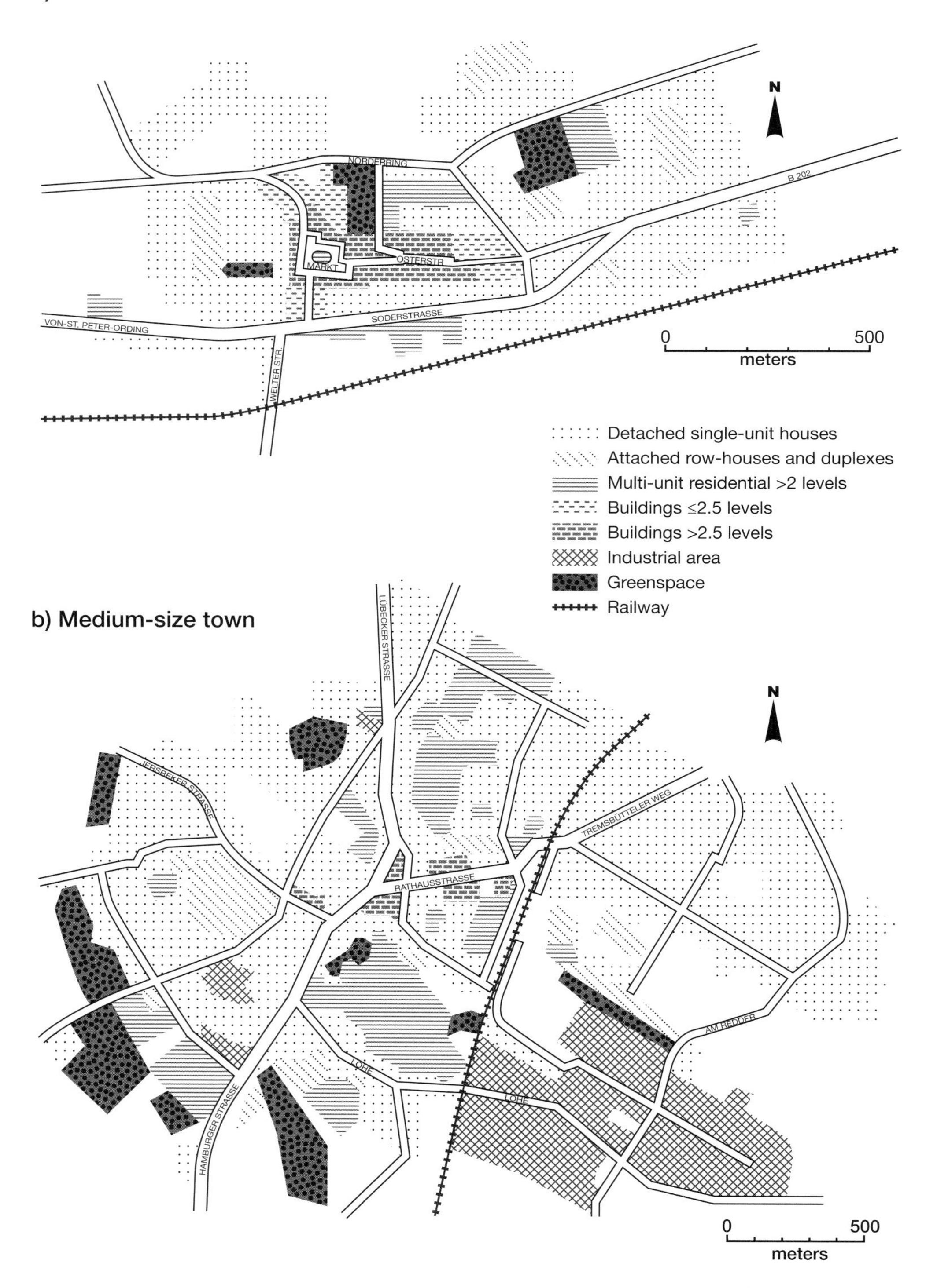

Figure 1.12 Anatomy of a small town and medium-size town. (a) Town of Garding, with estimated population 2,000; surrounded by cropland; far from a large city (Jebb, 1987). (b) Town of Berteheide, estimated population 11,000; surrounded mainly by pastureland; 27 km from Hamburg (Ekert, 1987). Schleswig-Holstein, Germany.

sediment, manufacturing outputs, commercial wastes, and solid waste. Boating activities and shipping/warehouse materials add to the polluted mix. A moderately large water pollution plume is characteristic, and a medium-size industry may also create an air pollution plume. Much is concentrated in the harbor in addition to boats, fishing/shellfishing, recreation, and ferry travel.

Corridor-Dominated Towns. Elongated, even linear, towns often spread along streams, rivers, coasts, and roads. A stream-dominated town often has a dam and pond providing hydropower to run one or more mills. A farm/cropland town and a cattle/sheep town are usually at a crossroads, which includes a primary route to markets. A forestry/logging town is located downslope and near several logging areas, and by a major route to markets. Such towns are conduits with strong flows/movements and directionality. The barrier or filter effect means that flows/movements across the town are slow.

Prominent Site/Feature Towns. A prominent feature at a site is typically in the town center or near the town edge. Towns normally have one or more special feature, such as a historic building, castle, famous former resident, beautiful walkway, historical event, or sacred site. The feature is highlighted and attracts visitors.

Thus, a desert oasis town with a specific water source is surrounded by tree shade, food production, and uncommon species (e.g., Alice Springs; see Appendix). A company town is dominated by one industry (e.g., Palmerton) (White, 2012). A prominent building such as a fort/castle or religious structure attracts tourists and residents alike (e.g., Tomar). A marketplace attracts vendors and shoppers from the town and surrounding communities (e.g., Tetbury). A mine pit may be dramatic (e.g., Falun).

An edge-of-town feature may be an attractant, such as an industrial center, shopping/commercial center, or institution. Alternatively, it may be a repellent, such as a mine, wetland, dump, sewage treatment facility, quarry, or steep slope.

Town Molded by Surrounding Flows. Suburban towns, hillside towns, and towns in sections are strongly affected by flows/movements from the surroundings. For instance, a suburban town with straight boundaries suggests competition with adjoining suburbs. The town is denser on the city side, and more parkland/semi-natural land exists on the distant side, especially where three or more towns converge. A major highway passing through or close to town creates a barrier, noise, habitat fragmentation, and roadkill (e.g., Socorro, Concord). Commuter railway, trains, station, and car park mold the conditions of a town.

A mountain or hill town is strongly affected by gravity and things moving downslope. Water, from scarce source to major flooding, dominates the flows. Wind and eroded soil follow suit. But vehicles, walkers, and wildlife are also strongly affected by the slope.

A town closely surrounded by low-density houses may have an indistinct boundary. Low-density houses come in at least three forms: sprawl (e.g., houses surrounded by lawn), dryland enclosure houses (e.g., enclosures for farm animals), and homegarden or vegetable-garden houses (e.g., productive house plots mainly for vegetable/fruit production).

Town with two sections may be dissected by stream, wetland, canal, flood zone, ridge, road, or railway. Some towns have three, four, or more sections separated by green wedges and transportation corridors.

Other town anatomies are distinctive including retirement, company, planned, native people, high-income wealthy, low-income poor, historic, and abandoned/ghost towns.

Spatial Indicators of Ecological Conditions

The presence of one object type often suggests the presence of another. A dump attracts wildlife (bears) including dense populations of common species (rodents, gulls), so such wildlife may live near a dump. A tiny area of high-income homes in town is usually upwind or by water. Some components (cemetery, dump, small airport, fairground) in the newer residential area or town edge, were formerly located out of town, but town expansion has put them in town today. Many small land uses in the adjacent zone are there because of the proximity of people. Apartments are often near parks and ballfields. Housing often locates near protected lands (Prados, 2009b; McDonald *et al.*, 2009; Radeloff *et al.*, 2010; Roca, 2013).

Sacred sites, as well as sacred landscapes or mountains, may be important organizing forces, and of considerable ecological interest (Pungetti *et al.*, 2012). Springs, ponds, special trees, or particular rocks may be sacred sites, while common sacred landscapes are the habitat areas for wildlife, such as wolves, certain raptors, or special hole-nesting birds. Other sacred sites may be those of special historical or cultural importance.

Spatial and other patterns often suggest ecological conditions. In the small city of Dusseldorf, Germany, 38 common habitat types were identified (from a total of 75 types) (Godde *et al.*, 1995). The species richness of five plant and animal groups was plotted against the degree of design/planning/management/maintenance (Forman, 2014). The diversity of all five groups decreased with more design/maintenance.

Many simple spatial measures may be useful in evaluating the ecological effects of town/neighborhood layouts (Nolen, 1927; Kendig *et al.*, 1980; Friedman, 2014), such as: (a) percent of equal-sized rectangles; (b) percent of blocks with back alleys; (c) percent of house plots with adjoining back-line corridors; (d) distance from residences to town park; (e) percent low rise, duplex, single-attached, and single-detached units; (f) density of tiny neighborhood parks; (g) spaces for stormwater basins; and (h) location of wildlife corridors.

Town anatomy changes over time (Reps, 1965; Francaviglia, 1996; Albrecht, 2014), even rapidly (Greenberg and Schneider, 1996) (Chapter 3). Thus, over 25 years two southern Japan towns changed noticeably in vegetation types, orchards, working paths, people by age, farmer/fisherman ratio, and farm products (Nakagoshi and Ohta, 2001).

The street network is readily measured by road widths, short no-outlet streets, road density, size hierarchy, connectivity among neighborhoods, access to the central public space and main road, and overall neighborhood network form. Road ecology aids in ecologically interpreting such patterns (Forman *et al.*, 2003; van der Ree *et al.*, 2015).

The especially old and venerable trees represent the effects and nurture by generations of town residents. Such trees provide shade, but more importantly, are often scarce habitats for the biodiversity of mosses, lichens, fungi, beetles, and vertebrates nesting in cavities. The old trees may represent an historical or cultural event, and often residents treasure and have a personal linkage with venerable trees in town.

Where is the hottest place in town? How wide are strips affected by traffic or railway noise? Where in town is the highest microhabitat diversity, and therefore probably the highest biodiversity? Where are flood-prone zones? What do dog and cat densities suggest about different built areas (Forman, 2014)?

Ecologists use numerous useful assays for understanding patterns and processes, including species richness (number), evenness, competition, plant traits, and ecological traits (e.g., epiphytes, guilds, fruits). Bird foraging traits include migratory strategy, nesting substrate, and body weight. Moth traits include larval host specificity, overwintering stage, flight period, body size, and host plant distribution.

For urban planners, common measures of built areas may highlight population density, building density, impervious cover, and vegetative cover (Joyce Rosenthal, 2013 personal communication). Ecologically evaluating small communities might consider: single and double tree-lined blocks radiating out from greenspaces; distance and number of blocks between residences and public greenspaces/parks; percent of blocks more than some distance away from a park; and connection between town center and a green corridor or wedge extending outward. How important are blocks with rows of house plots suitable for a narrow back-line green corridor? The preceding assays are simply examples to highlight the ease of ecological research in towns and villages.

Villages and Hamlets

Most of the preceding concepts could also refer to villages. As suggested at the beginning of the chapter, the spatial arrangement of a village is easy to understand, compared with a town's diversity of buildings and complex layout. Villages, as smaller population centers with at least one common building and usually a common green, are often simply described as linear or enclosed/compact/squared (Sharp, 1946; Roberts, 1982; Wood, 1997; Arendt, 2004). Villages are generally too small to sustain a factory.

Two main village types are widely recognized: linear and compact (Sharp, 1946; Arendt, 2004). *Linear villages* (e.g., Selborne, Rangeley; see Appendix) are thoroughly affected by and dependent on the surroundings (Figure 1.13) (Johansson *et al.*, 2008). Although some linear villages line a river, most contain a road, rather than being sliced by it. A gentle curve or sudden turn or rise makes a place, as well as a good location for a special feature, common building, or village center. Features, such as a prominent house, pond, or cluster of distinctive shops, may punctuate the roadside and the village.

In contrast, an enclosed or *compact village* (e.g., Petersham) is more oval in shape and has a more distinct center, often with a village green, small market building, and water source such as a well with pump (Wood, 1997). Traffic must move slowly through a village.

Figure 1.13 Linear village with rather stable population of 119 and distant church the only common building. All houses adjoin both road and pastureland, mostly elongated fields, as in medieval times. Land slightly slopes to Baltic Sea on right; ditches on left in the groundwater emergence zone drain water away. Livestock in and around barns in early April; tank in foreground for livestock waste. Handicrafts made, but no manufacturing. Planted coniferous woods in distance. Stenasa, Oland, Sweden. Photo courtesy and with permission of Lars Bygdemark, photographer. I appreciate the assistance of Margareta Ihse and Ulf Sporrong. (A black and white version of this figure will appear in some formats. For the color version, please refer to the plate section.)

A planned village appears quite different and is uncommon and excluded here. Instead of the normal organic development of the village over time, the arrangement of houses, shops, and other features is somewhat regular and formal, with occasional flairs or complex patterns. So-called eco-villages, utopias, communes, and collectives, mostly of a few hundred people and mainly spiritually centered, are also excluded (Litfin, 2014). These rare places are largely planned, have distinctive social attributes, and are apparently quite environmentally sustainable.

In the 1940s England had 10,000 villages and hamlets (Sharp, 1946). Virtually all grew "organically" from an earlier farmstead. Planned villages were rare and planned formality almost absent. Considerable variability was present. For example, they were either a row or an agglomeration of homes with or without a village green, simple organic in form or with a rectangular road network, and with a single or multiple local points and meeting places (Arendt, 2004). While not directly linked to ecological attributes, the most beautiful villages in Europe have been recognized (Vaz *et al.*, 2013a).

Various indicators of an effective village (settlement) size have been suggested (Arendt, 2004): a five-minute walk from distant houses to the center; about 1 km^2 (100 hectares; 250 acres) including greenspace; 250 houses; or about 600–650 people. Villages are usually large enough to support a few shops for most daily needs. Without such characteristics, a village needs a good access road to a nearby town with shops for all daily and most weekly needs.

The enclosed or compact village green is somewhat isolated from the surrounding farm or natural land (Wood, 1997; Arendt, 2004). Settlements in New England (USA) generally developed with scattered farmsteads tied together for walking, and then became centered and more tightly integrated by a meeting house. Later development near the center provided enough people to support a few shops. Continued growth was mostly "organic," i.e., occurred without an overall centralized village plan.

A *hamlet* is normally village-like in appearance, but smaller and with no common building (Arendt, 2004). In the USA a hamlet is typically one or two rows of houses on small plots facing each other across a road (Arendt, 2004). All backyards adjoin farm fields, pastures, or natural land. A break in a road's sight line is characteristic of a hamlet. Although some hamlets are new, most are old, maintaining their integrity rather than growing and metamorphosing into villages.

Houses in hamlets and villages are close to the road, whether house plots are short or long. Thus, in the rear space, vegetable gardening is common. Also, many or all house plots adjoin farmland and/or natural land where wildlife are common.

Land Surrounding Towns

Adjacent Zone and Surrounding Land

Together these two areas surrounding a town are analogous to the zone of influence referred to with Von Thunen bands and central place theory (see above) based largely on economic dimensions.

Adjacent Zone

A town's outside *adjacent zone* is the adjoining area of intense environmental interactions (flows/movements in and out) with the town. This zone often contains scattered farmsteads and houses, as well as mixed-use buildings especially near well-traveled roads. A relatively irregular road network close to town usually includes both paved and gravel roads (see book cover). While a stream network remains in place, stream-corridor vegetation ranges from good to nearly absent. Culverts and bridges are at road–waterway crossings. Recreational paths and sites are common in the adjacent zone. Active farm fields on good agricultural soil with scattered patches of semi-natural vegetation, are characteristic of an adjacent agricultural landscape. The straight lines of roads, paths, ditches, pipes, fences, powerlines, and property boundaries mix and conflict with curvilinear topographic patterns. Water, wildlife, and people movements are channelized by this rectilinear pattern.

The width of an adjacent zone seems to be commonly two to a few kilometers (about 1.5–3 mi) wide. That distance includes an array of effects, such as the distance: many residents walk outward on a path; common foraging by animals; wells affecting groundwater; strip development extending out main roads; most nonnative plants spread; bird feeders attracting birds; pet dogs' explorations; most town noises are audible; penetration by most direct town lights; town industry pollutants are conspicuous; location of cemetery; dump location; firebreak is burned; seeds are dispersed by wind or animals; and aroma of spreading manure on fields reaches the town. The adjacent zone normally includes most of the scattered houses outside the compact town built-area.

Special features may pockmark the close-to-town adjacent zone, and create considerable habitat diversity. Common examples are: about five to six primary roads radiating across the area; an edge commercial-shopping-area car park (the town's largest impervious surface); quarry for construction gravel; golf course; wetland; flood-prone area; farm stand for selling local agricultural products; hedgerows; fences; and agricultural and roadside ditches (Forman, 2014). Few towns have an adjacent zone with a: mobile-home neighborhood; mine; or sprawl associated with highway or strip development (Davies and Baxter, 1997; Gutfreund, 2004).

Outer Surrounding Land

Farther out, the *outer surrounding land* or sphere of influence, while less predictable, is extremely important to the town. This area extends outward until there are no significant interactions with the town, or interactions with a neighboring town predominate. Scattered low-density houses, more near roads, and perhaps an occasional hermit's abode are typical. Road density is low, and more of the roads are gravel or simply scraped soil. Bridges, culverts, signs, and numerous other small anthropogenic structures embroider the landscape, providing countless diverse microhabitats. Natural areas, hills/ridges/mountains, and recreation areas are magnets for some town residents. Villages and hamlets exist in the town's sphere of influence. Streams and pond undegraded by the town are present. The stream, river, or reservoir water source for the town is commonly piped in from the surrounding land.

This outer surrounding land is effectively a convoluted ring extending from the adjacent zone outward to about one-half the distance to neighboring towns.

Additional common features of the surrounding land include: sand/gravel extraction/quarry area; mine; primary radial roads; natural land (woodland, forest, grassland, desert) widespread or predominant; clean water-supply source; unpaved roads common; waterbodies; and protected natural and recreational areas and sites.

Agricultural Land and Natural Land

Farmland Structure

An agricultural landscape contains diverse important structures (Chapter 10). Town, village, and farmstead for inhabitants. Cropland, pastureland, and fragmented woodland. Groundwater, streams, irrigation canals, filled or drained wetlands, and water points. Farmstead barn, grain or hay storage, outbuildings, farm pond, dump. Farm animals such as chickens, sheep, cattle, horses. Fields, fences, hedgerows, ditches, furrows. Tractors, plows, harvesters, trucks, all-terrain vehicles. Rodents, flocks of birds, large herbivores, pest insects. Diverse other features characterize the agricultural landscape, such as feedlots for cattle, tall grain-elevator buildings, slaughterhouses, and farm equipment stores.

Agriculture uses some 40 percent of the global land surface, including 70 percent of the freshwater used, and produces about 35 percent of the greenhouse gases affecting global climate change (Thornbeck, 2012). Essentially all of the preceding objects and groups have interactions with farm towns and villages. Their arrangement within the adjacent zone and surrounding land strongly affects the communities.

In the southern Kenya grasslands, wildlife populations were compared in a mobile pastoral area and an adjacent area that was subdivided for development and settlement (Western *et al.*, 2009). The number of homes (huts) remained approximately the same in both areas before, during, and after settlement. In the first area with livestock grazing shifting seasonally from place to place, grass production continued little changed, and wildlife populations increased somewhat. In the second area following settlement and continuous livestock grazing, grass production dropped, and wildlife populations decreased. Settlement (sedentarization) led to more intensive human land use, and a threat to wildlife.

Natural Land Structure

Natural land includes tundra, but around towns and villages is mainly forest/woodland or desert (Chapter 11). Forest/woodland has ample precipitation to grow trees and support flow in visible streams, whereas water in desert grows low dispersed plants and generally flows slowly through the soil. Flash floods especially occur in deserts.

Natural forest, tree plantations, and agroforestry perhaps have somewhat similar effects on towns and villages. Forest/woodland provides cover and often has many recreation sites. Weekend/holiday houses are particularly located in forest (Prados, 2009a; Roca, 2013). Indeed weekend/holiday houses often aggregate near protected areas

(Radeloff *et al.*, 2010; Forman, 2014). People's use of such houses commonly degrades surrounding habitat to various degrees.

Wildlife populations also benefit from woodland cover. Relatively distinct wildlife routes or corridors through forest may lead to towns and villages. Some, though not strong, evidence exists that poor residents of the communities gain significant income from surrounding biodiversity (Vira and Konteleon, 2013). More income is derived from biomass. Ecosystem services related to forest/woodland (Roe *et al.*, 2013) are considered in Chapter 4.

In deserts, towns and villages share the same locations favored by wildlife, rare species, and recreationists. Basically scarce sites with surface or near-surface water attract virtually all uses in the desert. Consequently, continuous conflicts related to water characterize desert towns and villages.

Spatial Models

Ecologically, as well as for people, towns and villages are key components of rural and remote landscapes (Jackson, 1952; Smailes, 1957; Young, 1999; Kosek, 2006; Thornbeck, 2012). Numerous spatial models are available to understand areas surrounding a town (Forman, 2014). Consider the rings of dachas, community gardens (allotments), newer weekend/holiday houses, farmland, and woodland around Russian cities. These might be modeled as a "Russian doll" (Ioffe and Nefedova, 2000; Vaz *et al.*, 2013a). Different zones can often be recognized around towns, but they are few and irregular. As mentioned above, the two areas, adjacent zone and surrounding land, are particularly useful for understanding towns.

A "donut model" with only the town and its surroundings (Forman, 2008; 2014) is perhaps simplest. Or consider the town as a stable, expanding, or shrinking nucleus, with radial transportation corridors extending into the land. Or model the topography (land forms, exposures, soil types) and habitat diversity (Anderson *et al.*, 2016). Or the land-use patterns (Friedman, 2014). Or the energy infrastructure, powerlines, pipelines, turbines, and transmission substations.

Or determine the outer boundary of the town's zone-of-influence by estimating the distances that primary inward and outward flows go. That is how an urban region around a city is determined (Forman, 2008). The result may be circular in flat terrain. Or the zone-of-influence may appear oblong, extending plume-like in the direction of a main road or river, or elongate in a narrow valley.

A study in the Italian Alps related the species richness of several groups to many variables of the complex topography: three climate variables, four water habitats, five grassland habitats, twelve forest habitats, five human-created habitats (shrub plantations, orchards, arable fields/gardening, cultivated gardens/parks, developed areas), and two environmental variables (Dainese and Poldini, 2012). Plant species richness best correlated with human land use, showing a negative linear correlation with percent of agricultural and urban land. Plant diversity also correlated positively and linearly with habitat diversity. On the other hand, bird diversity best correlated positively and linearly with habitat diversity, and secondly with percent forest cover. Orthopteran richness (grasshopper and cricket species) and butterfly richness did not correlate well with any of these broad-scale variables. The insect diversity may be more sensitive to fine-scale variables related to natural vegetation and human land uses.

An experimental study, evaluating whether total habitat area or habitat fragmentation is more important for biodiversity, used arthropod community responses to the amount and fragmentation of clover habitat separated by mowed areas in Ohio (USA) (With and Pavuk, 2011). Overall, habitat area had a greater and more consistent effect on increasing species richness than did habitat fragmentation. The greatest increase in species richness occurred where more suitable habitat reached about 40 percent of the area, thus suggesting a useful threshold.

The spatial area or extent at which landscape structure best predicts a species population response has been called the "scale of effect." A modeling study indicated that dispersal distance is the best predictor of scale of effect (Jackson and Fahrig, 2012). Movement behavior was second best, and reproductive rate least important. A general guideline based on the model results suggests that, to maintain an animal population, the radius of a landscape should be four to nine times the median dispersal distance of a species (or 0.3–0.5 times the maximum distance). Movements of species that avoid gaps in habitat connectivity resulted in a smaller scale of effect (0.2–0.5 times) than species that move in fragmented habitat.

To predict the species richness of terrestrial vertebrates in the large Willamette Basin, Oregon (USA), a biological land-use model evaluated three key variables: (1) habitat preference, (2) area requirement, and (3) ability to disperse between habitat patches (Polasky *et al.*, 2008). Also, a spatially explicit economic model was used that incorporates site characteristics and location to predict economic returns for various land uses. To achieve high biodiversity (ca. 240–255 species) plus high economic return (ca. 20–27 billion dollars) in the area, an "efficiency frontier" is identified. This is the point or zone above which change in land uses is most pronounced. In this case, the amount of agriculture and forest wood-extraction decreases, and rural residential use and conservation land increase. Thus, molding the landscape to achieve both high biodiversity and high economic return ironically creates both better and worse spatial patterns on the land. Less food is produced, less wood produced, more sprawl occurs, and more land is protected for conservation.

Hybrid Nature–Euclid Patterns

This final section introduces: (1) universal, natural, and euclidian patterns; (2) hybrid patterns; and (3) ecological implications of spatial patterns.

Universal, Natural, and Euclidian Patterns

Three sets of spatial patterns, universal, natural, and euclidian (anthropogenic), are especially prevalent in landscape ecology (Forman, 1995; Turner and Gardner, 2015; Rego *et al.*, 2018). *Universal patterns* are apparently present in all landscapes. Common examples include patches, corridors, matrix, edges, abrupt boundaries, networks, hierarchies, aggregations, smooth curves, wedges, parallel lines, perforated area, and size variability. A study of 25 landscapes at several spatial scales identified other potential candidates (Cantwell and Forman, 1994).

Common *natural patterns* include oblong/elongated, irregular, richly textured interior and edge, curvy/curvilinear, convoluted with lobes and coves, dendritic networks, fractal patterns, gradient/gradual spatial variation. A circle (e.g., hill, bog, hammock, kettlehole, sinkhole) is a less common natural pattern in the landscape. Design and planning could increasingly use these for ecological benefits.

Euclidian geometry or anthropogenic patterns are also common in the landscape, such as straight line, right angle, rectangle, square, grid, diagonal, "T" intersection, cross, rectilinearly linked nodes, and irregular rectilinear network. Less common are dendritic straight lines (canal system, drainage ditch system) and circle (traffic circle/roundabout, central pivot irrigation).

Hybrid Patterns

Hybrid nature–euclid patterns combine the curves or irregularity of nature with the straight lines of euclidian geometry or human construction. Thus, the processes producing these patterns are complex. How the processes affect patterns, and how the patterns affect processes, are apparently unstudied. Hybrid patterns may be especially important outside towns and villages in natural land, where strong natural and human processes interact (Figure 1.14). Hybrids may be less common in intensive agriculture which has mainly eliminated natural patterns, or alternatively, hybrids may characterize this phase. Irrespective, interesting mitigation approaches may exist to deal with hybrids. The following illustrate hybrid nature–euclid patterns.

Corridors

a. *Linear–curvilinear route.* Road, railway, or trail following hilly/mountainous topography followed by a straightened stretch or in flat terrain. Curving waterway with channelized stretch containing rocks/concrete added along streambanks that accelerate streamflow, reduce stream habitats and fish, and increase downstream flooding.
b. *Constrained stream.* Bridge in road causeway crossing a stream and floodplain, with rock fill just upstream such that streamwater must flow at a determined point, under the bridge. Natural stream migration across the floodplain is eliminated for an upstream stretch and a longer downstream stretch.
c. *Side-sliced corridor.* Road, railway, pasture fence, or crop-field edge removes stream corridor or riparian vegetation on one side.
d. *Fragmented corridor.* Disconnections (breaks, gaps) in a strip reduce movement along, but may increase movement across, the strip. Roads, railways, and powerlines often cross natural corridors. Crossing may be enhanced or inhibited, but movement along is inhibited. Crossings may increase erosion/sedimentation, chemical pollutants, and noise in the corridor.

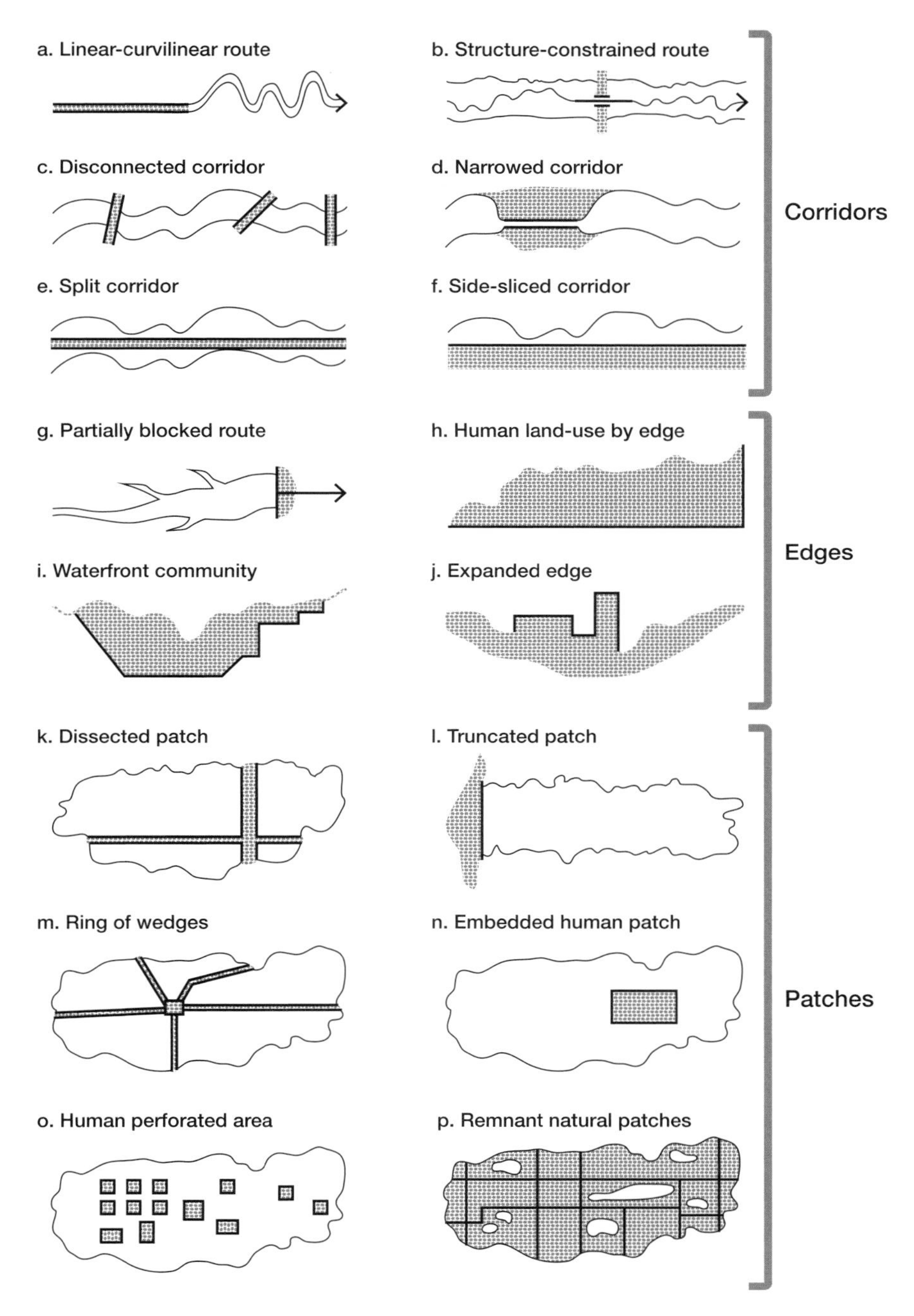

Figure 1.14 Hybrid nature–euclid forms in landscapes. Each form combines a nature-produced irregular or curvilinear line and a human-created straight line (often with a 90°right angle).

e. *Split corridor*. The corridor is dissected lengthwise by road, railway, powerline, or pipeline, for example along a valley. A stream corridor may be crossed many times with bridges, and is noticeably degraded by the human structure alongside.
f. *Narrowed corridor*. Towns and other streamside communities often narrow a stream for development on one or both sides using rock/concrete fill. One or more bridge typically connects opposite sides.
g. *Dam and pond/reservoir*. Dams may be for hydropower or flood control, and the dammed water for clean water supply or recreation. Streamside industries commonly use the water for hydropower, cooling, and waste disposal, as in many towns.

Edges

h. *Human land-use by edge*. Agriculture and development are often surrounded by both roads and natural land.
i. *Waterfront community*. Coastal, lakeshore, and riverside towns have the shoreline on one side and development or roads on the opposite side.
j. *Expanded edge*. Filled wetland and landfilling of urban waterfronts increases high real estate value land for development.

Patches

k. *Dissected patch*. A natural patch is sliced into sections by straight human corridors. Generally, the base of a "T" intersection is the newer corridor.
l. *Truncated patch*. Removing part of a natural area for human activity such as agriculture, development, or transportation leaves a straight edge.
m. *Ring of wedges*. Radial roads extending out from a town leave a series of wedge-shaped natural areas adjoining, even projecting into, the town. Both residents and wildlife use the wedges to move in and out of town.
n. *Embedded human patch*. Large or small anthropogenic spaces in a natural area degrade the resulting donut-like area (Forman, 1995).
o. *Human perforated area*. A natural area with many small human patches, while not fragmented, is severely degraded ecologically. Low density housing, dispersed patch clear-cuts, and tropical shifting agriculture clearings are examples.
p. *Remnant natural patches*. Hills, ponds, and wetlands remaining in a housing development or agricultural land with a rectilinear road network are characteristic.

Ecological Implications of Spatial Patterns

Form and function principles introduced above are best known. Rounded patches best protect internal resources. Convoluted patches with lobes and coves best facilitate interactions with the adjoining area. Linear and network features are structured to facilitate long internal flows.

Many landscape ecology studies correlate ecological phenomena like biodiversity, water quality, mineral nutrient flows, insect abundance, bird nesting, and vegetation cover with a spatial pattern(s) or arrangement. For instance, the movement of lizards was correlated with the distribution of trees and other features in a residential area (Kolbe *et al.*, 2016). A correlation study found that wildlife movement was greater along a straight edge, and across a convoluted edge, between woodland and grassland (Forman, 1995).

Alternatively, in Nova Scotia (Canada), a small-field experiment to examine cause-and-effect compared movements of meadow voles (*Microtus pennsylvanicus*) across different edge types between mowed and tall grass areas (Nams, 2012). The voles crossed at concave edges twice as often as at convex or straight edges.

The land of towns and villages is extensive and bulging with human activities. It is difficult to do experiments demonstrating cause-and-effect where large areas and human activities are involved. Therefore, ecological phenomena are typically correlated with spatial pattern, which provides valuable evidence and conclusions. Science moves ahead when later rigorous studies of any sort support or add to our existing understanding, or produce different results.

2 Flows and Movements

… the whole country seemed, somehow, to be running … I felt motion in the landscape; in the fresh, easy-blowing morning wind, and in the earth itself, as if … herds of wild buffalo were galloping, galloping …

Willa Cather, *My Ántonia*, 1918

Flows/Movements around Towns

How does a town or the land work? Flows and movements basically provide the answer. People move, vehicles move, water flows, air moves, heat flows, birds fly, terrestrial animals forage and run, and seeds are transported. The distance, amount, and rate of such flows/movements, relative to the spatial patterns present, tell how a town or land works.

This topic is explored from four perspectives: (1) background principles in brief; (2) in and out of towns; (3) edge of towns; and (4) corridors and connectivity.

Background Principles in Brief

Land mosaics, indeed towns, as seen from an airplane window, are open systems created and maintained by energy flow (thermodynamic heterogeneity) (Forman, 1991). Today's solar energy and yesterday's fossil fuel energy mainly sustain and change the patterns around us. Some objects move (movement of matter) through the mosaic pattern by *passive diffusion*, that is, from areas of higher to lower concentration due to a concentration difference. Much more prominent is *active (diffusion) movement*, which uses energy to go in any direction, including from lower to higher concentration.

Three types of energy-based flow/movement are central. (1) *Mass flow* or transport uses outside energy. Thus, objects are moved by wind due to air temperature differences, and carried by flowing water due to gravity. (2) *Locomotion* by animals and people uses energy from food originally derived from plant photosynthesis. And (3) objects moved by *motor vehicle transport* use fossil fuel. Each of the three flows acts as a vector carrying objects, such as pollutants, seeds, and mud.

In natural land virtually all routes of objects are *curvy*, and some convoluted (Forman, 2012). Groundwater flow and animal migration routes often are gentle curves, whereas a meandering stream or slowly foraging animal traces a convoluted route. Human routes however tend toward *straightness* (Figure 2.1), which provides two advantages.

Figure 2.1 Roman-era aqueduct for carrying clean water to town. Tomar, Portugal. See Appendix. R. Forman photo.

A straight line is most efficient for getting from here to there. It also minimizes transport-caused degradation of the surrounding land.

Three types of mass flow are important in wind and water flows (Forman, 1995). *Streamline flow* more or less parallels a surface, like air across an airplane wing. *Turbulence* usually has strong up-and-down flows called eddies, which are familiar around rocks on the surface of streams and rivers. And *vortex flows* have somewhat cylindrical movement, such as a tornado or whirlpool.

Two other types of air flow may be present in the land. *Cool air drainage* down a hill or mountain may occur on a still night, because cool air above is heavier than the warm air below. And *breezes* move from a warmer to an adjacent cooler place, such as an onshore breeze in early autumn coming from a warmer sea onto a cooler land.

Rainfall hitting the land usually has three trajectories. Some soaks into the soil as *infiltration*. Some flows as *stormwater runoff* on or under a surface. And some moves upward in *evapotranspiration* to the atmosphere. Evaporation is from non-living surfaces such as soil, and transpiration from living surfaces, especially plants.

Plants mainly "move" as seeds and spores, carried by wind or animals. Indeed, vegetation in a town edge spreads seeds, spores, and pollen over the surrounding land.

Animal movement is more diverse and may be distilled into four major categories. Some species *defend* a relatively small *territory* around the nest or den. Most animals have a larger *home range*, within which they daily *forage* for food. When young animals mature, they leave the home range, look for a mate, and establish their own home range at some distance, a process called *animal dispersal*. Finally, some animals *migrate* cyclically, moving to one habitat and then back to the first, taking advantage of suitable environmental conditions in both locations, which are often far apart.

Finally, ecosystem flows mainly refer to the *flow of energy* through a food chain. Also, mineral *nutrients cycle within or flow through* an ecosystem (e.g., the biogeochemistry of phosphorus and nitrogen). The amount of energy and total biomass typically decreases from plants to herbivores to predators to top predators, as heat is progressively given off (á la the second law of thermodynamics). Somewhat analogous to the cycling of nutrients, system flows often highlight the inputs and outputs to a "black box," which may be the entire system or any component in a connected system.

In and Out of Town

A town functions as both source and "sink" (Pulliam and Danielson, 1991). As a *source*, countless objects are constantly dispersing outward from town, many with a destination or target. As a *sink*, arriving things collect, and some are disposed of while others persist. For some arriving objects, the town is a target or destination, and for some it acts as a magnet-like attractor.

Consider briefly the characteristic objects flowing in and out of towns.

Entering Towns

People: motor vehicle traffic, cars, school buses, trucks; commuters/workers; visitors; shoppers from surroundings, villages, and nearby town.

Species: herbivores/scavengers; large predators; pollinators of town plants; native and nonnative species from farmland; native species from natural land.

Water: clean water supply; groundwater; floodwater; sediment.

Materials/energy: dust from farmland; heat; fire; gravel/sand for construction; goods for commercial shopping; materials for industry.

Leaving Towns

People: same as above, except omit visitors and add recreationists.

Species: native town species; nonnative town species.

Water: stormwater; stormwater pollutants; sewage wastewater or residue; groundwater; road salt; industrial water pollutants; commercial water pollutants; sediment.

Material/energy: solid waste; industrial air pollutants; goods/products for city; agricultural products for city.

Some of these objects both enter and leave (e.g., trucks in Rumford and Astoria; see Appendix). Seemingly more things enter than leave. Towns are especially a sink for diverse species, and a source of different water-related flows. Larger towns have internal public transport, whereas villages have none (e.g., Sandviken, Petersham).

The concepts of attractants and repellents or avoidance are especially useful in understanding landscape flows/movements (Forman, 2014). Familiar *attractants* include food sources, water sources, cover against predators, pheromone sites, more suitable temperature places, and favorable smells. For many people and some animals, towns and villages are attractants. But *repellents* are also widespread, such as noise, vehicles, people, big dogs, unfavorable smells, chemical pollutants, large predators, and even towns/villages themselves as places to avoid.

Consider a town as a round or oval node with a center. Flows radiate in and out along radii or radial routes between the town center and outside of town (e.g., commuters of Frome, Concord). *Inward flows* converge, and concentrate. From town edge to center, house-plot sizes shrink, traffic grows, and things accumulate, with diverse ramifications. Yet concurrently, *outward flows* from town diverge, and disperse. Interactions, some positive, some negative, between inward and outward flows occur (consistent with countercurrent theory). Indeed, interactions between converging and diverging flows are highest near the center. Perhaps the town edge is an optimum place to reduce conflicts and smooth the flows.

Heterogeneity within a town and in its surroundings also plays a key role in the in-and-out flows. In town, a set of some four to six radial roads is normally superimposed on a fine-scale mosaic of tiny patches and corridors. Many movements/flows are channeled along the radial roads, but other movements such as wildlife avoid the radii, and depend on house-plot characteristics and scarce greenspaces.

Outside the town the land is also heterogeneous, but at a broader scale with diverse land uses. The arrangement and habitat quality, ranging from good to poor, in the town's adjacent zone noticeably affects the rate of convergent movement toward town, as well as divergent flows out of town.

The *straightening* of features in town and the surroundings strongly affects routes, and especially rates, of movement in and out. The gentle heterogeneous curves of nature are linearized. Ditches, pipes, roads, powerlines, fences, canals, railways, and property boundaries in and near town are mainly straight (Figure 2.2). Straight routes facilitate, indeed increase the rate of, most movements/flows. Yet herbivores generally prefer *curvy heterogeneous lines* for avoiding detection by predators. Nature recreationists also generally prefer curvy, more heterogeneous and biodiversity-rich, routes. Curvy channels are normally better for protecting wetlands, reducing incoming floodwater, and cleaning water pollutants.

Finally, think about a town's infrastructure (Forman, 2012; Pollalis *et al.*, 2016) – water infrastructure, landscape infrastructure, energy infrastructure, food infrastructure, transportation infrastructure, waste infrastructure. Diverse flows never end, a large essential component of towns.

Figure 2.2 Boardwalk in snow connecting car park across stream to neighborhood park in town center. Concord, Massachusetts, USA. See Appendix. R. Forman photo.

Edge of Towns

The narrow *town edge* as the outermost strip of a town contrasts with the town's surroundings. In addition, the town edge typically differs from the newer residential area just inside. The edge between built area and surroundings may be straight, curvy, or convoluted. It is a quite heterogeneous strip, commonly containing farmhouses, commercial shopping center, manufacturing center, cemetery, water supply facility, large-plot housing, medium-plot housing, mobile-home neighborhood, and/or other features.

Edge Attributes and Flows/Movements

Like a semi-permeable cell membrane, the town edge functions as a *filter* or permeable barrier, through which some objects move readily, some slowly, and some not at all, with crossing rates changing over time (Forman and Moore, 1992). The edge strip slows most crossing movements.

Typically *forest edges* are visibly different from the forest interior for several meters or a few tens of meters from the boundary, though subtle effects extend much further into the forest (Forman, 1995; Lindenmayer and Fischer, 2006). Dense shrubs or small trees (mantel) in an edge provide cover and attract wildlife (Forman, 1995). Indeed, *edge species*, those only or mainly in edges, may be common where edges are vegetation-dense or heterogeneous.

People or animals may move along an edge (Forman, 2014). Predators, such as hawks feeding on treetop squirrels, or mountain lions (*Puma concolor*) feeding on dogs wandering outward, are characteristic. Scavengers also like moving along edges. But herbivores may avoid movement along an edge, because a heterogeneous strip increases the chance of meeting a dangerous predator.

A convoluted boundary is basically composed of alternating lobes and coves (Forman, 1995). These may be wide or narrow and long or short. Looking more closely at an individual lobe or cove typically reveals a finer-scale series of lobes and coves along its boundary. This illustrates the *self-similar fractal* pattern, whereby a giant zoom lens would reveal similar patterns at a series of scales. A fractal edge is likely to be rich in microhabitats and species, but not a good route for movement.

A *straight edge* may have considerable movement along it, while few animals move along a curvy edge. Thus, deer and elk (*Odocoileus hemoinus, Cervus canadensis*) in New Mexico (USA) move in large numbers along a straight forest–grassland boundary (Forman, 1995). In contrast, hardly anything moves along a *convoluted boundary* with *lobes and coves*, either following the convolutions or going from lobe tip to lobe tip. Perhaps a foraging predator would move from lobe tip to lobe tip so it could easily see into the intervening coves.

In contrast, for crossing boundaries relatively few deer or elk seem to cross a straight boundary. Instead, large numbers of animals move in both directions across convoluted boundaries between woodland and grassland. The coves in a convoluted edge seem to offer protection for animals cautiously moving in and out. The New Mexico deer and elk mainly leave a convoluted woodland boundary into a grassland at the tips of wooded lobes (Forman, 1995). Entering the forest from the grassland, the herbivores may enter most commonly along the side of a lobe, rather than at the lobe tip or cove end.

Built Edges and Flows/Movements

Edges can be described as hard or soft (Figure 2.3) (Forman, 1995, 2014). *Hard edges* are relatively straight and abrupt. *Soft edges* come in three forms: (1) a gradient, gradually changing from one side to the other; (2) convoluted with lobes and coves, as just described; and (3) a patchy strip, with tiny habitat patches of one or both sides present. Soft boundaries are normally crossed much more readily than hard ones.

A study of birds along a hard edge between forest and suburb (with typical house plots containing several trees and shrubs) of Brisbane (Australia) provides useful insight (Catterall *et al.*, 1991). Twenty locations were sampled, each including a: (1) forest site (>250 m from suburb boundary fencing); (2) forest edge (0–15 m from boundary); (3) suburb edge (sampled along front spaces/yards/gardens of outermost houses); and (4) suburb sites (250–750 m from boundary). Bird species and densities differed strongly among forest sites, suburb sites, and combined edge sites. Birds in edge sites were abundant in both the forest edge and suburb edge, i.e., in both the outer 15 m of forest and the

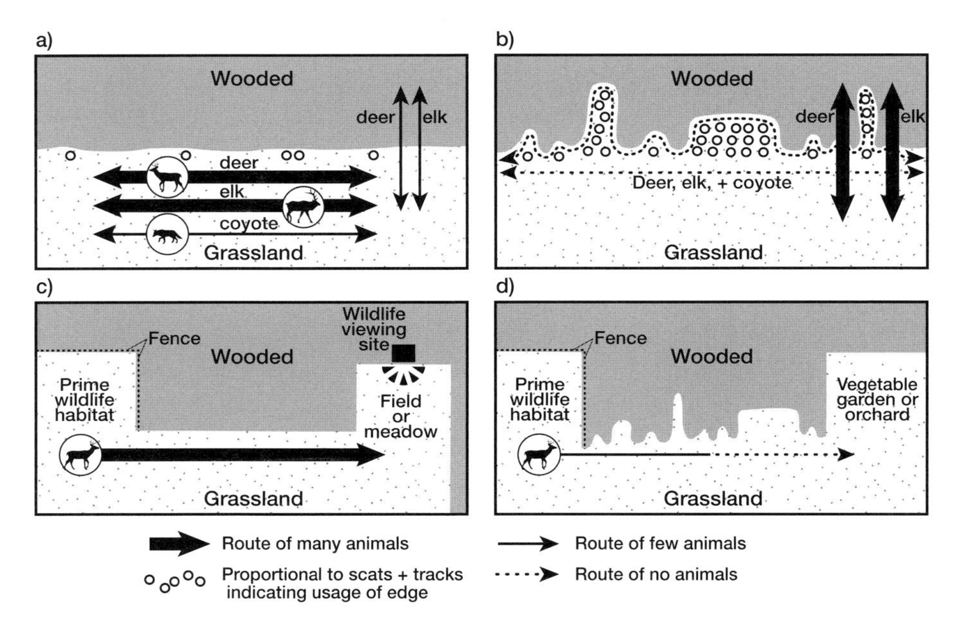

Figure 2.3 Wildlife movements relative to a hard straight edge and soft convoluted edge. (a) and (b) based on preliminary data by edges of pinyon-juniper woodland and grama-sagebrush grassland south of Tres Piedras, New Mexico, USA. Mule deer (*Odocoileus*), elk (*Cervus*), coyote (*Canis*). Adapted from Forman (1995). (c) Results from (a) and (b) used to stimulate wildlife movement patterns to a site. (d) Results used to inhibit movement to a site.

outermost strip of town-edge house plots. The edge species, a rather distinctive group, were mostly large and aggressive birds.

The forest species were mainly small birds feeding on foliage insects. Within the suburb, no effect of distance from boundary, or of vegetation structure, was evident for the low density of forest birds present. Finally, few forest birds were observed flying into the suburb, and few suburb birds flew into the forest. The edge effect, composed of outer forest and outer strip of house plots, plus the edge bird species present, functions as a strong filter between forest and built area. Many other studies have compared species densities to understand the role of edges separating contrasting types of land use (Gascon *et al.*, 1999; Rodriguez *et al.*, 2001).

A closer look considers the effect of housing density on edge crossings by birds (Sewell and Catterall, 1998; Parsons *et al.*, 2003). One study sampled a woodland–town edge, with samples located 20 m (66 ft) into the woodland side and 20 m into the outer row of house plots (Hodgson *et al.*, 2007). The presence and movement of birds was recorded for high-density housing (20–25 house plots/hectare; 1 ha = 2.5 acres) and low-density housing (1–2 house plots/ha). About half the bird species seemed to prefer edges by high-density housing, and half in the edges by low-density housing. No difference in bird crossings of edges by high-density and low-density housing was evident.

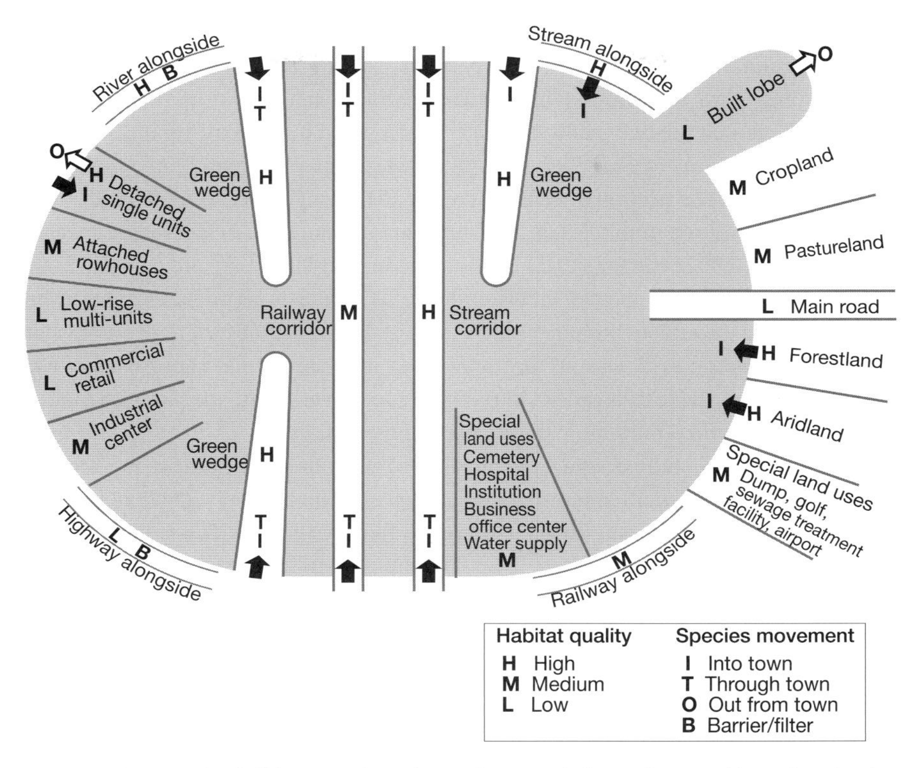

Figure 2.4 Expected wildlife routes in and out of town relative to form and boundary land uses. Town = shaded area; adjacent zone beyond.

However, crossings by bird feeding types differed. At high-density housing edges, omnivores crossed more than nectavores (feeding on flower nectar) which crossed more than insectivores. Low-density edges were crossed about equally by the three bird types. Also, more diverse vegetation in the house plots increased the crossing frequency between woodland and built area.

A broader view of the types of land uses in and adjoining a town edge provides further insight into wildlife crossing (Figure 2.4). Varied housing densities, commercial shopping and manufacturing centers, and various special features such as hospital and cemetery are commonly in a town edge. Adjoining the town may be cropland, pastureland, natural land, or special features like golf course and small airport. A river, stream, highway, or railway may go along the town edge or enter town as a corridor. The town may have a built lobe projecting outward or a green wedge projecting inward. These land uses vary widely in habitat quality, from high quality to wildlife avoidance.

Also, we can estimate the relative importance of each portion of a town boundary for species crossing into town. Natural land or stream adjacent to town, stream or railway corridor slicing through, and green wedge extending into town are obvious important routes for entering town (Figure 2.4). Several edge sections, based on adjoining inside-and-outside land uses, would seem to be secondary routes. The least likely places for species to enter towns seem to be across or along a major highway, and at some commercial shopping centers, and perhaps a manufacturing center. A built lobe may be valuable for wildlife moving outward from town.

Note that the two avian studies above suggest that characteristics within the outer row of house plots affect species-crossing frequency (Catterall *et al.*, 1991; Hodgson *et al.*, 2007). House plot characteristics such as heterogeneity, bird feeder, building location, and shrub/tree number and arrangement can increase, or decrease, species movements (Owen, 1991; Gaston *et al.*, 2013; Forman, 2014).

Finally, some species entering do not consider town to be the target, but are simply moving through to the other side. A river, stream, railway, or highway bisecting the town provides a corridor route through. River and stream are major routes, while highway, even with wide roadsides, is rarely a route. Very few species move along highways, especially in a built area, unless carried by vehicles.

A Dutch study found that movement of butterflies is most inhibited through built areas, while small mammals and large mammals are somewhat inhibited (Knaapen *et al.*, 1992; Forman, 2014). Compared to these groups, forest bird movement across a built area is least inhibited. Thus, perhaps normally butterflies could not cross through a town, whereas forest birds would have more success crossing the town's built area. Entering a green lobe and continuing on across town looks promising for certain species. Using a row of greenspace stepping stones is another possible route, if present.

Corridors and Connectivity

People live in a web of *corridors* (strips). The major types range from intensively used, relatively straight roads, railways, and powerlines, to streams, rivers, and semi-natural wooded corridors of varied width. Greenways combine linear recreational use and nature protection, and connect communities with the surrounding landscape (Jongman, 2004; Ahern, 2004; Benedict and McMahon, 2006).

Structural characteristics of the corridor strongly affect the rate and success of movement (Forman, 1983, 1995). Key internal attributes are a sharp environmental gradient from side to side, considerable lengthwise heterogeneity, and with or without an internal stream, road, or other entity. Externally, corridors are wide to narrow, continuous to discontinuous with gaps (or narrows), straight to convoluted, and having higher or lower vegetation than in the surroundings. Corridor *width* and *connectivity* are typically the two most important factors affecting flows, movements, and other functions.

Functions of Corridors

Six major functions of corridors mainly center on flows and movements (Figure 2.5) (Forman, 2015). As a *conduit*, corridors facilitate the movement of numerous natural and human objects/materials. As a linear feature through which objects flow, the corridor minimizes or prevents recycling and recovery (Evans, 2012). Stream and river corridors contain the waterway and adjacent strips (see Figure 1.2). Water, floating algae, seeds, sediments, and pollutants move downslope, while fish, invertebrates, butterflies, amphibians, reptiles, birds, and mammals move in both directions (Chapter 6). Wildlife

Green corridor patterns

a) Slicing through heterogeneous land

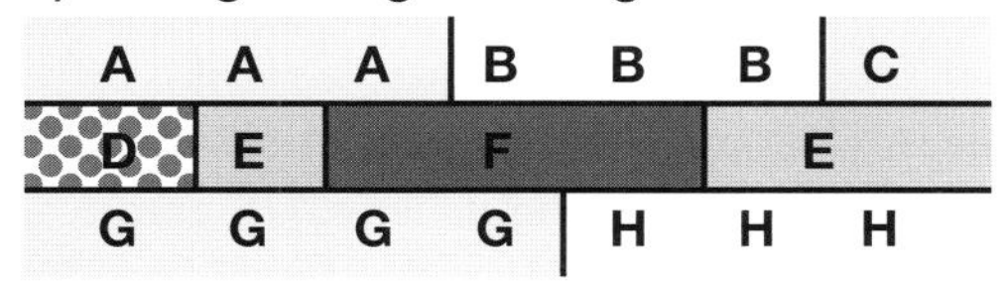

3 land uses north side
3 habitat types in corridor
2 land uses south side
7 of the 18 possible corridor environments illustrated

b) Edge species

Edge of A | Edge of B | Edge of C
Corridor habitat edge vegetation
Edge of G | Edge of H

Green corridor as center of a rich strip of edge species and vegetation

Functions of green corridors

c) Conduit

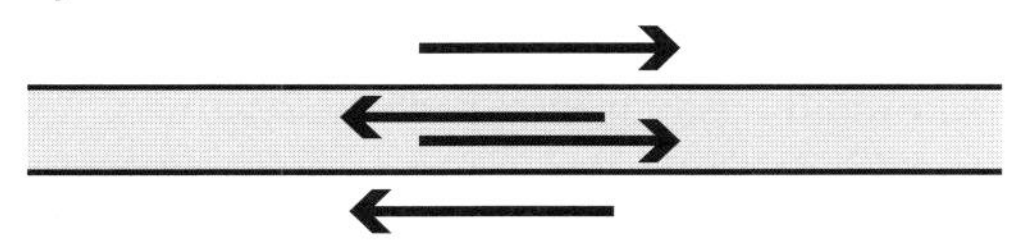

Movements along within corridor
Movements along close to corridor

d) Attached-node linkages

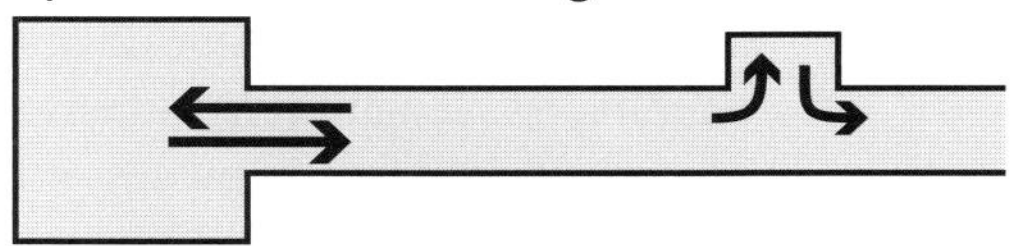

Large flows from, + small flows to, large patch
Small node as rest stop for wildlife moving along corridor

e) Filter and barrier

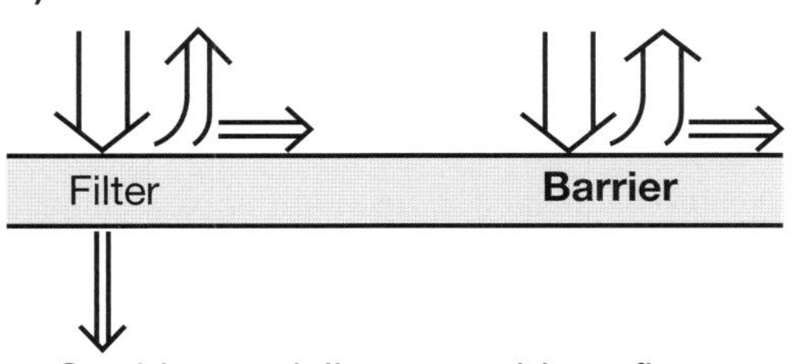

Corridor partially permeable to flows across the landscape
Filtration rate changing over time

f) Source

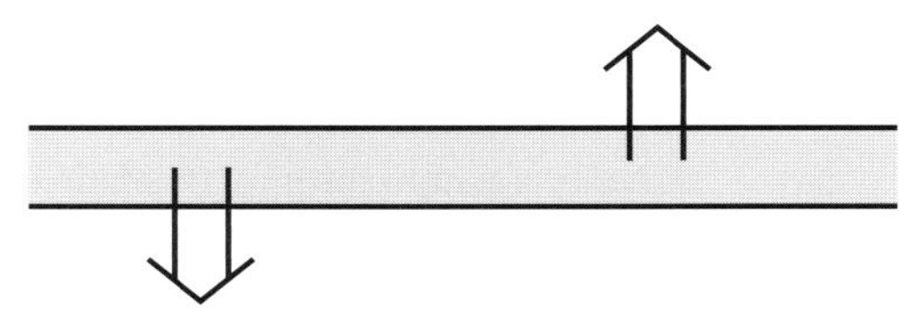

Positive + negative effects of corridor on surrounding land

g) Collector

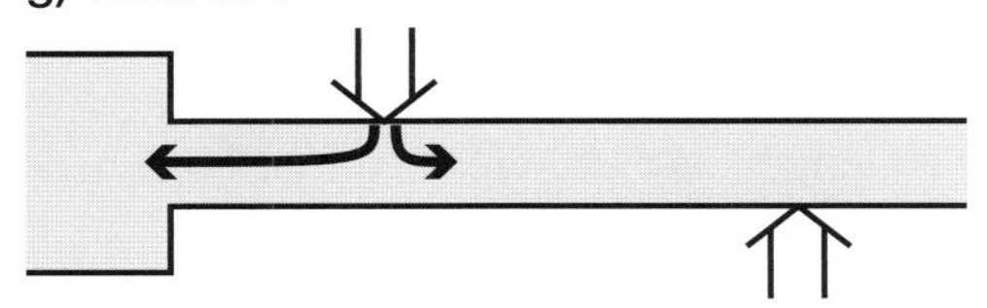

Catches objects moving across land
Drift-fence effect funneling wildlife to large attached node
Sink that absorbs flows

Figure 2.5 Spatial patterns and key processes or functions of green vegetation corridors.

move both within and alongside a hedgerow. Recreationists and wildlife move along greenways in urban regions. Walkers from town visit the surrounding land on paths/trails. Abandoned railways are converted to walking trails (e.g., Vernonia, Concord, Millstone/East Millstone; see Appendix).

Corridor interactions with attached nodes highlight the typical connections of a corridor with large nodes at its ends, and often small nodes along its length (Figure 2.5) (Bennett, 2003). Large nature-protected areas are connected by a wooded corridor with adjoining small woods along its length. Towns are connected by a road which goes enroute through villages. In these cases, flows go along the corridors the entire distance, or part way to an attached node. At each node, objects move between corridor and node. The quality of both corridor and node strongly affects the amount of interaction.

As *filter or barrier*, the corridor intercepts objects moving across the land. None cross if the corridor is a barrier. More typical is the filter effect, whereby some objects cross the corridor and some do not. Corridors crossing a town edge often contain considerable movement (e.g., tractors from farmland passing into Tetbury, Askim, Clay Center). Wildlife underpasses or overpasses increase the rate of crossing a highway corridor (e.g., Plymouth, Concord). The quality of a corridor includes the presence or absence of narrows and gaps/breaks, which increase the chance of objects crossing. Even if an object crosses and continues onward, the corridor slows the movement. For instance, in crossing a hedgerow a deer pauses to check for predators on the other side.

As a *catcher of objects moving* across the land, the corridor itself is enriched (Figure 2.5). This of course is most effective where the corridor is perpendicular to the direction of movement. More important perhaps is that the caught objects may then move along the corridor, and enrich an attached node (drift fence effect). If the node is large or a nature reserve, this helps sustain its biodiversity. The town of Ponoko, Alberta (Canada) has a stream-corridor park used by recreationists that has small attached nodes and leads to a large attached greenspace (Friedman, 2014).

Nonnative weeds were studied in corridors attached to three different semi-natural habitat types (Hobbs and Atkins, 1988; Panetta and Hopkins, 1991). The corridor from habitat #1 had both incoming seeds (seed rain) and weeds present almost only in the first few meters. Corridor #2 received a fairly constant seed rain from 0 to 90 m, peaking at 70 m from the habitat, whereas the weeds in the corridor were mainly at 35–55 m. Corridor #3 had most seed rain at 0–2 m and at 50 m from the habitat, but plants were primarily at 0–35 m and 65–70 m from the habitat. The data suggest that vegetation in corridors results from both catching species along its length, and by spread from a large node commonly at the corridor end.

Corridors as a *source of effects* on surrounding land have both negative and positive roles. In England, rabbits in hedgerow burrows go out and feed on farmers' delicious crops. Yet farmers also benefit from predatory beetles (carabids) which move out of hedgerows to feed on field insects, some of which otherwise could multiply to pest proportions and damage crops. Wine growers sometimes maintain strips or areas of meadow in a vineyard, from which pollinators go out and pollinate the grapes, increasing production. Traffic noise from busy roads inhibits sensitive birds for hundreds of meters outward. The list of corridor effects on the surroundings goes on.

Finally, the heterogeneous habitat within corridors supports an assortment of typical edge plants and animals (Forman, 2014). These are normally all generalist species and often a mix of native and nonnative species. Across a corridor is a sharp environmental gradient from one side to the other. Along the corridor is often an environmental gradient, for instance of moisture on a slope or distance from a waterbody. Also, patchiness along a corridor due to patchy soil, past disturbance, and so forth is common. For example, a natural corridor radiating outward from a town often has a gradient of human use, as well as disturbance, from high to low with distance outward. In effect, corridors channel flows/movements, block them, share them with attached nodes, and share them with the surrounding land.

Networks

Connecting corridors together forms a network (Figure 2.6). A *dendritic (tree-like) network*, such as a stream system, has familiar attributes including directionality, curviness, and converging/diverging corridors (Forman, 1995). The main alternative on land is a *rectilinear network*, such as a road or hedgerow network, again with easily seen characteristics including connectivity, circuitry (loops), mesh size, linkages per node, and corridor density. Both network types also have some characteristics in common, e.g., corridor width, hierarchy levels, and pattern of attached nodes.

Flows/movements in networks depend on all the preceding spatial attributes, perhaps especially corridor connectivity, width, and corridor density. Regional walking and bike trail networks in a town's surrounding land are good examples (e.g., Huntingdon, Concord, Tetbury). Heavy bicycle use interconnects towns around Xinguara (Para, Brazil) (Browder and Godfrey, 1997). Network resistance mainly relates to having to cross intersections and nodes, and move through narrows in corridors (Forman, 1995). High circuitry means many loops or alternative routes to avoid blockages. Water flows down a converging dendritic network, while a fish or deer may move in the opposite direction, up diverging narrowing network corridors.

Another useful concept is the *ecological network*. Here a main goal is establishing connectivity rather than corridors, and often no network is visible. An ecological network refers to a system of nature reserves with interconnections that make a fragmented land coherent and support more biological diversity (Jongman, 2004; Opdam *et al.*, 2006). Ecological connectivity refers to the ability of a species to move between locations (Mayor Farguell, 2008). While continuous corridors are normally considered to be the best way to move between sites, other options exist. Consider a cluster of stepping stones, a row of stepping stones, or a medium-quality habitat between sites. The movement route may be straight or quite convoluted, but the species of interest can successfully cross. Even though no corridor connects the sites, connectivity is sufficient to be part of an ecological network of functionally interconnected nature reserves.

Corridors play a key role in *habitat conservation*, which effectively has three top priorities, i.e., sustain: (1) large natural areas or patches; (2) connectivity among the large patches; and (3) waterbodies with protective vegetation strips alongside. The

Figure 2.6 Infrared image of small town with diverse corridors. Main roads through town; short green corridors in town; northwest/southeast walking trail on former railway bed; small freshwater stream at top; hedgerows in farmland; multi-lane highway on right; wide branching stream in saltmarsh lower right; tiny drainage ditches against mosquitoes in saltmarsh. Southeastern Massachusetts, USA. US Department of Agriculture, Soil Conservation Service photo. (A black and white version of this figure will appear in some formats. For the color version, please refer to the plate section.)

vegetation strips of number (3) are corridors. Number (2) is probably best achieved with corridors, though a cluster of stepping stones may be almost as good. Large green areas interconnected by green corridors, appropriately called an *emerald network*, is normally considered to be the optimum design (Forman, 2004a, 2014). With distinct corridors providing connectivity, the network is visible to all and thus potentially easier to monitor and sustain.

Flows/Movements through Land Mosaics

We begin with (1) background mosaic-flow principles, and follow with (2) some land-use pattern effects.

Background Mosaic-Flow Principles

Solar radiation enters the atmosphere which reflects and absorbs some of the energy. The solar radiation which penetrates the atmosphere, plus diffuse sky radiation, encounters the earth's diverse surfaces. Some energy is reflected and the rest absorbed (Forman, 2014). Energy is then emitted from the surfaces in the form of heat, which has three major effects: heat the air; heat the soil or surface material; and accelerate evapotranspiration of water molecules (mainly by plants) to the air. The relative proportion of these three functions varies by surface type (Forman, 1995). Thus, for an asphalt/tarmac car park most emitted heat raises air temperature, whereas for a meadow or woodland most emitted heat drives evapotranspiration.

Not surprisingly, different habitats and land uses have different air temperatures, mainly due to the relative amount of energy reflected (albedo) and energy absorbed. However, since warm air flows toward cool air, a horizontal movement called *advection* occurs. This affects temperature in the adjoining land use. Wind also blows heat horizontally. Imagine living just downwind of a large black car park in summer. Oven-like heat seems to broil you. Plants in a wetland receiving such heat evapotranspire at a high rate, which pumps water out of the ground. That lowers the groundwater surface (water table) which in turn dries out the plants. The horizontal flow of heat between sites by advection and wind causes these changes.

Particles, aerosols, and gases, including air pollutants, are also transported across the land mosaic by both wind and advection. Since heat rises toward cold outer space, upward moving air occurs over warm sites, especially in summer afternoons (Forman, 2014). Cool sites, such as wetlands, dense forests, and waterbodies, are most susceptible to deposition of the circulating particles (particulate matter) and aerosols (particles in droplets) (Forman, 1995). Air pollutants thus tend to accumulate in particular cool locations in the land. The town center, as well as commercial and industrial centers, are pollutant sources with upward moving air, while water-related areas in and around town receive some pollutants.

Hills and mountains provide the third dimension with several important ecological patterns and flows. Vegetation zones associated with progressively cooler temperatures upslope may be evident. North- and south-facing slopes differ noticeably in temperature, and therefore plant growth and vegetation. With rainwater infiltrating into the groundwater, most hills and mountains effectively contain an *underground pond* perched up high. Water is heavy, and the large downward hydrologic pressure means that wetlands, seeps, and springs commonly appear at the base of the hill. This former groundwater up high is both a benefit and a problem for a town next to the hill.

Windspeed is often 15–20 percent higher on the hilltop. Also, moist wind may be forced upward by the mountain where cooling causes precipitation (e.g., Icod de los Vinos; see Appendix), leaving a dry wind flowing onward and then downslope on the other side creating dry vegetation. On a still summer night without wind on a hill or mountain, cool air, which is heavier than warm air, flows downslope, i.e., *cool air drainage*, pushing the warm air below upward and out (Forman, 2014). For towns next to hill or mountain without development on top or on the town-facing slope, this cool air drainage is a free ventilation process, both cooling and cleaning the town air.

Rugged terrain provides concentrated high habitat heterogeneity. Biodiversity, which typically correlates with habitat heterogeneity, is also high. Wind turbulence is widespread. The diverse north and south slopes, hilltops, valleys, and so forth provide both protective cover and breeding sites for numerous animal species. Convoluted movements are common.

The *patch-corridor-matrix model* of landscape pattern highlights the *matrix* as the background habitat or land use within which patches and corridors exist (Forman, 1995). The matrix of course is heterogeneous but it is a fine-scale heterogeneity. The matrix may be continuous or subdivided, perforated/porous, or sliced with corridors. It has long fetches or runs, where wind accelerates, sheet flow of water over the surface may occur, seeds may disperse far, fires spread long distances, and predators can see far (Brady *et al.*, 2011). Scattered patches may represent danger to be avoided by wildlife of the matrix.

Patch boundaries may also indicate change. Normally patches with concave boundaries are shrinking, and those with convex boundaries expanding (Forman, 1995).

A built-area matrix in The Netherlands was found to be most inhibitory to crossing by butterflies, intermediate for small mammal and large mammal crossing, and most readily crossed by forest birds (Knaapen *et al.*, 1992; Forman, 2014). In contrast, a rural area was most easily crossed by large mammals and least suitable for forest bird crossing (intermediate for the other two groups). This suggests that for a rural town, forest birds in remote forest would have difficulty moving across the agricultural land. But upon reaching the town, the forest birds could readily spread across it.

Several principles apply to animals moving across heterogeneous land mosaics. *Traplining*, i.e., following the leader, is efficient for groups where, like bees foraging for nectar, a leader has learned the route or destination. More generally, movement across a mosaic decreases with more boundaries crossed (*boundary-crossing frequency*), more corridors crossed, more patches crossed, and lower-quality patches. Some animals may preferentially move through corners of rectilinear patches, where they can forage or scent-mark the spot for other animals to notice their presence (Forman, 1995). Wildlife, perhaps especially predators, often move through *convergency points* or coverts, where three or more habitats or land uses converge.

Use of a land mosaic depends on an animal's home range fitting with the scale of spatial pattern present. Thus, a small-home-range species might only move within a patch, and a large-home-range species might move well beyond the landscape where it raises young. Home-range size relative to scale of pattern is important in considering the success of wildlife within and surrounding a town.

A *metapopulation* is a population with individuals living in different locations or patches, which are connected by occasional movement of individuals between patches (Opdam, 1991; Morin, 2011; Forman, 2014). Over time, the individuals on one or more patch may disappear (*local extinction*), and later individuals *recolonize* the patch, a process called metapopulation dynamics (Figure 2.7). Local species extinction is most likely on small patches and isolated patches. Recolonization is most likely

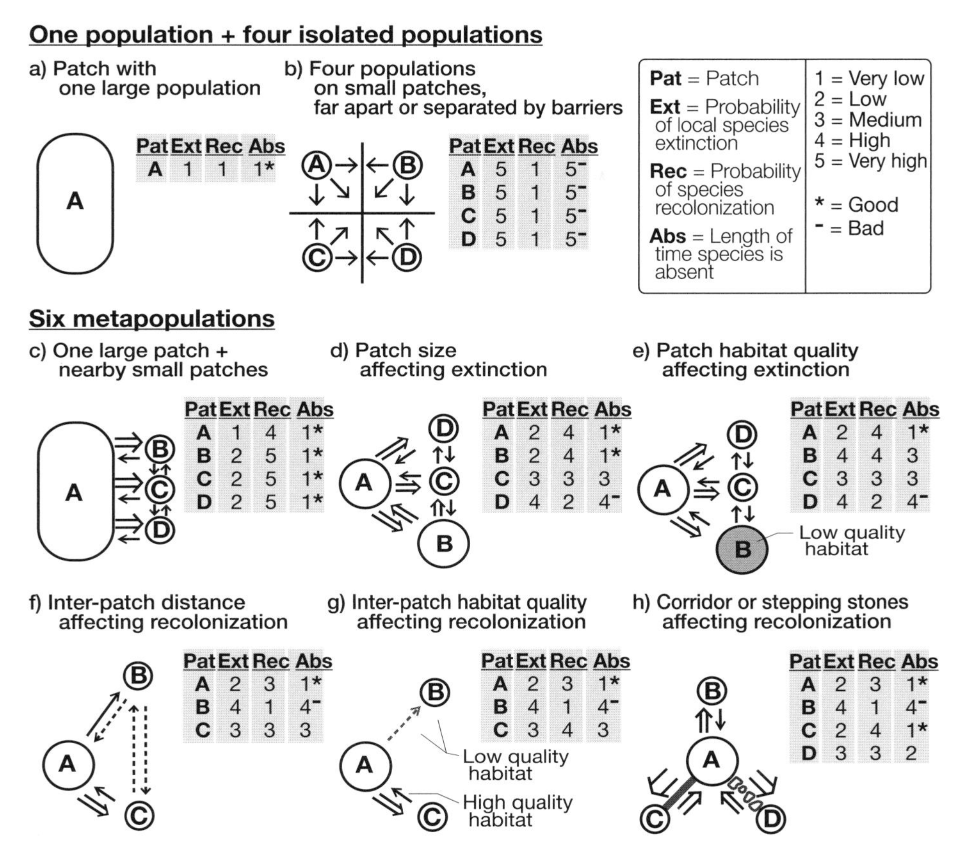

Figure 2.7 Metapopulations, local extinction/recolonization, and length of time species are absent from patches. Adapted from Forman (1995).

where patches are separated by a high-quality matrix, and where stepping stones or a corridor are present. In effect, patch attributes mainly affect local extinction, and matrix characteristics affect recolonization. Metapopulation dynamics is especially important in the area surrounding towns, as roads, housing, and other human activities fragment large natural populations in large habitats to small populations on small patches.

A useful spatial-movement principle emerges from geography's "near things are more related than distant things," plus animal behavior's rule that animals preferentially move toward more suitable habitats (Forman, 1995). Thus, we can predict that animals in a patch will preferentially move to nearby and high-quality habitats. Mapping habitats in and near a town then permits estimates of the major or potential wildlife routes.

Although some wildlife species live only in a single habitat, *multi-habitat species*, which regularly use two or more habitat types, are apparently most common. Some 90 percent of the common African large mammals, and more than 75 percent of the common terrestrial vertebrates in New England (USA) (DeGraaf and Yamasaki, 2001), are multi-habitat species. For such species the different types of habitat must be present, and be close enough for movement between the habitats. Furthermore, the route connecting the habitats must be suitable, not threatening, for movement.

Mammals that migrate seasonally between winter and summer habitats have the same requirements. Such animals also require *escape cover*, where they can quickly avoid danger, next to both habitats and especially the connecting route. Sheepherders moving sheep between low-elevation winter range and high-elevation summer range (transhumance), provide the protection along the connecting route (Farshad, 2001; Sow, 2001).

Habitat selection, when animals choose their nest or den site, determines where in the land mosaic the appropriate arrangement of habitats is located. Animals apparently choose a location based on two spatial scales: (1) the immediate surroundings which offer safety and can be defended, and (2) the broader area containing the habitat types suitable for its home-range foraging. The combination and arrangement of habitats in and around a town strongly determine what wildlife will be locally present, and where they will move.

Main roads radiate from towns across the land while fine networks of small roads, especially near a town, fill in some of the spaces. Road ecology, introduced in Chapter 12, highlights animal movements relative to roads and traffic (Forman *et al.*, 2003; van der Ree *et al.*, 2015). Wildlife approaching a two-lane highway with traffic may remain back from the road because of traffic noise, and perhaps vibration and danger. Road width is especially important for small animals. Some animals attempting to cross are road-killed while others reach the far side. Wildlife underpasses, increasingly incorporated in new and existing roads, are quickly used by animals to safely cross (Clevenger and Ford, 2010; Clevenger and Huijser, 2011; van der Ree *et al.*, 2015). Less common overpasses are even more effective. Culverts for water flow are widely used by small animals to cross under roads (Chapter 12). Small mammals and beetles move along roadsides, scavengers fly along looking for roadkill, and predators forage along rarely used, especially non-paved, roads. But overall, few animals move along roads or roadsides.

Finally, *landscape genetics* studies the distribution and flow of genes relative to land pattern, and is rapidly growing because genetics tells much about population processes (DeYoung and Honeycutt, 2005; Sannucks and Taylor, 2008; Sannucks and Balkenhol, 2015). In sampling the tissue of an organism, typically two components of genetic variation are estimated: the (1) amount of genetic variation (or diversity) present, and (2) spatial distribution of the genetic variation (i.e., genetic structure or differentiation). The amount of variation suggests the relative capacity to survive or thrive in diverse environmental conditions or with major change, as well as the threat of inbreeding problems. The spatial arrangement of genetic material provides insight into sources of successful reproduction, parental lineages, and gene flow across the land.

The degree to which species dispersal indicates gene flow is illustrated by predators that crossed a California multi-lane highway (Riley *et al.*, 2006). Genetic analysis found very little gene flow across the highway. This indicates that, although a few predators crossed, little breeding occurred on the other side of the highway. Also see Chapter 8.

Some Land-Mosaic Pattern Effects

A large-city study in Berlin provides quite interesting results applicable to greenspaces, such as parks, within and adjacent to towns. In summer, small urban greenspaces up to 30 hectares (75 acres) are about 1°C cooler than the surrounding built area (von Stulpnagel *et al.*, 1990; Forman, 2014). Mid-size parks average about 3°C cooler, and large ones, greater than 500 ha, about 5°C cooler. Furthermore, the cooling distance extends outward from a small greenspace for tens of meters, and from a medium-large greenspace for several hundred meters. Cooling distances vary somewhat with windspeed and on upwind/downwind sides. Probably the cooling distance is shorter out from lobes or corners. These numbers may differ for greenspaces in and adjacent to towns, but the idea of cooling towns with greenspaces is promising. A research frontier beckons.

Greenspaces, particularly semi-natural ones, are sources of species dispersing across the built area. The zone of summer cooling around a greenspace, highlighted in the preceding study, may enhance growth and survival of colonizing species. Cooling distances probably also facilitate animal and human movement between greenspaces.

The spread of species from a town depends on pattern in the nearby land mosaic. If a relatively homogeneous pattern, such as a grain field or pasture extends far outward, the distribution of seeds or animals or people dispersing outward from town would be a negative-exponential curve (or distance-decay curve), for example, decreasing with the square of the distance. Thus, numerous seeds are deposited close to the town boundary, some farther out, and very few at a great distance. That produces a curve with a steep initial decrease that progressively tends to flatten out.

But the usual land pattern is a mosaic of patches varying in suitability for each species. Thus, a species adapted to, and enhanced by, built-area conditions would likely compete poorly in farmland and woods, but could grow in a village. Thus, plotting abundance of a town-adapted species with distance from town produces a convoluted curve, due to both the negative-exponential input with distance, and the patchy distribution of quality habitats in the surrounding mosaic. A woodland species dispersing outward from town would show a similar convoluted curve of success, but in this case small woods in farmland may act as a stepping stones, so more distant woodland receives a large number of colonizers. In effect, distance and habitat-quality pattern determine outward patterns of species spread from town.

Maps of patches and corridors in land mosaics are abundant, yet maps of movements and flows among the spatial elements are scarce. Such movements/flows are widespread and effectively describe how a landscape works.

Consider what some ecologists learned in an Ontario (Canada) farmland composed of crop fields, small woods, and hedgerows (Figure 2.8) (Wegner and Merriam, 1979). For small mammals, most movements were between woods and hedgerow, with rather little movement between woods and fields and between hedgerows and fields. Birds also predominantly moved between woods and hedgerow. Some birds moved between hedgerows and fields and few birds between woods and fields. So, the locations most used for these movements in the landscape are hedgerow–woods

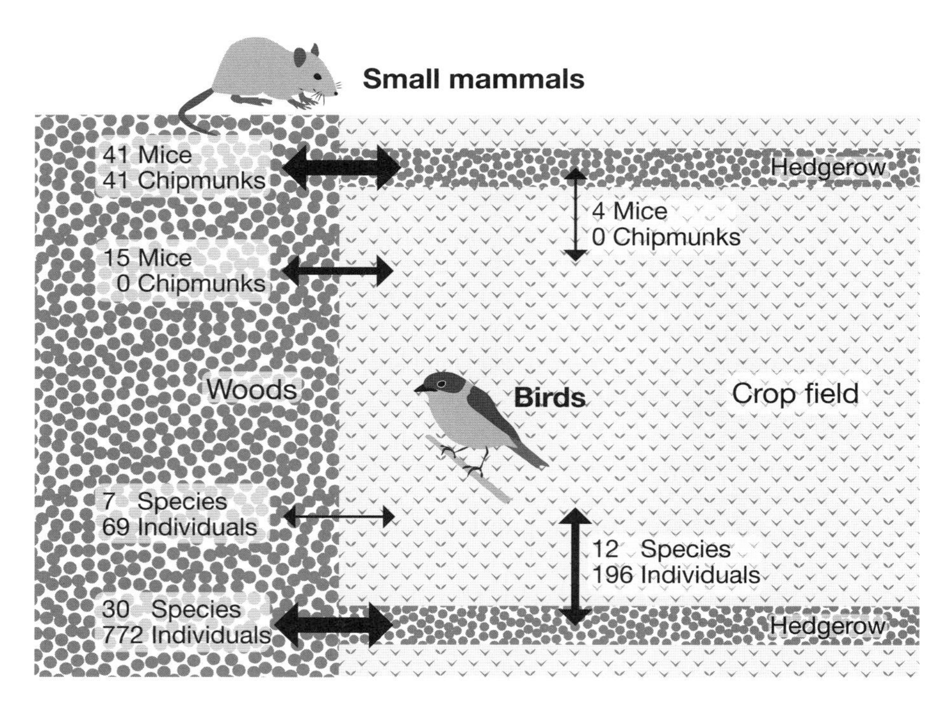

Figure 2.8 Map of wildlife movements between habitats in a heterogeneous landscape. Mouse (*Peromyscus leucopus*); eastern chipmunk (*Tamaius striatus*). Southern Ontario, Canada farmland. Adapted from Wegner and Merriam (1979).

connections. The least movement occurs at straight abrupt woods–field boundaries. Identifying such strategic points in the land is valuable for both species protection and land management.

A map of surface water flows from storm events around a stream with tributaries provides wonderful insight (Figure 2.9) (Diaz Pineda and Schmitz, 2011). Rainwater falling onto the southwestern portion of this map (outside the dashed-line ridge or divide) flows elsewhere, not to the stream. The remainder of the map has drainage to the southeast, as indicated by the two stream tributaries with acute angles pointing downstream. The mapped soil types have the common amoeba-like form.

The wide arrows indicate that the soil where they come from is porous, so considerable water has infiltrated into it. Most porous soils drain to lower watercourses. Thin arrows indicate low infiltration, and therefore more surface-water runoff. Long arrows indicate steep slopes. Thus, the greatest or fastest surface water runoff occurs at the long thin arrows, such as at one spot on the right and in the upper left. These are strategic points, because water quickly rushes into the stream, raising the stream/river level and potentially contributing to flooding. Also, rapid water runoff over the surface typically causes the most soil erosion. The eroded soil is mostly deposited in the watercourse just downstream of an arrow, thus smoothing the stream bottom and reducing habitat for large fish. Water, especially infiltrating into soil at the wide arrows, may reach the

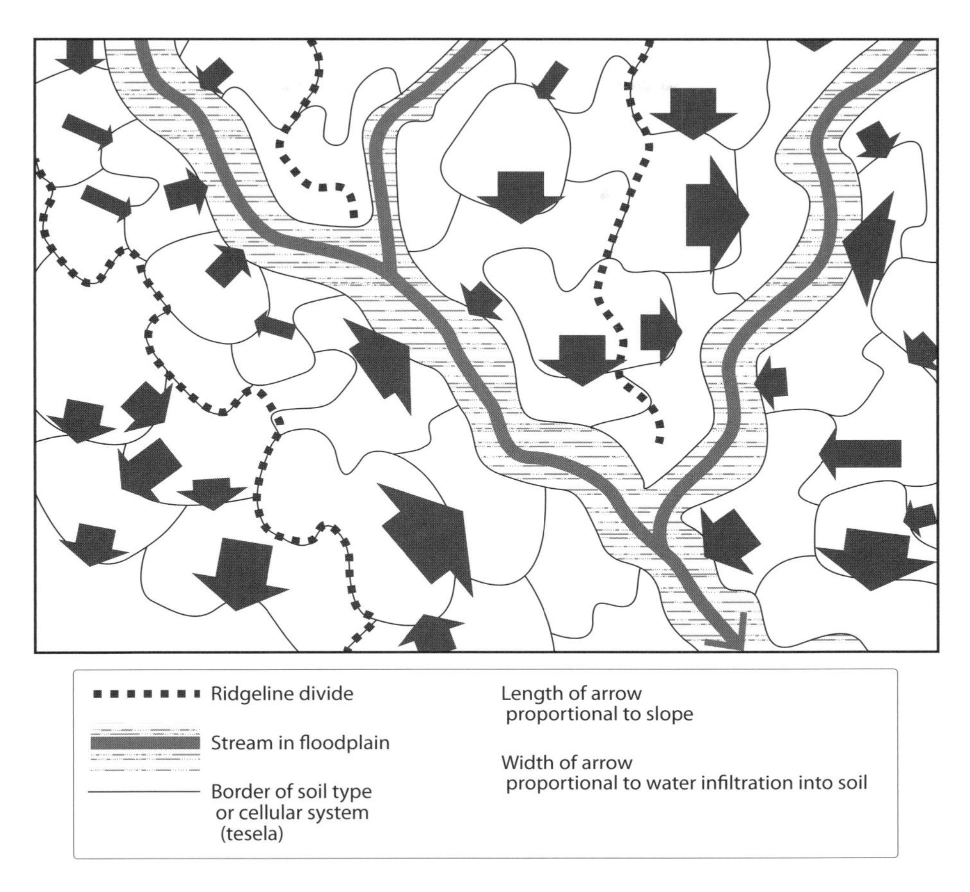

Figure 2.9 Map highlighting surface water flows in heterogeneous land. Greater slope and less water infiltration into soil indicates more water runoff. Adapted from Pineda and Schmitz (2011) and Solntsiev (1974).

groundwater beneath. In essence, a map of flows is a goldmine of useful information on how the landscape works.

In addition to specific locations catalyzing or altering flows, flows and movements are affected by many broad-scale processes such as the following. Regional winds and storms. A nearby city as a source of effects, and an attractant. A river corridor with rich wildlife attracting hunters from distant villages (Jones *et al.*, 2013). Major inputs to farms or factories, and outputs to markets. Road network with a vast array of water-flow culverts in varied states of repair. The cumulative light of towns and vehicles affecting movement and night feeding by bats, certain birds, and other animals, plus varied effects on plants (Rich and Longcore, 2006; Gaston *et al.*, 2013; Blackwell *et al.*, 2015). Hills and mountains partially filled with groundwater up high that creates springs, seeps, wetlands, and ponds in valleys. Thus, the richness of flows/movements in a land mosaic is driven by both local conditions and broad-scale processes, and towns contribute to both.

Many town residents may work nearby in the surrounding land (e.g., Casablanca), whereas other towns may have little interaction with the surrounding land (e.g., Akumal, Capellades). Also, people may work in other towns or a distant city (e.g., Frome, Ovcary, Capellades).

The centers of towns occasionally move, usually as a result of a new commercial area (e.g., Flam, Akumal). Indeed, rarely whole towns and villages move (e.g., Superior, twice). Some villages were relocated during the establishment and early management of Tiger Reserves in India (Botteron, 2001). Hill Village, New Hampshire (USA) moved a short distance and Tallangatta, Victoria (Australia) moved 8 km (5 mi) to avoid flooding from new dams (*New York Times*, May 21, 2016, page A5). Hibbing, Minnesota (USA) moved 3 km (2 mi) from a site believed to be rich in iron ore. Kiruna in northern Sweden, with 18,000 residents, announced that it was moving 3 km in order not to fall into a vast iron ore mine expanding beneath it.

Another way to visualize the effects of a town on its surrounding land is to consider the town as a source of flows outward. Mapping flows then produces a *plume* or zone of influence, which varies in shape. In flat terrain the plume would be rounded if over time flows went outward about the same in all directions. This is analogous to an airport wind-rose, where winds are about equally frequent from all directions. Much more common is a broad plume extending outward from a town in one general direction. An extreme is the highly elongate, or even linear, unidirectional plume.

Many different types of flows, or a longer period of measurement, lead to the more rounded or wider plume. Groundwater plumes tend to be fairly narrow compared with air pollutant plumes, which are affected by more atmospheric conditions, including turbulence along edges. Several factors contribute to creating elongated plumes, including a valley, downslope area, preponderant wind direction, built lobe or strip development, major highway, and stream/river flow direction. In general, flows/movements from a source location create a plume and initiate its general direction, while external factors mold its specific direction and shape. The plume in turn alters flows and movements in that portion of the surrounding land.

Increasing and Decreasing Rates of Flows/Movements

While the direction and route of flows and movements is a key step in understanding processes, the rate is also important. This can refer to the actual speed of an animal or particle, or to the amount that moves per unit time (Forman, 1995). Movements and flows in the landscape occur in three major places: (1) across a land mosaic; (2) along a terrestrial corridor; and (3) in a stream/river. For each of these, several spatial attributes are presented below that typically alter the rate of movement.

Some features of a route increase the rate, such as straightening and deepening a stream. And some decrease the rate, such as dogs and cars affecting arboreal koalas (*Phascolarctos cinereus*) having to move on the ground between separated trees (Prevett, 1991). Most of the attributes affecting flows are useful for land planning and management.

Figure 2.10 Elephant family on the move between known sites for food, water, and cover. Okavango Delta, Botswana. R. Forman photo.

For each wildlife and water attribute listed below, an increase or decrease in rate of movement is indicated (Forman, 1995, 2014). Wildlife here also refers to the chemicals, materials, and organisms transported by wildlife (Figure 2.10). Many attributes for wildlife also apply to people. Terrestrial corridors range from paths and hedgerows to riparian strips and railways. Streamwater transports plankton, seeds/spores, aquatic insects, sand/silt/clay sediments, mineral nutrients, and water pollutants. Floodwater also transports rocks, logs, and debris.

Wildlife Movement across a Land Mosaic

Increasing movement: high quality of habitats; from large green patch to nearby small one; entering or leaving a cove; leaving a narrows between low-quality patches; along a straight boundary; presence of row of stepping stones; an attractant.

Decreasing movement: many habitats to cross (high habitat-crossing frequency); many boundaries to cross; many corridors to cross; entering a cove; entering a section next to a low-quality patch; people or dogs/cats present; a repellent or danger.

Wildlife Movement in a Terrestrial Corridor

Increasing movement: shorter route; stream alongside; attached green nodes; wide corridor (some species), narrow corridor (other species); smell, food attraction.

Decreasing movement: high habitat diversity; many boundaries to cross; many habitats to cross; many other corridors to cross; many logs and rocks; strong wind or pollution from side; a low-quality habitat section to cross; predators or hunters present; adjoining low-quality habitat; busy road alongside; people present; dogs, cats present; gap or narrows; convoluted corridor boundary; repellent smell, light, heat, cold; curvy corridor.

Water in Soil, Stream, or River

Increasing flow: groundwater in sandy soil; subsurface water in sandy fill area; channelization/straightened; greater slope, greater depth; narrows.

Decreasing flow: groundwater in clay or patchy soil; surface sheet flow on well-vegetated slope; curvy route; many rocks or logs; heterogeneous bottom substrate; wide place in a corridor (Figure 2.11).

Many of the attributes above decrease the rate of movement/flow. This suggests that altering a natural pattern often reduces movement rate. In land planning and management, the goal may be to increase or to decrease movement rates.

Distance Effects of a Town

In essence, how far outward do town effects extend? Also see Chapter 11. We explore this question from four perspectives: (1) basic models; (2) some distance effects of flows/movements; (3) transportation and distance effects; and (4) land changes and distance effects.

Basic Models

Early English settlements in poor-soil cool-climate New England (USA) were often about 57 km^2 (6 x 6 mi, or average radius 3.3 mi; 1 mi = 1.6 km) (Platt, 2004). This area was considered suitable for agriculture, natural resources, and the people of a central village. Somewhat similarly then in population-dense farmland-rich Europe, there should be no other town (population more than 3,000) within 7 km of a town's edge (van Leeuwen, 2013). That represents the distance from small- and medium-sized towns over which town-related actors, activities, and values extend, effectively creating a local economy (or economy of place) (van Leeuwen, 2010; Ratajczak, 2013). The suitable or sufficient area around a town would doubtless differ in different climates. Also, normally the area around a town includes villages and hamlets with their surroundings.

Zones of influence or *von Thunen bands* were early recognized around towns (Losch, 1944; Cronon, 1991; Barreira, 2013). Typically, cropland was immediately around a

Figure 2.11 Small river from rainforest mountains during dry season. Floodwaters in wet season carry stones and finer sediment into the Caribbean Sea, where waves wash the material back to form a coarse-sand beach, ideal for nesting of huge leatherback sea turtles (*Dermochelys coriacea*). Village of Grande Riviere, Trinidad. R. Forman photo. (A black and white version of this figure will appear in some formats. For the color version, please refer to the plate section.)

town, pastureland surrounded the cropland, and woodland was the outermost zone. Larger towns and cities often had a narrow inner zone of high-maintenance gardens and orchards between the town and cropland. But varied topography meant that these zones were neither concentric nor symmetrical.

A *town-centered region*, based on the diverse positive and negative interactions a town has with its neighbors, provides a useful broader perspective (Forman *et al.*, 2004; Forman, 2008). For Concord, Massachusetts (USA; see Appendix), the functional town-centered region includes 17 towns and has a radius of 20 km (12.5 mi). Many patterns and processes, including bike and walking trails, rivers with floodplains, transportation patterns, and protected forest and wetland, help tie the local region together.

The distance effects of a town, i.e., the outward extent of significant effects from a town, are asymmetric. This results largely from wind, hill/mountain slope, transportation routes, or a stream/river. Also, nearby farmland, villages, towns, and a city increase or decrease the effects in different directions. Some effects could be readily measured in concentric rings or pie-shaped sections around town. Examples are sizes of farm fields, all-terrain vehicle (off-road vehicle, ORV) tracks in desert/grassland, and density of dispersed houses.

A *negative-exponential (or distance-decay) model* or curve describes the typical distribution of effects outward on a homogeneous surface. Thus, objects dispersing from a point decrease exponentially at increasing distance. Particles, seeds, animals, even people typically decrease curvilinearly in number with the square of the distance from a source (Forman, 2014). A sharp environmental gradient outward compresses the pattern, and heterogeneous surrounding land creates wiggles in the curve.

Examples of the negative-exponential pattern include villagers mainly working in a Chinese dike/pond/recycling/fish-production landscape adjacent to their village, with few villagers slightly farther out (Duning *et al.*, 2001). Agricultural workers go out from the town of Casablanca (Chile; see Appendix) to work in vineyards and crop fields in the adjoining valley, and with livestock and pastures on surrounding hills. Most work within 5 km (3.1 mi), some between 5 and 10 km, and a few as far as 20 km.

The outward dispersal distance also depends on the size of the source. In the city of Seattle, Washington (USA), coyotes (*Canis latrans*) mainly move from a shrub habitat up to 0.4 km into the built area, and from a small forest patch up to 1.0 km (Quinn, 1995; Forman, 2014). But from a large forest patch, these top predators may move more than 1.1 km into the built area.

Homing studies of movement by translocated animals, which are removed from a habitat and placed in a distant similar habitat, may provide useful insight into wildlife distance movements. Also, stream/river studies, which indicate how far downstream effects of a point pollution source extend (Batty and Hallberg, 2010), should be useful for understanding the distance effects of a town.

Some Distance Effects of Flows/Movements

Air pollution from town factories deposits particulate matter at progressively lower amounts further out. Most day-long recreationists will readily drive outward from a city up to one hour but not two hours, and hardly anyone goes three hours out. Weekend recreationists will commonly drive outward up to two hours, but a few go three hours. These patterns may apply to movement from towns, though numbers of such recreationists from a town are normally small.

Although apparently no synthesis of town distance effects has been done, an overview of distance effects from urban areas is useful (Hansen and DeFries, 2007; McDonald *et al.*, 2008). The following distance effects also often apply to towns, and are the average distances reported in the literature (n = number of studies) (McDonald *et al.*, 2009):

500 (km; 1 km = 0.62 mi) Air pollutant sulfates ($n = 23$)
100 Water quality ($n = 3$)
100 Nighttime light (affecting aesthetics) ($n = 7$)
50 Illegal hunting of wildlife ($n = 12$)
40 Water pollution ($n = 4$)

30 Illegal logging ($n = 8$)
15 Garbage dumping ($n = 14$)
10 Conflicts with wildlife and crop raiding ($n = 7$)
7 Firewood collecting (subsistence) ($n = 8$)
6 Nighttime light (biological effects) ($n = 3$)
6 Wildfire ignition ($n = 5$)
0.8 Noise (biological effects) ($n = 15$)
0.5 Trampling ($n = 12$)
0.2 Firewood collecting (recreational) ($n = 2$)
0.2 Nonnative species establishment ($n = 2$)
0.1 Edge effects on microclimate ($n = 3$)

Almost all of the distances from a city are sensitive to population size and activities, and thus are doubtless greater than for a town. Only the last few effect distances might apply also to low-population towns. The list of effects ranges from regional to local. All have negative environmental consequences and many negatively affect people. Yet positive results also exist for some, such as: illegal hunting providing animal protein for subsistence farmers (Roe *et al.*, 2013); nighttime light enhancing foraging by predators (Rich and Longcore, 2006); and firewood collecting (at low intensity) thinning a forest that may increase wood production and result in a richer herbaceous layer.

Many natural and human processes operate at spatial scales within and around a town. For example, female domestic cats may have home ranges of some 100 m radius, with male cats foraging out twice as far, even farther at night (Chapter 8) (Forman, 2014). But other processes often cover a considerably greater area, such as river flows, coastal dynamics, wildfire, large mammal home ranges, and migration routes. The *home range* concept, i.e., the area used in daily movement by an animal, is useful in understanding both ecological and human effects at the town scale.

A *human home range* is the area used from day to day for residence, job, shopping, school, park, entertainment place, and so forth (Forman, 2014; Friedman, 2014). A large home range indicates that the visited locations are far apart and that considerable transportation is required (Figure 2.12). A small home range indicates that the locations are close together, and time and cost are likely to be invested in maintaining a neighborhood rather than in transportation. *Mixed-use* areas, where residential, commercial, and other land uses are somewhat intermixed, can produce small human home ranges.

Town residents overwhelmingly shop for daily needs in the town, indeed mainly in the town center or in a town-edge shopping center near their homes. Most villagers mainly shop for daily needs in the village, though may also use a shopping center on a town edge convenient to the village. Both townies and villagers primarily use town commercial centers for weekly needs or weekend shopping. More specialized needs, such as special musical instrument, furniture, medical clinic, or attending a show or sports event, are commonly met by occasional trips to the city. Containing many multi-generation residents with deep ties in town, most towns arrange business and commercial shopping resources to provide both daily and weekly needs for town and village residents.

Figure 2.12 Transportation by pedi-rickshaws in front of building construction site. City of Tulear (Toliara), Madagascar. Photo courtesy and with permission of Mark Brenner. (A black and white version of this figure will appear in some formats. For the color version, please refer to the plate section.)

Transportation and Distance Effects

Streets, roads, and trails provide ready walking and biking access to jobs, plus some shopping for many town residents (Figure 2.12). Others commonly move by motor vehicle at low speed in town. Although motor vehicles at low speed may be relatively quiet, they tend to produce somewhat more pollutants which cover surfaces and enter waterbodies and groundwater.

Outside of town, mail delivery vehicles and public service maintenance/repair vehicles daily use roads about half way to the next town. School buses go out twice a day, commonly up to several kilometers beyond town. Large towns often have an internal bus system with somewhat infrequent service. Regional buses connect the town with villages, other towns, and a city. Passenger trains that stop at a station in town connect it to a city, and sometimes to a few other towns. Train noise and vibration degrade nearby town and land conditions (Kanda *et al.*, 2007; Guarinoni *et al.*, 2012).

Traffic noise is an annoyance, and at high level a stress, for people. A busy multi-lane highway, either through or close by the town, may cause these problems for residents within hundreds of meters. The noise problem is worse with a high usage by trucks (especially if more than 7% of traffic; Forman *et al.*, 2003). Such traffic should be diverted away from schools, hospitals, and analogous facilities in town. Unless a trail under or over a highway is present, crossing the highway is hazardous for walkers.

Research is increasingly clarifying traffic noise effects on wildlife (Chapter 12) (Forman *et al.*, 2003; Reijnen and Foppen, 2006; van der Ree *et al.*, 2015). For instance, in the outer suburbs of Boston, Massachusetts (USA) the presence of grassland birds of conservation interest was studied near multi-lane highways with more than 50,000 vehicles per day on average (Forman *et al.*, 2002, 2003). Both the presence of the birds and their regular breeding (during a five-year period) are inhibited for approximately 1,200 m (0.75 mile) outward on both sides. This 2,400 m wide strip would be poor for a nature reserve.

Near a two-lane highway with 15,000–30,000 vehicles per day, bird presence and regular breeding are reduced within 700 m (0.4 mi) on each side of the road. Two-lane highways with a speed limit exceeding approximately 45 mph (72 km/hr) also have the highest rates of wildlife roadkill, including wildlife–vehicle collisions (Forman *et al.*, 2003). A through two-lane road with 8,000 to 15,000 vehicles per day has no significant effect on bird presence, but no breeding occurred within 400 m (0.25 mi). The through two-lane road characteristically connects the centers of two suburban towns in the area. Finally, the good news is that a local collector road with 3,000–8,000 vehicles per day has no significant effect on either the presence or regular breeding of the birds of conservation interest. Noise effects on lower-traffic roads remains poorly known (Reijnen and Foppen, 2006).

Many studies of rural or remote roads indicate an avoidance zone alongside, where certain wildlife are scarce or absent (Forman *et al.*, 2003; van der Ree *et al.*, 2015). Examples include reindeer, elk, deer, coyote, small mammals, birds, amphibians, and snakes. A broader view of road and traffic effects indicates that most effects (Chapter 12): of road and roadside activities occur within 20–30 m (66–99 ft) of the road; of materials and chemicals within 50–100 m; on water and aquatic ecosystems within 100–1,000 m; and of traffic disturbance (especially noise) within 300–>1,000 m (Figure 2.13). Overall, the effects of road types and traffic levels surrounding towns and villages remain a key research frontier.

Land Changes and Distance Effects

Major processes changing land provide further insight into distance effects around towns. Firewood collection providing energy for heating and cooking, e.g., in East Africa, progressively extends further out from population centers, because collection rate exceeds tree growth in the savanna (Roe *et al.*, 2013). Wood removal degrades a widening area with reverberating effects: soil erosion by wind and water; loss of soil organic matter, especially fine particles, and fertility; reduced agricultural production; loss of shade; loss of wildlife habitats; increased heat; reduced air quality; and increased aerial particulate matter. At a site, more firewood collection means more environmental degradation and thus less wood production, and the larger the degraded area the worse it gets. So, firewood collectors keep moving outward until the time and cost of transporting firewood back to town becomes excessive. Both people and land are then in bad shape.

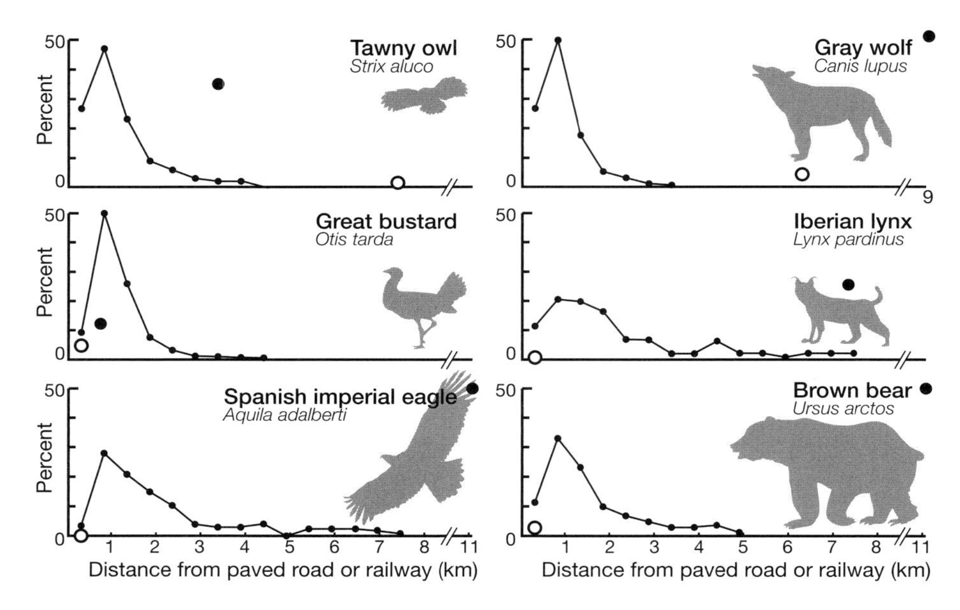

Figure 2.13 Distribution of notable predators relative to transport infrastructure in Mediterranean Europe. Curves represent the proportion of a species distribution within 500 m-wide zones from road or railway. Solid and open circles represent the highest and lowest prevalence respectively in a zone (ratio of number of cells with species present divided by number of available cells). Mainland (peninsular) Spain. 1 m = 3.3 ft. Adapted from Torres *et al.* (2016).

Historical land uses around a town are informative. Around a present coastal village (Grande Riviere, Trinidad; see Appendix) former plantations of cacao and other products extended some 13 km (8 mi) into the surrounding rainforest. The plantation trees were basically abandoned when economic conditions eroded, so today's rainforest there is quite distinctive.

Accessibility to a location is a key determinant of land cover change (Nagendra *et al.*, 2003). Near two medium-small sub-Saharan cities (Sikasso, population 160,000; Bobo-Dioulasso, 480,000) in West Africa, the locations of four major land-changing processes (1986–2009) were studied: abandonment, deforestation, crop expansion, and urbanization (Brinkmann *et al.*, 2012). In both cases, land abandonment and deforestation were much greater more than 5 km (3.1 mi) from a road. Urbanization was greater less than or equal to 2 km from a road in both cases. All four processes were about equal from 0 to more than 5 km from water. Abandonment, deforestation, and crop expansion relative to distance from the urban area were higher close to Sikasso (<3 km), and lower at greater distance. In contrast, near Bobo-Dioulasso the three processes were about equal with distance, except for more deforestation beyond 5 km.

If these results also apply to towns, the four major land-change processes mainly correlate with the distance from roads and urban areas, rather than from water-bodies. Land abandonment and deforestation were highest at greatest distance from roads. Deforestation was most variable relative to distance from water and urban area.

Urbanization was highest close to roads. Overall, crop expansion was about constant with distance from all three features, roads, water, and urban areas for both cities.

The location of houses and housing developments relative to distance from city, road, existing house, and so forth, as well as models for planning, have been much studied (Forman, 2014). Some distances found or used in these studies may be useful in considering distance effects from towns. Disturbance zones of 50, 100, 200, and 400 m around a house site are identified in the Rocky Mountain area (Theobald *et al.*, 1997; Compas, 2007). On the North Carolina Coastal Plain (USA), the average "leapfrog" distance (Haase and Lathrop, 2003) between existing development and new development is 200 m in coastal areas and 500 m for inland areas (Crawford, 2007). At a finer scale, the best predictor of the presence of nonnative plants (in a forested landscape with scattered houses along roads in Wisconsin, USA) was the number of houses (0 to 7) within a 1 km radius of a forest site (Gavier-Pizarro *et al.*, 2010a).

Consider the percentage of the protected areas within different distances of a town (average population 25,000) (McDonald *et al.*, 2009). Almost all towns in "urban/high-population-density nations" are within 50 km (31 mi) of a protected area, and noticeably closer to protected areas than are towns in other countries. Rural nations, with either many parks or few parks, have about the same distances from town to protected area. In "urban/low-population-density nations," towns are farthest from protected areas. This last case is worst for people wishing to visit a protected area, and best for protecting the protected area.

Extensive protected area around a town or village may affect movement of residents in the surroundings (e.g., Petersham; see Appendix), but serve as a magnet for recreational use by others. Also, a nearby protected area may provide subsistence resources for villagers and most farm laborers using a close-by protected forest (Gunuma Gede Pangranga UNESCO Biosphere Reserve, Java, Indonesia) (Nakagoshi *et al.*, 2014). Six villages just outside the forest boundary have 77 to 374 households each. The distances from village to main road vary from 1.0 to 13.7 km (0.6–8.5 mi), and distances to forest edge vary from 2 to 413 m (6–1,360 ft) (plus one at 3,570 m). About 59 percent of the villagers live less than 1 km from the forest edge, 24 percent are 1–2 km away, and 17 percent more than 2 km distant. Thus, villagers use footpaths, not the road, into the forest.

Residents in all six villages collect firewood in the forest plus a limited amount of bamboo for construction. Many villagers collect fruits and a few collect other resources such as gum. Over twelve months (2009), there were very few cases of illegal agricultural fields or illegal logging, 136 cases of illegal hunting, and no forest fires recorded in the forest. Overall, slightly over half (53%) of the population is less than 20 percent dependent on forest products. Yet 10 percent of the villagers are more than 80 percent dependent on forest products from the protected reserve.

Finally, key wildlife and water movements are mapped around Lake Placid (see Appendix), a small town in lake country on a sandy ridge of south-central Florida (Figure 2.14). Bears (*Ursus*) reside within a kilometer or two of the town edge and 6 km from the town center. Primary bear routes are just outside the town edge except in the northeast, where the animals move through a low house-density area of town. Panther

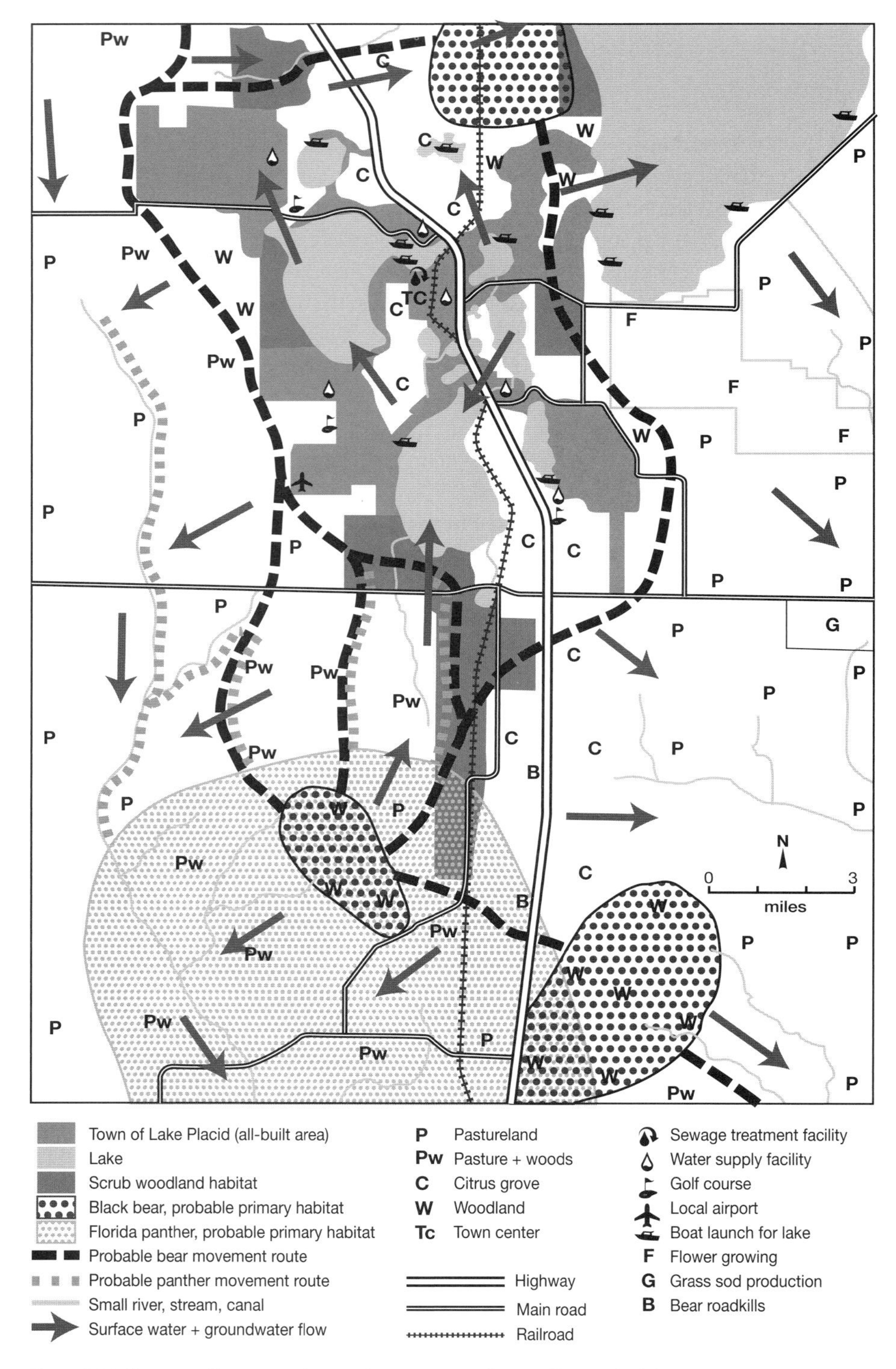

Figure 2.14 Routes of bear, panther, and water around a small town in lake country. Black bear, *Ursus americanus*; Florida panther, *Puma concolor*; bear patterns partially based on Murphy *et al.* (2017a, 2017b). Sandy soils in central vertical third of figure, with clayey and organic soils to east and west. Fires mainly move in and downwind (east) of even small scrub areas. Lake Placid (population 2,200), south-central Florida, USA. With appreciation and aid of Michael Binford, Mark Brenner, Hilary Swain, and the Archbold Biological Station. (A black and white version of this figure will appear in some formats. For the color version, please refer to the plate section.)

(*Puma*) habitat is about 8 km south of the town edge, and panthers move at least to within a kilometer of town. A famous biological station just south of town conserves rich habitat and nationally known biodiversity, while catalyzing conservation thinking throughout the town and region (Swain and Martin, 2014).

Rain adds water evenly over the town. Some surface water enters town from the south (Figure 2.14). Surface water and groundwater flows within the town are mainly northward. In contrast, water flows just outside town are southwestward, eastward, and southeastward. Local pollutants somewhat degrade the lakes and groundwater in town. Outside town for a considerable distance, agricultural fertilizers (perhaps especially phosphorus) eutrophicate streams, canals, and groundwater, and pesticides degrade the aquatic communities.

In summary, no direct measure yet exists of the distance a town affects its surroundings, and vice versa. However, we can hypothesize that distances typically are greatest, both for environmental and human effects, in the directions of main roads radiating from town.

3 Change

... six hours by train ... the scenery seems as patterned as wallpaper: a peasant, a field, a road, a village ... Move back in time, and it's the same: a peasant, a field, a road, a village ... history ... has to do with trajectory, progress ... wallpaper.

Peter Hessler, *Oracle Bones: A Journey Through Time in China*, 2007

Time and Change in Towns

Every town and village area is festooned with clues to the past (Watts, 1975; Lewis, 1982; Zelinski, 1992; Hart, 1992), and also the future. As suggested above, many dramatically different landscapes of the past reappear in similar form, one after another. The present feels brief, as most visible features are also clues to what lies ahead. The future takes shape before our eyes.

Four perspectives highlight time and change in towns: (1) geological foundations and the Anthropocene; (2) past-present-future at Scratch Flat; (3) glimpsing village and town history; and (4) climate change and towns.

Geological Foundations and the Anthropocene

The far-far-far-back geological era (Paleozoic, 541 to 252 million years ago) started with a proliferation of marine animals in the globe-covered sea. Lands appeared and coalesced into a supercontinent (Pangea). Plants, amphibians, reptiles and winged insects spread across the land, and trees formed "coal and oil forests." Next, in the far-far-back era (Mesozoic, 252 to 69 million years ago), the supercontinent broke up into several continents, on which dinosaurs, small mammals, gymnosperms (especially coniferous seed plants), and then flowering plants proliferated. The far-back period following (Tertiary, 69 to 3 million years ago) provided an abundance of modern plants and very large mammals.

Then came the Ice Age (Pleistocene, 3 million to only 12,000 years ago) with climate changes producing a series of major glaciations and intervening warmer phases. The huge mammals flourished and then disappeared. Modern humans appeared and spread, as a stone-age culture.

The recent 12,000 or so years since the last glaciation (the Holocene) included the development of human civilization. Extensive portions of the land had a sequence of

Figure 3.1 Row of stone moai from an earlier civilization, southeastern Easter Island (Isla Pascua), Chile. Moai statues were "walked" from a quarry along a forest road to a coastal village. R. Forman photo. (A black and white version of this figure will appear in some formats. For the color version, please refer to the plate section.)

major climatic changes, beginning with post-glacial cold, then warm, then warm-and-dry, and finally today's global pattern. Vegetation changed accordingly during this sequence. Today's land of towns and villages shows many traces of these earlier periods.

The *Anthropocene* refers to the time during which humans have significantly impacted the planet. Some scholars recommend starting the phase 7,000 years ago, when agriculture became global. Or at ca. 1500 AD, when explorers headed off around the globe, altering places and changing Europe (Figure 3.1). Or the late 1700s, with the beginning of widespread burning of coal and its effects. Or the late 1800s, with the onset of the industrial revolution. Or the 1940s, with a global World War II and the dawn of the nuclear age. Or the 1950s with widespread combustion of coal and oil (Voosen, 2016; Worster, 2016). Sometimes called the "great acceleration," numerous features of human civilization with global effects have accelerated since the 1950s (Voosen, 2016; Alberti, 2016). The land of towns and villages affects, and is affected by, these global changes. No previous species has changed the globe.

Past, Present, and Future at Scratch Flat

A remarkable study of 15,000 years on one square mile (2.6 km^2 or 640 acres) in a "typical" town (Littleton, Massachusetts, USA) provides insight into change from past to future (Mitchell, 1984). Today the site, called "Scratch Flat," 48 km (30 mi) west of Boston, has mostly good soil, a winding stream, a pond, meadows, hayfields, pine and oak woods (*Pinus*, *Quercus*), roads, busy highway, industrial buildings, a few old farmhouses, and suburban house developments. Scattered outcrops reveal rocks from the far-far-back era mentioned above, when today's West Africa separated from this area of North America.

Geology textbooks clearly describe the 1.6-km (1-mi) thick, very heavy Ice Age glacier, which molded the land surface here and then melted in place. The glacier left a hill (drumlin), mounds (kames), a blanket of sandy loam soil, and a lake, which in turn drained away leaving a pond and a winding stream depression. Walking around on Scratch Flat, these features are easily seen, and together signal former glacier. Indeed, digging with a shovel near the stream reveals beautiful beach sand from the glacial lakeshore.

Although these clues to the post-glacial past are easily seen, a quite different interpretation comes from certain local Native People (Indians). A monster snake arrived that ate people, but was chased away by a more ferocious monster. In rapidly escaping, the snake carved a curvy depression (where today the stream runs) (Figure 3.2), and disappeared down into a cave (where today's pond is). The features themselves are quite clear. Either or both explanations for the patterns present could be correct. The preponderance of scientific evidence indicates that the features reflect the glacial stage, when Scratch Flat was covered by ice as thick as 50 tall trees atop one another.

More exploring here reveals small areas of spongy peat moss (*Sphagnum*) with dwarf black spruce trees (*Picea mariana*), some barely head-high (Mitchell, 1984). Such vegetation covers extensive tundra areas of Siberia, Canada, and Alaska. This second phase of the town's past comes into focus as a spear point characteristic of this period, now in a museum, was found in the area. Indeed stone-age "Paleo-Indian" People hunted here, perhaps for wooly mammoths or other huge mammals, in this post-glacial stage.

Pollen extracted in cores from the bottom of peat-moss wetlands and ponds of the region reveal changes in the vegetation since the Ice Age. Several tree species from just south of here replaced the spruces and formed a forest canopy, indicating a warmer climate. Then another overlapping set of species from the southern USA and northern Mexico dominated, indicating a warm dry climate. The mammoths and other "megafauna" disappeared, while woodland caribou, elk, moose (*Rangifer, Cervus, Alces*), and other familiar animals roamed the land here.

Slightly more recent Indian sites of about 2000 BC–1630 AD, including a hunting camp and a rock shelter, have been identified in the area by archaeologists. With lots of arrowheads collected over time, the meat-eaters of the time consumed the range of herbivores common today, plus elk and heath hen (*Tympanuchus cupidus*), which later were hunted to extinction by European immigrants. The land also provided wild plant foods for the hunter-gatherers. Indeed, two Scratch Flat residents recently lived four

Figure 3.2 Scratch Flat, Littleton, Massachusetts, USA. Stream created by a giant serpent according to a Native Indian People's tradition, or in glacier-molded land according to science. See text. R. Forman photo, with aid of John H. Mitchell. (A black and white version of this figure will appear in some formats. For the color version, please refer to the plate section.)

days on wild plants, fish, and a few turtles they caught. But to the modern palate the diet was too tasteless to continue for a fifth day.

In addition to gathering food, Indians in the region began cultivating food, especially the "three sisters": corn, beans, and squash (*Zea*, *Phaseolus, Cucurbita*). On a small plot, they probably cleared most vegetation, made small soil mounds perhaps a meter apart, buried a fish in each mound, planted several corn seeds atop the mounds, and planted bean and squash seeds close to the mounds. But the Indian population dropped before 1620 due to disease, probably contracted from early European explorers, fishermen, or beaver-fur (*Castor*) collectors.

Yet another phase appears with European settlement only 16 km (10 mi) away (in Concord; see Appendix), that brings radical change to Scratch Flat: confrontation between the native and English communities, with killings; large wood structures built too heavy to move seasonally; forest eliminated; livestock introduced; proliferating stonewall networks to control the livestock; hayfields and meadows needed for the livestock; a culture of land ownership and property boundaries; and a new god putting strict controls on people, yet seemingly uninterested in nature. The native Indian population noticeably shrinks as the white population grows.

When did the present appear? Perhaps it arrived in 1922 when electric wires spread along the road (Mitchell, 1984)? Or, when sounds on the land changed from mainly birdsong to traffic noise, chainsaw, lawn mowing, and hammering? Or more recently, with the addition of motorized grass-trimmers, leaf-blowers, snow-blowers, snowplows, and construction equipment beeping in reverse? Or, when new people moved in thinking it was a nice place to live, rather than because of specific features of the land important to their daily life?

More key clues have been appearing since the 1950s. A major highway sliced through. An industrial "park" moved in. A shopping center was proposed, and the town citizens rose up and voted it down, but it was later built. A pine forest was cut for fashionable expensive houses on large plots. A dairy farm was not doing well. A long-time landowner began selling house plots. An array of high-tech somewhat-secret companies settled in. Although representing the present, aren't these really conspicuous clues to the future?

All towns have a past with several stages readily detectable on the land by observant residents, who regularly get out and explore. At least one early stage is continuous natural land with a symphony of birds. Stages of the past leave intriguing footprints everywhere. The present is brief, and relatively clear. The future is long, and quite uncertain. No matter, reading the land for clues reveals much about the upcoming future.

Glimpsing Village and Town History

General Characteristics

Cities developed about 5,000–6,000 years ago in many parts of the world, so towns and villages surely preceded these large population centers with extensive specialization (Pacione, 2005; Benton-Short and Short, 2008). Areas with the earliest cities include Mesopotamia (today's Iraq), Egypt, and the Indus Valley (today's Pakistan). Early cities also appeared in Huang Ho Valley (today's China), Greece, Rome, and Maya land (today's Middle America). Damascus could be the oldest continuously inhabited city.

With some 2,000 to 30,000 residents, a town normally has a significant portion of people with non-agricultural occupations (Russo, 1998). Thus, towns developed where residents could make a living, especially where a mill or factory could make products for residents and/or for transport elsewhere (Bell and Bell, 1969; Harris and McKean, 2014). A stream provides hydropower, cooling, and waste disposal. A dammed pond may provide fish and recreation. Indeed, streams, as well as lakes and the sea, have long been the major mechanism for cleaning towns, carrying away human waste, solid waste, industrial and commercial waste products, and polluted stormwater (e.g., Tavistock, Concord: see Appendix). The process has long polluted local waterbodies. Towns also must have a water supply and a transportation linkage to another town or city.

The general trajectory of a small town often looks like the following (Osborn and Whittick, 1963). Population increases. Intensification of buildings and people in the center. Outward residential spread. Industry on the edge of town (sometimes also remaining in the center). Lengthening of daily trips. Congestion. Shortage of space in the center. Commercial growth on the edge.

Oddly, published histories of town and village commonly overlook or minimize prominent and conspicuous environmental issues. Consider stormwater runoff, flooding, impervious-surface-area effects, heat buildup, loss of soil surface for growing food, solid waste disposal and treatment, pollution of local water bodies, pollution of town air, degraded biodiversity, wildlife movement and pests, and so forth. Also, human waste was commonly dumped onto the street. Only recently in the history of towns did "water closets" appear. These contain a seat under which piped water enters, a person sits down, and then with the push of a lever, the liquid departs as wastewater, quite a novel invention for towns. But where did the wastewater go? Even disease and human health are often avoided in town histories.

Another important omitted feature where horsepower was important are horses and their effects. In the pre-motor vehicle era, imagine a town's stables, manure buildup, seeds and sparrows on the manure, mud, dust, wagons and carriages, repair shops for wagons and horse hardware, hitching posts, water troughs, and horse feed. Lots of horses were coming and going, with effects doubtless permeating ears, eyes, nose, throat, buildings, streams, and more.

This overview of a town trajectory is for a growing town that effectively disappears by growing into a city. Other trajectories are more common. A village may persist as a village, mainly stabilized by a limited population tied to its surrounding land. If the surroundings degrade, as in a prolonged drought, the village may disappear into a hamlet. In central Italy, gradual population increases and decreases characterize the history of five communities (Figure 3.3). As all the communities now decrease, grassland and shrubland patches decrease, while forest patches grow in size and forest cover increases.

A town may also persist long term, but major population growth or loss can eliminate it. Indeed, its typical dependence on one or a few industries makes a town especially susceptible to population gain or loss. Indeed, in the general town trajectory, pulses of both population gain and loss are the norm. Towns that persist without becoming a city (e.g., Casablanca) generally have limited surrounding resources or are in an isolated location. Extreme cases are villages that move to a nearby location (e.g., Superior, Flam), frequently due to mining activity.

Characteristics over Time and Space

Early Greek towns often had a central way or spine along which key buildings were located (Bacon, 1974; Russo, 1998). The town layout was relatively heterogeneous, often fitting somewhat with topography, and contained relatively small regular geometric spaces. In contrast, early Roman towns had a quite regular geometry, reflecting the top-down military control on town and road construction. Fortified by thick walls and gates, the town was centered by an open square (forum) rigidly ringed by mainly government buildings. Most residents lived inside the wall, where a dependable water supply was present and vegetable gardens could be maintained.

A central marketplace in towns, typically upgraded from a former village green (Figure 3.4), provides a hub to bring and sell food and goods, as well as to shop for

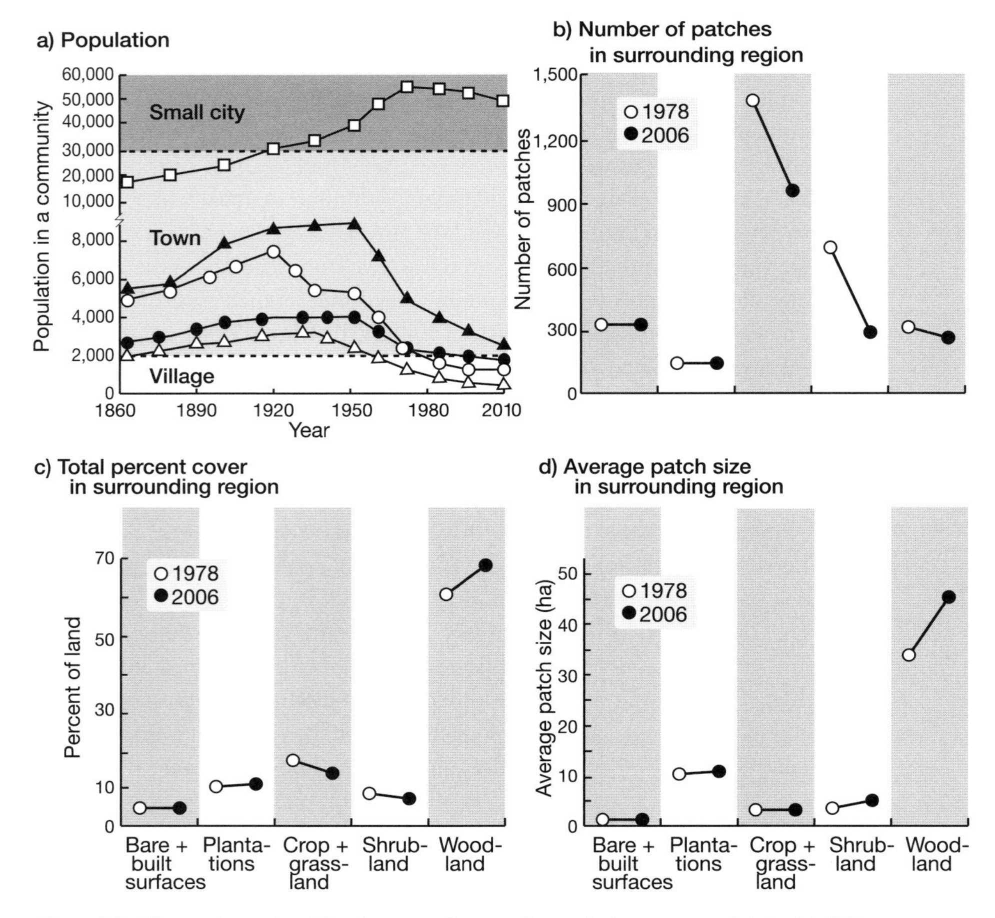

Figure 3.3 Change in regional land uses as five rural population centers shrink. (a) 150-yr population change of today's three villages, one small town, and one small city. (b), (c), and (d) Land use changes of surrounding region during the last shrinkage phase, 1978 to 2006. 16,800 ha (65 mi^2) mountainous area in Ascoli-Piceno province, central Italy. Plantations mostly old chestnut (*Castanea sativa*); cropland only slightly greater cover than grassland; woodland mainly dominated by oak, chestnut, or beech (*Quercus*, *Castanea*, *Fagus*). Based on Bracchetti *et al.* (2012).

residents' needs (Platt, 2004). Most marketplaces also serve social, civic, celebratory, and even religious functions. Typically today's central public space or plaza or square in a town evolved from the early marketplace as a central organizing space (Bacon, 1974).

In European towns diverse fortifications evolved in subsequent centuries. A prominent building or two – a castle, a religious structure – near the center became the norm, sometimes even in small towns (Bell and Bell, 1969; Bacon, 1974). These towns expanded by accretion, and often extended linearly along a road or shoreline. They were internally renovated, sometimes based on tension between powerful forces in town. The focus seemed to be on buildings, residents, and daily needs, with infrastructure

Figure 3.4 Village green of Petersham, Massachusetts, USA with gazebo on right and general store on left. Green and gazebo used for community activities and festivals. General store, one of five village shops, provides soup and sandwiches, gifts, and small selection of groceries. See Appendix. R. Forman photo. (A black and white version of this figure will appear in some formats. For the color version, please refer to the plate section.)

seemingly of secondary interest. Thus, traffic congestion, waste buildup, uncoordinated ditches/pipes, plus mud and dust along travel routes, persisted as severe problems.

Some towns had mainly attached buildings around a central green and a transport route (Chapter 1), creating inward-looking places with backs of buildings facing the adjacent farmland or natural land (Sharp, 1946; Roberts, 1982). The central space formed an effective core as the town expanded outward with both attached and detached buildings. Such towns seem compact.

Other towns inherited a village of mainly scattered homes and some central shops (Reps, 1969; Wood, 1997). In this case, infill or densification was the primary early building process, which perhaps usually led to a small central area of attached commercial buildings, but mainly detached residential homes across the town. Other than the main street, probably the road network was rather irregular in this town type.

Many interesting exceptions to these general patterns of town and village history exist. Three cases briefly illustrate the point. First, towns near a city, which usually expands outward, seldom shrink, but keep growing along with the city. Second, some towns were basically religious centers, so the structure and change over time are tightly tied to the nature and persistence of the leaders. For example, Brodgar, a Neolithic (3200–2300 BC) community on Scotland's Orkney Islands, was a ceremonial center apparently with some 5,000 people farming around it. Third, Upper Canada Village, Ontario (Canada) and Sturbridge Village, Massachusetts (USA) were formed by collecting numerous historic buildings in the respective region, and moving them to a site to produce a heritage educational experience for visitors. These are essentially artificial villages with no residents, but form interesting large outdoor museums.

Villages tend to form close to resources promising economic opportunity. Informal squatter settlements near Nairobi (Kenya) close to a large wildlife protection area seem to fit this model (Muriuki *et al.*, 2011). In the landscape, composed of large dense forest, open forest, shrubland, and rock/lava areas (and little cultivated cropland), 11 villages formed (1962–1979). Ten of the 11 were adjacent to or close by dense forest. Within years, a few hundred tiny cultivated fields and two large cultivated fields appeared near the villages. After two decades (1999) and the appearance of 11 more villages, extensive cultivated land covered the area around all the villages. No dense forest remained near the squatter settlements. Apparently, the dense forest offered some benefits to early colonizers, but cultivatable land for crops quickly became an economic foundation.

Buildings in three villages in the Lublin area of eastern Poland provide an interesting picture of changing village and living conditions (Banski and Wesolowska, 2010). A linear agricultural village, Jakubowice Koninskie, within commuting distance of a city, now has 30 percent of the buildings belonging to people from the city (Figure 3.5). In 1970 only nine houses had piped water, and today all have it. New houses are larger and have piped water, piped natural gas, and central heating. Furthermore, these features have spread to other older houses. New houses have concentrated along the main road, often by infilling between existing farmhouses and homes. Also, new houses have appeared along secondary perpendicular and parallel lanes, adding some width to the linear village.

The second village, Krasne, was formerly for farming and now is mainly a recreational/holiday place by a small lake. Eighty percent of the summer homes are less than 20 years old, and half belong to city residents. House plots are tiny, and excessive development has somewhat degraded the lake. Some farm buildings now provide recreational resources.

The third village, Antoniowka, along a transportation route remains as an agricultural village (Figure 3.5). Over six decades it has lost one-third of its original buildings and two-thirds of its population. No school, shop, or milk collection point remains today. Some new residents, not involved with agriculture, have built homes in infill spaces, or remodeled farmhouses, so the still shrinking village retains its compact but linear form.

These three villages are in a rural area (Lublin Voivodship), where on average buildings have the following attributes (in 2002) (Banski and Wesolowska, 2010): 3.8 rooms per dwelling; 0.98 people per room; 85 m^2 (925 ft^2) usable area per dwelling; 60 percent with public water supply plus 20 percent on-site well water; 61 percent with bathroom; 58 percent with hot water; 59 percent with flush toilets; 9 percent with piped

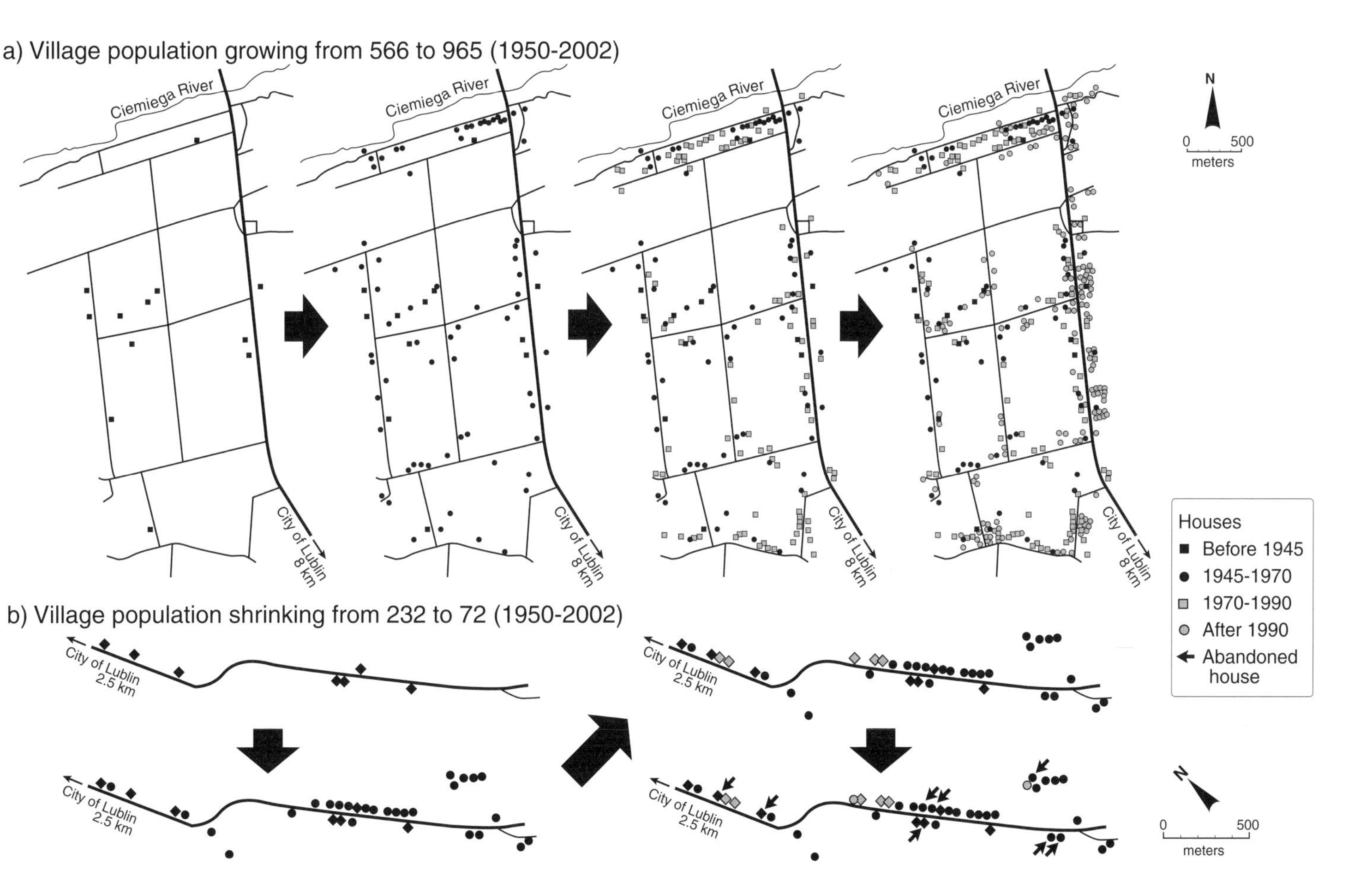

Figure 3.5 Villages growing and shrinking. (a) Jakubowice Koninskie and (b) Antoniowka, eastern Poland. 1950 to 2002: population (a) 566 to 965, (b) 232 to 72; dwellings (a) 126 to 279, (b) 63 to 40. Note that population changes from 1950 to 2002, whereas houses are mapped from before 1945 to after 1990. Based on Banski and Wesolowska (2010).

natural gas; and 54 percent with central heating. Providing these features for daily life in isolated villages with rather few residents and essentially no village government is an accomplishment.

Climate Change and Towns

Town effects on climate change, and vice versa, remain a barely studied frontier. Heat liberated by urban areas has a negligible role in heating the global atmosphere, whereas CO_2 emitted by urban areas is a major contributor of greenhouse gas (Oke, 1987; Alcoforado and Andrade, 2008; Erell *et al.*, 2011; Forman, 2014). Although individually towns and villages are small, cumulatively they cover an unknown but significant area of the land surface. Having barely any heat island effect (Gartland, 2008; Erell *et al.*, 2011; Forman, 2014), towns probably liberate negligible heat to raise global air temperature. The cumulative CO_2 liberated in towns by combustion in vehicles, buildings, industries, and so forth may be a small but significant input to global greenhouse gas levels.

However, the levels of greenhouse gases CO_2 and CH_4 (methane) liberated in raising cattle in the land of towns and villages are considerable. Most likely the amount exceeds that from the towns and villages, and perhaps approaches the amount from transportation or industry worldwide (Chapter 10).

Another way to think about towns and climate change is to consider the well-documented, ecologically related effects of human-caused global warming, and which effects in turn may affect towns and villages. While CO_2 and CH_4 are the most discussed greenhouse gases, others include N_2O, HFCs, PFCs, NF_3, and SF_6. Here is a representative list of greenhouse gas effects on much of the globe (IPCC, 2007; Heller and Zavaleta, 2009; Molina *et al.*, 2014):

1. Global air temperature increasing.
2. Extreme heat waves more frequent and longer.
3. More extreme-precipitation events.
4. Increased flooding in some regions.
5. Ice sheets and glaciers melting faster.
6. Many plant and animal species ranges moving toward poles, and up mountainsides.
7. Spring arriving earlier, with earlier plant phenology and animal behaviors for many species.
8. Some latitudinal migrant birds leaving later in autumn, and not flying as far equatorward.
9. More warm-adapted species in local natural communities.
10. An increase in forest pests, which are normally limited by cold temperature.
11. Oceans acidifying.
12. Sea level rising.
13. Coastal flooding.
14. Salt spreading landward in groundwater intrusion, up streams/rivers, and by wind-driven deposition.

Figure 3.6 A community of cliff dwellings abandoned in the 1200s after a 25-year drought. Prolonged drought usually means former residents do not return. Long House, Mesa Verde, southwestern Colorado, USA. R. Forman photo. (A black and white version of this figure will appear in some formats. For the color version, please refer to the plate section.)

15. Wildfire-burned areas larger, and the fire season longer.
16. Deserts heating up.
17. Increasing diverse negative effects on human health.

Unknown, but worrisome, is the growing risk that, above some threshold level, almost any of these climate change effects will accelerate and cause widespread degradation or destruction.

In the near absence of studies relating these effects to towns and villages, only hypotheses are possible, which we leave to readers and researchers. The numerous small communities, plus their relatively even distribution over large portions of the globe, are probably important in some of the above climate change cases.

Also, towns or villages affected by climate change could in turn affect surrounding lands. For instance, one study found that cities affected by drought lasting up to a decade were resilient, whereas those affected by multi-decade or longer drought (Figure 3.6) effectively collapsed, with extensive population abandonment (deMenocal, 2001).

Major population loss, or even a major economic drop, in a town would normally reduce the number of people working, using, and caring for surrounding lands.

In thinking about how to reduce the effects of, or plan for, climate change at the local level, a study of 3,101 counties across the USA provides insight (Brody *et al.*, 2008). For each local county, three variables were calculated:

(1) *Risk* of severe climate change effects, based on: (a) coastal proximity; (b) previous extreme weather casualties; and (c) expected temperature change.
(2) *Stress* variables contributing to climate change: (a) population not using light-transportation commuting to work by walking, biking, bus, trolley, rail, subway, or ferryboat; (b) CO_2 emissions per capita; (c) importance of carbon-intensive industries, i.e., agriculture, forestry, mining, construction, manufacturing, transportation, warehousing, and utilities.
(3) *Opportunity* to take action on climate change based on: (a) existing solar energy usage; (c) population with higher education; and (c) number of environmental non-profit organizations present.

Local counties with high risk, low stress, and high opportunity were then considered most likely to take action to reduce greenhouse gas emissions and climate change effects (Brody *et al.*, 2008). The Chicago, Seattle, and San Francisco areas, plus most of the northeastern USA within 200 km of the coast (Keenan and Weisz, 2016), fit these three criteria. Also, climate change action was considered more likely if other neighboring counties showed evidence of addressing climate change issues.

A survey, based on 112 scholarly articles, of recommendations for climate change adaptation strategies for biodiversity management identified 104 strategies (Heller and Zavaleta, 2009). The most frequently cited adaptation strategies were: (1) increase connectivity; (2) mitigate climate change with planning exercises; (3) mitigate other threats; (4) study responses of specific species to climate change; (5) practice intensive management to secure populations; (6) translocate species; and (7) increase the number of reserves. None of the 104 adaptation strategies mentioned towns or villages, despite their conspicuous abundance across the land.

Finally, in pondering towns and climate change, it may be helpful to link the issue with some of the most used and recommended landscape solutions for sustaining biodiversity in the face of climate change and urbanization (Forman 1995, 2008; Opdam and Wascher, 2004; Lovejoy, 2005; Millar *et al.*, 2007; Heller and Zavaleta, 2009; Keenan and Weisz, 2016). Consider the following land-protection solutions for biodiversity at the landscape scale: (1) large patches or areas of natural or semi-natural vegetation; (2) corridors and connectivity for movement across the landscape; (3) widely distributed small patches across the surrounding matrix, supporting dispersed uncommon species and providing stepping stones for movement; and (4) protected areas covering the range of topographic sites and rock and soil types present. Land protection slows the rate of climate change, reduces greenhouse gas emission, provides resistance against severe events, and creates resilience to bounce back from destruction (Nassauer, 2011; Pickett *et al.*, 2015; Turner *et al.*, 2016).

No one yet knows how towns and villages would best fit into this landscape pattern for biodiversity (Chapter 13), especially considering climate change and urbanization.

Focusing in on the options for wildlife and people – move, genetically adapt, behaviorally acclimate, or die – highlights great uncertainties (Beever *et al.*, 2017). Clearly rich research frontiers lie just ahead.

Characteristics of a Trajectory

For this topic, let us explore: (1) stability, variation, resistance, resilience; (2) legacy effect and complex resilience; (3) disasters, hazards, threats; and (4) trajectory for planning and management.

Stability, Variation, Resistance, Resilience

Over time, stable natural systems and their interacting components vary, and may fluctuate, within limits. Resistance and resilience strengthened by adaptability from previous disturbances sustain the system.

Human goals are related to these characteristics (Wu and Wu, 2013). In engineering, commonly a goal is to sustain or increase stability (even constancy). Indeed, the persistence of a university stabilizes a university town (e.g., Starkville, Mississippi, USA) (Fallows and Fallows, 2018). Efficiency, optimization, predictability, and/or persistence may be additional goals. In contrast, ecological systems over time are often stable, variable, diverse, resistant (within limits), resilient (able to recover), and adaptable. For instance, mangrove and screw-pine trees (*Rhizophora*, *Pandanus*) are resistant to tsunamis, as water simply pours through the network of stems and roots (Das and Vincent, 2009). A resilient forest may resprout and grow fast after logging. So, for a town with both socioeconomic and ecological dimensions, these types of ecological system characteristics are important in setting goals.

Variation or vary means the system's trajectory, although perhaps stable, is not constant, and *fluctuate* indicates wide variations in the trajectory. *Within limits* means that variation and fluctuation remain as part of the system's trajectory. *Adaptability* is a pliable capacity permitting a system to become modified in response to disturbance, yet persist (Forman, 1995). *Resistance* means the ability to withstand disturbance without exceeding the variability limits (Figure 3.7). *Resilience* refers to the ability to rapidly recover from a disturbance, even an extreme disturbance that exceeds the limits. Resilience prevents the system from being converted to another system, and hence disappearing. Finally, *persistence* indicates that the system lasts or endures "tenaciously," suggesting that it is difficult to exceed the threshold or limit.

Typically, the components of a natural system are plants, animals, microbes, soil, water, and air. These interact in countless ways. A richness of species and their interactions, plus spatial heterogeneity, are generally considered to increase adaptability, resistance, and resilience. Socioeconomic systems also include people, buildings, roads, and other anthropogenic components.

Over time a persistent trajectory can be called *stable with variation*. Using dictionary concepts, the system is on a steady trajectory that is likely to continue, is changeable

Figure 3.7 Mayan temples rising above the rainforest, and courtyard with stelae. Former city of Tikal, Peten, northern Guatemala, which depopulated along with climate change about 850 AD. R. Forman 1962 photo.

in form, and is able to react to disturbance by sustaining or re-establishing its key characteristics.

Thus, the ecological system is not constant, generally not in equilibrium (maintained by opposing forces), and generally not wildly fluctuating. Uncertainty is present, but also predictability within limits. If a disturbance causes change beyond the threshold limit of variability thus creating a new form, a highly resilient system may still rapidly recover. But if the new form becomes stable with its own characteristics and variation, the original system is gone.

Many dampening or *stabilizing forces* to sustain a system's trajectory are familiar (Forman, 1995):

1. *Hierarchy*. Flows, normally from a predominant top level to the bottom level, maintain all levels.
2. *Negative feedback*. One element has a positive or stimulatory effect on the second element, which in turn has a negative or inhibitory effect on the first element; in this way both elements vary and persist.
3. *Inertia of large size*. As a stabilizing mechanism, it is harder to degrade or eliminate a big object (e.g., major highway, large forest) than a small object.
4. *Self-sufficiency*.
5. *Diversity, heterogeneity, and redundancy*. For example, maintain a diversity of employers and jobs, avoid homogeneity, and "don't put all eggs in one basket."
6. *Mosaic stability*. Interactions among neighboring elements dampen disturbance-caused fluctuations.
7. *Network connectivity and circuitry* (alternative routes).
8. *Long-term stable components*. A large percent of residents with three or more family generations in town results in slow change (e.g., Socorro; see Appendix).
9. *Major stable outside force*.

The variation of a system over time is usefully described by the general tendency or trend of a trajectory, plus three variables measuring oscillation around the trajectory (Forman and Godron, 1986; Forman, 2014). The *general tendency* or trend can be indicated as increasing, decreasing, or level. Disturbances or perturbations cause oscillations in the trajectory. In essence, oscillation varies by: (1) *frequency* (infrequent, frequent); (2) *amplitude* or size (small, large, extreme); and (3) *regularity* (regular, irregular).

In general, it is easy to adapt to disturbances that are frequent, small (even large), and regular. Infrequent, extreme, and irregular disturbances are much harder to adapt to. Other combinations can be expected to produce intermediate levels of system adaptability, and associated resistance and resilience. Infrequent events commonly catalyze changes (e.g., St. George's; see Appendix). An extreme amplitude oscillation exceeds the threshold limit.

In a study of land uses in England and Wales over seven decades (1930–1998) (Swetnam, 2007), four patterns of stability were recognized: (1) stable (with small variations); (2) cyclical; (3) dynamic (with large variations); and (4) stepped (pulses of sequential change). Grassland and arable cropland were the leading land covers/uses for all four stability types, with one exception. Woodland, reflecting logging and regrowth, was most common for the dynamic stability type.

Legacy Effect and Complex Resilience

The *legacy effect*, or lag time between disturbance and major effect, is often overlooked in our ecological understanding and models (Foster *et al.*, 2003; Johnstone *et al.*, 2016). In essence, the effects of a disturbance or change may remain for a long time, often unnoticed, and then be recognized as a key factor or variable. In Central America, today the rainforest is different where the Maya had home gardens with tree foods, and then abandoned the spot more than 11 centuries ago (Ross, 2011). Understanding some present pattern or process requires a close reading of long past events.

For example, woodlands in Africa's Sahel region that grew up in somewhat moist conditions can no longer replace themselves, because now the climate there is too dry for successful seedling and sapling growth. The loss of species from a fragmented landscape may occur long after the original habitat was reduced and fragmented (sometimes called an extinction debt). A study of Ontario (Canada) wetlands found that the present species and diversity best correlated with land changes 30 years earlier (Findlay and Bourdages, 2000). In Massachusetts (USA), wetlands are protected by a strong 1970s law, but some wetlands in a national park are due to groundwater blockage by a road constructed in 1806, rather than being of natural origin (Watters, 2016). The legacy effect is much commoner than we realize.

A complex concept of *resilience* (see above) emphasizes the ability of a system, based on adaptation and self-organization, to absorb change and disturbance without changing its basic structure and function, or shifting to a qualitatively different state (Holling, 1996; Alexander, 2013; Wu and Wu, 2013; Alberti, 2016; Davidson *et al.*, 2016). Using part of this concept, social resilience has been described as the ability of a human community to withstand and recover from external environmental, socioeconomic, and political shocks or perturbations (Adger, 2000). In the societal case, separating the idea of resistance (withstand) and resilience (recover from) is useful, since the concepts are different and both are ecologically important.

This approach also indicates that a system passes through multiple stable states separated by thresholds or regime shifts (Folke, 2006). However, if a system is replaced by a different stable state, that is simply a different system with its own characteristics. Normally preventing the loss of a system by transformation to a different one is desirable (Lindenmayer and Fischer, 2006). A persistent meadow rich in wildflowers, grasses, insects, and soil that is turned into a decades-long intensive pasture with livestock, or indeed a housing development, is no longer a meadow. A mined-out mining town becomes a tourist town (see book cover). The Scratch Flat history described above highlighted a sequence of eight persistent land systems, each with quite distinctive characteristics.

Although lots of resilience definitions muddy the water, the concept seems to work fairly well in general systems theory and somewhat in systems-related aspects of ecology, but perhaps becomes diffuse in social contexts (Alexander, 2013; Davidson *et al.*, 2016). At present this resilience concept remains in the realm of theoretical

Figure 3.8 Damaged building and former building site remaining in town center 12 years after a severe hurricane. Note successional vegetation. St. George's, Grenada, W.I. See Appendix. R. Forman photo. (A black and white version of this figure will appear in some formats. For the color version, please refer to the plate section.)

exploration. Therefore, here we use the familiar concept of *resilience* as found in the dictionary and used in many disciplines, i.e., recovery or to bounce back.

Disasters, Hazards, Threats

In considering the stable-with-variation trajectory of a town or ecological system discussed above, extreme disturbances appear as the main threat to persistence. So now we focus on these extreme disturbances which produce, and are commonly called, disasters. But also hazards and threats that may lead to disasters are considered.

In essence, a *disaster* is a calamitous event occurring suddenly. Typically, it causes some loss of human life, extensive damage to built structures, and extensive ecological degradation (Figure 3.8). A disaster may be natural or human-caused, though often natural disasters are worse because of human development, for example, in a flood-prone area or on a steep slope in an earthquake area. In contrast, a *hazard* is something that causes danger or high risk. Often overlooked, a *threat* is an indication or warning of probable trouble. These represent a gradient. The disaster has happened, the hazard is an imminent problem, and the threat is likely to develop into a problem.

Figure 3.9 Volcan de Fuego eruption sending ash skyward and hot lava downslope. Antigua, Guatemala. R. Forman 1963 photo.

Disasters

Almost all towns face the risk of extreme weather events, fires, epidemics, and industrial accidents (Hardoy *et al.*, 2001). Most towns are subject to floods and many to earthquakes. Extreme droughts and famines are a particular risk in rural areas. Tsunamis and hurricanes (cyclones, typhoons) are a coastal risk, though the latter may move well inland. Hurricanes involve heavy rainfall, high wind velocity, and high-tide coastal flooding, with rain-caused flooding normally the most damaging. War/bombing/civil strife normally is targeted to cities, but towns, villages, and their wildlife are often severely damaged or burned as casualties of war (Barrow, 1991; Gaynor *et al.*, 2016). Global warming exacerbates several of the disaster types. Also note that people may be killed by floods, earthquakes, or industrial explosions, but annually many times more people die of preventable/curable diseases.

Disasters last for different periods (Forman, 2014). Usually bombing lasts minutes to hours. Earthquakes and flash floods, minutes to days. Hurricanes and desert dust-or-sand storms, hours to days. Floods and volcanic ash-or-lava deposits/flows, last days to weeks (Figure 3.9).

But resilience can be high, as illustrated in a city park (Halifax, Nova Scotia, Canada) that lost 70 percent of its tree canopy in a hurricane (Burley *et al.*, 2008). Two years later, both the seeds in the soil and the carpet of seedlings present were primarily native early-succession tree species, with few nonnative species. Also soil properties were similar to those in a non-urban control site. Rapid recovery indicates high resilience. Resilience following "superstorm Bob" that recently hit New York and New Jersey integrates socioeconomic, engineering, and ecological processes (Keenan and Weisz, 2016).

Disasters in rural communities are especially damaging because market and government forces provide little insurance or stabilizing effect (Zimmerman and Carter, 2003). Natural ecosystems, especially their geomorphic, soil, and land-cover components, can provide some stabilization (Vira and Kontoleon, 2013). Little evidence yet exists that high biodiversity mitigates the effects of landslides, hurricanes, and dust storms, whereas some evidence is available that genetic variation reduces the effects of flood and fire risk (Ash and Jenkins, 2007). Natural forest cover, presumably with high biodiversity, may increase flood protection (Vira and Kontoleon, 2013). High biodiversity may help regulate fire frequency and severity. Mangrove swamps reduce the risk to communities from tsunamis and coastal storms (Das and Vincent, 2009).

Some unrelated disaster cases add insights. Paradise, a California town of 26,000, was eliminated by a 2018 wildfire; history emphasizes that houses don't fit in a fire-adapted woodland. Following deforestation in Haiti, a 2010 earthquake caused widespread destruction in villages, towns, and cities (Sharma-Laden and Thompson, 2014). A sea-level village by a sandy cove on Fiji, if hit by a cyclone, would probably rebuild as is, because residents know the convenient linkages with both sea and land work well (Barnett and Margetts, 2013). In 2013, a train carrying oil through the town of Lac-Megantic, Quebec (Canada) derailed and exploded, destroying the town center and killing 47 people. In ca. 2003, an earthen dam in Martin County, Kentucky (USA), holding black sludge or slurry from washing coal, broke and poured a wave of 300 million gallons (1.1 million m^3) down the valley (Grisham, 2015). Parts of Oklahoma (USA) seem to be an epicenter of fracking (injecting diverse wastewater deep into the ground to extract oil), and only years later earthquake frequency there has accelerated enormously. In 1966 two hydrogen bombs, accidentally dropped near Palomares, Spain, where vegetables are grown, did not detonate but broke open spilling plutonium, and 50 years later radioactive soil and probably plants cover a significant area.

Finally, over decades, Kyoto, Japan has been impacted by cyclones/typhoons, floods, and major fires (Sakamoto, 1988). Also new developments have wiped out trees, and new trees have been planted in other neighborhoods. Despite these extreme disturbances and changes, the cover of the dominant elm-like trees (*Ulmaceae*) has fluctuated around a rather constant level. The author notes that the residents' appreciation of trees and their benefits stabilizes the tree cover.

Hazards and Threats

Hazards, which cause danger or peril, were highlighted in six towns (population 2,600 to 18,000) in the US Mid-Atlantic region (Greenberg and Schneider, 1996). In each case, an entire neighborhood was considered to be in danger from a nearby hazard. Hazards for the towns were: chemical company; hazardous waste site (major superfund); former pesticide plant/facility; two petrochemical complexes; and county incinerator. Blight and crime-drugs, as internal hazards, were also mentioned in one town.

Eight small cities (34,000 to 125,000 population) in the region had still more adjacent hazards, again a danger for an entire neighborhood: chromium site; landfill; industrial lagoon; former chemical company; numerous small waste sites and abandoned factories; global landfill; international airport; natural gas tanks; and major highway. As hazards, these could become or cause disaster events. Such hazards seem especially risky for towns, which have limited resources for mitigation or to prevent worsening conditions.

Irradiation or radioactivity is an especially dangerous and well-known long-term hazard. The 1986 meltdown of the Chernobyl nuclear power facility (Ukraine) spread radiative material over a distance of thousands of kilometers. Around the close by town of Prypyat, numerous effects on people and ecosystems have been documented and a large area has been fenced off. Numerous other smaller radioactivity release cases exist. At the village of Ozyorsk, Russia (just east of the Ural Mountains), a nuclear-waste tank exploded at a plutonium production facility (Silverman, 2015).

The town of Chimayo, New Mexico (USA) apparently has received radioactivity from two sources (Kosek, 2006). Doubtless over time, radioisotopes have spread in the air from a nearby nuclear research facility (Los Alamos), and Chimayo employees there have inadvertently carried more radioactivity back to their homes and town. Finally, consider Concord (see Appendix), a suburban Boston town, which had a company for decades dumping radioactive and other toxic waste into the soil. The waste spreads, and the hazard worsens. A map of hazards in and near many towns and villages would be eye-opening.

Threats, while less imminent, warn of trouble ahead. Lists of threats to ecosystems, human health, and well-being are published (Sutherland *et al.*, 2008; Stanley *et al.*, 2015). Some examples of issues representing threats illustrate the point: drones; nanotechnology; nature deficit disorder; increasing demand for biofuels/biomass; ocean acidification; and novel pathogens developed by biotechnology. Often threats are little known or overlooked or ignored.

Trajectory for Planning and Management

Above we describe the characteristics of an ecological or natural system over time. Now consider the goal for planning, managing, conserving, designing, or establishing policy for a town or village area, or agricultural or forestry land (Chapter 13). Commonly the objective has been mainly socioeconomic, especially to enhance jobs, housing, transportation, and economic development. In the present case, the objective is both socioeconomic and ecological, and so in part we use the characteristics of an ecological system that is *stable with variation*. Success also depends on effective communication, so we use concepts close to those in the dictionary, and simplify.

Stability or persistence thus is a viable goal for society, but combined with modest variation rather than wide fluctuation. Variation provides some adaptation benefit for resistance, or recovery from, disturbances. Fluctuations normally are too disruptive of the socioeconomic dimension, and thus dampening or stabilizing mechanisms may be needed. Predictability within limits is desirable. Maintain disturbances, at least small ones, which increase adaptability. Maintain the integrity of key system components. Certainly, avoid fluctuations that exceed the limits or threshold, thus potentially changing the system to something else. Avoid constancy or equilibrium, and avoid homogeneity and low diversity in town planning.

In many economic models, the shortage or loss of a resource is an example of a disturbance in ecology. Although economic conditions change, at least three time-tested solutions may prevent the economic system from exceeding the variability limit or threshold, and crashing. (1) Another resource may be substituted for the one in short supply. (2) Or technological development spurred by the shortage may fulfill the need. (3) Or if those don't work, serendipity remains, i.e., something will appear to solve the problem. On the other hand, an engineering goal may provide for constancy or stability, predictability, and efficiency. Both economic and engineering goals vary from the ecological approach of stability, diversity, and variation providing adaptation.

Changing Towns

We begin this section with (1) shrinkage and expansion, and follow with (2) town growth.

Shrinkage and Expansion

The majority of towns worldwide are apparently slowly losing population (Figure 3.10). Yet many stabilizing mechanisms are available for towns. For example, in the western USA (Albrecht, 2014) most residents have a higher level of self-sufficiency than urban residents. Thus, water conservation, money from tourists, "creative class" jobs (requiring complex knowledge or skills), and interest in renewable energy have increased, affecting many towns. Government has put some limits on overexploitation of natural resources surrounding towns. Amenities are plentiful, including views, good air quality, generally favorable climate, diverse waterbodies, and proximity to recreation resources. Adding these together should provide stable trajectories with variation for towns. Ecological lessons from shrinking cities should also be useful (Haase, 2008).

In general, towns seem to have low resistance, but high resilience (Sewell, 2009). Disturbances from floods to loss of a major employer cause considerable disruption. But many town and village residents are highly adaptable. Commonly they have experienced a series of severe floods, windstorms, droughts, and other disturbances. These past events have increased people's *adaptability*, the capacity to deal with disturbances, and provided some resistance against the next disturbance.

Town residents are also *resilient*. In most cases, rather than waiting for distant government to get things going again after a severe disturbance, instead many residents

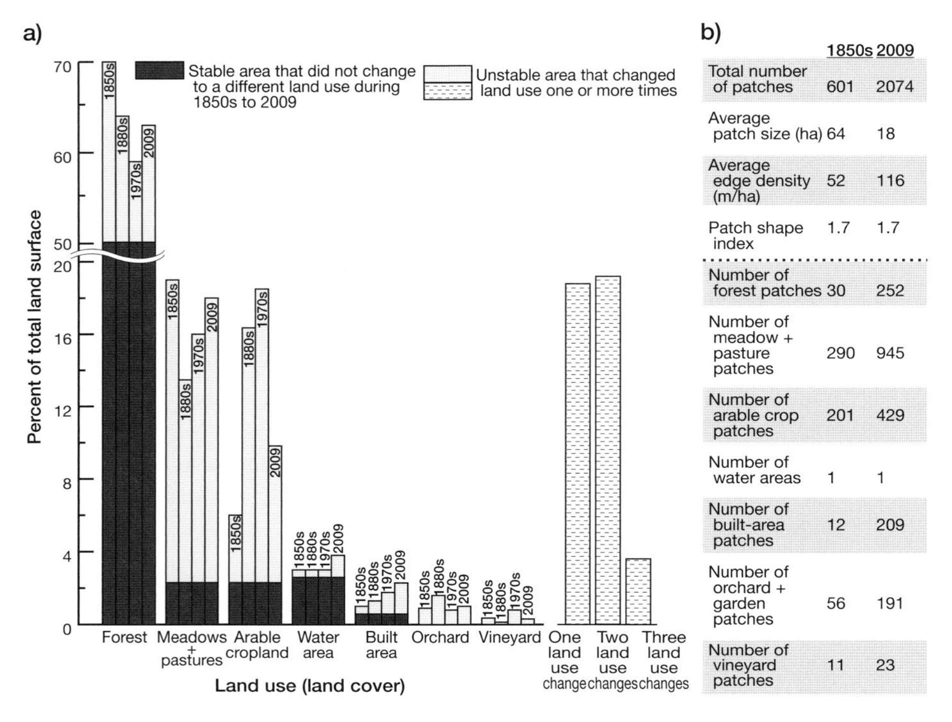

Figure 3.10 Change in land uses and patch patterns over 150 years in a forested landscape with villages. Banat Region by the Danube River in Romania. Land use (land cover) change areas are mainly close to villages. Based on Romportl *et al.* (2014).

quickly dig in to re-open, re-store, re-build, re-plant, and repair the area. Thus, the community bounces back rapidly. Furthermore, without the air pollution, impervious surfaces, traffic effects, and dense population in urban areas, vegetation and animals usually recover quickly from disturbances in towns and villages.

Villages of course also shrink and expand, which have ecological consequences (Browder and Godfrey, 1997). A study of the flora in ten villages in Bohemia, Germany before and after slight growth found that the types of plants present changed somewhat (Pysek and Mandak, 1997; Wittig, 2010). Species indigenous or native to the region decreased slightly with village growth. Plants (archaeophytes) that had colonized the region before 1500 (when world explorers returned with a richness of new species) decreased a bit more. In contrast, plants (neophytes) that recently colonized the region increased by 50 percent. Most of the neophytes are cosmopolitan generalist species that disperse well and colonize diverse suitable habitats (Muller, 2010).

A shrinking farm town (population 11,000 to 6,000, 1970s to 1990s) in South Korea shows continual change in the place (Hong, 2001). Over this 25-year period, forest types, crops grown, cover of plantations, amount of firewood harvested and used, amount of leaves/branches used, and charcoal produced all changed. But also various cultural attributes and attitudes of the people changed during the rapid population loss in town.

Droughts and Floods

Weather of course changes, often rapidly and markedly. Short-term floods, including flash floods, occur along with long-term droughts. Prolonged stress periods and prolonged benign periods also occur. Flood-prone areas such as in southern Manitoba (Canada) and southeastern Texas (outside Houston) have been repeatedly inundated by extensive persistent floods. People, plants, and animals there must have a range of adaptations to flooding.

With weather fronts or patterns commonly covering a large area, many towns and villages experience similar weather conditions. Indeed, communities subject to the same weather often thrive concurrently, but also degrade together, especially in droughts. A striking example is a 25-year drought in the southwestern USA during the 1200s (Figure 3.6). Numerous Native People's villages and hamlets were abandoned, with survivors mainly moving into a small number of remaining communities over an area 1,000 km in diameter. Somewhat analogously, today's global climate change affects essentially all towns and villages.

In the US Midwest, villages (<2,000 population) are most susceptible to closing up (Joan I. Nassauer, 2015 personal communication). Commonly the post office closes, churches are closed or consolidated, the majority of shops become empty storefronts, and wind-eroded soil particles fill the air and cover the place (see Figure 4.9). In north-central Texas, droughts seem to recur about every five years, which means that the economy suffers, population decreases, and towns and villages deteriorate together, though adaptation could be high. After a drought the economy may partially rebound, but the other problems persist. In Kansas (USA), people leaving the shrinking villages and hamlets mainly relocate either to towns (especially county seats) or to the outer suburbs of cities such as Topeka and Kansas City.

Effect of Surrounding Communities

Towns also shrink or expand according to changes in surrounding towns and villages (Figure 3.11). Thus, a town may shrink if the villages in its zone of influence shrink, as during a prolonged severe drought. Or a town may shrink if a neighboring village grows to town size, and effectively becomes the central draw for other villages in the original town's orbit. Or a town may shrink if neighboring towns grow. In all of these cases, the population loss mainly results from people who shop and are locally employed, rather than to young or elderly people.

Conversely, a town may expand as more or bigger villages use it as a hub (e.g., Askim; see Appendix). A town often grows if its neighboring town loses population. If a town and its neighbor have the same type of industry as a major employer, the two industries may collaborate to both towns' benefit. But more likely they compete, and eventually one employer and its town shrink. A town with several industries or major employers tends to be stable. On the other hand, a town with only one or two major employers is likely to fluctuate according to the employers' successes and failures.

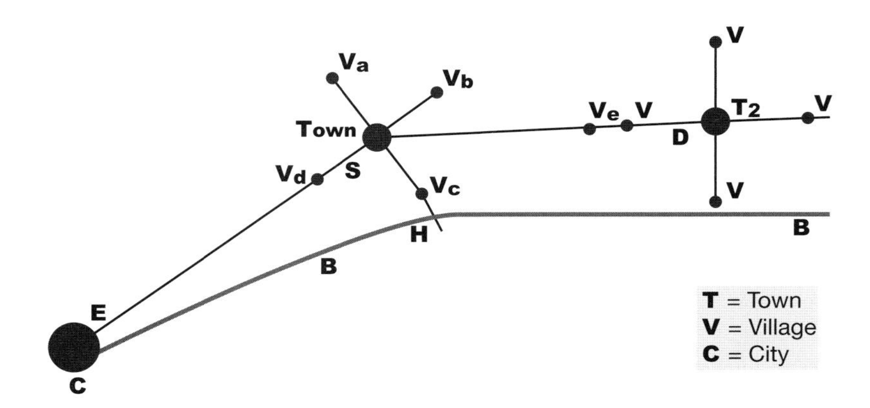

Town may grow if;	Town may shrink if;
Population in villages around town, **V_a**, **V_b**, **V_c**, **V_d** grows.	Population in villages around town, **V_a**, **V_b**, **V_c**, **V_d** shrinks.
Another village, **V_e**, uses town as a center.	A village, **V_b**, grows into a competing town, also serving villages **V_a**, **V_c**.
Neighboring town, **T_2**, shrinks.	A neighboring town, **T_2**, grows, and serves villages, **V_b**, **V_c**, **V_e**.
Major-employment industries in town and **T_2** collaborate.	Major-employment industry(s) of **T_2** outcompetes that in town.
Major-employment industry(s) in town outcompetes that of **T_2**.	A bypass highway, **B**, increases business at highway interchange, **H**, and village **V_c**
Commuters to city increase in town.	A major commercial retail center, **D**, is built on the town side of **T_2**.
Strip/ribbon developments, **S**, expands in direction of city.	A major commercial retail center, **E**, is built on the town side of the city, **C**.

Figure 3.11 Surrounding communities causing town to grow or shrink. Hypothesized patterns vary with distance between communities, and with rate of change.

Town Growth

Typically, towns welcome slow growth, as long as it provides a balance of new people and activities consistent with the existing town. Also, residents do not want growth to alter the "specialness" of the town or village.

Town Expansion and Densification Patterns

Analogous to urbanization by cities, town growth ("townification" or "townization") includes both external expansion and internal densification. Overall, densification seems ecologically better than expansion, though for both, certain spatial processes can minimize problems.

In considering the spatial patterns of *town expansion*, we learn from outward urbanization models (Figure 3.12) (Forman, 2008, 2014; Liu *et al.*, 2003, 2010). Thus, outside towns there seem to be four major ways growth occurs: (1) by bulges (patches sequentially expanding around the town edge); (2) in nearby villages (small population

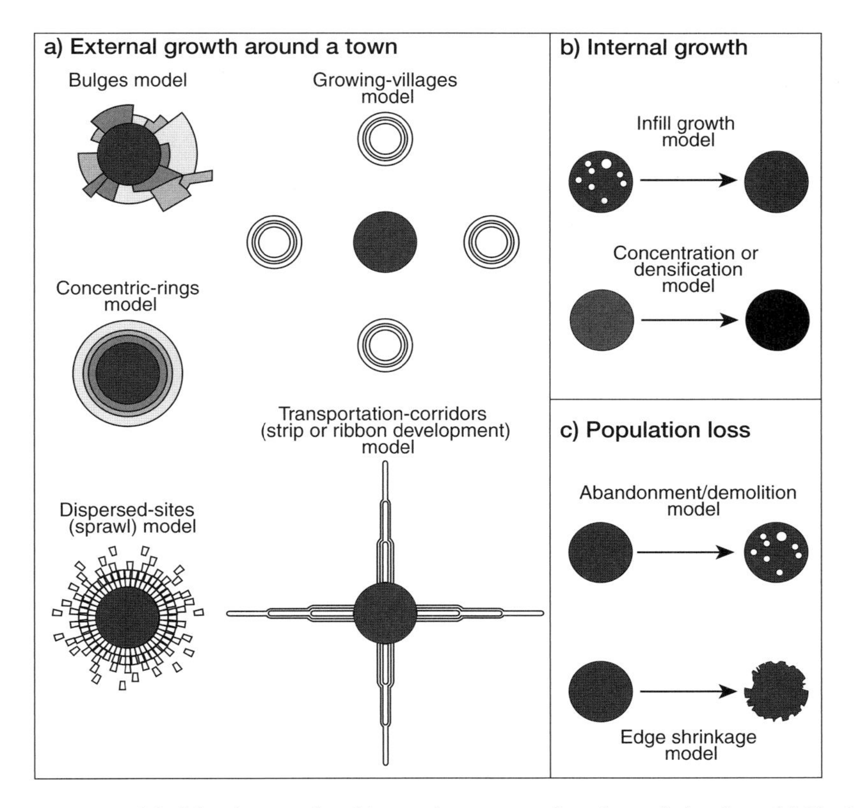

Figure 3.12 Models of external and internal town growth and population loss. (a) Shading indicates three stages of town (central circle) expansion. Adapted from Forman (2014); mainly based on analyzing 38 small-to-large cities worldwide (Forman, 2008).

centers); (3) along transportation corridors (strip or ribbon development); and (4) in dispersed patches (sprawl).

A study of four analogous processes degrading 14 environmentally related variables around 38 cities (250,000 to >10 million population) produced clear results (Forman, 2008). The first two processes (by bulges and in satellite cities) are much better environmentally than the last two (strip development and sprawl). Very little difference in impact was found between the first two, and also between the last two. Applying these urbanization results to town expansion, growth by bulges from a town edge and/or growth of nearby villages are environmentally best (Catalan *et al.*, 2008). To minimize environmental degradation, avoid growth along transportation corridors and in dispersed patches.

A fifth unusual pattern, or modification of the bulges model, is effectively concentric growth. Perhaps applicable to towns and villages, this is expansion house-by-house (rather than housing developments) around the edge.

Densification may include both infill (of unbuilt spaces) and upward growth (increasing the number of building levels on a site) (Figure 3.12). Whale houses in

Nantucket's village of Siasconset, Massachusetts (USA) progressively evolved over time by creatively adding "warts" (attached rooms) (Forman, 1991a). Residential land is sometimes converted to commercial land. Former industrial land commonly changes to residential, commercial, and greenspace areas. Greenspace may change to residential or commercial area. Each of these conversions may improve or degrade the town's overall ecology, including soil conditions, vegetation cover/biodiversity, wildlife habitat/movement, water quantity/quality, and air.

Residents Leaving, New Residents Arriving

Towns shrink or expand primarily by residents leaving and new residents arriving (emigration and immigration), not changes in deaths and births. Immigration/emigration in turn is related to the population's age distribution or stages in life. In general, young people leave numerous towns and villages for the city, while retirees from the city move out to certain towns. For instance, Australian towns have a higher proportion of elderly residents (>60 years old) and a lower percent of residents 15–35 years old than in cities or the nation as a whole (Davies, 2014). Typically, the younger cohort heads to the city for more opportunities in education, employment, and a diverse lifestyle, seemingly a characteristic pattern worldwide.

In Australia the elderly leaving cities mostly head to towns within 300 km of a major city, perhaps to facilitate future contacts with their former community. Also, in this dry continent, they overwhelmingly move to towns with natural amenities, mostly in forestland or near the coast. In smaller numbers the elderly also move from villages to towns. Elderly newcomers might have a smaller ecological footprint than the departing younger residents.

Why do elderly people leave the city? Mainly it is related to a decrease in income, or to be near family or a primary caregiver, or to have institutional care (Davies, 2014). Also, a town's amenities are important. Although small towns or villages may have amenities, the elderly head for larger towns which have good health facilities. Some move mainly for a change of pace and a new social network with cultural activities. A vibrant town center and main street, where the elderly can shop, meet people, and perhaps participate in festivals, is a major draw. Also, towns with an active engaged community plus a stock of affordable housing attract retirees. Employment opportunities and manufacturing diversity are not of much interest. In essence, a distinct set of towns is likely to benefit from population growth by retirees.

Constant Change

A closer look at a town or village reveals that, with a resilient population, it may be constantly changing in noticeable ways. Consider a group of three towns (populations from 3,500 to 12,000) in Hiroshima prefecture (southern Japan) over three decades (1962–1990) (Nakagoshi and Ohta, 2001). Orange production increased, then decreased, and then orchards were abandoned. An aggressive plant (kudzu, *Pueraria lobata*) spread across the former orchards and also pine forest. Upland fields were established and crops planted. Organic fertilizer (litter collected from forests) was replaced by expensive chemical fertilizer. Small and medium pine forest grew into tall pine forest, which turned into tall oak forest. Tall and medium pine forests changed to medium and low forests, presumably by cutting. Farmland fields were abandoned, and forest succession replaced the fields. Forest fires changed the vegetation pattern, and favored pine forest. Seven vegetation types continued to change in percent cover. Farm villages were established on poor sites, but then had aging and fewer farmers. The number of fishermen increased, and then stabilized. The density of working paths decreased. Favorable side-businesses for farmers' families increased. The population of all age groups decreased. Wood and litter used as fuel decreased. Good diversity (balance) of agricultural products was followed by little diversity. Income from non-commercial products was low, and then high. Rapid economic growth was followed by slow growth. The population of all age groups decreased.

Clearly, there was variability in these changing patterns from place to place and time to time. Nevertheless, such conspicuous, mainly large continuous changes describe towns with, e.g., fluctuating agriculture, vegetation, activities, economy, and population. Continuous changes produce valuable adaptations in people and towns.

In short, the history of a town portrays the never-ending changes in and near it. Concord (see Appendix) changed from a 1635 village in the forest to an eighteenth- and nineteenth-century agricultural town to a twentieth-century suburban town (Donahue, 2004). Meanwhile, pulses of little changes involving land use, buildings, roads, vegetation, water, and people never stopped.

Town Edge and Strip Development

Adjacent to a town's edge is usually productive agricultural or natural land, such as crops, pastures, woodlands, and/or desert, whereas adjacent to urban built areas often abandoned land is conspicuous in places (Forman, 2004a). A study adjacent to a large town with a long history of silver mining (Kutna Hora, Czech Republic, population 20,300; see Appendix) found that 17 percent of the area within approximately 2 km of town was unused abandoned land (Lipsky and Kukla, 2012). Apparently, the percentage was higher close to the town edge or outskirts. Abandonment had occurred before 1990 (38%), during 1990–2000 (33%), and 2000–2010 (29%). The greatest unused area was abandoned orchards and gardens (37%). Abandoned arable (crop) land, grassland, and mining area each were 10–15 percent. Abandoned industrial, infrastructure, floodplain, and pond/wetland areas were less common. Many ecological attributes are favored by land abandonment, while some ecological and human attributes are unfavorable (Lipsky and Kukla, 2012).

Adjacent to a town edge is a prime place for development. At least a dozen types of structures and land uses (e.g. cemetery, fairground, local airport, sewage treatment facility) are often built or created there (Chapter 1), each with distinctive ecological effects.

Strip or ribbon development on a town's radial road usually extends only a short distance. Typically, it is composed of scattered houses, with a small number of commercial retail establishments and sometimes small manufacturing. Unlike urban strip development which, combined with the road and traffic, is a major barrier to wildlife movement across the land, town strip development normally is a porous filter to movement. With less commercial activity along the road, there is less water pollution, truck delivery, waste removal, night light, and traffic.

Development Outside Towns

This important topic is considered in: (1) landscape change; (2) town development outside cities; (3) house-plot size; (4) development in the land of towns and villages; and (5) scenarios for future growth.

Landscape Change

Towns change, but in the context of a changing landscape. Much is known about the patterns, causes, and types of landscape change (Forman, 1995; Lindenmayer and Fischer, 2006; Franklin *et al.*, 2007; Turner *et al.*, 2013; Thompson *et al.*, 2016). But most studies focus on land uses (covers) such as forest, grassland, cropland, and wetland, and surprisingly either ignore or only peripherally include conspicuous towns and villages in analyses.

Examining a mosaic of patches and corridors on the land over a century (1884–1981) emphasizes that some types of elements are relatively stable, while others disappear rapidly (Agger and Brandt, 1988; Forman and Godron, 1986). Thus, in an agricultural land of towns and villages in Denmark, the investigators determined the fate of 1,566 spatial elements (Figure 3.13). These represented 11 groups of patch/corridor types. Three types were most stable, i.e., remain unchanged over the century: gravel pits (barrows); water elements (water courses, ponds, wetlands); and roads (gravel and paved). The least stable element types were on well-drained soil: footpaths, hedgerows, dry patches, dikes, ditches, and other linear corridors. Overall, most of the stable elements were patches, whereas almost all the unstable elements were corridors.

Also, during the century some elements, especially hedgerows and dry patches, appeared in new places. On the other hand, small elements of many types disappeared. Semi-natural vegetation elements in this intensive cropland were especially present along boundaries, such as between farms, along roads, and along county boundaries.

Shortly after this study, apparently a law passed requiring farmers to leave a strip of vegetation along all farm boundaries, and a wider strip along county boundaries. That then produced a large ecologically valuable habitat network which continues to enhance wildlife movement. The network helps tie the towns, villages, farmsteads, and land together.

What are the primary driving forces leading to landscape change? Studies in Quebec (Canada) and the Limpach Valley (Switzerland) suggest that there are three broad driving forces: key geomorphology-related characteristics; demand for resources (or goods and services); and technological change (Domon and Bouchard, 2007; Burgi *et al.*, 2010). The first two driving forces exist in towns and the third to a limited extent. Only the first applies well to villages.

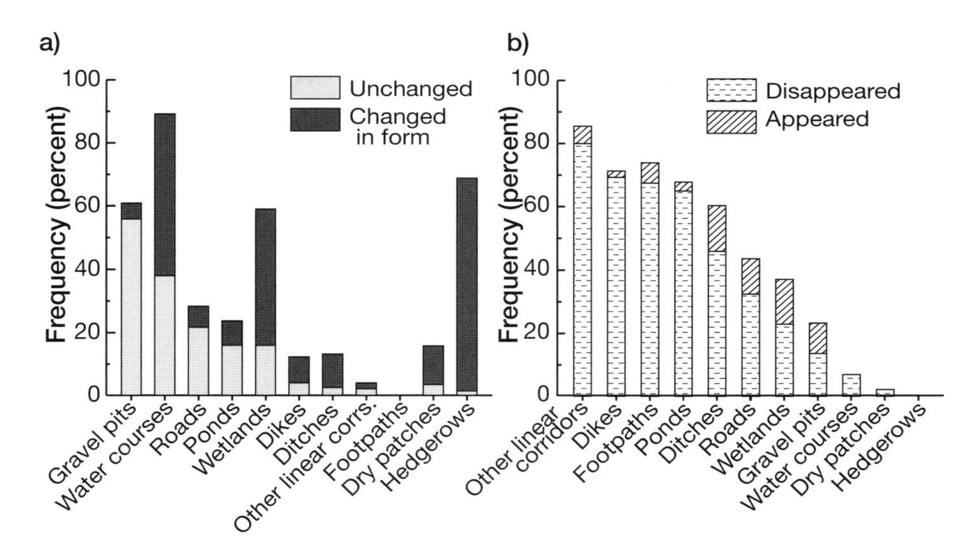

Figure 3.13 Changes in small elements of the landscape during a century of agricultural intensification. The fate of 1,566 elements in 11 categories from 1884 to 1981. Based on sampling 249 1-km^2 plots evenly distributed across eastern Denmark (Agger and Brandt, 1988).

Town Development Outside Cities

Russia is a nation of large cities and isolated towns and villages (Ioffe and Nefedova, 2000; Vaz *et al.*, 2013a). Moscow residents buy weekend/holiday houses in growing towns and villages within about three hours of the city, and some 20 percent of village houses are used seasonally by city residents. But towns also grow by people arriving from villages and towns in more remote areas. Irrespective of population size and people's origin, vegetable gardens and small orchards are abundant in and around towns and villages near Moscow. Such vegetable and fruit production near the city is for families, and not equivalent to market-gardening production for city markets as near Barcelona, Valencia, and London.

In an area of towns and villages outside the city of San Jose (Costa Rica), birds were recorded before and after a 16-year period of extensive housing development (Biamonte *et al.*, 2011). During this landscape change, 32 resident species disappeared, especially habitat specialists, and 34 migratory species are no longer detected, with their stopover habitat mostly eliminated. In addition to this biodiversity loss, the abundance of both resident and migratory birds has decreased. Wildlife patterns are ever changing.

Low-density settlements around Rome seem to correlate with two urban patterns, a decrease with distance (urban-to-rural gradient) from the big city's edge, and population aggregation or increase around satellite cities (Salvati *et al.*, 2012). This creates a polycentric pattern of sprawl. Low-density housing development is most likely to occur on former arable crop land, heterogeneous agricultural areas, and "dispersed urban fabric." A small amount of such development occurs on vineyards, forests, and olive groves, whereas little occurs on orchards, low-vegetation areas, and parks.

Patterns of settlements outside cities in the USA seem to fall into four categories (Clark *et al.*, 2009): clustered, and continuous with the urbanized area; clustered and isolated; unclustered and isolated; and linear (along roads). Towns, villages, and other buildings outside cities reflect both the pre-city settlement pattern in the land, and later urban expansion patterns from the city. Towns near a major city typically grow until they are absorbed into suburbia, and effectively disappear as a distinct town independent of urban forces.

House-Plot Size

In all the town expansion patterns above, the average residential house-plot size, and hence the total amount of land developed, can vary from small to large (Theobald, 2001; Gordon *et al.*, 2005). Growth by bulging outward and in nearby villages is adjacent to existing built areas. Using small to medium house-plot sizes, this growth pattern maintains compact population centers. Large house plots produce ecologically degraded land, and if the plots are dispersed outside of towns, the amount of degraded land is considerable. Yet by using smaller plots, aggregating them, and building communities rather than mainly housing, the towns around Toronto ended up with twice the housing density – half the sprawl – as around comparable towns in the USA (Sewell, 2009).

A key exception to the negative effect of large house-plot development is where the plot is mainly used for productive local food or other societal value (Forman, 2008). Homegardens, especially in the tropics, may often produce a fifth of a family's food-related needs (Arifin and Nakagoshi, 2011; Arifin *et al.*, 2014). That has the added benefit of providing stability for the family during difficult times regionally or nationally.

Development in the Land of Towns and Villages

Development in a rural mountain area (Macon County, North Carolina, USA) from 1906 to 1960 was categorized as rural (Kirk *et al.*, 2012), based on measuring housing density in forest and non-forest areas (Hammer *et al.*, 2004; Theobald, 2005). Then development from 1960 to 1975 was exurban, and since 1975 development has been mainly suburban around two small cities present.

In a large area close to protected forestlands (Southern Rocky Mountain Ecoregion of Colorado, USA), 8 percent of the total area was estimated to be modified by human land use, such as agriculture and housing. But 13 percent of the area was wildlife habitat modified or degraded by this human activity (Leinwand *et al.*, 2010). Agriculture modified the most land area. Residential (exurban) land use had a lower impact, but was especially significant by being close to protected areas, dispersed in pattern, and rapidly increasing. It would be interesting to compare the ecological effects of the long-term towns and villages versus the effects of the recent low-density exurban housing spread.

Around 57 large national parks in the USA, population density is relatively high and associated housing widely distributed (Davis and Hansen, 2011). The housing, much of which is rather recent, consists of both rural and exurban types (Figure 3.14). These

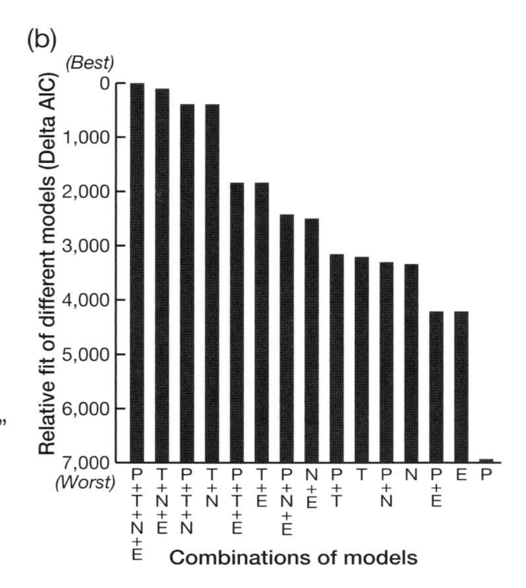

Figure 3.14 Relative fit of different spatial models for patterns of growth in the Greater Yellowstone region. Initially new houses during 1970–1999 in 2.6 km^2 "sections" were separately correlated with 45 natural and socioeconomic variables. Primary variables listed in (a) were mainly selected from the highest ranked 29 variables (all $p < 0.0001$). Based on the 20 Montana, Wyoming, and Idaho (USA) counties surrounding Yellowstone and Grand Teton national parks. Adapted from Gude *et al.* (2006).

national parks are basically ringed by development, which does not bode well for wildlife movement or biodiversity in the parks (Soule, 1991).

Over a decade (1998–2007) in Northern Ireland, most construction was on productive agricultural land (McKenzie *et al.*, 2011). Limited building occurred on many other habitat types, some with valuable biodiversity, which covered less area. Considerable development was of buildings next to existing buildings, an ecologically valuable approach illustrated by recent patterns in parts of Germany and Israel.

Many scholars have suggested that low-income residents, especially in villages, have a greater dependence on biodiversity for their income than other residents (Roe *et al.*, 2013). Some studies support the hypothesis and others do not (Vira and Kontoleon, 2013). Overall, at present the evidence is weak.

In a sense, Rachel Carson's 1962 book, *Silent Spring*, ushered in the environmental age. The book movingly described pesticide spraying against insects in towns of New England (USA). The persistent DDT pesticide spread widely in the environment, and the spraying spread widely across North America and elsewhere. Consequently, key bird and insect populations decreased over extensive areas including in towns and villages.

Scenarios for Future Growth

Scenarios for the future often focus on landscape change, outlining alternative futures (Steinitz and McDowell, 2001; Thompson *et al.*, 2016). Options commonly include no change, what may occur, and what may be best or hoped for. The assumptions in such scenarios or predictions are of course critical. Thus, an economist concludes that, with 100 million more people in the USA, the "Heartland" (mainly the Midwest), which today continues losing population, will thrive economically with vibrant growing communities (Kotkin, 2010). This scenario may be a combination of what the available socioeconomic data and trends indicate, plus the author's preferred future.

Four scenarios have been outlined and compared for the future of forest in a densely populated state (Massachusetts, USA) (Thompson *et al.*, 2014, 2016). Three scenarios are indicated as undesirable: recent trends continued; mainly economics-driven, with minimal government regulation; and biomass harvesting for energy and forest clearing for agriculture. The optimum recommended scenario highlights a multiple-use future forest, as a source of low-cost carbon storage, renewable energy, local wood products, clean water, and wildlife habitat (Foster *et al.*, 2010, 2017). Such a forest would also provide ample biodiversity and recreational opportunity. Linking such scenario analyses to the arrangement and changes in the towns and villages present might add considerable insight.

Change is the norm, constancy the exception, and therefore particularly interesting.

4 Human Dimensions

Human life swings between two poles: movement and settlement … one trades mobility for security, or in reverse, immobility for adventure.

Lewis Mumford, *The City in History*, 1961

While nature is at the heart of ecology, people are central to towns, villages, and their interactions with farmland and natural land. To understand the essence of the human dimensions, we start with (1) living in a town or village. Then we explore (2) culture and character, (3) economics, jobs, and commercial/industrial areas, and (4) social patterns and policy/planning.

Living in a Town or Village

Why would someone prefer to live in a town or village rather than a city? Clearly, some people love being in cities while others thrive in smaller communities. Consider the advantages and disadvantages of towns and villages: (1) a good place to live; and (2) problems living here. Then try (3) peering into a village.

A Good Place to Live

Towns are often in an attractive setting. Town neighborhoods lie close to outside rural open space, and residents have convenient access to nature (Jackson, 1952; Friedman, 2014; Arendt, 2015). Many residents are tied to the surrounding land, where important supplies and services are present for economic activities. An accessible diverse resource base surrounds many towns (Berry, 1977; Sargent *et al.*, 1991; Kosek, 2006). Wildlife frequently enter residential areas, and livestock are commonly seen. A water tower often stands tall as a visual icon, and a dependable source of water pressure for every building. Towns provide food and fuel for travelers.

The town form and its internal features are appealing (Francaviglia, 1996). As a small area at the human scale, typically one can walk throughout the town, usually with many north–south and east–west streets easy to navigate. Sounds, noises, smells, and scenes are quite different from those in urban areas. The town center is anchored by a central public plaza or square, either a green space or containing a prominent building. The mixed-use center with commercial shops and residential housing provides ample

Figure 4.1 Black cat and witch metalwork as cultural symbols over small city of Ghent, Belgium. R. Forman photo.

gathering and meeting places. Some streets are narrow, and large old trees embroider the old residential neighborhoods nearby. Perhaps a single traffic stop light changes color at the main crossroads in town, reflecting limited vehicle traffic and the importance of pedestrians. A waterbody such as a stream is usually present, often with some hardened streambanks which accelerate water flow.

A long history breeds *tradition*. Many residents have a strong sense of place (Jackson, 1952; Wuthrow, 2013; Friedman, 2014; Arendt, 2015). Residential life often includes a supportive family or relatives present. Many residents, and many city dwellers who return for a visit, grew up here and fondly remember the place in earlier days. Or their parents or ancestors were from here. Some residents take comfort in eventually being buried here.

Many cultural traditions remain conspicuous (Figure 4.1). Vernacular and heritage buildings are visible. Cradle-to-cradle life-cycle thinking (Simonen, 2014; Anton *et al.*, 2016) may apply to town roads, houses, even developments. Normally growth is incrementally outward from the town edge. Most change is slow. Pride is manifest in historical buildings, a museum, and special sites. Towns have a higher proportion of elderly residents with experience and wisdom than do cities. Former leading residents are honored.

Figure 4.2 Children playing on an engine in central town park. Town of St. Malo, Bretagne, France. R. Forman photo.

This is an authentic place where residents feel a self-identity (Wuthrow, 2013). Cooperation and sharing are conspicuous and contagious. Little rituals become motifs, and social networks vibrant.

Towns have a local government with relatively little bureaucracy between residents and leaders (Sargent *et al.*, 1991; Young, 1999). Residents know or recognize other residents (Hardy, 1886; Faulkner, 1957; Wilder, 2013). Residential neighborhoods are mostly close to town center (Figure 4.2). Small housing units are common and housing is relatively affordable. Resources are available for children and the elderly. Pets are widely valued. An annual fair and periodic festivals stimulate social interactions and economic benefits (Gibson, 2014). Some residents value *traditional living*, in touch or harmony with nature, seasons, and history (Berry, 1977; Vaz *et al.*, 2013b; Nakagoshi and Mabuhay, 2014). Towns have a relaxed lifestyle.

Local jobs, shops, and businesses thrive in perhaps most towns (Jackson, 1952; Mumford, 1961; Friedman, 2014; Arendt, 2015). Home-grown initiatives and local homemade products are common. The three legs of sustainability (economic, social, and environmental), plus local culture and local government, are valued by most residents. The town center business section is typically several blocks of two- or three-level buildings, at a human scale (Francaviglia, 1996). Commercial retail shopping, including many mom-and-pop shops, is sufficient for daily requirements and some specialized needs. Many residents can walk or bike to work. Vehicle traffic is limited and slow. Manufacturing pollutants are limited or often limited to specific locations. Conveniently, the town center portrays the condition of a town.

Villages have many of the same features as towns (Arendt, 2004, 2015). A village is a place, simple in form and initially readily understandable (Chapter 1) (Sharp, 1946; Roberts, 1982; Wood, 1997). The village green or public space is a magnet for events and games, and the common building is the center for meetings. Handmade products may be important. Food is stored, providing stability for difficult times (Mumford, 1961).

Remember the adage: it takes a village to raise a child. Village families nurture children, and care for the elderly. Almost all homes border farmland and/or natural land. Livestock are typically visible, and wildlife from the surroundings are common in town.

Towns and villages are great places to live.

Problems Living Here

Towns by the hundreds, indeed by the thousands anchor the land. Yet, as J. B. Jackson (1952) poignantly observed, "you have never stopped there except to buy gas." Even ecologists, closely attuned to the land, traditionally think of towns as simply places to get food or fuel. Pickup trucks commonly rule the roads in town, and even common pigeons and house sparrows (*Columba livia, Passer domesticus*) may be glimpsed.

A town often has little or no zoning, so a home may be adjacent to a vehicle service station or a pizza place staying open late. Car parks are small, and vehicles parked along streets mean the traffic moves slowly (Arendt, 2015). The choice of jobs is limited and salaries often low. Some towns are effectively divorced from their surroundings by land managers from afar, making jobs even scarcer (Kosek, 2006; Nakagoshi and Mabuhay, 2014). This is tough for a town's hunters and fishermen, and also for the often less influential local nature and environmental groups.

Over time, solid waste has been buried by wetlands or waterbodies, and along town edges (Mumford, 1961). A waterbody in town gets lots of liquid wastes, and has a bottom strewn with a surprising array of discarded objects.

The infrastructure for piped services, such as stormwater, sewage, natural gas, and more, is limited. Toxic chemicals may persist in the water supply, yet chemicals and materials used in town receive limited monitoring.

Employment for town residents is a complicated issue. A single large company or industry over time typically leads to a *company town*, where the company exerts control over many aspects of the town's and residents' life beyond employment. When the company eventually closes in town, a convulsion occurs. Many residents remain unemployed or depart, and many local shops close. A few medium-size manufacturers, or many small/medium ones, provide a diversity of employment opportunities, and some stability, when a company closes (Faulkner, 1957; Forman, 2014). Yet a government or corporate entity controlling economic activities in town may hire workers from elsewhere rather than town residents.

Residents worry about their job and economic future (Sargent *et al.*, 1991; Wuthrow, 2013). Having grown up in town, they also worry about the younger generation. Limited opportunities will send many youth away (Turker *et al.*, 2014). Will they return? When? Indeed, some town residents worry about the future of the town itself.

Problems in home, neighborhood, and workplace characteristic of small cities (Hardoy *et al.*, 2001; Bell and Jayne, 2006; Friedman, 2014) are also prevalent in towns: pollutants, pathogens, shortages affecting health, and physical hazards. Unregulated or little regulated cesspools and pits, dumps, sewage disposal, dogs, smells, and erosion are often widespread. More elderly residents mean more need for medical care.

Market days, festivals, and local fair days, especially with livestock-related activities, produce noise, smells, and rodent pests (Brown, 1986; Powe *et al.*, 2007). Long rows of stalls sell food, other organic items, and diverse goods, all resulting in rubbish. Crowds of people eat, discard solid waste, create sewage, and overuse parking facilities.

Disasters are rare, but especially devastating in a town or village (Greenberg and Schneider, 1996; Hardoy *et al.*, 2001). Consider the recent Lac Meguntic (Quebec) explosion in town, Mt. Etna villages below lava flows, severe earthquakes of Nepal/Chile/Pakistan, Java/Fukushima tsunamis, flooded towns around Houston (Texas), major wildfires in fire-prone areas, and regional conflicts over time. Disasters also strongly affect ecological and biodiversity conditions, both positively and negatively (Vira and Kontoleon, 2013).

Villages have some of these problems, plus others (Sharp, 1946). Newcomers remain outsiders for a long time. Houses are often named for their previous resident. Many jobs may be mainly limited to agriculture (or forest industries). Schooling opportunities are limited. Public transportation is usually limited or absent. Shops for daily needs are commonly present, but with limited diversity/selection and competition. On-site waste or wastewater handling varies from good to bad.

Comparing the pros and cons of town or village life leads to the conclusion that these places are highly desirable. Comparing the features of towns/villages with those of urban areas indicates that many people would relish small population centers, while many others greatly prefer cities. No doubt towns and villages will remain as key features of the land for generations. Communities that care for and capitalize on their natural features are especially promising (Vaz *et al.*, 2013b).

Peering into a Village

Consider a village at the junction of two streams and surrounded by farmland (Figure 4.3). Kaihsienkung (Kaixiangong) (China; see Appendix) in 1935 had 1,500 residents on 184 hectares (461 acres) (Fei, 1939). The surroundings were mostly rice paddies, with some wheat, rapeseed, and vegetable fields. Fish, shellfish, and aquatic plants in the streams were also harvested. Small groves of mulberry trees supported caterpillars that make silk, so a small silk factory was present. Both local jobs and the economy were diverse.

Water was central. In the floodplain water was everywhere, as paddies were connected by irrigation canals bringing in water, plus drainage ditches getting rid of water. For good rice production, every paddy needed just the right amount of water, and not too much. Villagers collaborated to achieve this.

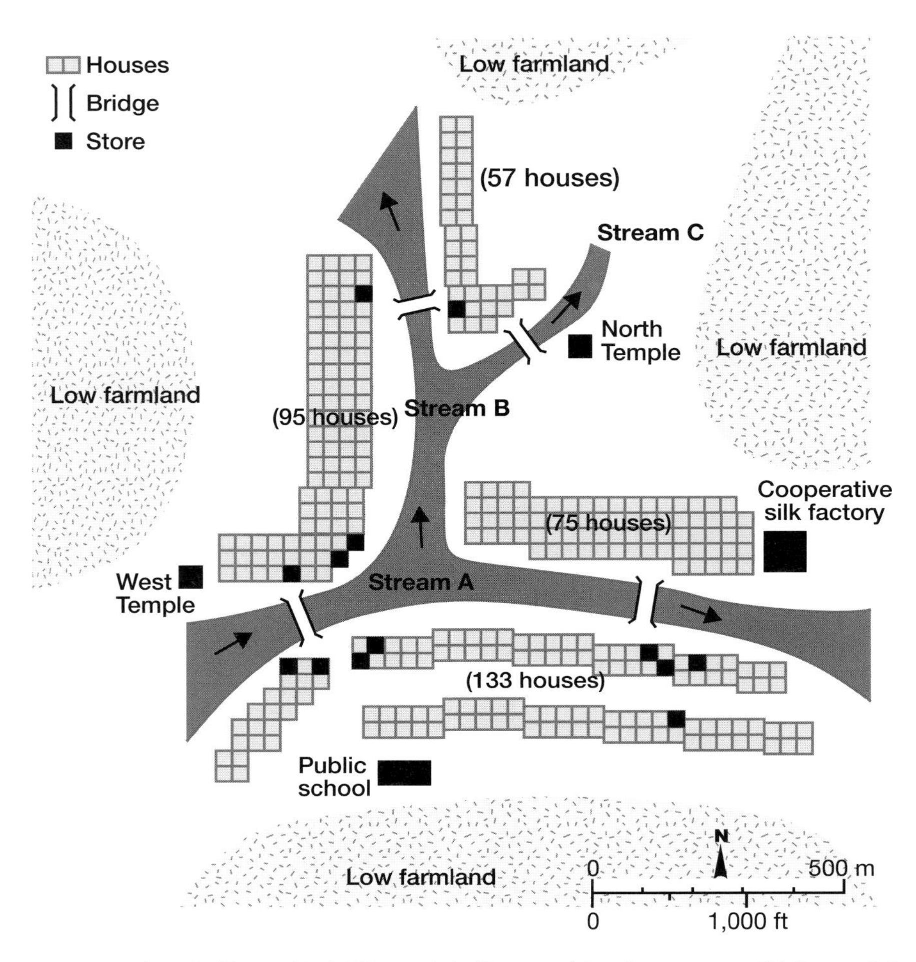

Figure 4.3 Plan of village with 1,458 people in Yangtse plain of streams, small lakes, and rice fields. Village center on slightly elevated land surrounds the bridge and stores in lower left. Nearby villages are walkable, many about 1.5–3 km apart. Boats are used for transport, selling goods, and fishing. Wet-dry monsoon climate. Kaihsienkung village, China, in 1935 near southeast end of Lake Tai. See Appendix. Based on Fei (1939).

Most of the houses were aggregated around the stream junction, where narrow bridges tied the village and its fields together. Trees were small and sparse. A school was added at a central easily walkable and monitored site. Because everyone depended on the stream water, the recently built silk factory and its pollution were at the most-downstream site. Streams and adjoining groundwater provided household water supply, as well as some waste removal.

Weaving was common in the homes, and 7 percent of the households contained craftsmen or professionals. Ten households were immigrants from elsewhere. All were well known and reasonably well integrated in the community. Most of the outsiders worked as craftsmen or in the tiny local shops, but none as a farmer.

Many houses had a tiny vegetable garden growing gourds, cucumbers, and the like, and a few had a hut for storage, even for keeping sheep. Human manure was the most important fertilizer, so it was preserved in pots which were buried and later used for farming. Clean clothes were hung out to dry. House maintenance required regular oiling of the wood and rearranging of roof tiles.

Culture and religion were stabilizing forces in the community (Fei, 1939). Education, thrift, repairing/recycling, and preparing for disasters were the norm, as *guiding principles*. Women and children commonly met in homes, while men often met or gathered in teashops. Local festivals, mostly religious, occurred occasionally, but village meetings were infrequent. Village elders met as needed at the silk factory to discuss matters of common importance.

Roads across the floodplain provided for local movement, while boat traffic provided long-distance travel. Every day local boats from neighboring villages with access to tillable soil moved slowly along the banks with goods to sell. Village women negotiated for vegetables and diverse objects of daily life.

The villagers were *adapted to disasters* that reoccur, and pondered readiness for new ones. Floods and droughts were fairly frequent though at irregular intervals. Locust outbreaks occurred less frequently. The introduction of new machinery was a perennial threat to the way things were done. The *most urgent economic problem* was broader-scale industrial reform. Technology and organizational changes in silk production had lowered the price of silk precipitously. Consequently, the village silk factory may soon be vestigial. Village jobs and its economy were in jeopardy.

So, like so many villages today, this 1930s village in China had created a local diverse economy based on surrounding resources and its own small industry (Figure 4.3) (Fei, 1939). (Ironically, villages usually do not have a factory.) The village structure fit into the land, though it had eliminated most floodplain vegetation and probably degraded a relatively small stretch of stream habitat. Villagers lived and worked together, and were daily tied to other communities. Yet big, widespread natural, as well as societal, forces shaped their thinking, their lives, and their uncertain future.

The author observed that the village's future remained stunted or uncertain, because of a scarcity of incoming new residents and related diversification (Kirkby *et al.*, 2006). He noted that viable small towns can be effective constraints on excessive migration to cities, and concurrently sustain a productive and vibrant countryside.

Culture and Character

Culture and Natural Heritage

Culture may be described, based on dictionary concepts, as the intellectual, aesthetic, and moral traditions passed by a group from generation to generation. Such traditions include learning, art, music, dance, morality, and ethics (Figure 4.4).

Nature is closely linked to almost all components of culture (Berry, 1977; Kosek, 2006; Kienast *et al.*, 2007; Wuthrow, 2013; Jackson, 2014). For instance, an old forest may provide medicines, foods, and diverse products for clothing, hats, hunting, fishing,

Figure 4.4 The Little Theatre in town of Pompeii, 77 AD, which was covered by hot ash from eruption of nearby Mount Vesuvius. In a theatre, culture is alive and passed through generations. South of Naples, Italy. See Appendix. R. Forman photo.

dance, musical instruments, and decoration, especially for native peoples. Nature's objects also provide meaning. The seashell on a desk or mountain painting on the wall of a skyscraper office often has deep meaning to the person there. In addition, the object evokes different meanings to others who see it. The nature–culture link is strong, and ever changing.

Town and village residents commonly live near *woods* along a stream or river, near wooded patches at the edge of town, and with extensive natural land further away. Each of these pieces of *nature has meaning*, but different meaning for diverse residents and groups of people. For some, the woods are dangerous, places to be controlled, pushed back, turned into productive agriculture, or converted to much-needed housing. To others, the natural places are exhilarating, places to be protected, treasured, viewed with wonder and awe, worshipped, bathed in, or perennially explored. The nature around us is part of the town's heritage experienced by preceding generations, with legends and stories, art and ethics passed onward to today's residents. Nature evokes powerful sentiments and is tightly linked to culture in villages and towns.

○ = Biomass
● = Biodiversity

Conservation mechanism	Poverty-reduction benefit	Very poor	Moderately poor	Better off	Land-owners
Locally managed marine areas	High	○	○	○	
Mangrove conservation	Medium	○	○		
Community forestry	Medium	○	○	○	
Agro-forestry	Medium		○	○	○
Commercializing non-timber forest products	Low	○		○	
Payments for environmental services	Low				○
Nature-based tourism	High		●	●	
Agro-biodiversity*	Medium		●	●	
Protectetd-area jobs*	Low		●	●	

Group that benefits

Figure 4.5 Beneficiaries, resources, and amounts of poverty reduction provided by different conservation mechanisms. Based on >400 documents for rural areas worldwide. * = based on <10 studies. From Leisher *et al.* (2013).

Consider a town with extensive nearby forest, where growing and harvesting trees, as well as hunting and fishing, are important to residents (Figure 4.5) (Kosek, 2006). For wood production, both adults and children have repeatedly seen and are somewhat knowledgeable about a series of basic *forest-ecology dimensions*. These often include tree species, seeds, seedling growth and competition, sapling dimensions and thinning, growth curves and when to cut, alternative types of cutting, logging processes, big logging trucks, mills and markets, soil erosion/sedimentation, water infiltration, downslope water supplies, old growth, and what residents do for each of these processes and phases.

The tradition of hunting animals in the same forest adds several more *cultural and wildlife biology dimensions*, including: safety, careful use and maintenance of equipment, understanding how animals use and move in the land, traditional food gathering, preparing and transporting a carcass, minimizing wastage, being challenged by the hunt, getting temporarily lost, sharing the experience with a family member or friend, and understanding the idea of living off the land. Fishing has analogous traditions. Towns and villages often bulge with such smart, knowledgeable, and adaptable people, steeped in first-hand experience in addition to book learning.

Indeed, the nature-and-human relationship gives rise to, and is a result of, *value systems* (Kienast *et al.*, 2007). One's value system underlies both perception and valuation. It may lead to, for example, openness or resistance to change, self-enhancement, in-town or broader-scale actions, or nature conservation. Some might value and attempt to conserve rare and threatened species. Others value ecological functions, such as providing clean water, protecting against soil erosion/sedimentation, sustaining harvestable species, or developing resistance or resilience in the face of expected disturbances.

Culture gives meaning to objects around us, as well as for life itself. Consider the objects dominating an agricultural village of 400 people (Peterson, Iowa, USA), and what the objects mean for residents (Hokanson, 1994). Geological processes helped locate the village, where a small river winds through a heterogeneous mosaic of soils, mostly rich for agriculture. A road to elsewhere was built a century and a half ago. Cornfields (maize) surround the village.

Meanwhile large grain-storage cylinders tower over the center of the rather compact village. The prominent towers, with "Peterson Co-op" painted in big letters, say to everyone for miles: "This is Peterson, our home, and we work together cooperatively."

Tidy streets lined with sidewalks (unexpected in a village) create a regular *grid*, a small version of the big grid of roads dividing up the land in all directions. These grids are clear and easily navigated, readily maintained and controlled, and comforting. Only outsiders from afar call them monotonous, boring. Those are different people.

About four *paved and gravel roads* radiate outward from the village. They connect with many other gravel and dirt roads, as well as lanes that tie isolated farmsteads together and to the village. Roadsides are embroidered with flowers and alive with butterflies and bees. Slowly moving through these ways by vehicle or feet keeps one close to the land, where soil irregularities are felt and the richness of species, indeed species interactions, is experienced. In addition to getting from here to there, that's the meaning of dirt roads.

Farmsteads dispersed across the land outside Peterson appear as homes with farm buildings and a small group of trees. Farmsteads mean independence, living with the land, a productive place, and freedom from the ills of dense housing in urban areas, even the nearby village. The trees provide shade and attract both familiar and unusual birds. A vegetable garden sits close to the home, just like those of preceding generations. That means getting one's hands in the soil, working with plants, pests and weather, and the comfort of food security (Berry, 1977; Jackson, 2014).

Beyond the waving foreground of productive plants, an *endless horizon* extends in all directions (Hokanson, 1994). That means think big, no closed in claustrophobia here. Indeed, big sky arches overhead, a blue dome often peppered with white cottony objects. The ever-changing clouds, sun, moon, and stars represent the soothsayer to be watched, the predictor of what's just ahead.

Winds in many forms endlessly sweep the village. What do they mean, or portray? Clouds build, move, and disappear. Rain falls. Snow falls vertically or horizontally, and piles up in known places. Dust eroded from the fields appears and dissipates. Bitter cold winds from Canada alternate with hot oven-like winds from the west. Dry winds desiccate the area. Leaves seemingly blowing in all directions remind us of the turbulence around buildings, trucks, trees, and shrubs. Small twisters and periodic tornados stalk the land. Livestock all pointing in one direction provide a weathervane to be read. Grasshoppers, unknown insects, and even birds fly with the wind, but sometimes are simply blown by it. Windbreaks and snow fences standing still remind residents of the pervasive power of wind in their lives.

Heterogeneity, uncertainty, and surprises are also a key part of the Peterson village. The population may decline. Droughts, pest outbreaks, and damaging winds recur, so residents generally adapt to and prepare for such disturbances. The arrangement of soil types, depths, and fertility varies widely and at many spatial scales. Fields appear heterogeneous and may be managed accordingly, at greater time and expense. Rectilinear roads require considerable surface work, maintenance, and repair, in part because they are at odds with nature's curvilinear patterns and processes. The river floods to different levels each year, periodically inundating crop fields and house plots. Furthermore, during a big flood the river migrates, or tries to migrate, to a new location in the floodplain, in the process washing out human structures. The village history records many such "natural" patterns or events, both large and small, along with the residents' responses and adaptations to change. Nature and culture here are tightly knotted.

Both spatial and temporal dimensions of culture are important in understanding the ecology of towns, villages, and the land (Wuthrow, 2013). In a broad context, *culture* (and religion) helps to see one's place in the universe, and geographically on the continent. A town or village resident perceives life as being far from a city, and the urban environmental and social ills therein. A newspaper or television program enters the home picturing something bad there. It is easy to put that out of mind as not part of one's life.

Local issues become all-engrossing. As suggested at the beginning of this chapter, settling into a place long term means less movement, less uncertainty, less adventure. Blinders may form, so new ideas and technologies emerging afar are easily missed, or ignored, or avoided. Over time the town or village gets out-of-date and becomes a backwater, increasingly ignored by the nation or broader society.

The *time dimension* is especially important in understanding heritage, culture, family, and town (Wuthrow, 2013). Over generations, town stories and legends develop, and are passed along. Festivals, fairs, and such events pockmark the decades, creating shared memories (Brown, 1986; Bradley and Hall, 2006; Dufty-Jones and Connell, 2014). The core of a town's *heritage* are the historic homes and buildings, famous former residents and what they did, and the places where important events occurred. Today's markers of such features reinforce the memory. Knowing these is to be culturally savvy. Loyalty to friends, to local groups, and to the town evolves. Family stories and values emerge, and are passed on to children. Traditional family values result. *Adaptive value* develops by knowing both town heritage and family values. Learning the past and understanding the present greatly increase one's ability to deal with the uncertain future.

Local change is often unwelcome, unless it is small or gradual (Wuthrow, 2013). Conservative political and moral perspectives may reverberate from generation to generation.

Aesthetics and Architecture

Aesthetics

A sense of place developed from familiarity and caring occurs in both ugly and beautiful locations, as well as everywhere in between. But *aesthetics*, or in simplest terms, beauty, increases the proportion of people with a sense of place. Imagine attempting to identify the most beautiful towns of Europe (Akgun *et al.*, 2013) or elsewhere.

Aesthetics is a much analyzed concept, for example, focusing on combined foreground-midground-background, prospect-and-refuge, relative preponderance of natural and cultural objects, a particular feature, or simply form. Measuring aesthetics is complex, especially because different people see and judge things differently. Nevertheless, a few aspects of aesthetics seem useful as an introduction to considering the architecture and character of towns.

Clean air with minimal agricultural dust and factory pollution is of course the ideal lens to view a town. In town, streetscapes including street, sidewalk, street-side plantings, and adjoining front yards/gardens and buildings are key determinants of better or worse aesthetics. Tall structures such as steeples, minarets, grain-storage structures, high-rise apartment buildings, even wind turbines affect aesthetics. Highlighting, or at least respecting, a special feature usually enhances aesthetics. A long view through town or a scenic vista from a high point is typically an enhancement (Figure 4.6).

The central public plaza seen by so many people is sometimes beautiful, but more often mainly functional. Perhaps the greatest missed opportunity is the town center's main street, often simply a long clutter of colors, shapes, sizes, signs and objects, each competing for the eye. Thus, the whole may appear as an unaesthetic or ugly mess.

Outside a town, farmsteads traditionally used local materials in construction, fit buildings to sun and wind patterns, and tailored farm buildings to traditional animal care and farm-usage patterns. Too often today farm buildings seem to be generic pre-engineered structures erected on-site (Thornbeck, 2012). From roads, the farmstead structures may resemble a small factory or even a military barracks. However, grazing cows or sheep in fields between road and farmstead may replace that feeling with a sense of peace and beauty.

In the town of Concord (see Appendix), a field next to the busiest road, carrying more than 50,000 vehicles per day, has grazing cows and scattered bird boxes for bluebirds, all visible beyond a row of apple trees. This beautiful arcadian image is purposeful, to remind both outsiders and residents of the agricultural heritage of this town.

Some people have *weekend/holiday houses* in or near a natural area, in part to repeatedly experience the beauty of nature or exercise in it. In Sweden, 23 percent of the people who own, rent, or use someone else's weekend/holiday house (second home) use a house in a small rural settlement, and 10 percent in the countryside with only a few other houses in sight (Figure 4.7) (Muller *et al.*, 2013).

Per year these users have about 20 one-day visits and 40 overnight visits, suggesting that most use is during a weekend. When at the house on weekends, 14 percent never or seldom visit a natural area, while 86 percent often or very often visit one. In contrast, on holiday (commonly 2.5 to 4 weeks), 6 percent do not visit while 93 percent

Figure 4.6 Beach exposed for diverse recreation at low tide around a coastal tourist town. Saint-Malo, Bretagne, France. R. Forman photo. (A black and white version of this figure will appear in some formats. For the color version, please refer to the plate section.)

do visit a natural area. The top three activities while at the house for either weekend or holiday use are: (1) recreational walks; (2) nature walks; and (3) gardening. Other frequent activities while at the weekend/holiday house are: sunbathing; jogging; alpine skiing; biking on roads; outdoor swimming; and picnic/barbecue. Overall, both weekend/holiday houses and their use have reverberating environmental effects in natural landscapes.

Architecture

The architecture of buildings is a key to both town aesthetics and character (Rifkind, 1977; Bailey, 1982; Longstreth, 1987; Arendt, 2015). Long ago the Roman architect Vitruvius pinpointed the five key elements of *good design*. In a treatise highlighting the importance of context, the first two are: (1) effect of a structure on the surroundings, and (2) effect of the surroundings on the structure. Later he lists the other three, i.e., the structure should: (3) be beautiful or inspiring; (4) constructed well and of good materials; and (5) work effectively for people using the building. This combination of context and building attributes applies nicely to structures in a town or village.

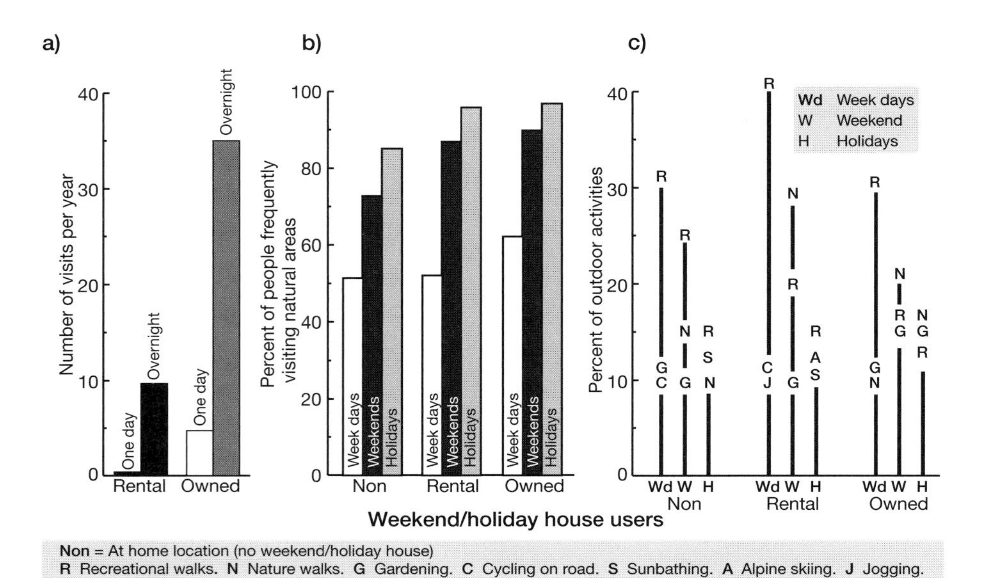

Figure 4.7 Outdoor activities of weekend/holiday house users in Sweden. (a) A quarter of the weekend/holiday houses are within 8 km of home, and a quarter >93 km away. (b) Other people surveyed seldom or never visit natural areas. (c) Other common outdoor activities are mountain biking, Nordic walking, cross-country skiing, recreational fishing, picnic/outdoor barbecue, and outdoor swimming. Adapted from Muller (2013).

Most architects relish designing distinctive *buildings* and mainly work in urban areas. Thus, rather few town buildings have been designed for the site by an architect (Vaz *et al.*, 2013a). More likely they have been built by building contractors using widely available generic architectural plans. Also, many houses may have been hand-designed and built by a former or current owner. Most of the buildings in the town center and older residential area have been altered and generally enlarged over time.

Such built areas and histories often contain numerous microhabitats supporting diverse animals and plants (Catry *et al.*, 2009; Forman, 2014; Adams, 2016). Decaying buildings provide many microhabitats. Some buildings may have been built on another site and moved to their present location.

The net effect of these construction activities is that towns and villages commonly appear distinctive or idiosyncratic, with many vernacular buildings. Such homes and buildings may reflect the personalities of the owners or former owners, even the values of rural life. But buildings in the town center are the most seen (Reps, 1969; Lewis, 1972; Jakle, 1982; Francaviglia, 1996). A prominent building on the central plaza, usually a government or religious structure, sets the architectural tone for the town. At least equally important are the multi-aged commercial, residential, civic, cultural, and sometimes manufacturing buildings squeezed together on the adjoining main street.

Although often only noticeable to knowledgeable people, two architectural features are especially important, both highlighting Vitruvius' primary good design principles.

First, some buildings *fit with the neighborhood*, which in turn seems to enhance a new building. Such new structures do not eliminate old trees, or degrade or disrupt surrounding buildings and house plots. They often look distinctive but they fit. A building that does not fit probably should not have been built, or should be in another place or another town.

Second, some buildings, especially old traditional ones, *fit with environmental conditions*. They mesh with the topography, avoiding steep or erodible slopes, ridgetops and hilltops, and sites subject to severe environmental conditions. They are tailored to solar angles, gaining more solar energy in cold periods, and less during hot times. They fit with the seasonal and daily winds, protecting against or diverting strong wind, cold wind, or hot wind. They fit with hydrologic conditions, avoiding a high water table, floodwaters, and disruption or pollution of a waterbody. They are built on soils with suitable engineering characteristics, such as bulk density, pore size distribution, compressibility, shear strength, and plasticity (Chapter 5) (Forman, 2014), and are appropriately ringed with porous fill for good drainage. Judge a town by how many structures fulfill these criteria of fitting in the neighborhood, and fitting with ever-changing environmental conditions.

Many *properties of buildings* that affect aesthetics, material, and construction are familiar (Francaviglia, 1996). Colors, texture, material, sizes, shapes, angles, ornamentation, density, and arrangement of features. Wear and degradation. Care, maintenance, and repair. Building plot sizes. Architectural styles, and their consistency or combination with nearby buildings. Many types of buildings were built at different times, reflecting changing styles, needs, and budgets. A high diversity of homes reflects construction by different owners. In contrast, a series of similar buildings, analogous to military barracks, was built at one time, often either by a major company for its employees or by government.

Finally, *farm buildings* tend to be conspicuous in isolated farmsteads, prominent in villages, and on the edge of towns. Normally the traditional farm buildings fit well with changing environmental conditions. This keeps farm animals "happy," enhances energy efficiency, and provides effective working conditions for the farmer.

Long ago during winter nights in northern Europe, farmers brought their livestock into the first level of the home, while the family slept in the upstairs loft. The livestock and the consequent decaying manure provided free heat for the home. But too often today farm buildings are generically designed and manufactured elsewhere, to be erected almost anywhere in the farmstead (Thorbeck, 2012). Once specialized agricultural buildings, tailored to cattle or dairy or poultry or pigs or grain or hay, now are somewhat hard to differentiate except by farmers.

Character

For travelers through farmland, farmsteads may be the prime feature portraying the character of an area, in this case *rural character* (Thornbeck, 2012). Farmsteads often reflect ancestral family roots or even immigrant origin. The farmhouse and cluster of trees and diverse farm buildings are generally the opposite of monotonous. The traveler

Figure 4.8 University of Virginia central campus and distinctive architecture originally designed by Thomas Jefferson. Small city of Charlottesville, Virginia, USA. R. Forman photo.

immediately sees these places as beautiful, ugly, welcoming, forbidding, vibrant, or worn out. A farm stand selling fresh locally grown food and flowers by the road adds a dimension. Rural character comes in many forms.

As with rural character, so too with towns – towns in forest have a forest character, in desert a desert character. The setting or topographical location of a town strongly molds its character. The town or village fits into or stands out from the surroundings. Indeed, many towns advertise aspects of their character upon entering. Welcome to "Our Town," home or center of X----, Y----, and Z----. A prominent natural feature or cultural/historical site may be highlighted. The specialness of a place often emerges from evocative distinguishing features (Figure 4.8) (Sargent *et al.*, 1991). A stream, river, lake, estuary, or sea alongside strongly molds the character of, indeed pervades, the place. One senses that people either live in housing or in neighborhoods with character. Community values are on display.

The aggregation of town center buildings strongly portrays a *town or village character*, based on many characteristics (Francaviglia, 1996; Arendt, 2015). Building types, styles, heights. Separate or attached buildings. Different housing types and densities. The scale of structures relative to a person's height or eyes. Tall structures, such as minarets, church steeples, and cylindrical grain-storage structures, affecting the perception of a town. To understand the pattern of buildings forming a town's character sometimes requires detecting the sequential history of fires and rebuilding.

Also, *streets mold character* (Arendt, 2015). Are lots of people out walking or shopping, or does the place seem empty? Do two cars meeting in the center make a

Figure 4.9 Village with two-thirds of the storefronts on Main Street closed. All places visible in photo are abandoned except for a cafeteria on far right. North of Emporia, Kansas, USA. R. Forman photo, courtesy of Joan I. Nassauer. (A black and white version of this figure will appear in some formats. For the color version, please refer to the plate section.)

traffic jam, or are there lines of vehicles moving slowly past street-side parked cars and pedestrian crosswalks? Are sidewalks shaded from hot sun? Is there a tangle of wires overhead on posts, or have utilities been put underground? Are tiny structures like bike-racks, fire hydrants, and traffic calming structures abundant, or scarce? Do streets and roads form an even grid, a hierarchy of widths, a network of curvy lines, or a combination?

Street *trees*, greenspaces, and bits of nature are major determinants of town character (Thorbeck, 2012). The central plaza, small parks, and school areas may be prominent, minimal, or largely hidden. What tree species are prevalent? Are there rows of elms (*Ulmus*) arching over roads, pollarded plane trees/sycamores (*Platanus*) with large leaves and yellowish trunks, beautiful flowering flanboyans/poincianas (*Delonix regia*), cherries (*Prunus*), magnolias (*Magnolia*), or various nuisance trees? Evergreen or seasonally deciduous? Do they have distinctive forms, such as tall palms, pointed conifers, fig trees with rampant hanging roots, or multi-stemmed screw-pines (*Pandanus*)? Old trees especially reflect the town's history (Clare and Bunce, 2006). A town or village seemingly covered with trees may project monotony, or intriguing richness.

Empty storefronts and abandoned houses with ecological succession scream "bad economy" (Figure 4.9) (Francaviglia, 1996). Lots of busy mom-and-pop shops, restaurants, and apartments above suggest a vibrant community, even in the evening. Are unbuilt spaces well maintained, or left as scruffy vacant plots with trash? How compact or spread out is the town center? Is the neighborhood all residential or clearly mixed use?

Icons of small towns in popular culture and national media – e.g., highlighting hardworking residents, honest shopkeepers, accessible local government, and a good life – reinforce the notion of town and village character (Francaviglia, 1996; Qian *et al.* 2012; Dufty-Jones and Connell, 2014). The character of a town, shaped by both natural and built features, evokes a Pavlovian response, a gut reaction. Character is a great umbrella concept that instantly integrates an array of town characteristics, often into a single word.

Economics, Jobs, and Commercial/Industrial Areas

Towns may thrive under economic conditions ranging from mainly market forces to primarily active government controls. In *market or growth conditions* (e.g., "let the market determine it"), human consumers are central, preferences and tastes are the main driving force, and the resource base is virtually limitless, since technology or substitution can overcome shortages. In contrast, *regulatory conditions* depend on government activities and other limitations, sometimes with planning, to head off problems or crises. In both situations, accounting for the use of natural resources as inputs to, or outputs from, production is important.

Meanwhile, *environmental economics* has focused on the by-products of production and wastes of consumption (Stavins, 2012). *Ecological economics*, on the other hand, generally combines both economic inputs and outputs – natural resources as input capital, plus by-products and wastes as output (Costanza *et al.*, 1997; Forman, 2008; Kareiva *et al.*, 2011). These broad perspectives, generally familiar at a national level, provide context for a town focus (Grek *et al.*, 2013; Vaz *et al.*, 2013a).

Local and spatial factors affecting a town's economy provide valuable foundations for understanding a town and its environment. The primary way economic growth occurs in a town is from exporting goods or services to regional or broader markets (Daniels *et al.*, 2007; van Leeuwen, 2010; Ratajczak, 2013; Vaz *et al.*, 2013a). The secondary base for potential growth are the businesses providing day-to-day goods and services to the local market (Fallows and Fallows, 2018). The unlimited regional market in the first case permits economic growth, whereas the limited number of local residents commonly constrains growth. At the same time, some towns, e.g., in China, are tied to national or global markets, and thus the industries do not relate to local features or resources (Kirkby *et al.*, 2006).

Neither free-market forces nor government intervention alone may sustain towns (Vaz and Nijkamp, 2013). Self-organization, self-reliance, diversification, effective economic use of surrounding resources, highlighting special features of a town and its surroundings, and regional linkages may be important for success.

Here we highlight: (1) shopping, business, and manufacturing centers; (2) beginning and end of a major employer; (3) jobs; (4) economics and the surroundings; (5) place branding and festivals; and (6) ecosystem or nature's services.

Shopping, Business, and Manufacturing Centers

The descriptive types of towns (Chapter 1) – mining, tourist, farm crop, shopping, manufacturing, fishing, and so forth – encapsulate the predominant sources of income and types of jobs present. People and businesses often cluster in communities to take advantage of *agglomeration economies*. Thus, costs for a business usually drop if accountants, attorneys, suppliers, and so forth are nearby.

Shopping or *market towns* as hubs for retail shopping by town residents plus people from surrounding villages and one or more neighboring town, illustrate useful patterns (Powe *et al.*, 2007). In England each market town of 2,000 to 30,000 people serves several to many surrounding villages. Villagers shop weekly in the town, and may do daily shopping in a commercial center on the town edge, thus requiring considerable transportation time and cost. Services, from government to schools and professions, become more important in the market town economy and employment. Often the pressure for nearby housing growth increases.

Some 1,200 towns of this size exist in England, of which about one-fifth are officially designated as market towns commercially serving surrounding areas (Powe and Hart, 2007). Their general economic structure is similar to that of the United Kingdom as a whole, but with slightly more manufacturing and less financial, business, and real estate activity. Services are diverse (van Leeuwen, 2013), with five types being particularly characteristic: banks; solicitors (lawyers); ladies' clothing shops; doctors' surgeries (offices); and supermarkets in town. Yet one-third of the market towns lacks at least one of the categories (supermarket and ladies' clothing shop the most missed), and 1.5 percent of the towns have none of the five services, suggesting how diverse market towns are.

English market or shopping towns were mainly so designated to support the surrounding agricultural areas (Brown, 1986). Livestock were brought to the weekly market to sell and buy in the town center marketplace. Grain, produce, and numerous goods were sold in stalls and shops. Although today livestock movement is limited, the market generally remains as a major weekly economic and social event. Fairs and festivals also occur periodically. Industries and trades are highlighted. Shops frequently change. The transport of people and goods to market is important. Every week a surge occurs in food consumption, waste accumulation, water use, vehicle fuel usage, greenhouse gas production, and town–village social interactions.

Many towns worldwide have a single predominant economic base, such as farm crops, mining, or a university. Yet most towns are multifunctional, having some two to four important economic bases (Powe and Hart, 2007). Despite the different bases present, a clear town identity is usually characteristic. This identity may result from a conspicuous natural feature, a politically influential economic base, or indeed from conscious town branding (e.g. advertising a key feature to remember).

The town center is usually a key commercial hotspot, though a town-edge retail center may leave the town center as a "dead zone." Some towns retain a factory in the center, though most such factories have closed or moved to an industrial center on the town edge. In the tradition of homemade products in villages, tiny craft-making and manufacturing shops are often dispersed across a town. Distinct business or office centers containing a cluster of companies with regional business are common in cities, but uncommon in towns. If present, a business or office center generally occurs on a town edge. A retail commercial center or shopping mall is commonly located on the edge of town close to a main radial road. In large towns shopping centers may also occur in the newer residential area.

A closer look at people's in-town movements for commercial uses is illustrated by a survey of 968 residents in four English market towns (Alnwick, Morpeth, Wymondham, and Downham Market) (Powe and Gunn, 2008). Residents living near the town center are more likely, and retirees less likely, to patronize the main food shop and a pub than are other residents. In contrast, recent in-migrant residents from an urban area use these two services less than do longer-term residents. No other differences in usage were found for long-term versus recent residents.

The location of work does not affect the use of the services, except for the main food shop. The aggregation of similar services seems to increase usage. Thus shoppers of the main food store also use other town center food shops, and residents out in the evening use more than one entertainment type.

A mixed-use town center with residential and commercial activity provides retail shopping for the daily needs of all town residents, except those choosing to use another, usually town-edge, shopping center (Figure 4.10) (van Leeuwen, 2013). Normally towns have few wealthy residents, so a high-income neighborhood is tiny or absent. In villages, one or a few local shops typically serves the daily needs of villagers. Residents in farmsteads and other dispersed houses outside population centers mainly depend on village shops or town-edge shopping centers for daily food and household needs. With a limited clientele, sometimes villagers help subsidize village shops (see Figure 3.4). Weekly shopping for diverse items then draws dispersed residents, villagers, and town residents into town shopping areas, providing an important impetus for a town's economy. Less frequent shopping for specialized uses and needs, such as a medical facility, sports event, furniture, or show usually requires travel to a city.

Beginning and End of a Major Employer

The entry and exit of a major employer is a key indicator of a town's economy. This may be a government facility or non-profit operation, but is usually a manufacturing or commercial company. In market economics a company often grows, maintains itself while competing with other companies, and finally closes or moves away. Growth normally produces more jobs, and a sustained period of successful competition maintains jobs. A major company's exit often means considerable unemployment. A public utility sustained in part by government has persistence and helps maintain a more predictable economic level.

A major company in town with an important competitor in a nearby town suggests that one or the other will eventually disappear, although merger is an option. Replacing a lost major company with a new one may involve a similar number of employees that

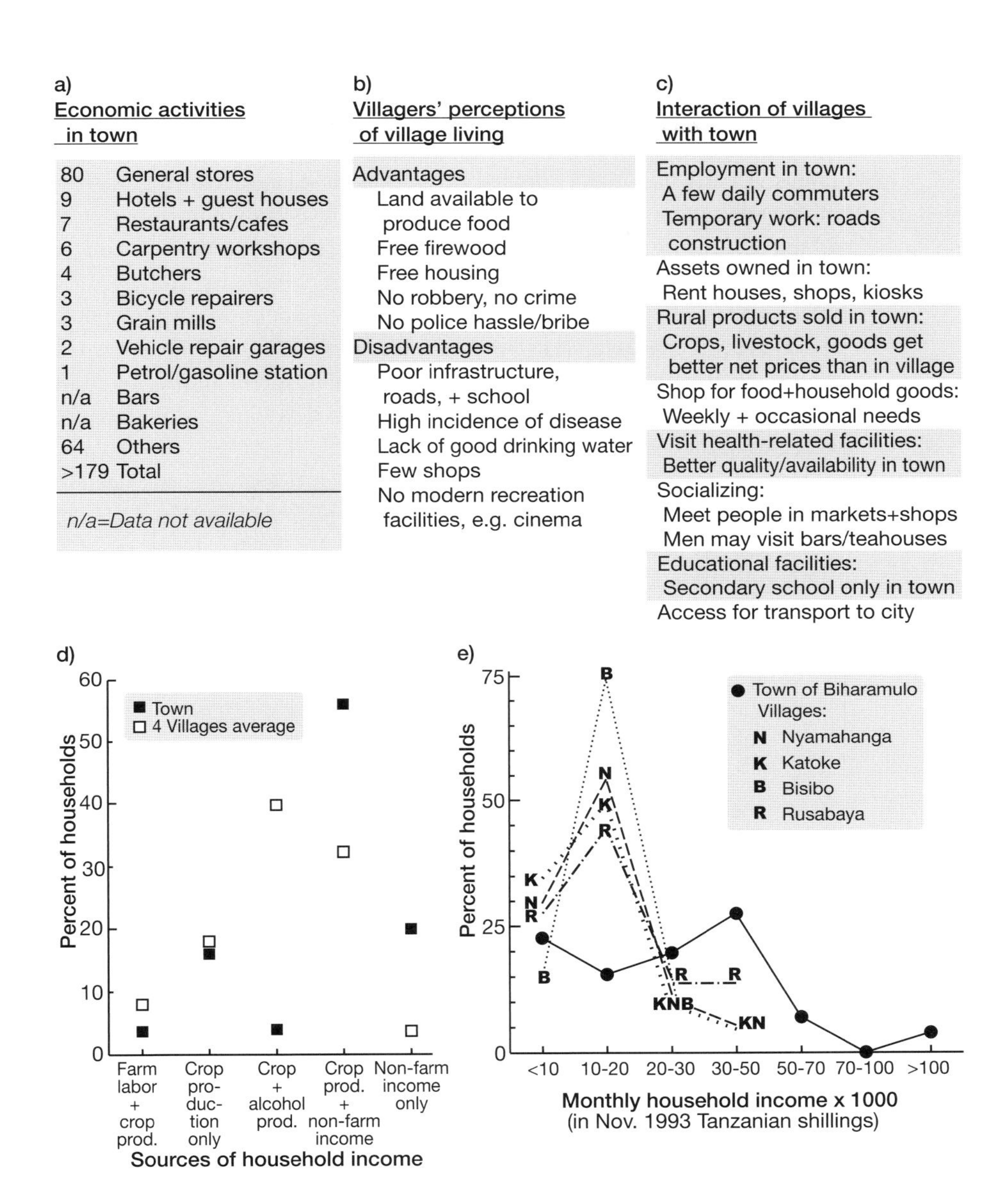

Figure 4.10 Employment, income, and interactions of town and surrounding villages. Bihiramulo, northwestern Tanzania (1993 data), population about 20,000; large villages/small towns: Nyamahanga 2,130; Katoke 2,278; Bisibo 2,161; Rusabaya 2,059. First three villages 7 km from town and fourth village 10 km; each a 2–4 hr walk. Adapted from Baker (2006).

sustains the town economy. But if the lost and new companies are quite different types, a temporary convulsion is normal, with departing employees being replaced by different new employees. The company town, dominated by a single large industry, causes huge unemployment and town disruption when it closes (Green, 2010; White, 2012). In contrast, several medium-size industries provide stability when one closes.

Company entry and exit also emphasizes the dependency of a town on its region (Leibovitz, 2006; Grek *et al.*, 2013). Large companies provide products to the region. Entrepreneurship is typically high in a region, along with a diversity of economic milieus. A particular major company has access to a large gross regional product, along with market potential and demand. Variables, such as employment rate, education level, and the number of small companies in town, correlate with the entry and exit of major employers.

A village (<2,000 people) may have shops along about one or two blocks of its main road or street (Francaviglia, 1996). A small town of 2,000–5,000 people often has commercial development along some four to six main-street blocks of the town center. Larger towns typically have compact commercial development on many blocks around the central public space.

The town center or main street symbolizes, even fuses, time and place (Francaviglia, 1996). It displays the past in layout, old buildings, and signs. Yet it also highlights the present activity or inactivity of local residents, shopkeepers, entrepreneurs, and visitors. All towns have unique interesting stories. The town center reveals a distinctive combination of shops, a place of celebrations, government and religious buildings, the public transport available, pedestrians in daily clothing, as well as their apartments and houses. This is the economic and social heart of the town.

Jobs

The desire for more local businesses and local jobs seems to be widespread in towns. Even in a small desert town (San Ysidro, New Mexico, USA), with small farms, limited mining, through traffic, highway widening, and chronic water-related problems, the prime goal of residents remained increasing local businesses and employment (Sargent *et al.*, 1991).

In many rural landscapes the traditional natural resource-based economies of agriculture and forestry, as well as manufacturing, are increasingly capital – rather than labor-intensive (Albrecht, 2014; Gibson, 2014). That means fewer jobs in town, a widespread barometer of economic conditions. Empty storefronts in the town center are conspicuous indicators of fewer jobs as well as economic decline.

Production agriculture and forestry persist as only one of the important components of the economy of towns and villages. Farm landscapes become scenery for tourists and real estate developers, and forests as recreational places. Rural communities often face slowly declining employment and population. Concurrently, the proportion of aging residents increases. Towns with few amenities and few residents with higher education are increasingly disadvantaged. These are pervasive long-term trends.

Meanwhile tourism, creative industries (Florida, 2002), road transport, residential construction, and community and government services grow in importance (sometimes called post-productivism). Information technology somewhat reduces the importance of location (Albrecht, 2014). Local resources providing amenities become more important in attracting and maintaining town residents (Figure 4.11).

Figure 4.11 Location where all-terrain vehicles (ATVs or ORVs) roar around making tracks. USA desert.
Source: BraunS / E+ / Getty Images.

Town and village residents largely escape urban competition, smog, crime, and traffic. They value being closer to nature, to family and friends, and having the ability to sink roots into a community.

Several local features lead to more jobs (Albrecht, 2014). The local production of food, energy, and other usable products provides *added value* by being close to users and consumers, and thus having low transportation costs. The attraction of entrepreneurs to town, e.g., largely tied to computer information technology or to the establishment of small start-up companies, adds jobs, which commonly lead to further growth. Workers may be more productive in a small company, and a cluster of entrepreneurial companies is probably better able to withstand or adapt to economic and other disruptions. Towns are more likely to have broadband internet access than villages, so telecommuters with a small home office and distant business may cluster around a town with amenities. Still, sustaining a business requires meshing somewhat with local community values.

The common reasons for population decline or growth are diverse (Daniels *et al.*, 2007):

Decline in Town Population

Decline of manufacturing; loss of natural resource base; regional population loss; shift of trade patterns to regional centers; changes in transportation patterns and routes; loss of major employer;

erosion of local businesses; seasonal jobs; loss of community service capacity; failure of leadership; inadequate planning for change.

Growth of Town Population

New technologies; development of natural resources; metropolitan population spillover; growth as a regional center; new transportation patterns; town center revitalization; tourism; recreational resources; good leadership; planning for change.

The shrinkage of a town or village need not lead to disaster and fall. Instead of the subtle oxymoron "sustainable development" (sustainability implying strong ecology; development implying continued growth), *sustainable shrinkage* seems more appropriate for a globe that has exceeded its ecological capacity or footprint (Callenbach, 2014). Shrinking or stable population, and especially consumption, peaceably sustains the resources we depend on.

How? Provide the right incentives. Switch from consumption to maintenance. Build to last. Control shrinkage instead of it controlling us. Treasure a compact community, and conservation. Rediscover our social and cultural roots. Encourage population stability. Restore nature. Redefine a healthy economy. All of these involve trade-offs of costs and benefits, which are evaluated against a worsening ecological footprint. Put hope on a pedestal – a stable or shrinking economy can contain opportunities for entrepreneurs, and jobs for all kinds of people.

Economy and the Surroundings

Central place theory, highlighting towns and cities as sources of things flowing outward to the surroundings, was one of the earliest land models (Christaller, 1933). Conceptual boundary lines separating the surroundings of adjacent towns suggested zones of economic competition. Many things flow outward from a town, including stormwater, sewage wastewater, water pollutants, road salt, industrial pollutants, town species, commuters, groundwater, school buses, and recreationists. But the town–surroundings interaction goes both ways, with equally diverse and important things entering the town: dust, water, floodwater, wildlife (e.g., deer), predators, gravel/sand construction materials, native species, traffic, groundwater, wildfire, commuters, school buses, shoppers, and the list goes on. Thus, towns and their surroundings are normally tightly interlinked (Vaz *et al.*, 2013a).

Many of the examples are *externalities*, whereby a pattern or process affects a surrounding area or a broader public without (adequate) market compensation (Platt, 2004). Such externalities may cause problems or provide benefits. Government policy may need to address problems.

For these examples, changing policies and practices in the surroundings would strongly affect a community. For instance, in northern New Mexico (USA), towns with residents whose ancestors used the surrounding relatively natural lands, today find the land largely controlled by government (Kosek, 2006). Government policies have gradually changed in major ways, including increases in logging, water supplies, fishing and hunting, fire management, export of economic benefits elsewhere, conservation of rare species and biodiversity, wilderness protection, and exclusions of local residents. Sheep grazing, local jobs, and the local economy increased, and then decreased with government control. Imagine the complications and conflicts from such a changing set of practices surrounding the towns!

China's countryside is a contrast, where the government places strong emphasis on agricultural production, but exerts relatively little effort on environmental protection (Zhao, 2013). The diversity of livelihoods is minimal, especially with a typical per capita average of only 0.08 hectare (<¼ acre, or 90 x 90 ft) of arable land. Overall, the net result is a static low-income population and considerable land degradation.

Linkages among towns is another approach to strengthen a local economy and employment (Vaz *et al.*, 2013a). Common ways towns collaborate are to facilitate commuters for jobs, weekly shoppers, trucks carrying goods, competitive games, festivals, and entertainment. Companies in nearby towns may be linked by complementary characteristics, whereby companies benefit from the linkage. Of course, if the companies are competitors, it is likely that one or more will prosper while another shrinks, leading to unemployment in that town. Another model is the town closely linked economically with a nearby city.

In the US "Heartland," where most towns and villages are gradually losing population, a rosier future of resettlement has been suggested based mainly on socioeconomic perspectives (Kotkin, 2010). Consider the suggested components for stimulating growth in the area. New technologies and infrastructure. More diverse population, especially of Latin and Asian immigrants. New ideas and cultural expression. Manpower to run specialized automated enterprises. Communities rooted in traditional values. A flexible business culture. More of the elderly working part-time. Internet access readily available. Companies attracted to large towns, creating an economic force. Coal and natural gas powering rural areas. Growth in biomass, wind, and solar energies. A characteristic work ethic. A tolerance of ethnic, racial, and religious diversity. Upward and social mobility. More children and a growing population, fueling economic growth.

An Achilles heel potentially dooming this vision is the near absence of environmental dimensions, especially puzzling for a rural region. Environmental protection and restoration for water quantity, water quality, fish, wildlife, vegetation and habitat, soil erosion and deposition, air quality, and biodiversity seem to be a key. Also, sharply reducing greenhouse gas production. Strong regional plans (supported by government, the private sector, or non-profit organizations) can outline and enforce the best areas for protecting water supply, food production, industrial centers, biodiversity, housing, transportation, soil protection, groundwater recharge, renewable energy capture, and recreation. These overriding environmental dimensions could be sustained, thus providing the essential framework for a resettling of the Heartland.

Festivals and Place Branding

Local approaches to improve employment and economic opportunities are highly diverse and tailored to the distinctive characteristics of a town or village (Connell and McManus, 2011). Some places have advertised themselves as weekend destinations or telecommuter places to live (Gibson, 2014). Another direct local economy approach is to construct a "flagship" tourist destination, such as Plimoth Plantation (Massachusetts, USA) or the Waltzing Matilda Museum (Winton, NSW, Australia). A tiny village of 288 people (Cumnock, NSW, Australia) developed a plan with good media coverage to attract people and also address housing affordability – new families moving to the area could rent a farmhouse for one dollar per week. Sometimes advertising big things – the world's largest pickle, wedding dress, or bottle of beer – is an effective magnet.

The much more widespread and diverse approach has been to organize and host *festivals* (Figure 4.12) (Bradley and Hall, 2006; Gibson *et al.*, 2010; Gibson, 2014; Kavaratzis and Ashworth, 2015). In a single year, more than 2,800 festivals took place in non-metropolitan, mainly agricultural areas of three Australian states (Gibson *et al.*, 2010; Gibson, 2014). Three-quarters of the festivals highlight sports, community, agriculture, and music, though more than 27 other types occur. Several festival types built on, and reinforce, the farming backbone of a town – agriculture, food, wine, gardening, animals/pets, and environment festivals. Many festival types highlight some distinctive characteristic of the town, while sometimes a community essentially creates a new idea as a festival focus.

The Australian festivals produce an estimated 175,000 full- and part-time jobs, compared with 136,000 agricultural jobs in the same areas. Festivals particularly stimulate the local economy, by providing income to the local general store, other shops, vehicle service stations, local caterers, and so forth (Gibson *et al.*, 2010). Festivals also facilitate diverse interactions, mostly positive but some negative, between visitors and residents. And they cause diverse environmental impacts, from excess human wastewater to soil compaction/erosion, stormwater runoff, pollution of local waterbodies, and noise.

Typically, festivals are organized by small non-profit organizations interested in cultural or social issues, rather than direct income generation for the town (Gibson *et al.*, 2010). Economic benefits, however, provide wider support for continued festivals in town. Creative industries focused on film, music, design, and fashion have been especially important in some towns (Florida, 2002).

Place branding goes beyond simple advertising of an event, but rather attempts to provide a continuing boost to employment and the town economy. Branding is a set of activities and methods to create a desirable image of a place (Lucarelli and Berg, 2011; Kavaratzis and Ashworth, 2015). Culture is the powerful force in place branding. Thus, common approaches highlighting a town or small city emphasize distinctive events such as festivals, associating the town with a well-known person, or indeed creating the idea of a signature or flagship place. Natural features are also widely highlighted in place branding.

Rural tourism and creative industries of course have their limits, and information technology only partially overcomes the constraint of remoteness. A somewhat coordinated regional approach, where individual communities have a mix of different economic solutions, offers promise.

Highlighting the culture of a particular town, or typically one aspect of a town's culture, is widespread in place branding (Jensen, 2007; Ashworth and Kavaratzis,

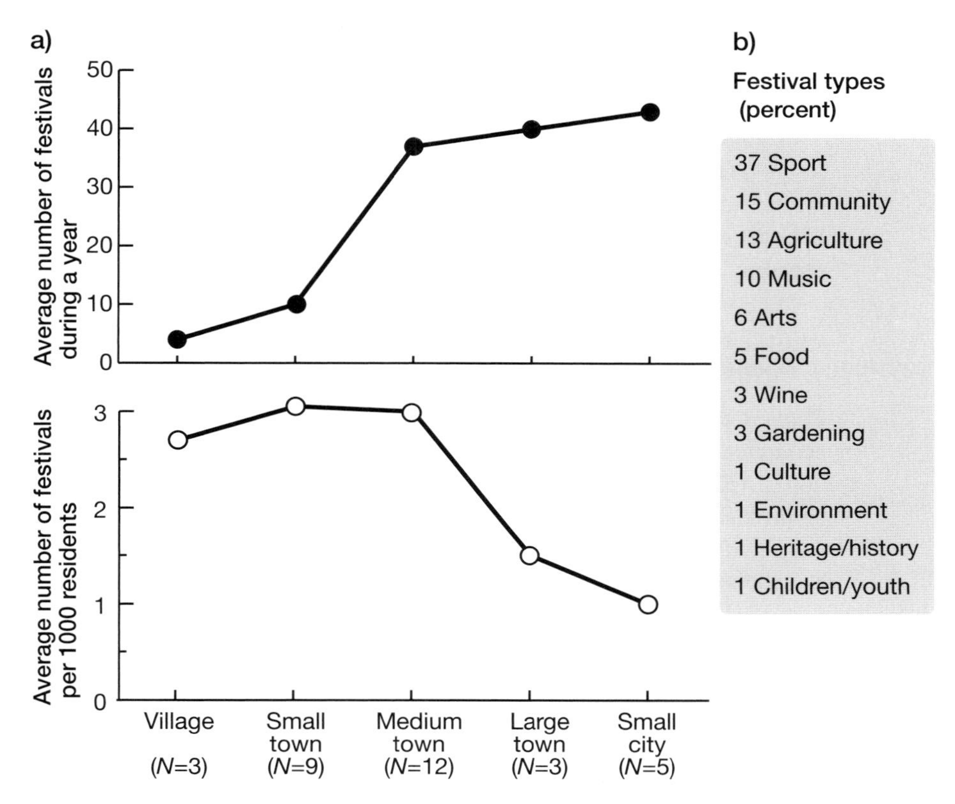

Figure 4.12 Diverse festivals in rural communities of southeastern Australia. Data are for the 32 non-metropolitan communities in Tasmania, Victoria, and New South Wales with the most festivals in 2007. Village = <2,000 population; small town 2,000–5,000; medium-sized town 5,000–20,000; large town 20,000–30,000; small city 30,000–50,000. Less common festival types are: Christmas/New Year; life style; outdoor; science; religions; seniors; innovation; education; animals/pets; bees; cars; collectables; crafts; air shows; dance; theatre; gay/lesbian; indigenous; and new age. Adapted from Gibson (2014).

2014). Yet successful branding that brings hordes of visitors gradually alters the town's culture itself. In place branding, culture may be used for varied goals, i.e., to provide saleable or consumable experiences, act as an economic resource, attract tourists and indeed creative residents, or highlight a distinctive locality (Kavaratzis and Ashworth, 2015). Events highlight a place's heritage and potential for creative and cultural production. Such events though must be strongly tied to the place's culture, and overcome differences between visitors and local residents.

Associating a place with a famous former resident or historical event is familiar (e.g., Salzburg and Mozart, X'ian and the terracotta warriors army, Barcelona and the architect Gaudi, Concord and the beginning of the American Revolution). In contrast, the town of Parkes (NSW, Australia) built an Elvis Presley museum even though the rock 'n' roll star never visited the continent. Overall, the linkage between festivals and features of host communities is weak (van Aalst and van Melik, 2011). Normally the architecture of a town is aesthetic and consistent with the idea or policy being highlighted. Creative and entrepreneurial people tend to be attracted to such places.

In addition to culture, nature is also a strong focus of place branding. Consider Jackson Hole (USA) and the Teton Mountains, Livingston (Zambia) and Victoria Falls, Zermatt (Switzerland) and the Matterhorn, Alice Springs (Australia) and a desert oasis. In each case the natural feature is strongly advertised, and visitors help sustain the local economy and employment.

Ecosystem or Nature's Services

Ecosystem services have been described as the benefits that people obtain from ecological functions (de Groot *et al.*, 2002, Millennium Ecosystem Assessment, 2005; Adler and Tanner, 2013; Roe *et al.*, 2013). Such benefits have been grouped as provisioning, regulating, cultural, and supporting services. By including values from hydrology, weather, and geology (without organisms), a more appropriate concept of *nature's services* refers to the human benefits provided by natural patterns and processes (Daily, 1997; Forman, 2014). The public apparently relates better to nature's services.

An important narrowing of the concept used by many scholars emphasizes that ecosystem services are *non-market values*, i.e., not compensated by market forces. Farmers produce food and are compensated by the market of buyers and sellers. Trees are cut for wood products, also subject to the market. In contrast, hillside or park vegetation reduces stormwater runoff, thus reducing flooding and big downslope mudslides. The vegetation provides a valuable service to society not compensated in the financial market of buyers and sellers.

Examples of urban ecosystem services, some of which apply to towns, include (Gomez-Baggethun and Barton, 2013; McDonald, 2015): water flow regulation and runoff mitigation; urban temperature regulation; noise reduction; air purification; moderation of environmental extremes; waste treatment; climate regulation; pollination and seed dispersal; recreation and cognitive development; animal sighting; and ecosystem disservices. Using diverse models, economic valuations of many ecosystem services have addressed economic, sociocultural, and insurance values, though the process is at an early stage. So far, towns and villages have received scant attention in ecosystem services studies.

Long lists of ecosystem or nature's services are often made, a recognition that gigantic nature pervades almost everything. While the basic ecosystem service idea is powerful and important (Millennium Ecosystem Assessment, 2005; Roe *et al.*, 2013; Geneletti, 2016), the long lists of services seemingly everywhere make it unlikely that government or society will soon take effective action on ecosystem services (Anderson *et al.*, 2009; Holland *et al.*, 2011; Roe *et al.*, 2013; Crouzat *et al.*, 2015). Perhaps nature's services are most usefully highlighted when, say, the most important two or three can be quantified, and effectively addressed by society.

Social Patterns and Policy/Planning

Social Patterns and Linkages

Sidewalk behavior provides valuable insight into a town's social patterns. When meeting a stranger in town, do most residents greet the person and even provide helpful insight? If so, there's probably an open collaborative social structure (Wuthrow, 2013). Alternatively, when residents meet, are perfunctory greetings or eyes averted the norm? We are too busy to pause, or we can't wait to get away. Such a town may have some mix of intimidation, deceit, drama, or intrigue. Intuitions and emotions are central to responses. Sidewalk encounters with conversation, even brief, suggest that neighborliness rules.

Who are the *town residents*? Usually a small group of relatively wealthy residents includes large landowners, successful business people, and some professionals (Wuthrow, 2013). Such residents play an important role in decision-making. A larger group of people with some advanced education provide services, especially related to schools, medical facilities, business, and government. Typically, wage-workers are the largest group of town residents, including contractors, manufacturing employees, office assistants, and agricultural employees. A fourth group, usually relatively small, are the retired/semi-retired residents. Low-income residents form the fifth group, which is usually small and heterogeneous, but may be quite distinct.

Social networks, however, tie people together into interacting groups, typically creating institutional norms or rules of behavior (Young, 1999; Fallows and Fallows, 2018). Such rules tend to cluster around the economy, political systems, class/inequality, kinship/family, religion, education, or leisure/recreation. Sharing, cooperation, rituals, and self-identity also may underlie social networks (Wuthrow, 2013). Group activities and behaviors come and go over time, though tend to be relatively stable in towns. Changes typically result from solidarity around a common cause, or from rearrangements related to shrinking or growing numbers of people. Or indeed from new ideas.

Living in a town means meeting and interacting with almost everyone over time, very different from city life. Leaders are from here, and generally want only incremental growth or change. Differing faiths are usually present. Contentious issues commonly divide neighbors, but usually only temporarily. Caring for children, the elderly, and low-income residents is widespread in most towns. A community spirit develops in, and supports, even celebrates, a town. Many residents expect to be buried here (Figure 4.13), and have their sons and daughters or grandchildren live here.

A small city (or large town) with 35,000 residents in India provides spatial insight into the locations, interactions, and movement of people within a town (Kamra, 1982). Nabla served some 265 surrounding villages and was connected to them by bus and rail. The area of this compact city was only 4.7 km^2 (1.8 mi^2), so walking, bicycles, and rickshaws were sufficient for most internal movement; few cars were present. The place was a government administrative center (with a big jail), a center for services and shopping, and a market town for the surrounding grain-agriculture land. Local manufacturing produced household goods, agricultural implements, and food products. Together

Figure 4.13 Tidy treasured cemetery on elevated well-drained site in edge of town. Hanga Roa, Easter Island (Isla Pascua), Chile. See Appendix. R. Forman photo.

these helped sustain a relatively self-sufficient population. The small city with a town center commercial strip was also a crossroads with an old fort and gates.

Comparing the place of residence and place of work for residents provided insight into potential social interactions in neighborhood, commuting, and work areas. In Nabla, of 200 people surveyed, 22 percent lived in the inner zone (city center), 25 percent in a middle zone (older residential area), and 50 percent in the outer zone (newer residential area) (Kamra, 1982). Almost half of the inner-zone residents worked there. Half of the middle- and outer-zone residents worked in the outer zone. Only 10 percent of the city residents worked in the middle (mainly residential) zone, while 20 percent worked outside the small city.

Furthermore, half of the inner-zone residents had their "two most intimate" friends in the inner zone. The friends of middle-zone residents were mainly in the middle and outer zones, while friends of outer-zone residents were overwhelmingly (71%) in the outer zone. Neighborhoods in these zones were fairly distinctive with relatively different groups of people. People from elsewhere moved into all three zones, and some movement of residents among zones occurred. Thus overall, in this small area-and-population city, commuter transportation was limited, and residents regularly met and interacted in their neighborhoods.

New residents in a town neighborhood commonly provide hybrid vigor. Combined with competitive and collaborative interactions among neighborhoods, this suggests continuous often slow neighborhood change. Disturbances of course provide pulses of rapid change. People adapt to the changing social norms, or they leave (Young, 1999). Analogously, people adapt to the changing environmental conditions in neighborhoods.

Studies of sprawl towns outside North American cities highlight lots of environmental issues, most with public health implications (Frumkin *et al.*, 2004). Where social capital or connections among people is strong, trustworthiness and reciprocity become the norm, an important binding force. But towns tend to have a higher proportion of elderly residents, low-income residents, and people with special needs than do cities. Combined with limited medical resources, that creates reverberating problems. Mental health issues are a particular challenge in towns. With more auto driving, people's physical activity drops, leading to overweight and diverse health problems.

Air *pollution* problems generally correlate with the types and location of industries, either in the town center or on the edge of town. Pollution of water bodies used for drinking or recreation mainly results from human wastewater and industrial effluents, though in some places from stormwater runoff. Both the pollution sources and the local waterbodies are at specific locations, amenable to solutions. Traffic noise from a busy highway, especially with many trucks, annoys people. Vehicle-caused injury and death in towns seems targeted to pedestrians and bicyclists, rather than vehicle occupants. Yet often the problem is readily addressed with relatively inexpensive traffic-calming solutions.

Utopian communities or eco-villages centered around a distinctive social or religious idea are a special case (Litfin, 2014). A spiritually centered belief or philosophy binds the typical hundred or few hundred people together for a period. Apparently, most eco-villages are characterized by, or have a goal of, relative longevity, small size, limited resource consumption, limited waste, economic prosperity, and a ripple effect. Social goals tend to be cohesiveness, an embodied vision, happiness, and satisfaction. The per capita ecological footprint is reportedly low, though comparison with other rural villages and the characteristic walking-and-public-transport cities apparently remains to be done.

Policy and Planning

Villages and towns fit within a *hierarchy* of legal and political jurisdictions superimposed on heterogeneous land (Young, 1999; Platt, 2004). Common governing hierarchies highlight: nation; province or state; county, parish, municipality, township, or town; and private or public ownership. Boundaries of jurisdictions are often invisible, though sometimes marked on the land or recognizable by different adjacent land uses. Most of the overlapping jurisdictions establish their own policies, regulations, and practices.

Villages normally have no official government and are limited in both population and funds. Thus, government policies mostly made in cities are controlling. Yet villages, being small and remote from the cities, are typically of tangential interest to centralized government, and often policies are little enforced in villages. Rural poverty, social inequity, and unsustainable natural resource use are commonly conspicuous (Zhao, 2013). Villagers tend to self-regulate, though are also sensitive to the local government of their nearby town. Solutions for improving villages and the land link the independence of villagers and dispersed residents with flexible regulations of towns and larger government jurisdictions, but are seldom sustained.

Towns, almost by definition, have a *local government* (Chapter 1), which also serves the dispersed people surrounding the town, though typically not the nearby villages. Town governance, often focused on housing, transport, energy, schools, water/wastewater, public health, and disadvantaged residents, commonly works to sustain, or increase, its economy and its residents (Brown, 1986). Local policies may enhance diverse aspects of town, such as providing for tourists, retirees, affordable housing, business investment, clean water, or sustainability (Hart and Powe, 2007). Towns both collaborate and compete with other nearby towns, all of which are subject to the governmental hierarchy, e.g., of province/state and nation.

Free-market forces combined with some government intervention may or may not help a town prosper (Vaz *et al.*, 2013a). But towns, which have a local government, some funds, and some distance from cities, can prosper by adding a mixture of organization and self-reliance. Indeed, many towns are well situated to plan for sustainability, or to protect nature, or to stimulate environmentally sensitive agriculture or forestry (Chapter 13). Some forward-looking towns plan for climate change, by reducing energy use, protecting or building up soil with organic matter, using less water, recycling materials, reducing vehicle use, maintaining growing forest, and much more.

Much about society, economics, and engineering depends on remaining in place, stationarity. Yet flows, movements, and change are pervasive, even predominant for policy and planning. To maintain an object or pattern in a flowing and changing land is costly, and hence a challenge for towns. A rural area in Mali, where livestock are moved between pasturelands for the dry-and-wet seasons (transhumance), must plan and use the land to fit with the livestock movement, which is critical to the economy (van der Zee and Zonneveld, 2001). Somewhat analogously, planning for weekend/holiday-house development near natural land in Europe depends on the seasonal movement of people (Prados, 2009b; Roca, 2013). Planning patterns that fit with such flows, movements, and changes result in lower maintenance and repair budgets, and therefore are likely more sustainable.

Some villages depend in part on *income from nature or biodiversity* in the surroundings (Figure 4.14) (Roe *et al.*, 2013). For example, wildlife provides considerable meat for some villages in Namibia (Jones *et al.*, 2013). Deer hunters in Pennsylvania and Michigan (USA) annually fill their freezers with deer meat that provides feasts for weeks or months. Although nature's resources may increase income and help avoid risk for some subsistence villagers, income and biodiversity are poorly correlated (Vira and Kontoleon, 2013). Protecting natural lands around a community against overuse or development is a chronic challenge, or losing effort. Thus, planning to conserve nearby biodiversity may be a solution for some to escape poverty, but paradoxically, where conservation fails, a route to more poverty. Nevertheless, environmental policies for increasing farm production, wood product harvest, soil protection, clean water supply, biodiversity, poverty reduction, and sustainability can create jobs at the local level.

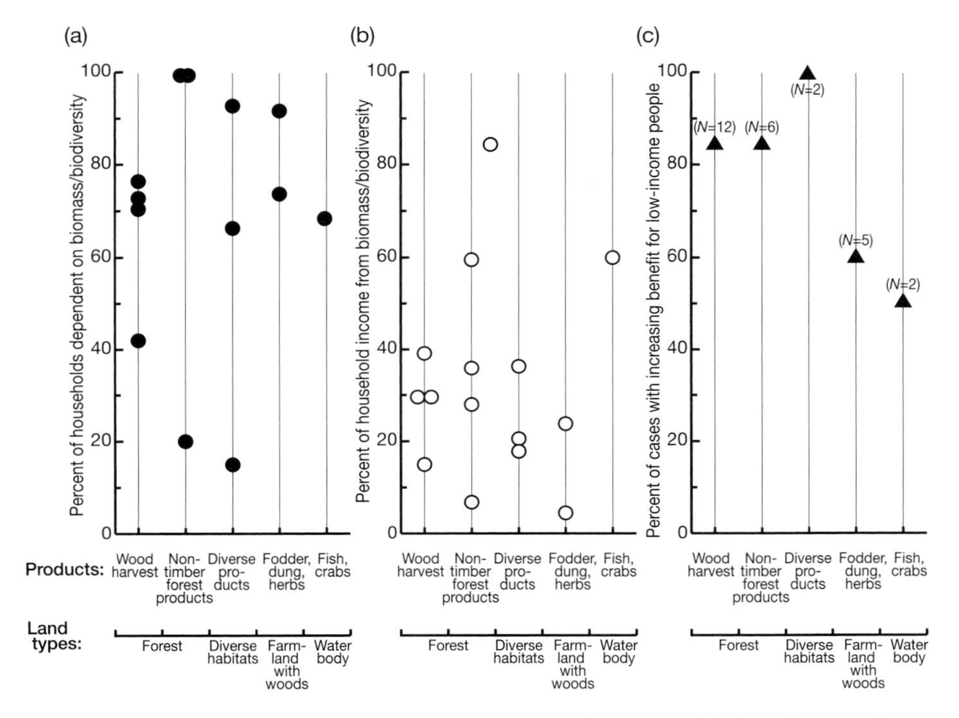

Figure 4.14 Dependence of rural low-income residents on biomass and biodiversity. Based on (a) 13 studies, (b) 14 studies, and (c) 27 studies in Asia, Africa, and Latin America. Biomass and biodiversity apparently not differentiated in many studies. Wood harvest includes forest, timber, and fuelwood. Farmland with woods includes wild foods, wild herbs, fodder, manure, dung, and leaf flitter. Diverse products include common pool resources. In (c) the alternative is increasing benefit for the wealthy. Adapted from Vira and Kontoleon (2013).

With a limited population in town, not surprisingly planning and policies effectively result from consultations and compromise by diverse entities. Local leaders, volunteers, the professions, local citizens, university specialists, non-profit organizations, and local, state/province, and national government agencies may contribute to planning and policies.

Studies of several local communities in New Mexico and New England (USA) provide insight into rural planning (Sargent *et al.*, 1991). The emergent prime objective seems to be to enhance a community's long-term viability, by respecting the carrying capacity of the surrounding natural and agricultural environment. In these areas, town planning covers both the concentrated built area and the dispersed homes in the surroundings.

Yet, as towns gradually empty of population, policies and planning become increasingly important. Often the *shrinking population* mainly results from broader forces, such as economic globalization, changing social networks, and/or new technologies (Tacoli, 2006; Vaz *et al.*, 2013). The change degrades but does not obliterate the town's traditions and heritage, or its culture. One solution is to recognize and conserve the town as essentially a large museum. Highlight not only old buildings and other structures, but also environmental memory, romantic childhoods, enchanting stories, and so forth.

I recall visiting Abanades (Castilla y La Mancha, Spain), a charming village previously of 1,000 people, now with 89 residents (see Figures 1.3 and 8.8). Upon entering, a sign said "Be careful of our children," and a nice playground was prominent in the village green. I later learned that the population of children was zero. The village was a *de facto* museum, attractive, historical, and supporting a rather well-maintained history museum. Over lunch, leaders indicated that they would welcome new families but not new large enterprises. Funds and caring for the village largely come from recent residents, who now live in Madrid 1.5 hours away and regularly return with their children on weekends and holidays.

As cities urbanize outward, land values increase and the peri-urban or *exurban area* effectively migrates further out. Nearby villages and towns are swallowed up by the urban tsunami (Forman, 2008; Vaz *et al.*, 2013a). Policies and plans generated in cities increasingly focus on this nearby visible changing exurban land, with reduced emphasis on rural economies and employment. Planning may use urban growth models that project exurban development.

A study outside Washington, DC compared a pattern-based model that largely mimics changing historical spatial patterns, versus a spatially explicit econometric model that highlights landowner profit-maximizing choices subject to market and regulatory constraints (Suarez-Rubio *et al.*, 2012). The pattern-based model was successful at describing patterns at the relatively broad (county) scale of many towns and villages, whereas the spatial econometric model was more successful at projecting development at both the broad scale and the local landowner scale. Neither model projected changes that were sensitive to policy changes. More distant towns outside the exurban zone are sensitive to policies made in cities, but balance that with molding their own policies and destinies.

Studies of small cities suggest that big fix solutions are rare (Bell and Jayne, 2006), a conclusion also applicable to the relatively conservative smaller communities. The residents' sense of place, value of small city/town living, walkability for daily needs, perceived economic improvements in more local businesses and jobs, and small changes desired in aesthetics and design together argue for gradual rather than transformative change.

Inventories for this planning emphasize natural, cultural, and human resources. Keep the farmland in agriculture. Protect natural areas. Plan drainage basins for adequate and clean water in river, streams, and lakes. Support and stimulate local business and employment. Invest internally. Plan for equity among the diverse groups present. Maintain local control of resources. But only establish policies that lead to small gradual changes.

Part II

Ecological Dimensions of Towns

5 Soil, Chemicals, Air

If our roots die, our lands will slip away.

Marcelo Romero, interview by Jake Kosek, Truchas, New Mexico, 1999

This chapter is the first of four digging more deeply into the ecology of a town or village. The chapter is divided into four sections: (1) town soil; (2) town chemicals; (3) air in towns; and (4) sound and noise effects.

Town Soil

Imagine living in the center of a town before about 1850. During rainy times mud covers the streets. During dry times dust blows down the street, temporarily blinding everyone and depositing particles everywhere. Except in tropical rainforests, street trees in the town center are scarce. Except in cold climates, the commercial buildings have overhangs, awnings, slanting porch roofs, or arcades to provide shade and repel the rain. Sidewalks are often raised and less muddy, and partially lined with horizontal bars to tie horses. No tall poles with wires tower overhead.

In this horsepower era, horses, wagons, carriages, and sometimes other livestock seem to move endlessly along the streets, with familiar squeals and rattles and clopping and neighing and voices (Figure 5.1). Horse and horsepower traffic compact the soil, so little rainwater infiltrates downward. Instead small channels of draining water curl across the street surface. Soil is eroded by water as well as wind. Ruts from wagon wheels cut into the compacted soil. Horse manure arrives daily, accumulates, and is sometimes removed.

Retail commercial buildings are concentrated in the town center, with residences and a few government buildings also present. These attached or detached buildings have been built one-by-one, displaying different widths, heights, and styles. A valuable town well is often present in an open area. Fences keep horses out of yards by houses. People shop, work, eat, and drink in the buildings, get water at the well, and endlessly deal with the muddy/dusty streets. Soil was prominent in the life of villages and towns, which grew into today's towns and cities.

Yet some of these characteristics remain. Unpaved dirt roads typically appear in the nearby area surrounding town, and often in the outer part of town. In remote areas, many villages and towns have unpaved streets throughout residential areas, even in the center.

Figure 5.1 Former Roman road in town with large flat stones, some long ago gouged by wheels of noisy wagons mainly moving at night. Residential houses with interior courtyards, plus occasional shops, are next to walkways, with lead piping for the water supply. Pompeii, AD 77 (restoration) before it was covered by hot ash from a nearby Mount Vesuvius eruption. South of Naples, Italy. See Appendix. R. Forman photo.

Informal squatter settlements close to cities often have mostly unpaved dirt streets. For instance, 40 percent of the total surface of recent Tijuana development (northwestern Mexico) is exposed soil (Biggs *et al.*, 2010).

What exactly is *soil* in a town? Geologists, engineers, and land planners commonly consider soil to be the unconsolidated or loose earth material above bedrock, thus highlighting the mineral dimension (Way, 1978; Berke *et al.*, 2006; Marsh, 2010; Forman, 2014). On the other hand, agriculturalists, soil scientists, and ecologists commonly consider soil to be the upper portion of this material that is modified by biological activity (Jenny, 1980; Craul, 1992; Wall *et al.*, 2012). Both perspectives are important in towns.

So, we describe *town soil* as the unconsolidated material above bedrock that is composed of a lower mineral zone, plus an upper biological zone in which mineral particles are mixed with abundant organic matter and organisms. Minerals and organisms exist in both zones, but biological activity is concentrated in the upper level. *Fill* is soil

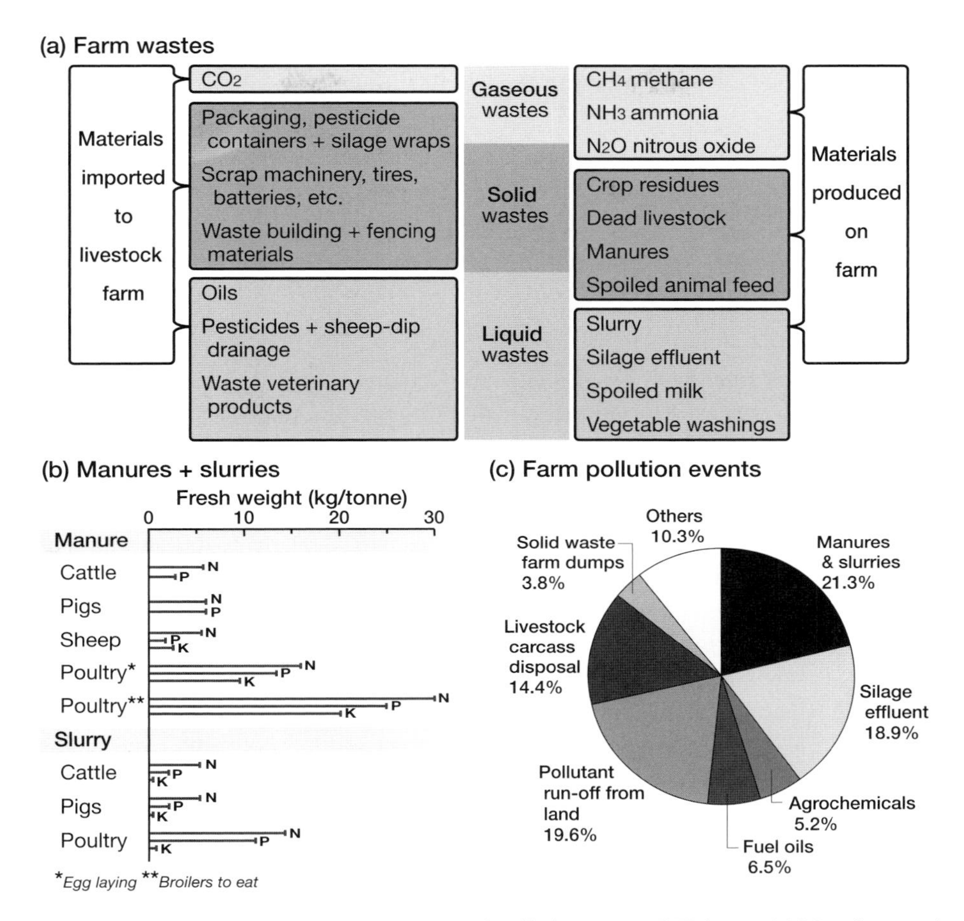

Figure 5.2 Livestock farm wastes, manures, and pollution events. UK farms. (a) Most imported materials are inorganic; most materials produced on farm are biological. (b) N = total nitrogen (N); P = total phosphate (P_2O_5); K = total potassium (potash [K_2O]). Percent dry weights: manure = cattle 25, pigs 25, sheep 25, poultry* 30; poultry** 60; slurry = cattle 10, pigs 8, poultry 25. 1 kg/tonne = 2 lbs/ton. (c) Data for Scotland. Based on Warren *et al.* (2008), Burchett and Burchett (2011), Tivy (1990).

added in a depression or hole. Fill for construction is usually sandy or gravelly, but fill for growing plants contains considerable organic matter.

Many types of soils are common in towns (Wheater, 1999), including in a: ballfield, vegetable garden, streamside, quarry, mine deposit heap, golf green, railway, roadside, dump (tip), cemetery, swamp, and crack in pavement. Farmland towns are mostly on former farmland with its residual characteristics derived from fertilizers, wastes, and pollutants (Figure 5.2).

Irrespective, essentially all *soils* are composed of mineral particles, water and dissolved nutrients, air, non-living organic matter, and living organisms (roots, soil animals, and microbes) (USDA, 2001). Not surprisingly, soils perform many functions (Wall *et al.*,

2012; Forman, 2014; Adams, 2016): structural support for engineered objects; water drainage; decomposition of organic matter; providing water and mineral nutrients for roots; and accumulation or breakdown of diverse chemicals including toxins.

Primary Characteristics

Town soils seem different from urban soils and from agricultural and various natural ecosystem soils (Russell, 1961; Craul, 1992; Wall *et al.*, 2012; Forman, 2014). First, we consider the nine primary or most distinctive characteristics of town soils.

1. *Soil types finely patchy due to construction and usage*. The distribution of sites where fill was used for the construction of buildings, roads, and other structures creates a much finer mosaic pattern on the land than a map of soil types produced by geological processes (Hole and Campbell, 1985; Craul, 1999; Blume, 2009). Fill often originates from outside town, and thus introduces different species into town. Many boundaries of the soil types formed tend to be relatively straight and sharp. This fine-scale patchiness is also characteristic of urban areas, but differs from the broad-scale patchiness of agricultural and natural soils.
2. *Vertical profiles modified by small additions of fill*. Adding small amounts of mostly sandy fill, and sometimes human artifacts, mostly in house-plot projects, adds to the vertical layering or stratification in town soils. Stratification in cropland soil due to plowing is quite different, and urban soil has both small and large areas of fill, associated with highways, high-rise buildings, and so forth.
3. *Impervious surfaces, mostly small, covering ca. 10–50 percent of area*. Much of the heat emitted from town roofs, roads, car parks, driveways, and sidewalks is absorbed by adjoining vegetated areas commonly on porous fill soil. This limits air temperature buildup. Much of the stormwater runoff from impervious surfaces infiltrates into the adjoining vegetated areas, resulting in limited flooding. In contrast, cities have a much higher percent cover of impervious surface (and less plant cover), and consequently, more heat buildup and stormwater runoff/flooding problems.
4. *Human wastewater added in numerous locations throughout town*. In many villages and towns human waste or wastewater from homes and buildings mainly enters an outhouse, pit, or covered cesspool in the ground, or is washed or drains to a waterbody or low place. In many other towns, wastewater is piped directly from buildings to septic systems in adjoining grassy spaces. Some towns, like most cities, have a sewage treatment system covering much of the area, but which includes leaks and combined sewage overflows (CSOs) during heavy rains.
5. *Compaction and elevated pH limited but widespread in small sites*. The compaction of soil particles by repeated walking, athletic field use, vibration, and other processes is generally limited to small areas distributed throughout the town. Compaction degrades several soil processes (Craul, 1992). Analogously, in town an increase in soil pH (more alkaline), mainly due to water flowing over concrete or mortar between bricks, generally occurs in small but widely distributed locations (Gilbert, 1991). In contrast, cities normally contain large areas of concrete and/or brick buildings.

6. *Entire town enriched by outside particles, pollutants, seeds, and organisms*. As a small place, a town surrounded by farmland or natural land has seeds and other species continually blowing across the town and accumulating in its soil. Dust particles from farmland also pour in, and if downwind of industry, pollutants likely cover the town. In contrast, large agricultural and urban areas have similar inputs, but some only penetrate the upwind edge portion of the area.
7. *Concentrated chemicals of a mainly residential community*. Diverse chemicals widely used in buildings, house plots, and commercial areas enter the soil. A wider range of chemicals comes from food, vehicle usage, and household products sold and used. Many of these compounds reach the soil by wear, spillage, or decomposition. On the other hand, urban areas have a considerably higher proportion of industrial chemicals in soil. Also, unlike towns, urban soils commonly have a hydrophobic surface crust of accumulated hydrocarbons, mainly from extensive vehicle combustion of fossil fuel (Craul, 1992; Siegardt *et al.*, 2005).
8. *Organic matter highly variable due to diverse vegetation cover*. The amount and chemical composition of soil organic matter, derived from dead leaves, fallen wood, and dead roots, varies with vegetation density and diverse species composition across the mainly residential area. On the other hand, cities have much less vegetation and diversity per unit area, and leaf litter disappears rapidly by decomposition in warm air or by truck removal.
9. *Concentrated buried structures and artifacts in the top 2–3 m*. This upper layer of a town's soil contains a complex of foundations, pillars, walls, and drainage structures, both in use and as remnants of former use. A network of diverse pipes both used and remnant, permeates the soil. Countless small human objects enrich the soil. In contrast, in cities this buried layer is frequently greater than 10 meters deep, and often contains underground walkways and transportation routes (Clement and Thomas, 2001).

In short, town soil is quite distinctive, contrasting sharply with urban, agricultural, and natural soil.

Key Natural and Human Processes

Geological Processes

Numerous geological processes, from mountain building to glaciation, volcanic activity, rock weathering, erosion, and sedimentation, have produced the soils in towns (Schaetz and Anderson, 2005; Gutierrez, 2013). Almost all towns are located in the following six land-surface types resulting from these processes.

Large flat plains, with silty and clayey soils high in pH and rich in calcium (Ca), carbonate (CO_3), and other mineral nutrients, supporting plant growth, agriculture, and septic systems (e.g., Casablanca, Chamusca, Emporia; see Appendix). Within the large plain, sandy floodplains of eroding streams or silty floodplains of meandering rivers may be present.

Second, moderate slopes commonly have silty and sandy soils, and eroding streams are common (e.g., Pinedale, Socorro). Third, steep slopes have rocks or gravel on cliffs, or silty or sandy soil covering bluffs (e.g., Icod de los Vinos, Stonington). Rock quarries and semi-natural vegetation are common.

Fourth, hilltop towns on almost any type of soil, and with wind velocity some 10–20 percent higher than in the surroundings, used to be common for providing defense (e.g., Capellades, Kutna Hora). Except for limestone hills, the top is usually acid and nutrient poor. Fifth, in valley bottoms sandy soil and a small stream are common, or alternatively, a river with a silt-covered floodplain (e.g., Tavistock, Huntingdon/Smithfield). Agriculture usually thrives.

Finally, on a coastal plain near the sea (or large lake), the flattish area may be covered with sands, silts, or organic soils (e.g., Sooke, Akumal, Swakopmund). Sandy soils are often acid and poor for farming, whereas silt from river floodplain or delta is rich in nutrients and prime for agriculture. Aquifers and biodiversity are often especially important. Flooding occurs on coastal plains, and marshes and swamps with anaerobic organic soils are common.

All six land-surface types support towns and only the steep slopes normally do not support agriculture. Soil erosion by water occurs in all types, and sedimentation leaves the eroded particles further downslope or downstream. Erosion by wind also may occur in all locations, but is most characteristic of hilltops and plains, particularly in dry climates (Figure 5.3) (e.g., Page, Alice Springs). The wind-eroded soil particles are deposited downwind in low areas or by upward-protruding structures. Normally heavy sand particles are dropped locally, silt particles go greater distances, and clay particles may be carried thousands of kilometers. Within towns, construction sites, abandoned plots, and soil-covered fields are often sources of eroded soil particles.

Imagine being a town planner, and starting by looking at the town's geology and soils (Daniels *et al.*, 2007). In general, the town's geology combines the underlying rocks and the surface topography (particularly slopes). Most town boundaries do not coincide with geological or even ecological boundaries. Much of importance lies deep in the soil: a fault line suggesting earthquake and landslide; an aquifer providing a water supply, and threatened by contamination; groundwater recharge area, where impervious surface should be minimized; minerals or sand/gravel for mining; and a shallow depth to bedrock, thus unsuitable for wastewater disposal.

A flat shallow soil typically is wet with poor drainage, and a poor location for most development. Sandy soil areas are bad for septic systems because the effluent moves rapidly to, and may pollute, local surface waterbodies or groundwater. But clayey soil areas are also bad because the effluent, instead of draining downward, puddles up near the surface and stinks. Highly erodible soils are unstable. Woodland and other natural areas in town are often on relatively poor soils which remain around the development on better soils. Yet woodland is effective in reducing soil erosion, stream sedimentation, and flooding. Geological features, such as a cave, fossil site, rock outcrop, and unusual strata, often warrant protection.

Ecological and Human Processes

Some processes mainly affect the top of the soil. Precipitation is added. Solar energy heats the upper soil, even to 180 cm (6 ft) under an impervious surface (Wessolek,

Figure 5.3 Red sand dune >160 m (500 ft) high in extensive desert area with dunes reaching twice that height. Dune 45, Sossusvlei, Nambia. R. Forman photo. (A black and white version of this figure will appear in some formats. For the color version, please refer to the plate section.)

2008). Sun and wind cause evapotranspiration, pumping water to the atmosphere. Dead leaves, branches, and logs fall, producing the litter layer. Frost and fires occur. The water table below rises and falls. Ecological succession changes the vegetation and depth of roots. Physical, chemical, and biological conditions in the soil all change with these ecological processes.

Other ecological processes occur within the soil. Organic matter is decomposed by bacteria, fungi, and other organisms. Water infiltrates downward. Evapotranspiration sends water skyward. Mineral nutrients are washed from upper to lower soil layers. The water table (top of the groundwater-saturated zone) below rises and falls. Specialized bacteria convert nitrogen gas from the air to ammonium and nitrate absorbed by plant roots. Earthworms and other large invertebrates move up and down, mixing organic and inorganic materials, and increasing the permeability for water, as well as for O_2 and CO_2 movement. A whole food web of soil animals is active. Roots absorb mineral nutrients, grow, die, and provide organic matter to the soil. These many processes show how a soil functions, how it works. Town soils are dynamic.

Human processes affecting soils are equally compelling. Particulate pollutants from afar, e.g., dust, soot, and heavy metals from agriculture and industry, easily cover the entire surface of a town (e.g., Superior, Capellades, Chamusa; see Appendix). Fluctuating floodwaters from upstream cover, scour, and redeposit soil elsewhere (e.g., Vernonia, Flam). Nearby construction sites, industries, power facilities, and busy roads pour pollutants onto the soil. Human waste or wastewater is added in small amounts widely across most towns. Thus, in numerous outhouses, cesspools, septic systems, and other locations, organic matter, nitrogen, phosphorus, and bacteria including pathogens enter the soil. Hydrocarbons from vehicle combustion of petroleum are spread over the soil.

Buildings alter soil conditions by shading, reflecting energy, and emitting heat. Building foundations, sidewalks, and roads absorb energy and transfer heat sometimes deep into the soil. Footpaths and ballfields may have compacted soil tens of centimeters deep, while heavy construction equipment compacts soil a meter or two deep. Train and vehicle usage produce vibrations compacting the soil. Topsoil, fertilizers, and pesticides are added. Soil itself is often removed for construction. Numerous structures and human artifacts are buried (Clement and Thomas, 2001). Human effects, mostly external, strongly alter internal soil processes. Almost all soil processes are altered by covering a soil with an impervious surface (Wessolek, 2008; Niemela *et al.*, 2011). Wind and water erode soil without adequate vegetation. As suggested at the beginning of this chapter, productive land disappears if we do not take care of the soil.

Soil Patterns and Processes

This section unravels the following key subjects for town ecology: (1) vertical and horizontal patterns; (2) soil texture and associated properties; (3) structural properties; and (4) water- and air-related soil properties.

Vertical and Horizontal Pattern

In natural soils of moist climates, soil stratification refers to layers called A, B, and C. The upper *A horizon* is a zone where rainwater moves downward leaching or washing out mineral nutrients, whereas the middle-zone *B horizon* is where some of the nutrients accumulate. The A and B horizons are the main area of biological activity and commonly called the *topsoil*. The bottom *C horizon*, or *subsoil*, is typically broken rock material derived from the underlying bedrock (sometimes the A horizon has been called topsoil, and B plus C the subsoil).

A closer look at the A horizon is ecologically valuable. At the top is a *litter layer* of little-decomposed organic matter, normally mostly dead leaves. Below that is the *humus*, also all organic matter, but in this case it has decomposed into a somewhat homogeneous black soil. Below the humus is the *A_1 layer*, which is quite blackish and composed of both mineral soil and organic matter. Many fine roots are present in the A_1 layer, indicating that absorption of water and mineral nutrients by roots mainly occurs here. Below the A_1 is the A_2 layer, which is not blackish and has most of its nutrients leached out and downward to the B horizon.

Towns adjoin, and have spread onto, cropland, pastureland, woodland, or desert. Not surprisingly the vertical soil profiles in these locations differ markedly. Plowed cropland

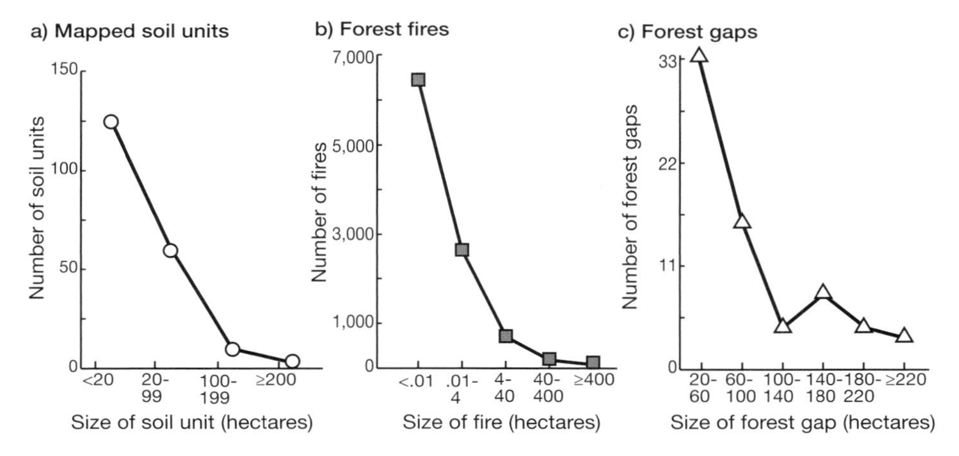

Figure 5.4 Size and number of soil, fire, and forest gap patterns in the landscape. (a) Soil units near Homer, Alaska. (b) USA forest fires in 1985. (c) Forest gaps (chablis) on Barro Colorado Island, Panama rainforest. Adapted from Hunter (1990).

(1) has a soil profile with a distinct plow or till line, above which the soil is relatively homogeneous. Pastureland (2) often has a grass-produced thick black rich A_1 layer in a moist area, and progressively less in dryer areas. Woodland (3) soil commonly illustrates the characteristic A, B, and C horizons just described. Desert (4) soil is mainly dry mineral particles, often with considerable wind-deposited material.

Most towns developed from farm villages on good soil, so the town has mainly spread on good agricultural soil (e.g., Tetbury, Casablanca, Ovcary). Farmland weeds sprout from buried seeds, and residential plants thrive on formerly plowed and fertilized soil. Within a town, some locations have recent fill added directly onto bedrock (5). Old fill and the beginning of an A layer (6) appear in other places. Also compacted soil (7) sites are present, and sometimes wetland soils (8).

A map of soil types shows the horizontal soil patterns essentially produced by geological processes. For a particular soil type, normally numerous tiny patches and progressively fewer larger patches are present (Figure 5.4). Interestingly, a similar pattern exists for other spatial ecological features.

Soil types have been named and described in many classification systems used over time and in different nations, though often town soils have been omitted or just called town soil. As a first step to understand town soils and enhance communication, perhaps the best strategy is to simply name a soil type by the last major human disturbance or use of the land. Thus, towns have ballfield soil, sandy fill, garden soil, and car park soil, in addition to soils of more natural conditions, such as wetland, floodplain, rainforest, and limestone grassland soil.

Soil Texture and Associated Properties

Soil contains mineral particles or grains of several sizes, though is mostly sand, silt, and/or clay. Of these, sand particles are large, silt medium, and clay small. Thus, *soil texture*, referring to the proportions of sand, silt, and clay (as percent by weight), is a key soil characteristic with many implications.

A widely used system portrayed by a *soil triangle* describes basic soil types based on soil texture (Jenny, 1980; Craul, 1992, 1999). Draw the triangle; it's easy. At the upper point of the triangle, sand or sandy soil is overwhelmingly composed of sand-size particles, with tiny amounts of silt and/or clay. A silty sand or clayey sand is mainly sand with somewhat more silt or clay. Similarly, near another triangle point, silt or silty soil dominates, and slightly further from the point appears a sandy silt or clayey silt. Near the third point are clay and clayey soil, and so forth. Near the center of the triangle is a *loam soil* with somewhat similar proportions of sand, silt, and clay. Sandy loam, silty loam, and clayey loam soils appear just outside the central loam in the directions of the triangle points. Nearly all soils seen worldwide are easily named and understood from a soil texture perspective.

Consider some common soils. Vegetable gardens usually are best grown on loam or silty soil, sometimes on sandy loam or clayey loam. Crop agriculture is also best in the loam and silt part of the soil triangle. Indeed, septic systems work best in silty and loamy soils. Fill for construction is normally a sandy soil, or even gravel composed of larger-sized particles.

So, towns, which have mainly grown on agricultural land, typically have predominantly silty or loamy soil. Consequently, plants usually grow well in town, and even vegetable gardens without nutrient additions produce lots of food. Street and yard trees are typically widespread in town, except in the commercial town center and in desert towns. Spontaneous plants colonizing abandoned house plots, construction sites, and cracks in pavement often grow luxuriantly in towns and villages.

Coastal towns are more likely on sandy soils (or clay if on limestone) (e.g., Sooke, Keflavik, Stonington). Towns in woodland may be on almost any soil type, though woodland normally indicates a moist or wet soil. Desert towns, are typically on wind-deposited sand or silt, or on diverse water-deposited sediments (e.g., Superior, Rock Springs, Swakopmund).

It is useful to understand *soil compaction*, introduced above, in a bit more detail, since it affects so many soil properties related to texture. Compacted soil from repeated walking, athletic field activity, and many other human activities compresses soil particles together. Sidewalks may be limited in towns, so compaction by walkers along roadsides is often common. Silty, clayey, and organic soils (of primarily organic matter) are most sensitive to compaction. Also, soils such as loam with a mixture of particle sizes compact readily. Sand with equal-sized particles compacts little. Compaction near the surface is most detrimental to plant growth. But soil compaction affects many other soil properties to be introduced below. It reduces soil porosity and structure, and degrades drainage, aeration, organic matter, and soil animal communities (Craul, 1992; Sieghardt *et al.*, 2005).

Structural Properties

In view of the buildings and roads throughout town, we first briefly consider engineering attributes of soil for construction. *Engineering soil structure* is the capacity of soil to sustain heavy weight. Four attributes are central.

Figure 5.5 Shallow complex pipe infrastructure in sandy fill under town center street. Scandinavia. (A black and white version of this figure will appear in some formats. For the color version, please refer to the plate section.)
Source: Georgeclerk / E+ / Getty Images.

1. *Bulk density* refers to the dry weight of soil material per volume of soil (Craul, 1999; Evans *et al.*, 2000; Sieghardt *et al.*, 2005). Sandy soils are lightweight (lots of space between particles) and clay soils are heavy.
2. *Compressibility* of soil is the capacity to resist volume change (Way, 1978). Soils change volume by having an external load, shrinking or swelling when wet (e.g., clay), or frost expansion.
3. *Shear strength* is the resistance to a diagonal stress or force that disrupts the continuity of soil pores (Way, 1978). Large and small *pores* result from dead roots, earthworm movement, and other processes. High shear strength is provided by similar sized and shaped mineral particles, as in certain gravels and coarse sands.
4. *Plasticity* is the capacity of a soil to become and remain deformed without volume change or rupture (Way, 1978). Plasticity reflects particle slippage and collapse at grain-to-grain contact points, mainly in clay soils.

Thus, sand and sometimes gravel are the usual soils used in fill under and around buildings, roads, and other structures in town (Figure 5.5). In fact, concrete, some asphalt/tarmac, and glass are almost all sand. Other than water and air, sand is the most used natural resource worldwide.

Towns have a special problem because in good farmland of silt or loam, deposits of sand are often scarce. The most likely local source of sand is certain floodplain or streambank locations by a stream or river which previously deposited the sand (e.g., Issaquah). Alas, digging out sand from floodplains and streambanks alters water flow, water quality, valuable habitat, fish, and bird populations.

Thus, residents of two settlements, Kuptembwo and Kwa Rhonda (Nakuru region, Kenya), have increasingly stopped farming to dig sand that is trucked to a nearby city (McGregor *et al.*, 2006). Socioeconomic benefits go to the sand diggers, but apparently not to their nearby settlements. Meanwhile, conspicuous environmental problems expand. Similarly, sand quarry mining by the town of Hotoro Arewa (near Kano, Nigeria) provides high incomes, and has drawn town residents away from farming (Binns and Maconachie, 2006; Maconachie, 2007).

Coastal towns mine beach sand for construction, a resource also in high demand for recreation. Rock from rock quarries can be pulverized to gravel, even sand size, but the process is expensive. With little sand available, just outside of Bangkok townhouses are built on locally mined clay fill subject to shrinkage and swelling (Hara *et al.*, 2008). But larger condominiums and roads with heavy truck traffic are built on sandy fill, mined from a floodplain 100 km upriver.

So, towns and cities use sand for fill, upon which roads and buildings made of sand are constructed. Sand becomes scarce, a growing if little recognized problem.

Water- and Air-Related Soil Properties

Several soil properties affect both the engineering and biological functions of soil in town. The following are particularly important.

1. *Drainage* refers to movement of water over the soil surface (surface runoff) and within the soil (subsurface runoff) (Craul, 1999; Marsh, 2010). Soils are often described in categories such as excessively drained, well drained, poorly drained, and very poorly drained. The terms refer mainly to suitability for plant growth. The widespread use of porous sandy fill in towns means that subsurface water often drains toward the nearest fill area. Furthermore, most fill areas are elongate or linear along foundations and roads, which effectively provides porous channels where subsurface water flow accelerates. Subsurface clay areas and compacted soil act as barriers. Soil water drainage patterns in residential towns are complex.
2. *Infiltration* refers specifically to the rate of water penetration by gravity into the soil from its surface (Craul, 1999; Sieghardt *et al.*, 2005; Marsh, 2010). Infiltration water flows mainly through soil pores, and recharges the groundwater if it reaches the saturated-zone water table. Unlike cities with much lowered water tables, towns usually have water tables only slightly lowered, so groundwater recharge is probably common in town (Chapter 6).
3. *Permeability* is the capacity of a soil to permit water or air movement through it (Craul, 1999; Marsh, 2010).
4. *Aeration* refers to air movement through soil, mainly O_2 from the atmosphere downward and CO_2 upward and out. With good aeration, oxygen levels remain relatively high and carbon dioxide levels low deep into a soil. With a relative abundance of trees in towns, roots are deep, and as they die aeration and organic matter increase.

5. An *oxidation-reduction threshold* or balance is present at low O_2 levels (Craul, 1992). Both positive and negative ions "shift," so many chemicals in oxidized form become converted to their reduced form (insufficient O_2), or vice versa. When oxygen drops below the threshold, several mineral nutrients become less available to roots and other soil organisms, and the soil may change color. Also, some elements such as arsenic and chromium become toxic to organisms.

These water- and air-related soil properties are combined with biological and chemical properties, introduced below, to understand the structure and dynamics of soils.

In essence, engineering soil structure is largely controlled by bulk density, compressibility, shear strength, and plasticity of soil. Water infiltration mainly depends on *pore size distribution* (arrangement of large and small pores in the soil), permeability, and compaction, while air movement of O_2 and CO_2 especially relates to pore size distribution. Diverse human activities in town affect the different characteristics above which determine engineering suitability, water flow, and aeration of soil.

Life in the Soil

Black soil organic matter is presented first as a key to all life below the surface. Next microbes and plant roots are explored. Then soil animals, from vertebrates to large invertebrates and tiny invertebrates, highlight the dynamic nature of soil.

Organic Matter, Microbes, Roots

Soil Organic Matter. Litter and humus layers are entirely composed of *soil organic matter*, i.e., dead tissue derived from living organisms. Almost all comes from above-ground leaves and wood, though, especially in grasslands, some comes from the death of roots. Below the humus, the A_1 soil layer also contains abundant organic matter (Figure 5.6). In places the soil surface in town is mulched by adding organic material such as sawdust, wood chips, or peat moss. These three soil layers, litter, humus, and A_1, are full of living organisms. Roots, microbes, and soil invertebrates are the main components in abundance.

Soil organic matter plays several important roles (Connor *et al.*, 2011; Forman, 2014). It lightens the soil, and increases both water infiltration and water retention. It enhances aeration and facilitates root penetration. Organic matter increases populations of both soil animals and microbes, thus accelerating its own decomposition.

Dead leaves are in part decomposed by bacteria and fungi. Yet perhaps more important are the soil animals or fauna in the soil, some of which chew the litter into smaller pieces, or digest the material and pass it out in partially decomposed form. These actions of the soil animals transform litter into forms much more readily decomposed by microbes. Humus is formed and its decomposition continues with microbial, chemical, and physical processes. Ultimately the carbohydrates and other organic compounds in living cells are broken down into their basic molecules of CO_2, H_2O, and various mineral ions in the soil.

With a town's highly heterogeneous above-ground patterns, the amount of soil organic matter varies sharply from place to place. Where woody plants are common, the

Figure 5.6 Pile of topsoil with considerable dark organic matter near stacked firewood. Organic matter holds both water and nutrients effectively. Driggs, Idaho, USA. R. Forman photo, with aid of A. W. Forman.

organic matter has more lignin which decomposes slowly. Near concrete structures and buildings of brick and mortar, soil pH is higher. Thus, most microbial decomposition is by bacteria, rather than fungi which are abundant in more acid soil locations.

Diverse human-made organic materials are also decomposed by the soil animals and microbes. Organic waste from the homes present, including paper products, food scraps, and some plastic items are readily decomposed in compost piles or bins (McGregor *et al.*, 2006). Periodically turning over the material provides adequate oxygen for good decomposition. Placed in the town dump, the waste may be quickly covered and become anaerobic (without oxygen), so decomposition is much slower. Lumber and wood materials are effectively decomposed by fungi in acid conditions. Herbicides and insect pesticides are widely used in town, as well as hydrocarbons, organic acids, and alcohols. Some decompose rapidly, others slowly. Some of the industrial chemicals are resistant to microbial breakdown, and thus accumulate. Small contaminated soil brownfields in town occur this way.

Microbes. Let us look more closely at these all-important microorganisms. Normally *bacteria* are by far the most abundant (Jenny, 1980; Craul, 1992; Schaetzl and Anderson, 2005; Wurst *et al.*, 2012; Wall *et al.*, 2012). Many bacteria decompose organic matter, but others recycle nitrogen or oxidize elements such as sulfur and iron. Many form spores resistant to heat and desiccation. Some can reproduce in 20 minutes, though the norm is considerably slower. Some accumulate around root tips, presumably enhancing the absorption of nutrients. *Aerobic bacteria* operate with oxygen, and *anaerobic bacteria* without oxygen. Rapid bacterial growth in soil can use up the oxygen, especially in wet soils.

Fungi are the second big group in most soils. Branching one-cell-thick filaments may total hundreds of meters in length. Stalked fungi or mushrooms are most familiar, but these are only the reproductive structures which produce spores dispersed in the wind. Some fungi, called *mycorrhizae*, penetrate or are attached to root tips creating a symbiosis with certain plant species. The plant receives mineral nutrients from the fungus, while the fungus gains carbohydrates from the plant. Also, some fungi are parasites, diseases, and antibiotic-producing species.

Nevertheless, fungi decomposing organic material is a main story in towns. Many can breakdown the somewhat resistant cellulose from plant cell walls, and in acid conditions can decompose the highly resistant lignin from wood. In town, fungi tend to be most abundant in the soils covered with trees.

Protozoa are the other large group of microbes in soils. These mobile single-celled animals mainly consume bacteria. Most protozoa produce resistant cysts that can blow in the wind and persist. And many species thrive in higher pH soils on limestone, or near concrete, brick mortar, and gardens with lime added.

Uncommon or specialized microbes doubtless thrive around places with specialized organic materials. Consider the local medical facility, veterinary clinic, cemetery, sewage treatment facility, commercial centers, and various manufacturing sites. Indeed, think of the networks of buried pipes. Every pipe system has leaks into the soil where microbes usually thrive.

Roots. We briefly introduce big familiar living things in soil, roots. Assuming the attached green plant above is thriving, roots metabolize and grow. That means they absorb considerable oxygen from the soil. Oxygen is important for nutrient absorption and to prevent root rot. Even a brief period without oxygen, such as in wet conditions, can damage roots.

Roots absorb water from pores in the soil. A continuous stream of water molecules in a plant extends from root tip to leaf surface. Therefore, when water evaporates from a leaf in *transpiration*, that same amount of water is absorbed by the root from the soil. Direct sunlight, hot air, and wind accelerate transpiration, and hence water loss from the soil. The requirement for both water and oxygen in the soil strongly determines where and how well town plants grow.

But roots are quite multifunctional. They anchor the tree against wind. Store carbohydrates and other organic substances produced in leaves. Produce hormones for plant growth. Absorb mineral nutrients, especially nitrogen, phosphorus, and sulfur, used in the synthesis of many organic substances. Bind soil particles together. Continuously die, creating soil organic matter. Roots increase water infiltration as well as aeration, so keep the soil filled with roots.

Roots, however, are sensitive to toxins such as those washing in from commercial and industrial areas. High levels of ammonia in solution, copper, lead, or zinc basically stops water uptake and root growth. Trees next to streets often grow poorly due to many stresses and disturbances. Unlike a typical tree where the root system and tree crown extend outward similar distances, street-side tree roots typically extend short distances except in one or two directions, where they may extend well beyond the crown. These extensions are in the direction of favorable soil water conditions. Keeping a street tree thriving requires water, not too much, not too little, but just the right amount for years and years.

Soil Animals

Soils are lively places. Vertebrates, large invertebrates, and tiny invertebrates coexist, the latter two groups in huge numbers. The soil animals perform many functions in the soil (Wall *et al.*, 2012; Mehring and Levin, 2015). Soil conditions also affect animals above the soil, such as a farmland bird, yellow wagtail (*Motacilla flava*), which is decreasing in English crop fields (Gilroy *et al.*, 2008). The best predictor of yellow wagtail presence is the penetrability of the field, presumably to provide feeding access to soil invertebrates.

Vertebrates. Soil vertebrates commonly have burrows, especially in the heterogeneous microhabitats and soils in towns and villages (Schaetzl and Anderson, 2005; Forman, 2014). This results in an uneasy coexistence with human pets, especially dogs and cats, which chase or feed on these animals with burrows nearby. Soil vertebrates include: frogs, salamanders, turtles, snakes, rabbits, hedgehogs, weasels, wombats, moles, voles, and burrowing owls. These burrowing (fossorial) mammals, reptiles, amphibians, and occasionally birds may feed underground on roots or litter, or on large invertebrates such as earthworms. Many soil vertebrates emerge to forage on the ground surface. Dry soils and locations with suitable sandy soil texture are generally preferred, partly because these tend to have adequate oxygen. Some animals use burrows formerly dug by other species. In desert towns, some shrubs are on soil mounds in which a group of soil vertebrates coexist, mostly emerging at night to feed.

Large Invertebrates. However, large invertebrates play the key role in most town soils. Six groups are prominent worldwide (Russell, 1961; Jenny, 1980; Wall *et al.*, 2012; Wurst *et al.*, 2012): (1) earthworms (annelids or oligochaetes); (2) slugs and snails (mollusks); (3) beetles (coleopterans); (4) ants and termites (Formicidae, Isoptera); (5) millipedes and centipedes (diplopods, chilopods); and (6) spiders (arachnids). Large soil invertebrate types vary in different habitats, and seem to be especially sensitive to soil moisture. Although data on these animals in towns is scarce, soil studies below in a crop field (wheat, *Triticum*), a pasture, and a forest suggest useful patterns (Russell, 1961). Note that repeated crop production often greatly decreases soil organic matter. On the other hand, organic matter commonly remains abundant with the use of manure or chemical fertilizer, or in wet rice or irrigated-desert production (Connor *et al.*, 2011). No indication of variability is present in these studies, so only preliminary trends are suggested.

The number of soil invertebrates is much higher in the pasture soil than in the crop and forest soils which have similar totals. This pattern also applies to the tiny invertebrates (springtails, mites, nematodes). Within the large invertebrate groups, earthworms and millipedes/centipedes are highest in the crop field. Since most towns spread over cropland, and others over pasture and natural land, these results may provide clues to the main soil animals in town soils.

Earthworms thrive in moist soil that is not too hot, and are particularly abundant in limestone areas and where soil pH is relatively high near concrete, brick mortar, and vegetable gardens with lime added (Craul, 1992; Broll and Kaplin, 1995; Schaetzl and Anderson, 2005; Forman, 2014). Earthworms play a key role by burrowing, which increases soil pores, drainage, and aeration. They ingest mineral soil, organic matter, and microbes, and then add digestive chemicals and other microbes. The animals excrete tiny aggregations (casts) of these materials. The casts improve soil aeration, as well as water infiltration and the capacity to hold water. By moving up and down in the soil, earthworms mix the organic material and nutrients from upper layers with the lower mineral layers. This mixing effectively deepens the A layer permitting deeper root growth.

Nonnative earthworms (*Lumbricus* spp.) have noticeably changed some earthworm-free deciduous forests of the northern US Midwest by consuming the litter layer, effectively reducing the cover and richness of herbaceous plants, and increasing the dominance of sedges and grasses (Loss *et al.*, 2012). Of four ground-nesting songbirds present, ovenbird (*Seiurus aurocapilla*) density correlated significantly and negatively with soil earthworm biomass. More nonnative earthworms mean fewer forest ovenbirds, which have a loud song that I like, something like: "*Teacher-Teacher-Teacher-Teacher*." Some cities have a high density of nonnative earthworms which alter soil characteristics. Towns may also be subject to nonnative earthworm effects.

Slugs and snails thrive in moist cool soil. These mollusks create pores that increase aeration and infiltration, and mix the upper organic and lower inorganic soil materials (Russell, 1961; Owen, 1991; Craul, 1992). Furthermore, the gut of slugs and snails has an enzyme (cellulase) which breaks down cellulose and other organic substances of plant cells, thus playing a particularly effective role in litter decomposition. Many mollusks feed on other large invertebrates in the soil. Similarly soil *beetles* and their worm-like *grubs* increase soil aeration and water drainage, and many species feed on other soil invertebrates.

Ants and termites (which are unrelated) provide many of the same soil functions as for mollusks (Russell, 1961; Owen, 1991; Craul, 1992; Schaetzl and Anderson, 2005). Moreover, by being numerous, ants can move huge amounts of soil material, and most are predators. Termites also contain cellulase in the gut, so they focus on decomposing wood, especially in the tropics. *Millipedes* are chewers important in the early decomposition process, while *centipedes* are predators on other soil invertebrates. *Spiders* mostly forage through the organic layers of the soil as predators. Other important soil-burrowing large invertebrates include pillbugs or woodlice (isopods) and bees and wasps (hymenopterans) (Russell, 1961; Owen, 1991; Wheater, 1999).

Figure 5.7 Spruce and pine (*Picea, Pinus*) growing on acidic mine waste of a former silver mine. People looking at a 40 m wide (130 ft) sinkhole or cave-in apparently 200 m deep, highlighting the depth of former mining and hazard of mining near a community. Kutna Hora, Czech Republic. See Appendix. R. Forman photo. (A black and white version of this figure will appear in some formats. For the color version, please refer to the plate section.)

Tiny Invertebrates. Looking still more closely at soil, we often see the tiny invertebrates, mainly mites, springtails (collembolans), even nematodes (Russell, 1961; Craul, 1992; Broll and Keplin, 1995; Schaetzl and Anderson, 2005; Wall *et al.*, 2012). Generally, *mites* are oval and hard-bodied, and just visible to the human eye. Different species feed on litter, fungi, algae, and other mites (detritivores, fungivores, herbivores, predators). *Springtails* are cylindrical and can "jump." *Nematodes* are tiny roundworms without segments. Both groups also include species with different food sources. All three tiny invertebrate groups thrive in somewhat acid soils where few large invertebrates survive (Figure 5.7). Places in town with peat moss or conifer species or acidic chemicals in the soil may have abundant tiny soil invertebrates.

In very dry and very wet conditions, the story of a town's soil is different. Desert towns are often covered with wind-deposited sandy or silty soil, and normally have little soil organic matter. Woody plant roots often penetrate down deeply to a buried moist or water layer.

At the other extreme, wetland areas in a town usually have a thick layer of organic matter. But being frequently inundated by water means that oxygen in the soil is limited, and mainly anaerobic organisms survive or thrive. With little soil oxygen, roots are shallow, many mineral nutrients are largely unavailable or toxic, and soil animals are limited. Every town may have a mosaic of dry soil, wet soil, and intermediate conditions.

Town Chemicals and Their Sources

Extremely diverse chemicals compose all objects including us, and support all ecological and human processes. The first law of conservation of matter emphasizes that matter (materials or chemicals) can neither be created nor destroyed, only changed in form. So, chemicals in town either leave, remain, or are converted to a different chemical form.

Some chemicals combine to make a different stable chemical, such as combining NO_X (nitrogen oxides) and VOCs (volatile organic compounds) in the presence of sunlight to form O_3 (ozone). Some complex organic compounds breakdown or degrade into a simpler stable compound called a "degradate," such as the pesticide DDT breaking down to DDE which has different and significant effects. In addition, some elements, such as nitrogen, cycle in an ecosystem, whereby different microbes convert the element into different molecular forms in a cycle.

Inorganic substances and organic substances represent the major division of chemical types. Organic substances contain C (carbon), or, more precisely, contain C, H (hydrogen), and O (oxygen). Organic compounds, such as sugars, fats, and proteins, mainly result from photosynthesis, but numerous other organic compounds are synthesized by industry.

Most of the common chemicals and their sources in soils characteristic of towns are highlighted in the following categories: (1) rock, mining, and fill chemicals; (2) soil chemicals; (3) industry and brownfield chemicals; (4) water supply, river, stream, and pipe chemicals; (5) road chemicals; and (6) vehicle chemicals. We explore the atmospheric input in a later section on air in towns.

Rock, Mining, and Fill Chemicals

Si (silica) is the prime element on land, and, combined with oxygen, SiO_2 (silicate) is the major mineral. SiO_2 is a hard or resistant mineral, the main constituent of volcanic rocks, metamorphic rocks, sandstone, shale, sand, silt, and some clay. Glass is basically silicate. However, in limestone areas, Ca (calcium) and CO_3 (carbonate), combined as $CaCO_3$ (calcium carbonate), are the primary components of the land (e.g., Emporia, Valladolid). Limestone and some clays are basically $CaCO_3$. SiO_2-based soils are acidic (low pH), and $CaCO_3$ soils are somewhat alkaline (high pH).

Normally porous fill for drainage is basically SiO_2. Rubble fill from brick and concrete construction is rich in Ca and Mg (magnesium). Fill from wetland or the bottom of a waterbody has considerable S (sulfur). Topsoil fill to enhance plant growth is rich in N (nitrogen), P (phosphorus), Ca, Mg, humic acids, and other organic compounds.

Excluding sand mining described above, most mining is for heavy metals, especially As (silver) (Figure 5.7), Au (gold), Cd (cadmium), Co (cobalt), Cu (copper)

Figure 5.8 Large copper-mining operation providing jobs for, supporting, polluting, and perhaps threatening a nearby town. "Copper moon" rising at Santa Rita Mission, New Mexico, USA. (A black and white version of this figure will appear in some formats. For the color version, please refer to the plate section.)
Source: Alan Dyer / Stocktrek Images / Getty Images.

(Figure 5.8), Fe (iron), Li (lithium), Ni (nickel), Pb (lead), Sb (antimony), Sr (strontium), and Zn (zinc) (Baker *et al.*, 2010; Gell, 2010). Smelters then "burn" the mined rock material at high temperature to purify and extract specific elements.

The common environmental problems from mining are: (1) altered and polluted groundwater; (2) piles (heaps) of often toxic mine-rock waste; (3) contaminated stormwater runoff, such as acid mine drainage; and (4) toxic dust.

Toxic dust is a special problem because of the extensive movement of above-ground machinery and trucks. Also, smelters usually produce considerable dust. For instance, a coal mine in northeastern India created dust rich in iron, lead, cadmium, and aluminum (Rout *et al.*, 2014). Such dust affects the respiratory health of nearby residents, and upon deposition degrades aquatic and soil ecosystems. Most mines have a town or village(s) nearby that is periodically blanketed with mine dust (e.g., Rock Springs, Superior, Falun [Figure 5.8]; see Appendix).

Soil Chemicals

Tests for soil quality focus on physical, biological, and chemical indicators, including the chemical elements (Schindelbeck *et al.*, 2008): C, N, P, K, Mg, Ca, Fe, Al, Mn, Zn,

Cu. New York State has established standards for the maximum amounts of "minor or trace" elements: As, Ba (barium), Be (beryllium), Cd, Co, Cr (chromium), Cu, Hg (mercury), Li (lithium), Mn (manganese), Mo (molybdenum), Ni, Pb, Sb, Se (selenium), Sr, Ti (tin), V (vanadium), and Zn.

A so-called Ellenberg fertility index, which includes the amounts of various soil chemicals for suitable plant growth, has been widely used in Europe to evaluate soil quality (Pellissier *et al.*, 2010). It also facilitates estimating the best fertilizer composition for farmers to use.

Towns and villages in farmland are typically bathed in dust containing mineral nutrients and often pesticides (e.g., Casablanca, Clay Center, Chamusca). N (nitrogen) and P (phosphorus) blown into town from surrounding farmland stimulates plant growth in house plots, vegetable gardens, and other vegetation. Industrial pollution covers towns (e.g., Ovcary, Rumford).

Clay is not only a tiny particle, but also is chemically important in soil because of its *cation exchange capacity*. In an acid soil the clay particle mainly has hydrogen ions (H+) attached to it. At progressively higher pH levels, hydrogen ions are progressively replaced by other positively charged ions, including Ca (calcium), K (potassium), PO_4 (phosphate), and NO_3 (nitrate). These other ions are elements important to plant growth, and hence their abundance is an indicator of soil fertility. In effect, clay particles can hold mineral nutrients.

Yet soil organic matter is even more effective in holding mineral nutrients. Thus, typically residents add humus or other soil organic matter to gardens, not only to improve moisture conditions, but to increase and sustain soil nutrients for plant growth.

Pit toilets or latrines and cesspools add organic matter, microbes, N, P, and other elements of the human body to the soil. Septic systems using water as a waste-transport mechanism also add water to the soil.

Soil also contains several important gases (Forman, 2014): CH_4 (methane), CO_2 (carbon dioxide), O_2 (oxygen), CO (carbon monoxide), H_2S (hydrogen sulfide), HCN (hydrogen cyanide), NH_3 (ammonia), ethane, ethylene, propylene, and radon. Uranium or phenols may be present in local areas.

Nitrogen deposition in the USA and many other regions ranges from 1 to 39 kg N per hectare per year, a major increase over several decades (Pardo *et al.*, 2011). This added nitrogen alters the nitrogen cycle, including NH_3 (ammonia), NO (nitric oxide), and N_2O (nitrous oxide), as well as various ecosystem processes. Excess nitrogen produces diverse responses of species, from diatoms and lichens to shrubs and trees.

Pb (lead) in town soils may reflect lead paint use, or former or present traffic exhaust. A study of 20 sites (two samples each) apparently in three agricultural towns outside Kano (Nigeria) discovered that the concentrations of almost all elements (Cd, Co, Cr, Cu, Fe, Hg, Mn, Ni, and Pb) exceeded, often considerably, the maximum safe levels for agricultural production (Maconachie, 2007). These high levels may have resulted from airborne air pollutants. But also some farmers in that area make fertilizer by collecting mainly organic material from a solid-waste dump, and mixing it with plant material from home and usually some livestock manure. The excess heavy metals may have come from the dump.

Radioactive elements are present in some rocks, and therefore soils. Radioactive radon gas from certain rocks and soil sometimes appears in cement $CaCO_3$ products, such as foundations of homes and buildings. Radioactivity from medical use and diverse research is sometimes present in soil and water. Radioactivity seems likely to have spread throughout the town of Chimayo, New Mexico (USA, population 3,200) from a nuclear research facility an hour's drive upwind (Kosek, 2006). Biomagnification, the increase in concentration through the food chain, may also occur. Certain bacteria might help remediate some radioactive waste (Geissler *et al.*, 2010). But apparently no level of radioactivity, which may lead to mutations, illness, and death, is safe.

Industrial use and disposal is also an important source of radioactivity. For instance, 60 years of metals manufacturing including army munitions at a small site in the town of Concord, Massachusetts (USA; see Appendix) left the bulk of its radioactive waste, especially depleted uranium, in the sandy soil and groundwater. Cleanups over two decades continue. But some contamination has percolated down to bedrock, and is to be encased semi-permanently in a huge buried concrete cylinder. A recent smaller plume of non-radioactive dioxane (a VOC used in solvents and degreasers for cleaning machines, probably containing 1,4-dioxane) has apparently now extended under a river in the direction of an adjacent town's water supply. Pumping the contaminated groundwater plume for 40 years is planned.

Industry and Brownfield Chemicals

The range of products produced in industries of course is enormous, yet the byproducts or pollutants released into the environment fall into relatively familiar groups (Batty and Helberg, 2010; Gell, 2010): salinity, acidity, and alkalinity all have major environmental effects on aquatic and other ecosystems. Fatty acids and alcohols are common byproducts released. SO_2 (sulfur dioxide) and PM (particulate matter) result from coal-burning factories and power facilities. Ca and CO_3 emanate from a cement factory.

Brownfields are relatively large areas of contaminated soil, usually resulting from former industry (Hollander *et al.*, 2010; Haninger *et al.*, 2017). Common soil contaminants are (Roast *et al.*, 2010): the inorganic metals, Cd, Cr, Cu, Hg, Ni, Pb, Zn; and the organic substances, benzo(a)pyrene, pentachlorobenzene, pentachlorophenol, tetrachloroethene, and toluene. Common contaminants in English brownfields are (Bambra *et al.*, 2014): As (arsenic), Asb (asbestos), Pb (lead), petroleum hydrocarbons, and radiation (radioactivity). People living near these brownfields have higher rates of morbidity/mortality.

Certain plants (hyperaccumulators) absorb unusually high amounts of heavy metals such as nickel or selenium (Rascio and Navari-Izzo, 2011). Yet for a variety of reasons, few brownfields are likely to be cleaned up (phytoremediated) by plants (Forman, 2014). However, diverse bacteria, each affecting a different metabolic pathway, may help remediate an industrial brownfield site which usually has many contaminants (Killham, 2010).

Water Supply, River, Stream, and Pipe Chemicals

Streams or a river bring sediment and agricultural chemicals from upstream farmland. Muddy water containing pesticides and nutrients from fertilizers is typical. Alternatively, the incoming flow may contain stormwater pollutants from an upstream town, or industrial pollutants from upstream industry (e.g., mercury to Concord; see Appendix).

Towns typically have small buried dumps near streams, wetlands, and other waterbodies. Often begun by individual farmers for farm and household waste, gradually the dumps became used by neighborhoods. When they caused too many problems, they were closed and covered. The town-wide pickup and removal of solid waste by trucks is a relatively recent phenomenon, replacing the many small burials of solid waste.

In a town water supply, chlorine is a common disinfectant, and fluoride a teeth protector (Frumkin *et al.*, 2004). Sometimes disinfection byproducts appear, including halomethanes and haloacetic acids. Well water sometimes has high levels of rock minerals, such as arsenic. Road salt (NaCl) often leaches downward into the groundwater and pollutes well water. Subsurface runoff also may carry NaCl to ponds and streams.

The output from a sewage treatment facility normally goes to a river, stream, lake, estuary, or sea. The pipe system typically has sewage leaks, and may add heavy metals such as Fe, Cu, or Zn. Primary treatment of sewage releases considerable organic matter, bacteria including pathogens, N, P, and low levels of heavy metals and other materials from buildings and pipes. Secondary treatment releases considerable N and P. Tertiary treatment ends up with very low levels of N and P.

Road Chemicals

Asphalt/tarmac road surfaces are in essence gravel and oil, with a complex of resistant hydrocarbons, whereas concrete roads are $CaCO_3$ with iron reinforcing rods. Repeated vehicle use, especially by trucks, gradually wears the surface material into particles that blow or are washed away. Gravel and earthen roads have loose surfaces so vehicles lift considerable dust into the air during dry periods. Sometimes oil-based fluids are spread on earthen roads to reduce dust.

When a paved road surface is replaced, some contaminated wastes of society may be included in the recycled asphalt (Forman *et al.*, 2003). Road surface wear then spreads the chemicals into the surroundings, a potentially serious problem by a water supply or rare species habitat.

A dozen main chemical pollutants come from roads and roadsides (Figure 5.9) (Forman *et al.*, 2003). These chemicals result from roadbed and road surface wear, but also from sanding and de-icing agents, herbicide, pesticide, and fertilizer use, and roadside maintenance practices.

Diverse items with diverse chemistries are discarded from vehicles and line most roadsides. In cold climates, road salt is periodically spread along roads. Usually this is NaCl, sometimes contaminated with Cyn (cyanide). The main alternatives used are CMA (calcium magnesium acetate), CaCl (calcium chloride), or MgCl (magnesium chloride).

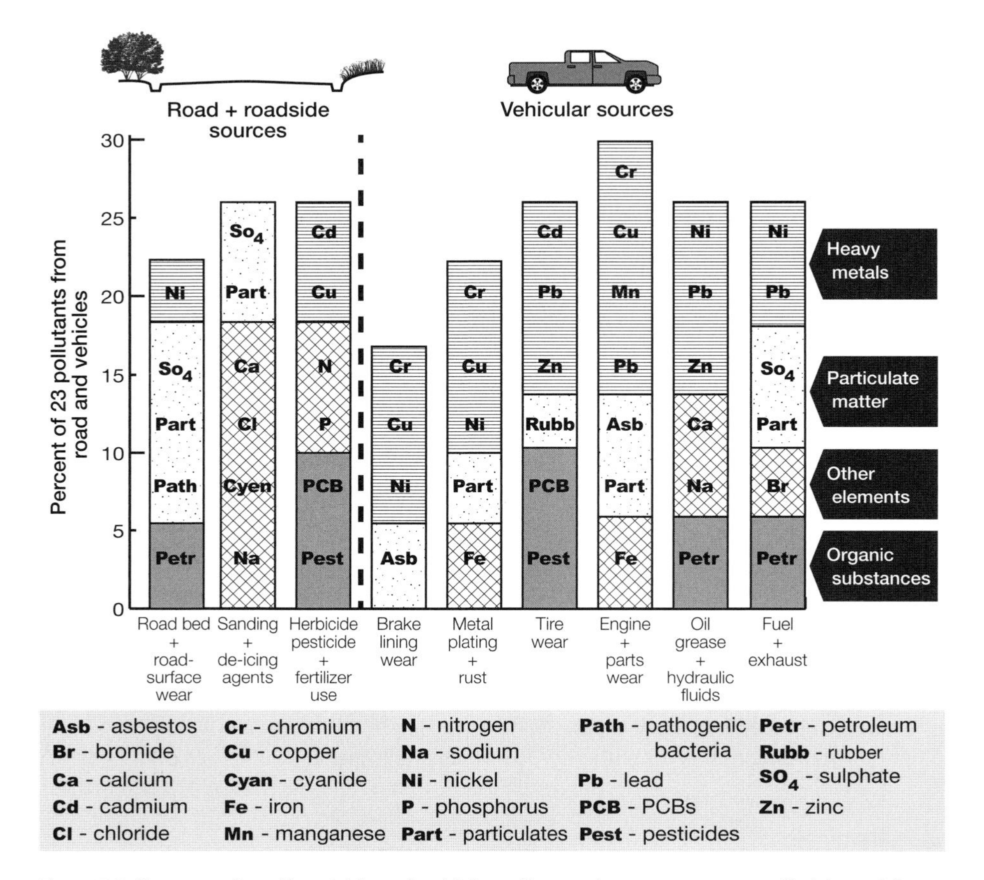

Figure 5.9 Sources of road/roadside and vehicle pollutants in stormwater runoff. Adapted from Forman *et al.* (2003), and based on Kobringer (1984) and Federal Highway Administration (1996).

Bridges are particularly important because chemicals leach out of the bridge, and wash or blow from the road surface and passing vehicles, directly into a stream or river. Stormwater from a Florida bridge carries the following pollutants to the water below (Forman *et al.*, 2003): NO_3- and NO_2-nitrogen, NH_3-nitrogen, (ortho)phosphorus, Al (aluminum), As (arsenic), Ca, Cd, Cu, Fe, Hg, Ni, and Zn.

Vehicle Chemicals

Transportation by truck, rail, boat, or wagon brings goods of all sorts from outside into town. Commercial retail shops are the major recipients. Industries may also be major recipients of outside materials, but this stream of chemicals often does not involve many town residents.

Major pollutants from vehicles are (Figure 5.9) (Forman *et al.*, 2003): Asb (asbestos), Ni, Cu, Cr, PM (particulate matter or particles), Fe, Cu, rubber, PCBs, Zn, Pb, Cd, petroleum, SO_4 (sulfate), Br (bromine), Na, Ca, and Mn. These 18 chemicals come mainly from six vehicle sources: (1) brake lining wear; (2) metal plating and rust; (3) tire wear; (4) fuel and exhaust; (5) oil, grease, and hydraulic fluids; and (6) engine and parts wear. Vehicle wear and oil-based fluids are the primary vehicle pollution sources.

Pollutants spread outward different distances from busy roads (Frumkin *et al.*, 2004). Black smoke and NO_2 (nitrogen dioxide) concentrations may be especially high in the first 100 m outward. Also, vehicles commonly leak or spill N, P, Cd, Zn, PAHs, and mono-aromatic hydrocarbon compounds (Forman *et al.*, 2003). Near busy roads, the pollutants doubtless affect the health of people in town neighborhoods, as well as terrestrial and aquatic ecosystems and their species.

Localized less frequent inputs from vehicles also occur (Forman *et al.*, 2003), i.e. spills of oil, gasoline, industrial chemicals, and other substances. Vehicles have accidents with other vehicles as well as with roadside structures, often leaving a spot of chemicals.

Chemicals are sometimes used as additives to gasoline (Forman *et al.*, 2003). Common ones include: tetraethyl lead; MMT (methyl cyclopentadienyl manganese tricarbonyl); manganese oxides; MTBE (methyl tertiary-butyl ether); and the heavy metals, Al, Cu, Fe, Hg, Ni, Pb, Zn. Petroleum-related compounds in gasoline may include PAHs, benzene, toluene, thylbenzene, xylene, and MTBE.

Gasoline service stations normally have buried storage tanks for fuel. Over time, they often leak. MBTE, a gasoline additive, not infrequently leaks into the surrounding soil and water, where it can spread rapidly (Frumkin *et al.*, 2004).

Chemicals in Ecological and Human Use

Today in the USA more than 60,000 chemicals are in use, many without careful evaluation of potential health risks. No human health data are available for 85 percent of them. Furthermore, essentially all chemicals can be freely used except for pesticides and pharmaceuticals.

Many pollutants from varied sources have been widely shown to be toxic or hazardous to human health (Hardoy *et al.*, 2001): Ar, Asb, Cd, Co, Cr, Cyn, Hg, Mg, Ni, Pb, and NO_3 (nitrates). Also, organic compounds, such as waste oils, PCBs, organic solvents, benzene, and chlorinated solvents, may be toxic or hazardous. Some of these bioaccumulate up the food chain in ecological systems.

We begin with (1) chemicals around us daily, followed by (2) sustaining chemical elements.

Chemicals around Us Daily

These appear in six groups: (1) retail shops and food; (2) buildings; (3) air indoors and in vehicles; (4) town spaces and house plots; (5) solid waste; and (6) emerging contaminants.

Retail Shops and Food

Generally trucked into town, highly diverse goods line the shelves of shops. Shoppers then buy, transport, and use the goods mainly in or around the home. Some goods originate locally and others are shipped from afar. Lots of paper and plastic packaging may be associated with the goods.

Diverse food is displayed for sale in small markets or small-to-medium supermarkets or outdoor sites. But many food items for sale are considered health risks (Reeves, 2014), including: cola drinks; high-fructose corn syrup; salt-laced vegetable juices; fried foods, pastries laden with trans-fats; and hormone-treated beef. Many food additives may lead to the metabolic syndrome of overweight, diabetes, high cholesterol, and heart attacks. Foods also may come with varied types of packaging.

Some goods are used outdoors, for yards, gardens, lawns, pets, and even livestock. And some are used in the construction, maintenance, and the embellishment of buildings. Chemicals are well dispersed in a town.

Buildings

The typical life span or life cycle of a house or small bridge in the USA is 50 years, before it needs a major overhaul or replacement (Yudelson, 2008). A non-residential building has a 75-year life span, and a school 60 years. Chemical objects and materials also have a life span. The use of recycled materials, such as plastics, nearly doubles the typical life span. Agri-fiber products, including strawboard, wheatboard, cotton, and wool, use agricultural products for construction uses, and are completely recyclable.

Buildings, in which town residents live and work, have lots of chemicals bad for health, and bad for the environment. For example, it is wise to avoid urea-formaldehyde resin in composite wood products, such as particleboard and composite wood (Yudelson, 2008). The author recommends using cabinets without formaldehyde. Low VOC-content paints, coatings, adhesives, and sealants. Also, low VOC-emitting carpeting and backing. Bamboo, cork, wood, or other natural-product flooring should be safe. Use low-VOC furniture. Avoid asbestos insulation. No PCBs (polychlorinated biphenyls). Use fluorescent or LED lighting, and minimize use of bulbs with high mercury content. Use CFC (chloro-fluoro carbon) refrigerants. And, according to the author, certainly avoid items containing toxic dioxins, Cd, Hg, Pb, di-ethyl hexyl phthalate, and natural rubber latex.

Finally, here is a set of reportedly harmful chemicals typically used in building products and materials (Harvard University Office of Sustainability, 2017 website). Obvious substitutes are not yet available for many of these, and the human health and environmental effects of many are little studied. This is a worrisome list of hazardous chemicals close to many of us daily:

(1). Halogenated flame retardants, including PBDE, TBBPA, HBCD, Deca-BDE, TCPP, TCEP, Dechlorane Plus, and other retardants with bromine or chlorine. (2). Heavy metals, specifically Cd, Hg, Pb. (3). Perfluorinated compounds (PFCs), including PFOA and PFOS. (4). Phthalates.

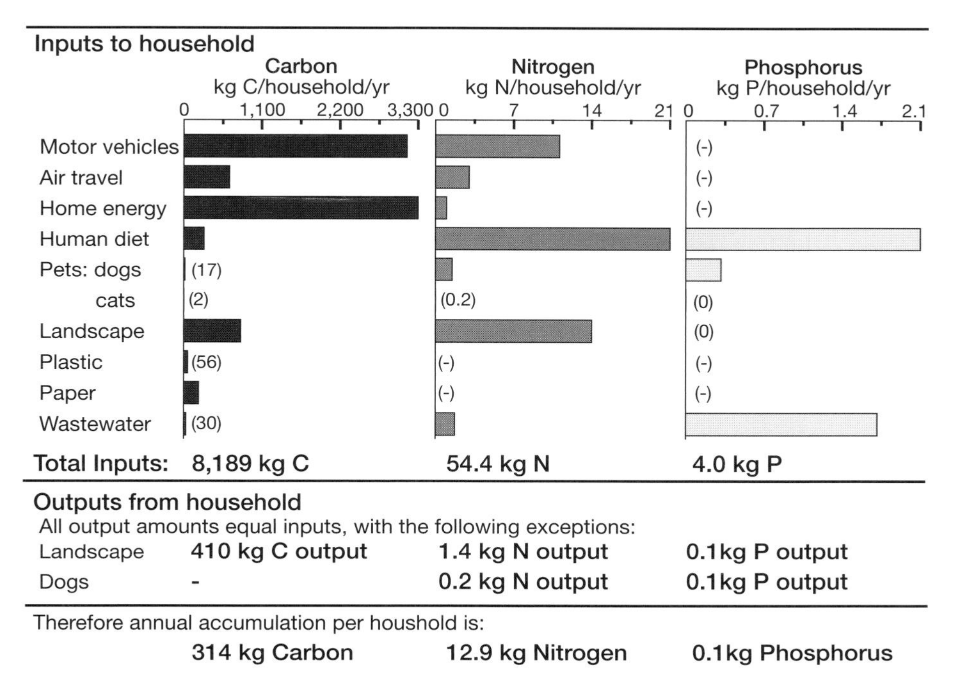

Figure 5.10 Carbon, nitrogen, and phosphorus flows through a household. Inputs and outputs calculated for 360 households (including their house-plot landscapes and off-site activities) along an exurban-suburban-urban gradient in Anoka and Ramsey counties, Minnesota, USA. Average lawn area = 1,457 m^2, finished floor area 138 m^2, number of household residents 2.6; 1 m^2 = 10.8 ft^2. Landscape input excludes pet waste; wastewater input includes detergents, etc.; some dog food input cycles though the landscape contributing N and P. Adapted from Fissore *et al.* (2011).

(5). Formaldehyde. (6). Chlorinated building materials. (7). Polyethylene and chlorosulfonated polyethylene (includes PVC). (8). VOCs, such as in adhesives, sealants, and paints. (9). CFCs and HFCs. (10) Asbestos. (11). Petrochemical fertilizers and pesticides, herbicides, fungicides. (12). Wood treatments containing creosote, arsenic, or pentachlorophenol. (13). Chloroprenet (neoprene).

Finally, in view of the array of chemicals used in town, it is useful to consider that the chemical elements are flowing from one place to another. The inputs and outputs of three especially important elements, C, N, and P, through a household or house plot in a year may be almost equal (Figure 5.10). In contrast, a major gain or decrease in an element would reflect major change in the household.

Air Indoors and in Vehicles

Fuel for cooking and heating are primary indoor pollutant sources, both as smoke and fumes (Spengler and Chen, 2001; Hardoy *et al.*, 2001; Benton-Short and Short, 2008). Biomass combustion of dung, crop residues, and wood is most polluting. Kerosene is somewhat cleaner, and natural gas and electricity result in the cleanest air. In the industrial indoor workplace, dust (often toxic), lead, and benzene from glues are often problems. Small industries, such as in towns, are generally little or not regulated or monitored for indoor air quality, so pollutant exposures are probably common (Barten *et al.*, 1998).

Indoor tobacco smoke is a well-known health hazard. Paints with high (>50 grams/liter) VOC content are also to be avoided (Yudelson, 2008).

Inside cars, trucks, and buses are several chemical hazards (Frumkin *et al.*, 2004): tobacco smoke from smokers, CO (carbon monoxide), and VOCs (benzene, styrene, toluene, 1,3-butadiene, formaldehyde). Inside diesel buses, black carbon and $PM_{2.5}$ (fine particulate matter) are hazards.

Town Spaces and House Plots

Most towns use and store a wide range of chemicals, many toxic, for maintenance and other functions. Indeed, Pinedale (see Appendix) maintains a small secure building for such chemicals on the edge of town.

Pesticides are chemical hazards on lawns (Law *et al.*, 2004). Spills of petroleum products are temporary localized but significant problems, in part because gasoline can move quickly through soil.

Nitrate (NO_{3-}) levels on residential lawns outside Baltimore, Maryland (USA) are higher in higher housing-density sites, though NO_3- availability does not correlate with homeowner management practices (Raciti *et al.*, 2011). Nitrate levels are also higher on lawns in previous agricultural land, suggesting that some nitrogen remains from previous farming, which may have used manure for fertilizing soil.

Solid Waste

As mentioned above, along streambanks, wetlands, and the edges of other waterbodies, small buried dumps are common in towns. Anaerobic decomposition of buried organic matter gives off CH_4 (methane), H_2S (hydrogen sulfide), and CO_2 that diffuse upward into the air (e.g., the smell of CH_4 in Cranbourne; see Appendix). Flooding sometimes opens a dump, scouring out buried material. During especially big floods, some stream- or river-side industries release accumulated chemical contaminants into the floodwater.

Solid waste traditionally built up in towns which had little or infrequent solid-waste removal (Yudelson, 2008). This included construction waste, waste from building operations, food waste, and other wastes including medical waste. A stream running through town was the main town-cleaning mechanism. Such a system of course polluted long stretches of streams and rivers with a rich array of contaminants. Ironically, solid waste may be considered as unwanted, yet valuable, nutrients (Pollalis *et al.*, 2016).

An overview of solid waste in the UK in 1990 emphasizes the predominance of wastes from resource extraction, agriculture, construction, and industry (Cullingworth and Nadin, 1997): 25 percent from mining and quarrying; 18 percent agriculture; 16 percent construction and demolition; 16 percent other industrial; 8 percent sewage sludge; 8 percent dredged spoils; and 3 percent commercial. Only 5 percent of the solid waste came from households.

Worldwide, solid waste is growing rapidly and outpacing solutions (Hoornweg *et al.*, 2013). In the USA, food waste is ca. 20 percent of solid waste (Pollalis *et al.*, 2016). In covered dumps (tips) anaerobic digestion slowly decomposes the material. Composting of food waste with oxygen greatly accelerates decomposition (e.g., a large composting facility in Issaquah; see Appendix). To reduce inputs to dumps, prolonging their life, and encourage composting and other approaches, in Halifax, Nova Scotia (Canada), no organic material can be added to a dump. Thus, food, paper, cardboard, and other organic materials are decomposed or combusted more rapidly than if buried in a dump. Such an approach should be easier to achieve in towns where more space is available.

Five major options exist for residential solid waste in the USA, from most to least wasteful (Friedman, 2014).

1. *Landfill* (dump or tip). Mainly slow anaerobic decomposition, which produces CO_2 and CH_4, both greenhouse gases, along with small amounts of N, O_2, toluene, vinyl chloride, and benzene, all basically produced from organic biodegradable waste.
2. *Incinerator*. Combustion of solid waste producing heat for heating systems or electricity generation for a town, but best with high-energy materials such as paper and plastic (which however are recyclable). Advanced filters needed to minimize pollutants.
3. *Anaerobic digestion*. Anaerobic microbes decompose organic material, giving off CH_4 (methane), which is burned and its energy collected for use. Some greenhouse gas CH_4 is given off in the process.
4. *Recycling*. Commercial and residential waste is transformed into other usable products, which reduces the natural resource extraction needed to make products. Transportation uses fuel and produces greenhouse gas.
5. *Composting*. With oxygen present, aerobic microbes decompose organic matter giving off CO_2 and H_2O, and the residue chemical elements can be used as fertilizer. CO_2 from aerobic decomposition, like that in natural systems, normally is not considered to be a human-caused greenhouse gas.

Emerging Contaminants

Newer concerns arise with so-called *emerging contaminants* (Batty and Hallberg, 2010). These are little known, human health effects are little studied, and environmental effects are probably even less studied.

The major types of emerging contaminants are (Boxall, 2010): (1) metabolites; (2) transformation products or "degradates" (organic stable or semi-stable chemicals resulting from the degradation of a more complex chemical); (3) dioxin-like compounds; (4) human medicines and pharmaceuticals; (4) veterinary medicines; (5) nanomaterials; (6) personal care products; and (7) flame retardants. Fifty-three veterinary medicines and 23 human medicines, some widely used, pose particular environmental and health risks. These human-produced chemicals have been released to the environment, and are especially found in waterbodies.

Human-engineered nanoparticles (NPs) are expected to greatly increase and lead to many new products. In the environment, nanoparticles seem to aggregate, but their behavior and effects remain little known. Pharmaceutical compounds have been linked to both health and environmental effects. Nanoparticles today have applications in cosmetics, sunscreens, paint, pharmaceuticals, water treatment, and bioremediation (Boxall, 2010). The ecological effects, based on 12 species studied, include growth inhibition, altered appendage movement, and mortality of organisms.

Twenty-three "degradates," resulting from the partial breakdown of pesticides, biocides, and veterinary medicines, are of particular risk. Ecotoxic effects have been shown on various species, including fish and daphnia (a tiny aquatic invertebrate).

The liberation of these emerging contaminants into ecosystems means that natural selection by bacteria is doubtless already in progress. Bacteria strains more resistant to the chemicals have probably already evolved.

Sustaining Chemical Elements

The big picture of chemical elements shows that some are already, or about to be, in short supply for economic extraction globally (Sackett, 2014). We have been mining, using, and discarding solid waste at an excessive rate. A major portion of these elements is not in rock or in our waste piles, so mining urban dumps will not suffice. Rather, most of the material is in use, as our houses, vehicles, cellphones, and the array of products of modern life. Thus, above-ground mining of materials in use lies in our future.

A study more than 25 years ago concluded that 80 percent of the world's mercury reserves; 75 percent of its silver, tin, and lead; 70 percent of gold and zinc; and 50 percent of copper and manganese had already been made into human products (Halada, 2008). Amazingly, consumer products in Japan's urban areas were estimated to contain (Sackett, 2014): 16 percent of the world's gold, 22 percent of its silver, 19 percent of its antimony, and 11 percent of its tin. Today silver and zinc are in limited supply, as well as less-known elements used in modern life, such as tellurium, hafnium, indium, and neodymium.

Global phosphorus production is expected to decrease after 2030 (Sackett, 2014), while the human population and its need for phosphorus-based food production are both expected to grow on into the twenty-second century. More than three-quarters of the world's phosphate rock are in one nation, Morocco. Similarly, an estimated 97 percent of the "rare earth" oxide-mineral reserves are in China, apparently mostly in a small area near Ordos, Inner Mongolia. These rare rare-earth minerals are in every vehicle, every cell phone, and countless other high-tech devices. Over the past century the demand for minerals has increased exponentially.

Today only lead and two little-known minerals, ruthenium and niobium, are more than 50 percent recycled into new products. Some industries have instituted major waste-reduction and recycling policies, which have also provided significant economic benefits. In two Swedish cities, phosphorus-rich urine is separated from solids in toilets as a step toward recycling phosphorus for food production, and reducing its eutrophication effects on waterbodies (Figure 5.11) (Cordell *et al.*, 2009).

To decrease the risk of shortages and increase the amount of key minerals available, Japan instituted a policy with four approaches: (1) substitution of a more readily

Figure 5.11 Green pond with goldfish carp, apparently eutrophicated by excess phosphorus stimulating phytoplankton growth. El Alhambra, city of Grenada, Spain. R. Forman photo. (A black and white version of this figure will appear in some formats. For the color version, please refer to the plate section.)

available element for an endangered one; (2) regulation to avoid shortages and chemical hazards; (3) reduction, e.g., by increasing manufacturing efficiencies; and (4) recovery, including recycling and mining of elements in urban areas (Nakamura and Sato, 2011).

Above-ground mining of elements in the products we use is a key step for a life-cycle or cradle-to-cradle approach to chemicals and materials (Sackett, 2014; Simonen, 2014). Designing products for deconstruction at the end of their useful life greatly facilitates the reuse of elements (Polallis *et al.*, 2016). The array of items we use and treasure are not the end for useful elements. Our possessions are the valuable precursors to the next phase of human consumption.

Countless more chemicals exist in town. They are constituents of rock and soil. On the shelves of shops and restaurants. In storage areas of commercial establishments. In homes, where every home in a neighborhood has different chemicals. In diverse industries across town. In vehicles and rapidly changing high-tech devices. These are all chemical hotspots in town.

The growing array of hazardous chemicals without obvious substitutes is worrisome. But the burgeoning mix of emerging contaminants might be scary for both our health and our environment.

Air in Towns

Microclimate, including air movement, heat, and moisture, is the lead-off topic. Air pollutants and effects follows.

Microclimate

Air in the typical town adjoining agricultural land is a hybrid, with characteristics of agricultural, natural land, and sometimes even urban air. Towns in natural land have mainly an atmosphere produced by nature, while seacoast towns have primarily coastal air.

In essence an agricultural area, with a single crop such as wheat (*Triticum*), has a fairly homogeneous surface and reflects upward, e.g., a quarter of the incoming solar and diffuse sky radiation (Oke, 1987; Erell *et al.*, 2011; Forman, 2014). Basically, all the remaining energy (75%) is absorbed by the plants and soil. Moist bare soil may reflect 20 percent and dry bare soil 30 percent of the incoming radiation. Forest and heterogeneous urban areas reflect about 15 percent. An average town with more tree and other vegetation cover than urban areas might reflect some 12–14 percent of the incoming radiant solar and sky energy, and absorb the rest.

Farmland has virtually no impervious surface but does have bare soil, whereas towns have some 10–50 percent impervious surface, and most urban areas 50–90 percent. Considerable heat is emitted from the exposed farmland soil and especially from impervious areas. With moist soil, the farm crop and the town vegetation transpire considerable water to the atmosphere, which effectively cools the air. With dry soil, transpiration cooling is less. Little town or urban vegetation means little cooling. Overall, the maximum summer temperatures tend to be very high in urban areas, high in farmland, and moderate in towns.

Humidity, which relates to soil moisture and plant transpiration, is likely higher in towns than urban areas, and may be similar to that in farmland. However, in towns, horizontal windspeed and air turbulence (see below) resemble those in urban areas. Cropland has much higher windspeed and lower turbulence than in towns. Visibility, haze, clouds, and fog conditions resemble those of farmland, not urban areas.

Air pollutants in town are normally much lower than those in urban areas, where dense transportation and industry emit a wide array of pollutants. In some towns, industries pollute the air, but usually these are small or medium industries, not large industry. Farmland air varies from good, especially with moist soil, to bad during dry conditions when wind lifts clay, silt, and some sand particles into the air. Agricultural chemicals such as pesticides and phosphorus are attached to some of the particles.

So, for summer heat, town air resembles a hybrid between farmland and natural-land air. For humidity, it most resembles farmland air. For airflows, town air resembles that of urban areas. For pollutants, town air most resembles that of natural land. Finally though, since these towns and villages mainly adjoin agricultural land, natural land, or a large water body, and not an urban area, the town's atmosphere remains much cooler and cleaner and healthier than that of cities and suburbs.

Moisture in Town Air

Humidity refers to the concentration of moisture or water vapor in the air (Moran and Morgan, 1994; Erell *et al.*, 2011). In town, humidity tends to be higher in localized places, such as just downwind of a pond, wetland, woods, or old residential area with large trees. Some rainwater runs off from impervious and other surfaces in pipes and ditches, leaving a film of water on impervious surfaces that evaporates, creating a brief increase in atmospheric humidity. Water also infiltrates into the soil, some of which is then transpired by trees and other plants to the air, increasing the humidity. Urban areas with extensive impervious surfaces typically have drier air than towns, which have drier air than woodlands. Fountains and splashing streams add moisture to the air.

Dry air accelerates plant roots-to-air evapotranspiration. Moist air reduces evapotranspiration, resulting in moister soil, more microbes, more soil animals, more decomposition, and more mineral nutrient cycling. Desert towns often have moister air than in their surroundings, due to considerable evapotranspiration from irrigated areas in town.

Diverse Airflows

Airflows due to temperature differences conveniently fall into two categories, local breezes and regional wind. Breezes are gentle, while wind ranges from low to high. Winds affect all places, but towns and villages as small population centers are particularly affected by breezes.

Local Breezes. As warm air from a town moves upward toward cold outer space, air from the surroundings, a *breeze from the country*, is drawn into the town, effectively cooling the town (Forman, 2014). Even at a finer scale, a large greenspace next to the commercial town center results in a breeze from the park that cools the center.

A vegetation-covered hill or mountain next to a town provides *cool air drainage* by a different mechanism (Thurlow, 1983; Erell et *al.*, 2011; Forman, 2014). On still nights with no regional wind, the cool air on upper slopes is heavier than warm air in the town below. Therefore, the cool air drains downward, effectively pushing the town's warm air upward and out (see book cover). Keeping development off the hilltop and town-facing slope is important in order for vegetation to create sufficiently cool air (Forman, 2008). Cool air drainage not only cools the town air, but also the town's upward-moving warm air carries out the pollutants present. This is a free ventilation system, both cooling and cleaning a town's air.

Coastal towns by the sea or lake have onshore and offshore breezes, which are seasonal. After the cold season, the land warms faster than the adjacent waterbody. Upward moving warm air from the land draws air from over the cool waterbody onto the land as an *onshore breeze*. This process cools and cleans the town air. Conversely, after the hot season, the land cools rapidly, so upward moving warm air over the warmer waterbody draws cool air from the land to the waterbody as an *offshore breeze*. This both cleans and cools the town air. In short, all the breezes cool the town, and all clean the town air.

Regional Wind. These winds are created by temperature differences at the regional scale. Three types of wind occur on land. In relatively flat open terrain such as pastureland or crop field, airflows occur as *streamlines*, effectively parallel layers of airflow or wind above the surface (Oke, 1987; Erell *et al.*, 2011; Forman, 2014). Streamline airflow continues over gentle smooth hillslopes and other objects, analogous to air passing over an airplane wing. Wind turbines generating energy work well in open areas with streamline airflows (Figure 5.12).

In contrast, a steep slope or object with an abrupt upwind or downwind side, such as a house, truck, or individual tree, creates *turbulence*. Turbulent airflow seems chaotic, but typically has eddies with strong localized circular up-and-down motion. Finally, airflow perpendicularly crossing a long abrupt object, such as a dense hedgerow or an isolated long building, creates a horizontal *vortex* on the downwind side, analogous to turbulence in cylindrical form.

Winds remove objects and energy from surfaces. High-windspeed streamline airflow effectively blows away objects, turbulence removes more, and a vortex removes the most objects. Butterflies in a garden, heat in a building, dust from a construction site, seeds from a tree, and hats from heads are blown downwind.

Windbreaks, from hedgerows to solid walls, reduce downwind velocity, but also create a range of important wind patterns (Brandle *et al.*, 1988; Erell *et al.*, 2011; Forman, 2014). Assume that the windbreak is in an extensive open flattish area, such as a field or treeless golf course, and that wind hits the windbreak perpendicularly. In addition to location, two simple variables affect the windspeed at any nearby point: *windbreak height* (H) and windbreak porosity. Windspeed is typically reduced over a distance of 3–6 H on the upwind side of a windbreak. A "quiet" zone with very low windspeed extends about 8H downwind of the windbreak. Then from 8 H to 15–25 H is a zone of

Figure 5.12 Two 1.65 Megawatt wind turbines on edge of town. The turbines provide >90% of the energy for a hospital and a community college. Gardner, Massachusetts, USA. R. Forman photo.

somewhat reduced windspeed (wake zone). Thus, for a 2 m high hedgerow, windspeed is reduced in an upwind zone 6–12 m wide, in the quiet downwind zone between 0 and 16 m, and in the final zone from 16 to 30–50 m downwind of the windbreak.

The other key variable determining windspeed is *windbreak porosity*, from low to high. A low-porosity wall or dense vegetation shortens the distances of windspeed reduction just given. Perhaps more importantly, turbulence occurs on both the upwind and downwind sides of a dense windbreak, with strong turbulence on the downwind side.

In contrast, a highly porous windbreak has essentially no turbulence and reduces wind velocity for a long distance, but the reduction is slight. For most uses, a medium-porous windbreak is considered optimum, because turbulence is minimal and the amount and distance of windspeed reduction are both relatively large (Brandle *et al.*, 1988). The windspeed-reduction distances are shortened, if the upwind open area has turbulence-causing objects present, or if wind encounters the windbreak diagonally.

Small cities often have one or more *street canyons* where about equal-height buildings align both sides of a street (Oke, 1987; Erell *et al.*, 2011; Forman, 2014). Wind blowing over the parallel rows of buildings creates a "canyon vortex" over the street, which tends to ventilate the air. Such street canyons are uncommon in towns.

However, a relatively wide vegetation corridor is not uncommon in a town. If the green strip is on the upwind side of town and somewhat parallel to wind direction, it serves to channel wind into town, as in many small German cities. Indeed, Saarbrucken (Germany) considers agricultural areas and stream valleys leading to the city as "climate regulation zones," and thus mainly unsuitable for development (Beatley, 2000).

More common in and around towns and villages are isolated or small groups of buildings or trees. Regional wind causes turbulent eddies around these structures. Airflow over a series of equal-height buildings, where the distance between buildings is less than or equal to building height, is *skimming flow*, since rather little airflow occurs between the buildings (Oke, 1987; Erell *et al.*, 2011). Small pockets of polluted air between buildings may form with skimming airflow. If the buildings are farther apart, ample turbulence occurs (and streamline flow may touch the ground), creating an overall rough pattern of airflow (isolated roughness flow).

Energy, Radiation, and Heat in Town

Radiant energy or *radiation* moves through the air as electromagnetic waves which, upon encountering an object, release heat and warm the object. *Solar radiation* from the sun encounters the Earth's atmosphere of water vapor, nitrogen, oxygen, ozone, carbon dioxide, and other chemicals. Some of the energy is reflected outward, some passes on through to the Earth's surface, and some is absorbed by the atmospheric gases (Oke, 1987; Gartland, 2008; Erell *et al.,* 2011). The atmosphere then emits some of the radiant energy as *diffuse sky radiation*. Thus, both solar and diffuse sky radiation reaches the Earth's surface. This incoming radiation is about half composed of *short wavelength* energy (especially UV [ultraviolet] radiation, plus the visible purple-to-red wavelengths), and about half as *long-wavelength* radiation (especially infrared). The long-wave energy is essentially equivalent to *heat*.

Figure 5.13 Solar panels atop a three-level parking garage on a desert university campus. The rotating panels capture solar energy, and reduce heat buildup by shading the parking surface all day. Arizona State University, Tempe, Arizona, USA. R. Forman photo.

Some of the incoming radiant energy is reflected outward by the Earth's surface. The percent of incoming energy reflected is called *albedo*, and different surfaces have quite different albedos. Characteristic albedos are (Gartland, 2008; Erell *et al.*, 2011; Forman, 2014): fresh white-painted surface 80 percent; desert 30 percent; concrete 25 percent; mowed grass 23 percent; crops 20 percent; commercial city center 14 percent; asphalt/tarmac 13 percent; forest 12 percent (deciduous 16%, coniferous 12%, tropical rainforest 8%); and lake 7 percent. The albedo of most towns may be about 13 percent.

Basically, all incoming radiation not reflected is absorbed as short-wave and long-wave energy by the surface soil, vegetation, and structures (Figure 5.13). A *wavelength shift* then occurs, whereby short-wave energy is converted to long-wave energy, and emitted from the surface to the atmosphere as heat. That warms the air.

This framework also leads to the *greenhouse effect*. Long-wave energy emitted by surfaces tends to be filtered out by the common gases, water vapor and CO_2 (as well as CH_4 [methane], N_2O, O_3 [ozone], and the other greenhouse gases [HFCs, PFCs, NF_3, and SF_6]; Harvard University Office of Sustainability, 2017 website). Thus, with considerable greenhouse gas accumulation in the atmosphere, most of the long waves or heat emitted from the Earth surface cannot pass through, and are trapped in the lower atmosphere. Heat accumulates and temperature rises. This is quite equivalent to a greenhouse in which long waves cannot pass through the glass, heat builds up, and the plants broil or fry.

Surface Energy Balance or Budget. Six easy-to-understand variables describe the energy balance or budget of a town, or some part of it, over any time period (Kuttler, 2008; Gartland, 2008; Erell *et al.*, 2011; Forman, 2014). Two variables are inputs: (1) the incoming *solar and sky radiation*, and (2) human-produced or *anthropogenic energy* added, such as heat released from heated buildings and heat given off from vehicle combustion. Also, two variables are outputs: (3) *sensible heat* (which we can feel) is temperature-raising energy emitted from materials and structures to the atmosphere, and (4) *latent heat* (which we cannot feel and doesn't raise the temperature) is the energy required to convert liquid water to gaseous water vapor especially in plant evapotranspiration (and is given off to the atmosphere). Finally, two variables are a change in amount over the time period: (5) net *heat storage* in the town (gain or loss of accumulated heat), and (6) net *horizontal heat advection* (transfer of heat to or from an adjacent area).

The incoming solar and sky radiation input is by far the largest energy flow. The outgoing sensible heat flow is normally second largest, and the anthropogenic energy input is the smallest energy flow (Gartland, 2008). The overall town energy budget is positive during spring when solar energy is increasing, negative during the fall, and about zero for a year.

Heat Island. A city has an *urban heat island*, indicating that the city air temperature is higher than that of the surroundings (Gartland, 2008; Imhoff *et al.*, 2010; Erell *et al.*, 2011; Forman, 2014). Hamlets, villages, and small towns probably have no heat island, or have a tiny area with slightly elevated air temperature (Landsberg, 1981; Mills, 2004). A slight *town heat island* is probably common in large towns. The *heat intensity* of a heat island generally refers to the air temperature difference between the hottest location in town and that in a "comparable" location outside town (Gunawardena *et al.*, 2017). Unfortunately, the heat intensity concept is a bit mushy, since finding a comparable location in a heterogeneous and different environment is dubious (Stewart, 2011).

Overall, the heat intensity of a heat island is higher at night than in daytime. Also, it is typically greater in winter than in summer (Imhoff *et al.*, 2010). Heat islands are most pronounced in population centers with a high percentage of impervious surface cover, such as roads, sidewalks, and roofs. Areas with considerable vegetated greenspace have little or no heat island.

Regional wind normally reduces or eliminates a heat island, which is more likely to be present in calm clear air. Calculations indicate that a modest 3–5 m/s (7–11 mi/hr) regional wind is sufficient to eliminate the urban heat island (see below) of small cities (33,000–50,000 population) (Schmid, 1975). Heat islands are uncommon in desert towns where the surroundings are often hotter than the town with its irrigation, and winds are common (Erell *et al.*, 2011). Trees over grass seem to be especially effective in reducing heat intensity (Shashua-Bar *et al.*, 2009). In short, town heat islands are probably uncommon in small towns and villages, in windy regions, in hot-dry climate towns, and in towns with limited impervious cover (e.g., <25%) and extensive vegetation cover.

Greenspaces and Heat. A Berlin study of 42 greenspaces, mostly parks, of different size provides insight into how to cool cities and towns (von Stulpnagel *et al.*, 1990; Forman, 2014). The small greenspaces (<30 hectares or 75 acres) have air temperature about 1°C cooler than the surrounding built area. Medium-size greenspaces (30–500 hectares) cooled the air about 3°C, and large greenspaces (>500 ha or 1,250 acres) had air about 5°C cooler than the surrounding built area.

Measurements in the surroundings of five medium (125–212 ha) and small (18–36 ha) Berlin greenspaces indicated that the cooling effect extends outward from the medium-size ones some 500 m upwind and 1,500 m downwind (von Stulpnagel *et al.*, 1990; Forman, 2014). *Outward cooling distances* seem to be several hundred meters by medium and large greenspaces, and less around small greenspaces. The outward cooling effect extends farther on calm days, and is eliminated by strong wind.

A few studies in northern Europe, Montreal, San Francisco, and elsewhere have found somewhat similar, though more limited, results (Schmid, 1975; Forman, 2014; Bowler *et al.*, 2010). A Tokyo multiple-regression study found that woodland area was the best predictor of maximum daily summer air temperature (Tonosaki *et al.*, 2014). Trees are especially effective in reducing air temperature. Temperature also correlated with the size of greenspaces irrespective of how much was covered by trees. The results of these urban studies suggest that adequate tree cover and/or greenspaces in towns can essentially eliminate a town heat island.

Impervious Surfaces and Heat. Unlike cities with some 50–90 percent impervious surface cover, towns are usually 10–50 percent covered by impervious surfaces. Moreover, the surfaces in town are overwhelmingly small, especially roofs of houses, but also one-block-long sidewalks, basketball/tennis courts, car parks by community buildings, and so forth. Typically, the only large impervious surfaces are the commercial town center, one or two other commercial centers, an industrial center on the town edge, and one major highway.

Impervious surfaces reflect (albedo) considerable incoming solar and sky radiation (Gartland, 2008; Forman, 2014). Light-colored, dry, and smooth surfaces such as new concrete, reflect more than do dark, moist, rough surfaces. The radiant energy not reflected is absorbed by the impervious material, and later emitted as heat. White surfaces absorb less energy and emit less heat to the air than do dark roofs and asphalt/tarmac car parks. Impervious surfaces shaded by trees or other objects also emit less heat (Figure 5.13). Think how many benefits would be provided in town by a plant-covered surface, such as a green roof, with solar panels above it.

The small impervious surfaces prevalent in residential towns are mainly surrounded by vegetated areas with lawn, garden, shrubs, or trees. Therefore, much of the heat given off from the surfaces doubtless moves horizontally by advection into the surrounding vegetated areas. Trees extending over impervious surfaces provide triple benefit. They shade the surface (e.g., wall, sidewalk, and street) reducing its radiant energy absorption and heat emission. The underside of the tree canopy absorbs some of the heat emitted from the surface. And the tree foliage evapotranspires water vapor which effectively dissipates heat outward without raising air temperature.

Air Pollutants and Effects

Agricultural and natural land mainly surround towns and villages, and cover them with air containing natural and human-produced chemicals. Consider the typical mix of air in natural land: about 78 percent N_2, 21% O_2, H_2O vapor (0–4%), 0.04 percent CO_2, and traces of neon, helium, methane, and other gases. Soil microbial activity adds CO_2 and NOx; volcanic activity, SO_2 and PM (particulate matter); sea spray, NaCl; lightning-caused fires, CO_2, CO, and PM; various trees, HCs (hydrocarbons) including VOCs (volatile organic compounds) (Smith, 1981; Gartland, 2008); wetlands, CH_4 (methane); plant evapotranspiration, H_2O vapor; and wind erosion, PM, especially silt. These natural air components are not (unwanted) pollutants.

Some trees emit airborne VOCs containing phytoncides or essential wood oils (Figure 5.14) (such as alpha- and beta-pinene, 68-cineole, D-limonene, and humulene) (Reeves, 2014). These seem to be anti-bacterial, anti-inflammatory, and anti-carcinogenic. They also tend to lower blood pressure, reduce stress hormones, and increase white blood cells that defend against viruses. Therefore, take many walks in the woods.

Human activities in non-urban areas also add to the atmospheric mix: livestock and rice paddies, CH_4; pastureland and crop fields, PM; coal-fired power facility, SO_2 (e.g., Page, Kutna Hora; see Appendix); smelters, SO_2 and HMs (heavy metals) (e.g., Falun, Superior); refineries, HCs; and human-caused fires, CO_2, CO, and PM. Wind transports all of these into towns and villages. And of course, urban areas add a huge diversity of chemicals to the air that blow into some towns.

Ten air pollutants characterize the air where people live (Forman, 2014): five gases, CO_2, CO, SO_2, NOx, and HCs; four particulates and aerosols, PM, HMs, O_3 (ozone) smog, CFLs (chlorofluorocarbons); and toxic substances (TOX), a heterogeneous group of chemicals. Gases are lightweight and invisible. Particles are heavier, visible, and "dry," while *aerosols* are particles in droplets of water.

Fossil fuels – gasoline, diesel, coal, oil, and natural gas – produce somewhat different pollutants and are the predominant underlying source of air pollution in towns and cities. Biomass fuels, solid waste, surface materials, and manufactured products are also indirect contributors. Motor vehicles and industries are the primary direct producers of the air pollutants. PM, CO_2, and NOx originate from the widest range of major sources.

Both roads and vehicles produce an array of pollutants in towns (Forman *et al.*, 2003). While some pollute water, almost all become airborne due to wind sweeping across road surfaces, traffic movement lifting particles from the surface, and chemicals directly emitted by vehicles into the air.

Most of the pollutants produce both environmental and human health effects (Forman, 2014). The primary health effects are respiratory disease and secondarily cancer.

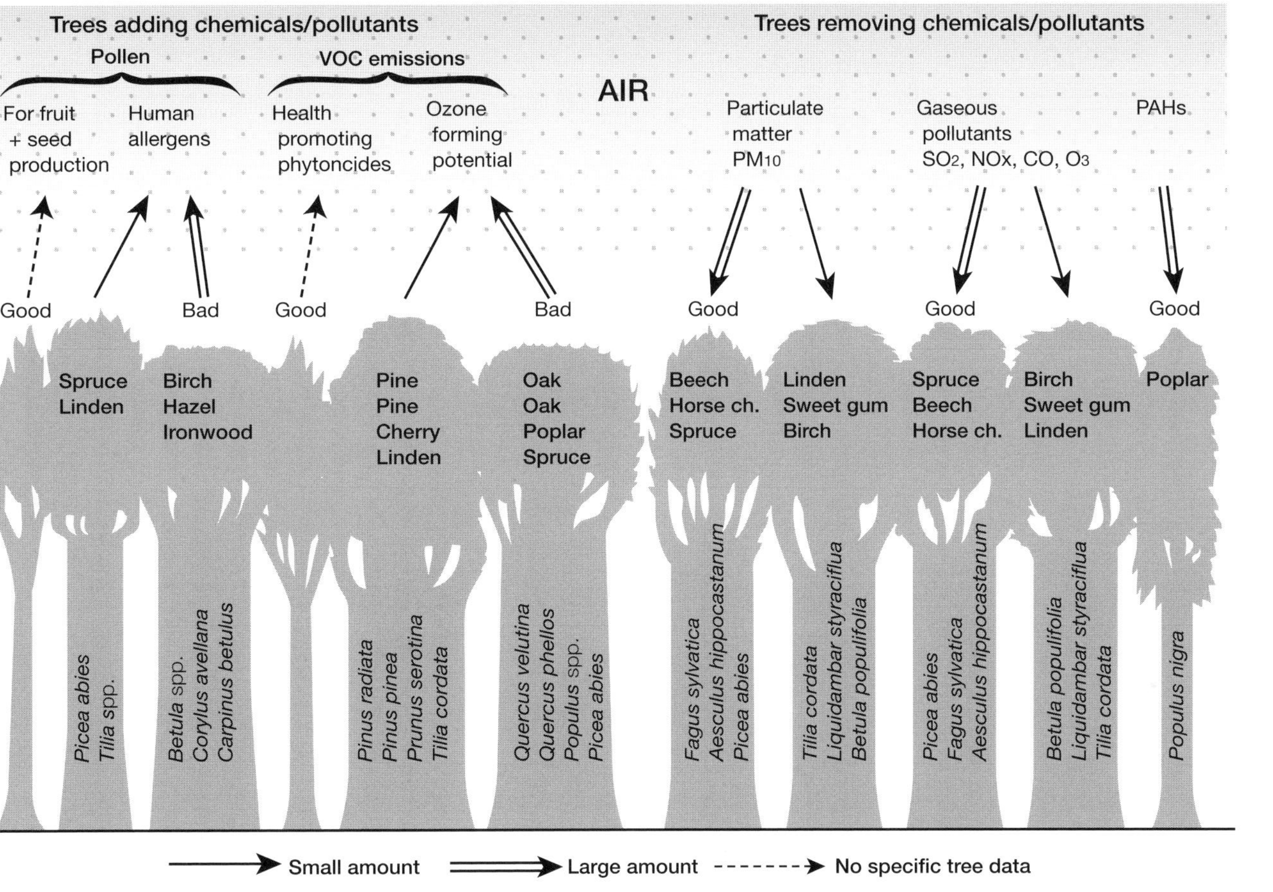

Figure 5.14 Effective and ineffective tree species for adding or removing chemicals/pollutants in the air. VOC = volatile organic compound. Phytoncide = essential wood oil, such as alpha- and beta-pinene, 68-cineole, D-limonene, and humulene (Reeves, 2014). PAHs = polycyclic aromatic hydrocarbons. Trees with medium amounts of emission/removal excluded. Based on Gartland (2008), Forman (2014), Grote *et al.* (2016).

On average, suburban towns have less air pollution than in the adjoining city. For example, in the USA, airborne NOx may be 99 percent lower, PM 60 percent lower, and SO_2 17 percent lower, though much variation exists (Marsh, 2010). In rural and remote towns with mainly residential areas, considerable PM may come from construction sites, unpaved roads, vacant plots, and burning of household/yard waste (e.g., Tomar, Swakopmund, Rock Springs). Commercial centers with extensive impervious surface may have considerable amounts of heat-related pollutants, such as NOx and HCs. Also, skimming air (see above) over detached buildings can result in pockets of air pollution between buildings (Erell *et al.*, 2011). Industrial centers of towns add distinctive pollutants from manufacturing processes to the town air.

Each town in an area has a somewhat different leading industry or group of industries (e.g., Askim's recent rubber factory, Kutna Hora's automobile manufacturing, Rumford/Mexico's paper mill). Therefore, each town is covered and affected by a different combination of air pollutants, making each town's chemistry distinctive.

In a region with widespread air pollution, such as the regional NOx (nitrogen oxides) covering East Asia, towns are more severely contaminated and more similar in chemistry. Similarly, in cropland, dust, with similar levels of particulates, N, P, and pesticides, blankets all the towns and villages with a similar chemical input. A town, such as Marcus Hook, Pennsylvania (USA), surrounded by polluting industries, is particularly impacted by the local inputs. Coastal towns are repeatedly covered with NaCl salt (e.g., Swakopmund, Stonington, Hanga Roa). Despite the array of potential air pollutants in towns, nature is resilient as trees and shrubs can partially clean bad air (Figure 5.14) (Nowak *et al.*, 2006; Escobedo and Nowak, 2009; Janhall, 2015).

Sound and Noise

Does a tree falling in a remote forest make a noise? If *noise* is unwanted sound (unwanted by humans) and nobody is nearby, the falling tree makes a sound, but no noise. That conclusion is relative to humans. But sound can also be loud, high or low pitched, or frequent enough to significantly inhibit or degrade other organisms and species. Such sound also is generally considered to be *noise*.

Sound, as radiant energy, typically moves as regular waves of pressure through air, water, soil, and human structures (Parris, 2015). A sound's pitch or frequency measured in hertz (Hz) is the number of wave cycles per second. Loudness or volume or amplitude, measured in pressure or intensity, is recorded in decibels (dB) on a logarithmic scale. "A-weighted" decibels (dBA) refer to loudness of sound perceived by people.

Overall, the primary sources of noise or noise pollution are road traffic (Figure 5.15) (e.g., Socorro, Lake Placid), industrial operations (e.g., Issaquah, Rumford/Mexico), construction activities, trains (e.g., Askim, Kutna Hora), and aircraft (e.g., Concord). These sources characteristically make loud, low-frequency noise. The first three typically make relatively continuous noise, regularly interrupted by quiet periods. Trains

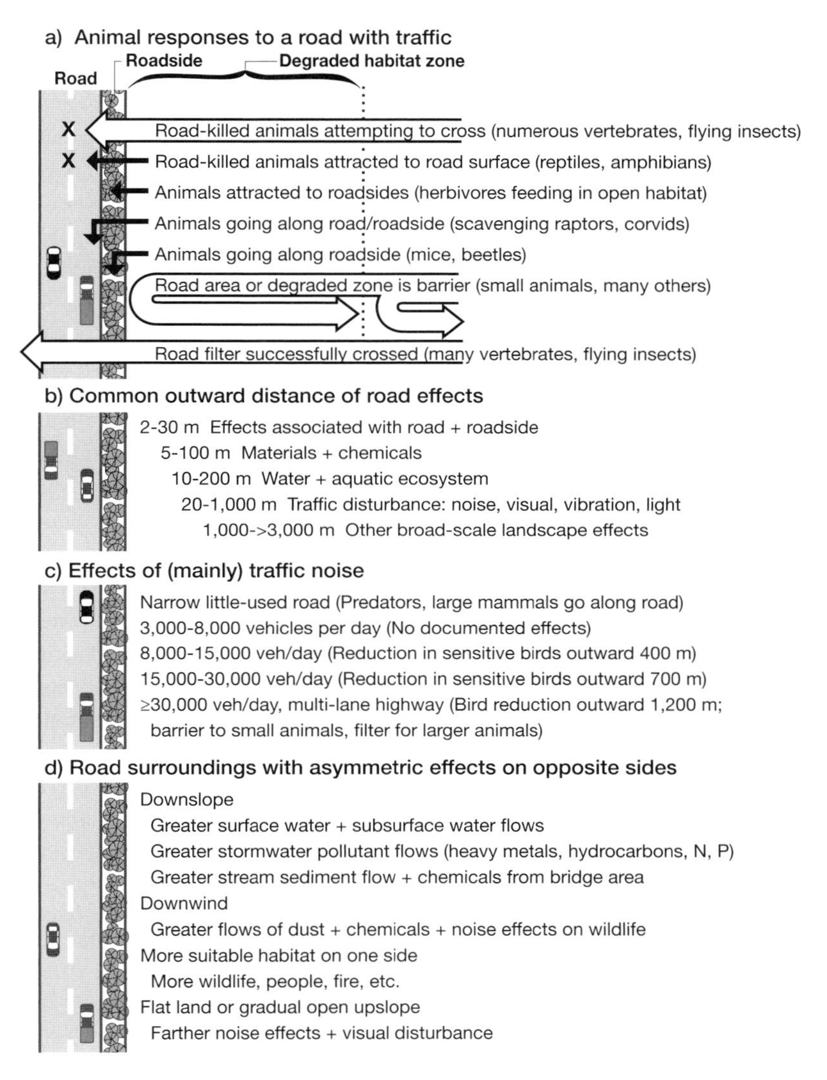

Figure 5.15 Wildlife responses to roads, and components of the road-effect zone. *Road-effect zone* = the convoluted strip significantly affected by the road and vehicles. Based on: (a) Forman *et al.* (2003), van der Ree *et al.* (2015b); (b) Forman (1995), Forman *et al.* (2003); (c) Forman *et al.* (2002), Reijnen and Foppen (2006); (d) Forman *et al.* (2003).

and aircraft create short-term loud noises. The muted sound (low intensity) of aircraft flying far overhead is probably audible in all towns and villages. Loud noise, such as pile-driving for bridge or harbor construction, also readily propagates through water and inhibits fish populations (Jefferson *et al.*, 2009; Halvorsen *et al.*, 2012). Drill-hammering and excavating a street near town center may drive rats from buried pipes to surfaces around buildings and houses. Although human activities commonly make loud noises, nature's loud noises are normally infrequent, such as thunder, strong wind in town, moose bellowing (*Alces alces*), and whale talk.

The low-frequency noise in towns and cities has sound waves disrupted by numerous large flat surfaces, especially walls (Slabbekoorn and den Boer-Visser, 2006). This pattern seems to have selected for shorter, higher frequency, louder, faster, and novel bird songs. For instance, common great tits (*Parus major*) in Europe now sing different songs in urban areas than in rural areas. Such a change benefits songbirds because, in a noisy place, intruders into a territory or potential mates can hear the birds singing (Slabbekoorn and Smith, 2002).

Wildlife and people are affected in three ways by noise (Parris, 2015), which may:

1. *Cause stress*. Physiological stress may make an animal avoid a habitat (Forman and Deblinger, 2000; Reijnen and Foppen, 2006). Alternatively, the animal remains in the noisy location, but with less efficient foraging or reduced breeding success (Kight and Swaddle, 2011; Blickley *et al.*, 2012). For instance, great tits near a busy Dutch highway lay fewer eggs and have fewer young fledged (Halfwerk *et al.*, 2011).
2. *Make it harder to hear other animals*. Wildlife communicate by sound or song to attract a mate, maintain contact with young, and warn of an approaching predator (Kight *et al.*, 2012; Proppe *et al.*, 2013). Some animals can adjust to noisy conditions by calling at quiet times or at a higher pitch (Brumm, 2004; Parris *et al.*, 2009; Barber *et al.*, 2010), though this seems to only partially compensate for noisiness (Parris and McCarthy, 2013). Noise also makes it harder to hear predators or, when foraging, to hear prey (Barber *et al.*, 2010).
3. *Cause hearing loss*. Both chronic loud noise and a single extremely loud noise can damage the cochlea in the inner ear (Kight and Swaddle, 2011). Furthermore, such noise may cause a temporary increase in an animal's hearing threshold, the point at which it can just detect a sound (Scheifele *et al.*, 2003).

We now explore sound and noise effects from the following perspectives: (1) soundscape; (2) local community noise; (3) airport and railway train noise effects near town; (4) road and motor vehicle noise effects near town; and (5) noise, birds, birdsong.

Soundscape

The mosaic of biological, geophysical, and anthropogenic (human-caused) sounds in a landscape is referred to as a *soundscape* (Pijanowski *et al.*, 2011a, 2011b; Liu *et al.*, 2013). *Biological sound* (biophony) is produced by all the organisms present. *Geophysical sound* (geophony) combines the sounds of wind, water, thunder, and so forth. *Anthropogenic sound* (anthrophony) is produced by people and all the human-made objects present.

Building on the concepts of environmental acoustics (focused on urban noises affecting humans; Hartmann, 1997; Bolteldooren *et al.*, 2004; Adams *et al.*, 2006) and bioacoustics (focused on sound affecting individual species; Bradybury and Vehrencamp, 1998; Marler and Slabbekoorn, 2004; Fletcher, 2007), the soundscape ties the spatial arrangement of sound and noise to landscape ecology patterns and processes

(Forman, 1995; Turner and Gardner, 2015; Rego *et al.*, 2018). Thus, measurements highlight the composition of sound, plus its temporal pattern, spatial arrangement, and interaction among sounds.

Consider walking through three adjoining soundscapes, town-edge commercial center, farmland, and large woods, and recording the sounds in each. Sounds in a commercial center include a distinctive sequence of: people talking, delivery truck, piped music, children playing, road-repair machine, sparrows chirping, and emergency vehicle siren. Along the farmland road are sounds of: tractor working, airplane high overhead, dog barking, corvids/crows calling, insects buzzing, wind across fields, distant highway traffic, and livestock talk. In the woods, one hears: several birdsongs, faint running water, wind in the treetops, creak of a tree branch, an overhead airplane, a rustling in the leaves, and a loud unknown animal call.

Anthropogenic sounds predominate in the first soundscape, a mixture of sound types in the second, and biological sounds in the third. The sequence of sounds recorded within a soundscape provides insight into both the spatial and temporal arrangement in the landscape. The changing combination of sounds along the route largely reflects the ecosystem processes and human activities present (Pijanowski *et al.*, 2011b). In effect, the soundscape is the sound dimension of landscape ecology.

More species present means a greater diversity of sounds or signals over a spectrum of sound frequencies (Farina *et al.*, 2014). Thus, acoustic or sound diversity is a valuable indicator of animal diversity. Birds, mammals, reptiles, amphibians, insects, and fish make sounds, e.g., to communicate, attract mates, defend territories, warn against predators, and socialize (Depraetere *et al.*, 2012). Water is also a good sound transmitter so, not surprisingly, sounds are a key component of freshwater and marine systems (Yan *et al.*, 2010; Trenkel *et al.*, 2011).

Acoustic data were recorded along transects in 15 adjoining 350-m-diameter (1,150 ft) sites in a 3.4 km^2 (1.4 mi^2) rural area on the island of Corfu, Greece (Mazaris *et al.*, 2009). The main biological sounds are wildlife, including birds and insects. Geophysical sounds are primarily sea waves, wind, and rain. And anthropogenic sounds are mainly transportation, agricultural activities, and livestock. Almost all sounds appear to be either in the foreground or background. Foreground sounds are short, sharp, and probably stimulate responses. Background sounds, on the other hand, carry information and provide landmarks. The foreground sounds do not correlate with readily identified spatial landscape patterns, whereas the background sounds correspond well with landscape pattern.

Overall, natural sounds seem to be decreasing, mainly due to the increasing noise of transportation (Jensen and Thompson, 2004; Dumyahn and Pijanowski, 2011b). Noise pollution from spreading road, rail, and air networks, plus growing numbers of vehicles, trains, and aircraft is the major cause. Anthropogenic noise is degrading soundscapes and therefore landscapes.

Local Community Noise

The major types of noise in towns are: noise from neighbors; traffic noise; and construction noise (Cullingsworth and Nadin, 1997). If one lives near a local airport, aircraft noise may be prominent. If a railway goes through or adjacent to town, ample noise occurs there. Commonly a bypass highway has been built just outside of town to avoid traffic congestion in town, and such highways are typically busy and noisy, often well into the night. Such bypass highways are also usually barriers to the movement of wildlife and walkers in that direction.

A busy commercial center is often noisy, with delivery trucks and shoppers' vehicles. An industrial center may have noise from manufacturing processes and from truck traffic bringing raw materials and transporting away products. Busy roads within town often have numerous cars, many trucks, and a few school buses, local public-transport buses, rubbish/trash trucks, and even occasional farm tractors or logging trucks from the surroundings. Neighborhood noise in residential areas that annoys some neighbors but not most is usually infrequent, short term, and moderately loud. Examples are hammering/construction, a party, or loud music.

Near the edge of town, farm equipment or logging equipment may make considerable noise. Loud noise usually results from highway motor vehicle traffic, trains, certain manufacturing processes, and other motorized equipment, such as building construction, road repair, drilling, and earth moving.

Neighbors often get used to some ongoing noises such as roosters calling or sounds from a religious building. Other chronic or repeated noise remains annoying, and may lead to various health problems. One response is to move away, and the other main response is to become acclimated to the noise.

Animals have similar responses. But most animals depend on acoustic communication to feed, warn of predators, successfully reproduce, interact with each other, and in essence to survive. Human-caused noise often blocks out or inhibits this critical communication, causing some animals to alter their behavior. For example, California ground squirrels (*Spermophilus beecheyi*) in areas with low-frequency noise from wind turbines make higher-pitched sounds than do animals without the noise (Rabin *et al.*, 2003; Evans, 2010). The wind turbine noise masks the normal ground squirrel calls.

Railway Train and Airport Noise near Town

Along railways (e.g., Askim, Concord, Kutna Hora), train noise is louder, briefer, and less frequent than road traffic noise (Forman, 2014; Silva Lucas *et al.*, 2018). The short duration noise of trains is accompanied by noticeable vibration, which apparently compacts soil and doubtless inhibits soil animals. A decrease in avian density, and probably a change in wildlife behavior, are apparently correlated with train noise level (intensity) and duration in The Netherlands (Dorsey *et al.*, 2015).

Noise (measured in decibels) increases exponentially with train speed (Proneldo, 2003). Noise from low-speed trains (<50 km/hr) mainly results from acceleration or deceleration, and from the wheels rolling on rough worn rail surfaces (Evans, 2010). Medium-speed (50–275 km/hr) train noise mainly results from wheels rolling on the rails, and high-speed trains (>275 km/hr) mainly make aerodynamic noise. Electric trains are much quieter than diesel and steam-driven trains, but most electric trains operate close to cities. Also, all trains have a loud horn or whistle directed forward.

A narrow strip of trees by the railway has little effect on reducing noise propagation. A forest strip 61 m (200 ft) wide next to a railway is reported to only reduce train noise from 79 to 73 decibels (dBA) (Marsh, 2010). Forest hundreds of meters wide is apparently needed to attenuate most noise from a passing high-speed train.

Trains, especially high-speed trains, produce noise over a range of frequencies (Dorsey *et al.*, 2015). Forest, perhaps especially due to its litter/humus layer, reduces the high-frequency noise (4,000–8,000 Hz) somewhat. But low-frequency noise (<1,000 Hz) is little reduced by forest.

Since loud trains repeatedly pass on the same tracks, the wildlife nearby must have *habituated* (become accustomed) to the trains and noise. Something unusual, such as an engine badly needing repair, a suddenly roughened rail surface, or a loud blast of the horn might send wildlife running away.

Aircraft noise at a local airport may be considerable, both in the downwind direction due to wind carrying the noise of aircraft activity, and in the upwind direction around aircraft noisily taking off. These noises are usually infrequent, loud, and short. Little is published about the wildlife and other ecological responses to such local airport noise. This noise regime near a town's airport may be analogous to that near a railway with infrequent commuter trains.

Road and Motor Vehicle Noise near Town

Motor vehicle traffic on roads and highways has a little recognized but huge impact on the environment (Forman *et al.*, 2003; van der Ree *et al.*, 2015). Effects include lots of roadkill, a barrier to wildlife movement, fragmentation of the landscape, pollutants entering waterbodies, production of unhealthful air pollutants, and greenhouse gas emission. Honking a vehicle horn makes brief loud noise.

But traffic noise also has a giant effect on wildlife (Reijnen and Foppen, 1995, 2006; Forman *et al.*, 2003; Parris, 2015; van der Ree *et al.*, 2015). Traffic noise is low frequency. Automobile traffic noise is overwhelmingly caused by the tire surfaces interacting with the road surface. Smooth tires and dense smooth paved roadways absorb little noise energy and thus produce considerable noise. Also, a cracked bumpy road surface is noisy. A vehicle's motor and drive-train and its vehicle aerodynamics make much less noise. However, a high proportion (e.g., >7%) of trucks in the traffic is another major contributor to traffic noise.

Traffic noise may often affect the ability of birds to raise young. For instance, when a predator approaches, adult birds give alarm calls to each other and to the young (Barber *et al.*, 2010). But with traffic noise, if the calls are not heard, the young may become predator food. Some birds and frogs respond to traffic noise by calling more loudly, at a higher pitch, in shorter intervals between noise pulses, or at a different quieter time period (Warren *et al.*, 2006; Parris *et al.*, 2009; Nemeth and Brumm, 2009; Evans, 2010). Wide wooded strips (e.g., >200 m) attenuate traffic noise (Fang and Ling, 2005).

Road noise can also produce physiological stress, which may lead to a weakened immune function or decreased breeding success (Kight and Swaddle, 2011; Blickley *et al.*, 2012; Parris, 2015). Traffic noise may also cause temporary or permanent hearing

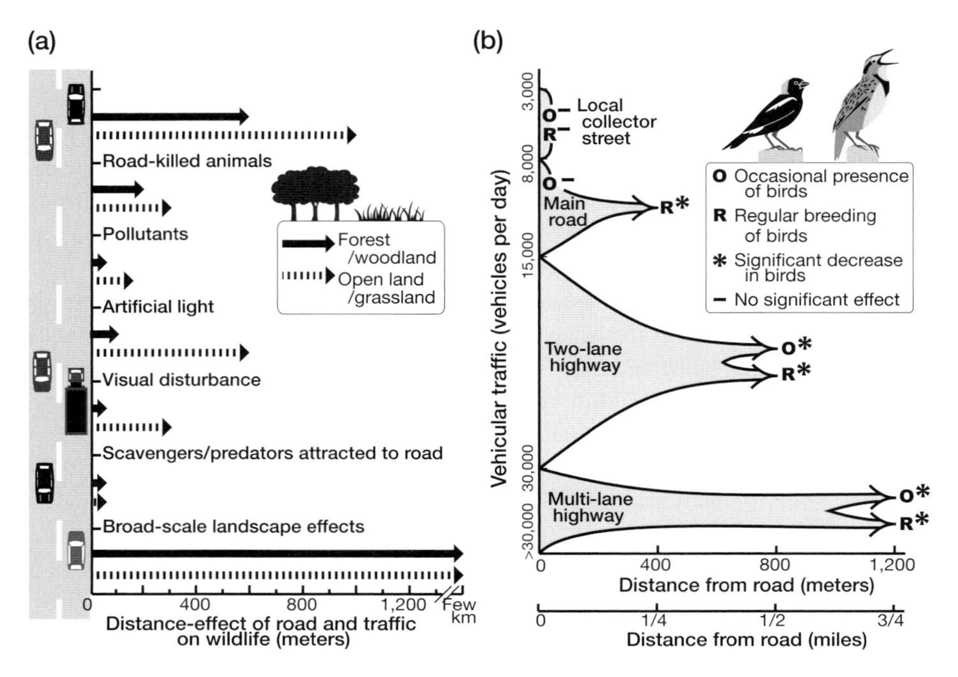

Figure 5.16 Traffic noise, distance from road, and traffic-level effects on wildlife. (a) Generalized diagram based on Forman *et al.* (2003), Reijnen and Foppen (2006), van der Ree *et al.* (2015), and a few rough estimates. Pollutants include exhaust and road salt; artificial light includes roadside and vehicle lights; broad-scale effects include disrupting a wildlife movement corridor, habitat fragmentation, and human access into a remote area. (b) Road traffic and grassland birds of conservation importance (Massachusetts, USA). Based on Forman *et al.* (2002) and Forman (2004b).

loss, or an increase in the *threshold of hearing*, i.e., the detectable sound level (Kight and Swaddle, 2011; Parris, 2015). Research is needed in comparing the decibel levels causing physiological responses with decibel levels in habitats near busy roads.

Noise from road traffic extends over large areas on both sides of a road, largely depending on traffic density (volume) and the presence of adjoining structures or hills. With an average road density (total road length per area) of about 1.3 km/km^2 in the USA (much higher in Europe), nearly 20 percent of the entire nation is estimated to be ecologically affected by traffic noise (Forman, 2000).

A study of grassland birds of conservation importance in suburban towns of Boston found no significant effect of a road, with an average 5,500 vehicles passing per day, on the occasional presence and the regular breeding of birds (Figure 5.16) (Forman *et al.*, 2002). But main roads with 11,500 vehicles/day showed a significant decrease in regular breeding extending outward 400 m (1/4 mile) on each side. Similarly, significant decreases in both the occasional presence and regular breeding extended outward 800 m from two-lane highways with an average 22,500 vehicles/day, and outward 1,200 m (3/4 mile) on both sides of multi-lane highways with more than 30,000 vehicles per day. Similar results for birds are documented in The Netherlands (Reijnen and Foppen, 1995,

2006). Such effect-distances are smaller for small terrestrial animals, and sometimes greater for large mammals (Forman *et al.*, 2003; Barber *et al.*, 2011).

An experiment, testing how important traffic noise is in affecting birds, played recorded traffic noise (sound levels about 37–57 dBA) along an 0.5 km (0.31 mi) forest strip ("phantom road") (McClure *et al.*, 2013). This was a known stopover site for migrating birds in southern Idaho forest (USA). Compared with a control, 13 songbird species were reduced more than 25 percent in abundance around the phantom road. With this traffic noise alternating between on and off periods, several migrating songbird species were almost totally absent during noise periods. The most sensitive species were reduced in both number and abundance by traffic noise.

Other ecological effects occur along roads, including road-killed animals, pollutants, artificial light, visual disturbance, and scavengers attracted to the road. However, traffic noise seems to account for most of the decrease in wildlife near roads, and almost all the long-distance effects (Figure 5.16) (Forman *et al.*, 2003; Benitez-Lopez *et al.*, 2010; McClure *et al.*, 2013). Apparently sound decreases about 3 dB with a doubling of distance from a line source such as a road, whereas the decrease seems to be about 6 dB from a point source such as a bridge construction site (Barber *et al.*, 2011).

Narrow strips of trees next to roads have little effect on reducing traffic noise propagation. Adjacent buildings, soil berms, noise walls (preferably covered with plants), and wide tree strips (e.g., 200 m) significantly reduce noise propagation (Fang and Ling, 2005). However, hard surfaces, such as building walls and some vertical noise barriers, simply reflect traffic noise to the neighborhood on the opposite side of a road.

Numerous auditory deterrents, such as whistles on the front of vehicles, have been used to reduce wildlife–vehicle collisions, especially with large mammals such as deer and kangaroos (Scheifele *et al.*, 2003; Valitski *et al.*, 2009; D'Angelo and van der Ree, 2015). Several problems limit the effectiveness of this technique, and at present there is no convincing evidence that auditory deterrents significantly modify animal behavior or reduce wildlife–vehicle collisions.

Busy multi-lane highways passing close to towns thus produce the greatest traffic noise effects on wildlife (e.g., Casablanca, Lake Placid, Socorro). Two-lane highways and other busy main roads with speed limits above 75 km/hr (45 mph) are perhaps equally damaging ecologically, because they combine considerable traffic noise degradation with the highest rates of wildlife–vehicle collisions and roadkill (animal mortality) (Rajvanshi *et al.*, 2001; Forman *et al.*, 2003; Laurance *et al.*, 2009).

Most species subject to roadkill can reproduce faster than vehicles can kill animals. But roadkill reducing populations of rare and uncommon species, especially slowly reproducing mammal, reptile, and amphibian predators, is particularly serious. This combination of traffic noise and roadkill effects applies to the four to six main roads typically radiating outward from a town.

A final major ecological effect of these radial main roads, and smaller roads as well, is the barrier effect disrupting connectivity for wildlife movement. This fragments habitat and natural land. Road width is especially important for many small animals, whereas traffic is a greater impediment for mid-size and large mammals (Oxley *et al.*, 1974; Forman *et al.*, 2003). Small fragmented habitats contain small populations of animals and plants. For both demographic and genetic reasons, small populations are more likely to go locally extinct than are larger populations.

Noise, Birds, Birdsong Change

Noise and bird species were recorded in six parks in Spanish and Portuguese towns (population 4,000–16,000) (Paton *et al.*, 2012). Avian species richness seems to strongly correlate, both positively with park size (1–20 hectares; 2.5–50 acres), and negatively with noise level (50–59 dB). In these towns, parks with greater than 54 dB noise have few bird species (four to nine), while the two quieter parks (50 and 52 dB) are much richer, with 13 and 42 species. Most of the town park birds are common, though the quieter town parks (50–52 dB) probably contain some uncommon species.

A study of eastern bluebirds (*Sialia sialis*) in diverse greenspaces of a single town, Williamsburg, Virginia (USA; population 14,700), relates number of fledglings hatched (productivity) and other measures of fitness to noise levels (Kight *et al.*, 2012). The surrounding terrestrial human noise is typically loudest in the 200–2,000 Hz range (Slabbekoorn and Peet, 2003), while the adult birds mainly sing in the 1,500–4,000 Hz range. Thus, consistent with a study of great tits (*Parus major*) (Halfwerk *et al.*, 2011), this study focuses on the overlap 1,000–5,000 Hz range.

Of 45 eastern bluebird nesting attempts, 43 successfully fledged young birds, yet nesting success significantly decreased with increasing noise. The surrounding environmental noise included biological sounds (other animals), geophysical sounds (rustling leaves, wind), and anthropogenic sounds (e.g., cars, voices, air conditioners, recreational activities, construction, and landscaping equipment). But, especially in loud low-pitched noise sites, sounds were overwhelmingly human-produced noise.

Considering temporal patterns, sustained noise during daytime in town may result in more birds singing at night, as illustrated by urban robins (*Erithacus rubecula*) in Europe (Fuller *et al.*, 2007). Noisy industrial processes run by daytime employees, or daytime traffic on a busy highway in town, could mask bird calls so the birds learn to call during quieter nighttime.

Changes in birdsong can also involve pitch and loudness. Thus, ambient environmental noise was recorded at 13 sites along a gradient from rural to periurban to urban around the small city of Salamanca (Spain; population 155,000) (Mendes *et al.*, 2011). Rural sites have higher frequency, less loud, and less variable noise than urban locations. At each site the songs of common blackbirds (*Turdus merula*) were recorded. At sites with either very high-frequency or low-frequency noise, the blackbirds had changed their vocalizations or songs. The rural and periurban blackbird songs are significantly different from the urban birdsongs.

Birdsong evolution seems to be affected by noise, which may alter communication, mate selection, and vocalization learning (Luther and Baptista, 2010). Thus, from place to place the white-crowned sparrow (*Zonotrichia leucophrys*) has considerable variation in the structure of phrases, syllables, and syntax. Regional differences in the sparrow's song are conspicuous, and one or more dialects can be recognized in each region. Since anthropogenic noise is a primary driving force, this pattern is effectively an example of cultural evolution.

Solutions to the widespread ecological effects of human-caused noise at present are limited or elusive. Quiet costs money. The best solution is to limit noise at its source. Second, separate noise from the wildlife and people responders. Thus, double- or triple-layer glass windows separate outdoor noise from indoor quiet conditions for people. Soil berms, noise walls (e.g., covered with plants), and extremely wide forest buffers help separate railway and busy highway noise from sensitive wildlife species.

6 Water Systems and Waterbodies

Reflection and water are wedded for ever.

Herman Melville, *Moby-Dick*, 1851

Water Flows, Water Supply, and Towns

Water falls from the sky, and then evaporates from land and sea back to sky. Peering more closely at this global *water cycle* (Welty, 2009), some rainwater infiltrates downward through the soil to the water table as the upper surface of groundwater. Some may runoff to a waterbody over the surface (surface runoff or flow) or under the surface (subsurface runoff). And in evapotranspiration, some evaporates upward from the non-living soil surface, or is transpired upward by plants. Water cycles globally, but in towns, flows from here to there are the story (Figure 6.1).

Towns, like farmland and natural land, are a particular type of land surface. Towns require a water supply of clean freshwater for drinking and household use. That water is pumped from the groundwater or flows from a surface waterbody (e.g., lake, river, stream). Most towns have a separate input(s) of water for industry. The water cycle focuses on *water quantity* or *hydrology*, i.e., the amount of water in different locations and the size of water flows. But *water quality*, the biological, chemical, and physical characteristics of water, are also important to towns.

In addition, global freshwater provides numerous values or ecosystem services to society (Ecological Society of America, 2001; Larsen *et al.*, 2013).

Towns and Water

Towns exist where good agricultural soil, suitable water conditions, and good building locations come together. Location within a drainage basin (catchment, watershed) tells much about a town's water conditions. Consider nine town locations:

1. Ridge or hilltop. No stream. Water supply typically a deep well. Susceptible to drought effects. High evapotranspiration with wind. Low temperature may mean frozen pipes (e.g., Petersham, Capellades; see Appendix).
2. Slope of mountain or hill. Water supply from stream or groundwater. Susceptible to flooding. Protection of headwater area minimizes flooding, and often protects uncommon terrestrial and aquatic species (e.g., Stonington, Astoria).

Figure 6.1 Visible stream with concrete-armored sides flowing through town center. Strings of lights give a festive flair and warn boaters of the low bridge. Southwestern England. R. Forman photo.

3. Base of mountain or hill. Water supply from spring, seep, or wetland, where groundwater flowing downward comes to the surface. High and relatively constant water table. Flood prone (e.g., Vernonia, Flam).
4. Center of small valley. Town on both sides of stream. Water supply from upstream or a tributary. High water table. Flood prone. Water quality often a problem (e.g., Rumford, Issaquah).
5. In an agricultural plain. Groundwater suitable for pumping, if not contaminated by agricultural chemicals, and if water table not lowered by irrigation. Susceptible to drought. Winding stream or river polluted by agricultural sediment and chemicals. Protection of groundwater recharge areas and cones of influence for town wells important (e.g., Emporia, Cranbourne).
6. In a desert. Water pumped from groundwater originating in nearby or distant hills/mountains. Cisterns for catching and storing scarce rainwater. Low water table. Desert surface runoff greatly exceeds infiltration, so sheets of water commonly create flash floods (Shanahan, 2009). Intense sun and considerable wind mean high evapotranspiration, but plant cover is low. Irrigation common in desert towns. Susceptible to drought (e.g., Alice Springs, Superior).
7. Riverside. Water supply from upriver or a tributary. High water table in floodplain. Water quality threatened by agricultural chemicals and sediment from upriver. Flood prone. Town wells often in floodplain near river (e.g., Tavistock, Swakopmund).

8. Lakeshore. Water supply from lake or nearby stream. Relatively constant water table. Water quality threatened by wastewater, industry, and boating/recreation (e.g., Sandviken, Rangeley).
9. Coast of sea or estuary. Water supply often pumped from the *groundwater emergence zone*, the narrow zone typically within 100–200 m of the coastline (Todd and Mays, 2005; Forman, 2010) where freshwater from the land nearly reaches (or does reach) the sea. Coastal storms cover town with rain, fog, and salt. Coast and nearshore areas used for recreation and fishing, or polluted by town or industry. Susceptible to flooding. Excessive groundwater pumping causes saltwater intrusion under town (e.g., Hanga Roa, Akumal).

Most of the locations have short-term problems (floods) with too much water, as well as long-term problems (drought) with too little water. Many towns, especially in dry areas, use considerable water for irrigation.

The land surrounding a town is often higher, at least on one side, so some surface water and groundwater flow into town. Indeed, the water table under a town in a valley or depression is usually high, and thus flooding is common in heavy rainfall events. Deforestation in the surrounding higher land often raises the water table under a town, due to the reduced evapotranspiration (Thaitakoo *et al.*, 2013). Yet covering the slopes with development greatly increases surface runoff and consequent flood levels downslope. On the other hand, irrigation in towns, especially in dry areas, usually lowers the water table.

Groundwater

Groundwater lies invisibly beneath us as the big supply of freshwater on Earth (Todd and Mays, 2005; Shanahan, 2009). Somewhat like an underground lake, groundwater may form an aquifer, filling a large porous rock or sandy area. In other places groundwater simply fills spaces between particles in the lower portion of the soil. A distinctive fauna of strange (unseen, little-known) "ugly" invertebrates often lives in the upper portion of groundwater (Gibert *et al.*, 1994; Forman, 2014).

The upper surface of the saturated-soil groundwater is the *water table*, which typically is only slightly lower in towns than in natural conditions (and much higher than in urban areas) (Figure 6.2). In general, the water table is higher in clayey soils and lower in sandy or porous soils (Winter *et al.*, 1998; Todd and Mays, 2005) (e.g., Lake Placid, Hanga Roa). Indeed, sandy soils are effective in absorbing rainwater and reducing floods, if development and impervious surfaces are kept off such soils (McHarg and Sutton, 1975; Morgan and King, 1987; Yang and Li, 2011). Rainwater infiltrating through the soil of natural or semi-natural land is the basic way to *recharge* the groundwater, i.e., add water and raise the water table (Welty, 2009) (e.g., Issaquah). In towns, groundwater commonly provides water to deep-rooted trees and other plants, as well as to wetlands, pools, and streams.

The water table may be lowered by human activities such as groundwater pumping, irrigation, diversion of streams, mining, electric-power generation, industry, and

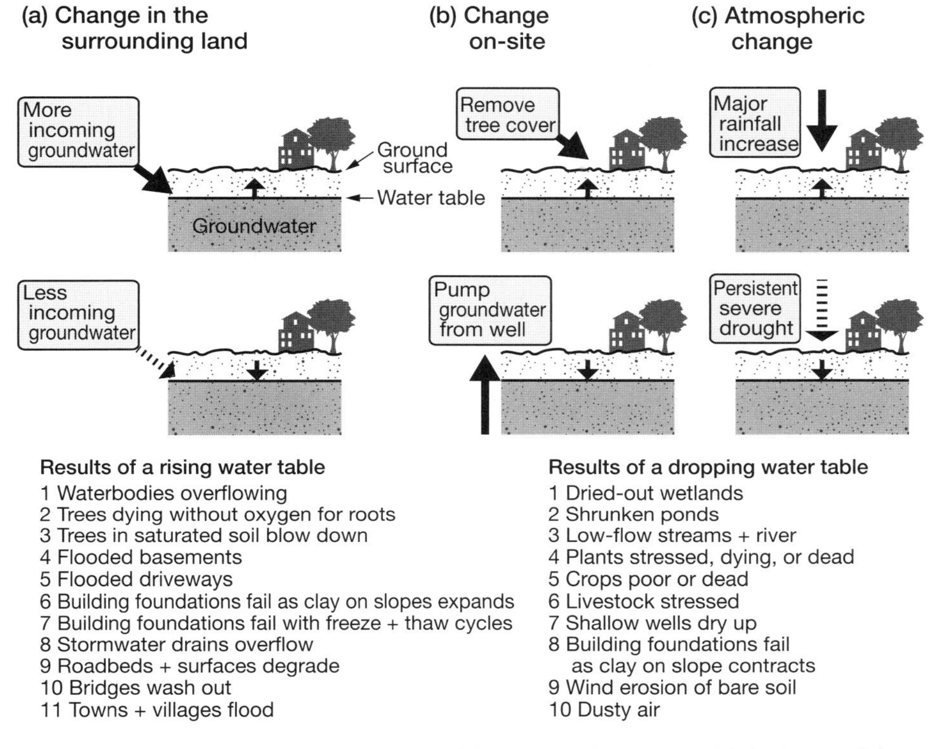

Figure 6.2 Causes and results of a water table rising or dropping. Water table is the top of the saturated-soil groundwater.

construction of channels (Todd and Mays, 2005; Shanahan, 2009; Westerhoff and Crittendon, 2009; Marsh, 2010; Marcarelli *et al.*, 2010) (e.g., Kutna Hora, Capellades). Typically, towns have a limited stormwater drainage system and no heavy industry pumping groundwater, and therefore a relatively high water table. A rising water table in town may cause septic systems to malfunction, damp or flooded basements, flooded driveways, and roadbed failure. A lower water table leads to dried out wetlands, ponds, and streams, as well as to the death of trees and shrubs.

Groundwater often originates from precipitation on nearby or distant mountains (e.g., Casablanca, Scappoose). Usually groundwater flows in the same direction as surface water in stream and surface runoff water. But since groundwater flows though porous material, the direction of flow is altered when the water encounters buried hard rock or dense clay (Todd and Mays, 2005). Mapping surface water flows and groundwater flows may produce different patterns, and in extreme cases, surface water may flow to one drainage basin while groundwater flows to a different basin.

Water quality, on the other hand, focuses on the biological, chemical, and physical characteristics of the water. As mentioned, the town water supply typically includes groundwater from beneath the town. Cities may pump groundwater for industry but normally not for clean water supply. Usually urban water tables are much lowered and heavily contaminated. Groundwater beneath cropland is also typically contaminated with agricultural chemicals, including pesticides and especially nitrate, which pollute a significant portion of rural America's wells.

The contamination of groundwater is serious because the water flows so slowly through soil. Pollutants from mining, industrial wastes, stormwater runoff, human wastewater, solid waste dumps, construction fill, farmland, pipe leakage, and chemical spills persist in groundwater for long periods (Todd and Mays, 2005; Shanahan, 2009). The main exception is groundwater in limestone, which is dissolved by water and full of cavities through which groundwater flows fast (e.g., Valladolid).

Water Supply

As a daily human need, the *water supply* of clean freshwater for a town may be pumped from the groundwater in relatively deep town wells and in shallow home wells (Todd and Mays, 2005). Town wells are often close to a stream or river where silty soil (alluvium) helps clean the water (Westerhoff and Crittendon, 2009) (e.g., Alice Springs, Swakopmund).

For water supply, generally villages use groundwater, small towns use groundwater and/or a stream, large towns/small cities use diverse sources, and cities build and use dammed reservoirs. Large towns typically get their water supply by piping in upslope water by gravity, or by pumping from elsewhere (e.g., Superior, Emporia). In dry areas, maintaining vegetated soil to catch rainwater and storing water runoff in soil and vegetation is a key to getting past dry spells that last for weeks (Rockstrom *et al.*, 2014).

Town wells have a *cone of influence*, with its point at the bottom of the pipe within the groundwater, and a gradually widening circle up to the ground surface (Winter *et al.*, 1998) (e.g., Concord). Intense water pumping tends to dry out surface wetlands, ponds, and streams within the large circle. Also, roads and buildings within the circle and nearby produce chemicals which move downward, and may contaminate the water supply being pumped. Protection of the cone of influence of town wells normally is a town priority. Towns, though not villages, typically have a water treatment facility which cleans and disinfects the water supply (e.g., Clay Center).

Irrespective of whether the water originates in town or from outside, except for desert towns, normally the water supply amount is very small compared with the amount arriving in precipitation (Forman, 2014). Freshwater is stored in water towers or water tanks at a higher elevation to provide water pressure to all homes and businesses (Figure 6.3) (Westerhoff and Crittenden, 2009). Two or more water towers/tanks provide stability for this crucial resource (e.g., Rocky Mount, Casablanca, Socorro).

Most components of the underground water system are constructed of materials designed to only last 25–50 years (Westerhoff and Crittendon, 2009). Thus, typically in towns the extensive pipe network carrying water is old and leaky. Occasionally one of the few water-main pipes or connections bursts, causing a local flood and cutting off water to numerous houses.

Figure 6.3 Water tower providing water pressure for buildings and houses in small city. Rocky Mount, North Carolina, USA. See Appendix. R. Forman photo.

In residential areas of the USA, most water use (59%) is outdoor use on a house plot (American Water Works Research Foundation, 2014 website). Other uses are: 11 percent toilet, 9 percent clothes washer, 7 percent shower, (6 percent leaks), 6 percent faucet, 1 percent bath, and 1 percent dishwasher. An analysis of water use in a suburban neighborhood of Zaragoza (Spain), a relatively dry area with house plots of about 93 m^2 (1,000 ft^2), found that 57 percent of average daily water use was in the house and 43 percent was for outdoor use (Salvador *et al.*, 2011). Approximately 60 percent of the households over-irrigated (based on irrigation requirements). A third of the plots received appropriate irrigation water, and 6 percent were under-irrigated.

Surface and subsurface runoff generally flows to a local surface waterbody, such as stream, river, lake, or pond. Any of these freshwater objects can serve as a water supply source. Sometimes a town dams a stream, creating a small reservoir as a water supply (Westerhoff and Crittendon, 2009) (e.g., Frome, Emporia). If agriculture, or deforested land with soil erosion, covers a significant portion of the drainage basin upstream of the reservoir, sediment washes into the waterbody. Then the reservoir fills with sediment, sometimes quickly, which progressively reduces the amount of water available (reservoir capacity).

Drainage basins (watersheds/catchments) for drinking water in the continuous USA are 77 percent covered by natural vegetation, 8 percent by agriculture, and 5 percent built area (Wickham *et al.*, 2011). Yet only 3 percent of this natural land is protected for conservation. Natural vegetation is decreasing and built land increasing in the drinking-water drainage basins, a worldwide pattern (McDonald, 2016).

During droughts, when the reservoir level is low, water from streambanks (upstream bank storage) is slowly released and helps maintain reservoir water supply (Winter *et al.*, 1998). Nevertheless, many towns have small or limited water supplies, and thus are at risk during droughts. One solution in dry regions is to introduce water via pipes or canals to serve many towns, but without drying out the water source area (Birkenholz, 2013). Since many towns are growing in population, providing for a greater water supply by enlarging an existing dam or using a second water source with good land protection, may be important.

Impervious Surfaces and Runoff

Above ground, precipitation pours water on the town and on surrounding areas including upslope. Water then flows over the ground and built structures as *surface runoff*. Also, water soaks into the ground as *infiltration*, where it has three trajectories. Some may continue downward to the water table and become groundwater. Some may be absorbed by plant roots, and later evapotranspired. And some may flow downslope as *subsurface runoff* in the soil but above the water table.

Almost no water infiltrates through *impervious surfaces* such as roofs, concrete walls, roads, and paved driveways. Yet an abundance of cracks and holes in streets and sidewalks characterizes towns, which often have limited budgets (Wessolek, 2008). Most towns seem to have about 10–50 percent impervious surface, which is much higher than in natural and agricultural areas, and noticeably lower than in urban areas.

Four suburban neighborhoods of single-family houses (11–33 house plots/hectare = 4.4–13.2/acre) on green plots outside Leuven (Belgium) have an average of 51 percent (range 39–60%) impervious surface (Figure 6.4) (Verbeeck *et al.*, 2011). The impervious area within a house plot (terrace, solarium, garden shed, garage, driveway, path to front door, and garden path) was high within about 15 m (50 ft) of the street and about 8+ m of a house. House-plot impervious area increased in the neighborhood over approximately 65 years mainly by adding pavement for parking near the street, plus house additions.

Similarly, in a 1.16 km^2 (287 acre) suburban area outside Leeds (England), impervious surface increased from 32 percent to 44 percent over three decades (1971–2004) (Perry and Nawaz, 2008). Three-quarters of the increase was due to paving in the front (garden) space of house plots (Figure 6.4). Such house-plot changes affect flooding (Warhurst *et al.*, 2014).

Towns vary in percent impervious surface cover from low (e.g., St. George's) to high (e.g., Capellades, Chamusca). Furthermore, most impervious surfaces in town are dispersed, small, and bordered by vegetation. House roofs, paved driveways, streets, sidewalks, and public buildings predominate. The few relatively small commercial and industrial areas in a town direct runoff of water both to vegetated areas and, via pipe or ditch, to a waterbody.

The impervious surfaces transferring water via pipe, road, or ditch directly to a waterbody is called *direct-connection impervious area* (DCIA) or drainage connection (Taylor *et al.*, 2004; Ladson *et al.*, 2006). Direct-connection water flows fast, increasing

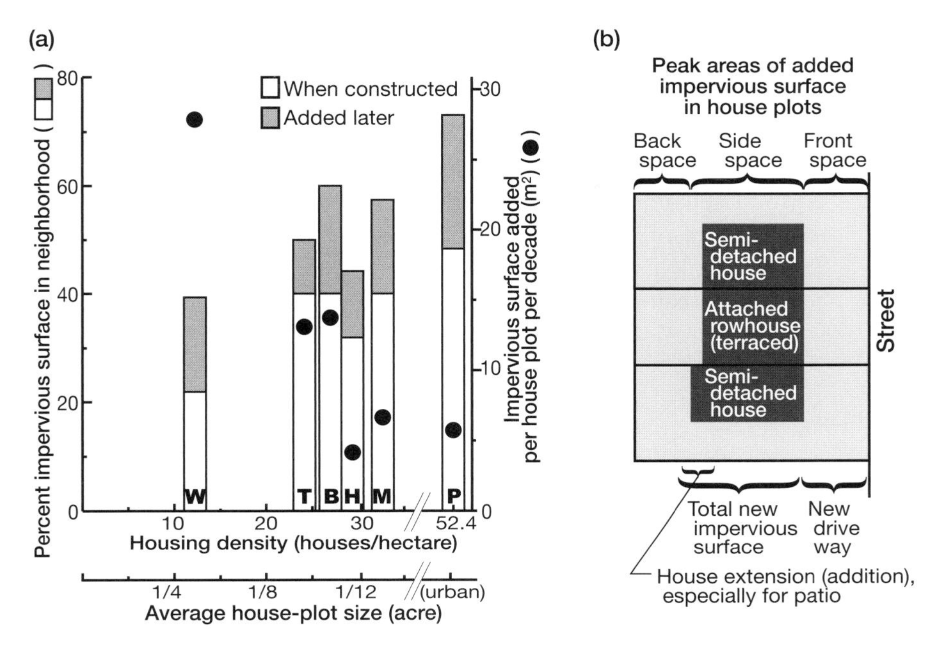

Figure 6.4 Impervious surface expanding over time on house plots. Based on one suburban UK neighborhood (Halton, Leeds) (Perry and Nawaz, 2008), four suburban Belgian neighborhoods (Withof, Ter Duin, Berkenhof, Mataki), and apparently an urban neighborhood (Pastourwijk) near Leuwen, Flanders, Belgium (Verbeeck *et al.*, 2011). Years of construction: W, 1952; T, 1962; B, 1949; M, 1930; P, 1923; year of Belgium study, 2008. H, 1971; end of England study, 2004. (a) Peak areas based on Belgium study; peak area in Halton is front driveway area. Percent impervious surface in neighborhood includes house plots, streets, and car parks. (b) Letters at base of histograms refer to the suburban neighborhoods.

the probability of flooding. But for much of a town, precipitation water on the numerous small surfaces quickly runs off into the adjoining plant cover, where most water infiltrates into the soil.

Suppose a typical town has about 30 percent impervious surface, overwhelmingly as *dispersed small surfaces bordered with vegetation*. Perhaps only 15 percent of the precipitation water is directly connected and channeled to waterbodies, an amount unlikely to produce severe or frequent flooding. Such linear flow-paths are scarce in natural areas, and predominant in urban areas (Hosl *et al.*, 2012). The 85 percent or so of precipitation water, which infiltrates into the soil widely across the town, supports a relatively high water table (by groundwater recharge). Infiltration water also supports a local waterbody and root absorption by plants. This pattern somewhat mimics natural conditions. However, the relatively high water table may contribute to frequent flooding in town.

Flooding is a particular problem for most towns, due to often being in low sites in the landscape and/or having a high water table. Flooding is reduced in a variety of ways. Hard-surface engineering solutions to reduce flooding are common, while ecological solutions are increasingly used. Engineering solutions generally channelize and accelerate water flow between walls through a town. Another approach is to divert water in channels around the town. Both approaches tend to produce flooding downstream of the town.

Ecological solutions in town to reduce flooding, on the other hand, generally focus on three natural processes, reduce water runoff, increase infiltration into soil, and increase evapotranspiration to the atmosphere. Increasing natural vegetation and evapotranspiration upslope of town may decrease flooding. Lowering the town's water table may also reduce flooding. Protecting the headwater areas of a stream system with natural conditions is particularly effective for protecting distinctive fish and wildlife populations. But it also provides clean water and reduces floodwaters.

Evapotranspiration

Evaporation from non-living surfaces such as soil and roof, plus *transpiration* from living surfaces, especially plants, gives off water (water vapor) to the atmosphere. In the typical town just mentioned, 30 percent of the horizontal town surface evaporates water, though rainwater also evaporates from wet vertical surfaces.

The vegetation component is more interesting. In mainly residential towns, tree foliage commonly covers a significant portion of the impervious surfaces (Gartland, 2008). Often the total leaf surface area is two to three times the horizontal surface area (leaf area index). Different tree species transpire at different rates (Pataki *et al.*, 2011). Nevertheless, trees generally transpire more water than the same area of shrubs, which in turn transpire more than do herbaceous plants.

Dry air, sun, and wind accelerate evaporation and transpiration (*evapotranspiration*). Since the sites of many trees in town are directly exposed to wind and sun, the somewhat savanna-like structure of towns helps produce a high transpiration rate. Furthermore, considerable irrigation of house plots and town-managed spaces, especially in dry areas, characterizes towns. Irrigation further increases transpiration, such that relative humidity may be higher than in the surroundings, except where irrigated agriculture is nearby.

Overall, the water cycle of a town seems similar to that of a natural wooded area. Impervious surface area and pumping groundwater in wells are both significantly higher in town. Yet typically the water table is only slightly lower. With mostly small dispersed vegetation-bordered impervious surfaces, infiltration of water into the soil is only slightly lower than in natural woods. Direct-connection water runoff is slightly higher. Flood frequency is slightly higher. Evaporation is slightly higher. Transpiration is slightly lower, or higher. Unlike the similarity to a natural woodland, the town's water cycle differs markedly from that of an agricultural area or an urban area.

Wastewater and Stormwater

To the east of the southern mansions are several streets of workmen's houses. These were very small and poor, containing only four rooms each ... All the streets appear to have had a channel of stone down the middle ... This channel is not deep, but rather a slight curved hollowing of the upper sides of the line of stone, which is about 22 inches wide. Probably therefore the street sloped down to the middle ... and thus occasional rain, and waste water from the houses, would be led off without making the street muddy. This is far the earliest example of street drainage known; and the system must have been general in Egypt at that age [12th Dynasty] for it to have been used in a labourers' town such as this. (W. M. F. Petrie, 1891)

Human Waste and Wastewater

Systems for Disposal and Treatment

Traditionally, wastes from villages and towns (e.g., in the USA) were carried away by streams, rivers, estuaries, or sea (Metosi, 2000). These waterbodies were the basic cleaning mechanism. This approach persists in many areas, while in other areas clean water is maintained and valued for both environmental and human health reasons.

Human waste (urine and feces) is inherently a resource. In dry areas, water is valuable. Fecal organic matter, somewhat analogous to humus (except for abundant pathogenic bacteria), can be incorporated into garden or farm fields, where it retains moisture and slowly releases nutrients. In highly productive food systems, including fish farms, the nutrients, especially phosphorus and nitrogen, are valuable as fertilizer for growing plants. For instance, human waste or "nightsoil" has long been collected in large jars and used to increase production (see village of Figure 4.3). Human waste has also been commercially dried, with bacteria spores mainly killed, to use as lawn fertilizer. But human health problems may occur downwind.

Human waste poses problems in the environment because of the high level of phosphorus, nitrogen, and organic matter, along with the presence of pathogenic bacteria (Westerhoff and Crittenden, 2009; Butler and Davis, 2011). These problems are minimal when waste is directly deposited into a hole in the ground, as in a *pit latrine* or *outhouse* (enclosed latrine) close to the home (Figure 6.5) (Anderson and Otis, 2000; Hardoy *et al.*, 2001; McGregor *et al.*, 2006; Butler and Davies, 2011). Several types of latrines exist, and solar, electrical, or other heat may be added to increase the rate of breakdown or decomposition of waste. A composting toilet mixes human waste with certain other organic matter, and keeps the material well oxygenated to facilitate decomposition (Van der Ryn, 1978). Inconvenience, odor, and personal health are generally considered to be problems with such waste-into-soil approaches, especially where people are concentrated.

The usual solution is to have human waste deposited directly into water, forming *wastewater* carried away in pipes or ditches (Butler and Davies, 2011). This water with waste is usually combined with water from bathtub, shower, basin, kitchen sink, dishwasher, and/or clothes washer, which together contain numerous chemicals from household use (Baker, 2009a; Butler and Davies, 2011).

The resulting wastewater may be piped directly to a *cesspool* just outside the building (e.g., St. George's). This is basically a large hole filled with gravel and covered over. During very wet conditions a cesspool may overflow. Alternatively, a pipe or ditch may carry and simply empty the wastewater into a nearby low spot in the landscape, where the material may accumulate. Such ditches and low spots produce odor and public health

Figure 6.5 Outhouse with hole in the ground in human waste system using no water and producing no wastewater. Waste may fill a tank to be emptied, or may be decomposed by heat from solar radiation. Edge of city of Tolear, Madagascar. Photo courtesy and with permission of Mark Brenner. (A black and white version of this figure will appear in some formats. For the color version, please refer to the plate section.)

problems. Also, rainwater may then carry the waste further, and pollute a stream or other waterbody.

The two main alternatives to these problematic solutions are septic systems and a sewage treatment system. *Septic systems* just outside of buildings are designed to treat or clean the wastewater on-site, incorporating the wastes into soil with oxygen, where they are broken down and rendered harmless (Marsh, 2010; Butler and Davies, 2011).

In contrast, *sewage treatment systems* collect and treat wastewater piped from numerous buildings, residential, commercial, and industrial (thus containing some industrial waste) (Butler and Davies, 2011). Sewage is viscous, flowing slowly by gravity through pipes with acute-angle intersections, and sometimes needs to be pumped (Clement and Thomas, 2001). Sewage wastewater is piped to a sewage treatment facility, which decomposes and renders harmless the waste for a community.

An on-site *septic system* has three key components (Anderson and Otis, 2000): (1) a large buried septic tank which collects wastewater from the building; (2) a drainfield (or leachfield) area with perforated pipes, through which wastewater flows; and (3) a thick layer of oxygenated soil beneath the drainfield, through which the wastewater percolates. The septic tank normally is pumped at regular intervals to remove heavier sludge at the bottom, and floating scum (oils, greases, waxes) at the top. The liquid wastewater flows through the perforated pipes which are rich in decomposing bacteria. The somewhat cleaner wastewater finally percolates down through the soil where biological, chemical, and physical processes further clean the liquid, before it passes onward to groundwater or a local surface waterbody.

In contrast, a *sewage treatment facility* may treat or clean incoming wastewater in three basic stages, primary, secondary, and tertiary treatment (Anderson and Otis, 2000; Kalff, 2002; Baker, 2009a). Wastewater is composed of water, dead organic matter, bacteria, mineral nutrients (especially P and N), heavy metals, and numerous chemicals from residential, commercial, and industrial buildings.

Primary treatment removes large solid objects and other solids that settle to the bottom of a wastewater tank, as well as oils, greases, and waxes that float. *Secondary treatment* removes the bulk of the dissolved organic materials (BOD) and suspended solids (TSS), and significantly reduces bacteria. Usually at the end of secondary treatment, chemical disinfection (typically with chlorine) kills almost all the bacteria, including pathogenic or disease-causing fecal coliform bacteria (e.g., *E. coli*) (Baker, 2009b). *Tertiary treatment* removes about 90–95 percent of the nitrogen and phosphorus present, and often another chemical disinfection occurs. Further advanced treatment is rare. Unfortunately, pharmaceutical compounds and, in dry climates, salt also often remain after tertiary treatment (Westerhoff and Crittendon, 2009).

Septic systems mainly work well in low-density housing areas, whereas sewage systems work well in high-density areas (Anderson and Otis, 2000). Septic systems distribute the water and nutrients into the soil throughout the community. But some portion of a town's septic systems typically work poorly. Therefore, pollutants, pathogens, and odors are present in scattered spots, and local waterbodies may become somewhat polluted (e.g., eutrophicated).

On the other hand, sewage treatment systems have an extensive pipe network carrying the community's wastewater to a single large facility, where the liquid is treated and rendered harmless (Anderson and Otis, 2000). Thus, water from across the community is channeled to a waterbody at one point downslope of the sewage treatment facility (e.g., Huntingdon/Smithfield, Concord). This means that wetlands, ponds, and streams around town tend to dry out in dry periods.

But leaks occur in the extensive sewage pipe system (Westerhoff and Crittendon, 2009), and nutrient-rich and heavy metal-rich sludge accumulates. Heavy rains cause overflows (combined sewage overflows) (Butler and Davies, 2011) (e.g., Astoria). The sewage treatment facility itself is an area rich in nutrients, and thus rich in wildlife and biodiversity.

A combined sewage/stormwater system has a single pipe system, through which both stormwater and wastewater sewage flow. If the pipe system leads to a sewage treatment facility, during rainy periods the large amount of stormwater may overwhelm the capacity of the facility to treat wastewater (Benton-Short and Short, 2008; Baker, 2009b). Then combined sewer overflows (CSOs) occur, resulting in untreated (raw) sewage passing directly into the environment from the sewage treatment system (Benton-Short and Short, 2008; Butler and Davies, 2011).

Lots of technologies and alternative systems for treating wastewater exist and may be useful for towns. *Grey-water recycling* uses the water from kitchen, basin, and bathtub (grey-water), rather than the clean freshwater supply, to flush the toilet (Butler and Davies, 2011). This recycles the water from kitchen and bath use, and greatly reduces the amount of clean freshwater used.

Wastewater recycling cleans wastewater by having it percolate downward through the soil until it reaches (recharges) the groundwater, which is then pumped up as drinkable water supply (Todd and Mays, 2005). For decades, Windhoek, the capital of Namibia, has used many advanced technologies such that 100 percent of its clean freshwater supply is recycled human wastewater (Law, 2003; du Pisani, 2006). I stayed in the city and the water tastes fine.

Years ago, the town of Arcata, California (USA) established a 15 ha (40 acre) *marsh/pond area to treat wastewater* (Gearheart, 1992; Forman, 2014). Basically primary, secondary, and tertiary treatment was accomplished by natural processes (plus two chemical disinfections). The area became an important wildlife area, rest stop for migrating waterbirds, and recreation area. Indeed, the outflowing water fed a commercial oyster production area.

For wastewater treatment by natural systems, the basic idea is to have the liquid flow in a sinuous (curvy) trajectory through wetlands (Vymazal *et al.*, 2006; Mitsch and Gosselink, 2007). Success depends on both the length and rate of flow.

Another partial solution for a town's wastewater is to use it to grow food in an aquaculture system, as some cities in South and Southeast Asia do (Costa-Pierce *et al*, 2005; Bunting *et al.*, 2005; McGregor *et al.*, 2006). The high nutrient levels stimulate a large production of plants, algae, and fish, either separately or together. Considerable food is produced locally, though pathogenic bacteria in the ponds are doubtless bad for the workers (Little and Bunting, 2005).

Villages, Towns, Cities, and Wastewater

So, consider what commonly happens to human waste in communities from village to city. Villages with a low population density usually have no sewage treatment system, so wastes enter the ground *on-site*, i.e., next to nearly every building (e.g., Selborne, Petersham). Outhouses, cesspools, pipes to low spots, and septic systems may all be present, or one type may predominate. Outhouses (Figure 6.5) often predominate in many villages of developing nations (e.g., Grande Riviere), pipes to low spots in some areas of southern Europe, and septic systems in middle- and upper-income villages in the USA (e.g., Petersham).

Small towns (population 2,000 to ca. 10,000) with a relatively low population density have a more complex arrangement (e.g., Hanga Roa, half cesspools, half septic systems). Most such towns use the village solutions just described. But the higher building density and the on-site waste solutions produce considerable odor and public health problems. Since home septic systems require a significant area of adjoining lawn or other herbaceous vegetation, they cannot be used effectively on small house plots or in high-density housing areas.

Thus, some small towns invest in sewage treatment systems, with the pipe system at least covering the town center and often the adjacent older residential area (e.g., Lake

Placid, Casablanca, Sooke). Given the high cost, the sewage system commonly has only primary, or primary and secondary treatment. Interestingly, the European Union has a policy or goal such that all communities over 2,000 people, i.e., all towns, should have a sewage treatment system covering nearly all the buildings in town.

Large towns of, say 15,000–30,000 people commonly have a sewage treatment system (e.g., Frome). This may cover only part of, or all, the town, and may provide secondary or secondary and tertiary treatment (e.g., Emporia). Septic systems and other on-site solutions are often present, and sometimes abundant. A package or *neighborhood sewage treatment system* may serve a distinct concentrated population, such as in a university, hospital, prison, or group of apartments (e.g., Lake Placid). Large towns occasionally invest in advanced or experimental sewage treatment systems (e.g., Concord, Wausau).

Towns and villages, as well as their various buildings, are located on different soil types. The on-site human waste approaches only work satisfactorily on certain soils (Chapter 5). On porous sandy soil, wastewater drains rapidly downward, and pollutes the groundwater or a nearby waterbody. At the other extreme, dense clay soils are quickly saturated by wastewater, which then simply runs off over the surface often to a waterbody that becomes polluted. Clay soils get soggy and smelly. Silty or loamy soils provide the ideal intermediate conditions, so that wastes percolate steadily – not too fast, not too slow – and decompose in the soil.

On-site human waste disposal approaches across a town may pollute a stream flowing through town, or other waterbody present. On-site phosphorus, nitrogen, organic matter, and pathogenic bacteria are carried by subsurface or groundwater flow to the stream. Typically, a stream is most polluted just downstream of town, though added water from a tributary may improve or degrade the stream's water quality. Except for leaks and combined sewage overflows, a sewage treatment system will minimize this stream pollution in town.

A sewage treatment facility at the end of the sewage pipe system is virtually always downstream, downriver, or downslope of the town. Here, after treatment, the facility basically pours all water from the pipe system into the stream. If the facility only provides primary treatment, the downstream water is heavily polluted with organic matter, pathogens, and nutrients. If the facility provides secondary treatment, the stream is polluted (eutrophicated) with the nutrients, phosphorus, and nitrogen. If the facility provides tertiary treatment, in principle it does not pollute the stream, though low levels of P, N, and heavy metals (e.g., from pipes) are typically present. Modeling the flows of a nutrient such as phosphorus through the food and waste of a small city (e.g., Gavle) highlights different places where improvements would be effective (Nilsson, 1995).

Stormwater

Impervious Surfaces and Stormwater in Town

Most towns began next to a stream or other waterbody and therefore today their town centers border the water (Figure 6.6). This means that the central commercial area, with typically 70–90 percent impervious surface area, is adjacent to the waterbody. The commercial area produces considerable waste which in early days was largely carried away by the waterbody, a practice in some towns today.

Figure 6.6 Coastal town on hillside where stormwater and human wastewater rush toward the harbor. Wastewater is collected for treatment. Astoria, Oregon, USA, with mouth of the Columbia River in distance. See Appendix. R. Forman photo. (A black and white version of this figure will appear in some formats. For the color version, please refer to the plate section.)

The main road through the early town tended to go along the edge of the stream or lake or other waterbody, which is often still the case. Thus, pollutants are also readily washed by rainwater from road to adjoining waterbody. These may include sediment, horse manure, hydrocarbons, heavy metals, road salt, and materials spilled from vehicles. Commercial areas, industrial areas, and roads have relatively large continuous impervious surfaces (a long fetch or run), on which runoff water accelerates toward a low area or local waterbody (Butler and Davies, 2011).

Road surfaces with adjoining ditches are fairly homogeneous along their length, so water runoff may go long distances. On the other hand, street-side ditches in residential areas can be designed to better mimic nature, providing less surface runoff (and flooding), more infiltration into soil, more evapotranspiration, higher biodiversity, and much better aesthetics (Beatley, 2000; Hill, 2009).

Residential areas with narrow streets have considerably less impervious surface area than those with wide streets characteristic of many suburbs (Frazer, 2005; Girling and Kellett, 2005). Since the highest buildings in towns are typically only two to four levels high, towns have no need for wide streets to accommodate huge city/suburb firetrucks.

Impervious surfaces are not only major sources of surface water runoff, but also sources of heat emission that raises air temperature in town (Chapter 5). Rough averages of percent impervious surface cover for different housing densities have been calculated, based mainly on North American residential suburbs (US Environmental Protection Agency, 1993; Arnold and Gibbons, 1996; Paul and Meyer, 2001; Forman *et al.*, 2003):

20% for 1-acre house plots (2.5 houses per ha)
25% for 0.5-acre house plots (5 houses per ha)
12% for cluster development (average 5 houses per ha)
30% for 0.33-acre house plots (7.5 houses per ha)
38% for 0.25-acre house plots (10 houses per ha)
65% for 0.125-acre house plots (20 houses per ha)

All of the house-plot sizes are large enough to have an outdoor private family space. The 0.25-acre plot size is typically sufficient for a septic system, whereas the 0.125-acre size may not be sufficient, depending on soil, slope, and other factors.

Industrial areas average about 75 percent impervious surface cover, commercial areas 85 percent, and shopping centers 95 percent. The extensive impervious surface cover of industrial and commercial centers usually has many tarmac/asphalt and concrete surfaces of different sizes and shapes. The boundaries between surfaces often have cracks and holes over time, due to weathering and vehicle-caused vibration (Wessolek, 2008). Thus, some stormwater (perhaps commonly 3–11% of the total) infiltrates into the soil rather than runs off over the industrial and commercial impervious surfaces.

Smooth surfaces of course have the most and most-rapid runoff. Roughening a surface or adding objects on it slows and reduces runoff (Wessolek, 2008). After runoff, water that remains on a surface evaporates to the air, so a rough surface has more evaporation than a smooth surface. In a light rain most of the water is evaporated, whereas in a heavy rain almost all water runs off.

Ditches and pipes carry stormwater from impervious surface to a low area or local waterbody. Small towns typically have a limited drainage pipe system (e.g., Vernonia, Clay Center, Page), whereas large towns, hill towns, and those in heavy rainfall areas have extensive drainage pipe systems (e.g., Astoria, Rumford). In contrast to the extensive stormwater pipe networks with large collector and outlet pipes in cities/suburbs, towns normally have many short ditches and pipes for runoff. Also, town pipe networks typically have few branches.

In this way, stormwater in town is distributed to many local low areas, wetlands, and/or waterbodies. This limits floods resulting from peaks of stormwater runoff (Hosl *et al.*, 2012). Ditches and pipelines in town also serve as prime routes for some wildlife movement. Scavengers such as raccoons and foxes (*Procyon*, *Vulpes*) readily use stormwater ditches, especially at night.

The effect of a town's (or city's) greenspaces on water runoff and flooding remains little studied (Forman, 2014). Probably overall town flooding is reduced more by several small greenspaces than by one large park (SS>OL). Several small greenspaces have more edge facing the sun and wind, and thus greater evapotranspiration. SS reduce run or fetch of a built area, providing friction that reduces the acceleration of stormwater runoff. Several small greenspaces widely distribute opportunities for constructed basins receiving runoff from their surroundings. With towns typically having a relatively high water table and being flood-susceptible, many small greenspaces seem to make good sense.

A *hydrograph* is a plot/graph showing water flow rate or discharge, e.g., of a stream, over time (Welty, 2009; Marsh, 2010; Marcarelli *et al.*, 2010; Butler and Davies, 2011). Typically, the hydrograph shows streamflows before and after a storm event, compared with flow over time with no storm. At least four important patterns appear in a hydrograph: (1) *peak flow*, that is, the maximum level of flow, which may reach flood level; (2) *lag time*, that is, how slowly or quickly peak flow occurs after a rain event; (3) *flashiness* or frequent *flash floods*, i.e., many storm events with short lag times, a pattern that usually degrades the stream channel; and (4) *low flow*, the minimum level of flow (well after peak flow), which often indicates high water temperature and degraded conditions for fish (White and Greer, 2006).

A higher percent impervious surface cover produces a higher peak flow, shorter lag time, more flashiness, and lower low-flow conditions. More water running off quickly means that less rainwater infiltrates into the soil, and thus less subsurface flow to support a stream. Between rain events, a flashy stream may barely trickle from tiny pool to tiny pool, not a good situation for most fish.

Stormwater Pollutants

Rain washes surfaces clean, which dirties local waterbodies. Stormwater runoff in town flows over tarmac/asphalt surfaces picking up hydrocarbons (including PAHs) and heavy metals, both from vehicles and the surface material (Chapter 5) (Frazer, 2005; Baker, 2009b). Runoff over concrete and mortar between bricks picks up calcium carbonate, which raises the pH of water a bit. Particulate matter is carried from dusty surfaces, heat from dark surfaces, and garbage and dog waste from certain surfaces. Also, heavy metals are commonly leached from concrete bridges into streams.

More broadly, stormwater pollutants originate from many sources, including: rain itself; aerial deposits during dry periods; vehicle use (especially wear, exhaust, and leaks); industrial waste; commercial waste; construction sites; debris from residential houses; pet wastes; road salt (de-icing); agricultural plots and vegetable gardens in town; and maintenance of lawns, flower gardens, and parks. Phosphorus, the main source of freshwater eutrophication, typically accumulates in the soil attached to particles. Thus, bare soil areas and long-fertilized lawns are often important sources of phosphorus runoff during heavy rains (Baker, 2009b).

With this cornucopia of pollutants, the numerous outlets (ends) of relatively short pipes and ditches provide a good stormwater solution for towns. Many of the outlets may deposit stormwater and pollutants into low vegetated areas or wetlands, rather than directly into local waterbodies. Much of the water and pollution is thus absorbed by the soil and vegetation, and only a limited amount reaches waterbodies by surface or subsurface flow. Stormwater runoff channeled to vegetation and wetlands may infiltrate downward, thus cleaning the water and recharging groundwater. Then the water is available to pump in wells as water supply for residents.

First flush, the initial water runoff from a storm event, has the highest concentration of pollutants, so absorbing the first flush into soil and plants, rather than piping it directly to a waterbody, is especially important. Both the water and the pollutants in stormwater runoff produce significant effects on local aquatic ecosystems and fish (Paul and Meyer, 2001; Lee and Heaney, 2003).

Lots of other solutions for stormwater runoff exist. For example, *porous pavements* have tiny holes so some water infiltrates (Ferguson, 2005; Scholz and Grabowiecki, 2007). These are particularly useful where trucks do not go. *Permeable pavements* are usually composed of concrete or plastic blocks with cracks and/or holes through which considerable water infiltrates (Munchow and Schramm, 1998). These are especially useful for driveways and some walkways. Houses collect water from the roof in *rain barrels* for watering gardens, or in *cisterns* for water supply in dry times or climes.

Many types of *constructed basins, ponds, and wetlands* (and rain gardens) are designed to either slow down water runoff (*detention basins*) (e.g., Emporia), or to both slow runoff and treat the pollutants (*bioretention basins*) so they do not pollute waterbodies. Rain gardens usually refer to small constructed plant-covered depressions to receive stormwater on house plots (Dietz and Clausen, 2008; Mehring and Levin, 2015; US Environmental Protection Agency, 2018). Sometimes stormwater from roads drains into and irrigates adjoining land, such as town parks, or even residential house plots for flower gardens, lawns, and shrubs/trees.

Low-impact development (LID) refers to housing in a small drainage basin with a decentralized stormwater system (Richman *et al.*, 1997; Forman *et al.*, 2003; Hood *et al.*, 2007). Vegetation is maintained in the upper headwater area, vegetated ditches are used for much of the water flow, and big-pipe flows at the bottom of the drainage basin are minimized.

Finally, *split-flow theory* has been proposed to mimic stormwater in natural landscapes (Echols, 2008). Using a model, precipitation water is separated or split into its three following component volumes: evapotranspiration, infiltration, and runoff. Rather than only addressing large rainfall events, e.g., with detention and bioretention basins, the model also focuses on the numerous small rainfall events that in total often exceed the large events. By including small rainfall events as well as evapotranspiration and infiltration (along with large rainfall and runoff events), we better simulate the stormwater pattern of natural systems, including rate, quality, frequency, duration, and volume. Since town water flows, unlike urban stormwater, are not too different from flows in natural land, the relatively small modifications required for mitigation may be easily accomplished.

At a broad scale, a well-connected landscape is especially suitable for stormwater flows without disruption (Figure 6.7). Thus, in Jehikal, Tochigi (Japan), connectivity is rather high for stream valleys, floodplain rice-paddy agriculture, villages with roads, woodland, and upland agriculture. Each feature and the whole system are connected.

Wetlands and Towns

Both wetlands and ponds are in depressions in the ground. The wetland depression is filled with soil saturated with water, and normally flat on the surface. These slow-flowing waters play important roles but also contain pests and hazards. Abundant flies

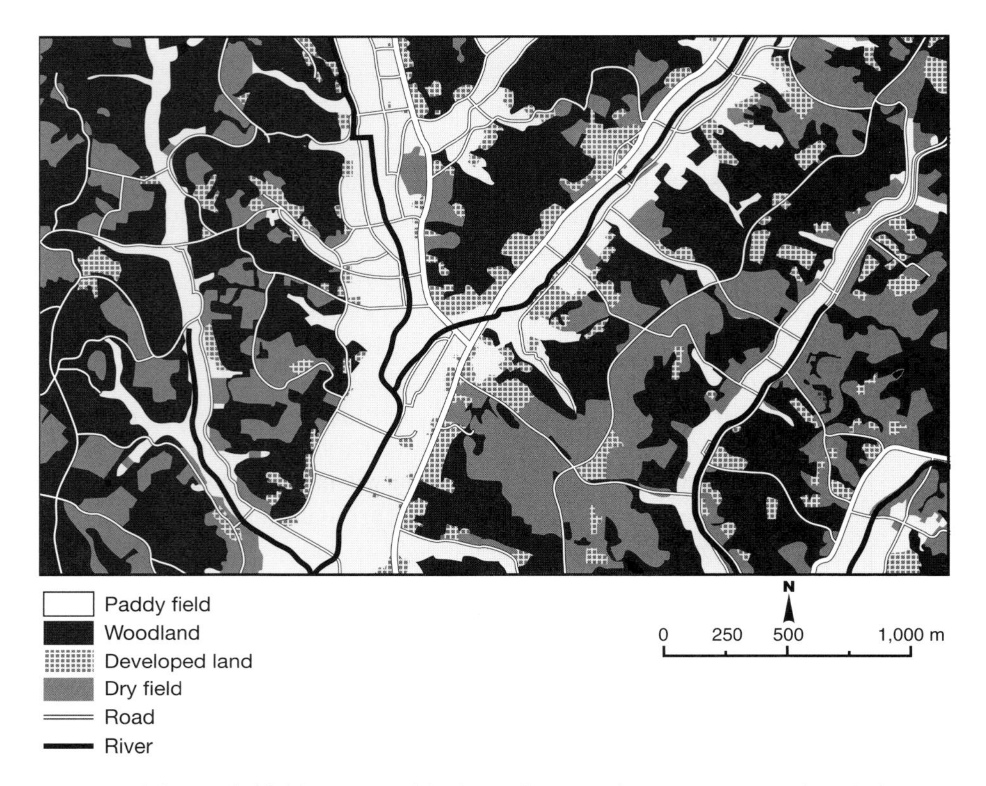

Figure 6.7 A fine-scale highly connected land mosaic supporting numerous people and also wildlife. Rice paddies in river and stream valleys of "Satoyama" landscape, Jehikal Town, Tochigi Prefecture, Japan. Adapted from Katoh *et al.* (2009).

and mosquitoes, snakes and crocodiles, seemingly bottomless muck, risk of drowning, and simply the unknown are characteristic.

Freshwater wetlands and ponds have slow-flowing water, with ample oxygen near the surface, less oxygen lower down, and often no oxygen (anaerobic conditions) in the lowest portion. The faster-flowing water of streams and rivers carries considerable oxygen.

Slow-moving waterbodies are also town treasures. Town wells are often located next to these dependable water sources. Stormwater drainage systems in town often lead to these waterbodies, where stormwater chemicals are readily absorbed or decomposed. Houses may line the shores of ponds for views and recreation, while houses alongside a wetland have nature and privacy. Economic values range widely, such as harvesting reeds or fish, and local recreational opportunity from nature appreciation and hunting to swimming and boating.

Exploring wetlands is a treat, including: (1) attributes, types, locations; (2) inflows, outflows, soil, chemicals; (3) habitats, productivity, species; and (4) wetland and town interactions.

Attributes, Types, Locations

Wetlands are areas of water-saturated soil with the water surface at or slightly above the soil surface for a prolonged period most years (Keddy, 2000; Kalff, 2002; Mitsch and Gosselink, 2007). Normally the soil is predominantly either peat or clay, with *peat* being accumulated organic plant material that decomposes slowly when saturated with water. Wetlands store a lot of carbon (Kayranli *et al.*, 2010), and support a variety of plants adapted to the saturated-soil conditions.

Three key characteristics of wetlands are used in determining the border of a wetland (*boundary delineation*): (1) hydrology (water quantity); (2) soil conditions; and (3) organisms (especially vegetation). In general, the boundary extends farthest outward based on wetland hydrology conditions. Wetland soil conditions extend outward an intermediate distance, and wetland vegetation extends outward the least. Wetland boundary delineation is important to protect wetland values, and emphasized where regulations prohibit or limit development within a certain distance of a wetland (e.g., Lake Placid, Concord).

Not surprisingly, many types of wetland occur worldwide, including swamp, marsh, bog, fen, mire, and moor (Mitsch and Gosselink, 2007). *Inadvertent or accidental wetlands* form as an unplanned result of human activity in the land, such as a road or building blocking water drainage (Palta *et al.*, 2017). All wetlands vary by depth and duration of flooding, and are affected by adjoining deep water and upland conditions. The species present are either only adapted to wet conditions, or to both wet and dry conditions.

Swamp, marsh, and peatland are the primary, widely recognized wetland types. *Swamps* are dominated by woody plants. Most tree roots require a considerable amount of oxygen, so tree swamps tend to be inundated or flooded for relatively short periods, e.g., one to two months per year. Persistent inundation often kills trees. Shrub roots on average tend to be shallower and apparently less oxygen-requiring, so shrub swamps are normally inundated a bit longer (e.g., two to three months) than tree swamps.

Marshes are dominated by herbaceous plants, often grasses or grass-like (graminoid) plants. Typically with relatively shallow roots, marsh plants are often inundated most years for long periods, e.g., three to several months. Inundation for a longer period each year commonly produces seasonal ponds, which temporarily dry up during the dry season. Herbaceous plants are often present in the ponds, but they may be marsh plants or aquatic plants.

Peatlands develop where nearly year-round inundation of soil occurs. Plant litter decomposes slowly and has accumulated over time to produce a thick black organic soil. Most peatlands, such as bogs and muskegs, are acid with a low pH. Peat moss (*Sphagnum*) and a limited number of plant species are adapted to acid wet soil conditions (Lindsay, 1995).

In areas with a high water table, groundwater mainly saturates the wetland soil (Shedlock, 1993; Mitsch and Gosselink, 2007). A groundwater-fed wetland has less fluctuation in water level than a wetland fed by surface and subsurface water runoff from the land. Cities typically have low water tables and wetlands have mainly been eliminated by drainage or filling (Forman, 2014). In contrast, towns often have higher water tables, are often subject to flooding, and commonly contain wetlands (e.g., Concord).

A large wetland adjacent to a town limits town expansion, but also may provide economic harvestable resources, from reeds to orchids, fish, waterfowl, and peat for fuel. Indeed, marsh people in the confluence of the Tigris and Euphrates rivers of southern Iraq lived in villages on constructed "islands" in the marshland.

Wetlands commonly occur on the edges of lakes (e.g., Lake Placid), ponds (e.g., Petersham), streams, and rivers (e.g., Concord), where the water table tends to be high. Such wetlands are important sources of organic matter for the open waterbodies (supporting detritus food webs), and for species that forage in the waterbodies. Town residents commonly degrade or remove this wetland vegetation alongside ponds and lakes to reduce pests, and to create views, docks, and waterfront facilities.

Towns are often bisected by a stream. Usually this means that the stream floodplain has been narrowed, i.e., partially filled for roads and buildings. Therefore, to minimize streambank erosion, especially during floods, the streambanks are *armored* (covered by large rocks or walled by concrete). Upstream and especially downstream of the narrows, small often elongated wetlands commonly appear in the floodplain (e.g., Chamusca). Floodplains in towns have a high water table so these wetlands often persist, even with vegetable gardens or community gardens in the floodplain.

Towns next to rivers may be partly or entirely in the river floodplain, and hence generally subject to severe flooding, or next to the floodplain where they generally escape the floods. Flooding not only inundates the floodplain, including wetlands, it may remove trees and shrubs, deposit tree trunks and debris, and cover the floodplain with sediment (commonly silt) rich in mineral nutrients. River floodplain or riparian wetlands are elongate and often curved (oxbow wetlands).

In some cases, town residents remove the wetland vegetation and replace it with food-producing plants that thrive in wet soil. Cranberries are grown on acid peatland in Massachusetts (USA). Crustacean farming occurs in wet areas in Thailand and elsewhere in Southeast Asia. Fish farming is productive in numerous small ponds on wet floodplains of China. Most widespread is the cultivation of wet rice in paddies on floodplains throughout East, Southeast, and South Asia (Figure 6.7).

Water Flows, Soil, Chemicals

Normally *inflows* to a town wetland are precipitation from above, slow-moving groundwater from one side (Shedlock, 1993), and surface runoff from adjoining surfaces (sheet flow in heavy rain). Floodwater mainly enters a wetland by surface runoff and the town stormwater drainage system of pipes or ditches. The duration of soil inundation by floodwater, i.e., the *flooding hydroperiod*, largely determines the vegetation present. Wetlands may be semi-permanently flooded, seasonally flooded, or intermittently flooded (Mitsch and Gosselink, 2007). Changing wetland water levels changes mosquito and fly populations. It also alters access into a wetland by people and pets, and may improve or worsen the functioning of nearby septic systems.

Figure 6.8 Swamp wetland and boardwalk with Atlantic white cedar and peat moss (*Chamaecyparis thyoides, Sphagnum* spp.) dominant. Deep anaerobic acidic peat soil beneath. Wellfleet near Marconi Wireless Station, Cape Cod, Massachusetts, USA. R. Forman photo.

Outflows of water from a wetland are simpler. Evapotranspiration overwhelmingly by the vegetation pumps water upward, especially in sunny, dry, or windy conditions. Groundwater slowly carries out a small amount of water, and floodwater leaves as surface runoff downslope.

A wetland with more water flowing in and moving through has more oxygen in the water than does a wetland, such as a bog, with less flow (Lindsay, 1995; Richardson and Vepraskas, 2001). Most of the flow through a wetland occurs near the surface, so oxygen level decreases with soil depth (Figure 6.8). Typically, most of the soil volume is inundated, anaerobic (without oxygen), and acidic (low pH).

Wetland *peat* results from the accumulation of plant leaves and other organic matter in mainly water-saturated anaerobic conditions (Lindsay, 1995). Decomposition by anaerobic fungi and bacteria is slow. Dark brown peat generally has less than 33 percent of its organic material decomposed into unidentifiable material, while *black muck* has more than 67 percent unidentifiable material and the rest identifiable fibers (Mitsch and Gosselink, 2007). A long period of decomposition gives off the gases, hydrogen sulfide (H_2S), with a rotten-egg smell, and methane (CH_4), an odorless, strong greenhouse gas.

Mineral wetland soils, commonly clay-like (gley soil), are permanently or semi-permanently saturated (Richardson and Vepraskas, 2001; Mitsch and Gosselink, 2007).

These appear black, grey, or blue-gray due to the chemical reduction of iron and manganese at low pH (acid conditions).

Pollutant chemicals may enter a wetland in all the water inflows, i.e., precipitation, groundwater, surface runoff, subsurface runoff, and the stormwater drainage system. Air pollutants from town and regional industries, transportation, agricultural dust, and so forth pour in in rain.

Fortunately, wetlands have lots of processes to treat or clean the pollutant-rich water: (1) settling (of particles, including those with phosphorus attached); (2) filtration (of particles); (3) adsorption (attaching chemicals to particles); (4) absorption (uptake which removes chemicals from water); (5) assimilation (incorporation of chemicals into living tissue of growing organisms); and (6) decomposition (breakdown of organic material by fungi at lower pH, and bacteria at higher pH). Town wetlands clean the town, a valuable ecosystem service.

Habitats, Productivity, Species

Many wetland species have adapted to the perennially wet conditions and could survive or thrive nowhere else. Many other species in wetlands have a broader genetic tolerance, and can compete successfully both in the wet conditions and in drier conditions outside the wetland.

Wetland plants have several characteristic *adaptations* (Cronk and Fennessy, 2001), such as: shallow roots; woody structures (pneumatophores) projecting well above the water level that absorb O_2; and intertwined or connected root systems so trees are less likely to blow over. Little soil mounds, often around former stumps, provide slightly drier conditions benefitting some species. Mosses and lichens (epiphytes) on wetland trees grow in a continuously humid environment.

The stagnant or slow-moving surface water facilitates mosquito breeding. Bats and swallows that relish mosquitoes and flying insects circle around wetlands. Waterbirds including herons, egrets, rails, and some shorebirds find ample food in the shallow water of wetlands. Some species, such as turtles, may survive or thrive in residential areas by moving along wet drainage ditches and resting in culverts under roads (Rees *et al.*, 2009). Where wetlands are inherently scarce, or have been largely eliminated by filling or drainage, e.g., around large towns, they are key habitats for uncommon or rare species. For some mid-sized and large mammals, wetlands serve as escape cover, where they can escape from predators, dogs, and people (Landers *et al.*, 1979).

A study of 45 wetlands in the Beaverhills area of Alberta (Canada) found that "indices of biotic integrity" best correlated with surrounding land characteristics (including road cover and density, plus disturbed land) within 100 m of a wetland for plants, and within 500 m for birds (Rooney *et al.*, 2012). Large wetlands and wetland complexes have lots of habitat heterogeneity supporting a wide variety of species. However, habitat diversity in small wetlands usually relates mainly to the sequence or zonation of water levels, and to vegetation differences along a gradient from wetland border to center (Keddy, 2000).

Water level strongly affects the presence and productivity of plant species. If the water level drops for a prolonged period, species from surrounding upland areas tend to invade the wetland. Under normal fluctuating water conditions, plant productivity is low during dry conditions, and also low during wet conditions (Connor and Day, 1982). Productivity is highest during intermediate water conditions or with alternating seasonal dry-and-wet conditions (Golet *et al.*, 1993).

Severe drought may dramatically change the wetland soil and vegetation in quite a different way. Normally a moving fire burns slowly or goes out when entering a wetland. But in a severe drought, the water level or water table drops far below the wetland surface, leaving a thick layer of dry peat (used as fuel in many far northern climes). Fire during a severe drought not only enters the wetland of dry vegetation, it burns deeply into the peat layer, which may continue burning for a long time. That kills the tree roots and hence the trees. When rains return and the water level rises, ecological succession begins the multi-year process of wetland vegetation return. This is a familiar pattern in the New Jersey (USA) Pine Barrens, where cedar swamps (*Chamaecyparis*) on peat line the streams winding through a fire-prone land of dry pines and oaks (*Pinus*, *Quercus*) with scattered villages (Forman, 1979).

Wetland and Town Interactions

A small town sometimes builds around wetlands with unstable sinking soil, and gradually fills in the wetland edge for constructing roads and buildings (Figure 6.9). Large towns, however, have more funds and are short on space near the town center. Constant mosquitoes are considered unsuitable for a large town, and may carry diseases. So, wetlands are more likely to be filled and/or drained. Also, farmers outside town have often drained most of their wetlands to create plowable land for crops (Millennium Ecosystem Assessment, 2005). Sometimes the center of a wetland is dug out to create a pond for views or recreation. Removing wetland vegetation to produce food, such as rice, fish, crustaceans, or cranberries, essentially eliminates the wetland and its natural values.

Town residents may value wetlands both for absorbing excess stormwater, and for absorbing and treating stormwater pollutants. In town, dumping rubbish in a wetland is generally unappreciated, but the same wetland on the town edge or just beyond town may be welcomed as a convenient dump site. Yet that puts in lots of toxic chemicals which spread in soil and water.

During high winds, a lakeside wetland reduces wave energy, minimizing damage in town. On the other hand, during high rainfall periods, a town area next to a wetland may readily flood, since a saturated-soil wetland can no longer absorb rainwater. If a dam is built by beaver (*Castor*) or people just upslope or upstream, the wetland water level may drop, drying out the upper soil and roots. If the dam is just downslope or downstream, the wetland water level rises, again altering the vegetation.

Some towns protect their wetlands as valuable nature reserves or for nature-based recreation (e.g., Concord, Swakopmund). These are the only places where certain species of interest live. A raised curving boardwalk may be built through a portion of the wetland that does not threaten the species to be protected (Figure 6.8). Urban wetlands may have higher bird abundance and species richness than nearby upland sites, but in rural areas the relative value for birds is less clear (McKinney *et al.*, 2011). Restoring

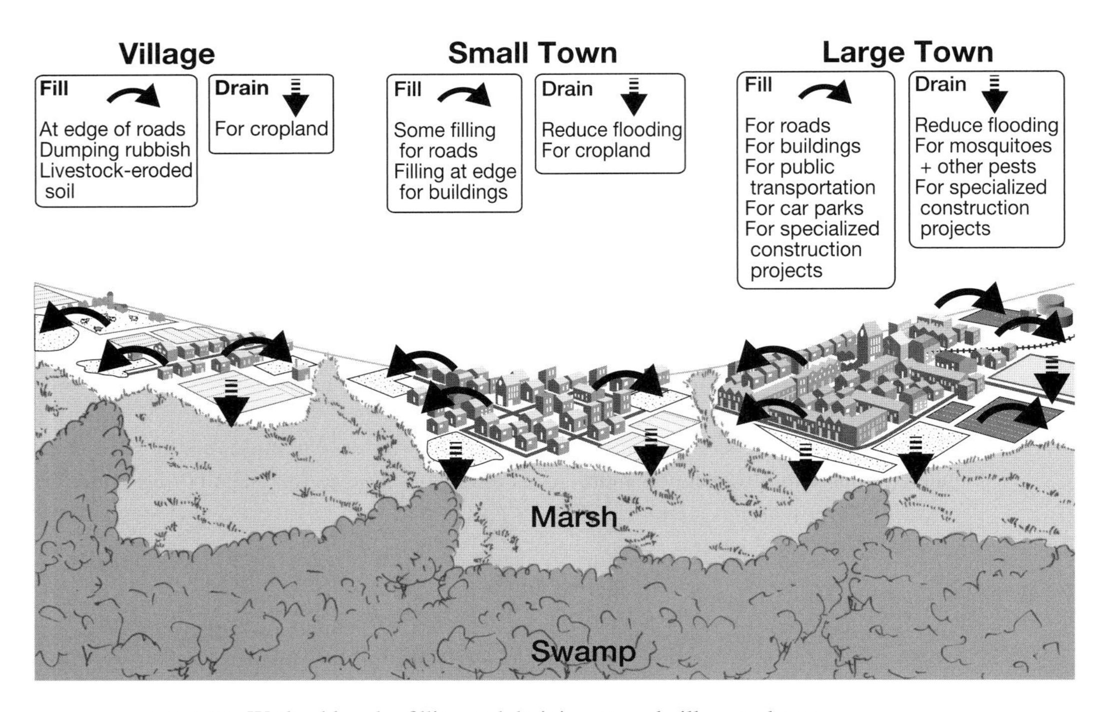

Figure 6.9 Wetland loss by filling and draining around village and town.

wetlands in farmland where they were drained results in rapid recovery of both invertebrate and avian biodiversity (Eglington *et al.*, 2008; Thiere *et al.*, 2009).

Normally people, pets, and livestock can only move slowly through a wetland, which usually keeps them near the edge and permits native wildlife to readily escape into the center. Livestock in a wetland browse out much of the low vegetation, and trample or compact the upper soil layer. Nature-focused people often love the challenge of moving through a wetland and discovering seldom-seen species. In contrast, most other people in a wetland are annoyed by pests, and worried about dangerous reptiles, deep muck, inability to move rapidly, and the unfamiliarity of everything.

Finally, *wetlands and global climate change* seem to be strongly linked (Mitsch and Gosselink, 2007). Wetlands remove the greenhouse gas, CO_2, from the atmosphere by plant photosynthesis, and sequester or store carbon in the plant tissue and particularly in the soil. Natural wetlands store an enormous amount of carbon in the saturated soil, and sometimes more in trees. Converting swamp vegetation to cultivated fields effectively gives off CO_2, a key greenhouse gas. However, draining a wetland effectively transforms its dense carbon-storage soil layer to much more CO_2 in the atmosphere.

Three other greenhouse gases are given off to the air by wetlands. Water vapor is not considered significant, considering the huge supply in the atmosphere. Nitrous oxide, N_2O, given off (from nitrification and particularly denitrification) is also not considered to be significant globally. But methane, CH_4, emissions from wetlands may amount to 20–25 percent of the global methane emissions. Furthermore, methane is more than 20 times worse than CO_2 for greenhouse gas effects. Most of the CH_4 emission comes from peatlands and freshwater marshes.

Ponds, Lakes, and Towns

Numerous types of ponds and pools can be found in towns. These are grouped into three contrasting categories: (1) freshwater ponds; (2) temporary ponds; and (3) constructed basins, ponds, pools, and wetlands. The final section explores (4) lakes.

Freshwater Ponds

A *pond* is a small surface waterbody with little water flow, and shallow enough for aquatic plants to grow across much of its muddy or silty bottom. Temporary ponds frequently dry out during dry periods, usually either seasonally or every few years (see below). Ponds are generally less than 1 or less than 0.1 km^2 in area (Kalff, 2002), and much more common than lakes (see Figure 5.4).

Ponds in and around towns may vary from small to large, and from rounded to highly convoluted in shape. Most are connected to a stream or river. This contrasts with ponds in urban areas which are generally moderate sized, little convoluted in shape, and often disconnected from stream or river (Steele and Hofferman, 2014). The urban pond patterns apparently result from society's removal, addition, physical reshaping, and locations targeted for development. These processes probably characterize large growing towns.

Town-area ponds may have an *inlet* where a stream enters, and an *outlet* where a stream leaves a pond. If no outlet is present, the pond water is probably sustained by groundwater (e.g., Lake Placid, Concord), though in areas with year-round rainfall, surface and subsurface runoff could support a pond. If no inlet is present, groundwater and/or runoff supports the pond water level.

If light reaches the entire bottom of a natural pond, photosynthesizing aquatic plants may cover the bottom. Various worms and other invertebrates live in the bottom sediment, which typically contains oxygen provided in part by plant photosynthesis. Another result of a pond being shallow is that winds create waves which effectively mix the whole water volume, a process that also provides oxygen to the bottom area (Figure 6.10). The mixing by waves also means that, unlike many lakes, a strong stratification of water temperature from top to bottom (and associated lake turnovers) ordinarily does not occur in ponds.

Normally the depth of water gradually increases from pond margin inward. This supports a *littoral zone* with *pond zonation* as a sequence of semi-aquatic and aquatic vegetation (e.g., Lake Placid, Petersham). Nearest the edge are emergent plants with roots in the pond bottom and stems extending above the water surface. Proceeding inward, next are floating aquatic plants with roots in the bottom and leaves floating on the surface. Deeper water has submerged aquatic plants that do not reach the water surface. The diversity of aquatic animal species is particularly high in this littoral zonation near the pond edge.

Figure 6.10 Former farmhouse by a pond enlarged with dam to provide power for a series of mills on stream. Could the photo be upside down? Harrisville, New Hampshire, USA. R. Forman photo. (A black and white version of this figure will appear in some formats. For the color version, please refer to the plate section.)

Where the land surface gradually slopes down to the pond edge, often some or all of a natural pond is surrounded by a narrow zone of wetland (rather than aquatic) vegetation, such as marsh or swamp (e.g., Petersham). Organic matter falls or is washed into the pond which promotes a rich aquatic food web. Protecting a town pond as a nature reserve means also protecting the ring of wetland vegetation. Encroachment by development often includes dumping, thus filling ponds from the edge (Figure 6.9).

The ecological condition of a pond is readily estimated by land uses around it. Residential housing by the shore often means that pond-side vegetation and emergent aquatic vegetation have been largely removed, the shore area is transformed for recreation activities, and wastewater may enter and pollute the pond. Vehicles bring pollutants and recreationists. Boats bring motors and fishermen with related impacts. Pipes and ditches bring stormwater and pollutants. Other uses targeting ponds include town wells next to ponds, town fire trucks pumping water from ponds in emergencies, ice skating in cold wintertime, and ponds as beautiful/aesthetic places attracting people.

Imagine a 20-hectare (50-acre) Florida pond with houses snuggled up forming a ring around the pond. Memorable events never end (R. Pickert, 2013 and 2018 personal communications):

- Early residents dug out a sinkhole wetland to create the pond. With water over your head near the shore and less than a meter deep in the center, imagine the sight in a severe drought.
- When alligators grow to 4 m (12 ft), a wildlife official comes to remove them, while residents watch. Then dogs and children are safer, and fish populations regrow.
- Later officials decided to dig a canal for small boats connecting the pond to a nearby lake. But then someone noticed the shallow-pond water level was a foot or two higher than that of the lake.
- Residents with big pontoon boats perform the ritual of circumnavigating the little pond clockwise, then counterclockwise, and finally triumphantly across. Then bored, they sell the boat to the next new neighbor.
- Fishermen occasionally arrive with a small boat to catch small fish. But the boat sometimes brings invasive plants, which clog the pond, reducing target fish populations.
- Occasionally a new house with septic system in the sandy soil is built. Phosphorus flows into the pond, stimulating phytoplankton growth and thus murky water.
- One hurricane-like storm catalyzed a huge invasion of walking catfish over everyone's yard. Neighbors attacked the invading army with garden tools, which resulted in weeks of summer heat bathed in a strong putrid odor.

Inputs to a pond may add house-plot and lawn chemicals, stormwater runoff chemicals, agricultural chemicals, industrial chemicals, and particulate matter from eroded and other surfaces. Many of the chemicals added are toxic to aquatic organisms or degrade the aquatic ecosystem in varied ways. A solid waste dump by a pond contaminates the pond long term. Adding particulate matter to a pond makes turbid water, which decreases light penetration and degrades fish populations. Progressively, sedimentation occurs and the shallow pond gradually fills with sediment.

Particulate matter entering a pond commonly also carries in phosphorus adhering to the particles. Most phosphorus may accumulate in the bottom sediments, where it might enhance growth of the rooted aquatic plants. But some phosphorus is absorbed by the phytoplankton in the water. The phytoplankton rapidly reproduce turning the pond water green, an example of *eutrophication*, i.e., nutrient enrichment catalyzing rapid algae growth (see Figure 5.11).

Intense eutrophication leads to a more serious ecological effect. As phytoplankton multiply, the dense green upper water shades out the water below, eliminating photosynthesis and oxygen production. The lower phytoplankton cells die and filter downward. Bacteria in enormous numbers decompose the dead algae cells moving downward. These aerobic bacteria use oxygen to decompose the cells, and the rapid bacterial growth quickly uses up all the oxygen. The bottom portion of the pond becomes anaerobic, and all the fish and other aquatic organisms requiring oxygen die. The pond is in bad shape because of that added phosphorus. Furthermore, many chemicals added are toxic to pond organisms.

Phytoplankton is composed of numerous algae species, though often dominated by a few green algae and some blue-green algae species. The aquatic plants rooted in the bottom, and appearing in zonation near the edge, usually include a fair number of species. A small shallow low-flow waterbody is an unusual habitat and composed of many microhabitats. Not surprisingly, a pond contains many locally scarce species and is often of conservation importance.

Animals in the bottom sediment include (benthic) worms and other invertebrates. Aquatic and semi-aquatic invertebrates are normally in huge abundance and rich diversity. They are not evenly or randomly intermixed, but rather well sorted out by microhabitat, such as at different levels in the water and associated with different plants from pond edge to center. Various beetles, flies, and tiny crustaceans may be particularly common. Like the invertebrates, fish species sort out by microhabitat. Fish species richness is often modest, though fish abundance may be quite high in a moderately eutrophic pond.

Terrestrial vertebrates from the surrounding land are strongly attracted to a pond. Many amphibians benefit from the proximity of water and land, while many reptiles similarly spend considerable time in water and readily move onto land. Small to large mammals are all attracted to ponds. Birds wade or dive for fish, amphibians, or reptiles, while others mainly relish the concentration of aquatic invertebrates. Still other species feed on flying insects around the pond.

Change is the norm in ponds. Sunlight and water temperature change cyclically, both daily and seasonally. Stormwater runoff and pond water level also change seasonally, but in quite irregular spurts and drops and vary from year to year. Human uses of the pond, such as for recreation, change diurnally, weekly (e.g., especially weekend use), seasonally, and over years.

The lifetime of a pond in a town may be determined by the gradual buildup of sediments, pollutants, and accumulations from added nutrients and plant/phytoplankton growth. Encroachment of development, both roads and buildings, accelerates pond shrinkage, perhaps a prevalent pattern in towns. Occasionally a whole pond is filled in town and covered over by another land use. In surrounding agricultural land, pond removal may be by filling or by drainage. Being small, shallow, and with little flow makes ponds fragile, readily subject to gradual degradation or abrupt destruction.

Numerous pond types exist, the following being of particular interest in and around towns (Forman, 2014):

1. Oxbow pond, resulting from a former curved channel on the floodplain of a meandering stream or river.
2. Kettle pond, originating from melted blocks of glacial ice (e.g., Walden Pond on the edge of Concord; see Appendix).
3. Pond with rock or clay bottom (e.g., Stonington).
4. Blocked-drainage (or blocked–valley) pond, such as where a main stream fills with sediment blocking a tributary flow.
5. Karst pond, which fills sinkholes in limestone terrain with a moist climate (e.g., Valladolid, Lake Placid).

6. Beaver pond, created by beaver damming a stream (e.g., Petersham, Concord).
7. Pond in former gravel pit, often deep, with steep banks, acidic, and with low biodiversity.
8. Aquaculture pond using wastewater (Costa-Pierce *et al.*, 2005).
9. Constructed ponds for diverse purposes (see below).

Temporary Ponds

Temporary ponds and pools normally are small shallow waterbodies with no inlet or outlet that regularly dry out during dry periods, resulting in no established fish population (Colburn, 2004). Sometimes called intermittent, ephemeral, or vernal, most ponds are seasonal. They fill during the wet season and dry out in the dry season. Temporary ponds are present worldwide (Crosetti and Margaritora, 1987; Ross, 1987; Lake *et al.*, 1989; Taylor *et al.*, 1990; Mahoney *et al.*, 1990; Serrano and Toja, 1995; Bazzanti *et al.*, 1996).

With no inlet, temporary ponds fill by surface runoff and/or subsurface runoff. Or in sandy soils, the ponds are typically in equilibrium with groundwater, with pond levels rising and falling as the water table changes (Serrano and Tonja, 1995). Puddles and pools produced by individual storm events or occasional rainy periods do not have regular alternating wet and dry phases, and generally are not categorized as temporary ponds.

Different types of temporary ponds occur in and around towns (Colburn, 2004). Some occur in rock depressions, while others occur near streams which overflow their banks, filling the ponds. Temporary kettle ponds originated where blocks of ice from a glacier persisted in the soil. Some temporary ponds only dry out every few years (Colburn, 2004).

Species in temporary ponds are unusual by thriving in extremely wide environmental fluctuations. The species must survive both inundation and desiccation, repeatedly occurring. Consequently, many uncommon or rare species may be present in temporary ponds, and often many of the animals present are not found in other types of waterbodies. Consequently, some temporary ponds (vernal pools) are a conservation priority.

Temporary pond plants use the wet-period water to grow, though roots in the pond-bottom soil may have limited oxygen available (Ross, 1987). Reproduction, such as by flowers, fruits, and seeds, occurs rapidly before the severe dried-out conditions arrive. The length of the water or inundation period (hydroperiod) varies from pond to pond and year to year (Eyre *et al.*, 1992). Since some of the species are better adapted to inundation and others to dry conditions, the species typically sort out along a gradient from pond edge to pond center (Welborn *et al.*, 1996). Also, a successional sequence of species, especially invertebrates, occurs with pond changes from wet to dry conditions, and vice versa (Lake *et al.*, 1989; Eyre *et al.*, 1992; Bazzanti *et al.*, 1996).

In contrast, animals may move to escape a stress period. Because of the dry period, temporary ponds are basically fishless. Without fish predators present, many amphibians and invertebrates can survive and thrive (Semlitsch, 1987; Welborn *et al.*, 1996). Some amphibians and reptiles successfully reproduce during the inundation period, and then bury themselves in the soil or migrate elsewhere during dry conditions

(Semlitsch, 1987; Ireland, 1989; Massey, 1990; Taylor *et al.*, 1990). The amphibians and reptiles that migrate, often hundreds of meters or more than 1 km, generally move along moist ditches and through culverts to another pond or wetland. Blockages by roads or development are serious, and many animals are killed by vehicles on roads. Road density near a temporary pond especially affects species that move across the land (Veysey *et al.*, 2011).

Invertebrates thriving in ponds without fish in the northeastern USA include many species of micro-crustaceans (Colburn, 2004). Also abundant are aquatic insects, including water beetles (Coleoptera), true flies (Diptera), water bugs (Hemiptera), and caddisflies (Tricoptera). Temporary ponds in Australia, southern Europe, and the southeastern USA contain many micro-crustacean species, especially cladocerans and copepods, which have diverse strategies for surviving the dry season (Crosetti and Margaritora, 1987; Mahoney *et al.*, 1990; Taylor *et al.*, 1990). In California ponds (vernal pools) and Carolina Bays (southeastern USA), unique plant communities are present containing many endemic species (found nowhere else) (Ross, 1987). In contrast, vernal ponds in the northeastern USA have mainly typical wetland plant species. The dry phase of temporary ponds not only eliminates fish, but also many beetles (Coleoptera), dragonflies (Odonata), and other insect groups (Colburn, 2004).

Not surprisingly, the area surrounding a temporary pond strongly affects pond conditions and the species present (deMaynadier and Hunter, 1995). Trees around the edge shade the pond keeping water temperature down (Colburn, 2004). Trees provide abundant leaf organic matter as the base of a pond's mainly detritus food web, and also drop branches providing microhabitat heterogeneity. In addition, trees pump considerable water upward, which lowers the pond level. Amphibians and some reptiles often use the adjoining habitat as part of their life cycle.

Being small and shallow, temporary ponds are easily filled by society, which causes pond extinction and loss of species. A survey of temporary ponds in and around a town is warranted, and the most valuable ones, along with their surrounding area and connections, warrant protection. For example, within the town of Concord (see Appendix) are 2 state-certified vernal pools (temporary ponds) and approximately 12 others (see Figure 7.14). In the area surrounding the town are 32 certified and more than 100 other vernal pools.

Town residents and farmers also dig large holes to create temporary ponds for diverse uses. Farmers may pump water into the depression, often near the farmstead, to provide water for livestock in dry periods. Temporary ponds may provide water for irrigation, though streams, rivers, and groundwater are the main sources for irrigation (Figure 6.11).

Constructed Basins, Ponds, Pools, Wetlands

Small *constructed basins* or depressions with ponds, pools, or wetlands provide many benefits to society (Zhang *et al.*, 2009; Zhi and Ji, 2012; Forman, 2014). These structures are particularly appropriate for towns, which often have ample space, adequate water, and limited funds. Almost all constructed basin types have a shallow edge for safety, at least a narrow zone of aquatic or semi-aquatic plants, and relatively steep internal sides. Some water flows through the ponds and wetlands to keep them oxygenated

Figure 6.11 A water-irrigated terraced landscape providing food for the former Inca community of Machu Picchu high in the Andes. Water entered from the forest at top in three streams/gullies; every terrace had not too much and not too little water. Peru. Photo by Steve Bennett (https://commons.wikimedia.org/wiki/File:Machu_Picchu_overview_Stevage.jpg), "Machu Picchu overview Stevage," https://creativecommons.org/licenses/by-sa/3.0/legalcode. (A black and white version of this figure will appear in some formats. For the color version, please refer to the plate section.)

(Lavielle and Petterson, 2007; Vollertsen *et al.*, 2007). Most common are various types of structures which treat or clean up pollutants to improve water quality (Mitsch and Gosselink, 2007; Davis *et al.*, 2009; Hurley and Forman, 2011). But first we introduce constructed food-producing ponds and detention ponds for reducing flooding.

Food-producing or *aquaculture ponds* are commonly constructed for the high production of phytoplankton-feeding fish, especially in China and Southeast Asia. The ponds are usually built on floodplains where a high water table sustains water levels. Also, the ponds are fertilized, for example with animal wastes, which provides phosphorus and nitrogen supporting high production. Fish ponds are perhaps most common, though ponds for harvesting rooted or floating plants also occur. *Wastewater aquaculture ponds* use human waste or wastewater as fertilizer for food production (Costa-Pierce *et al.*, 2005).

Detention basins or ponds (or catch basins) are built to temporarily hold stormwater, and release it slowly to flow onward toward a local waterbody (e.g., Lake Placid,

Socorro) (Mallin *et al.*, 2002). This process reduces peak flows (cuts off the tops of hydrographs; see above), thus reducing flooding. Very little evaporation or infiltration into soil occurs. If the water table is low, the detention pond often dries out between storms, whereas a high water table tends to keep water in the pond. Detention basins are commonly built along the lower side of a car park. In effect, detention basins slow water runoff, reduce flooding, but do little to treat or clean up water pollutants.

Bioretention or treatment structures reduce flooding, but are primarily designed to clean polluted water (e.g., Sandviken) (Davis *et al.*, 2003; Hsieh and Davis, 2005; Knight, 2008; Davis *et al.*, 2009; Kazemi *et al.*, 2009). If the basin is above the water table, typically the bottom is a clay or other nearly impervious layer, on which a more porous soil mix is added (Campbell and Ogden, 1999). Nevertheless, bioretention basins, ponds, and wetlands are usually not used on chemically contaminated brownfields, because flowing water might carry chemical pollutants to a waterbody (but see biofilters in Hurley and Forman, 2011; Forman, 2014).

A key is to get the hydrology right, i.e., the incoming water and the basin water level. Water is evapotranspired upward, infiltrates downward, and flows onward. *Pollutants are treated* (cleaned) or removed by several processes (Braskerud *et al.*, 2005; Dunnett and Clayden, 2007; Hogan and Welbridge, 2007): settling, filtration, assimilation, adsorption, and degradation/decomposition. Typically, plants and animals rapidly colonize a bioretention basin if water conditions are suitable.

Subsurface-flow structures have no standing surface water, but instead water flows through porous material, such as sand or gravel, in the root zone (Vymazal *et al.*, 1998, 2006; Rousseau *et al.*, 2004; Mitsch and Gosselink, 2007; Saeed and Sun, 2012). These structures require little space, often have emergent wetland plants (e.g., common reed: *Phragmites*), often have a shorter lifetime, and are widely used in Europe. In contrast, surface-flow structures have surface water, typically with emergent (e.g., cattail, common reed: *Typha, Phragmites*), submerged, and floating plants (water hyacinth, duckweed: *Eichhornia, Lemna*), and have low-permeability soil beneath (White *et al.*, 2000; Day *et al.*, 2004). Surface-flow systems better mimic natural wetlands and are common in North America.

In ponds and wetland, suspended solids (particulates) with attached material such as phosphorus settle to the bottom. Aerobic bacteria with oxygen decompose organic matter, while anaerobic bacteria near the bottom do the same at a much slower rate. Other microbes oxidize inorganic compounds such as ammonia-nitrogen. The soil filters and absorbs many pollutants. Basically, bioretention basins are the flagship cleaning structures, and require relatively little maintenance.

Excluding park ponds built for aesthetics and/or recreation (e.g., Capellades), the types of treatment ponds, pools, and wetlands may be differentiated by the source of water pollution (Mitsch and Gosselink, 2007; Forman, 2014): town stormwater runoff, agricultural runoff, human wastewater, livestock wastewater, dump leachate, industrial wastewater, mine drainage, and river/stream pollution.

1. *Town and road stormwater runoff.* Averages of many studies suggest that stormwater treatment structures are most useful in retaining suspended solids, such as sediments

from construction sites (75% retention) and heavy metals (75% lead, 50% zinc) (Mitsch and Gosselink, 2007). Such structures retain lower amounts of nutrients and carbon (25% total nitrogen, 45% total phosphorus, 15% organic carbon). Bacteria numbers are also reduced. Severe storms tend to reduce basin efficiency and increase nutrient release. A concentration of some stormwater pollutants in the basin may stress the plants present, reducing treatment effectiveness. Deep ponds by the structure's entrance and exit, plus semi-separate sections of marsh may be most effective. *Swales* are wide shallow ditches covered with grass or other herbaceous plants that reduce stormwater runoff and partially clean the water (Thompson and Solvig, 2000). Street swales function alongside roads in town (Dunnett and Clayden, 2007; Hill, 2009). *Rain gardens* are tiny vegetated depressions designed to absorb stormwater, and thus are often important in house plots (Dunnett and Clayden, 2007; Dietz and Clausen, 2008).

2. *Agricultural runoff*. Sediment, nitrogen, phosphorus, and pesticide residues are the main pollutants (Moustafa *et al.*, 1996; Reddy *et al.*, 2006; Budd *et al.*, 2009). A constructed wetland in rural Ohio (USA) retained 59 percent of total phosphorus and 40 percent of nitrate/nitrite-nitrogen, but major storm events reduced the numbers considerably (Fink and Mitsch, 2004). Also, see the preceding runoff results.
3. *Human wastewater*. Constructed wetlands are particularly effective in controlling organic matter (BOD), suspended sediments, and nutrients (especially phosphorus and nitrogen) in basin wastewater (e.g., Emporia) (Vymazal, 1998, 2006, 2010). Heavy metals and other toxic materials are also often removed by the basin system, but may concentrate to high levels in the sediment or wetland fauna.
4. *Livestock wastewater*. Confined and concentrated farm animals, including cattle, dairy cows, pigs, and chickens, produce exceptionally high levels of organic matter, nutrients, and bacteria. Constructed wetlands apparently noticeably reduce organic matter, total phosphorus, and fecal coliform bacteria (pathogens) (Newman *et al.*, 2000; Schaafsma *et al.*, 2000). Nitrogen concentrations tend to be very high, but the basin system effects on nitrogen are less clear.
5. *Dump (tip, landfill) leachate*. An impermeable layer beneath a dump collects any water that percolates through the solid waste. This water generally has high concentrations of ammonia-nitrogen and chemical oxygen demand, but may also have high concentrations of diverse pollutants from heavy metals to toxic organic compounds (Kadlec, 1999; Maconachie, 2007). Constructed wetlands would be a treatment option.
6. *Industrial wastewater*. Industrial wastes, as byproducts of manufacturing processes, are highly heterogeneous and often highly toxic. The amount of pollutants or contaminants is typically far greater than the treatment or cleaning capacity of a basin pond. Thus, the basin is basically a "holding pond" (e.g., Emporia). Continued pollution input plus some evaporation increases the pollution concentration. Eventually the pond water may be transported away for disposal by truck, or drained (often illegally) into a river during a flood stage.
7. *Mine drainage*. Constructed wetlands are an option for acid mine drainage with a low pH and elevated levels of iron, sulfate, aluminum, and heavy metals resulting

from coal mines (Mitsch and Gosselink, 2007). Cattail (*Typha*) seems to grow satisfactorily in such conditions, and cattail marshes seem to reduce the concentration of iron and sulfate, as well as raise the pH a bit (Wieder, 1989; Tarutis *et al.*, 1999). Hydrologic conditions are especially important on a mine site. Pollutant input (loading) rate may be too high and residence time of pollutants in the system too short, to treat the polluted area. The wetland might accumulate enough material to eventually be minable.

8. *River/stream pollution*. In a slightly different approach, polluted river water has been pumped for several years to a constructed wetland area, analogous to an oxbow or Australian billabong, in a floodplain in two US Midwestern states (Kadlec and Hey, 1994; Mitsch *et al.*, 1995, 2005). Both sediment and nutrient levels were significantly reduced. For instance, consistent 50 percent reductions in phosphorus and 30 percent reductions in nitrate + nitrite-nitrogen were recorded at one site.

A closer look at the key elements in constructing successful basins is helpful. Five components essentially create the system (Mitsch and Gosselink, 2007; Forman, 2014): hydrology, basin form, chemical inputs, soils, and vegetation.

In this case, *basin hydrology* mainly refers to: the hydro period (length of wet season or water input), rate of water input, and depth of water in the basin. A long dry season usually means that the surface water is temporary, and thus many treatment processes only operate for part of the year. An excessively low water input may result in the basin water becoming anaerobic, whereas an excessively high input usually means that many pollutants flow on through without being sufficiently cleaned up. The depth of water affects the kinds of plants present, the oxygen level in the lower portion, and some of the cleaning processes.

Basin form includes width of the shallow edge, which in turn determines the width of emergent vegetation and habitat for aquatic organisms. An optimal length-to-width ratio for the basin may be 2:1 to 3:1, or longer (Steiner and Freeman, 1989). A variety of deep and shallow areas seems optimal. A deep area, especially by the input, increases sedimentation, provides fish habitat, and helps create variable water velocity. Shallow areas support emergent vegetation and enhance soil–water chemical interactions.

Chemical inputs, if high, may inhibit growth or survival of the plants in the basin. An excessive input-and-output rate means that many pollutants simply pass through without being treated. Pollutants need to be retained in the system long enough to be adequately cleaned by basin processes. The efficiency of constructed basins is usually measured by the ability to remove nutrients (P and N), sediments, and organic carbon (BOD) (Mitsch and Gosselink, 2007). Yet wetland soils and organisms also absorb heavy metals, many of which bioaccumulate (increase in concentration) in the food chain.

The *constructed-basin soil* in surface-flow constructed wetlands is often 60–100 cm thick, sufficient for good root growth, and is underlain by a relatively impermeable liner such as a clay layer (Mitsch and Gosselink, 2007). Silty clay and loam soils facilitate plant growth as well as various treatment processes. Generally organic soils are avoided because of low nutrients, low pH, and inadequate support for rooted plants. Soils with high iron and aluminum concentrations are often favored due to their ability to absorb phosphorus. For subsurface-flow treatment systems, sand, gravel, or another permeable medium is preferred, generally 15–30 cm below the surface, to facilitate horizontal flow (Cooper and Hobson, 1989). This directs the flow of pollutants into the lower root zone rich in microorganisms important for cleaning the water. Over several years, constructed subsurface-flow wetlands often gradually become clogged, decreasing in water penetration and treatment effectiveness.

Constructed-basin vegetation, generally not on a floodplain, is preferably planted with genetic stock adapted to the local climate. Planting plants on wet soil or in less than 5 cm of water enhances the chance of establishment, and few rooted emergent aquatic plants thrive in more than 30 cm of water (Mitsch and Gosselink, 2007). Too much or too little water is the main cause of plant death. Species selected must be both tolerant of the pollutant input, and effective in facilitating treatment. Relatively few species prosper in high nutrient and high BOD wastewater. Cattails (*Typha*), bulrushes (*Scirpus, Schoenoplectus*), and reed grass (*Phragmites*) are widely used. Floating plants (e.g., water hyacinth, duckweed, water lettuce: *Eichhornia, Lemna, Pistia*) thrive in deeper water. A constructed wetland for dairy wastewater found three large grasses (*Zizania latifolia, Glyceria maxima, Phragmites australis*) to be most effective for treatment (Tanner, 1996). Also, the removal of total nitrogen linearly correlated with total plant biomass.

Several considerations require thought in planning and constructing basin systems (Mitsch and Gosselink, 2007). Try to mimic local natural ponds or wetlands. Choose a wetland size and shape to be effective for both reducing water flow and pollution concentration. Expect disturbances and change. Address public health issues. Arrange constructed basins to fit into and enhance the broader arrangement of waterbodies (Hurley and Forman, 2011). And consider additional benefits and values such as wildlife habitat and aesthetics.

After constructing basin systems for treating polluted water, management tends to be minimal (Kadlec and Knight, 1996). Trash is periodically removed. The basins may be located to attract wildlife or serve as stepping stones for movement (Le Viol *et al.*, 2009). Wildlife effects on treatment efficiency may need monitoring (Anderson *et al.*, 2003). Mosquitoes may need to be controlled (Knight *et al.*, 2003). Other ecological and human values may need attention.

Lakes

Lakes generally have a number of inlets (stream mouths) and an outlet where a stream/river leaves the lake. Lakeshore towns and villages at an inlet or outlet normally have a bridge over the stream, plus ample fish and good fishing (e.g., Falun, Issaquah). An inlet town's pollutants tend to affect a small shallow portion of the lake, whereas an outlet town especially affects the initial portion of the outflowing stream/river.

Inlets, where most lakeside communities are found, are normally in coves of the lake (e.g., Rangeley). The surrounding land slopes down to the lake amphitheater-like. Often extending down the slope are a stream, gullies with intermittent flows, and ditches, pipes, and roads. In general, the gradual slope around a cove is more buildable than on

Figure 6.12 Students poling and paddling in huge Tonle Sap lake. Lakeside town, central Cambodia. Photo courtesy and with permission of Mark Brenner.

a point or peninsula. The cove provides some protection against storm waves for docks, boats, and shoreline buildings. Also, the presence of a beach and recreational activities is more likely (Figure 6.12). A wetland is more likely in a cove, yet a wetland in town is often gradually filled to minimize mosquitoes and provide more usable space.

Downward-flowing stormwater on the slope tends to be concentrated in the town area, sometimes creating a flood risk. If the town is mainly on sandy or rocky soil, the water table may be at about the same level as the lake surface (e.g., Lake Placid). But if the town soil is largely clayey or silty clay, the water table will be higher. That suggests more flooding and difficulties with septic systems.

If a mainly vegetation-covered hill or mountain slope adjoins the town, cool air drainage is likely (Chapter 5). Because the large waterbody heats up more slowly than the land in spring, and cools more slowly in autumn, the town's air is cool in spring and warm in autumn (e.g., Rangeley).

The water supply of a village or small town may be piped from the lake, a well, or a stream upslope. A large town is likely to have a small upslope reservoir on a stream. A solid-waste dump is also usually far enough upslope so that chemicals do not wash down and pollute the waterfront.

Human waste or wastewater is a particular problem because it could rapidly pollute the lake water, at least with phosphorus, which turns the water green with phytoplankton (e.g., Lake Placid). On-site waste/wastewater approaches, including septic systems, might only work effectively for some buildings away from the shoreline (e.g., Grande Riviere). In contrast, a sewage treatment system set well back on the slope requires pumping wastewater upslope, and might still pollute the cove unless tertiary treatment (see above) were included. Visit a lakeside town and ponder its human waste/wastewater solutions.

Stormwater runoff from slope and town reaches the cove's lake-water. Unfortunately, enroute the runoff picks up materials and chemicals from construction sites, road maintenance strips, logging activities, farmland, vegetable gardens, flower gardens, and lawns that cause water pollution. Constructed bioretention basins (see above), especially wetlands, alongside roads and on the downslope sides of fields and gardens reduce stormwater pollutant concentrations.

A study of 204 Danish lakes found that the total amount of phosphorus and of nitrogen in lake water strongly correlated with the percent of agricultural land in the drainage basin (catchment/watershed) (Nielsen *et al.*, 2012). Even more interestingly, percent agricultural land in the whole watershed is a better predictor of lake nutrient levels than is percent agricultural land in zones closer to the lake.

In addition, towns usually contain small- or medium-sized industries which produce economic products, plus polluting or toxic byproducts. If the industries are well back from the lake, pollution problems are reduced. Air pollutants may affect the town and lake. However, stormwater runoff, picking up industrial pollutants, funnels them to the town's cove (e.g., Sandviken, Falun).

An array of pollutants may reach the cove water. The mineral nutrient phosphorus (and to a lesser extent nitrogen) normally catalyzes a green bloom of phytoplankton (Kalff, 2002). Organic matter catalyzes a bloom of bacteria decomposers. Zooplankton feed on both phytoplankton and bacteria. But most of these microbes die and slowly filter downward in the water. Bacteria then explode in numbers, decomposing the dead cells. The bacterial decomposition uses up all the oxygen, which effectively kills the fish and other aquatic organisms. The cove ends up with a dense green upper layer and nearly a dead zone beneath.

To make things worse for the town's cove water, some of the stormwater pollutants, such as pesticides and various industrial chemicals, are toxic (Kalff, 2002). Consider the inorganic chemicals, arsenic, mercury, lead, and cadmium, or organic substances, dioxins, PCBs, DDT, and chlorinated organics. Toxicity for a species may be acute (rapid) or chronic (slow), the substance may be persistent, or it may volatilize to the air. Furthermore, some chemicals bioaccumulate in the food chain, becoming more concentrated in, and killing, predators.

A lake with lots of lobes and coves normally has an extensive littoral zone, especially in coves. The *littoral zone* is the shallow-water lake edge, which is particularly important for rooted aquatic vegetation and associated habitat for numerous aquatic animals, including small fish. Many waterbirds, amphibians, and reptiles thrive in the littoral zone. Small fish there escape the jaws of large fish.

Commonly a road goes along or close to the shoreline. Residential houses usually extend a considerable distance along the lakeshore in both directions from town (Figure 6.13). Except where natural vegetation is retained or protected, this line of

Figure 6.13 Multi-level houses of lakeside town in the dry season. Many rivers and streams flow into the large Tonle Sap Lake, so during the monsoon season water level rises several meters and doubles the lake size. Central Cambodia. Photo courtesy and with permission of Mark Brenner. (A black and white version of this figure will appear in some formats. For the color version, please refer to the plate section.)

house plots has removed the lakeside vegetation and pours chemicals into the littoral zone (e.g., Lake Placid). Lower fish densities and diversity are characteristic, mainly due to less lakeshore habitat heterogeneity and fewer "fish refuges," in addition to increased bank erosion and phosphorus concentration (Hickley *et al.*, 2004; Scheuerell and Schindler, 2004; Elliott *et al.*, 2007). Lake water opposite protected natural vegetation may be considerably clearer than opposite lakeshore houses (e.g., Lake Placid).

The town center with its commercial and residential activity is normally on the lakeshore (e.g., Sandviken). A waterfront has a concentration of docks, boats of different sizes, fuel spills, impervious surfaces, stormwater pipe outflows (Royal Commission, 1991), transfer and shipping of goods, recreational activities, restaurants with food and waste, noise, and of course people. Some of the waterfront lakeshore is usually *armored*, that is, lined with large rocks or other hard materials to minimize lakeshore erosion by storm waves.

As a result of these diverse activities, normally the littoral zone around the waterfront is severely degraded. Still, docks and other structures provide new habitats for different species. Birds benefitting from human activities, such as gulls, cormorants, and ducks, are often conspicuous. Lake water pollutants can be blown into town by a storm, or even carried short distances inland by arthropods (Raikow *et al.*, 2011).

Boating and fishing are the main ways people on the lake affect its ecological condition (e.g., Rangeley). Overfishing of a particular species not only reduces its population, but alters the populations of other species in its food web, some increasing, some decreasing. The effects of fishing on fish populations apparently decreases with both fewer anglers and distance from town (Hunt *et al.*, 2011). The use of lead weights on fishing lines may lead to lead poisoning of fish or predatory birds. Boats and motors can degrade a littoral zone or particular fish habitat. Fishers sometimes introduce new or nonnative species to a lake, and boats transported from one lake to another can introduce nonnative aquatic weeds (e.g., Lake Placid). Both large and small boats may cause fuel spills and produce considerable noise.

Streams, Rivers, and Towns

Streams

Since towns and villages are often located by streams, we explore the subject from several dimensions: (1) regional differences, town locations and layout; (2) roads, bridges, culverts; (3) reservoirs, pollutants, flooding; (4) stream corridor and stream.

Regional Differences, Town Locations and Layout

In forest and woodland landscapes, small streams (first to ca. third-order) generally have rather fast water flow, and are shaded by a cover of tree canopies (e.g., Sooke, Rumford/Mexico). The detritus of dead leaves and twigs, rather than photosynthesis, is the base of the aquatic food web. Dead branches and logs fall into the stream producing considerable microhabitat heterogeneity, including deep holes with fish, some big.

Towns are usually located on somewhat larger slightly curving streams (third to ca. fifth-order), which have less rapid water flow, and a bottom covered mainly with stones and sand, plus perhaps patches of silt. Over time the stream may migrate back and forth across a narrow floodplain. Mainly open overhead, the stream has ample light to support aquatic plants growing along streamsides. Downstream, still larger streams and rivers are meandering, with slow-moving water, considerable phytoplankton, and smooth bottoms mainly covered with silt.

In cropland landscapes, commonly with silt or loamy soils, extensive erosion from plowing, parallel channels, and rain is characteristic, leading to extensive sedimentation of silt in streams (e.g., Chamusca, Clay Center). Floods are common and leave a blanket of silt and nutrients over the floodplain. Towns are normally located on the streambank area and may still get flooded occasionally. The stream is often straightened or channelized, and even with a narrow strip of streambank trees, is mainly open to the sun. Town centers are especially flood-prone, because of high water table, stream channelization, stream narrowing, concentrated impervious surface, and stormwater from upslope development.

Prolonged drought creates low-flow conditions, which especially degrade populations of plants, fish, and *stream-bottom* (benthic) *macroinvertebrates* (large invertebrates important, in part, as food for fish) (Lake, 2011). Stream-bottom invertebrates are reduced by the effects of development, except in farmland where they are already reduced (Cuffney *et al.*, 2010; Walters and Post, 2011). Different agricultural insecticides affecting these stream-bottom species are used in different regions of the USA (Wohl, 2004).

In desert and dry pastureland, streams are invisible most of the time, as water flows slowly through the sand in gullies (e.g., Swakopmund, Alice Springs). Only visible is a curving strip of sand with scattered trees. The water tends to be somewhat salty and rich in heavy metals. Riparian vegetation on the floodplain is highly sensitive to soil texture and salinity (Briggs, 1996; Decamps and Decamps, 2001; Beauchamp and Shafroth, 2011). Surface flow from rains is intermittent, with periodic floods (Walters and Post, 2011). *Flash floods* occur, the sudden big water flows, often from rain elsewhere at a higher elevation (e.g., Alice Springs, Socorro). Thus, towns in such areas are located on higher surfaces, often well back from a stream channel.

Towns normally appear near good agricultural soil, by a stream floodplain, on a high place unlikely be flooded or eroded, and near a place such as a narrows where vehicles can cross the stream. Floods reaching the town are believed to be infrequent. The high water table near the stream facilitates food production, though erosion usually pours sediments into the stream.

The town develops on both sides of the stream, often originally connected by a ford or ferry, and now commonly by two or more bridges (e.g., Pinedale, Tomar, Gavle). A road along the streambank quickly appears so the town is at a transportation intersection. The town is split by a stream corridor, through which water and wildlife move. For example, bears, wolves, and other species use the corridor to enter and move through the small town of Jasper, Alberta (Canada). Sometimes an elongated island is present. In effect, the town is at a highly diverse spot with many convenient resources close by.

This slightly curved eroding stream (perhaps third- to fourth-order) is on a moderate slope with moderate water velocity. Often another criterion for establishing a town is to dam the stream, creating a pond to run a mill or factory. Pond water provides hydropower, cooling, and convenient waste disposal. If resources and technology are inadequate to dam the stream, often the town is located near the mouth of a tributary which can be dammed.

The stream seldom disappears into a pipe through town, as is common in cities, and thus is a central town feature. Some towns have an elongated park along the floodplain and/or on one side of the stream (e.g., Huntingdon/Smithfield). Small greenspaces, often elongated and connected to the surrounding open or wooded land, are scattered in town. Many result from a high water table or flood-susceptible location. Streamside wetlands in or adjoining town are typically eliminated by filling with soil.

The impervious surface area in streamside towns is generally less than 25 percent, yet even an impervious cover of 5–10 percent significantly degrades the assemblage of downstream bottom invertebrates (Cuffney *et al.*, 2010). Although only some town center and hillside house surfaces are close to the stream, wet ditches or concrete troughs (in small towns) (e.g., Hanga Roa, Grande Riviere) or drainage pipes (in large towns) carry stormwater directly to the streamside.

Irrigation canals and channels without a concrete bottom provide enough water to support a line of trees with shade and wildlife (e.g., Socorro). The vegetation along *moist ditches* is normally dense and species rich (Geertsema and Sprangers, 2002; Blomqvist *et al.*, 2003; Williams *et al.*, 2003). Wildlife moves along the ditches which change in vegetation and periodically flood. Large invertebrate numbers and types in wet ditches seem to correlate with nitrate level, pH, and sediment particle size (texture) (Vermonden *et al.*, 2009). In contrast to a buried-pipe drainage system for a built area, a surface water system, such as moist ditches, provides several ecosystem services (Larson *et al.*, 2013). These include water quality improvement, reduced flooding, wildlife habitat and movement route, aesthetics, education, even recreation.

Roads, Bridges, Culverts

With a main road parallel to the stream, many perpendicular streets in town lead slightly upward into residential areas away from the stream. These perpendicular streets serve as stormwater runoff surfaces carrying water to the floodplain and stream. If roofs, paved driveways, and sidewalks are connected to the streets, the town area is quickly emptied of surface stormwater, though concurrently the stream level quickly, but temporarily, rises. Surfaces are cleaned as pollutants pour into the stream. Some stormwater pollution is absorbed by floodplain silt and clay and diluted by stream flow. The worrisome time is during low-flow drought conditions, when pollutants inhibit fish and other stream organisms concentrated in small pools and already stressed.

A town's bridge remains subject to washout due to an extremely high flood level or failure of a bridge pillar/post. Often a levee extends, sometimes several hundred meters, upstream on one or both sides to channel water under the bridge, and protect the pillars or foundations on opposite sides of the stream. This fixes or *constrains the stream* in a specific location in the floodplain.

Consequently, the stream can no longer *migrate* naturally from side to side in the floodplain. The effects of constraining the stream are conspicuous upstream, but especially far downstream. Over time, annual floods no longer periodically flood far to the right side or left side. The extensive floodplain areas no longer flooded change to quite different vegetation, and aquatic habitats are lost. This is one of the largest effects of towns on streams.

However, a bridge pillar in the floodplain has a more localized environmental effect. Immediately around it the water velocity is higher which scours out sediment, producing a deep bottom hole with different and sometimes bigger fish. Some people like fishing from bridges. The pillar location is shaded by the bridge, may be noisy from traffic, and receives sediment and chemical pollutants from vehicle and bridge runoff.

Tiny stream and intermittent stormwater-channel flows pass through *culverts* under roads and railroads. Providing for effective movement of both water and fish during both high flows and low flows is optimal. The least effective culverts are round pipes, and the best are arch-shaped with a natural stream bottom (Figure 6.14) (Forman *et al.*, 2003).

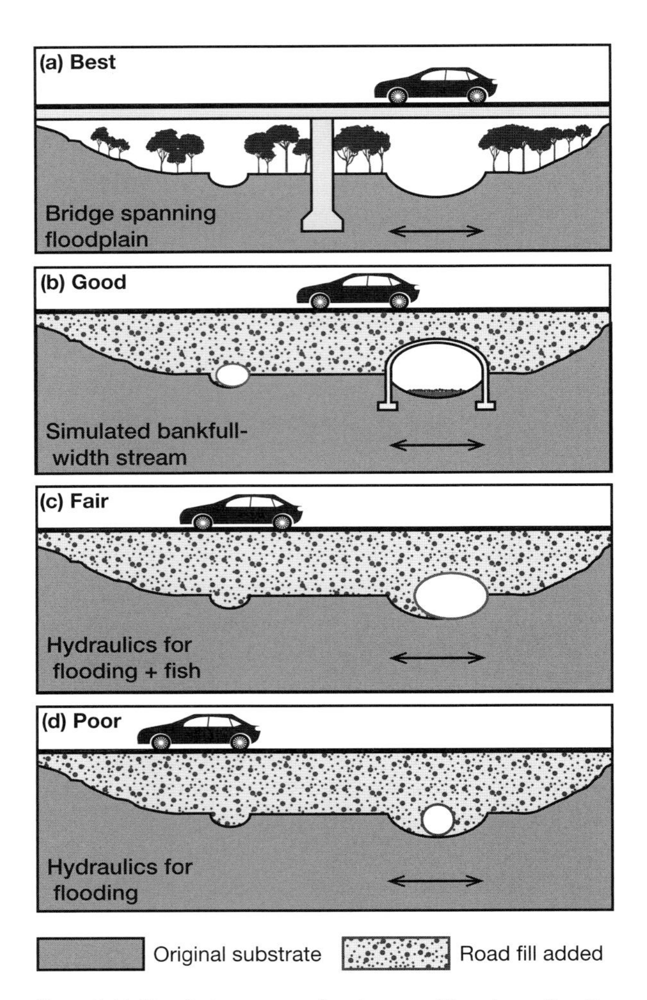

Figure 6.14 Road-stream crossing types, with a storm flooding example. Objectives are to combine effective passage of: high water flows (e.g., 100-yr flood); aquatic organisms (including fish); sediment and organic debris; and some terrestrial wildlife. For culverts, an arch-shape with a relatively natural stream bottom, and 120% of the natural bankfull-width stream is often recommended. (a) Based on Forman *et al.* (2003) and Gillespie *et al.* (2014). (b) From Gillespie *et al.* (2014).

Wildlife, from amphibians to badgers and bears, also move through culverts rather than crossing roads with traffic.

Towns may have hundreds, even thousands of culverts. The states of Oregon and Washington (USA) evaluated their culverts for the passage of salmon and found that about 30 percent inhibited passage. Simple modifications for incoming and outgoing water levels of culverts solved half of the problems. A third state, Vermont, apparently found about 90 percent of its culverts reduced or inhibited fish passage.

Fish communities in Great Plains (USA) streams upstream of a culvert blockage have both fewer species and an altered species composition (Perkin and Gido, 2012). If a species is blocked by a single culvert, the entire upstream network is no longer accessible to that species. Thus, blockages of fish by culverts of town roads is another extensive effect of towns.

A 2011 mega-storm ("Irene") in the hilly Vermont (USA) region killed six people, and isolated 13 villages/towns for multiple days by washing out roads. Stream habitat was extensively altered. Many kilometers of local roads washed out in part due to inadequate-sized culverts (averaging 54% of bank-full water flow) near development or houses (Gillespie *et al.*, 2014; Alison Bowden, 2012 personal communication). Yet roads with five recently upgraded *culverts for fish passage* did not wash out. As a general guideline, the culvert width should hold 120 percent of bank-full water flow (amount of water held between banks without spilling over onto the floodplain) for suitable hydrologic and fish passage.

Reservoirs, Pollutants, Flooding

Towns commonly dam streams creating pond-sized *small reservoirs or impoundments*. Perhaps most common are ponds for irrigation in dry areas (Kalff, 2002). Other uses of small reservoirs are for water supply, industry, fish farming, flood control, hydropower, and recreation (e.g., Wausau/Weston, Kutna Hora). Tiny ponds, e.g., behind earthen "check dams," are used for domestic purposes or watering livestock. A shortcoming of reservoirs is that unusual species appearing in a reservoir often spread in the stream up and down the valley (Rajvanshi, 2016).

Almost all of the impoundments are elongate and deepest by the dam, have considerable sediment buildup and turbidity, and fluctuate markedly in water level (Kalff, 2002). Phytoplankton are prominent, though rooted aquatic plants may be extensive. Since most stream fish do poorly in slow-moving waterbodies, fish species are often introduced. Summer water temperatures in the pond are high compared with those in the stream. Periodically, the dam may be opened to kill invasive plant cover and flush out sediments, which of course degrades the stream downslope. Lowered water levels during drought (Lake, 2011) and dam openings desiccate plants, oxidize organic matter muck, and kill fish.

Sedimentation, which reduces the amount of pond water, is often severe with surrounding cropland and in dry areas (Decamps and Decamps, 2001). Also, erosion of streambanks and pond shores, for example by livestock, adds considerable sediment. The original uses of dammed ponds are finite in duration, and little-used dams and ponds often remain, effectively fragmenting a stream into sections between warm-water ponds. A series of dams on a stream is particularly disruptive ecologically (Rajvanshi, 2016).

Dams and reservoirs across the stream floodplain inhibit the movement of floodplain riparian weeds as well as stream fish (Rood *et al.*, 2010; Marschall *et al.*, 2011). Even the gene flow of a widespread floodplain tree, *Populus fremontii*, in the dry southwestern USA appears to be dependent on the connectivity of stream networks (Cushman *et al.*, 2014).

Light or medium industry characterizes towns. In principle, manufacturing pollutants can be cleaned up or recycled for use, but in practice much of the pollution reaches the town's stream (Batty and Hallberg, 2010; Forman, 2014). The dam and pond may provide hydroelectric power for a major industry or for the whole town. Pond water in turn may cool the industrial equipment and processes. The byproducts or pollutants from manufacturing might be treated on-site, though, in the context of historical practice, the flowing stream is a magnet for getting rid of waste, particularly during flood times.

Lots of industrial and mining pollutants, almost all stormwater pollutants, and from all to almost none of the wastewater pollutants leave town in the stream (e.g., Kutna Hora) (Kalff, 2002). Even pharmaceutical chemicals, such as caffeine, ciprofloxacin, and diphenhydramine, from the town's residential areas enter the stream, apparently reducing algal growth and microbial respiration (Rosi-Marshall *et al.*, 2013). Heavy metals from industry generally reduce the abundance and diversity of stream macroinvertebrates, with an increase in pollution-tolerant groups (Batty *et al.*, 2010). Overall, stream pollution from towns ranges from serious to minor, even just for fish populations (Forman, 2014). Fully addressing stormwater, wastewater, industry, and dumping is a challenge for town budgets.

A stream's *base flow* is the water level determined by groundwater (Shanahan, 2009). Surface and subsurface stormwater runoff temporarily raises the stream level above base flow, even to flood level. Nevertheless, water flow rates especially depend on headwater areas and tiny upstream (first-order) tributaries (Kalff, 2002; Allan and Castillo, 2007; Forman, 2014). Vegetation removal there increases streamflow downstream and thus the "flushing rate" of ponds.

Towns seem to be particularly susceptible to *flooding*, particularly in certain types of locations. Familiar places are high water table spots, in deforested mountain gullies (e.g., Icod de los Vinos, St. George's), at the base of a hill or mountain (e.g., Vernonia), close to a stream (e.g., Flam, Sooke, Issaquah), a wetland or former wetland, just downstream of where a stream was narrowed, and just downstream of town (Forman, 2014). Floods keep returning. Flash floods appear rapidly, often unexpectedly (Walsh *et al.*, 2005a). Damage to the human side of town occurs, and degradation of natural features also. For instance, invertebrate abundance (fish food) is significantly reduced by flood events, especially in pools and on sand and boulders (McMullen and Lytle, 2012).

Yet both the town and nature are resilient. People go right to work cleaning up, repairing, and reconstructing the pre-flood conditions. Often this resilience means residents miss the opportunity make the place better before the next flood. Nature also quickly begins the ecological succession process, yet, unlike the typical human response, it brings a somewhat different set of species for the post-flood period. Tall floodplain vegetation rapidly grows providing friction against another flood.

Flood control mechanisms include altering the stream, changing land-use practices, and socioeconomic actions (Decamps and Decamps, 2001). Constructing dams, reservoirs, detention basins, levees (e.g., Huntingdon/Smithfield), channel modifications, canals, and bypass flood channels (e.g., Rock Springs, Socorro) are all mainly engineering approaches (Figure 6.15). Increasing natural vegetation cover (especially on steeper slopes), changing farming practices, decreasing direct-connection impervious area (DCIA), and maintaining vegetation around the headwaters and along tiny (first-order) streams are mainly ecological approaches.

Figure 6.15 Dry canal in desert and group from nearby research station ready to measure water characteristics. Near town of Socorro, New Mexico, USA. R. Forman photo. (A black and white version of this figure will appear in some formats. For the color version, please refer to the plate section.)

The engineering approaches emphasize sites and approaches amenable to rigorous equations and technical solutions. In contrast, the ecological solutions emphasize the big picture and the long term, e.g., the whole drainage basin, where every piece matters and disturbance and change are expected. Public education, flood forecasting, planning controls, land acquisition, and relocation of buildings are also advance actions for town residents.

Stream Corridor and Stream

A *stream corridor* includes the floodplain with its riparian vegetation and land uses, the adjoining hillsides, and vegetation, if present, along the upland top of the hillsides (Forman, 1995). Woody vegetation on the floodplain provides friction against flooding. The floodplain normally has high habitat diversity and therefore rich biodiversity. These result from many waterbody conditions, water table depths, sediment combinations, vegetation variations, and human effects. One analysis relates floodplain conditions to hydrology, bed load (sediments), connectivity, biodiversity, water quality, flood protection, and public attitude (Rohde *et al.*, 2006). Adding the hillsides and well-drained soils above the floodplain normally makes the stream corridor the most ecologically rich area in the landscape. Some floodplains have a canal and towpath along their edge (e.g., Tomar, Millstone/East Millstone) (Braithwaite, 1976; Forman, 2014). The stream corridor slicing through town serves as a source of species enriching biodiversity throughout the town. But also it provides pests and hazards such as deer, bears, and floods to the town.

Not surprisingly, town residents use the stream corridor in many ways. Old buried small dumps tell what earlier residents did with solid waste. Small vegetable gardens on the floodplain in town are often common. Some stream corridors are grazed by livestock such as sheep. Agricultural fields on a floodplain normally have high productivity associated with a high water table and ample mineral nutrients deposited by recent floods. In addition to natural processes, logging, fires, hunting, and other activities of town residents significantly mold the diverse vegetation in stream corridors (Briggs, 1996; Decamps and Decamps, 2001; Wohl, 2004; Sass and Keane, 2016).

A study in Georgia (USA) found that streamside activities tend to create deeper water, while opening the forest canopy over a stream creates warmer water (Barrett and Guyer, 2008). Overall, both effects favor reptiles over amphibians. Similarly, the stream-bottom invertebrates correlate with the amount of riparian vegetation present (Shandas and Alberti, 2009). With the series of linear or elongated habitats present in a stream corridor, a wide range of wildlife species regularly moves along the corridor and into towns (e.g., Sant Llorenc).

As a narrow waterbody with water perennially flowing except during severe droughts, a *stream* is a highlight of the town. Unlike urban streams which are heavily polluted and mainly invisible by being buried in pipes (Paul and Meyer, 2001; Booth and Bledsoe, 2009; Forman, 2014), a town's stream is daily seen or appreciated by residents. Thus, "stream daylighting," which transforms a buried water course to a visible engineered channel or semi-natural corridor (Bernhardt and Palmer, 2007; Forman, 2014), is essentially irrelevant in towns.

Key characteristics to understand a stream are its: (1) hydrology (water flow amount and velocity), (2) form/shape (width and straight to meandering), (3) internal structure (microhabitat heterogeneity and arrangement), (4) physical/chemical/biological properties of the water, (5) stream-bottom (particle sizes and aquatic invertebrates), and (6) fish (abundance/composition/diversity, and distribution).

The stream bottom often has a *pool-and-riffle* structure, i.e., pools with slow-moving water alternating with riffles of fast water splashing over stones. In the town center and by the bridges, almost always the streambanks are *armored*, that is covered with rocks or concrete for a short distance (e.g., Flam, Falun) (Vince *et al.*, 2005; Walsh *et al.*, 2005). Also, the stream is often narrowed and straightened a bit at these locations. These engineered changes constrain the stream in one location, and *stream migration* back and forth across the floodplain over time (e.g., Casablanca, Alice Springs) (Decamps and Decamps, 2001) is eliminated.

Stream-bottom macroinvertebrates are major sources of food for many fish (Cushing and Allan, 2001; Kalff, 2002; Lake, 2011). Some of the species such as worms live in the fine sediment, while often many more tiny crustaceans and larvae are attached to the bottom of stones. An abundance and diversity of stream-bottom (benthic) invertebrates indicates both good water flow and water quality. Stream macroinvertebrate abundance correlates with less surrounding impervious surface area, and with more vegetation cover, both at the scale of a local riparian site and the whole drainage basin (Riva-Murray *et al.*, 2010).

A diversity of fish populations especially thrives with stream habitat heterogeneity (e.g., Selborne, Sooke, Issaquah). Habitat diversity includes a range of water velocities, from slow, deep, and shallow water to small and large turbulence locations. The range of flow velocities creates an array of particle sizes on the stream bottom, for instance, patches of boulders, gravel, sand, silt, clay, and organic matter. Trees and shrubs overhanging the stream edge provide cooling shade, and create habitats by dropping logs, branches, and brush into the water.

Stream restoration for fish uses diverse techniques to create these heterogeneous conditions. For instance, logs and boulders, even flat-sided quarried rocks, are located in the central stream portion. Deep holes created by turbulent scouring around a large boulder or log are valued by large fish and by fishermen. Also, logs, branches, bundled brush, and rock piles are added along the edge. A strong well-maintained fence keeps livestock off the streambanks, where dense shrubs and trees overhang the edge. These are site solutions, yet stream conditions at a site also depend on conditions in the entire upstream drainage basin (Forman, 2014).

The movement or annual migration of fish up and down a stream may be blocked by a dam, a water mill, or stretch of polluted water (Lucas and Baras, 2001; Wolter, 2008; Raemaekers *et al.*, 2009). Fish ladders by stream blockages may facilitate movement (e.g., Tavistock), though water in a stream and a pond differs markedly. *Stream refugia*, where fish stay until conditions permit movement upstream (such as during floods), are often in tributaries, stretches of deep water, and stretches through natural forest (Carolsfeld, 2003; Lake, 2011).

Low flows during severe droughts often appear as a sequence of pools or deep holes connected or partially connected by a trickle of water (e.g., Capellades) (Lake, 2011; Walters and Post, 2011). Thus, pools and deep holes are particularly important for the survival of some fish and other aquatic organisms. However, major fish kills happen during drought, or from pollution events.

Town streams have normally had a "tough" history (Grossinger, 2012; Sass and Keane, 2016). Consider a forested stream a few meters wide in the center of Concord (see Appendix) (see Figure 7.14). In 1636 a dam, pond, and mill were constructed on Mill Brook, effectively blocking fish movement (Forman, 1997). Mill wastes washed downstream. The village's human waste, solid waste, agriculture-eroded sediment, stormwater, and later manufacturing wastes accumulated in the pond and washed downstream for almost two centuries. Livestock eroded and flattened streambanks, and residents made little dumps, quickly buried, nearby. The pond became so polluted and smelly that every year the dam was briefly opened to flush the accumulated waste out of town. The dam and pond were removed in 1826, but the stream was still the main mechanism or service cleaning the town (Carpenter *et al.*, 2003). Parallel ditches and pipes drained to the stream. Local manufacturing increased and the population grew. The town's stables or livery for horses was on the stream bank, and doubtless shoveled out every day. A small channel was diverted under the jail so prisoners' wastes went directly into the flowing water. After 1900, streets in the town center were paved, a streambank gasoline station opened, and stormwater runoff doubtless increased. Finally, in the mid-twentieth century, trucks became widely available for efficient removal of solid waste.

In the late twentieth century, 350 years after Mill Brook was dammed, government regulations plus local land-conservation protection noticeably cleaned up the stream. Today little-treated stormwater runoff keeps the town center stream somewhat polluted.

In agricultural areas, plowing and livestock keep town streams muddy. In manufacturing towns, pollutants from industrial processes predominate in the town stream. Imagine the effects of one or more pollutants, such as heavy metals, hydrocarbons, toxic organic compounds, wastewater organic matter, nitrogen, and phosphorus (see above), on a small stream (Paul and Meyer, 2001; Rabeni and Sowa, 2002; Miltner *et al.*, 2004; Walsh *et al.*, 2005b).

Three contrasting studies provide an overview. Adjacent built areas such as towns make streams deeper and more open, with less habitat complexity, smaller stream-bottom particle sizes, altered macroinvertebrate assemblages, and fewer sensitive specialist species (Violin *et al.*, 2011). In an agricultural region of southern Ontario (Canada), instead of broad land-use characteristics to understand fish and macroinvertebrate assemblages, finer-scale variables (related to field-tile drainage, non-agricultural use, transportation, septic systems, livestock farms, and water pumping/withdrawal) were the best predictors of the stream fauna (Yates and Bailey, 2010). In a southeastern Queensland (Australia) landscape, overall stream conditions (measured by fish, macroinvertebrates, water quality, mineral nutrients, and ecosystem processes) best correlated with the amount of forest cover, particularly close to a stream (Sheldon *et al.*, 2012). Towns affect almost all of these stream variables.

Rivers

Rivers flow in floodplains, with floodplain banks (hillsides) commonly on both sides. Towns by rivers are usually atop the floodplain banks to escape periodic large floods that may cover the floodplain (e.g., Emporia, Concord, Swakopmund). Floodplain banks with a sharp gradient of wet to dry habitats tend to be species rich.

Without human structures in the floodplain, a river *migrates* from side to side over time, so that the floodplain normally is highly heterogeneous with small curvy wetlands or ponds (oxbows or billabongs) (Dunne and Leopold, 1978; Decamps and Decamps, 2001; Rohde *et al.*, 2006). Furthermore, floodplains themselves are highly dynamic over time (Jones *et al.*, 2010; Jordan *et al.*, 2012; Sass and Keane, 2016). However, bridges, causeways (elevated roads and railways on constructed low ridges), and dam outlets constrain a river channel to a fixed location in the floodplain (see above) (Brooks, 1998).

Rivers (usually fifth-order or higher) are wider than streams, typically meandering, and open overhead, with slow-moving water, smooth silt-covered bottoms, food chains based on both phytoplankton and detritus, and many narrow adjoining wetlands (e.g., Socorro, Chamusca) (Kalff, 2002; Price *et al.*, 2007). Fish may be abundant, and river invertebrates seem to reflect the local surrounding abundance of invertebrates rather than long-distance dispersal from upriver (Sundermann *et al.*, 2011). Rivers repeatedly flood, depositing uprooted trees, brush, debris, and a cover of nutrient-rich sediment which is mainly silt. Flood-control levees may be present to help protect towns and villages along the river (e.g., Huntingdon/Smithfield).

Towns are connected by roads along the floodplain bank. In extremely wide floodplains, a few villages may persist on somewhat elevated spots in the floodplain. Characteristically, no road bridge crosses a major river by a village or small town, but rather crosses by a large town or city, which is a transportation crossroads (e.g., Wausau/Weston). Agriculture, especially rice, is productive on the nutrient-rich silty floodplain. Parks and other protected land may be common near the towns and villages (e.g., Huntingdon/Smithfield). Essentially, all river towns have fishermen plus boats of varied size. Docks may be present, particularly if protected by a diagonal rock jetty.

Town residents appreciate, but degrade, the adjoining floodplain bank. Pollutants from the town are mainly absorbed by the bank and nearby floodplain, and those reaching the river are quickly diluted by the volume of flowing water. This is quite different from much-degraded urban rivers (Forman, 2014). Mining and industrial towns pollute rivers (e.g., Rumford/Mexico, Wausau/Weston). Otherwise, overall the environmental effects of river towns, unlike towns by streams, seem to be small.

Coastal Zone

Finally, let us explore (1) coastline characteristics, (2) coastal towns, and (3) climate and other coastal changes.

Coastline Characteristics

The *coastal zone* is the irregular strip where land and sea interlink. Habitat heterogeneity, species richness, and in places productivity are extremely high. Furthermore, the place is highly dynamic, markedly changing with the tides, day and night, winds, rainstorms, erosion, and even cyclones/hurricanes or tsunamis. The coastline may be rocky, a sandy or gravelly beach, coral reef, stream/river delta, estuarine bay, salt marsh or mangrove swamp, or tidal muddy flat (Woodroffe, 2002). The connectivity of coastal wetlands (mangroves, salt marsh, seagrass channels, and tidal flats) is apparently important to some fish populations (based on catch-per-unit-effort) (Meynecke *et al.*, 2008).

An *estuary* with brackish water is the overlap or transition area between freshwater and seawater. A barrier beach or spit may be present protecting the estuary from high-energy sea waves (e.g., Sooke, Akumal).

A coastal microclimate typically extending several kilometers inland covers a town. Compared with a nearby area further inland, the coastal area has moister air, more mist or fog, less sunlight, more rain, more wind, and more aerial salt deposits. In autumn the town is warmer than the inland area, and in spring, cooler.

Both land and sea compose the coastal zone. The land may rise gradually from the sea across a coastal plain (Woodroffe, 2002). However, with a coastal range of hills or mountains, broad microclimate and vegetation zones are roughly parallel to the coastline (Forman, 2010). Short streams pour down to the coastline (Figure 6.16). Mountain rainwater also flows by subsurface runoff to the *groundwater emergence zone*, where

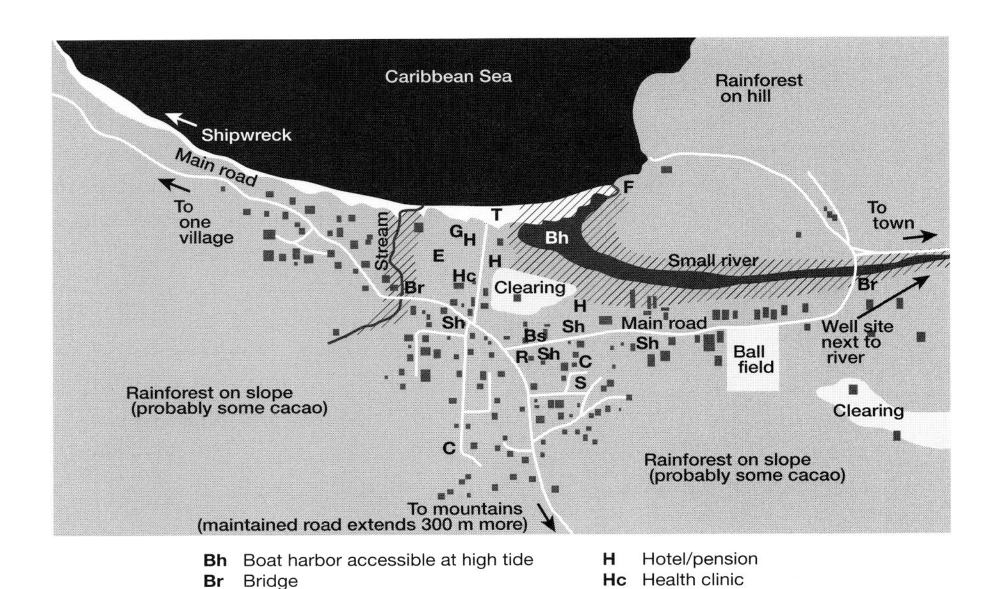

Bh	Boat harbor accessible at high tide	**H**	Hotel/pension
Br	Bridge	**Hc**	Health clinic
Bs	Bus stop	**R**	Recreation hall
C	Church	**S**	School
E	Ecotourist sea turtle center	**Sh**	Shop
F	Fishing dock	**T**	Turtle nesting
G	Groundwater emergence area	/////////	Flood-prone area

Some village characteristics:

- River floods from mountain rainforest carry pebbles/rocks to sea; waves wash them to beach
- Much stormwater and some human wastewater flows in open concrete troughs (50 cm wide and deep) toward river mouth or stream
- Buried water pipe along main road; 92% of households with continuous water supply
- 11% households on sewer system, 22% septic tanks, 67% pit latrines or outdoor toilet facilities
- Village area mainly open, with abundant flowering shrubs, butterflies, birds
- 32 village microhabitats noted in a visit
- Most house plots with fruit-producing trees/shrubs, sometimes attracting monkeys
- About 20 common village birds, e.g. Caribbean grackle, house wren, blue-gray tanager, palm tanager, kiskadee, swallows, hummingbird, tropical mockingbird
- Dogs, chickens, roosters widespread; cats scarce; essentially no livestock
- Constant sound of waves breaking, daytime birds calling; human noises include few trucks, limited traffic, and recreation hall music
- Scattered street and other lights attract flying insects, plus lizards and bats
- Only sidewalk extends from bus stop east to bridge
- Several partially built houses; a few abandoned houses with plots in ecological succession
- Many houses on concrete piers to minimize snake access
- 50% unemployment; jobs in ecotourism, agriculture, school, services; some home handicrafts
- Ecotourism focused on leatherback turtle nesting, Trinidad piping guan, and nearby national park
- Cooking mainly with gas; light mainly from electricity
- Limited river fishing, some offshore net fishing; foreign fishing boats and tankers often visible
- Men know the surrounding land, but don't use it much
- Apparently precipitation, river flow, and fish stocks decreasing; sea level rising
- Apparently villagers' main concerns: oil spill, waste disposal, climate change

Figure 6.16 Map and characteristics of a tropical coastal village strongly dependent on and affected by water. Grande Riviere, population 350, Trinidad. See Appendix.

freshwater reaches the ground surface or the heavier salt water, generally within a couple of hundred meters or so of the coastline.

In the coastal region, most roads go either parallel or perpendicular to the coastline. A broad rectilinear imprint, with finer-scale curvy patterns, characterizes the coastal region (Forman, 2010). Flows and movements of migrating birds, freshwater, vehicles, boats, fish, winds, and currents are also mainly parallel or perpendicular to the coastline.

Coastal Towns

Unlike coastal cities or coastal urban sprawl (Pilkey and Dixon, 1996; Ivanov, 2009; Forman, 2014) which produce intensive widespread pollution and disturbances, coastal towns and villages usually are in distinct coastal spots with relatively little degradation of adjoining waters (Figure 6.16). Rather few towns and villages appear by coral reefs, tidal flats, salt marshes, and mangrove swamps. Also, few towns are present on rocky points/headlands/lobes projecting into the sea.

Small coves are the most common locations of coastal towns (Barral, 2009; Mininni, 2010). Some coves, which are partially protected from strong winds and waves, have a beach. Many coves with a town or village also contain the mouth of a stream, often with small associated wetlands or tidal mudflats. The stream carries nutrients into the cove so saltwater fish are usually abundant (Figure 6.17). A diagonal jetty or groin of large rocks is often constructed to protect the harbor, docks, and boats against storms. In effect, coastal towns tend to have considerable habitat and resource diversity close by.

On a relatively flat site, a series of parallel town streets extends down to the waterfront, whereas in an amphitheater-like setting by a mountain or hill the streets tend to converge near the central waterfront (see Figure 1.1) (e.g., Astoria, St. George's). Either way, many stormwater drains pour sediment and stormwater pollutants into the cove. Water quality is reduced, which degrades fish populations, coral reefs, bottom seagrass beds, and beach recreation.

In Sabah (Malaysia) and Alabama (USA), productive seagrass areas are degraded by the addition of fine sediment (Freeman *et al.*, 2008; Anton *et al.*, 2011). The suspended sediment produces turbidity which cuts light penetration to the seagrass. In Malaysia nearby deforestation rather than adjacent development is the probable cause of degraded seagrass and associated marine animals.

Town waterfronts and harbors are the working town centers, which may support a fishing industry, port for diverse boat traffic, tourism, and some manufacturing (Faggi *et al.*, 2010). Boat anchors and some fishing techniques drag the bottom (e.g., Akumal). Excessive freshwater pumping in town sometimes leads to saltwater intrusion beneath, but probably this is uncommon or minimal, especially in the vicinity of a flowing stream.

Residential development usually spreads along the coast, and public access points facilitate coastal recreation (e.g., Akumal). Some of St. George's (Grenada) coastal buildings were damaged by a rogue wave, whereas an unbuilt coastal strip hundreds of meters wide mostly protects Hanga Roa (Easter Island, Chile) and the city Concepcion (Chile) against tsunamis. Overall, the two major town effects on the coastal environment are: (1) habitat loss due to development of the distinctive town site and destruction of

Figure 6.17 Fishing and shellfishing boats in harbor protected with rock jetties. Port town of Crystal City, California, USA. R. Forman photo.

nearby wetlands; and (2) degraded water quality and bottom conditions in the estuary/bay and sometimes the nearshore sea area.

Development along an estuary, of even as little as 3.6 percent of the estuary's border, was found to be the best predictor of degradation of the waterbird community (Deluca *et al.*, 2008). Flamingos in a bay by Celestun (population 3,000; Yucatan, Mexico) depend on the continuous protective cover of mangrove vegetation (*Rhizophora mangle*) (see Figure 8.3).

Indeed, 21 percent of a 574-hectare (2.1 mi^2) protected mangrove swamp (Altona, Victoria, Australia), with adjacent residential housing being built, was affected by human intrusions (Antos *et al.*, 2007). Walking and dog-walking were more common than cycling and motorbiking, and also extended furthest into the mangrove wetland. When adjoining housing is completed along 85 percent of the landward mangrove boundary, about half the protected area is projected to be disturbed by human intrusions. Such usage of the protected wetland disturbs migratory shorebirds and other species.

Climate and Other Coastal Changes

Climate change in the northern hemisphere leads to sea level rise due to far northern latitude ice melting (IPCC, 2007). Sea levels in the southern hemisphere greatly depend on the degree of West Antarctic Sheet ice melting. Towns within a few or several meters elevation of sea level can expect more saltwater inundation of low areas, storm surges, damage to coastal buildings and roads, more coastal erosion, saltwater intrusion contaminating wells, and warmer, perhaps more acid, coastal waters (Valiela, 2006). All of these alter marine ecosystems (such as coral reefs) and fish populations. Many species shift their ranges poleward. Since town centers are usually by the waterfront, they are most susceptible to sea-level rise damage. Terrestrial habitats and land uses by towns will likely be reduced or eliminated.

More extreme weather events produced by climate change also affect coastal towns. Seasonal cycles change in timing. Severe storms from the sea rearrange coastlines and degrade their human structures (Hauser *et al.*, 2015; Keenan and Weisz, 2016). Heavier precipitation events inland generate greater stream flows and flooding in or by the town. Streams overtopping their banks flood their floodplains and sometimes surrounding town areas (e.g., Sooke, Flam). Flooding streams produce turbid muddy waters and alter saltwater habitats in a town's cove. This in turn degrades floating plankton, bottom organisms, shellfish, and fish.

Hurricanes/cyclones are expected to become more intense and frequent, causing severe damage to some coastal towns. More extreme droughts from global climate change lower water tables, dry out wells, create stream low-flow conditions, reduce floodplain rejuvenation by periodic flooding, and favor warmer-water microbes in town reservoirs and water supplies. In effect, coastal towns that avoid making changes in advance are at great risk from climate change.

But climate change is only one of the many changes facing coastal towns (Valiela, 2006; Maslo and Lockwood, 2014). Wetland loss and degradation is the norm, usually little by little as buildings, docks, and pollution spread. The loss of salt marshes or mangrove swamps reduces protection against ongoing typical storms, habitat for numerous saltwater species, shellfish and fish populations, and recreational opportunities (Maslo and Lockwood, 2014).

Pollution of the town's cove and nearshore marine environment varies from limited to extreme. Stormwater pipes, troughs, and ditches pouring the town's surface water into the cove are nearly universal. Human wastewater seeping from numerous onsite systems or pouring from a primary or secondary wastewater treatment facility is common (e.g., St. George's, Astoria). This is a major source of pathogens and human disease, as well as leading to toxic shellfish.

Boats may pump out their ballast water and/or eliminate their human wastes into the town cove. Boats also widely disturb local natural communities and ecosystems. Motorboats drag anchors, spill fuel, and disrupt foraging by numerous waterbirds. Sound pollution by large and small boats inhibits communication of marine mammals and other coastal animals.

Thermal and industrial pollution creates wide plumes from their point sources in or by towns. Powerplants and desalinization facilities discharge heat, chemicals, and microbes into the cove. Lots of marine organisms are sucked into these facilities in the cooling water, and may emerge dead, as in conspicuous fish kills. Town industries commonly use the cove water for cooling and/or for byproduct waste disposal.

Some of the debris dumped into the cove sinks, gradually accumulating and decomposing, and some conspicuously floats. Such floating debris may also accumulate near the coast, reminiscent of the massive floating debris accumulations in parts of the ocean. Perhaps worse are coastal entanglements, the marine fish, birds, turtles, and mammals wounded or killed by plastic packaging, synthetic ropes and lines, and fishing gear. Residents of coastal towns all have stories about entangled dead or dying organisms near their town.

In effect, these many coastal changes, from climate change to spreading pathogens, thermal pollution, entanglements, and sound pollution, seem to be especially worrisome for towns. Lots of solutions have been proposed. For example, consider: hybrid natural and engineered structures (Hill, 2015); flexible, movable, modular, temporary structures, like nets/networks, dunes/sand, mangroves/pandanus trees (Pilkey and Dixon, 1996; Maslo and Lockwood, 2014); high habitat heterogeneity (Munguia *et al.*, 2011); wide wetland borders; periodically floodable parks; and so forth. Environmental change is the norm. Will coastal towns and villages change faster or slower?

7 Plants, Habitats, Greenspaces

> [S]cientists ... have identified a number of volatile organic compounds (vocs) from trees ... which may have antibacterial, anti-inflammatory, and even anticarcinogenic properties ... exposure to these airborne molecules from trees can reduce arterial contraction; lower arterial blood pressure; reduce the release of stress hormones ... lower ... vascular inflammation ... and increase the number of white blood cells that defend against viruses.
>
> So there you have it: taking a regular walk in the woods can help you avoid ... medication.
>
> Francois Reeves, *Planet Heart*, 2014

Town and village plants predominantly reflect residential areas, surrounded, and strongly affected, by farmland, forest, or aridland. In the pages ahead we explore town plants, habitats, and greenspaces from five broad perspectives: (1) town trees; (2) plant distribution and diversity; (3) species movement and change; (4) greenspace patterns; and (5) habitat diversity.

Town Trees

Trees in towns vary conspicuously in their (1) distinctiveness, (2) number and species, and (3) distribution, as explored below.

Town Distinctiveness

Several characteristics tend to differentiate town trees and shrubs from those in natural areas, agricultural areas, and cities. The woody plant species planted are strongly affected by species availability in the single nursery (or two) in or near a town. Town-planted trees, mainly along streets, are conspicuous, normally bought cheaply in quantity, and often grow to maturity (Figure 7.1). In contrast, species planted around houses typically express the individuality of residents, and include many less common species. Flower gardens with ornamental plants, though not formal gardens, are usually abundant in towns.

Many native species, even specialists, grow well in and often dominate towns and villages. Without the lowered water table, urban heat island, and air pollution of cities, planted trees in towns have a good chance of reaching maturity. Nonnative trees are present but usually not predominant, perhaps due to limited seed sources, as well as suitable growing conditions for native species. Also, residents, rather than

Figure 7.1 Street lined with umbrella nut pine (*Pinus pinea*) providing a midday shade corridor. Walls produce early and late shadows, reducing heat buildup. Small city of Sorrento/Sant'Agnello, Italy. See Appendix. R. Forman photo.

commercial landscaping companies, maintain most house plots. That normally results in a higher diversity of planted species, and especially of spontaneous plant species from surrounding land.

Shrubs are commonly widespread in villages and towns where walking is usually safe. Shrubs appear singly, in small patches, and in hedge or hedgerow corridors. Mature trees usually characterize the older residential neighborhood next to the town center. Unlike natural forest, these trees generally have wide full crowns, with small adjoining open spaces rather than competing trees. Unlike the extensive impervious surface cover in cities, the impervious surface in towns is much lower, and most surfaces are small. Therefore, much of the stormwater runs off directly into the adjacent soil rather than being piped away. In this way stormwater usually supports good plant growth in town. Probably reflecting limited town budgets over time, many of the older street trees have dead branches or holes in their trunks. Dead wood and cavities are also in house-plot trees. Numerous insects, mammals, and birds thrive in such microhabitats.

In farmland, town and village trees have ample phosphorus and wind-blown nitrogen that fertilize town plants, effectively promoting growth. Village houses may be surrounded by farmyard, or strongly influenced by plants and animals from the adjacent farmland. In desert towns, shrubs and small trees are common next to buildings with slanted roofs, which channel scarce rainwater to the roots. In forest, a "rain" of seeds, spores, and pollen covers the town.

Spontaneous vegetation appears in many spots throughout town, including in the middle of town center blocks, vacant plots, railways, dumps, industry waste heaps, and unused quarries (Kunick, 1990; Del Tredici, 2010). In difficult economic times or major disturbances, ecological succession with shrubs and trees present appears on vacant house plots (e.g., St. George's, Grande Riviere; see Appendix).

With little air pollution, epiphytes, including lichens and mosses, thrive on tree trunks. In the tropics, epiphytes such as ferns, bromeliads, and orchids often cover large tree branches. In short, the preceding array of characteristics make towns quite distinctive habitats for trees and other plants.

Tree Number and Species

How many trees are there in a town? A 1982 survey of 5 small towns, 5 medium-size towns, 113 large towns, 198 small cities, 23 medium cities, and 2 large cities in the USA found considerable variability in tree abundance within each category (Grey and Deneke, 1992). In part, this probably reflected a community's setting in either a forested or non-forested landscape. The median number of trees in small towns was 150, medium-sized towns 985, and large towns 4,000. The average number was much larger in all three cases, suggesting that a few towns had numerous trees, but many towns had few trees. The same pattern was present for cities of different size.

Towns of all three sizes spent little on annual tree care (about 0.3% of a total town budget). Small cities (25,000–250,000 population) spent considerably more on tree care (0.5%). In contrast, medium-sized cities (250,000–1,000,000) spent only 0.2 percent and large cities (>1 million) 0.01 percent, both considerably less than in towns. Small cities tend to have more trees per unit area than do large cities (Turner, 2010), and probably towns typically have a still higher density of trees.

A more detailed view of a large town, Manhattan (Kansas, in the agricultural grassland Midwest USA; population 29,500) in the 1980s, is informative (Grey and Deneke, 1992). The "older area" of town was laid out (platted) in the 1800s on the northwest side of a river, and subsequent major growth occurred to the north and west in approximately 1940–1960 ("intermediate area") and in the 1960s–1980s ("newer area").

Somewhat more than half (6,496 trees) of the town's street trees (>2.5 cm or 1-inch diameter) are present in the older area, which covers nearly a third of the town (Figure 7.2). This area corresponds to the town center plus the older residential area of most towns (Chapter 1). The intermediate area in the middle third of town contains 3,194 trees, and the newer area in the outer third of town has 2,334 trees. Basically, trees are concentrated in the older area.

The three areas have essentially the same diversity of trees (39, 38, and 40 species, respectively, in older to newer areas). Fewer uncommon tree species (each <1% of the total) were present in the older area than in the other areas (13, 20, and 20, respectively). The three areas appear very different due to differences in the most abundant species. *Tree health* (indicated by the average percent of trees rated poor or dead-and-dying

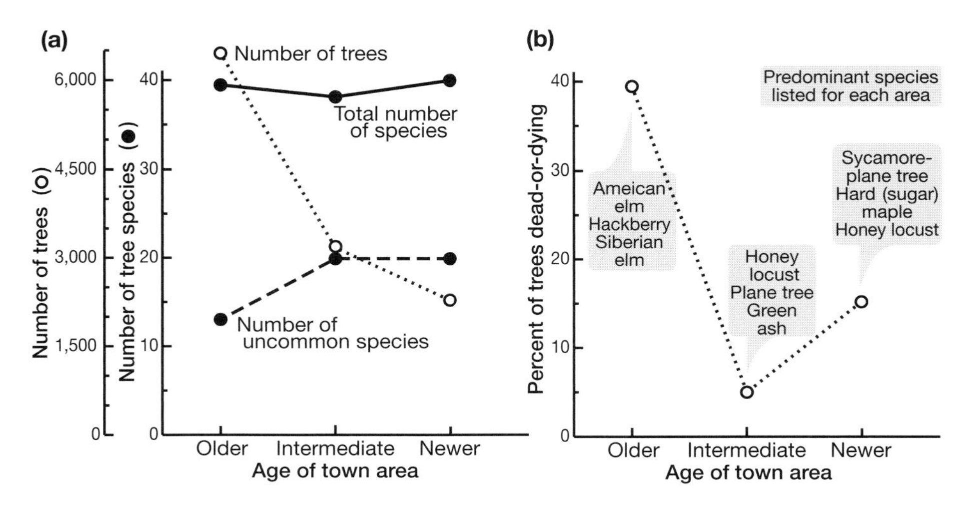

Figure 7.2 Street trees in town areas of different age. Manhattan, Kansas, Midwest USA. Based on Grey and Deneke (1992).

for the three leading species) is noticeably worse in the older area than in the other areas: 39 percent; 5 percent; and 15 percent. Unlike in most cities, recently planted trees in towns are mainly healthy and growing well.

In essence, the older residential area and town center have a denser street-tree cover, older and larger trees, different predominant species, and fewer uncommon species as in other neighborhoods. A much higher percent of trees in poor health or dead and dying results from an elm (*Ulmus*) disease and perhaps limited tree care. Lower-income neighborhoods have fewer street trees, but of the same species as other neighborhoods. In the older area of town, recent plantings are different species than the 40–60-year-old trees. The place is changing. Also, in the older area, a wider range of tree species was planted by residents on private property, mainly house plots.

For part of the town's history, local residents could plant street trees, which often resulted in poor species selection (Grey and Deneke, 1992). Several factors determine the common species planted by the town. Most important are the availability and affordability of saplings in the scarce nurseries in or near town, and recommendations from local university specialists. The size of a space for planting also strongly affects species selection. The relative sensitivity of species to SO_2, O_3, road salt, and even light pollution at night was sometimes considered. Except for the north and south sides of structures, temperature was not an issue in this flat terrain. Finally, species selection was also affected by human perception, such as tree shape, beauty, incense, a refreshing contrast, shade, protection, hazard, wood production, alien species, firewood, or symbol.

The ecological roles of trees in built areas are exceedingly diverse and important. Some 58 functions or roles have been highlighted for urban areas (Forman, 2014). These tree functions affect soil, water, air, plants, and animals, essentially all components of an ecosystem, in addition to city residents (McGregor *et al.*, 2006).

Tree Distribution in Town

Small areas of *semi-natural vegetation* with shrubs and trees are present in most towns (Clements, 1983; Airola and Bucholz, 1984; Ma *et al.*, 2001; Bertin, 2002). Typical places are a stream corridor (Groffman *et al.*, 2003; Burton *et al.*, 2009), a moderately steep slope, a green wedge projecting into town, or a portion of a park, cemetery, institutional land, or industrial land. Canopy, understory, shrub, and ground-cover layers are characteristic. Such small wooded areas contain many trees, essentially all of which arrived and grew spontaneously. Tree and shrub diversity is typically rather high and mainly of native species, including relative specialists (see greenspace section below).

The town commercial/residential center, in contrast, has relatively few trees (e.g., Rocky Mount, Wausau/Weston), many being somewhat stressed street trees. Shrubs typically are also in relatively low density and somewhat stressed. Large planted trees are usually present in the central public plaza (square). These often originated as *nursery plants*, horticulturally selected to accomplish certain goals, such as rapid growth, abundant flowers or fruits, low water requirement, cold hardiness, or pest resistance (Brickell, 2003; Cullina, 2009). Relatively large spontaneous trees are often scattered in the centers of blocks, i.e., the back spaces behind buildings (e.g., Valladolid, Huntingdon/Smithfield). Town center trees are stressed by diverse impervious surface effects, vehicles, and people.

Most of the town surface is covered by residential properties, usually house plots. *Street trees* providing shade are often limited in cool climates, where solar heat is appreciated (e.g., Rumford/Mexico, Wausau/Weston). Also, few trees typically line the streets in dry climes due to limited water, or where streets are narrow (e.g., Alice Springs, Socorro). On the other hand, trees are usually abundant in warm moist climates where shade is valued (e.g., Emporia).

Street trees provide aesthetics and shade (Phillips, 1993), and may provide harvestable products such as mangoes in the tropics and apple trees near Kutna Hora (see Appendix). Large street trees characterize the town's older residential area, and smaller trees the newer area (e.g., Lake Placid, Huntingdon/Smithfield). A particular block often has a single species of street tree, and a neighborhood may have been planted with one or two species at one time. Nevertheless, for a whole residential area planted at different times, native trees tend to be most abundant. The total tree diversity in towns is normally quite high.

A town's tree diversity may include as many nonnative as native species (DeGraaf and Witman, 1979). However, unlike cities, where nonnative trees are usually important in reducing both urban heat and pollutants, no particular ecological value is known for nonnatives in towns. Indeed, nonnative tree seeds regularly reach nearby semi-natural areas.

One or a few trees is commonly planted in the *front space* (yard, garden) of a house plot, where spontaneous trees are nearly absent. Trees in front spaces tend to be cared for to minimize the chance of a dying branch or tree falling on a person or vehicle. The management of front spaces can also affect street trees, as in a desert city (e.g., Albuquerque, New Mexico, USA) when front yards were changed to xeriscaping to reduce water use. Nearby street trees, which had prospered from house-plot watering, began to quickly die (William Pockman, 2015 personal communication). Front-space trees in a block are normally much more diverse, often including some nonnative species, than the street trees.

A few to many trees commonly appear in the *back space* or yard, commonly as a mix of native and nonnative species. The tree mix often reflects the choices of previous residents who may have liked native species, but more often liked the contrast or exotic feel of nonnatives. Trees near the house are mainly planted.

The *back corners* of a long house plot tend to be the most natural spots, with many spontaneous species. Stratification of foliage, with canopy, understory, shrub layer, and ground layer, may be present in back corners. Contrastingly, only one or two layers generally characterize the remainder of the house plot.

If the distance between house and back boundary line is large, e.g., more than 15–18 m (50 ft), a *back line corridor* often appears. Basically, vegetation along the back boundary tends to be little manicured by residents. Thus, many spontaneous plants appear, together with any planted species, along this narrow semi-natural corridor. The back-line corridor may connect with those of other house plots effectively forming a wildlife corridor cutting through a residential block.

Suburban towns often have considerable tree cover, such as 40 percent of a suburban residential area outside Milwaukee, Wisconsin (USA) (Dunn and Heneghan, 2011). A quarter of the trees here are street trees, 28 percent in front spaces, and 26 percent in backyards, where both planted and spontaneous trees grow. The species richness of front-space trees is triple that of street trees. Woods in three exurban built areas near Seoul (South Korea) have a thicker tree canopy (8–18 m above ground), thicker understory, and thicker shrub layer than in woods in the city center (Lee *et al.*, 2008). Although the number of woody plant layers is the same, suburban towns have noticeably denser foliage.

Unlike in cities, town spontaneous plants probably are mainly native species. Indeed, without the urban heat island, air pollution, and extensive area of cities, the considerable residential area of towns doubtless supports many specialist native species. Some species of conservation interest thrive in towns and villages (e.g., Lake Placid, Concord).

Venerable old trees (worthy of respect) may remain (Figure 7.3). These serve as unusual habitats usually rich in small species, such as beetles, mosses, and lichens that are uncommon or rare elsewhere in town (e.g., Icod de los Vinos, Selborne). Both tree species and size are important in supporting such species. The ancient trees are normally important in the town's history, or treasured by residents for other features, such as shade, sound, protection, or historical events. Old trees represent past land use and may be used to map historic areas (Clare and Bunce, 2006). These generally isolated trees are cared for and their life prolonged. An old tree or a group of old trees may remain within a semi-natural area. Here the tree harbors even more uncommon species, and provides seeds with the tree's genotype(s) for future generations of local trees.

A special case of old trees are the *sacred groves* or *trees* present in some towns (Castro, 1990; Bhagwat *et al.*, 2005; Bhagwat and Rutte, 2006; Malhotra *et al.*, 2007; Sheridan and Nyamnersu, 2007; Pungetti *et al.*, 2012). Over time these groves become special sites

Figure 7.3 Distinctive baobab tree (*Adansonia digitata*) treasured by town residents north of Tulear, Madagascar. Photo courtesy and with permission of Mark Brenner. (A black and white version of this figure will appear in some formats. For the color version, please refer to the plate section.)

that attract pilgrims and visitors. Overuse of a site, including considerable soil compaction, commonly occurs. Yet many such special sites sustain uncommon species. Protection of a sacred tree or grove normally protects other species of conservation interest.

Where a town's tree care or management is minimal or absent, dead trees, snags (dead branches), tree holes, and logs may be common, all features being of wildlife importance (e.g., Sandviken, Tetbury). This especially refers to street trees. Also many older residential area trees are in poor health or dead-and-dying (Figure 7.2). It is unknown how this differs by planted versus spontaneous, native versus nonnative, and generalist versus specialist species.

Plant Distribution and Diversity

The plants and where they are found in towns are now explored from five perspectives: (1) plant types; (2) plants in homegardens and house plots; (3) plants in different-sized population centers; (4) villages with useful trees; and (5) species types and town features.

Plant Types

Shrubs have a rather different role and arrangement than trees in residential house plots. Perhaps most common are rows of shrubs along the side plot boundaries, providing some privacy from adjacent house plots. Foundation shrub plantings often partially surround a house for insulation or aesthetic reasons. Especially in tropical towns, flowering shrubs, providing beauty, attracting birds, and minimizing grass cutting, are common in the front spaces of houses (e.g., Grande Riviere; see Appendix). Shrubs often separate sections of a house plot, and tend to spontaneously colonize the back corners and boundary line. Mature shrubs in residential areas are mainly planted, perhaps especially using nonnative species.

Vines, which generally benefit from ample light, are often on the edges of woods and in small woods (e.g., Rocky Mount). Some vines grow on trees and rapidly climb to the canopy (e.g., Kutna Hora). Others grow on fences or low walls. And some vines grow on the walls of buildings (e.g., Ovcary, St. George's). Any of these cases may be found in a town, though usually not commonly except for abundant vines in a tropical rainforest setting.

In the absence of severe air pollution (Linda and Peter, 2007), long exposed surfaces in town tend to support an abundance of tiny plants – lichens, mosses, hepatics (liverworts), and algae (Seaward, 1979; Forman, 2014). Quarries, rocks, cemetery stones, stone walls (e.g., Tavistock, Issaquah), tile roofs (e.g., Icod de los Vinos, Chamusca), and in moist climates, posts, benches, and buildings, may be decorated with different-colored patches of these tiny plants (e.g., Flam, Tetbury). Also, without the high pH urban soil due to rainwater running over widespread concrete or mortar, the typically more acid town soils, plus dead wood, support an abundance of fungi (e.g., Sandviken, Falun). Mushrooms pop up all over towns.

Flattened gray or gray-green lichen patches just above a surface (foliose) are usually the most conspicuous species on tree trunks and rocks in town. Flattened blackish, grayish, to whitish lichen patches partly embedded in the surface (crustose) of a structure are often more abundant (Figure 7.4). Foggy towns commonly have beard-like (fruticose) lichens (such as *Usnea*) on old trees and boards. Lichen abundance and diversity is especially reduced by SO_2 pollution and perhaps by dust (Loppi and Pirintsos, 2003; Spier *et al.*, 2010; Lisowska, 2011). Hundreds of studies have documented the scarcity of lichens in cities (Linda and Peter, 2007). Lichens on gravestones in town and village cemeteries are particularly interesting, because the stones contain dates of placement and may be of different types (e.g., Flam, Tetbury).

Bryophytes, which include mosses and liverworts, mainly have tiny green stems and leaves and reproduce by spores. Relatively flattened liverworts may grow on some tree trunks and shaded wet surfaces, such as by fountains. Mosses of many species appear in shaded spots in lawns, at tree bases, on rocks and rock walls, in woodlands, and in wet places (Figure 7.4). Two cosmopolitan disturbance-tolerant species, *Ceratodon purpureus* and the silvery-green *Bryum argenteum*, grow in the cracks of town sidewalks and other disturbed locations (Wheater, 1999). Large mosses often drape the trees and surfaces of rainforest towns.

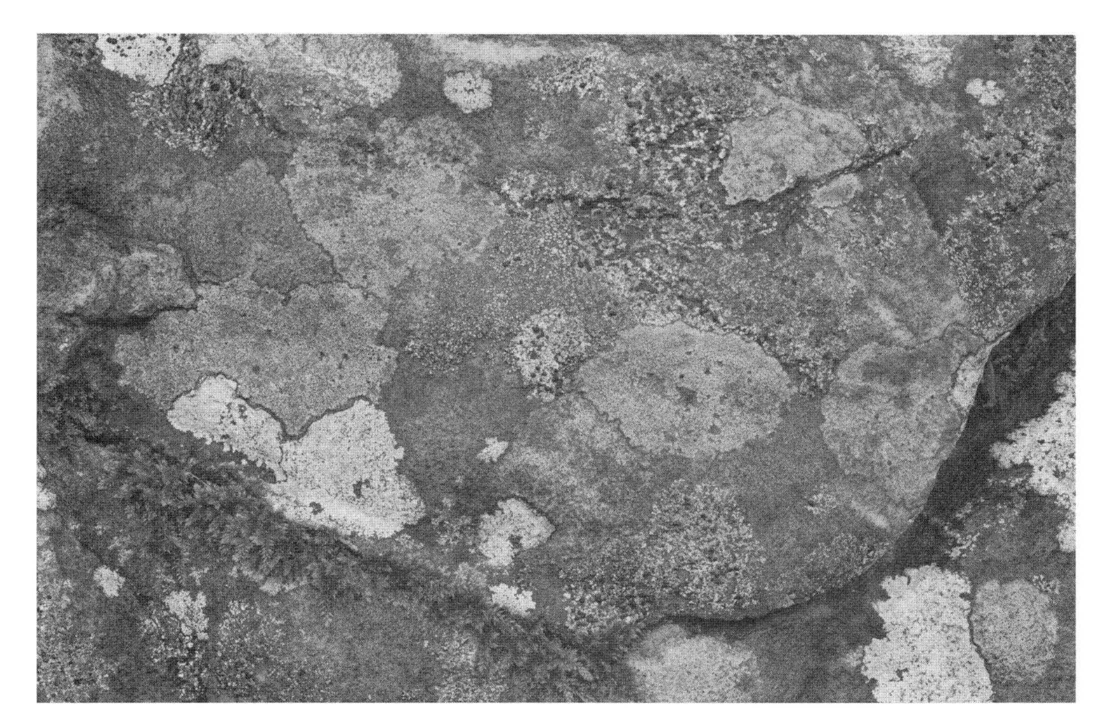

Figure 7.4 Rock surface with numerous lichen and some moss patches. About six, mostly crustose, lichen species; strip of creeping pleurocarpous moss with tiny stems and leaves in lower left. Ile de France, France.
Source: Jean-Luc Bohin / 500Px Plus / Getty Images.

Fungi decompose dead organic matter such as humus and wood, and mainly thrive in relatively acid conditions. Thus, in town, basically only areas that have been limed for plant production, areas next to concrete or brick-and-mortar structures, and a particular industry with high pH byproducts are unsuitable for fungi. The visible spore-producing structure, such as a mushroom, appears during a moist period. An important highly diverse type of fungi, the mycorrhizae, attaches to plant roots and is critical to growth of many plants.

Green algae and blue-green algae are basically in all town waterbodies. An excess of phosphorus in runoff from residential, commercial, or industrial areas turns a stream or pond green with phytoplankton algae. Some green algae thrive on moist tree bark, rocks, buildings, and even rainforest tree leaves. Coastal towns typically have brown and red algae on seawater rock structures, docks, posts, various walls, and boat bottoms. In effect, the preceding array of generally tiny plants in towns exceeds that of cities and farmland, and rivals the diversity in most forestland.

Plants in Homegardens and House Plots

Large house plots mainly covered with lawn and ornamentals in North America are typically a sign of sprawl, or high-income suburban residents (Thompson *et al.*, 2004;

Law *et al.*, 2004). Alternatively, large house plots in low-density housing can be mainly used for productive useful purposes such as growing food (Smith *et al.*, 2006). Such productive house plots or homegardens are widespread in moist tropical areas, where they help provide food security and many other values (Gajanseni and Gajanseni, 1999; Wezel and Bender, 2003; Kumar, 2006; Montagnini, 2006; Molebatsi *et al.*, 2010; Kumar, 2011). *Homegardens* have diverse productive useful plants over all or most of the house plot, unlike the typical small flower garden or vegetable garden within a house plot in most temperate areas. Homegardens with multi-layered vegetation somewhat resemble natural woodland, and support a rich mix of planted and spontaneous plant species (Albuquerque *et al.*, 2005; Kehlenbeck *et al.*, 2007). Rural South Africa home gardens, unlike peri-urban cases, are often arranged into six "micro-gardens": food gardens; medicinal gardens; ornamental gardens; structural species; open areas; and natural areas (Molebatsi *et al.*, 2010).

In Indonesia such house plots, called pekarangan, usually have considerable vertical and horizontal heterogeneity. These productive house plots markedly differ from upstream to downstream in the mountains-to-lowland area of the Bopunjur Region (Arifin *et al.*, 2014). The downstream area is flatter, wetter, more populous, and closer to markets than upstream. In upstream, middle-stream, and downstream communities, the average homegarden varies in area (188, 218, 562 m^2, respectively), plant canopy cover (sum of all woody plants >2.5 cm diameter) (167, 629, 1,733 m^2), number of plant species (27, 40, 44), and number of individual plants (1680, 4915, 1731). Thus, upstream productive home plots are smallest or lowest in all four variables. Downstream plots are largest and with the most woody plant cover, but have fewer plants than the middle-stream homegardens.

Ornamental species are predominant in the upper and middle homegardens, but only half as abundant in the lower house plots (Figure 7.5). Fruit-producing plants are important in all three areas, but highest in the downstream plots. Vegetable plants are relatively important in all three sites. Less common plants for spice crops (e.g., St. George's), medicinal plants, starchy crops, and industrial plants (products for industry) exist in all three locations. Other plant types, such as for fuelwood, handicrafts, and building materials are especially important in the downstream area.

Homegardens provide food, but also economic stability for residents when the national economy sinks. A study of 96 West Java homegardens (pelarangam) in four groups within each of four villages concluded that production provides up to 11.5 percent of a household's annual income and 12.9 percent of the residents' food expense (Kaswanto and Nakagoshi, 2014). Households range from one to ten people, averaging 4.6. Small productive house plots are smaller than 120 m^2 while large ones are 120–400 m^2. However, homegarden size seems to be related to whether residents also have other agricultural land.

Ornamental plants are most abundant in the West Java homegarden, averaging 47 percent of the total number of species present. Fifty-one percent of the ornamentals are in the bottom 0–1 m layer. However, 20 percent of these ornamentals grow more than 2 m high. Some village residents harvest resources in, and may degrade, nearby protected forest. The authors conclude that maintaining productive house plots is a way

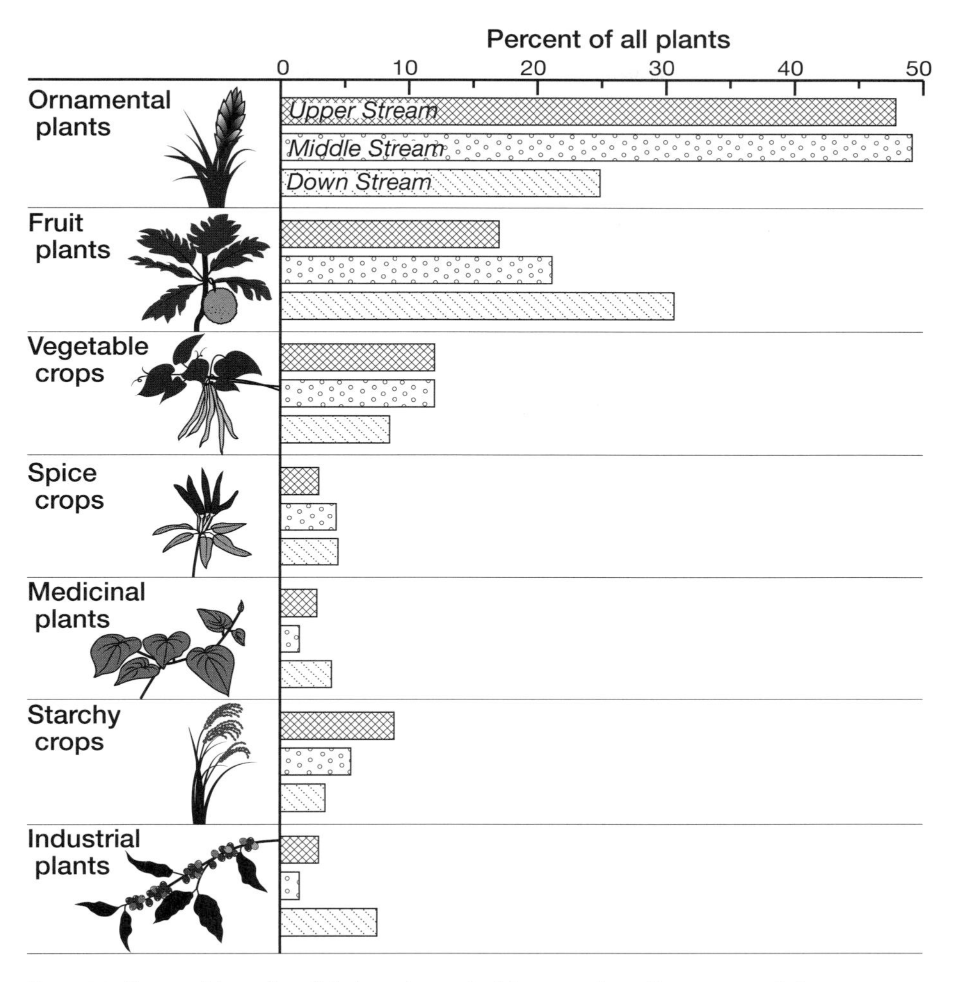

Figure 7.5 Composition of useful plants in tropical homegardens. Percentage of plants present in an average homegarden (pekarangan) in each of three elevational zones: upper stream (mountains), middle stream (transition), and downstream (lowland) area. Homegardens are productive house plots covered largely or entirely with useful plants, such as vegetables, spices, medicinal plants, and fruit trees and shrubs. Currently homegardens are most common in moist tropical regions, and may also contain farm animals. See text. Bopunjar Region, Java, Indonesia (Arifin *et al.*, 2014).

to reduce villagers' effects on the protected forest. Compared with nearby grassland farming, the heterogeneous multi-level homegarden stores more carbon and greatly improves local ecological conditions, from soil to wildlife (Arifin and Nakagoshi, 2011).

A rare alternative to the home garden and varied combinations of lawn, flower garden, and vegetable garden is the *native-plant house plot*, usually found in a town. For instance, in Florida a certified Florida Garden is a house plot completely covered with native species (e.g., Lake Placid). Soil organic matter, layers of vegetation, and diverse wildlife are all abundant. Maintenance involves periodic trimming, and weeding

Figure 7.6 Flowering shrubs plus food plants (banana, papaya, mango) covering front yard of village house. Grande Riviere, Trinidad. See Appendix. R. Forman photo. (A black and white version of this figure will appear in some formats. For the color version, please refer to the plate section.)

out of ever-invading nonnative plants. No mowing occurs, and herbicide and pesticide use is minimal or absent. Little or no watering is required since the species are adapted to the local climate.

In dry climates, many house plots in town may be *xeroscaped*, mainly to eliminate or minimize water use (e.g., Page, Socorro). Diverse forms of xeroscaped plots, especially the front spaces visible from the street, enrich a town. A xeroscaped space typically is a cover of pebbles, showy native desert plants, or local native and nonnative flowers. Various groups of rocks or cultural artifacts such as a wagon wheel or plastic pink flamingo are often added.

Generally, desert towns have a high diversity of shrubs and trees, which are scattered at low density (e.g., Alice Springs, Page). Many are next to buildings with slanting roofs, where roof runoff irrigates the roots and evapotranspiration may be less.

Compared with cities, towns have fewer walls and other visual barriers in front of residences, so more front spaces of house plots are designed for aesthetics, to be seen by people passing on the sidewalk or street. Flowering shrubs are commonly planted, sometimes over almost the whole front space, in addition to garden flowers and flowering trees (Figure 7.6) (e.g., Grande Riviere) (Owen, 1991, 2012; Smith *et al.*, 2005). Such a design seems alive with pollinators and birds.

Consider the plants in house plots (domestic gardens) of a low-income town, Ganyesa (population 19,000; Northwest Province, South Africa). A total of 346 plant species was identified in 60 house plots (Cilliers, 2010). Half of the species are cultivated exotic species, 21 percent cultivated indigenous species, 15 percent natural indigenous (native) species, and 14 percent are considered to be weeds, in the sense of unwanted exotic species. The town is in an area rich in endemic species (native nowhere else), and these house-plot weeds are prohibited by law and thus supposed to be controlled or eliminated.

Ornamental plants in these house plots are mainly cultivated exotic species, though cultivated indigenous species are common. The same pattern exists for food plants, though fewer species are used for food. Half of the medicinal plants present are cultivated indigenous species. Of all the human uses, natural indigenous species are most used for medicines. Trees for shade, hedge/windbreak, and wood are mainly cultivated exotic species.

In short, these house plots have 36 percent indigenous species (Cilliers, 2010). The predominant use of the flora is for aesthetics using ornamental plants, which are mainly exotics. Food and medicinal plants are about half natives and half exotics. The author suggests that residents with smaller families, higher income, and more modern houses with running water have gardens with higher plant diversity and more ornamentals. Lower-income residents have lower plant diversity and fewer ornamentals, but more medicinal and food plants.

Plants in Different-Sized Population Centers

Vegetation patterns are sometimes described or mapped for a town or village (Pysek and Pysek, 1990; Kovar, 1995). But how does species number or the density of species in a village or town differ from that in cities, and in the surrounding countryside? Furthermore, how does the percent of native species differ? A study of tree diversity in 54 villages, 38 towns, eight small cities (county seats), and four larger cities in the Jinzhong Basin (78% farmland and 7% forest) of central Shanxi Province (Central China) provides key insight (Wang *et al.*, 2009). Using 20 x 400 m transects (based on preliminary species-area curves), 6.4 hectares were sampled in each village (population 412 to 5,056), 7.2 ha in a town (3,208 to 20,924), 8.0 ha in a small city (51,200 to 84,300), and 8.8 ha in each larger city (population 95,000 to 267,400).

On average, a village had 33 tree species (total 178 species in all villages), a town 30 species (total 161), a small city 34 species (total 181), and a larger city 35 species (total 188) (Figure 7.7). Thus, the villages, towns, and cities have nearly the same number of tree species on average. These results are generally consistent with conclusions based on varied indirect evidence (Clements, 1983; Airola and Bucholz, 1984; Ma *et al.*, 2001; Bertin, 2002). Since the area sampled in each population center was similar, the actual numbers better represent the packing or density of species, rather than the species number in the entire population-center area.

The average number of indigenous species is 24 (village), 18 (town), 17 (small city), and 19 (larger city), respectively, suggesting that the village tree flora is mainly indigenous species. Towns and cities, with the same species number per unit area, have

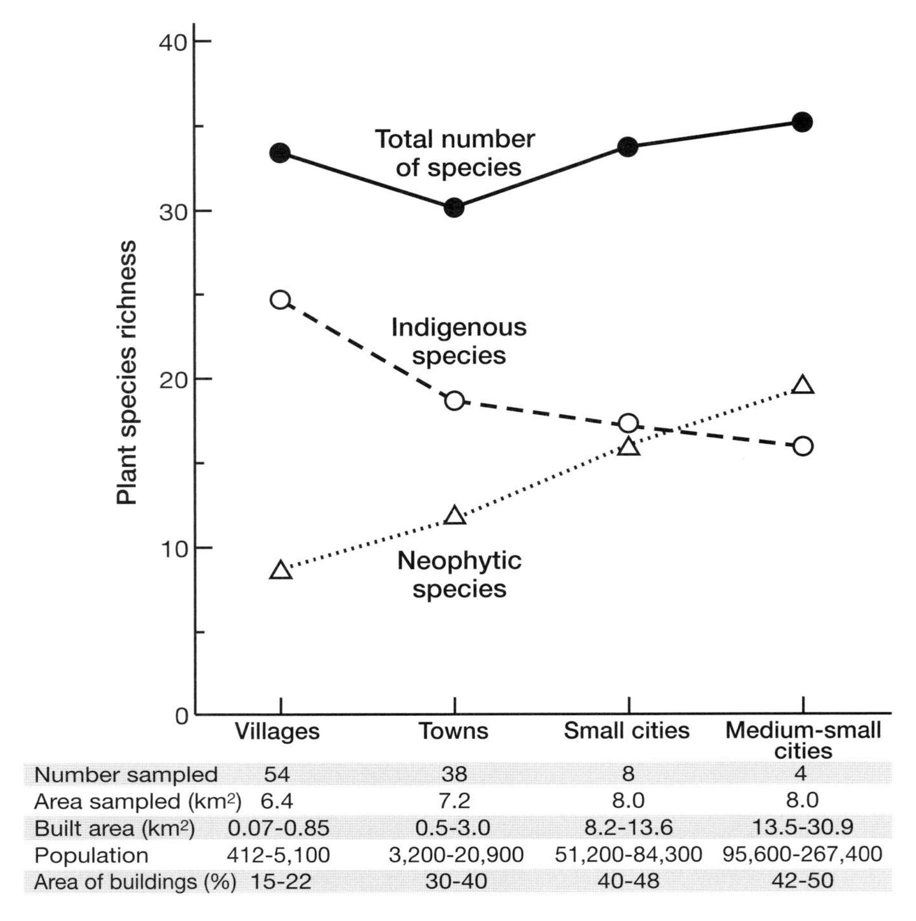

	Villages	Towns	Small cities	Medium-small cities
Number sampled	54	38	8	4
Area sampled (km^2)	6.4	7.2	8.0	8.0
Built area (km^2)	0.07-0.85	0.5-3.0	8.2-13.6	13.5-30.9
Population	412-5,100	3,200-20,900	51,200-84,300	95,600-267,400
Area of buildings (%)	15-22	30-40	40-48	42-50

Figure 7.7 Plant species richness in population centers of different size. Neophytic species presumably appeared over the centuries of intensive agriculture. Species richness calculated by the Gleason index (Gleason, 1926), D = S/ln A, (S = species number; A = area sampled), to correct for different areas sampled in the communities. Jinzhong Basin, Shanxi Province (China) (Wang *et al.*, 2009).

fewer indigenous species than do villages (Wang *et al.*, 2009). The average number of *neophyte* species (presumably species which arrived since 1500), is 9, 12, 17, and 16, respectively. Thus, for neophyte trees, towns are more similar to villages, and have fewer recent arrivals than do cities. Overall, except for indigenous species in towns, villages and towns are relatively similar in tree species and origin, and quite different from small and large cities.

Additional interesting insights emerge from examining human-dispersed plant species in Finnish villages (Hanski, 1982). Most of the village species fall into two categories, either common in the region, or rare and localized typically in one village. Species number increases with village size, and decreases with the degree of isolation of a village. Similarity in *species composition*, i.e., the particular species present, between villages decreases as villages are further apart. These results are consistent with patterns for species in natural patches documented in landscape ecology and conservation biology (Forman, 1995; Primack, 2004).

Villages with Useful Trees

A quite different and interesting picture of biodiversity emerges in a study of several villages near the Danube River in western Romania (Madera *et al.*, 2014). Twenty habitat types (biotopes) are recognized around the villages in the 21 km (13 mi) diameter area of generally small-field agriculture and woods. Widespread generalist plant species predominate, with a third (34%) of the flora in seven or more of the habitat types. Ten percent of the flora is only in three habitat types, and another 20 percent only in two habitats. A third (36%) of the species are limited to a single habitat type. Given the abundance of villages and prevalence of farming, a considerable number of the spatially limited plant species may be expected to locally disappear during the decades ahead.

But the villagers, mainly of Czech descent, have created a highly useful biodiversity in and adjoining the villages. Fruit trees predominate as street trees planted by residents, as a partial low canopy in productive house plots (small homegardens), and as orchards nearby (Bocek, 2014). Plums (*Prunus domesticus* and *P. insititia*) used for consumption, preserves, and distillates have the greatest cover and economic value.

Apricots and peaches are usually close to the home due to the care needed. Walnuts and mulberry along streets in Sfanta Elena (population 300) provide shade. Apples and plums dominate the street trees and orchards of Sumita (pop. 100). Peaches and plums predominate in Ravensca (pop. 100). Grnic (pop. 500) and Eibenthal (pop. 200) grow apples near their streams, while quince and medlar are scattered though the villages. Thus, trees with harvestable products predominate in all the villages, and the species are carefully arranged in microhabitats and residential locations.

Biodiversity here is enriched by the genetic varieties of fruits planted. At least eight plum varieties, 25 apple varieties, and eight pear varieties present help provide stability for the villages through extreme weather events and pest outbreaks. In this way the villages are less sensitive to global climate change.

Species Types and Town Features

Species are often categorized as common or rare, planted or spontaneous, and native or nonnative (Muller *et al.*, 2010; Forman, 2014). Variants of these types and other species types exist. The rare or common category highlights the widespread nature of some species, and the importance of conservation for other species. Planted versus *spontaneous species* (arrivals not purposely introduced) is especially informative ecologically, as it relates to available species, their genetic stock, and planting protocols, as well as to dispersal processes, competition, ecological succession, and sites without maintenance.

Native versus *nonnative species* has recently been of some interest in ecology, with nonnatives cast as between undesirable and evil. The subject is particularly relevant in Australia, which has had major economic problems, and still has significant ecological problems, with nonnatives. Studies have often focused on *invasive species* that spread rapidly and fit poorly in food webs and nutrient cycles. Later, when no longer spreading, the species may become *naturalized* by "fitting into an ecosystem" (Adler and Tanner, 2013; Bertin, 2013). Either in the invasive or naturalized phase, the species may thrive in a degraded habitat, such as a rail yard (Muhlenbach, 1979). Alternatively, it may alter somewhat a natural/semi-natural ecosystem.

Nonnatives are abundant in farmland where a small portion become pests. In cities, nonnative species play important roles (Forman, 2014), by reducing erosion, urban heat, and air pollution, and increasing ornamentals-based aesthetics and pollinator populations. Understanding the roles of nonnative species in towns and villages awaits researchers. Overall, somewhat analogous to teachers and professors evaluating students, it seems best to judge a species by what it does, not its origin.

Fewer sources of nonnatives exist for towns than cities. Few towns have an international airport, rail terminal, or shipping port where nonnatives arrive. Few towns have large trucking warehouses where some nonnatives multiply. Few towns have botanical gardens (e.g., Superior, Alice Springs) or an arboretum, where nonnative species often arrive in root balls and are cultivated. Towns have few plant nurseries (typically 0–2) where nonnatives arrive and are dispersed throughout residential and commercial areas (Ignatieva, 2012) (e.g., Lake Placid, Sooke).

Trucks servicing a town's commercial and industrial centers may be a prime carrier of nonnatives. However, residents traveling to cities and suburbs may also bring back nonnative plant species. Roadsides and railways are both habitats and conduits for species movement (Chapter 12). Nevertheless, the relatively little polluted and little heated town environment is more suitable for native species that may outcompete nonnatives.

The nearest suburbs and cities probably serve as one source of species in towns. For instance, 29 plant species were found growing at the bases of trees in seven major European cities (Muller and Werner, 2010). The most abundant species are: *Taraxacum officinale*, *Poa annua*, *Polygonum aviculare*, *Hordeum murinum*, *Chenopodium album,* and *Plantago major*. All except *Hordeum* are also at tree bases in Baltimore in North America. The abundance of these tree-base species in a town might reflect distance from a major city, or correlate with population size, or a town's stage of growing into a small city.

Town industries generally maintain piles of raw materials, stored products, and waste byproducts. In addition, the manufacturing process and smokestacks coat the soil with pollutants. In Britain each type of industry has its own collection of plants (Gilbert, 1991). For example, hundreds of plant species have been associated with woolen factories (Lousley, 1961). Since much of the wool was imported from Australia and South Africa, seeds in the wool germinated, but the plants did not survive the northern winter.

High-quality paper manufacturing used dusty rags containing seeds from Egypt and North Africa, which led to quite a different flora around the factory. A factory producing plant oils (soy bean, rapeseed, castor bean) had 85 unusual plant species around it (Palmer, 1984). Breweries and flour mills each had a distinctive surrounding flora

(Gilbert, 1991). A medium or large factory probably alters the flora in most of the town. A number of small industries, each with different inputs and outputs, probably also affects the plants present.

Specialized features or habitats present in some towns doubtless not only support a somewhat distinct set of plant species (Gilbert, 1991; Wheater, 1999), but also probably affect wider areas in town. Good examples are a: deep-mine pit (e.g., Falun; see Appendix); equestrian center (e.g., Cranborne, Sandviken); fort (St. George's); hazardous waste site (Clay Center, Concord); historic town meadow (Petersham); zoo/salmon hatchery (Issaquah); major ski jump (Falun); prison with dairy-herd pastureland (Huntingdon/Smithfield, Concord); institution (Emporia, Huntingdon/Smithfield); sewage treatment facility (Socorro, Lake Placid); community garden (allotment) (Frome, Scappoose); abandoned railway (Westbrook, Lake Placid); natural or quarry cliff (Capellades, Wausau); peat bog (Stonington); limestone cenote (Valladolid); and old-growth woods (Wausau/Weston).

The preceding insight on nonnative species and the richness of human-created habitats in town provide a useful foundation for the concept of *novel ecosystems*. The concept has briefly appeared many times over the past hundred years (Whitney and Adams, 1980; Rapoport, 1993; Hobbs *et al.*, 2006; Marris, 2011; Mascaro *et al.*, 2013; Alberti, 2016), and its meaning remains in flux. Novel ecosystems have been described as: (1) far removed from historical ones in environmental and biological attributes; or (2) characterized by new species combinations due to human action, but sustained without people.

Building from these, a third concept of novel ecosystems contrasts from the historical and hybrid (combined natural and human-influenced origin) by being irreversible. The ecosystem has lost the capacity to return to its preceding condition or historical state (Hobbs *et al.*, 2009, 2013). For this, a novel ecosystem represents an irreversible "regime change." Unlike many cities and severely eroded land, towns and villages may not have reached that irreversibility level. Thus, ecological succession in "ghost towns," "the lost towns and villages of Britain," or the archaeological tells of the Middle East may resemble patterns that existed in land surrounding the former towns.

Spontaneous species (adventives, self-sown), which are unplanted and colonize a site on their own, are widespread in both towns and villages (e.g., Sooke, Scappoose). Probably native species predominate though nonnatives are usually present, even common. A town site with spontaneous vegetation often appears species-rich, a magnet for diverse pollinators, a semi-natural habitat for wildlife, and even neglected, messy, or natural (Figure 7.8). Plant species richness is normally high, much higher than on planted sites. While naturalized and early successional species colonize former cultivated sites, native species may be scarce on such sites (Foster, 1992; Motzkin *et al.*, 2002; Ito *et al.*, 2004; Kobayashi and Koike, 2010). Spontaneous vegetation is not static, but ever changing in response to seasons, disturbances, and successional processes.

Ten villages were studied in Bohemia (Czech Republic) before 1982 and after 1995, a period of population increase and built-area expansion (Wittig, 2010). Indigenous species decreased only slightly in the villages, on average from 56 percent to 54 percent of the total flora over the period. *Archaeophytes* (nonnative species which appeared before 1500, when world explorers brought numerous new plants to Europe) decreased from 31 percent to

Figure 7.8 Profusion of successional flowering plants in front yard of abandoned house. Signs of house decay suggest potential animal habitats. Town of Dexter, Maine, USA. R. Forman photo. (A black and white version of this figure will appear in some formats. For the color version, please refer to the plate section.)

27 percent. Meanwhile, *neophytes* (nonnatives appearing after 1500) increased from 12 percent to 19 percent of the flora in the ten villages. Thus, over 13 years, increasing neophytes was the biggest plant change in the villages. Nonnative species normally keep increasing in built areas, especially new developments. Fighting or trying to control nonnatives seems only useful in the long term for large relatively natural patches in or near town.

Rare native plants also exist in some towns, usually as small populations (e.g., Lake Placid, Concord). The probability of their long-term survival in town is low, considering diverse daily human activities and major changes ahead, including climate change and possible growth from town to city. Usually, few residents know about the rare species, and little collective effort is made to protect them.

Some towns contain a botanical garden or arboretum which normally is an exceedingly rich concentration of native and nonnative species (e.g., Superior, Cranborne, Alice Springs). Just as for a plant nursery, the botanical garden uses considerable water. Pesticides are normally also widely used. Seeds and pollen from some of the species disperse into the town and surrounding area. Over time the native and nonnative plants become widely distributed in house plots and other sites throughout the town (e.g., Superior).

Streams partially bordered by riparian vegetation in a stream corridor are common in towns (Figure 7.9) (e.g., Pinedale, Sant Llorenc, Vernonia). The riparian vegetation is

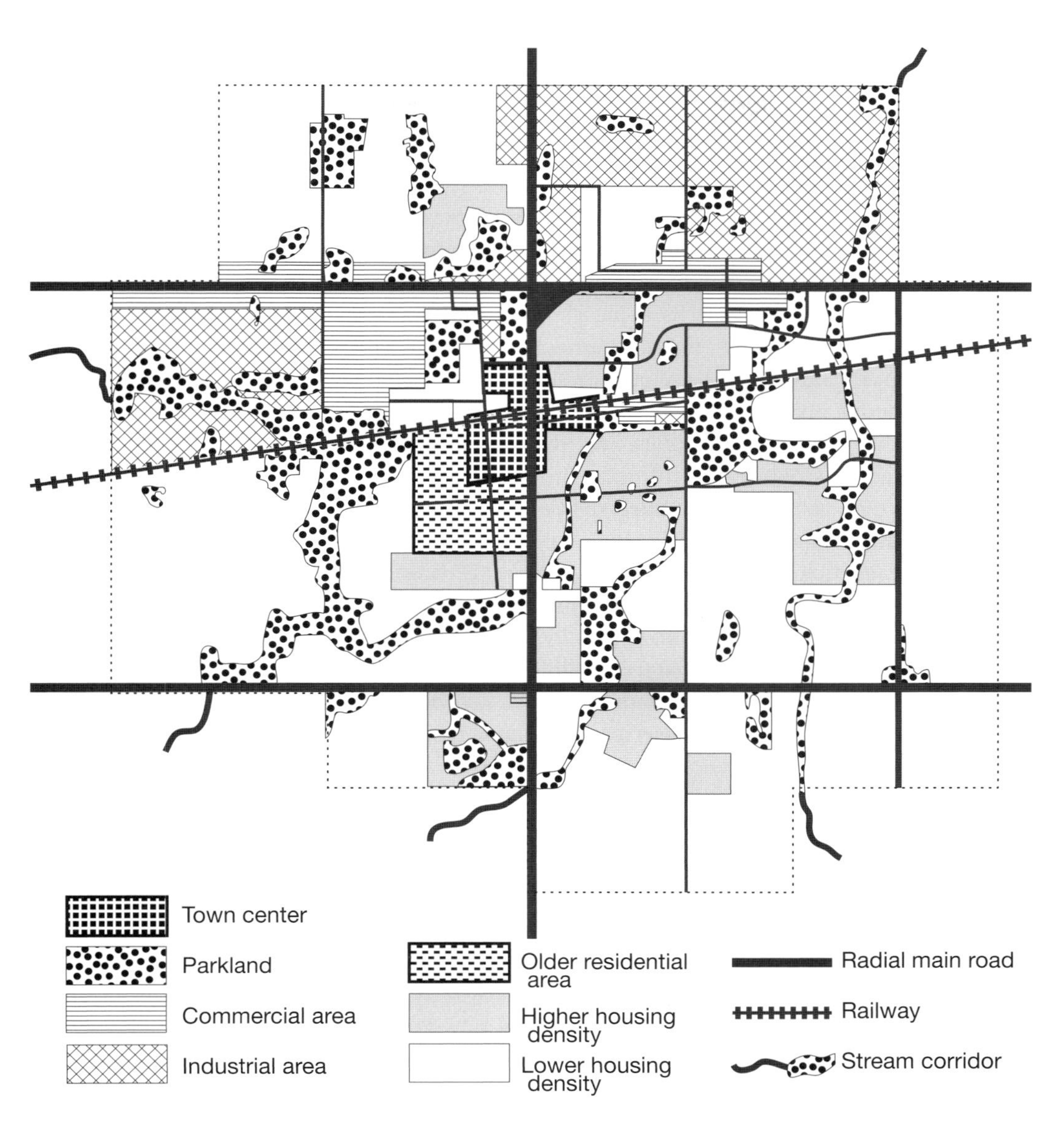

Figure 7.9 Green patches and corridors in a commuter town and commercial center for surrounding villages/hamlets. Note 1 long corridor, 8 short corridors, 2 large patches, 28 small patches, agriculture surrounding town, and highway and railway cutting through. Greenspaces are major factors attracting and sustaining residents. Stony Plain (population 15,000), Alberta, Canada. Adapted from Friedman (2014).

typically present in the outer residential areas, but largely eliminated in or near the town center. Stream corridor width has long been the primary focus of ecological research and conservation, though other variables such as adjoining land use and ecological connectivity are known to strongly affect biodiversity (Ward *et al.*, 2002; Moffatt *et al.*, 2004; Rodewald and Bakermans, 2006).

A study of 18 riparian floodplain corridors in suburbs with streams outside Sydney identified patterns that correlated with plant species richness (Ives *et al.*, 2011). Of 19 spatial, soil, stream, and adjoining built-area variables calculated, only two correlate significantly with total species richness of a sample site. More plant species grow (1) where the perimeter-to-area ratio of the vegetation patch containing the sample is greater, and (2) where the land surface adjoining the stream is less steep. Convoluted patches and flatter floodplains have more plant species. For species composition (the specific species present in a sample), riparian corridor width is the best predictor. Thus, landscape ecological patterns are useful in understanding species numbers and types at a site within a suburb's riparian corridor.

Old cemeteries are often near the older residential area in town. However, some cemeteries on the edge of town, in the adjacent zone, and especially around villages have an abundance of native species (e.g., Clay Center, Selborne). Thus, except for planted nonnatives, native species normally predominate in such cemeteries. Indeed, species richness may be quite high in old cemeteries, if near a large natural area and if maintenance activities are minimal (Barrett and Barrett, 2001; Forman, 2014). Also, microhabitat diversity, illustrated by the presence of ponds, diverse unpolished stone markers with lichens, old trees with holes and dead branches, a relative abundance of shrubs, and limestone or sandstone walls, also leads to high species richness. Many such cemeteries contain rare species (e.g., Kutna Hora, Tetbury) (Gilbert, 1991; Czarna and Nowinska, 2011; Paton and Sheahan, 2013).

Finally, to gauge the diversity of different groups of plants and animals, and for total diversity, seven plant and animal groups were measured in 27 protected areas of a medium-small German city (Brauniger *et al.*, 2010). As expected, size of greenspace was overall the best predictor of species richness for all groups. But also lichen species diversity correlated significantly with moss diversity. Carabid beetle (Carabidae) diversity correlated with butterfly diversity. Bird and snail diversity did not correlate with diversity of any other group. But impressively, moss diversity, butterfly diversity, carabid diversity, and total species diversity were all significantly correlated with vascular plant diversity. This suggests that vascular plant diversity has the widest influence on the diversity of other biological groups in the protected areas.

Species Movement and Change

Three perspectives explore this important subject: (1) natural processes and town processes, (2) flows and movements, and (3) vegetation change and disasters.

Natural Processes and Town Processes

Lots of natural processes related to plants are present, indeed conspicuous, in towns and villages. Wind blows, which erodes soil, transports leaves, disperses seeds and spores, and desiccates plants. Chemicals flow in plant photosynthesis, transpiration, and translocation between roots and leaves. Flowers are pollinated and seeds dispersed by animals. Plants compete with plants, are consumed by herbivores, and support epiphytes on their trunks and branches. Plants grow, die, and are decomposed. With limited heat buildup and air pollution in towns and villages, such natural processes may resemble those in natural/semi-natural areas more than in agricultural and urban areas.

Yet human activities or processes in towns interact with natural processes (Hough, 2004). Trees, shrubs, flower gardens, lawns, and hedges are planted and maintained. Scattered impervious surfaces produce a fine mosaic of dry and moist spots. People dig holes, compact soil, and add strips of usually porous fill. Fertilizers and pesticides are also often added to plants and soil. Planted palatable nitrogen-rich plants attract pest herbivores which over-browse vegetation. Bird feeders attract birds which disperse the commercial seeds provided.

Maintenance of vegetation has varied goals, such as aesthetics and debris removal, but most maintenance activity effectively combats natural processes. Consider the relative *maintenance costs* for a city's greenspace (Rotterdam, The Netherlands, 2002) (Hough, 2004). Based on the number of hours of annual maintenance per 100 m^2, one vegetation type in town is very costly to maintain: annuals require 150 hr/100 m^2. Four types are relatively costly (perennials, 39 hr; roses, 21; hedges, 19; and groundcovers, 9). And four types are inexpensive to maintain (shrubs, 3.5; lawns, 2; woodland, 2; and "naturalization," 1). Most towns have few or small parks which may be either woodland or lawn dominated, both at low maintenance cost. Shrubs can be combined with lawn at low cost. However, towns are mainly residential, so residents provide the intensive maintenance needed for abundant flower gardens with annuals, perennials, roses, and the like.

The balance between natural and human processes helps to produce the distinctiveness of a town. Tree cover may be the primary determinant. In wooded landscapes, the partial openness of towns makes them different, yet the abundance of native trees suggests a wooded savanna where natural processes predominate. In agricultural landscapes, towns often have many trees and, except when dust pours in, the town feels like an open savanna, with somewhat different natural processes important. In desert, intense sun and wind dominate so a desert town with scattered trees is a distinctive place contrasting with the surroundings.

Flows and Movements

Seed Dispersal, Pollination, Gene Flow

Town seeds are commonly dispersed by wind, water, vehicles, people, or animals (Murray, 1986; Black *et al.*, 2006). For instance, wildlife carry a set of attached seeds into town, and leave with a somewhat different set of seeds. Similarly, a set of seeds blows into town, and a somewhat different set blows out.

Relatively stable streamline airflow may predominate in open rural areas, whereas turbulence mainly characterizes built areas (Chapter 5) (Bornstein, 1977). Unlike urban areas, seed dispersal seems to be quite important in towns and villages, which have considerable surface area suitable for seed germination and plant colonization (Cheptou *et al.*, 2008). Also, towns have a higher proportion of spontaneous rather than planted plants, and native species rather than nonnatives.

An extreme windstorm carries seeds great distances (Swan, 1992; Nathan *et al.*, 2008; Dauer *et al.*, 2009). The tiny lightweight spores of fungi and mosses are carried across regions and continents, even intercontinentally, by winds. With town tree cover reducing windspeed, some of the incoming seeds and spores drop to the ground, later to germinate and produce plants (Baskin and Baskin, 2001). Meanwhile, existing trees and other plants disperse wind-blown seeds. The *seed shadow*, i.e., the area adjacent to a tree where almost all of the seeds fall to the ground, is asymmetric due to differing wind directions (Murray, 1986). The seed shadow size or radius depends on seed characteristics of a species, plus of course windspeed. Wind-blown seeds usually fall to the ground in a *negative exponential pattern* (e.g., decreasing with square of the distance), with numerous seeds close to the tree and progressively fewer at greater distance.

Where in town would be the most likely routes of wind-dispersed seeds? Wind whistles down long streets without overhanging trees. Traffic moving in one direction creates airflow that carries seeds, such as *Ailanthus latissimus* (Kowarik and von der Lippe, 2011), and even paper cups. Wind can accelerate across the occasional large open field, such as in a park, golf course, or airport, picking up seeds enroute. Similarly, seeds are picked up and carried along pipeline and powerline corridors cutting through towns. Seeds from riparian species are transported by water and wildlife along rivers and wide streams to and from towns. The edges of these corridors and open patches are likely to be rich in wind-dispersed species.

A stream through town carries lots of seeds. Seeds in mud attached to vehicles have moved long distances in rural or remote areas (Gilbert, 1991; Forman *et al.*, 2003). In the town's newer residential area, the *seed bank*, i.e., the seeds in the soil, may initially reflect in part the preceding land use, such as agriculture. Over time some of those residual seeds die, while new seeds are deposited on the ground.

People transport seeds in diverse ways (Black *et al.*, 2006). Active walkers in town redistribute seeds with tiny hooks which annoyingly attach to clothing. Some people bring seeds into town to sell at a farmers' market, or for other purposes (e.g., Alice Springs; see Appendix). Many nonnative species arrive inadvertently from suburbs and cities in vehicles, or with people in vehicles. A prime movement of seeds is by shipping wool, which is often full of seeds picked up by sheep moving through pastureland (Hanson, 2015). Within or around the town, seeds may be transported in great quantity with the collection and disposal of yard waste, such as fallen palm fronds in the tropics. People constantly disperse some seeds from food bought in markets, or collected in natural areas (Figure 7.10).

Animal dispersal may be most important in moving seeds around town (Tiffney, 2004; Lobova *et al.*, 2009; Martins *et al.*, 2009; Enders and Van der Wall, 2012). Migratory birds stopping briefly in the green patch of town bring seeds, redistribute seeds in town, and take seeds away. Both native plants and native wildlife species are relatively abundant in towns, so conditions are somewhat natural. Often shrubs are widespread, and in many areas a relatively high proportion of the shrub species produce seeds within a fleshy fruit. Wildlife readily consume the fruit, move locally, and defecate the undamaged seeds. In this way, fruit-producing plants are readily moved around town. The main

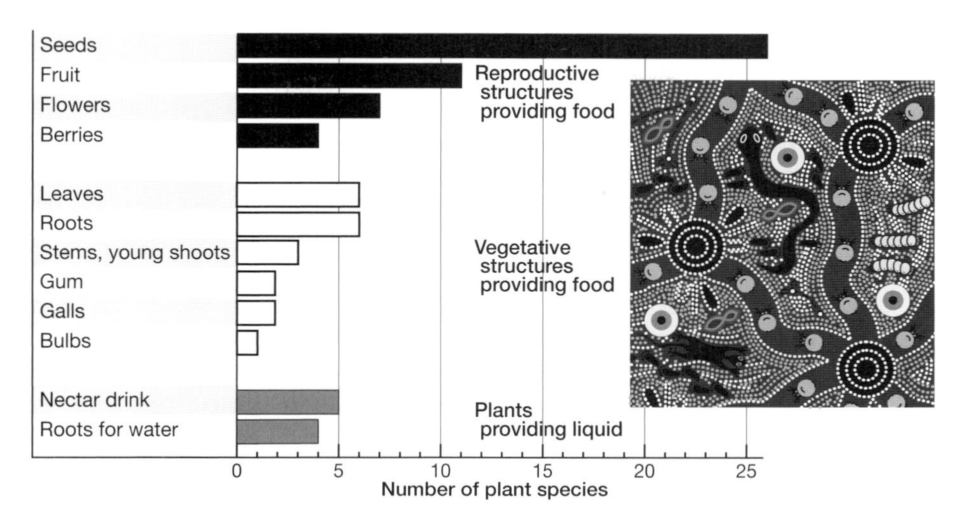

Figure 7.10 Plants providing food for nomadic and sedentary people away from towns and villages in Australia's Central Desert. Total plant species = 62 (some with two or more edible structures); 22 species used for medicines; and many land animals, insects, and grubs provide food. Aborigine people in area around the MacDonnell, Musgrave, Tomkinson, and Warburton ranges, and Alice Springs, Papunyus, Mt. Liebig, Amata, The Olgas, and Uluru (Ayers Rock). Rainfall 256 mm (11 in). Based on Isaacs (2002). Commercial dot painting of the land on right.

routes of birds and mammals in town, such as along hedges, from woodland to woodland, and between stream and nesting area, are the highways of animal-dispersed seed movement. Shrubs important for wildlife movement are often common along residential property lines and woodland edges.

A study in rural Spain found that bird dispersal of the two main fleshy fruit-producing plants, hawthorn and holly (*Crataegus monogyna*, *Ilex aquifolium*), was greater for plants within a forest, than if the plants were scattered in a field (Herrera and Garcia, 2009). However, in years with little fruit production, the birds readily fed on fruits of the isolated field shrubs and small trees. The field plants provide stability for the wildlife in tough times.

Pollination, the movement of pollen from the male flower structure to a female flower structure (which can produce a fruit), basically results from wind or animal movement. Streamline airflow tends to blow pollen somewhat linearly and horizontally to a distance, whereas turbulence produces multiple short routes and directions. The primary animal dispersers – bees, other bee-like insects (hymenopterans), bats, and certain birds – especially carry pollen along their main routes (Knight *et al.*, 2005; Osborne *et al.*, 2008; Stelzer *et al.*, 2010; Winfree *et al.*, 2011; Tarrant *et al.*, 2013). Pollen dispersal by animals is more prevalent in towns than urban areas, because animals are more abundant and the numerous habitats with flowers are closer to one another.

Flowers are widely distributed in town, and male flower structures produce pollen. Tree cover is especially important because trees produce huge numbers of flowers and pollen. Well above the ground, wind-borne pollen is readily picked up and carried considerable distances by wind. In some cases, male street trees which produce pollen are favored, such as ginkgo (*Ginkgo bilobata*) in Japan's built areas, because the female fruits make slippery sidewalks that stink.

Flower gardens represent another dense pollen source, yet in this case the pollen is usually produced by many different species (Corbet *et al.*, 2001; Garbuzov and Ratneiks, 2013). Flower gardens occur in parks, backyards, and most commonly in the front spaces between houses and street (Gilbert, 1991; Owen, 1991; Smith *et al.*, 2005). Flowering shrubs, often in lines along property boundaries, produce abundant pollen. Small patches of spontaneous plants, including wildflowers and weeds, are also common in towns. Finally, many roadsides and ditches receive limited maintenance and hence support numerous flowers. Adding these overlapping pollen sources together highlights the diversity and abundance of pollen in towns.

A study by volunteers of butterflies and flowers in more than 6,000 gardens fairly evenly distributed across France found that specialist butterflies had a longer proboscis (feeding tube) than did generalists (Bergerot *et al.*, 2010). Specialists fed on one or a few flower species, while generalists fed on many flowers.

To evaluate the differences in garden butterflies from remote areas to urban areas, the authors divided the data into eight groups based on the amount of urbanization in a municipality: <1% (986 gardens); 1%; 2–3%; 4–6%; 7–10%; 11–19%; 20–40%; and >40% (527 gardens). If 20–40 percent urbanized mainly represents small cities, most of the categories represent isolated homes, hamlets, villages, and different-sized towns. Butterfly richness and the number of specialist species decreased with increasing urbanization. Indeed, of 26 species in 9 butterfly families, all but 2 decreased with more urbanization.

Probably the routes of pollinators across town mainly follow corridors. In general, vegetation connection facilitates animal movement (Forman, 1995; Sarlov-Herlin and Fry, 2000; Bennett, 2003). Stream, railway, and main town roads with or without street trees are the primary long corridors. Short strips such as hedges, fences, and walkways may form fine-scale networks, especially in residential areas. Lots of animals, including bats, birds, and mammals, readily move along these corridors in town.

Gene flow occurs with both seed and pollen movement. Gene flow may lead to *genetic change*, the altered frequencies of genes in a population over time. Several attributes of towns may facilitate genetic change (Cheptou *et al.*, 2008). For instance, the characteristic small semi-natural patches contain many small plant populations. Over time small populations fluctuate in size, with some simply disappearing (local extinction). Also, small populations with inbreeding tend to have weak and sterile offspring. Indeed, both plant and pollinator populations are often small in towns.

Gene flow can also partially overcome the effects of habitat fragmentation. Thus, a genetic study of insect-pollinated trees (*Sorbus domestica*; Rosaceae) in Switzerland found that pollen and genes readily crossed closed forest, deep valleys, open land, and settlements (Kamm *et al.*, 2010). The pollinators, i.e., bees, bumblebees, and flies (Diptera), were considered to be generalists. The settlements were mainly small villages surrounded by orchards.

Little is known about gene flow and genetic change within towns and villages (Chapter 12). In an urban area, a moss (*Leptodon smithii*) and six herbaceous plants

Figure 1.1 Town on amphitheater-like slope descending to harbor with boats for fishing, tourists, or shipping goods. St. George's, Grenada, W.I. See Appendix. R. Forman photo. (A black and white version of this figure will appear in some formats.)

Figure 1.3 Village with most families returning from city on weekends/holidays to join 89 residents. Abanades, Castilla y La Mancha, Spain. R. Forman photo. (A black and white version of this figure will appear in some formats.)

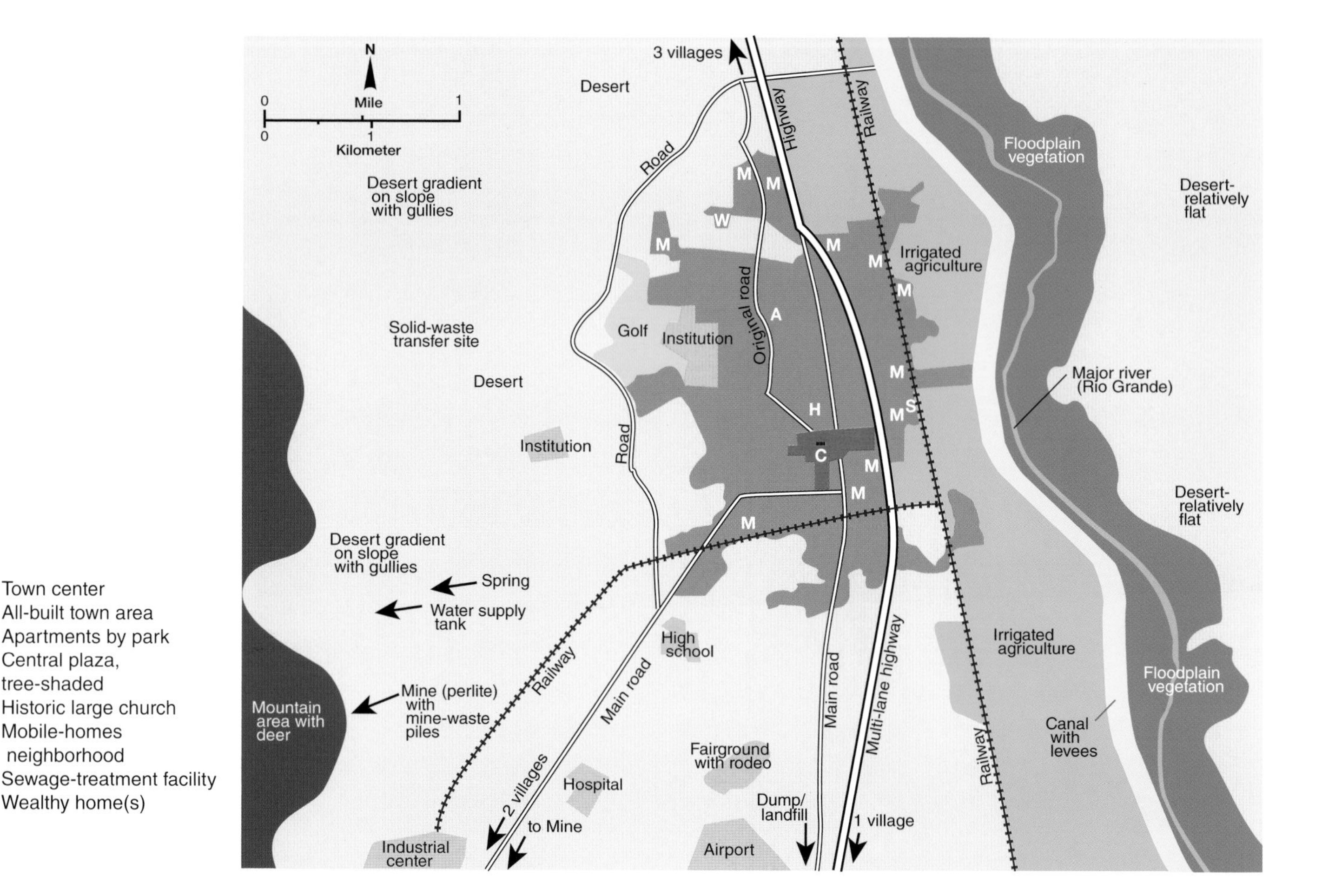

Figure 1.2 Desert town by major river and transportation corridors. Socorro, New Mexico, USA. See text and Appendix. (A black and white version of this figure will appear in some formats.)

Figure 1.13 Linear village with rather stable population of 119 and distant church the only common building. All houses adjoin both road and pastureland, mostly elongated fields, as in medieval times. Land slightly slopes to Baltic Sea on right; ditches on left in the groundwater emergence zone drain water away. Livestock in and around barns in early April; tank in foreground for livestock waste. Handicrafts made, but no manufacturing. Planted coniferous woods in distance. Stenasa, Oland, Sweden. Photo courtesy and with permission of Lars Bygdemark, photographer. I appreciate the assistance of Margareta Ihse and Ulf Sporrong. (A black and white version of this figure will appear in some formats.)

Figure 1.4 Main Street with several local shops of village in forested mountains and lakes. Rangeley, Maine, USA. See Appendix. Watercolor by H. Chandlee Forman; R. Forman, owner. (A black and white version of this figure will appear in some formats.)

Figure 2.6 Infrared image of small town with diverse corridors. Main roads through town; short green corridors in town; northwest/southeast walking trail on former railway bed; small freshwater stream at top; hedgerows in farmland; multi-lane highway on right; wide branching stream in saltmarsh lower right; tiny drainage ditches against mosquitoes in saltmarsh. Southeastern Massachusetts, USA. US Department of Agriculture, Soil Conservation Service photo. (A black and white version of this figure will appear in some formats.)

Figure 2.11 Small river from rainforest mountains during dry season. Floodwaters in wet season carry stones and finer sediment into the Caribbean Sea, where waves wash the material back to form a coarse-sand beach, ideal for nesting of huge leatherback sea turtles (*Dermochelys coriacea*). Village of Grande Riviere, Trinidad. R. Forman photo. (A black and white version of this figure will appear in some formats.)

Figure 2.12 Transportation by pedi-rickshaws in front of building construction site. City of Tulear (Toliara), Madagascar. Photo courtesy and with permission of Mark Brenner. (A black and white version of this figure will appear in some formats.)

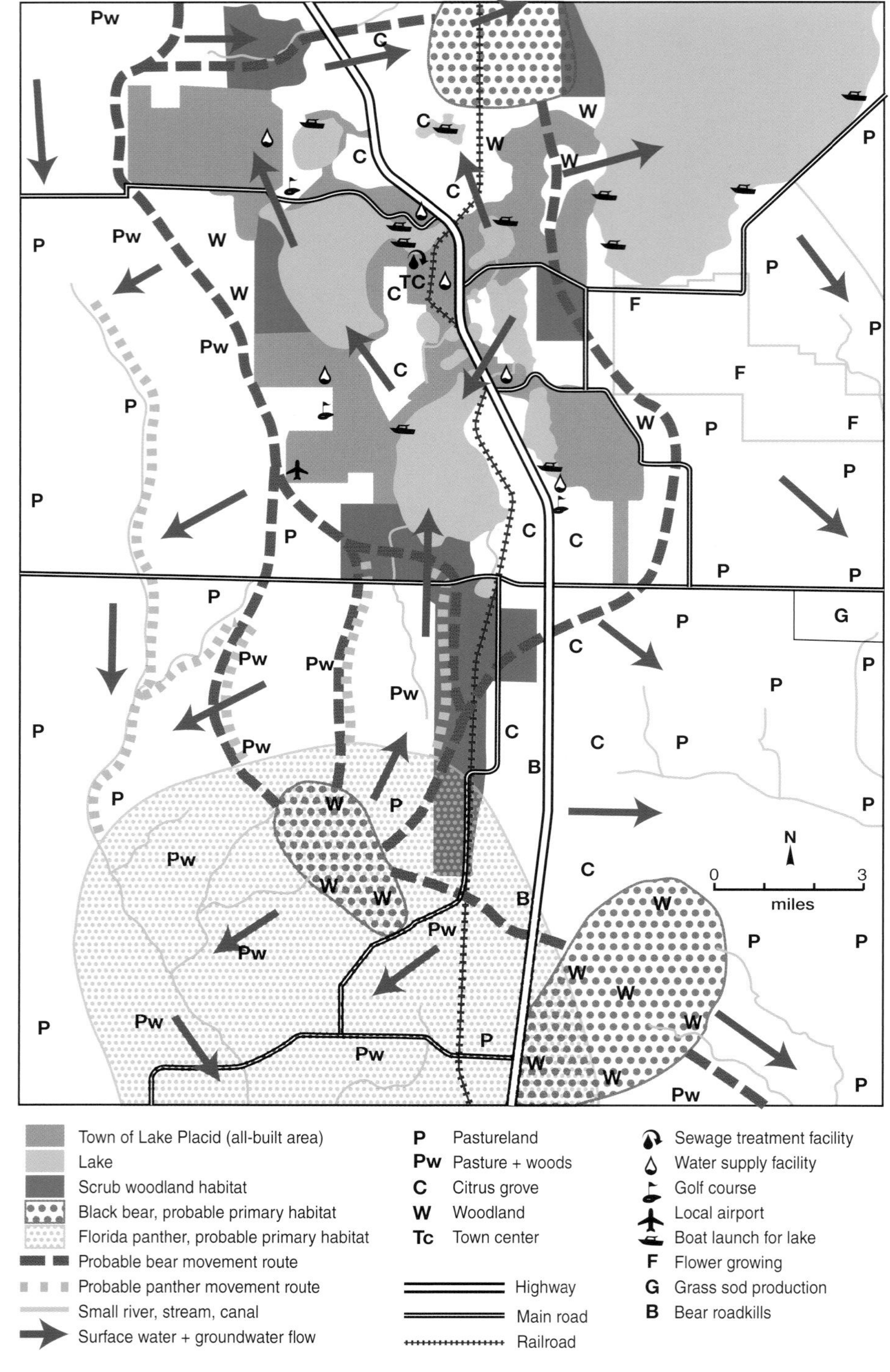

Figure 2.14 Routes of bear, panther, and water around a small town in lake country. Black bear, *Ursus americanus*; Florida panther, *Puma concolor*; bear patterns partially based on Murphy *et al.* (2017a, 2017b). Sandy soils in central vertical third of figure, with clayey and organic soils to east and west. Fires mainly move in and downwind (east) of even small scrub areas. Lake Placid (population 2,200), south-central Florida, USA. With appreciation and aid of Michael Binford, Mark Brenner, Hilary Swain, and the Archbold Biological Station. (A black and white version of this figure will appear in some formats.)

Figure 3.1 Row of stone moai from an earlier civilization, southeastern Easter Island (Isla Pascua), Chile. Moai statues were "walked" from a quarry along a forest road to a coastal village. R. Forman photo. (A black and white version of this figure will appear in some formats.)

Figure 3.2 Scratch Flat, Littleton, Massachusetts, USA. Stream created by a giant serpent according to a Native Indian People's tradition, or in glacier-molded land according to science. See text. R. Forman photo, with aid of John H. Mitchell. (A black and white version of this figure will appear in some formats.)

Figure 3.4 Village green of Petersham, Massachusetts, USA with gazebo on right and general store on left. Green and gazebo used for community activities and festivals. General store, one of five village shops, provides soup and sandwiches, gifts, and small selection of groceries. See Appendix. R. Forman photo. (A black and white version of this figure will appear in some formats.)

Figure 3.6 A community of cliff dwellings abandoned in the 1200s after a 25-year drought. Prolonged drought usually means former residents do not return. Long House, Mesa Verde, southwestern Colorado, USA. R. Forman photo. (A black and white version of this figure will appear in some formats.)

Figure 3.8 Damaged building and former building site remaining in town center 12 years after a severe hurricane. Note successional vegetation. St. George's, Grenada, W.I. See Appendix. R. Forman photo. (A black and white version of this figure will appear in some formats.)

Figure 4.6 Beach exposed for diverse recreation at low tide around a coastal tourist town. Saint-Malo, Bretagne, France. R. Forman photo. (A black and white version of this figure will appear in some formats.)

Figure 4.9 Village with two-thirds of the storefronts on Main Street closed. All places visible in photo are abandoned except for a cafeteria on far right. North of Emporia, Kansas, USA. R. Forman photo, courtesy of Joan I. Nassauer. (A black and white version of this figure will appear in some formats.)

Figure 5.3 Red sand dune >160 m (500 ft) high in extensive desert area with dunes reaching twice that height. Dune 45, Sossusvlei, Nambia. R. Forman photo. (A black and white version of this figure will appear in some formats.)

Figure 5.5 Shallow complex pipe infrastructure in sandy fill under town center street. Scandinavia. (A black and white version of this figure will appear in some formats.)
Source: Georgeclerk / E+ / Getty Images.

Figure 5.7 Spruce and pine (*Picea, Pinus*) growing on acidic mine waste of a former silver mine. People looking at a 40 m wide (130 ft) sinkhole or cave-in apparently 200 m deep, highlighting the depth of former mining and hazard of mining near a community. Kutna Hora, Czech Republic. See Appendix. R. Forman photo. (A black and white version of this figure will appear in some formats.)

Figure 5.8 Large copper-mining operation providing jobs for, supporting, polluting, and perhaps threatening a nearby town. "Copper moon" rising at Santa Rita Mission, New Mexico, USA. (A black and white version of this figure will appear in some formats.)
Source: Alan Dyer / Stocktrek Images / Getty Images.

Figure 6.5 Outhouse with hole in the ground in human waste system using no water and producing no wastewater. Waste may fill a tank to be emptied, or may be decomposed by heat from solar radiation. Edge of city of Tolear, Madagascar. Photo courtesy and with permission of Mark Brenner. (A black and white version of this figure will appear in some formats.)

Figure 5.11 Green pond with goldfish carp, apparently eutrophicated by excess phosphorus stimulating phytoplankton growth. El Alhambra, city of Grenada, Spain. R. Forman photo. (A black and white version of this figure will appear in some formats.)

Figure 6.6 Coastal town on hillside where stormwater and human wastewater rush toward the harbor. Wastewater is collected for treatment. Astoria, Oregon, USA, with mouth of the Columbia River in distance. See Appendix. R. Forman photo. (A black and white version of this figure will appear in some formats.)

Figure 6.10 Former farmhouse by a pond enlarged with dam to provide power for a series of mills on stream. Harrisville, New Hampshire, USA. R. Forman photo. (A black and white version of this figure will appear in some formats.)

Figure 6.11 A water-irrigated terraced landscape providing food for the former Inca community of Machu Picchu high in the Andes. Water entered from the forest at top in three streams/gullies; every terrace had not too much and not too little water. Peru. Photo by Steve Bennett (https://commons.wikimedia.org/wiki/File:Machu_Picchu_overview_Stevage.jpg), "Machu Picchu overview Stevage," https://creativecommons.org/licenses/by-sa/3.0/legalcode. (A black and white version of this figure will appear in some formats.)

Figure 6.13 Multi-level houses of lakeside town in the dry season. Many rivers and streams flow into the large Tonle Sap Lake, so during the monsoon season water level rises several meters and doubles the lake size. Central Cambodia. Photo courtesy and with permission of Mark Brenner. (A black and white version of this figure will appear in some formats.)

Figure 6.15 Dry canal in desert and group from nearby research station ready to measure water characteristics. Near town of Socorro, New Mexico, USA. R. Forman photo. (A black and white version of this figure will appear in some formats.)

Figure 7.3 Distinctive baobab tree (*Adansonia digitata*) treasured by town residents north of Tulear, Madagascar. Photo courtesy and with permission of Mark Brenner. (A black and white version of this figure will appear in some formats.)

Figure 7.6 Flowering shrubs plus food plants (banana, papaya, mango) covering front yard of village house. Grande Riviere, Trinidad. See Appendix. R. Forman photo. (A black and white version of this figure will appear in some formats.)

Figure 7.8 Profusion of successional flowering plants in front yard of abandoned house. Signs of house decay suggest potential animal habitats. Town of Dexter, Maine, USA. R. Forman photo. (A black and white version of this figure will appear in some formats.)

Figure 7.11 Diverse green roof on relatively steep roof angle. Vegetation luxuriance suggests an intensive green roof with soil depth >10 cm (4 in). Near Horgen and Zurichsee, Switzerland. R. Forman photo. (A black and white version of this figure will appear in some formats.)

Figure 7.13 Town park with picnic pavilion, large oak (*Quercus*), and stormwater from road runoff and pipe. Tree lost its top and resprouted. Emporia, Kansas, USA. See Appendix. R. Forman photo. (A black and white version of this figure will appear in some formats.)

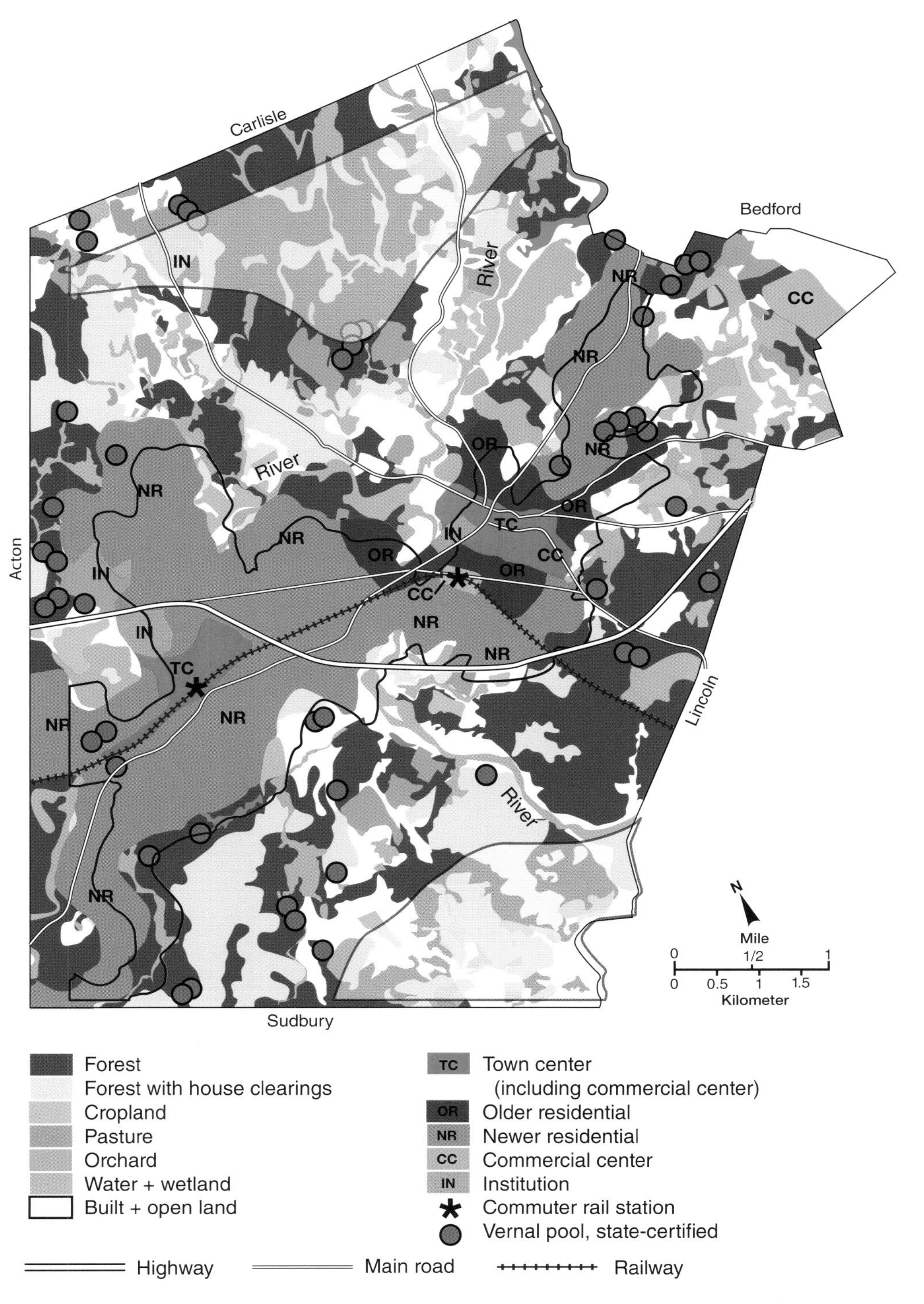

Figure 7.14 Land uses of town with two town centers and its adjacent zone. Concord, Massachusetts, USA in glaciated terrain. See text and Appendix. (A black and white version of this figure will appear in some formats.)

Figure 7.15 Community garden where families have small plots to grow diverse vegetables, fruits, and/or weeds. Background meadow with scattered trees and shrubs is affected by wind- and animal-dispersed seeds from adjacent woods. Concord, Massachusetts, USA. See Appendix. R. Forman photo. (A black and white version of this figure will appear in some formats.)

Figure 8.3 Population of flamingos in shallow bay protected by shoreline mangrove swamp close to town. Celestun, Yucatan, Mexico. Photo courtesy and with permission of Mark Brenner. (A black and white version of this figure will appear in some formats.)

Figure 8.9 Narrow stream corridor with channelized partially armored eutrophicated stream through village center (see Figure 1.3). Abanades, Castilla y La Mancha, Spain. R. Forman photo. (A black and white version of this figure will appear in some formats.)

Figure 8.14 Mid-aged house cat as cuddly companion and predator in residential neighborhood. Japan. (A black and white version of this figure will appear in some formats.)
Source: Mitsuharu Watari / EyeEm / Getty Images.

Figure 9.1 Two-level city-center building with arcades providing shade in warm climate. Note motorbikes and informal outdoor commercial activity. City of Siem Reap, Cambodia. Photo courtesy and with permission of Mark Brenner. (A black and white version of this figure will appear in some formats.)

Figure 9.4 Retail shopping commercial center and apartment buildings in forest (upper left) close to entrance–exit of north–south multi-lane highway. Town edge (far right) separated from highway by cemetery areas. Infrared photo; southeastern Massachusetts. US Department of Agriculture, Soil Conservation Service photo. (A black and white version of this figure will appear in some formats.)

Figure 9.5 Factory by dammed river which traditionally provides power, cooling, and waste disposal. Regulations now limit water pollution. Dover-Foxcroft, Maine, USA. R. Forman photo. (A black and white version of this figure will appear in some formats.)

Figure 9.8 Low-density non-sprawl houses on mostly fenced plots used for chickens, goats, and other food resources. Town of Maun, Botswana. R. Forman photo. (A black and white version of this figure will appear in some formats.)

Figure 9.11 Walkable compact village by road in border area of forest and pastureland. Note wooden walls and roofs, few windows, a few yurts, firewood, fences, and one electric pole on left. White Haba, Altay Xinjiang, China. (A black and white version of this figure will appear in some formats.) Source: Best View Stock / Getty Images.

Figure 10.11 Rock outcrops, slopes, fences, scattered trees, and wooded corridors providing habitat diversity that sustains biodiversity in pastureland. An egg-laying monotreme echidna (Tachyglossidae) discovered near vehicle. Southern New South Wales, Australia. R. Forman photo, with aid of David Lindenmayer. (A black and white version of this figure will appear in some formats.)

Figure 10.13 Farm tractor transporting soil, stumps, and other farm material though center of town. Askim, Norway. R. Forman photo. (A black and white version of this figure will appear in some formats.)

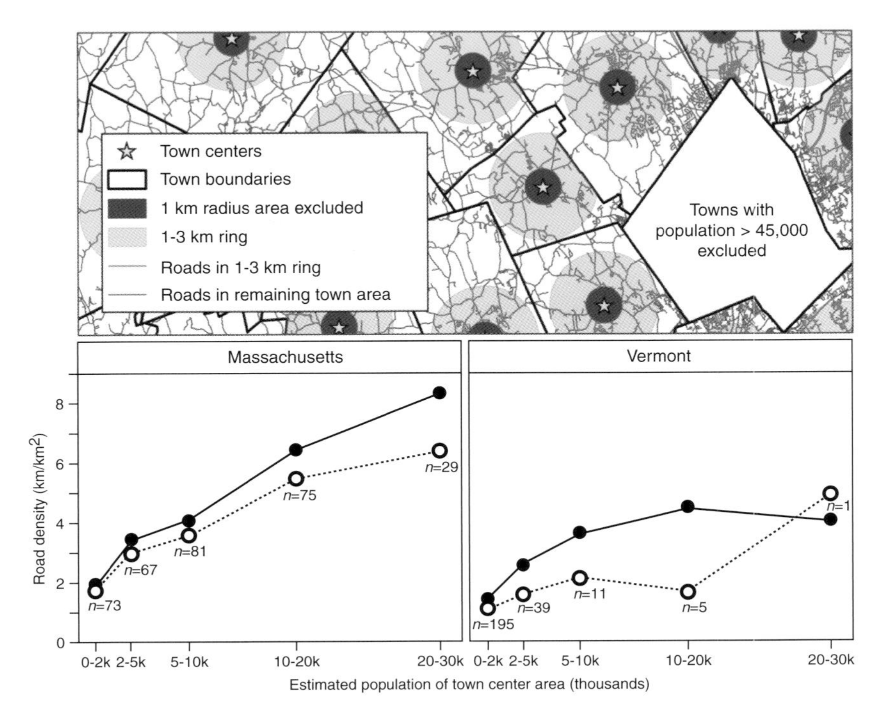

Figure 11.4 Road density in adjacent zone and outer surrounding land of towns in high and low population-density areas. In graphs, solid circles = 1–3 km ring; open circles = remaining town area. Population density: Massachusetts 336/km^2 (871/mi^2); Vermont 26/km^2 (67/mi^2). Figure courtesy of Joshua Plisinski and Jonathan Thompson, and with permission of Jonathan Thompson. (A black and white version of this figure will appear in some formats.)

Figure 11.5 Paper mill emitting steam, sulfur dioxide, hydrogen sulfide, and water pollution. Air and water pollutants degrade ecosystems downwind/downriver, while H_2S produces a strong smell, all effects extending for kilometers. Town of Rumford/Mexico, Maine, USA by Adroscoggin River. See Appendix. R. Forman photo. (A black and white version of this figure will appear in some formats.)

Figure 11.6 Fuelwood (firewood) collectors returning to town with large pieces of wood. North of Tolear, Madagascar. Photo courtesy and with permission of Mark Brenner. (A black and white version of this figure will appear in some formats.)

Figure 11.11 Desert town with no lawns, some abandoned buildings, mine and coal waste (far left), and hill providing cool air drainage. Superior, Arizona, USA. See Appendix. R. Forman photo. (A black and white version of this figure will appear in some formats.)

Figure 11.13 Transportation by two-humped camels in desert with moving sand dunes. Lobe of Gobi Desert, southern Inner Mongolia, China. R. Forman photo, courtesy of Jianguo Wu. (A black and white version of this figure will appear in some formats.)

Figure 11.14 Informal squatter settlement of recent arrivals in desert town edge. Government provides some widely separated public toilets and street lights, plus a straight earthen road through center with narrow perpendicular roads; residents create infrequent shops and occasional school rooms. Swakopmund, Namibia. R. Forman photo. (A black and white version of this figure will appear in some formats.)

Figure 12.1 Transportation by pedi-rickshaw, bicycle, and motorbike in built area. Abandoned building with arcade and colonizing plants on left. City of Tolear, Madagascar. Photo courtesy and with permission of Mark Brenner. (A black and white version of this figure will appear in some formats.)

Figure 12.8 Footpath connecting communities and other features across the landscape. Grassland with shrubland above and conifer plantations below. Skiddaw Mountain, Lake District National Park, UK. (A black and white version of this figure will appear in some formats.)
Source: John Finney Photography / Moment / Getty Images.

Figure 12.9 Inter-town bus being packed with goods, luggage, and people. Town north of Tolear, Madagascar. Photo courtesy and with permission of Mark Brenner. (A black and white version of this figure will appear in some formats.)

Figure 12.14 Bull bison or buffalo (*Bison bison*) walking slowly along road with kilometer-length lines of creeping traffic in both directions. Few species move along roads or roadsides. Yellowstone National Park, Wyoming, USA. R. Forman photo. (A black and white version of this figure will appear in some formats.)

Figure 13.3 Pedestrianized street for walking, biking, eating, and shopping in small city. Almost no motor vehicles. Residential and retail commercial area with car parks close by. Malmo, Sweden. (A black and white version of this figure will appear in some formats.)
Source: Werner Nystrand / Folio Images / Getty Images.

Figure 13.6 Street crosswalks for children and pedestrians in residential neighborhood. Maples (*Acer*) in autumn; northeastern USA. (A black and white version of this figure will appear in some formats.)
Source: christopherarndt / E+ / Getty Images.

Figure 13.7 Pickup truck converted for bus travel between villages and town. Town near Siem Reap, Cambodia. Photo courtesy and with permission of Mark Brenner. (A black and white version of this figure will appear in some formats.)

Figure 13.11 Netway-and-pod transportation system at London airport. Automated system (no driving) with electricity from wire buried in track running a small electric motor in pod, which carries ten people with luggage. R. Forman photo. (A black and white version of this figure will appear in some formats.)

Figure 13.12 Electric car with driver's door open next to house with green roof. Village of Flam, Norway. See Appendix. R. Forman photo. (A black and white version of this figure will appear in some formats.)

(*Taraxacum officionale*, *Primula elatior*, *Saxifraga tridactalites*, *Epicactus helleborine*, *Viola pubescens*, *Grevillea macleayana*) have had a change in genetic diversity (Evans, 2010). Four of these urban species have changed genetically over time. Three have had genetic divergence, producing relatively distinct populations. And three species have genetically diverged so that urban and rural populations differ. Another study of an urban weed (*Crepis sancta*; Asteraceae) found that the plant's seed dispersal process had changed genetically (Cheptou *et al.*, 2008). Indeed, short-term evolution had occurred in only 5 to 12 generations. These urban examples may be indicative of town and village processes, even though the places have such different characteristics.

Species Movement

Each species has a distributional *species range* describing or mapping the outward extent where the species lives. The range may expand in one direction, e.g., northward, or even in all directions, generally due to changing environmental conditions. Analogously the range may contract for the same reason. Towns are inside, outside, or on the boundary of a species range. Most species growing near their range boundary experience diebacks alternating with periods of local luxuriance or spread. Contraction of a species range may lead to the disappearance or extinction of a species in town.

However, near a contracting range boundary, a long-lived species such as a tree may persist as a remnant. When young it grew well in a favorable environment, but with a changed environment today seedlings and saplings of the species do not survive. Such remnant trees are common in towns and villages of the African Sahel region which has become noticeably drier over decades.

Species colonize a town area and also go locally extinct. The town area is strongly modified by residents, and native species from the surroundings repeatedly colonize the built area. Some die quickly, some persist without reproducing, and some successfully reproduce and spread. Nonnative species usually from outside the nation also colonize the town area, though less frequently. Also, nonindigenous species, those from outside the region but native to the nation, colonize.

Plant pests and diseases are endless problems at the core of plant pathology. A single example illustrates the shrinkage of a species distribution. The emerald ash-borer (*Agribus planipennis*: Coleoptera) is a conspicuous beetle that kills ash trees (*Fraxinus* spp.) (Prasad *et al.*, 2010). The species was introduced in 1998 in the North American Midwest where it has killed millions of trees, including in towns and villages.

The insect disperses short distances, whereas long-distance dispersal occurs by human activities. The moving front of beetles has expanded on average about 20 km per year. A larger local population spreads outward a greater distance. The main road network provides major dispersal routes for the beetle, as in Ohio where 84 percent of the infected trees are within 1 km of roads. However, species movement is not that simple. Consider the many other variables linked to the beetle spread and ash tree contraction: traffic level; distance from an infected zone; campground size and usage; wood products industry; size of wood and types of usage; and human population density.

Corridors are most likely the primary routes of plant species movement in and around towns and villages. At a broad scale, species cross the land and enter a town along roadsides, railways, pipelines, powerlines, streams, canals, and rivers. At a fine scale within residential areas, primary routes probably include back boundary lines and side boundaries of house plots, as well as hedgerows, hedges, fences, and walls. Connecting the town to the surroundings are walking paths, ditches, roads, and water courses.

Vegetation Change and Disasters

Ecological Succession

In general, *ecological succession* refers to a site's sequence of species, and often vegetation, over time. The sequence or process is mainly directional, i.e., from early to late successional stage.

Various cyclic processes at different time scales affect the sequence. Diurnal change may bring frost at night or intense heat by day. Weekly change may concentrate recreational trampling, lawnmower noise, and dumping on weekends. Some species are sensitive to monthly lunar or tidal cycles. And annual cycles bring bitter winters, early springs, hot summers, or hurricane (typhoon) season. Such cycles alter the species in the town's many successional sequences.

Ecological succession in towns differs considerably from that in natural land, agricultural land, and cities. In towns, many successional sequences occur, all strongly affected by the proximity of farmland and/or natural land. With limited funds, towns often have successional vegetation on unpaved car parks and neglected parkland sites. Also, with so many towns and villages losing population, abandoned house plots with successional vegetation are common (e.g., Grande Riviere, St. George's).

Often, with relatively clean air, town trees, rocks, and gravestones show a succession of lichens and perhaps bryophytes. Rural cemeteries by the edge of town are often unmown and plant succession contains uncommon species (e.g., Clay Center). Town trees contain dead branches, while fallen logs and standing dead trunks often persist, all with a succession of fungi, invertebrates, and vertebrates. Predominantly residential towns normally contain many relatively long house plots with little-manicured successional vegetation along the back boundaries. Commonly close to town edges are abandoned quarries, sand/gravel pits, and motorbike/off-road vehicle track areas, all with conspicuous but different successional vegetation.

Other common locations with ecological succession include periodically scraped roadsides, building construction sites, and mine-waste heaps/piles (Rapoport, 1993). Succession occurs on green walls and green roofs (Figure 7.11). Any neglected or low-maintenance site in town has successional vegetation. In essence, ecological succession is visible throughout a town. Perhaps unlike most other areas, the sequences of plants and vegetation are highly diverse from site to site in a town.

In addition to microhabitat conditions in town, the surrounding land has a powerful effect on succession. Entering wind, heat, pollutants, surface water, and groundwater affect town microhabitats. The biological role is equally important as a *seed rain* of species from the surroundings continually enters town sites. Seeds arrive by wind or are carried by birds and mammals. Different plant species arrive at different phases of succession. Surrounding natural or semi-natural land mainly disseminates native

Figure 7.11 Diverse green roof on relatively steep roof angle. Vegetation luxuriance suggests an intensive green roof with soil depth >10 cm (4 in). Near Horgen and Zurichsee, Switzerland. R. Forman photo. (A black and white version of this figure will appear in some formats. For the color version, please refer to the plate section.)

species to town sites. In contrast, agricultural land primarily disperses agricultural weeds, including many nonnatives, adapted to high light and nutrient levels.

Abandoned farmland around towns and villages leads to vegetation succession and often reforestation (e.g., Grande Riviere). Forest cutting with regrowth or replanting produces a successional sequence. Succession follows wildfire.

The rate of successional change is affected by varied town characteristics and environmental inputs. The relatively high water table of most towns facilitates rapid succession. The abundance of soil nitrogen blown in from surrounding farmland, from fertilizing lawns and gardens, or in the area around a sewage treatment facility (e.g., Concord, Lake Placid) accelerates plant growth and ecological succession. Excess nitrogen or phosphorus also tends to favor one or two dominant species which outcompete other plants, resulting in lower plant species richness. On the other hand, succession is normally slow on mine-waste heaps/piles, sand/gravel pits, and abandoned quarries (e.g., Falun, Issaquah).

In town, succession typically begins with seeds distributed over a site. The seeds may be present or stored in the soil as a *seedbank* from a former time. For instance, in an agricultural landscape, a town's newer residential areas often begin succession from a seedbank of agricultural plants. A species rain brings more and different seeds.

Succession on soil in town normally progresses from primarily herbaceous plants to shrubs and small trees to tall or mature trees, though several substages can be recognized. The plants of succession all or almost all spontaneously colonize and compete. Unlike in cities, succession in town often produces mature trees. Also, somewhat distinctive is the relatively rich abundance and diversity of wildlife, which comes from the nearby forest or agricultural land in addition to adjoining residential land.

Plantings, maintenance, and other activities by the town and its residents are often limited. In woods of suburban towns near Wilmington, Delaware (USA), a dozen types of human effects were concentrated along forest edges, near walking paths, and by vehicle tracks (Matlack, 1993). Such effects alter the species composition and successional sequence, indeed may effectively convert a natural woods to a semi-natural one.

Finally, ecological succession is particularly widespread in town after a major disturbance or disaster (e.g., St. George's). A stream's ten- or 50-year flood often erodes surfaces and washes away structures, leaving space for the colonization of successional species. A severe drought or fire or windstorm effectively does the same thing.

Disasters and Towns

If a major environmental or other disturbance occurs frequently relative to the length of human or species life cycles, the species is likely to have adapted to the change. In contrast, a so-called *disaster* is an infrequent sudden event causing great loss or damage (Forman, 2008). Disasters usually last minutes to days (Forman, 2014), and significantly degrade people, human structures, and natural resources in town (Campanella and Vale, 2005). Effects normally persist for a prolonged period. Illness, death, and property damage are common. Pipe, canal, and ditch systems are disrupted, causing many secondary effects. Transportation systems are disrupted. Industrial chemicals are spilled, and stagnant water persists. A town's local industry and some agriculture may be disrupted.

Disruption of a water supply is of immediate importance since freshwater is a daily need of people, and water is essential for wildlife and plants. Disruption of wastewater systems pollutes areas and leads to disease spread, while disruption of the solid-waste system leads to widespread accumulations of organic and other waste, along with the proliferation of pests. Transforming town surfaces by massive erosion or sediment accumulation (e.g., landslide, avalanche, lava flow, ash deposit, sandstorm) effectively removes the vegetation and provides new surfaces for a protracted ecological succession. Flood, tsunami, fire, hurricane/cyclone, and volcanic ash deposits significantly modify the town's surface, though some of the soil remains intact so vegetation succession may be rapid.

Earthquake, volcanic eruption, landslide/avalanche, tsunami, cyclone/hurricane, sandstorm, flood, and fire are familiar disasters, mainly of natural origin but in some cases accentuated by humans (Forman, 2008). Also, their effects are considered greater on a town with more people and structures than on a village. Yet mainly human-caused disasters also occur, such as a sudden major economic downturn, industrial accident, nuclear-facility radiation release, disease outbreak, strong government change, war/conflict, or massive immigrant arrival of environmental or war refugees. The more disasters a town has had, even of different types, the more adaptations are likely to be present.

The characteristics of different disasters affect plants and vegetation differently. A massive fire across town kills most plants, consumes some wood and soil organic matter, and leads to short-term erosion/sedimentation. An invading force or army may strip the food plants, eliminate livestock, cut trees for firewood, even burn the place. Bombing produces small craters often with water, mosquitoes, and moisture-favored plants (Gilbert, 1991; Forman, 2014). An industrial accident with air and/or water pollution kills the sensitive plant species, thus favoring species resistant to the chemicals spilled. Radiation released from a nuclear accident or bomb makes fruits and vegetables unsafe to eat, among other severe problems, and accumulates in the food chain making livestock and wildlife inedible (e.g., Prypyat, USSR, now Ukraine) (Forman, 2008). Net-like mangrove and screwpine (*Rhizophora*, *Pandanus*) shrubs or small trees may survive a tsunami and thus flourish thereafter. On brick/concrete rubble sites with low organic matter and very low nitrogen available, annual and perennial plants typically persist and legumes may be common (Dutton and Bradshaw, 1982; Gilbert, 1991).

Some pre-disaster measures can reduce disaster damage. For instance, a protected-land setback from the seacoast reduces tsunami damage (see Figure 13.14) (e.g., Hanga Roa). Prevent or remove structures on sites prone to disaster damage. A more proactive case would be to relocate a whole town or village to a less vulnerable site, which has been done around mines and for other reasons (Chapter 2) (Forman, 2014). Relocation can carefully remove vegetation in the new site and restore it in the former threatened site.

Other advance measures include creating effective evacuation options (Campanella and Vale, 2005). Firebreaks. Appropriate planting designs around buildings. Emergency storage water and supplies. The traditional stability measures of redundancy, alternative routes, diversity of resources, linkages with the surroundings, and so forth certainly apply to towns.

Alofi, a town on a 28 m (92 ft) wide seaside terrace in Niue, one of the smallest Pacific island nations, was hit by a category 5 cyclone/hurricane (Barnett and Margetts, 2013). Although wind velocity reached more than 300 km/hr, most damage was due to the sea when waves swept over the town. The town's response was slow due to varied factors, yet it included adaptations in anticipation of future cyclones. Buildings were made repairable by residents, and pipe networks were designed and rebuilt on, rather than under, the ground to facilitate maintenance and repair. Unlike the town residents who had never seen such a storm, the remnant coastal forest with adaptations rapidly resprouted and returned.

Kyoto (Japan) illustrates a case where over many decades, typhoons (cyclones/hurricanes), floods, and fires have all impacted the city (Sakamoto, 1988). Furthermore, trees have been cut down for developments and trees in other areas planted. Despite these major alterations, the number of the predominant elm-related (*Ulmus*) trees fluctuated around a rather constant number. The author concludes that the residents' appreciation of trees and their valuable roles has helped stabilize the tree cover.

Greenspace Patterns

Greenspaces in city or town are neither vacant nor open space (Wheater, 1999; Forman, 2008). Rather, like built areas of different type, each providing value, greenspaces come in many types, each providing value to society. Ballfields provide for active recreation, a natural area supports biodiversity, riparian trees protect streams and fish, a green corridor provides for walking and wildlife connectivity, and a grassy hilltop provides relaxing views. Greenspaces vary in the vegetation types or plant communities present (Kovar, 1995), but all greenspace types are recharge areas for groundwater (Chapter 6). Greenspaces in and around towns are of economic, social, aesthetic, and ecological value.

The number or *richness* of a variable highlights its diverseness or diversity. In contrast, the *dominance* or *evenness* of the types measured emphasizes their relative abundance, e.g., whether one or two types predominate (high dominance), or whether most of the common types are in similar abundance (high evenness).

We now consider greenspace patterns from two perspectives: (1) greenspace patches and corridors; and (2) greenspace in the adjacent zone outside town.

Greenspace Patches and Corridors

Greenspace in Villages

In mainly agricultural landscapes of cool-moist climate Britain, tree canopy areas are recorded in and around 21 villages (Sharp, 1946). Four roadside *linear villages*, 12 compact, squared, or *enclosed villages* (patch-shaped), and five planned villages (mainly planned and built at one time) (Chapter 1) are included. Individual trees, a short corridor, a small patch, and a large patch of trees are recorded in five locations around villages (Figure 7.12). The following median values indicate the distribution of trees in and around these villages.

Enclosed villages have the most tree vegetation, and planned villages the least. Separate individual trees in front and side yards/spaces of houses, and small tree patches in various locations, are most common in villages. Overall, the adjacent zone just outside a village and surrounding land further out differed little by village type, but these mainly reflected the surrounding terrain and agricultural land uses. Tree vegetation within the enclosed and linear villages provides shade, reduces windspeed, supports wildlife, provides aesthetics, and reduces flooding and erosion.

Patch-Corridor-Matrix Model

If one were dropped by helicopter at random in a large town or agricultural landscape, the landing would be in one of three possible types of places. It could be in a patch of some sort, or in a corridor or strip, or in the background built or agricultural matrix. The *patch-corridor-matrix model* is a valuable foundation of *landscape ecology*, effectively the ecology of the land mosaic, e.g., that one sees from an airplane window (Forman, 1981, 1995; Turner and Gardner, 2015; Rego *et al.*, 2018).

The model built on and replaced an earlier island-biogeography approach explaining the species richness on islands (MacArthur and Wilson, 1967). Unlike an island, a *patch*

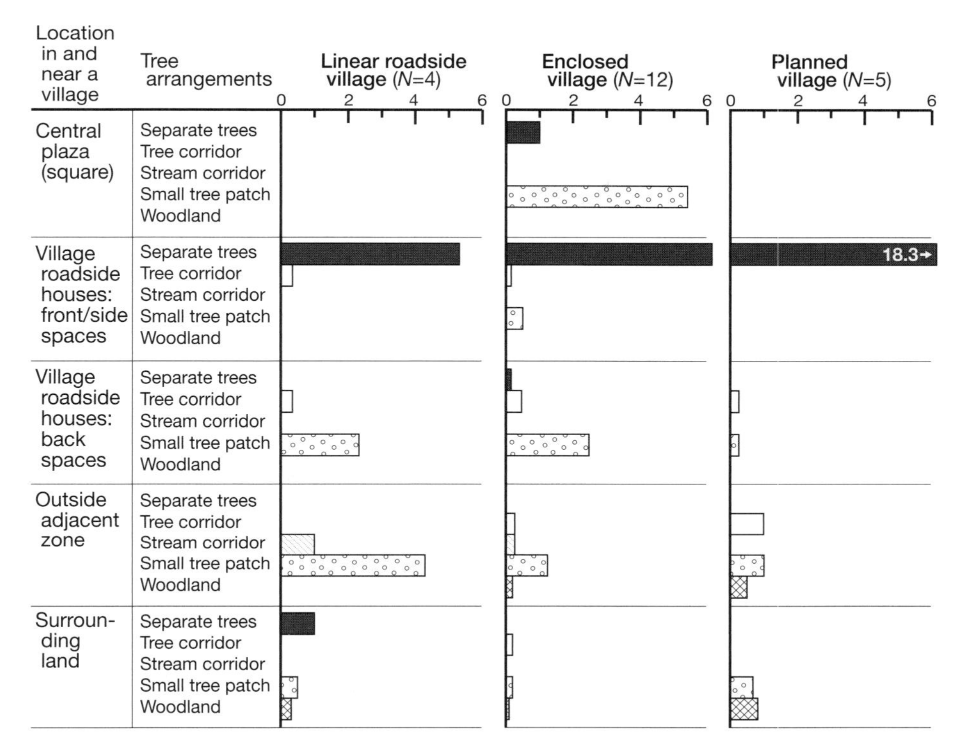

Figure 7.12 Tree distribution in different village types. From 21 mapped and drawn villages in agricultural landscapes in cool-climate Britain (Sharp, 1946). Average numbers plotted; considerable variation exists from village to village. Adjacent zone beyond the outer row of village houses, and surrounding land farther out (see Chapter 1).

Linear villages are normally along a road or coastline, especially at a bend (see Figure 1.13). Individual trees along the road provide shade for walkers in the village. Separate individual trees in front, side, and back yards/spaces provide benefits to homeowners. Trees in the adjacent zone or surrounding land beyond town benefit wildlife and livestock. Small vegetation patches, in varied locations from village green to surrounding land, are wildlife habitats, as well as stepping stones for movement across the landscape. Tree lines along the road are conduits for wildlife movement, and also "catch" wildlife moving across the landscape.

An enclosed village has homes clustered closer together, with more interaction among residents. A sidewalk provides safety for walkers and facilitates interactions. Thus the enclosed village of Petersham, Massachusetts (USA) (see Figure 3.4) has 1.76 km (1.1 mi) of streets with a sidewalk (some portions on both sides of street). Approximately 70 percent of the sidewalk length is tree covered, providing shade in the summer sun. Also 31 percent of the sidewalk length has shrub cover or 1–2 m high stonewall alongside (within 3 m). In cities, shrubs rarely line walkways for security reasons, but normally villages are quite safe for walking.

Planned villages have not developed "organically," and exhibit regularity, symmetry, even many urban features. Usually ecological dimensions are scarce, ironic given the location of villages in natural land or farmland.

on land is surrounded by a mosaic of habitats, each of which may affect the patch, is a source of different species, and is differentially suitable for movement of species between patches (Forman, 1995; Laurance, 2008; Collinge, 2009). Furthermore, the patch *edge-area-to-interior-area ratio* strongly affects the species richness of a patch. The patch edge contains an abundance of generalist species and the interior has many interior species.

Structure-function-change or pattern, process, and change characterize the landscape as a living system. The landscape below the airplane window has a structure, i.e., the spatial pattern. Numerous ecological characteristics have been correlated with spatial land-cover patterns, as illustrated by a study in western Georgia (USA) which found 152 significant correlations between spatial variables present (Styers *et al.*, 2010). The landscape or town functions or works, that is, flows and movements describe the landscape processes. And the spatial pattern or arrangement of the landscape changes over time. All living systems have these three attributes, structure, function, and change.

Before relating these concepts to towns, one more insight is important. What are the attributes of patches? Large to small, rounded to elongated, lobed to smooth, and so forth (Forman, 1995). Simple attributes. Attributes of corridors? Wide to narrow, straight to curvy, long to short, and so on (Bennett, 2003; Terrasa-Soler, 2006; Forman, 2014). Matrix attributes? Continuous to discontinuous, perforated or not, sliced or not, etc. These simple familiar types of attributes, describing the spatial patterns dictionary-like, facilitate communication among diverse disciplines. They also provide convenient handles for improving a landscape.

Although a town is much smaller than an urban or agricultural landscape, the landscape ecology concepts remain fundamental, especially in understanding greenspace patterns. Ignoring individual tiny vegetation spots such as a flower garden, green roof, or backyard (Forman, 2014), the most common greenspaces in towns are typically a park, school playground, cemetery, railway, and stream corridor. Less common types include pipeline, powerline, wetland, institution, dump, community garden (allotment), and so forth. The number of town greenspaces varies from relatively few (e.g., Hanga Roa, Rock Springs; see Appendix) to many (e.g., Falun, Sooke).

Railway corridors lined with vegetation often bisect a town (e.g., Askim, Cranborne, Emporia) (Borda-de-Agua *et al.*, 2018). Railway crossings may be frequent, though periodically blocked by a passing or stopped train. Also, a stream corridor often slices through town, usually with two or a few bridges for crossing (e.g., Pinedale, Sandviken). A town wetland may adjoin the corridor. Both stream and wetland may frequently flood. School playgrounds are mainly dispersed in neighborhoods. Cemeteries are few, dispersed, and on high ground, thus away from a stream corridor or wetland.

Parks in town are usually small and may adjoin any of the other greenspaces. Normally species richness increases with park size. But parks often contain planted and mowed grass-tree-bench areas for intensive visitor use (Figure 7.13), as well as less-used semi-natural areas. Species richness correlates even better (higher R^2) with area of semi-natural habitat than with park area (Forman, 2014).

The preceding characteristic combination of town greenspaces is rather low in *connectivity* (Forman, 1995). That is, wildlife and people commonly cannot move between

Figure 7.13 Town park with picnic pavilion, large oak *(Quercus)*, and stormwater from road runoff and pipe. Tree lost its top and resprouted. Emporia, Kansas, USA. See Appendix. R. Forman photo. (A black and white version of this figure will appear in some formats. For the color version, please refer to the plate section.)

greenspace patches without having to move through built areas. Yet towns tend to have more greenspace per unit area than do urban areas. This suggests that, despite the limited direct connectivity, abundant greenspaces provide stepping stones for movement by many species that can successfully move short distances through built areas (e.g., Alice Springs, Sandviken). Other greenspace-pattern examples in towns include: a large park surrounded by small ones (e.g., Emporia); semi-natural patches in the town edge (e.g., Astoria, Icod de los Vinos), and a trail network connecting town and surroundings (e.g., Tetbury, Issaquah, Tavistock).

The combination of greenspaces providing functional value to built areas is sometimes called *green infrastructure* (Benedict and McMahon, 2006; US Environmental Protection Agency, 2010; Yu *et al.*, 2011; Austin, 2014). Many but not all of the functional values involve linear movement along strips, so green corridors are a key to green infrastructure. Although green infrastructure focuses on natural processes, economic, social, recreational, and/or aesthetic values are often of primary importance.

The structure or spatial pattern of a landscape or town is more than simply the number of different types of patches, corridors, and matrix. Perhaps equally important is their specific *spatial arrangement* (Forman, 1995). Patches connected by corridors provide connectivity for wildlife and people movement. *Adjacency effects*, where an adjoining land use, such as car park or housing development, affects a greenspace (Hersperger and Forman, 2003), often has a major impact (Hamabata, 1980; Tilghman, 1987; Radeloff *et al.*, 2010). In suburban towns near Wilmington, Delaware (USA), the significant effects of surrounding residents extended more than 80 m (260 ft) into greenspace woods (Matlack, 1993).

The concept of an *ecological network* has evolved to be a functionally connected set of green patches (often with corridors), through which wildlife or people can successfully move (Opdam *et al.*, 2006; Pesout and Hosek, 2014). A connected set of large green patches or areas, with each large patch providing several key ecological benefits, in the landscape has been highlighted as an *emerald network* (Forman, 2008, 2014). This is an optimal spatial pattern for planning and nature protection.

A big green patch near a little green patch suggests much more movement from big to little patch, rather than vice versa. Wildlife movement often occurs outward from the lobe of a patch, and along a row of small stepping-stone patches. Indeed, a rich assortment of spatial arrangements indicates known directions of movement (Forman, 2014). This information is particularly valuable because, without knowing what's in or outside a greenspace, reasonably robust predictions can be made, even mapped, about the direction of movements and flows in the vicinity. In effect, spatial arrangement is a useful preliminary indication of how the landscape or town works.

The preceding patch-corridor-matrix concepts developed over three decades are central in understanding landscapes, including the land of towns and villages. Thus, a leading ecologist, Simon Levin (2015), points out that "landscape ecology has developed into one of the most vibrant branches of ecological science, with exceptionally strong links between theory and practice. Indeed, it is hard to think of any area in ecology where theory has had a greater impact on application, or where applications have done more to stimulate creative theory." This theory and application will catalyze town ecology.

Values of Town Greenspaces

The values of town greenspaces appear in three groups, i.e., values to nature, to society, and to health. *Nature's values* range widely, including protection of biodiversity, rare species, soil, mineral nutrients (including nitrogen and phosphorus), rare habitats, habitat diversity, clean air, suitable temperatures, unpolluted waterbodies, and other ecological values.

Greenspace services or *values for society* (Chapter 4) are equally broad (Frumkin *et al.*, 2004; Louv, 2005), including: nature-based recreation such as walking and fishing; cooling the town's air in summer; cleaning air; protecting soil against erosion and consequent sedimentation; protecting water supply; protection against disturbance damage; raising real estate values; constraining expansive development; providing outdoor social meeting places; delineating neighborhoods; decreasing violence and crime (Kuo and Sullivan, 2001); reducing CO_2 greenhouse gas emission and climate change; increasing aesthetics; and so forth. All can be of economic importance.

The quote at the beginning of this chapter highlights the *health values of greenspaces* (Kellert *et al.*, 2008; Ulrich *et al.*, 2008; Gaston, 2010; Chucas and Marzluff, 2011; McDonald, 2015). Some or all may also apply to animals and thus be important in ecology. Reduced blood pressure, cardiovascular disease, heart attacks, strokes, obesity, diabetes, and sedentary lifestyle seem to be generally related to time spent in greenspace, rather than in urban built space (Reeves, 2014). Less air pollution, lower air temperature, and beneficial chemicals from greenspace trees may be prime factors. In a hospital, patients with a window viewing vegetation recuperate faster than those viewing a brick wall (Ulrich, 1984). Even the birthweight of a newborn baby may correlate with the presence of trees near the mother's home (Donovan *et al.*, 2011, 2012).

Air pollution including smog contributes to high blood pressure, lung and cardiovascular disease, hardening (calcification) of the arteries, obesity, diabetes, stress (oxidative), dementia, and premature death (Frumkin *et al.*, 2004; Reeves, 2014). In the lungs and arteries, fine particles ($PM_{2.5}$), toxic free-radical chemicals, volatile organic compounds (VOCs), and polycyclic aromatic hydrocarbons (PAHs) enter the body and lead to oxidative stress.

A walking study in London is relevant to towns. Sixty adults with mild or moderate asthma walked two hours on a city street, and at another time walked two hours in a large park (Reeves, 2014). Concurrent environmental measurements found that the street air was 3.5 times higher than the park air in aerial fine particles ($PM_{2.5}$), and seven times higher in nitrogen dioxide. The street walkers had a significant decline in respiratory capacity. A study of walkers in forest and city in Japan had analogous results (Park *et al.*, 2010). The choice of route for walking or jogging, even for a few hundred meters, may have a significant health impact.

Excessive heat increases the concentration of several air pollutants including ozone, increases the penetration of particles into the lungs, increases heat stroke susceptibility, and increases mortality (Reeves, 2014). On a clear June day in Montreal, the air temperature in a woods was 23°C (74°F), a golf course 27°, a residential area 31°, and a heavily paved and built industrial area 40°C. The lowest and highest temperature areas were only 500 m (0.3 mi) apart. In short, the woods was coolest, the industrial area very hot, and temperatures varied widely at the scale of a land mosaic where people and wildlife live and move.

Unlike cities, towns of course do not have urban air laden with heat and pollutants, yet local industries or surrounding land may fill the air with moderate levels of heat, dust, and chemicals. Towns generally have relatively few greenspaces, yet the total greenspace area proportionately exceeds that of cities. The spatial pattern of town greenspaces greatly affects nature, society, and personal health in town.

A forest or park with dense trees provides several potential health benefits. Indeed, in Japan "forest bathing" (*shinrin-yoku*) is recommended (Li, 2010; Park *et al.*, 2010). Beauty, stillness (little wind), rustling leaves, bio-acoustics of birdsong and frog sound, and geo-acoustics of flowing water are typical properties.

But also trees produce "palliative" chemicals, the essential wood oils (VOCs, such as pinene, cineole, limonene, humulene), which may have anti-bacterial, anti-inflammatory, and anti-carcinogenic properties (see Figure 5.14) (Li, 2010; Reeves, 2014). Such chemicals may lower arterial contraction and blood pressure, reduce the release of stress hormones, lower vascular inflammation, improve immune function, and increase the white blood cells that defend against viruses. In essence, for good health, take a walk in the woods.

Greenspace in the Adjacent Zone

Consider first the natural conditions within a town. The essentially all-built area of Concord, Massachusetts (USA), a cool-moist climate outer suburb of Boston, is elongated, due to the results of glacial action plus the presence of three rivers (Figure 7.14) (*2015 Open Space and Recreation Plan*, 2015). Two town centers are present, one around the site of an original grain mill and the other around a train station. Vegetated land uses/habitats are recorded below for four key areas: (1) within town centers; (2) within older residential areas; (3) within newer residential areas; and (4) areas adjacent to the town.

The town centers are nearly devoid of green patches. Older residential areas contain forest patches and a few open lands (ballfields, etc.), while the much larger newer residential areas contain many forest patches, a good number of crop and open lands, and a few uncommon land-use/habitat types. On the other hand, land adjacent to the all-built town is a diverse mixture of forest patches, other developed land, forested wetland, waterbody, cropland, pasture, and open land.

Considering that Concord's town centers compose less than 1 percent of the town, and older residential areas perhaps about 10 percent of the area, the relative abundance of both forest patches and open lands in the older residential areas is striking. Crop and pasture fields however in town are limited to the newer residential areas. Town expansion is significantly constrained by forested wetlands and waterbodies, but also by protected forest patches and active agricultural fields.

Approximately half of the town boundary, particularly to the north and along rivers, adjoins areas of conservation importance (designated as Massachusetts BioMap2 Core Habitat). These are habitats with a concentration of rare plants, animals, and/or natural communities. The town and its surroundings contain 16 state-listed rare species, plus 7 species of conservation concern. Rare plants, birds, amphibians, a reptile, insects, and mussels are present in and around town.

Seven state-certified vernal pools (temporary ponds) are present in town, and 41 in the surroundings, as well as many potential vernal pools (Figure 7.14). Twelve bridges are in town and nine in the surroundings. A commuter railway with two active stations extends along about 60 percent of the town's length.

Most of the forested area in and surrounding Concord is a transition mix of tree species from the north and south. The uncommon semi-natural habitats are scattered: small-river floodplain forest, kettlehole wet meadow, sugar maple stand (*Acer saccharum*), pitch pine stand (*Pinus rigida*), peat bog (*Picea mariana*, *Larix laricina*), and circumneutral soil areas. Also, the town government is responsible for 15,000 to 20,000 trees, with annual planting of 150 trees chosen from a list of 30 species.

A landscape may be rich in rare habitats and species, or depauperate. But a town normally develops in a special location in the land, and thus may have previously had, or still contain, rare and uncommon species (e.g., Lake Placid) or natural communities. Concord, at the intersection of three rivers, has extensive protected land in and around it, so uncommon species have survived or thrived.

The edge of a town is mainly the line of outermost house plots of the essentially all-built town area. Sections of the town edge may have a commercial shopping center,

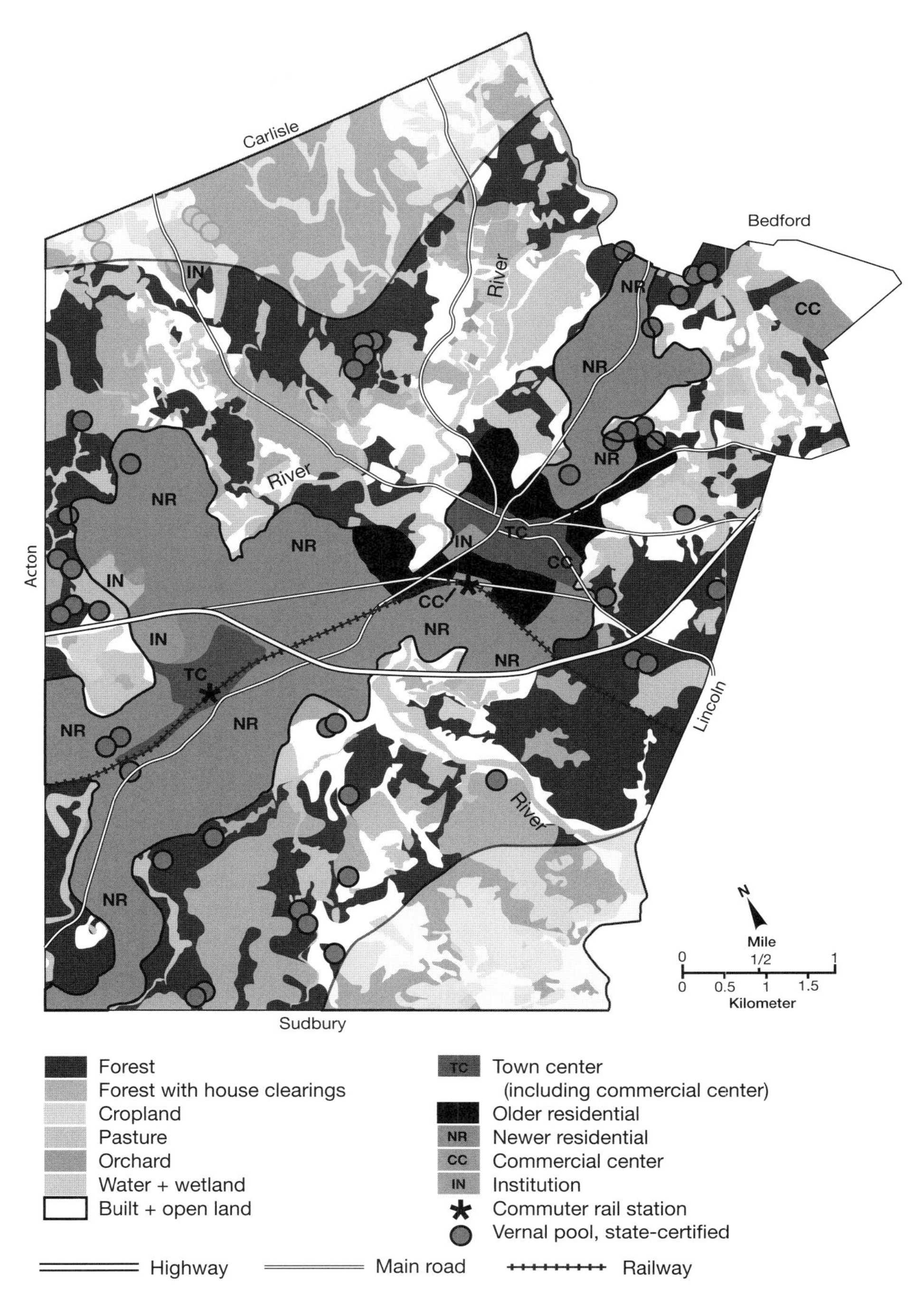

Figure 7.14 Land uses of town with two town centers and its adjacent zone. Concord, Massachusetts, USA in glaciated terrain. See text and Appendix. (A black and white version of this figure will appear in some formats. For the color version, please refer to the plate section.)

industrial center, cemetery, or other special use. In some sections, the *adjacent zone* (typically 2 to a few kilometers wide) outside the town edge contains a prominent corridor, such as a highway, railway, stream corridor, or river corridor (see Figure 2.4) (e.g., Casablanca, Chamusca) (Forman *et al.*, 2003; van der Ree *et al.*, 2015; Borda-de-Agua *et al.*, 2018). Such corridors often form a barrier or filter to movement in and out of town. Furthermore, the corridor may be a source of plants and animals, or of noise and chemical pollutants, entering the town.

More commonly, the adjacent zone contains large patches, such as woodland, farmland, aridland, wetland, hill/mountain slope, or lake/estuary/sea. These large adjacent areas strongly affect the town and its vegetation. Town plants, animals, and people readily move into and use the large adjacent areas (Hansen and DeFries, 2007).

A third type of adjacent zone is a specific town land use, such as a fairground, golf course, dump (tip), sewage treatment facility, waterworks (water treatment facility), plant nursery, campground, or airport. Each has a single intensive land use, yet all also serve as distinctive habitats, sources, and routes for species.

Consider a local *town airport* (e.g., Hanga Roa, Alice Springs, Wausau/Weston). Most of the typically fenced elongated area is mowed grassland, with the runway as grass, soil, or paved. Depending on climate, the grassland might be mowed a dozen times a year. Mowing and disturbance intensity is greatest near runways, so that much of the remaining area has the appearance of a low meadow. Diverse small to large vehicles cause disturbances. With the airport often on an elevated site, few wetlands are present. Plants, beetles, wasps, butterflies, and ground-nesting or feeding birds are commonly prevalent. Uncommon or rare species may be present, either benefitting from the continued disturbance or because the large grassland habitat is scarce in the area. For instance, grassland birds of conservation interest often thrive on town airports.

Some towns have a *fairground* with an annual event that attracts thousands of people (e.g., Emporia, Rocky Mount, Rock Springs). That means intensive trampling, food, waste, and wastewater. Furthermore, many fairs and other events may include livestock, including horses, which add significantly to disturbance of the vegetation. Smaller events often occur at different seasons. The fairground area typically receives little maintenance much of the year, and then is heavily impacted at the major annual event. Such a maintenance regime may discourage uncommon species and encourage weeds.

A *golf course* in the town border area may be about 54 ha (133 acres), the worldwide average (Hudson and Bird, 2009). Approximately 60 percent of the course is out-of-play area, and hence particularly appropriate for biodiversity and uncommon species. Many ecological studies indicate that the abundance of evergreen trees, and having one or more waterbody, noticeably increases wildlife diversity (Yasuda and Koike, 2006; Sorace and Visentin, 2007; Hudson and Bird, 2009). Maintaining native vegetation cover in and around waterbodies, and minimizing the width of intensively mowed areas, are particularly valuable for enhancing biodiversity. Many studies also emphasize the importance of nearby natural areas in the surroundings (Porter *et al.*, 2005; LeClerc and Cristol, 2005; Hodgkinson *et al.*, 2007). Combining high microhabitat diversity, varied maintenance regimes, and surrounding natural land tends to support a wide range of plants and animals on golf courses.

A *solid-waste dump* (tip, landfill) illustrates yet another important town feature in the adjacent zone. Small town dumps are often active, with solid-waste being continually added on top. A grassy weedy, much disturbed area typically surrounds the active dump. Larger towns and cities may have closed the dump and capped it with a thick layer of clay, or a protective liner, to keep water from infiltrating through the solid waste (Misgav *et al.*, 2001). Such a facility accepts waste and recyclable materials that are then trucked away to a large dump, incinerator, or recycling center elsewhere. A town's capped dump has the grassy weedy strip around it, but also normally has a grassland covering the dome of decomposing solid waste. This grassland may have tiny wetlands on the clay, and may contain some characteristic grassland species. The concentrated habitat diversity, along with low disturbance on a capped dump, tends to sustain a rich flora and fauna (Misgav *et al.*, 2001; Rahman *et al.*, 2012; Tarrant *et al.*, 2013).

Finally, the town's adjacent zone is also important because expansion normally occurs here. Commonly, existing primary streets of the outermost neighborhood are simply extended straight outward, forming a framework for a new street grid along which buildings are built. Except in relatively flat terrain, this new grid is incompatible with the adjacent-zone topography. Expanding compatibly with the land features, rather than trying to mold the land to the grid, is more sustainable (Arendt, 2004). That would reduce earth-moving plus long-term erosion, sedimentation, and flooding, and better protect natural habitats, biodiversity, and waterbodies. Fit the expansion to the land, rather than the land to a grid.

Habitat Diversity

Common and Uncommon Habitats

A 1995 study of Dusseldorf, a small German city, discovered 78 types of habitat for plants and animals (Godde *et al.*, 1995). Of these, 38 were considered to be common, that is, found many times across the small city. About a quarter of the habitat types recorded are characteristic of a city, and typically absent in a town. Listing the 38 widespread habitats in rough order from most designed and maintained to least designed and maintained, that is, left largely alone for ecological successional processes, provides valuable insight (Forman, 2014).

Five biological groups of organisms, including plants, butterflies, and snails, were measured in the common habitats. Interestingly, all five groups showed a similar pattern, i.e., habitats in the most designed/maintained portion of the list had few species. In contrast, habitats in the least designed/maintained portion were rich in species. This was true for both an individual habitat and for all habitats of a type. The pattern suggests that human design and maintenance reduces biodiversity. We are designing and managing against nature.

Town habitat patterns are different (Girling and Helphand, 1994). Of the 78 habitat types in the small city (Godde *et al.*, 1995), approximately 17 are common in a typical town. About 20 habitat types appear to be uncommon, i.e., only one or a few examples present. Many other types, e.g., low-rise housing, school yards, bridges, and construction sites, are often present in intermediate abundance.

In the town center, there seem to be four common habitats (building, road/street, sidewalk/walkway, base of post), and also about four uncommon habitats (railway, stream corridor, green wall, and town center as a whole). The older residential area contains 15 of the 17 common habitats, but only about four of the 20 uncommon habitats. Finally, the newer residential area contains 16 of the 17 common habitats, and about 14 of the 20 uncommon ones.

House plots, as one of the most common habitats across a town, come in many sizes and shapes. These vary with topography and location of road. Also, various government and development constraints are important, such as location of a septic system, the presence of large trees, setback distances from front, side, and back boundaries, and ratio of plot and house sizes (Kendig *et al.*, 1980). In addition, groups of house plots may form distinctive habitats (Belaire *et al.*, 2014).

Species richness or biodiversity normally correlates positively with habitat diversity, and especially the number of uncommon habitats present. Thus, the preceding patterns suggest that the town center, with few common and uncommon habitats, is species poor. The older residential area has many common habitats but few uncommon ones, again suggesting rather low biodiversity, and many generalist species of common habitats. In contrast, the newer residential area, containing many common habitats, and especially containing many uncommon habitats, is likely to be species rich overall. A particular small area in the newer residential area may be species poor, but each of the uncommon habitats contains a different set of species, which enriches the newer residential area as a whole.

Two habitats uncommon in most towns today – green roofs and green walls – probably originated in villages, remained in towns, and now are recommended in cities (Dunnett and Kingsbury, 2004; Lundholm *et al.*, 2010; Forman, 2014; Madre *et al.*, 2014). *Green roofs* with plants covering flat or low-slope roofs would work well in towns without urban heat and chemical pollution (Figure 7.11) (Forman, 2014). Both "extensive" shallow-soil (<10 cm or 4 in) roofs and "intensive" deeper-soil (10–30 cm) roofs would reduce water runoff, cool the air, cool the building, and prolong roof life. Unlike most urban environments, low green roofs with modest wind in town might be useful for food production, as well as sustaining meadow species diversity. Green roofs are rarely useful for migrating birds which need food plus cover for resting.

Similarly, *green walls*, which may exist by some gardens, are promising in towns (Buchan, 1992; Francis, 2010; Jim, 2010; Kontoleon and Eumorfopoulou, 2010; Forman, 2014). The cleaner cooler air, compared with cities, facilitates growth of wall plants with roots in the ground, in cracks and holes, in elevated containers or material, and directly in the air as epiphytes.

Habitat Clusters

Before examining the clustered habitats representing high habitat diversity, several habitat concepts are useful. Also, it is useful to remember that normally towns are not plagued by urban heat, particulate matter, and chemical air pollutants, so habitats as extreme as a car park or shopping center may support a richness of spontaneous species, most being native.

Useful Habitat Concepts

Spatial scale refers to the sizes of areas under consideration. Fine scale refers to small areas, and broad or coarse scale refers to large areas, as in the scale indicated on a map. The common habitats in town range in spatial scale from the base of a post to a whole neighborhood. Habitat patterns normally differ at each scale or level of scale, such as a single habitat at one scale having high microhabitat diversity at the next finer scale. Even a house contains lots of microhabitats, for instance, for ants, cockroaches, wasps, mosses, disease organisms, bats, and mice (Hardoy *et al.*, 2001; Adams, 2016).

Habitat diversity or heterogeneity usually reflects the combination of topographic heterogeneity or roughness, plus the results of human activities on the land (Cadenasso *et al.*, 2013). Both topographic structures and human buildings normally have a sunny side and a shady side, with contrasting heat conditions for the adjacent habitats. Habitat diversity increases with the number of habitats, as well as the number of types of habitat. Thus, many habitats of two types, checkerboard-like, or few habitats each being a different type, is not as diverse as many habitats and many habitat types.

A habitat differs from its edge to its interior, so for certain purposes we can recognize an *edge habitat* and an *interior habitat*. Usually the context or adjoining habitat(s) significantly affects a patch, which may appear like a clearing in a forest, or an oasis of trees in a desert, or indeed similar to the adjoining land. The more distant surroundings also affect a habitat, illustrated by the "rain of species" entering from agricultural or forest land surrounding a town. Clearly then, the *distance from a source* of species affects a habitat. Green corridors within a town, such as a hedge or stream corridor, are short. Unlike heterogeneous long corridors elsewhere, short town corridors generally can be considered as single habitats. A row or group of stepping stones may facilitate species or human movement in the absence of a connecting corridor.

Soil and water provide additional habitat conditions. *Soil patterns* roughly vary with topographic heterogeneity, thus increasing the diversity of vegetation. Also, soil may change gradually up and down a slope so that distinct boundaries or edges are absent, thus appearing as a continuum or continuous change in vegetation along an environmental gradient. In towns, however, human activities disrupt these natural patterns. Most conspicuous are roads and buildings, but typically in both cases the natural soils have been altered by earth-moving equipment, plus the addition of usually sandy fill. Fertilization also alters the soil and viable seeds within it (seed bank), and hence the subsequent succession of plant species (Thompson *et al.*, 2005; Pellissier *et al.*, 2010).

Where waterbodies are present, such as pond, stream, river, lake, estuary, or sea, a prominent environmental gradient is present. This land-to-water gradient is usually sharp, a narrow band with quite different vegetation from one side to the other. Thus, for example, broadly a pond may be considered a habitat, but at a finer scale, several microhabitats are squeezed together from dry land to pond bottom. Similarly, coastal tourist towns in Argentina are effectively a sequence of different irregular habitat strips squeezed together, effectively creating a biodiversity hotspot of plants and birds (Faggi *et al.*, 2010).

Clustered Microhabitats in Town

Habitat clusters or concentrations represent spots with high habitat diversity, and hence species-rich locations. Consider a multicolored checkerboard. Lots of edges and lots of interiors are packed together. Many adjacencies are also present, that is, adjacent red-blue habitats, red-green, yellow-green, purple-orange, and so forth. Adjacent effects imply that red affects blue, blue affects red (differently), red affects green, and so on. Some of these effects are negative, others positive. The edges are not only microhabitats, they are filters and barriers against flows between habitats. Habitat clusters are ecological hotspots.

A *community garden* or *allotment*, where residents mainly grow vegetables, seems to illustrate such concepts and more (Figure 7.15) (Breuste, 2010). Numerous quite different square or rectangular plots are packed together, some full of tomatoes, some with diverse delicious foods, and some bulging with weeds. But also narrow paths, each somewhat different and full of spontaneous plants, separate the plots. Furthermore, fertilizers and water are used in abundance. The result is a microhabitat cluster with a richness of luxuriant plants and lots of interacting animals (e.g., Tetbury, Scappoose, Concord) (Gilbert, 1991).

Now consider a green habitat completely surrounded by a yellow one, oasis-like. The green habitat would be different if half its perimeter were surrounded by yellow and half by red (Hersperger and Forman, 2003). Suppose four types, or eight, surrounded the green habitat. In effect, habitats vary by the number, and number of types, of adjacent habitats (Forman, 2014).

Clusters of habitats pack lots of plants and other species together. Each habitat has its own set of species or species pool, yet many of the species must be generalists since they are in close proximity to other habitats and environmental conditions. Wildlife and other animals in habitat clusters have a wide range of foods in proximity. That reduces travel time in foraging and provides a highly diverse diet.

In a town, each of about 16 habitat types (Godde *et al.*, 1995) seems to be a cluster of microhabitats (stream corridor, low-density residential block, older residential area, neighborhood, house plot/homegarden, community garden, plant nursery, semi-natural woods, farmstead, farmland edge, shrub area, pond and pondshore, industrial site, park, edge of woods, and quarry/old gravel pit). All but one habitat (older residential area) are primarily characteristic of the newer residential area. Not only are the habitats quite different, the individual microhabitats within each seem to be rather different, both within a habitat and among habitats. This suggests that adding these 16 habitat cluster types together produces an extremely high microhabitat diversity in a town.

Species Richness

High habitat or microhabitat diversity normally produces high species richness. For instance, a botanical garden with many habitats close together contains an enormous

Figure 7.15 Community garden where families have small plots to grow diverse vegetables, fruits, and/or weeds. Background meadow with scattered trees and shrubs is affected by wind- and animal-dispersed seeds from adjacent woods. Concord, Massachusetts, USA. See Appendix. R. Forman photo. (A black and white version of this figure will appear in some formats. For the color version, please refer to the plate section.)

number of plant species (e.g., St. George's, Superior, Alice Springs). Thus, the 16 habitat cluster types just mentioned are probably hotspots of biodiversity in town. Several other characteristics of a town also normally produce high species richness. Being adjacent to a semi-natural area, a low-maintenance or neglected location, or a wet area in a dry climate, is likely to produce high biodiversity.

Also, in a natural community, if the relative abundance of the several leading species is fairly even (high *evenness* or low *dominance*), the community is likely to be species diverse, since one or a few dominant species is not outcompeting and eliminating most of the uncommon species. A high level of evenness not only normally indicates high species diversity, it also may suggest a lower rate of species loss or local extinction.

Town locations with particularly low species richness are also of ecological interest, though little studied. A lawn, shallow soil, and site with intensive maintenance are likely to be species poor. Also, the town center, a busy road, and adjacent to a busy road may be species-poor locations. In aridland, a xeroscaped yard covered with stones may be species poor, though it also could contain a richness of desert plants.

Note that rare or uncommon species may or may not be in species-rich habitats. High habitat diversity increases the chance that there are some habitats present that support rare species. On the other hand, a highly unusual habitat with few species may contain some rare species that can survive or thrive nowhere else (Forman, 1979).

Whole Town as a Habitat

Tying together many of the above concepts, a town is a distinctive habitat surrounded by farmland, woodland, or aridland. Typically, a town originated at a somewhat unusual site in the landscape, often related to topography and water. Topographic heterogeneity and a waterbody provide rich habitat diversity. People over decades or centuries have altered and added to the natural features, producing a yet more diverse set of habitats. Uncommon habitats often harbor many species uncommon in the landscape. Still more important, the habitat types with clustered microhabitats as biodiversity hotspots greatly increase a town's biodiversity. These habitat types are largely absent in the surrounding land.

Furthermore, the surrounding agricultural, woodland/forest, and/or aridland landscapes are species sources producing a rain of species continually entering the town. A green wedge or patch-corridor network protruding into town greatly increases the flow of species into town. Lots of agriculturally related species blow in or are carried into town. Natural woodland or aridland species enter in abundance. The town edge is doubtless especially rich in species from both the adjacent zone and the surrounding landscape.

Nonnative species, particularly from urban areas and other towns, arrive. Spontaneous nonnative plants may be most common in the town center (Muller and Werner, 2010). However, residents widely plant nonnatives in residential house plots. Nonnatives, such as house sparrows, common pigeons (rock doves), and starlings (*Passer*, *Columba*, *Sturnus*), seem to be most frequent in the interior portion of a town. These species tend to survive or thrive in town centers, where buildings and trees with holes provide nest sites. Perhaps more important is the relative abundance of human-provided food, including spills, garbage, and bird-feeder seeds. Seed-eating birds from surrounding natural land are attracted to residential areas with feeders (Cowie and Hinsley, 1988; Brittingham and Temple, 1992; Fuller *et al.*, 2008).

In addition, spatial patterns in town are ever changing. Seasonal flooding and drying, windstorms, human construction, fires, and diverse other human activities are ongoing. Maintenance and repair budgets and activities go up and down over time. The human population grows along with development. In ten Bohemian (Czech Republic) "villages," plant species richness and the number of nonnative species increased during only a slight built-area expansion (Wittig, 2010). Towns and villages also often lose population, with corresponding abandonment of house plots and consequent ecological succession. Both native and nonnative floras change over time (Bertin, 2013; Bertin and Parise, 2014). Disasters occasionally happen, followed by massive town changes.

Yet some habitats or locations are more stable than others. A rural study of 13 habitat types over a century in Denmark found that some habitats change little, while others are readily changed or eliminated (see Figure 3.13) (Agger and Brandt, 1988). In brief, towns are evolving but remain extremely rich places biologically within their landscapes.

Scattered through this chapter are clues highlighting towns as ecosystems unlike any others on land. The prevalence of buildings and roads, among other things, of course separate towns from natural land and agricultural land.

Unlike cities, water conditions in towns typically feature little pumping of groundwater and most rainwater infiltrates into house plots (instead of being piped out of town). This results in a relatively high water table (also being located near a stream or other waterbody affects most of a town). Clean and cool air, plus ample soil water, support robust plant growth. Small patches of high pH soils (near concrete and mortar in town) provide habitat diversity with a richness of species. Many street trees are mature, and tree canopies cover a large portion of town. Shrubs are abundant in public spaces, native shrub and tree species are prominent across town, and specialist species grow well. Lichens are abundant on tree trunks and cemetery stones, and diverse mosses appear widespread. Flower gardens are abundant, and native species often dominate spontaneous vegetation.

Neighborhood house plots have many back-line semi-natural corridors, plus small multi-plot wooded patches in back corners. Stream, railway, and/or pipeline corridors, as well as green wedges, are major routes of wildlife and seeds into and out of town, where few major barriers to species movement exist. Finally, nearby surrounding farmland or natural land covers the entire town ecosystem with an endless rain of species.

8 Patterns of Wildlife and Other Animals

... the nobler animals have been exterminated here ... I listen to a concert in which so many parts are wanting ... thinking that I have here the entire poem ... I hear that it is but an imperfect copy ... my ancestors have torn out many of the first leaves and grandest passages.

Henry David Thoreau, *Journal*, 1856

In the time frame of species on land, villages and towns are newcomers. Some resident species have dispersed long distances and colonized a town, while migratory species briefly stop in enroute between distant locations. Still, most animals are town residents that came from the agricultural or natural land surrounding a town. Other animals enter and leave town within a day or even a minute.

Wildlife and biodiversity are now considered from five major perspectives: (1) understanding town animals; (2) wildlife of residential habitats; (3) wildlife of roadsides, waterbodies, and other town habitats; (4) dogs, cats, and farm animals; and (5) artificial night lighting and flying fauna.

Understanding Town Animals

The form of a village or town strongly affects the interactions between town/village and surroundings. In general, villages are either linear, usually along a road or edge of a waterbody (e.g., Selborne, Rangeley; see Appendix), or compact (enclosed) with buildings and streets surrounding a relatively central space (Chapter 1) (e.g., Vernonia, Petersham). Almost all building plots of a village adjoin the surrounding land.

Meanwhile, towns are typically ovoid (rounded to oblong) with four to six main roads radiating outward, sometimes with scattered buildings alongside (Figure 8.1) (e.g., Tetbury, Socorro). This pattern is usefully represented by a *hub-and-spokes model*. Although building plots along the town edge adjoin surrounding land, the great majority of town buildings are surrounded by other buildings and thus separated from the surroundings.

A town normally started as a point and then progressively increased in average diameter, with the commercial town center and adjacent older residential area (Chapter 1) near the central point. Because town area increases with the square of the radius (πr^2), relatively little space exists near the center and considerable space in the newer residential area. All-built town areas expand at the edge of the hub, and also at the base of some radiating main road spokes. Occasionally another radiating road develops. Town wildlife either reside in town habitats or enter across the town edge.

Figure 8.1 Horse riders along main street of town. Rio Lagartos, Yucatan, Mexico. Photo courtesy and with permission of Mark Brenner.

Biodiversity and Animal Populations

As described in earlier chapters, *biodiversity* normally refers to the variety or number of living systems or life forms present. These may be considered at three scales, the number or richness of habitats, species, and genetic types. The most common measure of biodiversity, also used in this book, is the number or richness of species. Infrequently, we consider the richness or diversity of habitats, as in the case of clustered habitats (Chapter 7). We rarely point out the diversity of genetic varieties, such as the varieties of fruit trees in a village (Chapter 7), or seed varieties, as in the following example.

Around villages and towns of the dry Sahel region of northern Nigeria and southern Niger, 25 crops (cultivars) are grown (Mortimore, 2013). Cereals, beans, earth nuts, root crops, melons, vegetables, sugar cane, cotton, and kenaf predominate. Yet the number of named and maintained crop varieties (land-races) in a wetter site (including irrigation) is 76. Much of this agro-biodiversity, measured by number of genetic varieties, is present in 2 indigenous cereal grains, pearl millet with 3 to 12 varieties at dry to wetter sites, and sorghum with 6 to 22 genetic varieties.

The biodiversity of "famine foods" used in five Kano area villages (northern Nigeria) during a severe drought (1972–1974) adds to the picture (Mortimore, 1989). In addition to the crop varieties used, 47 species of trees, shrubs, grasses, and herbs provided food. Other famine foods known to be locally used brought the list to 68 plant species. A later study of a 1992–1996 drought here identified 19 plants only used in times of scarcity.

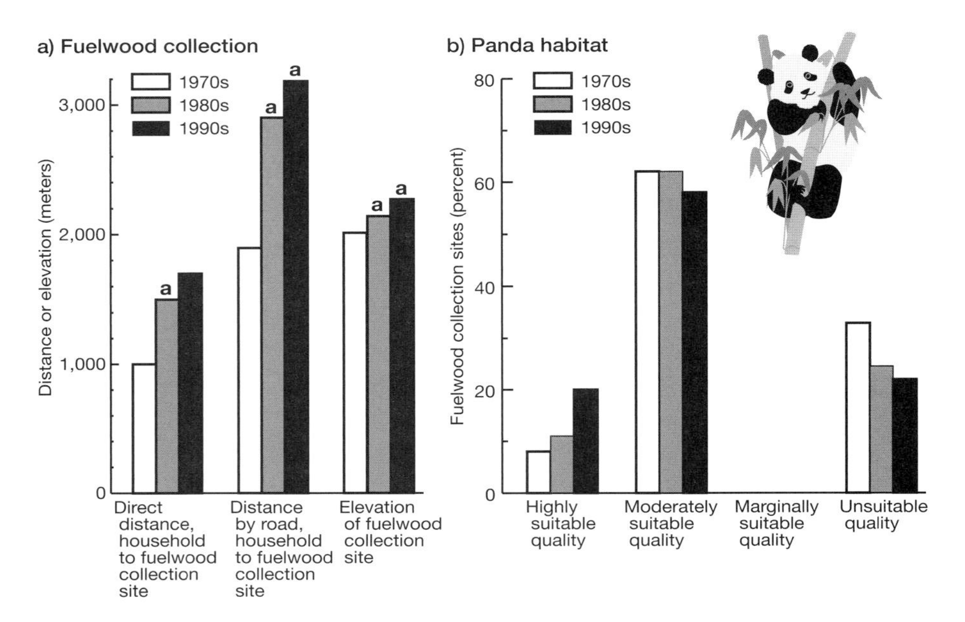

Figure 8.2 Changing fuelwood collection sites relative to households and panda habitat. Average values plotted; *a* = significantly greater than in the preceding period. Wolong Nature Preserve, Sichuan Province, China containing about 110 giant pandas (*Ailuropoda melanoleuca*), or 10% of the total wild population. Most households of fuelwood collectors (firewood cutters) are within the reserve. Based on He *et al.* (2009).

In summary, although habitat diversity and genetic variety add to our understanding, biodiversity here will be used in its commonest sense, i.e., the richness, number, or diversity of species. Also, *species-rich habitats* or areas are valuable from a variety of ecological and human perspectives. Species-rich habitats threatened by degradation or loss have been called "biodiversity hotspots."

Clearly species differ in many characteristics or attributes, for instance, big to little, dangerous or not, native or nonnative, reproducing or not, and edible or inedible. Rare or uncommon species, such as the giant panda in China (Figure 8.2) (see Chapter 11), are of particular interest in this chapter relative to towns and villages.

A species exhibits a set of traits (the species phenotype), each *trait* being a product of the species' genes and the environment (Evans, 2010; Adler and Tanner, 2013). Species traits or characteristics or attributes may be broadly grouped as morphological, physiological, behavioral, life history, and social.

Body size is a useful trait in part because the area of a species home range roughly correlates with body size. The *home range* is the area surrounding a nest or den that is used in daily movements, especially in foraging for food. Some animals, particularly birds, also have a smaller *territory*, the area around the nest defended against intruders. A third important spatial dimension is *animal dispersal* distance, that is, how far a subadult animal leaving the den or nest moves to establish its own den (with mate) and home range. Usually this is more than 1.5 times the home-range diameter.

Home-range diameters are usefully compared with town diameters. A bear's home range greatly exceeds a town area, so a bear in town is moving in and out (Figure 2.14) (e.g., Lake Placid). A songbird territory or house cat home range is much smaller than a town, and hence, except along the edge, the animal lives entirely in town. An animal from the surroundings frequently foraging in town has its den just outside town. The dispersal distance of a resident town species indicates its probability of dispersing to a nearby village or another town in the surroundings.

Consider the three sources of species in a characteristic town surrounded by farmland, which in turn is surrounded by natural land. The town may have some remnant native species from pre-town natural land, but is more likely to have remnant species from preceding agriculture. Second, species from both surrounding farmland and natural land continually arrive. Finally, both native and nonnative animals arrive in town with trucks, cars, trains, and people, and in nursery plantings.

Biodiversity can be measured as the number of species with any particular trait, though of course a huge range of possible traits exist. Broader, more integrative measurement approaches exist, including the number of "increasers" and "decreasers" (species increasing or decreasing in population size over time). Or, measure "town avoiders" that do poorly in town, "town adapters" that tolerate but don't depend on the place, and "town exploiters" that primarily live in towns (Macgregor-Fors and Schondube, 2011; Adler and Turner, 2013).

A study of spider species richness in Debrecen (Hungary) urban and non-urban areas highlights the value of understanding *habitat affinity*, that is, where a species or its relatives primarily lives or originated. Spider species richness is higher in urban than in suburban and rural sites (Magura *et al.*, 2010). The added urban species richness is related to more open-habitat species, suggesting the importance of the surrounding pastureland and cropland on the urban spider community. The pattern may be more striking for small towns in agricultural land. Thus, the number of species by habitat affinity provides insight into biodiversity measures.

A *population* of animals refers to all the individuals of one species in an area (Figure 8.3). Populations grow when birth rate plus immigration exceed mortality plus emigration. Natural and human limiting factors, from predators to toxic substances, especially influence mortality. A population also has an age structure, the percentage of individuals from young to old, and a sex ratio of males to females; both affect the population's reproductive potential. *Population density* or *abundance* (number of individuals of a species) depends on the carrying capacity, basically how many animals the available resources and limiting factors will support.

Particularly important in and around a town is a *metapopulation*, with individuals of a species distributed in separate patches, among which individuals can move (Opdam, 1991). A town area has lots of small habitat patches, and road and housing development often creates more. Local- or sub-populations of animals may temporarily disappear (go locally extinct) on a patch, followed by recolonization of the species. Metapopulation dynamics refers to the rate of species disappearance and recolonization on all the patches.

Also, *competition* between species for limited resources occurs. For instance, near Brisbane (Australia), the species richness and abundance of woodland-dependent

Figure 8.3 Population of flamingos in shallow bay protected by shoreline mangrove swamp close to town. Celestun, Yucatan, Mexico. Photo courtesy and with permission of Mark Brenner. (A black and white version of this figure will appear in some formats. For the color version, please refer to the plate section.)

small songbirds (passerines) correlate with the density of shrubs present (Kath *et al.*, 2009). But also richness and abundance are strongly and negatively related to the abundance of one bird species, the noisy miner (*Manorina melanocephala*). This aggressive species excludes most of the songbirds from habitats, except where the songbirds can use abundant shrub cover to escape. Indeed, in sites with many noisy miners, which compete successfully for open understory space, 13 of the 73 songbird species recorded are absent.

Lots of *species interactions* besides competition occur in towns. Predation, parasitism, and mutualism (where both interacting organisms benefit) are generally familiar. For instance, around Morelia (Mexico) the predation of eggs or nestlings in artificial bird nests is higher in urban and suburban areas than in outer shrubland areas (Lopez-Flores *et al.*, 2009). Birds are the main predators in all three areas. Snakes, however, may be major predators on ground-nesting birds in California shrubland (Patten and Bolger, 2003). Predators that control herbivore populations, consequently favoring plant growth or biomass, illustrate the "trophic cascade" concept. Effects at one feeding or trophic level affect two or more other feeding levels in the food chain.

These many briefly described concepts introduce the richness of wildlife and biodiversity dimensions in towns. However, few town or village studies on these concepts yet exist; a mammoth research frontier awaits us. Most of the town studies to date use species richness, abundance, and composition.

People, Biodiversity, and Faunas

People value and have preferences of animals for many reasons (Kellert, 1996). Indeed, in the USA, the most preferred animals are dogs and horses, and the most disliked are cockroaches, mosquitoes, and rats. Similar results are reported from Germany and Japan, though in the latter case, invertebrates generally have higher value. Dogs, cockroaches, and mosquitoes populate almost all towns. Rats are common in large towns where they clean up discarded food and other organic matter, as well as clean the insides of pipes with their hair. Horses are present or important in many towns and villages (Figure 8.1) (e.g., Pinedale, Petersham).

Information on animals and other ecological subjects is based on published science. Yet people in many villages and towns dispersed around the world do not read or speak the language of published science. The Zapotec-speaking community of San Juan Gbee (population ca. 1,300; Oaxaca, Mexico) recognize and name nearly 400 species of animals in and around the village (Hunn, 2008). These include about 12 snakes, 7 lizards, 1 turtle, 4 iguanas and crocodilians, 6 amphibians, 4 fish, 120 birds, 105 insects, 13 crustaceans, arachnids, etc., and 5 other invertebrates. Many of the snakes and invertebrates recognized by local residents are at the scientific genus level. In contrast, scientists have identified about 25 percent more types of snake and about 60 percent more bird species. Scientists using scientific literature from elsewhere as a foundation, and local people intimately familiar with animals around the village, recognize different levels of biodiversity.

Towns and villages have a special role in protecting biodiversity. Published national and state-listed rare species (e.g., Concord, Lake Placid) receive press coverage, funds, and protection efforts by agencies and organizations. The rest of the species of conservation interest are far from cities, but often close to towns and villages across much of the land. In effect, localities are responsible for protecting these species, many of which are rare and declining.

Detailed faunas doubtless have been occasionally described for towns and villages, but the information usually ends up in unreviewed and nearly unavailable reports. One list of animals in a 0.1 ha suburban house plot over 15 years (Leicester, UK), and some animals from a text on urban entomology (North America) provide insight into what to look for in towns. The garden study (Owen, 1991) recorded 2,204 species, including: 7 mammals, 49 birds, 3 amphibians, no reptiles, no fish, 3 flatworms, 4 snails, 17 slugs, 5 earthworms, 1 leech, 8 woodlice (isopods), 7 centipedes, 4 millipedes, 1 false-scorpion, 10 harvestmen (daddy long-legs), 64 spiders, plus hoverflies, butterflies, lacewings (and allies), ichneumons (wasps and allies), earwigs, bees, and ladybirds (ladybugs). Rare butterflies, moths, and wasps were found. Many short food chains were identified. The author indicated that no pests were found; all species were interesting.

The author also lists common birds in flower gardens of different continents (Owen, 1991). African gardens commonly have bulbuls, wax finches, robin chats, barbets, and sunbirds; Europe – thrushes, tits, robins, greenfinches, and warblers; and North America – robins, blue jays, leaf warblers, cardinals, and chickadees. She further notes that the American robin and the English blackbird are successful in towns. Flies, cockroaches, rats, house sparrows, starlings, and feral pigeons are in built areas worldwide.

No list of town species is complete. In the 15 years following the Leicester garden study, many more species were discovered (Owen, 2012). A story in Concord (USA) (see Figure 7.14) emphasizes the point. A tiny owl, the saw-whet owl (*Aegolius acadicus*), had not been found in town for decades. One night a birder with a tape recorder went into the town's largest woods and played a recording of the saw-whet owl. Almost immediately, a saw-whet close by called back. The birder then repeated the process in different parts of the woods, and at almost every place an owl promptly responded. Most likely the woods is full of the little mug-size owls, though scores of good birders constantly combing the town for birds had missed it.

The urban entomology text highlights the following groups as being widespread pests in built areas (Robinson, 1996): cockroaches, fleas, lice, mites, spiders, bugs, ants, flies, stinging hymenopterans (bees, wasps, etc.), termites, and moths/beetles in wool and other organic materials. Because these invertebrates are pests among dense populations of people, many pesticides are used in high doses and for long periods. Hundreds of insects have adapted and developed insecticide resistance.

Wildlife of Residential Habitats

Town habitats differ from town center to older residential area to newer residential area (Chapter 1). Within each town area it is useful to differentiate large habitats, small-patch habitats, linear habitats, and microhabitat clusters (Chapter 7).

Usually, few large habitats exist in the town center and older residential area, not surprisingly with little area in the central portion of a circle or town. But a variety of large habitats may be entirely or partially in the newer residential area (e.g., Huntingdon/Smithfield, Concord; see Appendix). Several small habitats are often present in all three areas of town, and each area contains a few different types of small habitat. Most linear habitat types are in all three areas of town, though the newer residential areas may contain some distinctive corridors.

Microhabitat clusters are concentrations of small habitats important in part as species-rich locations, though other species-rich sites also exist in town. Most of the microhabitat clusters are in the newer residential area. Typically, hardly any species-rich site exists in the town center while some are in the older residential area. Numerous types are typical in the newer residential area, which also often has diverse remnant habitats and species from the previous land use, usually farmland.

Few published studies of wildlife in towns and villages seem to be available, so most of the information following is modified from urban wildlife studies. Thus, 23 bird and marsupial/mammal species are characteristic of Australian towns and cities (Temby, 2004; Adams and Lindsey, 2011): Australian brush-turkey, Australian magpie-lark, Australian magpie, Australian wood duck, black flying-fox, common bush tail possum, galah, grey-headed flying-fox, house sparrow, laughing kookaburra, little corella, long-billed corella, Pacific black duck, rainbow lorikeet, red flying-fox, ring-tail possum, silver gull, spotted turtle-dove, sulfur-crested cockatoo, superb lyrebird, torresian crow, welcome swallow, and white ibis.

Abundant urban wildlife information is available for Europe and North America, much of which probably applies to towns. So, below are "characteristic town wildlife" by habitat based on Leutscher (1975), Landry (1994), and Boada and Capdevila (2000) for Western Europe, Burger (1999) and Adams (2016) for eastern North America, and my own experience living in both regions. Some of the animals listed are wildlife types, often equivalent to genera rather than species. No urban or town wildlife summary was found for South America, Africa, and Asia. A few species seem to be common in urban areas of different continents (Adams, 2016), especially: house sparrow (*Passer domesticus*), feral pigeon (or rock dove, *Columba livia*), European starling (*Sturnis vulgaris*), red fox (*Vulpes vulpes*), feral cats (*Felis catus*), white-tailed deer (*Odocoileus virginianus*), and raccoon (*Procyon lotor*). Towns also contain species of their agricultural, forest, or aridland surroundings.

Town as a Whole

Within the surrounding land a town appears as a patch, a relatively distinct habitat. Some 20 wildlife types appear to be widespread in Euro-North American towns. Five are mammals, 11 birds, 1 reptile, and 3 invertebrates. Seven types are commonly seen flying over a town. Some may live in town and others in the surrounding land. A handful of bird species are widely present in town centers and parks of European towns (Figure 8.4). Unlike the flocks of birds common in cities, town birds are more likely to be in ones or twos, as in most natural ecosystems.

The town center is a small area most unlike the surrounding landscape. House sparrows, feral pigeons, starlings, mice, rats, cockroaches, and other species thrive on the human-provided food and garbage (Gilbert, 1991). Dumpsters collecting food waste and bags of rubbish in a town center are favorite animal feeding locations (Goode, 2014; Adams, 2016). A city center study found that omnivorous and tree-nesting bird abundance correlate with the heterogeneity of buildings (Pellissier *et al.*, 2012). However, the abundance of insect-feeding animals is more related to vegetation, particularly shrubs in areas of dense medium-height buildings.

In towns and small cities of southeastern Australia, amphibian (anuran) species richness is low near town centers and in moist terrain with low-income housing, but high in high-income neighborhoods near town edges (Smallbone *et al.*, 2011). Around Tucson, Arizona (USA), roadrunners, a large running predaceous bird, are essentially absent in the city center, and highest in density in the urban border area (DeStephano and Webster, 2012). Probably many species common in the town edge or adjacent zone become less common in the town center.

Figure 8.4 Common birds in European towns.

Red fox (*Vulpes vulpes*), as one of the characteristic widespread species in towns, has acclimated or become habituated in the built environment. But red foxes are also widespread in surrounding areas. Foxes native to the forest and agricultural land tend to avoid towns when moving across the landscape (Storm *et al.*, 1976). Rainforest elephants in Xishuangbanna (southern China) feed widely through the forest, but when villages expanded in population and area, the elephants became increasingly aggressive. The animals invaded villages and kitchens, destroying gardens, crops, and pigpens, and injuring people (Wu *et al.*, 2014).

The predation of bird nests tends to increase in fragmented habitat. In an area of northern Finland (Lapland) with eight villages and two towns, nest predation was measured on a gradient from town area to village area to forest (Kaisanlahti-Jokimaki *et al.*, 2012). Village size had no effect on predation.

Avian diversity and feeding types (guilds) were analyzed in land around 14 coastal tourist villages and towns (population 3,400 to 42,700) in Argentina (Faggi *et al.*, 2010). Species richness is higher in grassy areas than in nonnative tree plantations. Mainly insectivorous rural grassland birds predominate in grassy areas, whereas forest mainly has seed-eating (granivorous) birds which are characteristic of the built areas.

Buildings as Habitats

In addition to many of the widespread species in towns, buildings typically contain at least a dozen types of animals of their own. Based on estimated characteristic town wildlife (see above), nine types typically breed on the outside of buildings. Older buildings and exterior ornamentation favor these species. Another dozen animal types are typical inside buildings. Pet dogs, cats, and fish are cared for by people in their rooms, where house flies also buzz about. Clothes moths thrive in storage spaces, feeding on woolens and other organic materials. Wasps and relatives often nest in walls and other enclosed spaces. The basement or cellar is home to house mice, ants, and cockroaches, which all use the kitchen and other places with food or garbage.

Small animals may be abundant on walls, especially old walls or surfaces with ornament. Moist surfaces, shady walls, deep cracks, and loose or irregular substrate favor most wall animals (Wheater, 1999). Walls of limestone, concrete, and mortar between bricks have a higher pH than those of silica-based rocks and bricks, and thus favor limestone species. A study of 216 moss samples from walls near Zurich (Switzerland) discovered over 60,000 animals present (Steiner, 1994; Wheater, 1999). Most common were nematode worms (19,054 individuals), tartigrades, rotifers, mites, and springtails, with a scatter of centipedes, spiders, and beetles. Hymenopteran bees commonly nest in walls, and ants may forage on walls.

A study of 27 green vegetated walls and 27 bare walls in north Staffordshire (UK) found that birds use the green walls year-round, gaining some visual predator avoidance and wind protection from the foliage (Chiquet *et al.*, 2012). Interestingly, birds could only perch atop the bare-wall building, whereas the entire surface of the green wall is available for bird use. Green walls next to fast-food restaurants or dumpsters with spilled food and garbage could provide good bird habitats (see Figure 9.3). Furthermore, lighted glass windows or bare walls may attract certain insects, such as aquatic species, that die on the surface, thus providing food for birds.

On buildings, feral pigeons and starlings commonly use ledges and window boxes. Various raptors, including peregrine falcons, kestrels, merlins, and red-tailed hawks, nest on building ledges somewhat protected from rain (Wheater, 1999). Many bird species, especially cliff-nesters, occasionally nest on roofs. In 1993 peregrine falcons (*Falco peregrinus*), basically cliff nesters, nested in high places of about a dozen towns (and many cities) in the USA and Canada (Cade *et al.*, 1996). Although town buildings are low, a water tower or other tall structure might be a suitable nesting site.

Buildings around Milan (Italy) built before 1936 are of brick and roof tiles, with many ledges and ornament used by birds, especially feral pigeons (Sacchi *et al.*, 2002). In contrast, newer urban buildings are mostly of smooth concrete and glass, providing little habitat for wildlife. Even two- and three-level buildings in town centers can provide rich habitat, which is also close to farmland and natural habitats as species sources.

Wall and roof holes, cracks, and broken windows permit many birds, some mammals, and numerous invertebrates to enter and live inside buildings. Furthermore, people in crowded cramped living conditions facilitate the transmission of airborne diseases (Hardoy *et al.*, 2001). Poor water supply, sanitary wastewater system, solid waste disposal, and food-borne conditions lead to an abundance of disease-carrying vectors, such as mosquitoes, worms, flies, cockroaches, and fleas.

Finally, architecturally LEED-certified new buildings may have green roofs, green walls, bioswales, and/or detention/bioretention ponds (Yudelson, 2008). These measures greatly increase the abundance and richness of animals in and on buildings, though studies are scarce.

House Plots and Microhabitats

Some 30 animal types, in addition to the widespread species mentioned above, seem to characterize house plots. Twenty-two are present essentially throughout the heterogeneous area surrounding a house (Figure 8.5). Six types, butterfly, moth, snail, slug, earthworm, and spider, are mainly in a flower or vegetable garden. A spot for yard waste favors two more types, sowbug/isopod and centipede. Vacant house plots commonly have house mouse, squirrel, rat, ant, garter snake, isopod, centipede, and termite.

The house plot provides the essentials for animals – food, cover, water, and nest sites. In addition to habitat diversity, the size and age of the house plot plus the surroundings are keys to the assemblage of birds present (Smith *et al.*, 2005). Older plots normally have mature trees attracting a rich fauna.

A house plot represents a patchwork of habitats or microhabitats, such as the house, lawn, shrub area, trees, rock garden, flower gardens, vegetable garden, hedges, walls, tiny pond, and brush pile/compost heap (Gilbert, 1991; Owen, 1991, 2012; Daniels and Kirkpatrick, 2006). Birds, including blue tit, blackbird, robin, house sparrow, starling, and great tit in Britain, tend to be extremely diverse and dense (Gilbert, 1991; Owen, 1991; Bland *et al.*, 2004). Site fidelity may also be high, such as the 41 bird species that continuously nested over many years in a large London garden (Goode, 2014). Huge

Figure 8.5 Common birds in eastern North America towns.

numbers of fly, beetle, lepidopterans (moths, butterflies), and spiders were recorded in the garden. Mammals in house plots there often include hedgehog, gray squirrel, and rabbit, with fox periodically moving through (Gilbert, 1991).

Biodiversity in house plots can be artificially increased to levels far higher than in surrounding natural areas by adding foods, diverse structures, and water features (Cowie and Hinsley, 1988; Johnston and Don, 1990; Brittingham and Temple, 1992; Gaston et al., 2005; Fuller *et al.*, 2012). For instance, three common bird species (house sparrow, European blackbird, starling) correlated with the density of bird feeders in urban Sheffield (UK), though the other common species, blue tit, great tit, common wood pigeon, and winter wren, did not (Fuller *et al.*, 2008). Because of the abundance of diverse flowers, pollinators are especially abundant in house plots (Stelzer *et al.*, 2010; Winfree *et al.*, 2011; Garbuzov and Ratnieks, 2013). Bees, wasps, butterflies, and moths are conspicuous. But hummingbirds, some beetles, and some flies also pollinate flowers.

Tree and Lawn Habitats

A town park is perhaps most characteristic of a place combining trees and lawn, though of course this combination often characterizes a golf course, institution, even many house plots. People parks are *lawn-tree-bench ecosystems*. Based on the characteristic town wildlife list, fourteen animal types, plus the widespread types above, are characteristic in parks. The European blackbird or American robin and the mourning dove or "calling" pigeon (wood pigeon and other species) are the familiar tree species. Bats often use a hollow tree. Squirrels use a hole in a tree or build a spherical twig and leaf nest. A tiny woods or copse often has another five types: woodpecker, turtle, ant, termite, tick. Five more wildlife types characterize the grassy lawn: mole or vole, American robin or European blackbird, Canada goose, song sparrow, and cricket.

A study around the town of Stevens Point, Wisconsin (USA) found that cooper's hawk (*Accipiter cooperii*) feeds and breeds in towns, suburbs, forest, and semi-open areas (e.g., Socorro, Lake Placid) (Rosenfield *et al.*, 1996). As essentially the only North American hawk that regularly forages inside the forest, the hawks also perch near bird feeders in town attempting to feed on songbirds or squirrels (Figure 8.6). Another study in six Kansas and Oklahoma (USA) towns highlighted the relatively recent breeding of Mississippi kites in towns and cities (Parker, 1996). The species breeds and forages gregariously, and is attracted to a potential nest site more than 6 m high on a tree with a nearby perch and other kite nests nearby.

A new lawn tends to be strongly dominated by the rapidly growing grass planted, with other plant species introduced in added topsoil disappearing after a few mowings (Gilbert, 1991). Over decades without fertilizing and watering, the initial dominant is gradually replaced by a diversity of species, with no predominant species present. In older lawns one species often assumes dominance in cover. Soil nutrient status (and probably water), but not slope, aspect, or pH, seems to be the major factor determining the dominance, diversity, and composition of lawns.

Herbicides are usually used to favor grass by eliminating non-grass (dicot) species, including those with flowers. Yet the opposite approach can be used to favor flowers

Figure 8.6 Hawk on bird feeder of residential house plot while other birds remain hidden, still, and silent. Female Cooper's hawk (*Accipiter cooperii*), suburban town, southern New England, USA. R. Forman photo.

in the lawn, and thus animals in the lawn (Gilbert, 1991). Some dicot plant families are resistant to quite different herbicides, and hence favored by limited use of these herbicides. Soil animals, birds, mammals, and pollinators may become frequent on such lawns.

The addition of peat or bark as mulch around plantings also introduces seeds and then plants. Six common species and 12 occasional species introduced in mulch were found in a London park (Gilbert, 1991). Plants are introduced into parks by providing seeds for birds, being attached to footwear, clothes, and dogs, and in other ways. Other than roosting or foraging geese and gulls, rather few birds are attracted to playing fields and other large lawns, except near woody plants (Wheater, 1999).

Semi-natural woodland in parks of course supports a dense and diverse fauna. But scattered trees and lines of trees are also used by some birds and other species. For instance, the bark of trees is habitat for ants and other invertebrates, which attract foraging birds and other animals. A study of tree-bark arthropods in Matsudo City (Chiba, Japan) found that tree species and diameter were the primary determinants of invertebrate abundance, diversity, and composition (Yasuda and Koike, 2009). The arrangement of trees (isolated, tree line, and tree cluster) was not a significant factor.

Institutions and town parks often contain all three arrangements of trees surrounded by lawn (Dexter, 1955).

A study of two animal and two plant groups in Osaka area (Japan) parks concluded that species richness of birds is mainly influenced by habitat area and distance to a species source (Natuhara, 2008). Ants are more related to site history. Some rare species of ants, ferns, and trees occur in small habitats, which therefore are important for total biodiversity. The great tit (*Parus major*), a woodland bird, can successfully breed by using scattered park trees, and thus has a larger home range than in rural woodland.

Sacred trees and groves, often associated with a town or village, are somewhat protected long-term, but may be gradually degraded by an excess of visitors. An old sacred grove of trees is likely to contain many species uncommon or rare in the landscape (Pungetti *et al.*, 2012). Close to a village in Ghana, one government-protected grove with some ecotourism contains the only population of beautiful mono monkeys (*Cercopithecus mona mona*) (Ormsby, 2012). Since the monkeys are associated with an idol or fetish placed in the grove by village ancestors, they are safe, even after sometimes stealing food in the village.

Or, around villages in northern Madagascar, taboos (*fady*) protect many animals which represent spirits of the earth (Mannle and Ladle, 2012). It is taboo to eat eel, lemur, turtle, tenrec, bat, and even green pigeon, and taboo to harm or kill chameleon, boa, fourline snake, hognose snake, and lemur. A harmed or killed taboo animal is believed to stay in the person and later seek retribution. Some sacred species may be associated with a religious building, a gravestone, or a particularly old tree, which effectively offers protection.

Wildlife of Town Corridors and Patch Habitats

Road, Railway, and Stream Corridors

Roadside

Town roadsides or verges vary widely but generally contain a wet ditch and some infrequently mowed grassy vegetation. A vacant plot between buildings may have grassy successional vegetation (e.g., Grande Riviere; see Appendix). Vehicle traffic provides disturbance, noise, and roadkill. Street lights prolong light into the night. Three of the species commonly use bridges, while five (bat, moth, beetle, lizard, and toad) are mostly around street lamps at night. Overall, 20 animal types listed as characteristic town wildlife (see above), plus the widespread types above, appear to be typical of roadsides (Figure 8.7).

The primary effects of roads (Figure 8.7) in and around towns are probably traffic noise, providing tall grassy habitat, roadkill (animal mortality), the barrier effect, altered stormwater flow, and chemical effects (Forman *et al.*, 2003; van der Ree *et al.*, 2015). Roadkill rates are typically highest on fast two-lane roads such as a town's radiating main roads. A 16-month study of roadkill on 12 km (7.5 mi) of roads in suburban towns of Tippecanoe County, Indiana (USA) recorded 360 mammals, 205 birds, 141 reptiles,

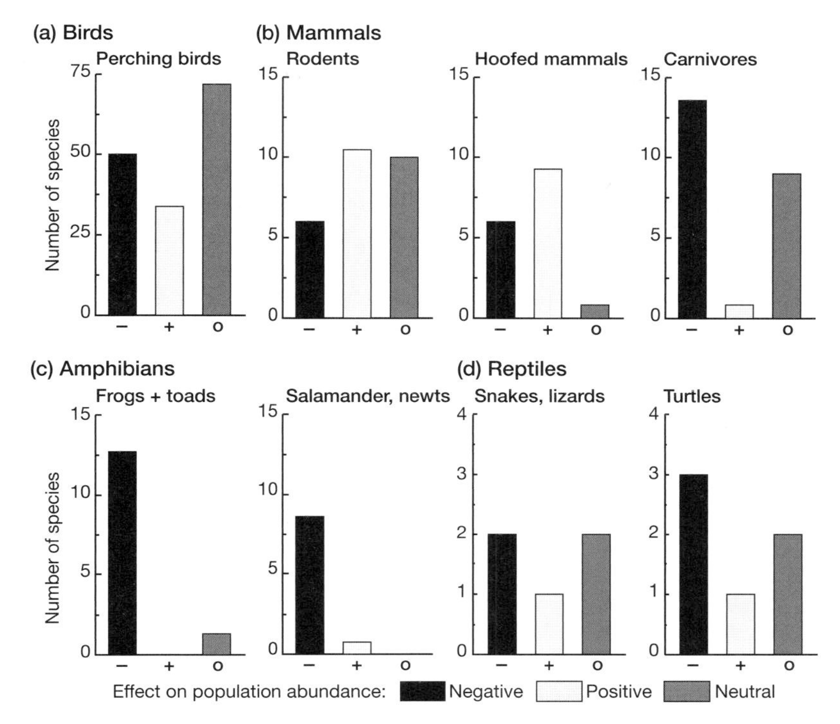

Figure 8.7 General response of different animals to roads. Results for: (a) 153 perching bird (passerine) species from many studies in several nations; (b) 67 mammal species based on numerous studies in several nations; (c) 23 amphibians from 16 studies in six nations; (d) 11 reptile species based on nine studies in three nations. Adapted from Rytwinski and Fahrig (2015).

and 9,809 amphibians (almost all frogs) dead (Adams and Lindsey, 2011). Traffic noise mainly correlates with the amount of traffic (plus the percent of trucks), so a busy highway usually affects sensitive birds and other species for hundreds of meters outward (Forman *et al.*, 2002; Reijnen and Foppen, 2006). Wildlife crossing structures such as underpasses (e.g., Plymouth, Concord) reduce both roadkill and the barrier effect (Forman *et al.*, 2003; van der Ree *et al.*, 2015).

Tall grassy roadsides, including a moist ditch and other microhabitats, provide habitat and connectivity for movement of some animals (Wheater, 1999). Air pollutants from roads/vehicles affect nearby arthropod populations as well as street trees (Gilbert, 1991). Under-road culverts provide escape cover and resting spots for many types of animals, while erosion and scouring are associated with bridges (Adams and Lindsey, 2011). Many small animals tend to avoid crossing a road with traffic, so the road acts as a barrier or filter.

Vehicles and road surfaces give off a range of chemicals, from heavy metals to hydrocarbons and road salt, blanketing roadsides and adjoining properties (Gilbert, 1991). Indeed, maritime plants have invaded British roads and towns due to road salt accumulations in roadsides, while some roadside trees and other species have died

Figure 8.8 White-tailed deer (*Odocoileus virginiana*) in field by woods and old tree. Near village of St. Mary's City, Maryland, USA. R. Forman photo.

due to the excess salt. Stormwater flows along roadside ditches and through culverts, carrying water, debris, and chemicals to a local waterbody.

In brief, roads and roadsides permeate and widely affect towns, perhaps having a greater ecological effect than in urban, agricultural, or natural land (Chapter 12). Wildlife–vehicle collisions with, e.g., deer (Figure 8.8), damage people, vehicles, and animals and are widely known about in a town.

Railway

The railway corridor is surprisingly heterogeneous with many different microhabitats (e.g., Askim, Concord) (Forman, 2014; Borda-de-Agua *et al.*, 2018). Wheater (1999) recognizes seven major habitats: the track area or cess (of coarse crushed stones or cinders between and around tracks); cuttings (mineral soil or rock banks); flats (level areas alongside); embankments (stony sides, alongside raised tracks); drainage ditches; rail (marshaling) yard (e.g., Falun, Emporia, Rock Springs) and unused siding; and masonry (buildings, platforms, tunnels, bridges, posts). These can be subdivided into more microhabitats.

Overall, the railway corridor often has an abundance of ants benefitting from the porous substrate, and lots of flower-pollinating bees, hoverflies, butterflies, and moths (Gilbert, 1991). The strip of cess/stones/cinders around the tracks commonly seems sterile – without plants and animals – because frequently herbicides are intensively applied (Forman, 2014). Foxes and rabbits are common railway animals. Foxes or coyotes may move along parallel to the tracks.

Cuttings or cut banks provide moist enclosed locations often supporting many types of animals, including songbirds, small mammals, and soil animals (Goode, 2014). Drainage ditches are widely used by wildlife, basically because moisture is present. Embankments on opposite sides of the tracks often have quite different vegetation and fauna, especially when they are north- and south-facing.

Unused railway sidings, often near former or current industries, are especially rich in animal species (Gilbert, 1991; Wheater, 1999; Goode, 2014). Also, some unused rail lines are converted into walking/biking trails, greenways, or linear parks through or close to towns (Flink *et al.*, 2001; Goode, 2014).

Finally, the masonry microhabitats support calcareous species, but individually generally have few species. On the other hand, along a railway the abundance of different masonry structures supports a rich flora, and a relatively little-known fauna. Railway bridges and walls in UK towns contain cavities and crannies sufficient for starlings, feral pigeons, house sparrows, and sometimes pied wagtails to nest (Gilbert, 1991). Also, grasshoppers, flies, and spiders populate masonry structures.

Unlike a main road or major highway slicing through town, a railway corridor provides habitat connectivity for wildlife to move into the heart of a town. A stream corridor and a railway corridor are likely the two main routes for wildlife through town (see Figure 2.4).

Stream Corridor and Pond

A stream corridor often slices through town (e.g., Pinedale, Sooke, Issaquah) and most towns contain one or more pond. A large pond generally has house plots along the shore, while a medium-size pond may have been constructed in a park, and a tiny pool in a house plot. In addition to the widespread species present, many characteristic animal types are recognized around these waterbodies. Thirteen seem especially common in a stream corridor: squirrel, deer, duck, redwing, garter snake, mink, butterfly, raccoon, great-blue heron, turtle, frog, fish, and dragonfly. A pond and pondshore have a similar but different list: duck, fish, frog, heron, bat, fox, raccoon, midge, gnat, swallow, turtle, redwing, mosquito, snake, bullfrog, dragonfly, and butterfly.

Unlike urban streams which are often eliminated by burial in pipes, town streams largely flow through a town intact. Especially in the town center, most are somewhat *channelized* (straightened) and *armored* (rocks or concrete added along streambanks) (Figure 8.9) (Biggs, 1996; Decamps and Decamps, 2001). The upstream portion is typically cleaner and more natural, being primarily affected by upstream surrounding land uses and the newer residential area. The downstream portion is more degraded due to the combination of upstream, town center, older residential area, and newer residential area impacts.

Three major effects on town stream corridors are characteristic. First, stormwater runoff from nearby impervious surface cover, particularly direct-connection impervious area (DCIA) (Chapter 6), alters both the hydrology and the form of a stream (Paul and

Figure 8.9 Narrow stream corridor with channelized partially armored eutrophicated stream through village center (see Figure 1.3). Abanades, Castilla y La Mancha, Spain. R. Forman photo. (A black and white version of this figure will appear in some formats. For the color version, please refer to the plate section.)

Meyer, 2008). Greater impervious cover, such as in the town center, increases flooding. Flooding in turn scours out the stream channel reducing stream habitat heterogeneity.

Second, surface runoff plus municipal, industrial, residential, and transportation discharges in pipes or ditches, adds a "witch's brew" of pollutants to the stream. Mineral nutrients, heavy metals, road salt, hydrocarbons, specialized industrial chemicals, organic matter, pathogenic bacteria, and rubbish especially pollute the water downstream.

Many stream characteristics are degraded by a percent impervious cover as low as 10 to 20 percent (Dunne and Leopold, 1978; Klein, 1979; Booth and Jackson, 1997; Paul and Meyer, 2008). Much of the residential area in towns may have 10–30 percent impervious surface cover. Hydrologic, geomorphic, fish, and other biological stream attributes are degraded. Channel width and depth are altered by increased water-flow velocity and flooding. Fine sediment is washed away leaving a coarser sediment stream bottom. Erosion occurs around bridges, and small pools frequently form. Stream temperatures rise due to runoff from sun-heated surfaces and fewer overhanging trees in the stream corridor. Warmer water holds less oxygen and, together with chemical pollution, many fish species disappear (Adams, 1994; Paul and Meyer, 2008).

Human wastewater channeled from buildings to stream or leaking from septic systems enriches the stream with organic matter, pathogenic and other bacteria, nitrogen, and phosphorus. A sewage treatment facility is likely to be near the downstream edge of town, where, if not fully cleaning wastewater, it creates zones of stream animals differentially sensitive to high-to-low levels of effluent (Gilbert, 1991).

Transportation adds heavy metals, hydrocarbons, and many other chemicals. Industry adds many more chemicals, typically including toxins. Residential areas add pesticides and fertilizers. The species composition of stream-bottom invertebrates changes sharply. Consequently, fish populations and other aquatic organisms normally decrease and species composition is altered in a town stream (Wolter, 2008).

The third big effect on stream corridors is an altered streamside or riparian zone. In town, streamside trees and shade decrease. The stream's floodplain is constrained by development, and may be largely eliminated in places. That reduces probably the richest zone of microhabitats in town. The biodiversity of floodplain animals is truncated. The buffer between development and stream that absorbs water and chemicals is squeezed.

In Orange County, California (USA), 52 bird species were recorded in riparian corridors through dense residential development (Oneal and Rotenberry, 2009). Of these species, 19 best correlated with local riparian-vegetation variables, 13 with broader landscape-scale variables, and ten with both local and landscape-scale variables. Woodland species, riparian vegetation species, and insect-feeding birds best correlated with local riparian vegetation variables. Shrubland species better related to broader landscape characteristics.

Another key role of a stream corridor is as conduit for species entering a town. The corridor through the town of Jasper, Alberta (Canada, population ca. 4,000) is a noteworthy route for bears, wolves, elk, and deer at different times (Douglas Olson, ca. 2001 personal communication). These animals create diverse, sometimes severe, problems for residents in town. Nevertheless, the people are largely used to ("habituated to") the large wildlife in their driveways, garbage bins, parks, schoolyards, and the like. Not surprisingly, interactions including predation occur between animals in the corridor. Also, in Washington State (USA), a mountain lion or cougar (*Puma concolor*) occasionally moves into a suburb, town, or village, where it may feed on a dog or in rare cases attack a person (Kertson *et al.*, 2011).

Ponds with slowly moving water are quite different habitats than a stream corridor. At the broadest scale the town formed around the side of a large pond (e.g., Lake Placid). Intermediate-size ponds may be natural, or as constructed ponds (e.g., Selborne, Petersham) in a park or institutional ground or by a company headquarters. Tiny pools are often created in house plots, or by government and commercial buildings. Since most towns have a rather high water table (Chapter 6), the large and medium-size ponds may be sustained by groundwater plus precipitation, while the tiny ponds usually have an artificial lining that prevents the water from draining away.

Fish, amphibians, herons, and waterfowl are classic pond vertebrates (Biggs *et al.*, 2005; Wheater, 1999). But invertebrates are numerous in slowly moving pond water. Emergent, floating, and submerged aquatic vegetation zones on the gentle slope edges of ponds (Chapter 6) as well as the pond bottom, provide microhabitats for diverse insects, spiders, crustaceans, mollusks, and worms. Towns may also have temporary ponds, as distinctive water and dry habitats often containing uncommon plants, plus amphibians and insects that thrive in the absence of fish (Colburn, 2004) (e.g., Concord).

Gene Flow and Wildlife

As yet few studies evaluate gene flow at the scale of towns and villages, yet one in Asturias (Spain) is quite informative (Figure 8.10) (Garcia-Gonzalez *et al.*, 2012). Genetic distances, that is how genetically different populations are, were analyzed (based on haplotype networks) between amphibians around a city and eight village areas. Genetic measurements were made of 199 midwife toads (*Alytes obstetricans*) and 275 palmate newts (*Lissotriton helveticus*), a colorful type of salamander. Gene flow correlated with road density (km/km^2) as a primary barrier or filter between village area populations.

For toads, road density was a significant barrier in seven out of eight areas separating communities, whereas for newts, only half of the separating areas acted as barriers to gene flow. Small low-traffic roads inhibit movement by the toads more than by the newts. Thus, wildlife gene flow is limited by roads between village area populations less than 10 km apart, and cumulatively limited over a distance of some 100 km.

Some rural species characteristic of the land of towns and villages also now live in cities subjected to an extremely different set of stresses and disturbances. Species which have reproduced in cities over many generations can be expected to be somewhat genetically different from the rural populations. So far, such genetic divergence has been demonstrated for surprisingly few plants and vertebrates (Figure 8.11). Nevertheless, the range of species types is noteworthy: one moss; two vascular plants; five amphibians; one reptile, four birds; and two mammals. It seems logical to say that many species in a city are genetically adapted to urban conditions, but that remains a hypothesis awaiting study (Hopkins *et al.*, 2018). Many city species are planted or introduced, and probably many more owe their existence to the "rain of species" arriving from the land of towns and villages. Perhaps few species have genetically adapted to towns and very few to villages.

Other Town Habitats

Rock Cliff and Quarry. A natural rock cliff often supports many uncommon or rare species in part because it is an unusual habitat and has rock faces and crevices facing in many directions relative to the sun (e.g., Flam, Rumford/Mexico) (Goode, 2014). Limestone cliffs in the Niagara Falls region (Ontario, Canada and New York, USA) provide habitats for a wide range of mammals, birds, reptiles, amphibians, and invertebrates (Larson *et al.*, 2000). A rock outcrop that protrudes modestly above the town may often support some uncommon species, but normally has a history of rather heavy human disturbance.

A quarry remaining from former rock extraction tends to be slowly colonized by species from the surroundings, and in early stages has few uncommon species (e.g., Stonington, Issaquah) (Wheater, 1999). However, with time, the flora becomes more

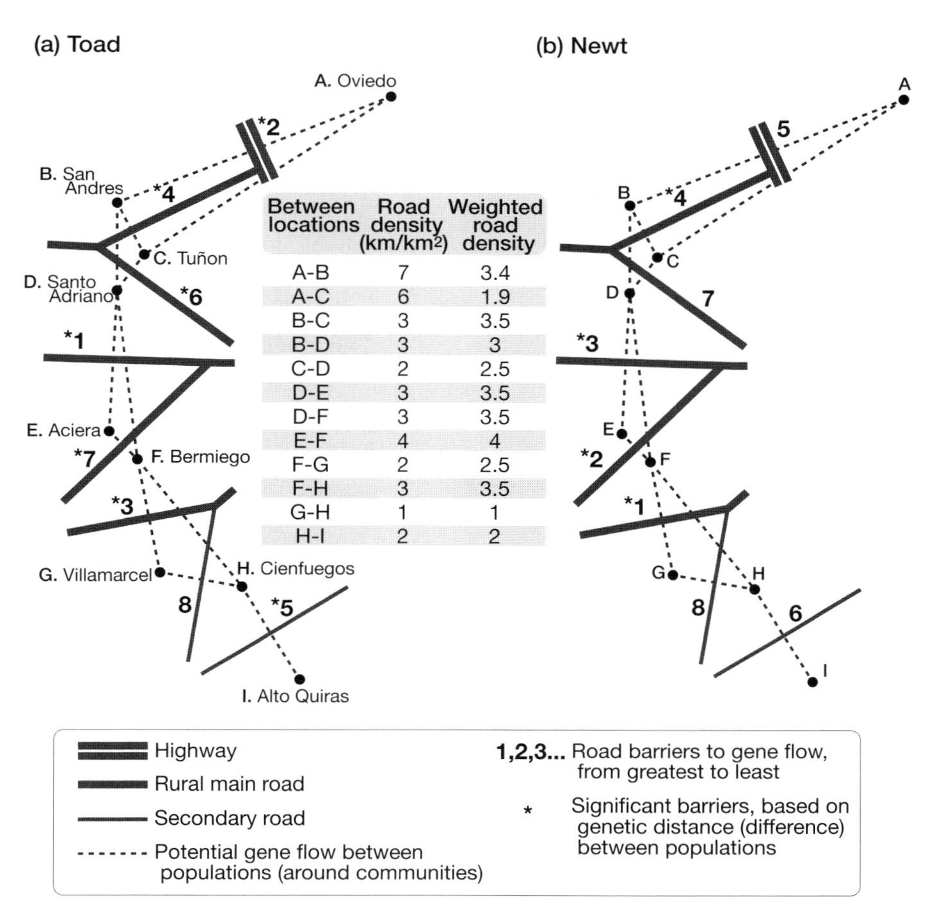

Figure 8.10 Road density between villages/hamlets as a barrier to amphibian gene flow across the landscape. Genetic distances measured for midwife toad (*Alytes obstetricans*) and palmate newt (*Lissotriton helveticus*). See text. Villages/hamlets with about 70–1,000 residents; Oviedo, a city. Asturias, Spain. Adapted from Garcia-Gonzalez *et al.* (2012).

distinctive as new species arrive and early colonizers persist in rocky microsites. Also, the fauna becomes richer, and caves add a whole new set of animals.

Cemetery. Burial grounds vary from old and overgrown to new and tidy lawn-covered (Gilbert, 1991; Goode, 2014). Some are adjacent to a religious structure in town and many are at least partly surrounded by a wall which supports some species. Old vegetation-rich burial grounds have large trees, many tree cavities, extensive vine cover, and many shrubs, all of which are scarce in carefully mowed cemeteries. Migratory birds relish the old cemeteries which provide ample cover and food for a brief stop. Not surprisingly, the lawn and burial stones cases have been called biological deserts, whereas the overgrown examples are rich in both plant and animal species. Intermediate cases of course exist, and may support a richness of butterflies, birds, and rabbits, plus

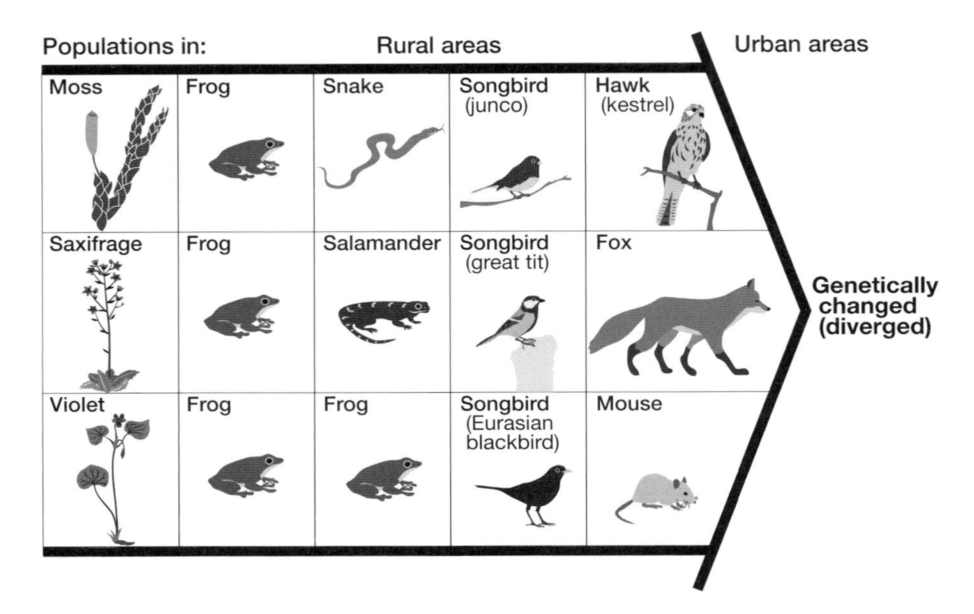

Figure 8.11 Fifteen plant and animal species of rural areas known to differ genetically in urban areas. Evans (2010): *Leptodon smithii* (Italy); *Saxifraga tridactites* (Germany); *Viola pubescens* (USA); *Leptodactylus ocellatus* (Brazil); *L. fuscus* (Brazil), *L. podicipinus* (Brazil); *Bothrops moojini* (Brazil); *Plethodon cinereus* (Canada); *L. labryrinthicus* (Brazil); *Junco hyemalis* (USA); *Turdus merula* (Europe); *Falco tinnunculus* (Poland); *Vulpes vulpes* (Switzerland); *Apodemus speciosus* (Japan). Marzluff (2012): *Turdus merula* (Europe); *Parus major* (Europe).

a foraging fox and raptor (e.g., Tetbury, Hanga Roa, Selborne). The old- or medium-age cemetery is usually in or near the town center, while newer cemeteries are near the town edge.

Dump (Landfill, Tip). Towns, as concentrated sources of food and garbage, and immediately surrounded by farmland and/or natural land, attract lots of wildlife (see Figure 9.3). With the town dump usually in the town edge or outside adjacent zone, it is well located as a feeding station for wildlife.

Many flying insects such as bees, wasps, and flies breed and feed on rubbish (Wheater, 1999). In the UK, cockroaches, house crickets, house mice, and starlings may benefit from the heat of decomposing organic matter. Foxes forage in the garbage and other materials. Gulls are especially common on many dumps worldwide. Near Sydney (Australia) silver gulls daily fly as much as 12 km to feed on a dump (Smith and Carlile, 1993; Wheater, 1999). At the gull nest sites more than 85 percent of food regurgitated (for young birds) is dump refuse, mainly meat and secondarily starchy material.

In the USA, mammals often seen feeding at dumps include bears (e.g., Rangeley), rats, free-ranging cats, dogs, coyotes, deer, raccoons, opossums, and bobcats (*Ursus* spp., *Rattus norvegicus*, *Felis catus*, *Canis familiaris*, *C. latrans*, *Odocoileus* spp., *Procyon lotor*, *Didelphis marsupialis*, *Felis rufus*) (Adams, 2016). Common dump birds include gulls, vultures, and ravens (*Larus* spp., *Cathartes aura* and *Coragyps atratus, Corvus corax*) (Horton *et al.*, 1983; Belant *et al.*, 1995). Hawks and owls are also common at dumps hunting rodents, rabbits, and birds.

Airport. A town airport with an unpaved or paved runway is mainly covered with infrequently mowed grass, though grass near a runway may be lawn-like (e.g., Socorro, Swakopmund, Stonington). This large grassy weedy area naturally attracts grassland birds and other species, which may be uncommon or rare in the landscape (Forman, 2014; Adams, 2016).

Periodically, efforts are made to reduce the number of large birds and tight-flocking small birds to decrease the chance of a bird–plane strike. Hawks, vultures, gulls, geese, ducks, and other waterbirds pose a hazard, as do flocks of starlings, feral pigeons, and doves (Cooper, 1991; Satheesan, 1996). Waterbodies and wetlands near airports are sources of diverse large birds. Especially hazardous to the many small planes and few large aircraft in local airports are large herbivores on the ground, such as deer and kangaroos. Conflicts in airport management may occur where rare species require protection and aircraft-hazard species are minimized.

Powerline (Electric) Corridor. Essentially, all towns have electricity mainly provided by above-ground electric transmission powerlines crossing the town edge. A powerline corridor often has extensive shrub cover, along with grassy and bare areas plus bare-soil tracks or roads. A rural Massachusetts (USA) study of 15 powerline corridors ranging from 15 to 78 m wide analyzed the abundance of seven shrub-nesting birds (King *et al.*, 2009). Four species were most common in intermediate-width corridors, and scarce or absent in narrow ones.

Raptors sometimes are electrocuted by powerlines. Also, across the USA, 12 species nest on powerlines, of which osprey and red-tailed hawk (*Pandion halietus*, *Buteo jamaicensis*) are apparently most common (Blue, 1996). In Germany ospreys have been increasing, and those nesting on powerline pylons produce more young than do ospreys on tree tops (Meyburg *et al.*, 1996).

A four-year study in the Sacramento Valley rural farming area (California, USA) recorded more than 5,000 hawks, which generally avoided human settlements, plowed fields, and most grain and row crops (Smallwood *et al.*, 1996). In contrast, they preferred wetland, riparian and upland vegetation, alfalfa fields, and rice stubble, all of which were rare habitats. The hawks also preferred trees with snags (large dead branches) and open views, as well as powerline (utility) poles with multiple horizontal cross-bars rather than a single horizontal bar.

Tower. Towers for radio and television, cellular phones, wind turbines, and religious structures (e.g., steeples, minarets), plus water towers and smokestacks, are probably habitats for few species (Forman, 2014). Owls may nest inside steeples, and peregrine falcons on smokestacks. Probably plenty of insects, spiders, and the like live or land on tall towers. But tall towers are best known for killing migratory birds attracted to their lights on top at night (Forman, 2014; Adams, 2016).

Miscellaneous Town Features. A sand-and-gravel pit is likely to be in the floodplain of a town's stream corridor, and may attract migrating waterbirds (Wheater, 1999). Industrial waste heaps or areas (Gilbert, 1991; Wheater, 1999; Hough, 2004) are considered in Chapter 9. A car park or paved area usually supports relatively few animals, though either abandoned car parks or those with abundant tree or shrub plantings may be somewhat species-rich (Wheater, 1999). A sewage treatment facility (Wheater, 1999) is considered in Chapter 6, plus community gardens (allotments) (Gilbert, 1991) and golf courses (Forman, 2014) in Chapter 7.

Dogs, Cats, Farm Animals

Towns and villages are epicenters of visible dogs and cats as the companion animals of owners. Also, unlike almost all cities, farm animals are often common in villages and present in most towns. Animal pests and diseases are also part of the picture.

Dogs, cats, and sometimes horses are the usual companion animal pets, as less affection and commitment are usually displayed with birds (e.g., canary, parrot), reptiles (e.g., lizard, turtle), fish, and invertebrates (Kellert, 2005). Other common mammal pets include mouse, gerbil, rabbit, and monkey. Companion animals normally provide many physical and mental benefits to owners, such as peace of mind and self-confidence (Serpell, 1996; Kellert, 2005). For instance, stroking a much-loved dog reduces stress in the person. The benefits to distressed people include improved emotional symptoms and faster recovery time. Pets typically provide therapeutic benefits to sick or disabled residents.

Some towns have a pet store (e.g., Icod de los Vinos; see Appendix). Pets apparently also teach *biophilia*, the affinity of people with nature (Wilson, 1984; Kellert *et al.*, 2008). We evolved with, depend on, and are inspired by nature.

Farm animals typically thrive around many village homes (e.g., Selborne, Petersham), as well as farmsteads in town edges (e.g., Chamusca, Concord, St. George's). Working horses or oxen for transport. Horses for riding. Dairy cows for milk. Cattle for meat, or religion. Sheep for wool and meat. Goats for milk and cheese. Chickens for eggs and poultry. Ducks, requiring a pond or paddy, for meat. Pigs and rabbits for meat. All of the farm animals also provide rich manure fertilizer.

Food and water are especially provided during droughts or extreme cold periods for animals around homes. In addition, homes provide protection against dangerous predators or getting lost. In some towns livestock are taken to market to sell and buy. The town may have a fenced or walled "pound" to temporarily hold escaped or loose farm animals.

Moreover, sacred animals are important in some towns. Sacred cows are protected in India. Monkeys are honored and protected around certain religious temples and sites. Also, various animals are protected around sacred groves and trees (Pungetti *et al.*, 2012).

We now consider the primary pets, dogs and cats, which may have important ecological effects near towns.

Figure 8.12 Walking with dog on leash along forest path in snow. Yellow labrador has discovered wild turkey (*Meleagris gallopavo*) tracks. Templeton, Massachusetts, USA. R. Forman photo.

Dogs

The world has some 400–700 million domestic dogs (*Canis familiaris*), with 75 million in the USA (Lenth *et al.*, 2008). Since dogs are associated with people and homes, and also need space to run, the density of dogs is particularly high in exurban areas and around towns, villages, and farmsteads (e.g., Hanga Roa, Socorro, Casablanca) (Odell and Knight, 2001; Maestas *et al.*, 2003). In such areas dogs are often taken to natural areas, even beaches, for exercise (Williams *et al.*, 2009). Unvaccinated dogs may carry dog distemper, parvoviruses, and parasites that can be passed to wildlife (Suzan and Ceballos, 2005; Fiorello *et al.*, 2006; Acosta-Jamett *et al.*, 2011; Davlin and VonVille, 2012).

Dogs serve many roles in society beyond being a companion (Figure 8.12). They are used: in hunting to find and retrieve game (e.g., Plymouth, Clay Center); to guide the disabled; to protect property and people; as partners for law officers; to detect chemicals; and as search and rescue dogs. Here we consider dogs relative to: (1) restrained, free-ranging, feral; (2) dog waste and urine; (3) disturbance of wildlife; and (4) dog-walking.

Restrained, Free-Ranging, Feral

Almost all town and village dogs daily spend some time outdoors (Forman, 2014). Some may be in a fenced area, from which they occasionally escape. Others, called *restrained* dogs, may walk leashed or unleashed with their owners, and generally obey the owner's instructions. Even without the owner present, a restrained dog generally follows the instructions learned. These dogs have an owner, a home, and food provided at home.

Free-ranging or free-running dogs are unsupervised, unfenced, unleashed, and spend all or most of the time on their own outdoors (Rubin and Beck, 1982; Gompper, 2014). Two free-ranging dog types are commonly recognized: those with an owner and food provided at home, and *feral* dogs (including strays) which have no owner and thus forage widely for food. *Dog packs*, usually of two to five dogs remaining together, may form where owner care and home feeding is limited.

Free-ranging dogs may attack wildlife, including small mammals, tortoises, wild turkeys (Daniels and Bekoff, 1989; Butler *et al.*, 2004; Lenth *et al.*, 2008; Hughes and Macdonald, 2013), and a wide range of other animals (Silva-Rodriguez and Sleving, 2012). These dogs will chase or attack a deer (*Odocoileus virginianus*), and in places are a major predator on fawns (Ballard *et al.*, 1999). Free-ranging dogs feed on and disperse rubbish from trash bins, which encourages pest rodents and insects (Rubin and Beck, 1982). Free-ranging dogs also often feed on dumps.

In the small city of Fort Collins, Colorado (USA), of 1,361 dogs observed in a study, only seven showed evidence of avoiding humans (Lehner *et al.*, 1983). Additional study concluded that 17 percent of the Fort Collins dogs were in packs, while 21 percent of the suburban dogs around Sacramento, California (USA) were in packs (Rubin and Beck, 1982).

In rural New Mexico (USA), feral dogs (in packs) with young had a home range of about 0.14 km^2 (14 ha; 35 acres) (Daniels and Bekoff, 1989). But after the pups reached about four months, the pack home range expanded 11-fold to an average of 1.6 km^2.

The home ranges of different breeds of dogs of course varies widely. In a small city (Berkeley, California, USA), the average dog home range was 3.5 ha (8.6 acres) (Berman and Dunbar, 1983; Forman, 2014). Beagle and spaniel home ranges were 0.01–0.02 ha, a fraction of a single city block. The largest home ranges were 5.4 ha (about five city blocks) for a collie, and 7.9 ha (seven blocks) for an afghan. All animals were fed at their owner's home. Because much of the area is composed of buildings or fenced-off areas, the "accessible home ranges" were much smaller, averaging 1.7 ha (4 acres) (Forman, 2014). The dog home range had a core area of dense movement, plus a larger travel area of occasional movements. Home ranges of nearby dogs overlapped travel areas but did not overlap a core area.

Free-ranging dogs in the city of Queens (New York) had an average home range of 4 hectares (10 acres) (Rubin and Beck, 1982). Occasionally restrained dogs had an average home range half that size, whereas totally unrestrained dogs daily traveled on average over about 10 ha. In contrast, a pack of free-ranging stray dogs had a home range of 25 ha in Baltimore, Maryland (USA), and 61 ha in St. Louis, Missouri (USA).

Dog Waste and Urine

Dog density in and surrounding towns is perhaps 20 to more than 100 times the density of native carnivores (Matter and Daniels, 2000). That density results in a lot of dog waste (Matter and Daniels, 2000; Adams, 2016), which is typically concentrated in house plots/yards, or near the house plot (since animals often defecate when changing behavior, such as beginning a run). In a dog-walking semi-natural area, dog waste is concentrated near the car park and along path edges near the beginnings of trails. In contrast, free-ranging dogs widely disperse their waste.

Dog waste contains lots of organic matter, bacteria including pathogenic ones (e.g., fecal coliform *E. coli*), phosphorus and nitrogen (Shueler, 2000). In poor-soil areas with concentrated dog waste, the organic matter and mineral nutrients may enrich the soil, creating tiny patches of luxuriant plant growth.

Various flies breed in dog and other feces, particularly houseflies (*Mus domestica*), at least in London (Erzinclioglu, 1981; Gilbert, 1991). More flies emerge from moist than dry dung. But five common fly species regularly visit the dog waste mainly to feed. Extrapolating from the London study suggested that dog feces could annually produce up to 130 billion flies in Britain, and a much larger number of flies would visit the dog waste. House flies and other dung flies could readily transmit pathogenic bacteria to people's food, skin, clothes, pets, kitchen surfaces, and dining tables.

Salmonellosis and campylobacteriosis from bacteria, toxicariosis from roundworms, or girardia may be present in dog waste. Thus, children playing outside, adults gardening, and pets may be at particular risk of infection by bacteria and parasites in dog waste concentrated on the ground (Shueler, 2000).

Rainwater also carries the bacteria pathogens from dog waste down into the groundwater, which feeds ponds, streams, and shallow drinking-water wells (Matter and Daniels, 2000). Or the surface stormwater may wash fecal coliform bacteria directly into ponds and streams (Shueler, 2000). Rather little is known about dog bacteria concentrations in the waters or their significance.

A drainage basin in northern Virginia (USA) with 11,400 dogs producing an estimated 2,270 kg (5,000 lb) of solid waste daily is considered to be a major contributor of bacteria into the main stream present (Adams, 2016). Similarly, a model suggests that, in a basin of up to 52 km^2 (20 mi^2) with waste from 100 dogs, drainage for two to three days would carry enough bacteria and nutrients to close a downstream estuarine bay to safe swimming and shellfish harvest (US Environmental Protection Agency, 2012). Overall at present, some researchers conclude that the overall effect of dog-waste bacteria in waters is limited (Hughes and Macdonald, 2013; Weston *et al.*, 2014), while others consider this to be a potential human health problem (Matter and Daniels, 2000; Shueler, 2000; Lee *et al.*, 2009; Adams, 2016).

Dog urine ends up on varied objects, including certain posts or tree bases scent-marked with urine for dog communication purposes (Gilbert, 1991; Matter and Daniels, 2000; McCormack *et al.*, 2011). On tree bases in towns with little or modest air pollution, the nitrogen in the urine apparently creates a distinctive flora of lichens (e.g., *Lecanora conizaeoides, Physcia tenella*), green algae (*Pleurococcus viridis, Prasiola crispa*), and even a moss (*Ceratodon purpureus*) (Barkman, 1958; Gilbert, 1991). Scent-marking by dog urine and feces along a semi-natural area trail negatively affects a wide range of wildlife, which then avoid the trail area (Kats and Dill, 1998; Miller and Hobbs, 2000). The scent of dog urine and feces typically lasts for hours, and may last for days (Kats and Dill, 1998).

In the Isoso (Bolivia) area, with rural villages and many forest-game hunters with dogs, more than 95 percent of the dogs carry both the canine distemper virus and canine parvovirus (Fiorelllo *et al.*, 2006). In natural areas of Mexico City, dogs regularly transmit parvovirus, toxoplasmosis, and rabies to many mid-sized mammal species (Suzan and Ceballos, 2005). Based on the frequent opportunities for contact between dogs and wild carnivores, plus the known susceptibility of wildlife to these viruses, domestic dogs represent a disease risk for native wildlife. Finally, various dog-borne zoonoses or zoonotic diseases are known, which could "jump" to humans (Weston *et al.*, 2014).

Disturbance of Wildlife

In a broad sense, there are three types of avoidance response wildlife may have to the presence of a domestic dog or a predator (Liddle, 1997).

1. The wild animal may sense the presence of a dog by sight, sound (barking), or smell, and the response may be physiological stress, or a change in behavior, such as flushing (running or flying away).
2. The disturbance makes the habitat less suitable, such as by widespread scent-marking or repeated flushing, so wildlife move to another suitable place (if such a place exists).
3. Direct damaging conflict occurs, with dog or wildlife injured or dead. If other wildlife observe this interaction, they will have learned how to respond to a dog in the future.

Alert or *agitation distance* is the distance between observer or dog and wildlife, when it experiences physiological or psychological changes (Miller *et al.*, 2001; Brearley *et al.*, 2012; Baker *et al.*, 2013; Knight and Gutzwiller, 2013; Hing *et al.*, 2016). These agitation changes may be stress, release of stress hormones, increased heart rate, or diverted attention.

Flushing or *flush distance* is the distance to the animal when it changes behavior. Flush distance has often been used in managing areas for wildlife. Flushing is usually a temporary response and the animal may return after the dog has left. Flush time refers to how long the animal remains away from the habitat, and distance moved is how far the animal went (Miller *et al.*, 2001).

In an area with many people, ungulates such as deer are less wary, i.e., have a shorter flush distance, of a person walking than in wilder areas, though this may relate to the scarcity of alternative suitable habitats (Stankowich, 2008). Hunted deer populations have considerably larger flush distances than non-hunted animals.

For many wildlife species flush distance may be modest when a person approaches, greater when a dog alone approaches, and greatest when a person and dog together approach (Figure 8.13) (Miller *et al.*, 2001; Taylor and Knight, 2003). Joggers, by looming faster in a bird's visual field and making more noise, cause greater disturbance than do walkers (Lethlean *et al.*, 2017). Approaching an animal directly tends to lengthen flush distance. Flushing an animal to a less suitable habitat typically means that the animal must forage for food longer that day to maintain its energy and metabolic balance.

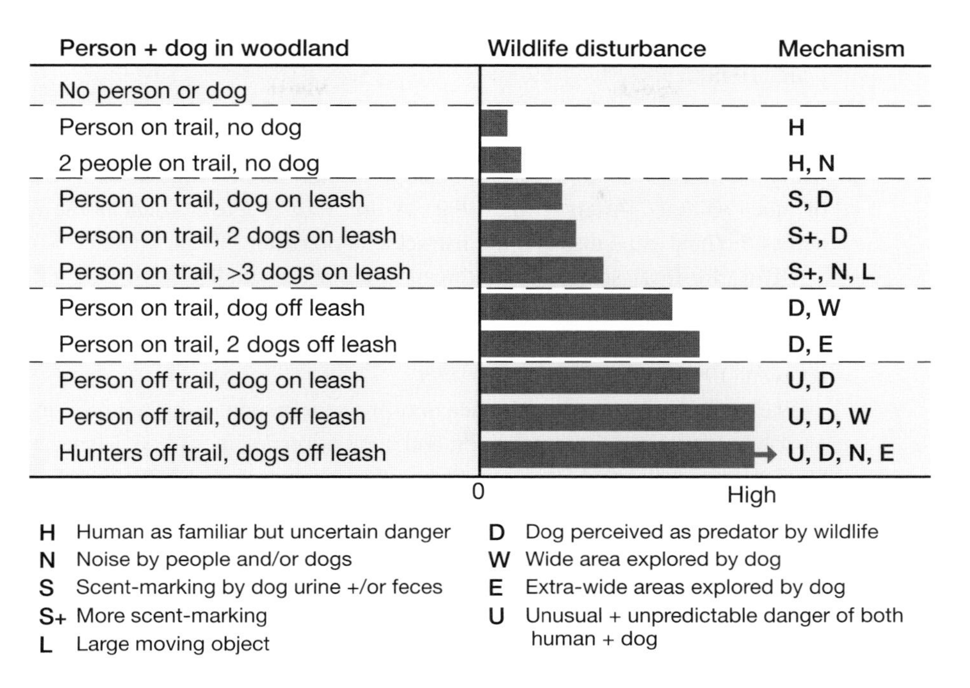

Figure 8.13 Wildlife disturbance in woodland affected by people, dogs, and use of trail and leash. Hypothesized level and mechanisms of wildlife disturbance.

Repeated appearances of a dog may mean that the wildlife becomes *habituated* to dogs, that is, accustomed to and less scared of them (Stankowich, 2008). In this case the flush distance is shorter. Such a process could bring herbivores closer to a dog trail, and increase the chance of a conflict (George and Crooks, 2006). However, some researchers doubt this result of habituation, because wildlife perceive dogs to be predators (Hill *et al.*, 1997). Irrespective, unpredictable dog movements in a seldom-visited area create long flush distances (Miller *et al.*, 2001; Taylor and Knight, 2003; Stankowich, 2008; Silva-Rodriguez and Sleving, 2012).

Scent-marking with dog urine or feces causes some animals to avoid a trail area (Kats and Dill, 1998; Miller and Hobbs, 2000). The strip of scent-marking by on-leash dogs might be about 10–20 m (33–66 ft) wide. However, the strip of scent-marking by trail-walkers' off-leash dogs might be several times that width. Such trail scent-marking is also likely to deter wildlife from crossing a trail. Such a scent-marked trail network tends to subdivide a natural area into semi-separate suitable sections for wildlife.

Unleased dogs may especially affect ground-nesting/denning wildlife in a broad swath along a trail. Birds seem to react as if dogs were a predator, even though generally dogs are good chasers but poor hunters. For breeding by ground-nesting and other birds, dog presence may increase the risk to eggs and young by other predators watching, such as corvids (crows, etc.). On trails near the small city of Boulder, Colorado (USA), the abundance of some bird species, nest occurrence, and nest success decreases near trails, a pattern interpreted as the effect of walkers, not dogs, in recreational activities (Miller *et al.*, 1998).

A remote little-used trail or road is a favorite route for predators to forage along (Kohn *et al.*, 1999). But a dog scent-marked trail is likely to be unsuitable. Wild carnivores typically spend considerable time at the edge of their home range investigating the smells of other animals (Allen *et al.*, 1999). A dog trail is like a linear home range cutting across the landscape that unduly distracts the predator.

Off-trail effects are also of interest. Around the small city of Ballarat (Australia), the iconic and uncommon koala feeds on foliage high in the trees. But when trees are scattered, the slow-moving animal comes to the ground where it is subject to dog attack (Prevett, 1991).

Walkers without dogs can induce anti-predator responses by wildlife, including long alert (vigilance) and flush distances, and population declines of some species (Hill *et al.*, 1997; Blumenstein and Daniel, 2005). Indeed, many recreational off-trail activities alter the behavior and displace birds from their habitats (Riffell *et al.*, 1996; Gutzwiller *et al.*, 1997). In a coastal protected area, the greatest dog impacts resulted from off-leash dogs, and those that penetrated farthest into the protected area (Antos *et al.*, 2007).

Dog-Walking

A high dog density around towns and villages means that a lot of dogs need exercise. Some is gained by walking sidewalks and streets in town, and some is focused on trails in nearby greenspaces and semi-natural areas. Experiments show that dog-walking provides diverse benefits to both human and canine health (Bauman *et al.*, 2001). Social benefits and conflicts alike occur in designated dog areas of parks (Manning *et al.*, 1996). Dog-walkers gain exercise by walking outdoors, and most seem to concentrate on their dog, other dogs, and people met. Still some are able to concentrate on nature, even though wildlife are scarce because of the dog. Some people, including children, cyclists, and horse riders, as well as some dog-walkers who fear dog-on-dog conflicts, take evasive action to avoid meeting dogs on a trail.

Thus, a small number of basic options has evolved for managing semi-natural areas relative to dog-walkers: (1) dogs prohibited; (2) dogs only on leash (lead); (3) dogs off-leash, but on voice-and-sight control; and (4) dogs free-ranging. Voice-and-sight control means that a dog is always in sight of the walker and promptly obeys instructions (rangers or wardens can easily test this, with diverse consequences for failure). As needed, the four options can be zoned spatially and temporally in a natural area. Thus, particularly valuable ecological sections, or times of key species breeding, can appropriately have dogs prohibited.

While dogs have various places to run, a wild animal basically has no other place to go than its habitat. Translocated to another promising looking place leaves it with only about a 10 percent chance of survival, due to the existing predators and competitors present, plus the inefficiency of foraging in an unknown heterogeneous place. Training and restraining a dog not to chase wildlife is an appropriate criterion for dog-walkers.

Voice-and-sight control largely awaits evaluation and many seem skeptical of its effectiveness. The effectiveness of *leash rules* (regulations, bylaws, laws) (e.g., Vernonia), which require dogs to be on a leash (lead), are beginning to be evaluated (Weston *et al.*, 2014). One leash-rule study in three habitat types of Edmonton, Alberta (Canada) parks concluded that there was no significant difference in avian diversity (and also probable breeding birds) between leashing dogs or free-ranging dogs, but the study did not record whether dog-walkers obeyed the leash rule (Forrest and St. Clair, 2006). Compliance with dog-leash rules or regulations is often low.

In effect, off-leash or free-ranging dogs along trails in a natural area negatively alter animal behavior (Steven *et al.*, 2011). This includes significant effects, for example, on small mammals (Lenth *et al.*, 2008; Shannon *et al.*, 2014), deer (Lowry and McArthuur, 1978; Ballard *et al.*, 1999; Miller *et al.*, 2001; Pelletier, 2006), and a predator, bobcat (*Lynx rufus*) (George and Crooks, 2006; Lenth *et al.*, 2008; Reed and Merenlender, 2011).

Off-leash dogs affect a much larger area of wildlife habitat than do leashed dogs (Figure 8.13) (Stankowich, 2008; Silva-Rodriguez and Sleving, 2012). People with dogs have a greater effect than people without dogs (Miller *et al.*, 2001; Thomas *et al.*, 2003; Lenth *et al.*, 2008). The negative effect on wildlife of a person with dog off-trail extends even further (Miller *et al.*, 2001; Taylor and Knight, 2003), probably because the effects of both person and dog are unfamiliar and unpredictable to wildlife.

Free-ranging dogs produce other ecological effects, including disturbing nearby grazing livestock. Streambanks and stream bottoms around a stream-crossing trail are often degraded. Dogs swimming and chasing objects in a pond degrade the pond margin by repeatedly running in and out of the water. Chasing frogs, fish, and other animals in the shallows of a pond may significantly degrade the pond-side vegetation and water clarity.

Three studies provide particularly useful insights into dog-walking effects in natural areas. The first, on birds and mammals outside a small city (Boulder, Colorado, USA), evaluated the effects of dog-on-leash, dog alone, and person alone, both on-trail and away-from-trail, on two grassland songbirds (vesper sparrow, western meadowlark; *Pooecetes gramineus*, *Sturnella neglecta*), one forest songbird (American robin, *Turdus migratorius*), and mule deer (*Odocoileus hemionus*) (Miller *et al.*, 2001). For all species, the presence of dogs created a greater alert distance, flush distance, distance moved, and total area disturbed (area of influence), when a dog was off-trail rather than on-trail. For grassland birds, person with dog-on-leash had the greatest impact, i.e., the largest flush distance, distance moved, and total disturbed area. Person alone had an intermediate-level effect, and dog alone had the smallest effect. For the forest bird, there was no difference in flush distance, distance moved, and total disturbed area between person alone and person with dog-on-leash. For deer, all four variables measured were greater for dog alone than person alone. In short, both dog and person effects are greater off-trail than on-trail. Also, person walking was most disruptive for grassland birds, and dog alone for deer.

Second, using five diverse methods from pellet plots to remote-triggered cameras, the effect on mammals of hikers, mountain bikers, and equestrians on trails was evaluated in areas permitting versus prohibiting free-running off-leash dogs (Lenth *et al.*, 2008). In areas with dogs, deer are less active (fewer recorded) within 100 m of a trail (see Figure 13.2). Small mammals including squirrels, rabbits, chipmunks, and mice are scarce within 50 m of a trail, and prairie dogs (a mainly subterranean small mammal) within 25 m of a trail. Mid-size bobcats (*Lynx rufus*, a small wild cat) are less frequent, and foxes more frequent, in areas with dogs than in areas without. Even in no-dog areas, deer are less active within 50 m of trails used only by people. Thus, although trail-using recreationists reduce deer activity near trails, the presence of free-ranging dogs has a larger impact, and on a wide suite of wild mammals.

Third, birds were sampled within 50 m of trails in 90 woodland sites outside Sydney to evaluate the effects of walkers with dog-on-leash, walker with no dog, and a control (no walker or dog) (Banks and Bryant, 2007). Compared with the control, bird species richness dropped 35 percent and abundance 41 percent along trails with dog-walkers. Walkers alone generally produced less than half of the dog-walkers effect. Ground-dwelling birds were most affected, with half the species absent along dog-walker trails. Dog-walker trails have 76 percent fewer bird individuals within 10 m of the trail compared with the control. Along walker-only trails (no dogs), no difference in bird richness or abundance was present with one versus two walkers together. In short, trail walkers with dog-on-leash have a major impact on the woodland avian community, though the area disturbed (influence zone) is relatively narrow.

What do such results suggest about protecting wildlife from dogs in semi-natural areas? Placing the car park and main trail entrance by low-quality habitat of the protected area is a key early step. Since animals tend to defecate when changing behaviors, such as riding in a vehicle and getting out to exercise, much of the dog waste is found in the first kilometer or less of a trail. This suggests having an initial designated section with kilometer-long dog-walking trails adjacent to the car park, so that dog waste is concentrated and conveniently removed (see Figure 13.2). Also, urine scent-marked locations are concentrated here.

A resource survey and map of habitats, wildlife, and other valuable resources of the entire protected area is a key step for dividing up or zoning the protected area, according to the relative ecological value of different sections. Note that any wildlife survey where dogs have recently been will likely show few wildlife species and individuals, especially sensitive ones. Thus, one goal will be to improve habitat and restore wildlife.

Several *trail design* guidelines are important. Avoid the ecologically most sensitive or valuable areas. Place trails mainly in edge portions of the natural area which are generally of lower quality (see Figure 13.2). Also, by avoiding or eliminating trails that penetrate the interior where dogs have the greatest impact, an edge trail protects interior areas for existing and future interior species. Since scent-marking of dog-used trails creates a corridor of relatively unsuitable habitat avoided by some wildlife, design the trail network so as not to subdivide suitable habitat into small sections for wildlife.

Then tailor trail usage to the high, medium, and low ecological-value areas. Free-ranging dogs off trail cause the greatest habitat degradation, so these are prohibited in the entire protected area, except perhaps in the small section by the car park. No dogs are permitted in high-value areas. Dogs on leash may be suitable for low-value areas. Use diverse techniques to keep people on trails. Removing a dog's waste and placing it in an appropriate container is important for dog-walkers.

Finally, where dog populations are high and dog-walkers numerous, to adequately protect natural areas, create dog parks or equivalent away from protected natural areas (Walsh, 2011; McCormack *et al.*, 2011). Creative designs might have partially separate and connected sections, which appeal to different types of dog and to diverse dog-walkers.

Many dog-walkers in natural areas feel that their dog has little effect on wildlife (Barnard, 2003; Taylor and Knight, 2003; Sterl *et al.*, 2008). If the dog-walker sees no wildlife, it may be because the dog disturbed the animals which have moved away, not the absence of wildlife. Or if wildlife are seen, he or she concludes that the dog is not disturbing animals (Sterl *et al.*, 2008). Yet as just described, wildlife are stressed, and the available habitat (for feeding, resting, and breeding) is significantly reduced, by the presence of a dog.

Cats

Domestic or house cats (*Felis catus*) in homes are usually cuddly companion animals for people. Yet a conflict arises since cat predators catch and kill certain wildlife, as poignantly portrayed in a one-woman Broadway (New York) play, *The Belle of Amherst*, about a noted nineteenth-century poet. Emily Dickinson likes birds; her sister likes cats.

In addition to house cats that spend more time indoors than out and are fed by the owner, three types of free-ranging cats are recognized (Campos *et al.*, 2007; Forman, 2014): (1) outdoor free-ranging cats that spend more time outdoors than indoors, but return for food, shelter, and care (e.g., Socorro, St. George's); (2) stray cats recently lost or escaped from the owner; and (3) feral cats born with no owner or owner's home, and surviving on their own.

Estimates of cat numbers for a town or nation are extremely rough because of varying methodologies, assumptions, and human attitudes toward cats. Britain may have about 9 million free-ranging cats (Woods *et al.*, 2003), and 11 million cats are estimated to live in Brazil (Campos *et al.*, 2007). Some 70–85 million owned domestic cats might live in about a third of the households in the USA, and a hard-to-estimate 30–90 million feral domestic cats may range freely outdoors (Marks and Duncan, 2009; Jarvis, 2011; Dauphine and Cooper, 2011; Adams, 2016; Woinarski *et al.*, 2017). Australia has some 7 million cats (Woinarski *et al.*, 2017). Cats seem to be considerably less common in some cultures and regions. Domestic cats retain their characteristic of being hunters, irrespective of whether they live indoors or outdoors, and whether owners provide food or not (Figure 8.14). Thus, cats are often maintained to catch mice or deter rats around a residence.

A cat's ecological footprint ("pawprint"), the total area required to provide resources supporting an individual or population, is probably rather large considering the animal's size. Being a meat-eater is most important. But also many cats are provided grooming, veterinary care, travel, and toys.

In considering the ecological effects of cats around towns and villages, we exclude three places where cat effects have been much studied: islands, cities, and Australia. On islands, such as in the Galapagos and Hawaii, introduced domestic cats have multiplied

Figure 8.14 Mid-aged house cat as cuddly companion and predator in residential neighborhood. Japan. (A black and white version of this figure will appear in some formats. For the color version, please refer to the plate section.)
Source: Mitsuharu Watari / EyeEm / Getty Images.

and caused declines in many native animals, even eliminating a species from an island (Konecny, 1987; Fitzgerald, 1988; Atkinson, 1989; Hess *et al.*, 2004; Nogales *et al.*, 2004). Australia is covered with diverse marsupials which evolved without mammal predators (Molsher *et al.*, 1999; Woinarski *et al.*, 2017), and cat predation is a serious ecological problem (New Zealand and South Africa are also special cases where cat predation is a problem; Tennent and Downs, 2008; Morgan *et al.*, 2009).

In cities, the density of indoor cats is much greater than in rural areas (Forman, 2014). Outdoor urban free-ranging cats may be about double the density of that in suburban and rural areas (Lepczyk *et al.*, 2003). In cities outdoor cats are seldom seen, apparently because, with so many people, cars, and noises during the day, city cats primarily forage at night (Haspel and Calhoon, 1991; Barratt, 1997). Outdoor city cats mainly feed on widespread garbage and discarded food, though some also consume food put out by residents. Urban cats, as inherent hunters, may also feed on prey populations with little effect, which are mainly common urban wildlife species such as house mouse, small rats, and house sparrows (*Mus musculus*, *Rattus* spp., *Passer domesticus*). As opportunist hunters, some city cats also periodically consume individuals of common and even uncommon native wildlife species. Yet many other stresses exist in cities, and scavengers, such as rats and raccoons, are typically denser than cats (Jackson, 1951; Childs, 1986; Courchamp *et al.*, 1999).

We now consider cats and towns from three perspectives: (1) cat density, food, and home range; (2) cat predation; and (3) cats, rare/uncommon species, and spatial town pattern.

Cat Density, Food, and Home Range

Cat density is much higher (20 to >100 times) than native predator density in areas of dispersed houses (Kays and DeWan, 2004; Sims *et al.*, 2007). In southern Michigan (USA), a third of the landowners in both rural and suburban areas have free-ranging cats, and the average number of such cats/landowner (2.59) does not differ significantly (Lepczyk *et al.*, 2003). Furthermore, cat density around the houses along roads in rural and suburban areas does not differ significantly (1.2 versus 1.4 cats/ha).

Along a 9.2 km x 30 m wide transect through a rural (agriculture and unpaved roads) and suburban (buildings, gardens, paved roads) area near Piracicaba (Sao Paulo, southeastern Brazil), free-ranging cat density averages 147/km^2 (391/mi^2) over the year (Campos *et al.*, 2007). Cats are more abundant in the suburban than the rural areas. In suburbs cat density exceeds dog density, whereas in rural areas cat and dog density may be about the same.

Various types of food are used by cats: cat foods indoors; cat food provided outdoors at the home; garbage and other discarded edible material; and prey animals. Prey include common urban, suburban, or rural species, plus less common native wild animals. The relative proportion consumed varies by cat type. House cats primarily indoors and outdoor free-ranging house cats that mainly return home to feed are provided with a high proportion of meat products by the owner. Stray cats probably feed mainly on garbage products with an occasional prey animal captured. Feral cats normally scavenge garbage products and are also predators on prey animals. However, stray and feral cats sometimes feed on food put out by residents for outdoor house cats. A study in rural Florida found that many people put food out for stray and feral cats (Levy *et al.*, 2003). Providing food for cats at home seems to have only a limited effect on cat hunting and predation, perhaps in part because hunger and hunting are controlled by different parts of the brain (Warner, 1985).

The home range of cats is relatively little documented. In residential areas of the city of Brooklyn (New York), the average home range of owned female cats was 1.8 ha (4.5 acres) and male cats 2.6 ha (Haspel and Calhoun, 1991). Two residential neighborhoods had essentially the same home-range sizes, though cats with more food available tended to have smaller home ranges (Konecny, 1987). In the suburbs of Champaign-Urbana, Illinois (USA), male and female owned cats average 2 ha (5 acre) home ranges (Horn *et al.*, 2011). Feral cats are more active and travel longer distances at night (Haspel and Calhoun, 1991; Barratt, 1997), when people, vehicle traffic, and noise are less. For instance, ten house cats living on the edge of a suburb had an average diurnal home range of 2.7 ha (7 acres) versus a larger nocturnal home range of 7.9 ha (Barratt, 1997). All cats remained within 100–200 m of the suburb, except at night when some roamed further into the agricultural surroundings (see Figure 13.2).

Cat home ranges in farmland mainly relate to the scattered homes or farmsteads present, and may vary considerably in size (Liberg, 1980; Barratt, 1997). Near horse-farm buildings in Georgia (USA), the cat home ranges average 1.1 ha (2.8 acres) for males and 0.64 ha for females (Kitts-Morgan *et al.*, 2015). Farm cats spend considerable time around buildings, as well as ecotone edges, especially between marsh and woods, a combination of habitats that provides a range of foods throughout the year.

Cat Predation

A village in Bedfordshire (UK) farmland had 173 houses and 78 cats, and all cats were fed by owners and spent some time roaming outdoors (Churcher and Lawton, 1989). Over a year, the 1,100 animals carried back to houses by cats and recorded by residents were 65 percent mammals and 35 percent birds. All the mammals were relatively common garden or backyard species (Figure 8.15): wood mouse 17 percent of the total captures; field vole 14 percent; common shrew 12 percent; bank vole 7 percent; pygmy shrew 4 percent; rabbit 3 percent; and other mammals 8 percent. Similarly, almost all birds caught were common species of house plots (gardens): house sparrow 16 percent; song thrush 4 percent; blackbird 3 percent; robin 2 percent; other birds 10 percent. Some sampling evidence suggested that more than a third of the annual house sparrow mortality may be due to cats. The least common species caught over 12 months were a few weasels, pipistrelle bats, and a kingfisher.

Mid-age cats were the best predators. Old cats seemed to get lazy, and kittens mainly chased frogs and butterflies. Cats on the village edge adjoining farmland caught more animals than did cats in the middle of the compact village. Also, the village edge cats caught proportionally more small mammals and fewer birds.

Domestic cats have been hypothesized to be an important potential threat for extinction of mainland vertebrates (Loss and Marra, 2017). Also, cats might act as a "mesopredator," keeping more damaging predators, such as rats, under control (Fitzgerald, 1988; Courchamp *et al.*, 1999). But little evidence exists for the second hypothesis relative to cats, and anyway cats seldom control rat populations.

Not surprisingly, cat density at least in rural and suburban areas, correlates with both housing density and human density (Biro *et al.*, 2005; Sims *et al.*, 2007; Thomas *et al.*, 2012). A survey of cat owners across Great Britain reported that 986 cats living in 618 households brought 14,370 prey animals to houses during the five-month growing season (Woods *et al.*, 2003). The main types of animals caught are: mammals 59 percent; birds 24 percent; amphibians 4 percent; reptiles 1 percent; fish less than 1 percent; invertebrates 1 percent; and unidentified items 1 percent. These include at least 44 wild bird species, 20 wild mammals, 4 reptiles, and 3 amphibian species. Ninety-one percent of the cats catch at least one animal, and the average is 11.3 animals/cat. Many of the animals may have been juveniles caught after nesting/denning season. For houses that provide bird feed (Lepczyk *et al.*, 2012), cats catch more bird species, but fewer individual birds and amphibians/reptiles, than at other houses. Fewer mammals are caught if a cat is kept indoors at night, and if it has a bell on its collar.

Diverse studies have shown differences in cat predation between rural and suburban areas (Coman and Brunner, 1972; George, 1974; Haspel and Calhoun, 1991; Marks and Duncan, 2009). In the Brazil rural-and-suburb study mentioned above, which analyzed cat scats (feces) for animals caught during a year, 70 percent is of animal

Figure 8.15 Mouse with baby mice in farm building, a predominant small mammal around houses and farmsteads. US Department of the Interior, Fish and Wildlife Service photo.

origin, 16 percent plant origin, and 14 percent non-food items (Campos *et al.*, 2007). Invertebrates are overwhelmingly the predominant animal material (63% of the total). Other animals are: mammals (especially rodents) 21 percent, birds 13 percent, reptiles 2 percent, and fish 2 percent. The consumption of mammals is estimated to be 2.9 kg (5 lbs) per cat per year. In the Brazil study, cat and dog numbers, as well as diets, do not differ significantly between winter and summer samples. Niche breadth is about the same for cats and dogs, and niche overlap is almost complete (Krebs, 1999).

Overall, small mammals are the main food of cat predators, songbirds are the second-most caught animal, and other animals are taken in very small numbers (Churcher and Lawton, 1989; Woods *et al.*, 2003; Baker *et al.*, 2005; Loss, 2013; Kitts-Morgan *et al.*, 2015). An interesting exception are invertebrates being most abundantly caught in a rural-suburban area of southeastern Brazil (Campos *et al.*, 2007).

Several studies provide data on the average number of prey animals caught annually by a cat: (1) 12.8 in English village (Churcher and Lawton, 1989); (2) 18.3 for the UK (Thomas *et al.*, 2012); (3) ca. 3.9 for rural and suburban southeastern Brazil (Campos *et al.*, 2007); (4) 14.6 for Britain during the five-month growing season (therefore a higher number for the year) (Woods *et al.*, 2003); (5) 15.6 (corrected from 4.5 prey carried to house) around three Australian cities (Woinarski *et al.*, 2017); (6) 34.6 in rural, suburban, and urban areas of southern Michigan (USA) during the five-month growing season (higher number for the year) (Lepczyk *et al.*, 2003); (7) 36.0 for the urban UK (Sims *et al.*, 2007; Adler and Tanner, 2013); (8) 21.0 for the city of Bristol (UK) (Baker *et al.*, 2005); and (9) 18.3 for the town of Reading (UK) (Thomas *et al.*, 2012). Some of these studies are based on the number of prey carried to the house, while others analyzed animal remains in cat stomachs or scats. Since only some of the prey killed are brought home (perhaps some 30%; Kays and DeWan, 2004; Blancher 2013; Woinarski *et al.*, 2017), some of the mortality numbers should be higher.

Suppose that every cat consumed on average one mammal or bird per day, and that home-provided food, scavenged garbage, and other edibles are insignificant as food. Consumption would be 365 prey per cat per year (Loss, 2013). In contrast, the numbers above suggest that perhaps 15–20 prey killed per cat per year may apply in both rural and suburban areas. These data for outdoor cats probably only refer to mid-aged cats; young and old cats typically catch few to no vertebrates.

Cats, Rare/Uncommon Species, and Spatial Town Pattern

No rare species, and individuals of very few uncommon species, were consumed by cats in the predation studies above. Over a five-month growing season, more than one bluebird and one ruby-throated hummingbird (*Sialia sialis*, *Archilochus colubris*), species of conservation concern, were killed by about 800 to 3,100 cats, in rural-suburban southern Michigan (USA) (Lepczyk *et al.*, 2003). In the English village study, weasels, pipistrelle bats, and one kingfisher were caught by 77 cats in a year (Churcher and Lawton, 1989).

Thus, apparently cats overwhelmingly consume common or abundant birds near buildings, and predation of rare and uncommon species is rare. Local population declines of course may occur, as in the Bristol and English village studies, but common species are likely to rapidly recover. Overall (excluding islands and Australia), the present evidence indicates that the effect of cat predation on uncommon species populations and conservation is minimal or negligible.

Mid-age cats (not kittens and old cats) kill huge numbers of prey animals annually (Pimentel *et al.*, 2005; Blancher, 2013; Loss *et al.*, 2013; Woinarski *et al.*, 2017). An unknown but probably large portion are young mammals and birds, many of which will otherwise die from limited food supply, environmental stresses, other predators, and migration (Balogh *et al.*, 2011).

Apparently virtually all of the mammals and birds killed by cats are common species associated with the massive numbers of people and houses on the land. Not surprisingly diverse control efforts to reduce cat predation are proposed and tested (Nogales *et al.*, 2004; Thomas *et al.*, 2012; McCarthy *et al.*, 2013; Adams, 2016; Marra and Santella, 2016). Common options include: (1) destroy on site, which generally means a poison bait, but that is non-selective and kills many more animals than cats; (2) trap, remove, and euthanize, a process only feasible in local areas, due to the extensive manpower required and damage to other animals caught in the traps; (3) trap, neuter, and release, which similarly is manpower intensive and seems only locally useful; (4) keep cats indoors, which can have an effect but is extremely difficult to sustain; and (5) attach a small bell(s) or other device to a cat's collar that alerts many potential adult prey in sufficient time to escape, though effectiveness remains little studied.

Domestic cats are also a key host species transmitting the protozoan parasite of tomoplasmosis, which causes mental disease, even death, of humans (Lilly and Wortham, 2013). Resistant cysts (oocysts) pass in cat feces to water and soil, typically close to home, where the cysts may persist for months, even years. Typically, a few percent of pet and stray cats contain oocysts of tomoplasmosis, which can also kill wildlife.

The spatial dimensions in the preceding analysis together provide useful insight into the role of cat predation. In villages, almost all houses are within 100–200 m of the surroundings, the approximate radius of a cat's diurnal home range (Laundre, 1977; Liberg, 1980; Haspel and Calhoun, 1991; Horn *et al.*, 2011; Kitts-Morgan *et al.*, 2015). In the common case of farmland surrounding the village, cats readily feed on common farmland species in open fields, especially near woods (Toner, 1956; Pearson, 1964; George, 1974). Even foraging farther at night, the cats normally only catch common farmland species. Farmsteads in villages, town edges, and especially in the surroundings often contain many cats that feed around buildings and on dense populations of farmyard animals, especially small mammals (Toner, 1956; Laundre, 1977; Churcher and Lawton, 1989). Normally no large woodland with interior conditions exists nearby so uncommon wild animals are scarce.

In a suburban residential area of Chicago adjoining a forested river corridor, the richness of 20 native migrating bird species correlated with both fewer outdoor cats and fewer outdoor dogs (Belaire *et al.*, 2014). Interestingly in contrast, the richness of four nonnative resident birds correlated with more outdoor cats and more outdoor dogs (see Figure 9.10).

In suburbs and towns, houses are close enough so that domestic cats almost entirely forage within the residential area (Figure 8.14). Almost all prey killed are common house-plot/garden/backyard mammals and birds (e.g., especially house mouse, wood mouse, and house sparrow) (e.g., Lake Placid), and very few uncommon species.

Yet cats on town edges roam in their home ranges into farmland, wetland, and woodland to a distance of some 100–200 m by day, and if left outside at night somewhat farther (Liberg, 1980; Marks and Duncan, 2009; Horn *et al.*, 2011; Kitts-Morgan *et al.*, 2015). If farmland adjoins the town edge, such cats may catch common farmland species. If woodland is adjacent, predation in the woodland edge is largely on common generalist edge species.

However, if cats reach the woodland interior habitat (Forman, 1995; Turner and Gardner, 2015), the uncommon specialist species, such as many insectivorous birds, may be subject to cat predation (see Figure 13.2). Indeed (excluding certain island and Australian habitats), the one place where abundant cats have access to abundant uncommon species, and thus cause a potentially important ecological effect, is where a row of houses adjoins natural land.

Artificial Night Lighting and Flying Fauna

The night half of the year is sleep time for people and most wildlife familiar to people. But night is wake-up time for the other set of wildlife that, by avoiding the sun, have a different palette of resources available for foraging and carrying out their daily activities. I've led numerous nightlife walks, from Venezuela and Costa Rica to New Mexico and Canada, and never failed to discover these interesting nocturnal animals. Ecologist and night-walker Ross Wein (2006) comments: "Reflection comes easiest after midnight … if you sit still in the woods long enough, the best of nature will come to you."

Town Lights

Almost every town is covered at night by *skyglow*, the reflection by the atmosphere of a town's cumulative artificial light. Skyglow forms an overarching *light skydome*, commonly visible from kilometers (miles) away. Indeed, skyglow affects organisms and natural communities well beyond the town or city source (Rich and Longcore, 2006). Skydomes are lighter in air with particulate pollution or low clouds.

The size and brightness of these skydomes have greatly increased worldwide since about 1950, especially in wealthy nations, along with population growth and sprawl. The 100 million street lamps are the largest source of artificial light globally (Stone *et al.*, 2015). Thus, for instance, a medium-small city (Kiel, Germany; population 240,000) has a streetlamp for every 12 residents. Night-sky artificial brightness, which prevents the complete transition from cones to rods in human-eye vision, now covers more than 44 percent of people in the USA (Cinzano *et al.*, 2001).

Artificial lighting (i.e., made by people) is mainly used for four purposes (Eisenbeis and Hanel, 2009): (1) pedestrian and driver safety on streets and public places; (2) commercial use, such as advertising and buildings used by the public; (3) event lighting, as used at a fairground or football stadium; and (4) lighting house plots for safety, security, or prestige. Most of the ecological studies to date focus on streetlamps by roads. Research frontiers await us.

Towers with warning lights often protrude above towns – water tower, smokestack, radio/television tower, communication tower, electric powerline tower, and wind turbine (e.g., Page, Rumford/Mexico). Towers have red or white constant or blinking lights on top (Gauthreaux and Belser, 2006; Forman, 2014). A rare high-rise building in a large town has lighted windows (Forman, 2014; Adams, 2016). Migrating birds especially are affected by towers and high lighted windows (Cusa *et al.*, 2015).

The town center with a high density of artificial lights is normally at the center of the light skydome. Relatively bright light remains all night, affecting trees, attracting flying animals, and enhancing foraging by scavengers and predators. Progressively through the night the town center changes to an irregular checkerboard of light and dark spaces.

Residential areas in early evening appear somewhat grid-like with rows of streetlamps separated by generally muted house lights. Streetlamps are particularly useful for pedestrian and driver safety at road intersections and curves. Skyglow and energy consumption are reduced if streetlamps emit no light above the horizontal, and only illuminate the area needing added light (Blackwell *et al.*, 2015). Housing areas may range from a continuous row of lights to separated house lights embedded in a dark matrix of largely house-plot vegetation. Bird strikes on indoor lighted house windows may occur at dusk.

Insects, and consequently predators, are attracted to both the streetlamps and outside house lights of residential areas.

A commercial retail center, like the town center, has numerous lights, forms a bright lighted patch, and noticeably contributes to the light skydome, but by midnight the light level is noticeably reduced. An industrial center with mainly daytime activity appears as a dark patch at night with scattered safety and security lights (e.g., Rumford/Mexico; see Appendix). Sports fields and stadiums typically have bright lights.

Moving vehicles with red and bright white lights vary from dense to sparse in the town center and retail commercial areas, especially reflecting alternating periods of commuting, and may continue through the night at low densities in residential areas. Often a stream corridor and perhaps a railway corridor slices through or alongside town. These are relatively wide dark strips through the light of town (e.g., Askim, Concord), serving as routes for nighttime wildlife movement, especially into town (e.g., Alice Springs, Sooke, Pinedale). A large park in town is a large dark space where streetlamps may line some walkways.

Finally, the outside adjacent zone and farther out surrounding land appear as a dark matrix with widely scattered house lights. The light skydome of a town may be conspicuous, and somewhat farther out appear the lights of villages.

Artificial Light Characteristics

Artificial night lights partially obscure the sky and stars, but the focus here is ecological light pollution. Three night-light types are particularly important: (1) direct glare; (2) chronic increased illumination; and (3) temporary unexpected light (Rich and Longcore, 2006). Each of these can vary in duration, intensity (illumination), and quality (wavelengths). A surprising array and number of animals is mainly or entirely nocturnal: owls, bats, moths, most small carnivores, most rodents, many reptiles, most frogs and toads, and the predominance of marsupials. In some town and village areas, the number of nocturnal species rivals or exceeds the diurnal species.

A range of street and other light types is available, though a particular town or village usually only uses one or two of the types. Each has a distinctive spectrum of wavelengths: (1) incandescent; (2) fluorescent; (3) low-pressure sodium vapor light (orange-yellow, with wavelength peaks in the red, yellow, and green); (4) high-pressure sodium vapor light (orange-pinkish, with peak in yellow); (5) high-pressure sodium-xenon vapor lamps; (6) mercury vapor light (bluish-white, with peaks in blue, yellow, and red); (7) metal halide; and (8) light-emitting diodes (LED) (a broad-spectrum white light that apparently widely interferes with animals; Stone *et al.*, 2012). The various types of light have different, yet little studied, effects on animal behavior, including of frogs and toads (Buchanan, 2006), salamanders (Wise and Buchanan, 2006), insects (Eisenbeis, 2006), and moths (Frank, 2006).

Three characteristics of artificial light are of primary ecological importance: (1) illumination or brightness; (2) wavelength spectrum or color; and (3) daylength or photoperiod (Rich and Longcore, 2006; Blackwell *et al.*, 2015). *Illumination* (illuminance) is the amount of light encountering a surface area, more precisely light intensity is the number of photons striking the surface per m^2 per second (Rich and Longcore, 2006).

Illumination or brightness is commonly expressed in lumens or lux or foot-candles, with intensity corrected for the wavelength spectrum detected by the human eye. Since animals see different wavelengths, illumination could be corrected for the wavelengths perceived by another animal, but this is rarely done. Characteristic illumination levels (in lux) are: 103,000 full sunlight; 50,000 partly sunny; 1,000–10,000 cloudy day; 400–600 bright office; 100–300 most homes; 10 lighted car park; 0.1–0.3 full moon; and 0.001 lux for a clear starry sky. Artificial light in this typical list ranges from 10 to 600 lux, though floodlights are brighter and lights from houses less bright. Sudden changes in illumination level often alter animal and human behavior, movement, and communication.

The *emission or wavelength spectrum* refers to the range of light wavelengths emitted by a light source. As in a rainbow, long wavelengths are red and orange, intermediate ones yellow and green, and short wavelengths blue and purple. Ultraviolet wavelengths are extra-short and invisible to humans. But all non-primate animals detect and process different colors or ranges of wavelength (Blackwell *et al.*, 2015). For example, in general birds can see some of the extra-short ultraviolet wavelengths, but see poorly in low-light conditions.

Furthermore, great tit (*Parus major*) songbirds in The Netherlands, that night-roost under LED (light-emitting diode) streetlamps of different color, sleep well and are healthier under greenish and reddish light (Monahan, 2017). Under white light at night, the birds remain more active, and lose sleep. In Britain some neighborhoods have LED streetlamps emitting in the 450–460 nm wavelength (blue) range (Gaston *et al.*, 2012; Blackwell *et al.*, 2015). Studies apparently suggest that wavelengths lower than 500 nm produce more deleterious effects on both animals and humans, including disrupting circadian rhythms and metabolic processes.

The third major effect of artificial light is *daylength or photoperiod* which in normal conditions simply reflects the seasonal tilting of the Earth relative to the sun, and the consequent shortening and lengthening of days. But artificial light at day's end or beginning effectively lengthens the light time or daylength, with three broad ecological consequences (Blackwell *et al.*, 2015). (1) Circadian rhythms of animals and plants are stimulated, thus changing developmental and health patterns, including altered growth, reproduction, and disease resistance. (2) Diurnal and seasonal behaviors of animals are disrupted, including foraging for food, breeding, dispersal, and migration. (3) Light cues for detecting predators, avoiding vehicles, and habitat selection are altered. Also, plants close to lights often grow longer into the autumn than usual without going dormant, and hence may be damaged or killed by winter frost.

Species Responses to Artificial Light

Street Lamps and the Insect Fauna

The flight-to-light by insects at night primarily depends on the local zone of attraction within the illuminated area (Kolligs, 2000; Eisenbeis and Hassel, 2000; Eisenbeis and Hanel, 2009). However, changes in the far background, such as turning on floodlights

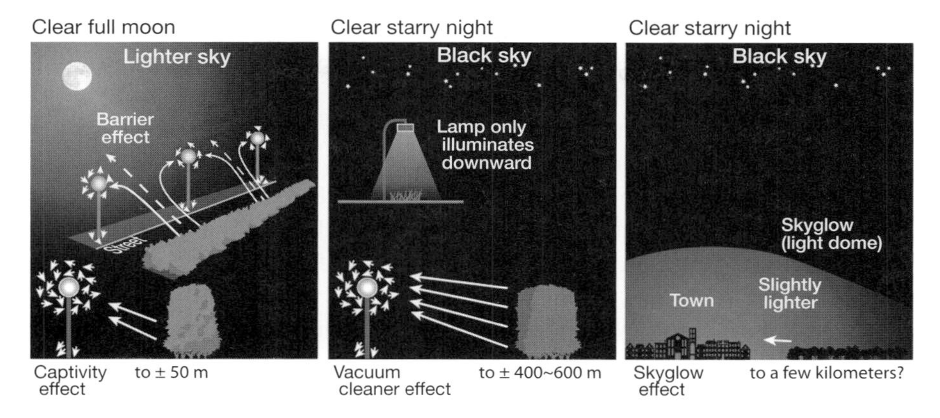

Figure 8.16 Artificial night-light effects on flying insects and different sky conditions. Street lamps illustrated. Vegetation = forest ecosystem with flying insects. Arrow = animals moving; X = insect death by predation or exhaustion. See text. Mainly based on Eisenbeis (2006).

across town or a cloud obscuring moonlight, also affect the insects attracted to a light (Bowden, 1982; Eisenbeis and Hanel, 2009). Many animals are especially active before, during, and after full moons. Yet others are most active at new moons.

Some insects fly directly to the lamp's hot glass cover, and die. More frequently, insects orbit the lamp seemingly endlessly, until caught by a predator or falling exhausted to the ground (Figure 8.16). Some escape to the cover of dark, rest, and may return to the light. The so-called *vacuum cleaner effect* in effect sucks insects out of their dark habitats, reducing populations and sometimes causing local extinctions.

Since flying insects are so important at night around artificial lights, an interesting study of night insects around a village (residential houses with some garden ponds), an isolated farmhouse (far from waterbodies), and a road (in an open landscape of fields and vineyards near a village) is highlighted. Insect traps installed below 19 streetlamps were monitored from June to September in and near Sulzheim (Rheinhessen district, Germany; population about 1,000) (Eisenbeis and Hanel, 2009). Streetlamp types included high-pressure (hp) mercury vapor (80 watts), hp sodium vapor (70 and 50 w; monochromatic 589 nm wavelength), and hp sodium-xenon vapor (80 w) with and without a UV filter. In total, 536 trap samples collected more than 44,000 insects which were identified into 12 orders.

Samples under mercury lamps collect the most insects (average 141 per trap per night), with sodium-xenon lamp samples averaging 107, and sodium lamp samples 64 per trap per night. The UV-filter lamp and control (no lamp) samples are much smaller (15 and 6, respectively). This suggests that high-pressure mercury lamps are most disruptive of the insect community, whereas hp sodium lamps are much less disruptive, attracting only 45 percent of the number of insects drawn to mercury lamps. For moths only, the pattern (25% attracted) at sodium lamps is even better.

All 12 orders of insects are present under streetlamps at each of the three habitats sampled, village, farmhouse, and road. However, the abundance of different insect groups differs sharply at the three sites. Residential village samples have predominately beetles (Coleoptera, 31%), plus moths (Lepidoptera, 16%), aphids (Aphidinae, 14%), flies (Diptera, 10%), and caddisflies (Tricoptera, 8%). Farmhouse samples are also mainly beetles (39%), plus moths (19%) and bugs (Heteroptera) (13%). In contrast, roadside samples in farmland overwhelmingly have flies (Diptera, 68%), plus a few beetles (7%) and moths (7%).

Least insect activity (smallest samples collected) occurs during full moons, and greatest activity near new moons (Eisenbeis and Hanel, 2009). Most activity (measured at 22.00 h [10:00 in the evening]) occurs above 19°C, and no activity occurs below 17°C (62°F).

The average number of insects caught per night was highest under high-pressure mercury lights (141), and progressively less under high-pressure sodium-xenon lights (107), high-pressure sodium lights (64), and high-pressure mercury lights with ultraviolet filters (15) (which did not meet government regulations for roadway illumination). Thus, high-pressure sodium vapor street lights attract fewer insects (is less ecologically destructive) than the mercury and sodium-xenon lights, highlighting the importance of light wavelength on degrading insect populations in nearby ecosystems.

The total number of insects approaching a street light is estimated to be three times the number caught in the light traps (Eisenbeis, 2006), though of course the proportion varies by insect group. A rough estimate of the death rate is 33 percent, that is, of the insects approaching the street light, a third die from predation, contact with the lamp, and other factors. Further estimates of mortality under high-pressure sodium lights there assume one streetlamp per ten people and 60 insects killed at a streetlamp per night for the 120-day summer. This suggests that artificial night flight from streetlamps may kill 720,000 insects in a village (population 1,000) and 7.2 million insects in a town (population 10,000) each year.

Rare "red-listed" insect species are also captured in light-trapping studies (Kolligs, 2000). Lights along a riverside or at a bridge doubtless attract many aquatic insects, such as mayflies (Scheibe, 1999). In fact, artificial light affects many aspects of an aquatic insect community in an urban Columbus, Ohio (USA) study (Meyer and Sullivan, 2013)

A village edge next to dark agricultural land harbors more moth species than does a more-lighted medium-housing density area with parks (Eisenbeis and Hanel, 2009). Numerous moth species, even in the relatively night-dark land of towns and villages in The Netherlands, seem to be decreasing over time. Nevertheless, dark areas at night contain a much richer insect fauna than do lighted areas (Scheibe, 1999).

Bats and Lights

Many nocturnal bat species survive at low population levels following a decades-long exponential increase in night light. Bats are mainly affected by the illumination or brightness of artificial lights, as well as attraction to insects around lights (Zurcher *et al.*, 2010). Unfortunately, both bats and insects near the lights may be hit by moving vehicles.

A Mexican urban area has four bat species using the night air differently (Rydell, 2006). Above the buildings are large fast bats. Between the buildings are medium-size fast bats. Under street lights (lamps) are small fast bats. And near ground level are broad-winged, slow, but highly maneuverable, bats. These species effectively use four different habitats, delineated by spatial area and night-light conditions. A similar vertical division of habitat by large-and-small slow-and-fast bats occurs in rural areas of northern Europe (Abbott *et al.*, 2015).

A high density of insects around street lights attracts many bats (Stone *et al.*, 2015). Commonly, fast-flying bats use open areas and are attracted to street lights, while slow-flying bats are often in cluttered areas and avoid street lights (Stone *et al.*, 2012, 2015). Street lights may lead to more predation and collisions with moving vehicles.

Night lights around a bat roost may increase the risk of predation, reduce the growth rate of juveniles, and reduce fitness (Boldogh *et al.*, 2007). Also, such night lights delay the emergence of bats from the roost in the evening (Stone *et al.*, 2015). This results in less foraging time, and the potential of missing the period of peak insect abundance.

Many bats regularly commute between roost and choice feeding area by following along stream corridors, hedgerows, and edges between land uses or habitats. Night light introduced along the route may disrupt flight patterns (Hale *et al.*, 2015; Stone *et al.*, 2015). A suboptimal route is then used, potentially exposing the animals to less-continuous vegetation, increased predation risk, higher wind velocity, and busy roads to cross. In the extreme, the bats abandon their roost.

Lights, Birds, and Other Animals

Artificial light has a major effect in astronomy by reducing the resolution or contrast between stars and blackness (Mizon, 2002). Diverse studies highlight the light effect on animals (Schmiedel, 2001; Rich and Longcore, 2006), but its effect on natural processes is little known (Blackwell *et al.*, 2015; Hopkins *et al.*, 2018). For plants, artificial light affects the efficiency of photosynthesis, direction of growth, and the timing of many developmental processes including bud expansion, flowering, and leaf fall (Briggs, 2006; Irwin, 2018).

Artificial light acts as a stimulus affecting animal physiology, behavior, and movement. The major effects on vertebrates and insects (Rich and Longcore, 2006; Eisenbeis and Hanel, 2009; Gaston *et al.*, 2012, 2013) are disturbance of: (1) biological rhythms; (2) foraging for nutrition; (3) mating behavior; (4) successful reproduction; (5) orientation and attraction/repulsion; (6) communication; (7) competition and predation; and (8) migration. Many of these responses involve the physiological production of hormones and altered timing of activities. The effect of street lights is of particular interest because they typically extend over much of a town (Rich and Longcore, 2006; Horvath *et al.*, 2009; Gaston *et al.*, 2013).

Adding artificial light at night interferes with the "magnetic compass" guiding birds during migration (de Molenaar *et al.*, 2006). The high intensity glare from floodlights, beacons, and lights on towers distracts and disorients birds, especially in cloudy and rainy weather. Birds striking towers (e.g., Astoria, Swakopmund, Superior), such as water tower, radio/television, smokestack, and a rare high-rise building, often die (Rich and Longcore, 2006; Forman, 2014; Adams, 2016). Also, at dusk in a residential area, even local birds sometimes fly toward indoor lights and hit the window glass of houses.

A study of nesting success of western burrowing owls (*Athene cunicularia hypugaea*), which nest in underground holes, in the small city of Las Cruces, New Mexico (USA;

population about 66,000) provides insight (Botelho and Arrowood, 1996). Significantly more nestlings (baby birds) were produced and more birds then successfully fledged in a human-altered area than in a natural area. Artificial lighting near nests in the human-altered area attracted numerous insects and bats, so the owls regularly perched under the lights feeding on insects, and sometimes even on the bats.

Light effects on other animals include illumination, light quality, and photoperiod. Several useful articles include literature reviews for different animal groups: terrestrial mammals (Beier, 2006); birds (Montevecchi, 2006; de Molenaar *et al.*, 2006); migrating birds (Gauthreaux and Belser, 2006); reptiles (Salmon, 2006; Perry and Fisher, 2006; Perry *et al.*, 2008); amphibians (Buchanan, 2006; Wise and Buchanan, 2006); fishes (Nightingale et al., 2006); and insects (Eisenbeis, 2006; Frank, 2006).

Tall buildings are scarce in towns, so bird strikes on lighted windows during migration are normally not a problem (Forman, 2014; Cusa *et al.*, 2015; Adams, 2016). But warning lights on towers, including radio, television, communications, cell-phone, wind turbine, and airport towers, may attract migrating songbirds (Rich and Longcore, 2006; Adams, 2016). Dead migrating birds beneath these lighted towers are common. It appears that red lights are more disruptive of flight pattern than are white lights (Gauthreaux and Belser, 2006). Also, constant lights are more hazardous than flashing lights. Thus, flashing white lights on towers seem to be least disruptive.

Seven animal responses to artificial night light are widespread, with examples known from two or more major animal groups. Night light:

1. *Alters foraging behavior and decreases or increases feeding success* by mammals (Beier, 2006), bats (Rydell, 2006), reptiles (Perry and Fisher, 2006), and salamanders (Wise and Buchanan, 2006).
2. *Increases the risk or avoidance of predation* by mammals (Beier, 2006), reptiles (Perry and Fisher, 2006), frogs and toads (Buchanan, 2006), salamanders (Wise and Buchanan, 2006), and moths (Frank, 2006).
3. *Alters the circadian rhythm* (biological clock) in mammals (Beier, 2006), birds (de Molenaar *et al.*, 2006), and moths (Frank, 2006).
4. *Alters reproductive behavior or decreases breeding success* by birds (de Molenaar *et al.*, 2006), frogs and toads (Buchanan, 2006), and moths (Frank, 2006).
5. *Disrupts longer distance movements*, such as corridor use, dispersal, homing, and migration, by mammals (Beier, 2006), birds (Gauthreux and Belser, 2006), salamanders (Wise and Buchanan, 2006), and moths (Frank, 2006).
6. *Increases mortality by striking objects*, such as moving vehicles, lighted windows, and lighted towers, for mammals (Beier, 2006) and birds (Gauthreux and Belser, 2006).
7. *Lengthens photoperiod which alters behavior* in reptiles (Perry and Fisher, 2006) and salamanders (Wise and Buchanan, 2006). Plants respond to both photoperiod (length of daily light) and light quality (different wavelengths, as in the types of light listed above) (Briggs, 2006). Responses include seed germination, stem elongation, leaf expansion, flowering, fruit development, bud dormancy, and leaf senescence and drop.

Other effects of artificial night light on animals probably include inhibited communication, reduced habitat, altered spatial arrangement of habitat, and loss of biodiversity.

The *zone of attraction* around a lamp is the area where the lamp's illumination (radiant energy) exceeds that from the sky (Eisenbeis, 2006). Thus, the attraction zone is smaller, and fewer insects attracted, on moonlit nights, or with a bright town skyglow, or under certain cloud conditions. Different studies suggest that the attraction radius is: (1) 10–17 m (1 m = 3.3 ft) for moths around a street lamp; (2) 50–700 m, 35–519 m, or 57–736 m from a 125-Watt mercury vapor lamp; and (3) maximum 130 m from a greenhouse light. The author summarizes the attraction zone distance as about 50 m in a moonlit sky, and about 400–600 m in a dark (starry) sky.

Finally, this exploration of the night side of towns, villages, and wildlife suggests that what we see by day is but a portion of the fauna and activity present. In conservation we work hard to protect species. But are those mainly the daytime active species we easily see? Should we also focus on the dark-side species? Would reducing artificial light be a key solution?

In addition to the varied specific solutions for night lights just mentioned, a few general perspectives seem useful. Light intensity (illumination) and spectral (wavelengths) output of night lights affect animals differently, and are readily controlled (Beier, 2006). Eliminating lights, dimming lights, and varying their timing are also easily accomplished (Shaw, 2014). Lighting especially rich in long wavelengths, lower intensity or illumination level, and shorter duration should greatly benefit the insect fauna of ecosystems. And the spatial arrangement of night lights can be planned to minimize ecological impacts (Beier, 2006; Shaw, 2014). Perhaps especially important is to maintain blackout areas with no development and motor vehicles, in protected nature reserves and wide green corridors (Minnaar *et al.*, 2015).

Part III

Town and Land Interactions

9 Commercial, Industrial, and Residential Areas

Towns … were little more than a mile across in any direction … new industrial sites … and … new residential areas were developed on the edge of town.

Jonathan Brown, *The English Market Town*, 1986

The preceding chapters focus on specific important components of towns, including flows, people, air, water, vegetation, and wildlife. Now we begin to weave the threads into patterns. The quote above usefully highlights spatial patterns, the size of towns and how they develop. Spatial arrangement provides a foundation for understanding flows and movements. Also, all the following patterns and processes change over time.

We now highlight four major parts of town: (1) commercial town centers; (2) edge-of-town commercial centers; (3) industrial centers; and (4) residential areas and the town. The fourth focus begins to tie commercial/industrial/residential areas together into a whole town.

Commercial Town Centers

Spatial Arrangement and Change

We start with the town center, usually the oldest and most important commercial center (Figure 9.1). This is followed by commercial centers on or near the edge of town. Business office centers in some large towns are only briefly mentioned, since they are more characteristic of urban and suburban areas (e.g., Issaquah, Concord; see Appendix) (Snep *et al.*, 2011; Forman, 2014; Serret *et al.*, 2014).

Towns in the eastern USA vividly illustrate a typical changing commercial *town center* (Francaviglia, 1996):

1850s–1860s. Technology and architecture developments changed the appearance of the traditional dust-and-mud commercial center.

1880s. Poles for electric power lined the streets or back alleys.

1880s–1890s. Planted street trees began to spread in town centers for shade and aesthetics, partially replacing awnings on buildings and wooden porch walkways that connected buildings together.

1880s–1890s. The old water towers of wooden staves were replaced by larger metal water towers with the town's name painted in big letters, showing pride, dependable water pressure, and fire safety with pipes to fireplugs.

Figure 9.1 Two-level city-center building with arcades providing shade in warm climate. Note motorbikes and informal outdoor commercial activity. City of Siem Reap, Cambodia. Photo courtesy and with permission of Mark Brenner. (A black and white version of this figure will appear in some formats. For the color version, please refer to the plate section.)

1890s. One or more electric streetlamps replaced gas or oil lamps, which had persisted since the 1830s.

1890s. Road paving began to appear, replacing oiling or sprinkling of water in town centers.

1890s–1900s. Electric streetcars on tracks ran on main streets, and in some cases connected towns.

1920s. Most commercial areas in town centers were paved with asphalt replacing bricks and stones, and connected by paved roads to other towns.

1920s. Cars and trucks were widespread in town, along with wagons, carriages, and horses, and strips of shops extended along main roads, with parallel parking alongside.

1920s–1930s. Many building facades were replaced by industrial materials, and neon-light signs lighted the evenings.

1940s. Secondary side streets in town centers were largely paved.

1960s–1970s. Building modernization, the connecting of one- and two-level town-center buildings, and historic preservation spread. Most town centers still had few street trees, especially to the north where rainwater, snow, and ice persist.

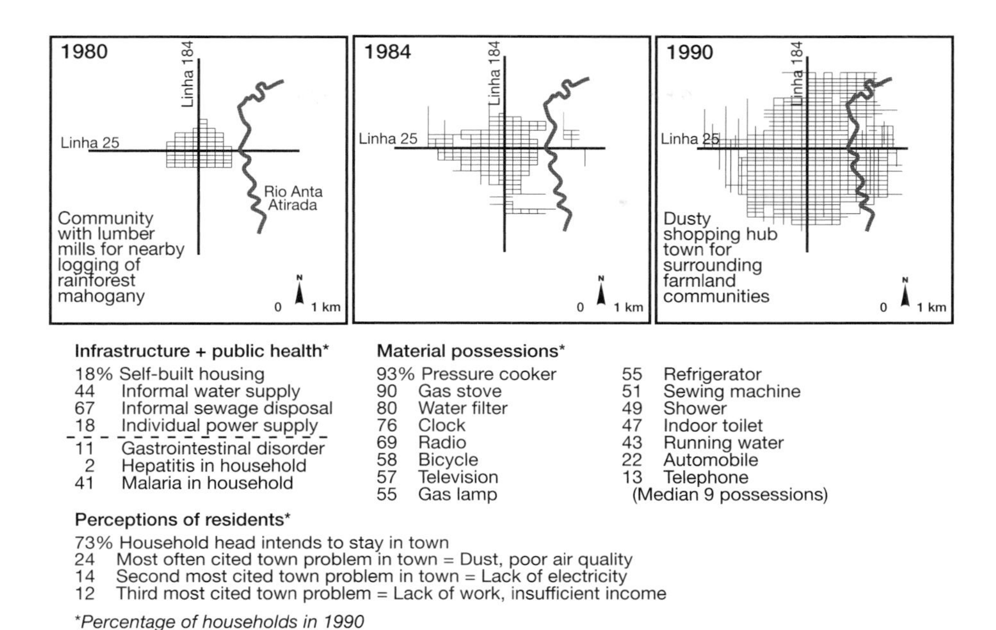

Figure 9.2 Growth and attributes of households in a changing frontier town. Town infrastructure, possessions, and perceptions in 1990. Informal water supply = well without pump, local stream, open spring. Informal sewage disposal = no septic tank or sewer line, and reliance on outhouse (Figure 6.5) or outdoors. Individual power supply = no electricity, and energy from gas/kerosene stove/lamp. Rolim de Moura, 140 km from main road, Rondonia, Brazil. Adapted from Browder and Godfrey (1997).

Such, though different, town center changes have occurred in all regions. Indeed, 20 Scottish towns over 13 decades (1691–1821) showed an ever-changing mix of growth, status quo, and population-loss periods (Harris and McKean, 2014). Rapid population growth, and sometimes loss, alters Amazon villages and towns (Figure 9.2) (Browder and Godfrey, 1997). Empty storefronts in the town center symbolize economic and population downturns. In contrast, some towns change slowly, such as Gurupa (Para, Brazil), a long-term Amazon River town, or Petersham, a rural Massachusetts (USA) village with strong ecological linkages and cultural traditions.

A study of 24 towns (population 5,000 to 20,000) in four European areas (England, the Netherlands, Portugal, Poland) analyzed shopping and jobs relative to residents in or outside a town (van Leeuwen, 2013). Not surprisingly, most town residents' shopping purchases (approximate average 70%) are in town, whereas residents from the surrounding areas do 32 percent of their shopping in town. Also, 35–82 percent of town residents work in town, while 25 percent of surrounding area residents commute to and work in town. Relatively few town residents (9%) work in the surroundings. Other studies suggest similar patterns (Brown, 1986; Hart and Powe, 2007).

Town centers differ markedly from city centers and village centers or greens (Gilbert, 1991; Thayer, 2003; Forman, 2014). Towns are commonly located by waterbodies, such as a stream, river, lake, or salt water. Usually, the town center was located to minimize flood risk. However, subsequent land-use changes upslope, upstream, or upriver may increase flooding problems. Towns bisected by a stream corridor and/or a railway corridor have their town centers adjacent to, or on both sides of, the corridor(s) (Williams, 2006; Grossinger, 2012). All towns have a main road passing through the town center, and most town centers are at the intersection of two or more main roads radiating outward from town (Bohl, 2002).

Thus, the town center is usually oval or elongate in the direction of the first or primary main road. The town's *central plaza* or square, normally with irrigated lawn, flowers, and scattered trees, is in the town center. The central plaza may or may not have a prominent government or religious building, but is an important meeting place and used for various celebrations and festivals (Francaviglia, 1996; Wood, 1997; Arendt, 2004; Friedman, 2014; Dufty-Jones and Connell, 2014).

Buildings, Streets, Infrastructure

The commercial retail shopping and offices dimension is conspicuous (Powe and Hart, 2007). But the town center is really mixed use, i.e., combining commercial and residential uses, and often with civic buildings, cultural resources, eateries, small car parks, and even local craft or manufacturing shops (e.g., Rock Springs; see Appendix) (Brown, 1986; Francaviglia, 1996). Unlike villages, infrastructure and utility wires and pipes may be mainly buried or inconspicuous (e.g., Selborne, Sandviken). Buildings often have shops on the ground level, professional offices or storage on the second level, and apartments on a higher level (if present) (e.g., Icod de los Vinos, Sandviken). Residences, restaurants, and a theater in the town center normally mean that the area has people active well into the evening, with lighted areas all night.

Commercial, residential, and government buildings, many often three-level and attached, line a few or several blocks adjacent to and near the central plaza (Arendt, 2004; Williams, 2006; Harris and McKean, 2014). In hot seasons or climes, these buildings, perhaps with awnings or arcades, provide a *shade corridor* as a cool route for shoppers (e.g., Alice Springs, Valladolid). Most shops, most restaurants, and most professional and business offices are in this town center. In towns, stores are mainly local mom-and-pop shops, whereas in small cities many national/international chain stores are conspicuous (e.g., Gavle). The mass of buildings is highly heterogeneous, varying in heights, architectural styles, materials, walls, roofs, and condition. Yet almost all could potentially support green roofs or solar panels (Friedman, 2014).

The number of levels in town-center commercial buildings often indicates town population size (Chapter 1), e.g., mostly two-level buildings in small towns, and four-level in large towns. In many regions, construction cost per level (or floor area) decreases from one to four levels, and sharply rises in higher buildings because of steel construction. Also, inside elevators are expensive, so two-level town buildings seldom have an elevator. Outside Beijing, apartment buildings are commonly seven-level, because elevators are required in higher buildings. The number of levels in town-center buildings doubtless correlates with many features of environmental import, including the amount of people, rubbish, pests, dogs/cats, impervious surface, noise, and light.

Buildings represent both habitat and hazard for animals (Forman, 2014). Bats, common pigeons, starlings, and house sparrows typically find holes and nest or roost especially in the upper parts of these buildings. Scavengers such as mice, rats, and raccoons may reside in the lower parts. With a relatively high water table in town, basements/cellars and underground car parks may be scarce, or wet. Termites and ants may be abundant, even degrading the buildings. In moist tropical towns, epiphytes such as mosses, lichens, ferns, bromeliads, and orchids may be common on roof tiles, in roof gutters, on posts, or even attached to wires. Leaks and drainage systems of buildings often attract fungi, algae, and diverse invertebrates. Flying animals sometimes hit lighted windows. Scavengers and predators quickly learn where they can catch and feed on animals in buildings.

Streets and various other spaces subdivide the town center often into small irregular rectangles (Williams, 2006). Generally, small car parks are distributed by streets and behind buildings, both for employees and shoppers (e.g., Sandviken, Huntingdon/Smithfield). To reduce stormwater runoff some are gravel (e.g., Sooke). Shops and restaurants need truck deliveries and waste removal that may occur by alleys behind the commercial buildings.

Town center streets are major routes of local residents, but also serve people from surrounding villages, as well as through-travel by cars and trucks. Some shopping streets, with bike-racks and nearby car parks, may be *pedestrianized* for walkers and closed to traffic (e.g., Sandviken). A local transport system of small buses or large vans goes to the town center (Friedman, 2014). A row of iron posts 0.7–1.0 m (2–3 ft) high may separate sidewalk pedestrians from vehicle traffic on narrow streets (e.g., Icod de los Vinos). With lots of pedestrians and traffic, plus street-side parking in front of shops, traffic moves slowly in the town center. This is especially noticeable on weekend market days or festival days (e.g., Sooke, Tavistock) (Brown, 1986; Francaviglia, 1996; Dufty-Jones and Connell, 2014).

Both grocery stores/markets and solid-waste transfer stations are sources of discarded food, packaging, and other organic materials (Evans, 2012). A study of birds around 27 solid-waste stations and four grocery stores found both similarities and differences (Figure 9.3) (Washburn, 2012). For grocery stores, about 85 percent of the birds are common (rock) pigeons, with a few percent each of gulls, European starlings, crows, and house sparrows. In contrast, solid-waste transfer stations have about 39 percent gulls, 23 percent European starlings, 12 percent crows, 10 percent common pigeons, and 8 percent house sparrows. Thus, five species are common at both places, but food waste at grocery stores overwhelmingly attracts common pigeons, whereas solid-waste stations mainly attract gulls and European starlings. Diseases can be transmitted by scavenging birds to other wildlife and humans. Nuisance wildlife of diverse sorts causing problems are generally more abundant in suburban and exurban areas than in either urban or rural areas (Kretser *et al.*, 2008).

Shops, restaurants, and indeed market and festival days produce piles of rubbish with wasted food and food scraps (e.g., Frome). Garbage bins and dumpsters therefore are abundant usually behind shops, where waste-removal trucks empty them (e.g., Capellades, Lake Placid, Falun) (Friedman, 2014). But lots of scavengers, from dogs and cats to raccoons,

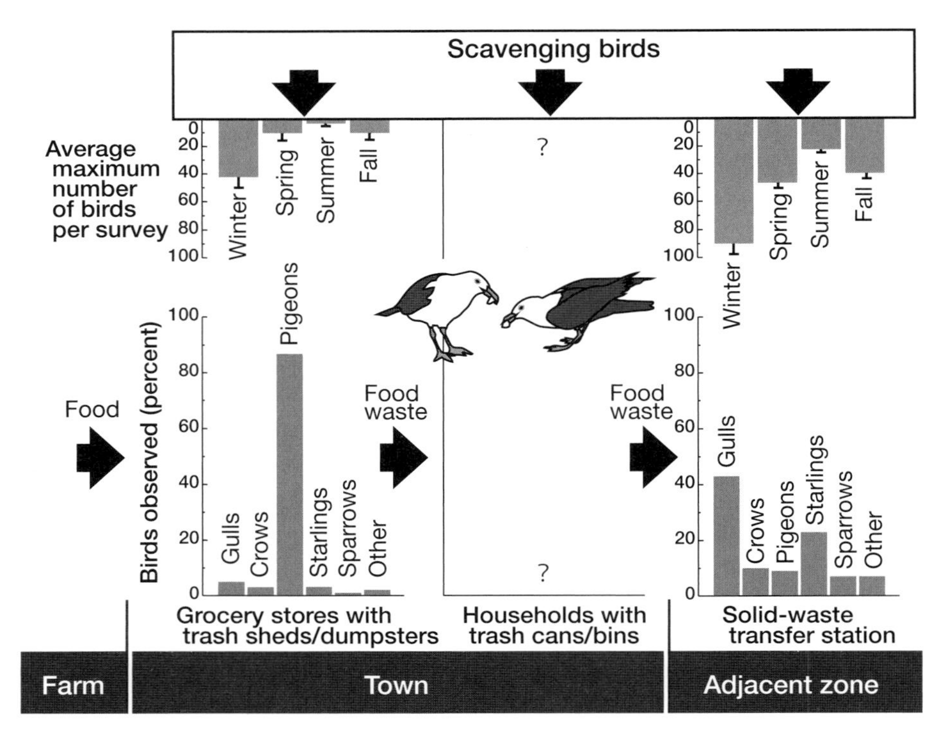

Figure 9.3 Scavenging birds and the town food and food-waste system. Birds sampled over 15 months at four grocery stores and 27 transfer stations across the USA. Transfer station receives local solid waste which is transported elsewhere for disposal. Gulls (*Larus* spp.), crows (corvids, *Corvus*), common rock pigeons (*Columba livia*), European starlings (*Sturnis vulgaris*), house sparrows (*Passer domesticus*). Adapted from Washburn (2012).

gulls, sparrows, and rats, feed on and around these many rubbish containers. Also, food scraps and diverse seeds are dropped on streets, and are fed on by birds and other animals when traffic is low. These species help clean the town center. Streets function as conduits for the movement of seeds, leaves, dust, and trash, which are blown by wind or by traffic movement in one direction (Kowarik and von der Lippe, 2011).

Tiny green areas in the town center are particularly conspicuous because of their relative sparseness and their diverse shapes. With space at a premium, lawns and bare soil areas are normally few and small. The central plaza often is the only relatively large green site, and thus serves as a "hub" for movement of animals. The scattered trees provide shade and the vegetation evapotranspires water molecules upward, so on a hot day the plaza is cooler than its surroundings.

A railway corridor through town normally has strips of vegetation, along which animals move into and sometimes through town (Gilbert, 1991; Forman, 2014). However, intense herbiciding of the track area leaves almost no plants between the vegetation strips. In agricultural towns, a grain-storage structure(s) may tower over the town, and attract lots of mice and other herbivores (see Figure 10.2) (e.g., Casablanca,

Askim) (Friedman, 2014). In addition, a train station may be located in the town center, thus acting as a partial barrier or filter for wildlife moving along the railway corridor.

A stream corridor slicing through town normally contains narrow strips of riparian vegetation on each side, though often in the town center they are very narrow or absent. Terrestrial animals moving into or through town in the stream corridor probably pass the town center when people and traffic are scarce, especially at night. If the water is relatively clean, fish can readily move through the town center. But old pipe and ditch systems for wastewater, stormwater, and former or present industries in the town center usually mean the water is not very clean.

These former and current pipe systems are major habitats and runways for other animals in the town center. Rats run alongside or through (partially cleaning) underground pipes. Cockroaches use the stormwater drainage pipes, which interconnect the bottoms of buildings, to find food sources in each building. Raccoons go down street-side storm drains to dry nest sites, and emerge to forage above ground at night. Empty storefronts during economic downturns indicate built habitats for many animals (e.g., Askim, Rumford/Mexico).

Microclimate and Water

Some street-sides also have street trees, though in the area of town center commercial buildings, these may be small (recently planted or in poor health) or scarce (Friedman, 2014; Harris and McKean, 2014). In cold climates, tree shade makes rainwater, snow, and ice last longer on sidewalks and streets, which is not favored by shop owners. Some hot-climate towns have many street trees for shade, while others try to save scarce water by not irrigating the street trees. Where treelines produce shade corridors, birds and people do tend to move along them more readily. Bird nests in town center trees are scarce, with so much movement of people, traffic, noise, and light.

The hottest place in town is usually the town center with its extensive horizontal and vertical surface area. Solar and sky radiation are absorbed in short and long wavelengths, and then longwaves are emitted from surfaces, which heats the air (Chapter 5). Trees, awnings, and arcades (Figure 9.1) are the most common outside solutions for this heat.

But also air in the town center is typically rich in particulate matter (Williams, 2006). People moving lift particles from sidewalks. Traffic contributes particles and also lifts particles from streets. Local craft manufacturing and other industry emit particles and other pollutants. Building construction sites contribute particles, and seemingly endless road-repair projects add to the richness of town center particulate matter. Nevertheless, probably the typical particulate concentration in town center air does not significantly decrease solar radiation input or increase precipitation.

Town center noise results from the same combination of sources. Trucks particularly add to the noise level. In addition, music from shops and residences increases the noise level. The effects of town center noise on noise-sensitive birds and other animals remains a research frontier.

Artificial night lighting is a different issue (Chapter 8). The town center is usually the brightest place in town, and is lit all night long. This is the center of the town's skyglow or light dome, commonly visible for kilometers beyond the town (Chapter 8) (Eisenbeis and Hassel, 2000; Rich and Longcore, 2006; Eisenbeis and Hanel, 2009). At night the light dome may attract night insects and other animals, apparently an unstudied subject.

But a row of streetlamps serves as a barrier to the movement of some insects, which are diverted to one of the streetlamps (see Figure 8.16). A single streetlamp creates a "vacuum cleaner effect" sucking insects out of their natural habitats. The streetlamp is also a major killer of the night insects, with (high pressure) mercury lamps twice as bad as sodium lamps. Animals, from lizards and toads to bats and rats, are drawn to the town center streetlamps to feed on the feast of beetles, moths, aphids, flies, caddisflies, and more.

Finally, water also ties the town center together. Rainwater falls on a roof, flows to a roof gutter, accelerates down a downspout, rushes to the street, disappears into a street-side drain, and is piped or channeled in a ditch to the nearest local waterbody. With a heavy rain and water flowing this way from many roofs, flooding may occur.

This problem is reduced in many ways. A green roof covered with plants, a scarcity in towns, slows the water flow to the gutter. Or, without a gutter or downspout, water drops directly to the soil below, where much is absorbed. Or, water from the downspout is directed to water the plants (rain garden) around the building. Or, water from the downspout is directed to a small constructed basin, pond, or wetland (Chapter 6), which reduces the velocity and amount of water flowing onward. Or, the pipe or ditch leading from the stormwater drainage system beneath the road carries the water to a larger constructed basin, pond, or wetland. All of these solutions reduce flood risk.

Indeed, they also all reduce pollution in the nearby waterbody. Rainwater flowing down a wall picks up particulate matter blown there, and for a concrete or brick-and-mortar wall, picks up $CaCO_3$. Rainwater flowing over a sidewalk and street picks up spilled or discarded items, dog waste, and particles from the sidewalk surface, vehicles, and the worn road surface. Flowing water also picks up heat from a wall, sidewalk, and street. These pollutants are then carried by the street drainage system and pipe/ditch to the waterbody. Disrupting that rapid flow with a constructed basin/pond/wetland system maintains clean water in the waterbody.

In short, the commercial and residential town center is a rich concentration of resources that provide habitats for a set of species much less common elsewhere in town. Furthermore, main routes and alternative routes of the ever-moving animals present are relatively easy to discern. Similarly, wind, water, and pollutant flows are widespread and readily delineated in the town center.

Edge-of-Town Commercial Centers

A newer commercial center in or near the edge of town normally has no central public plaza, no railway or train station, no residential units, and no shoreline/coastline (e.g., Astoria, Westbrook, Huntingdon/Smithfield; see Appendix). A main road or highway adjoins or is near the center, providing ready access for shoppers, as well as trucks bringing goods and removing wastes. An extensive paved area is present for shoppers' vehicles, but is essentially empty after midnight and on holidays. Such car parks

Figure 9.4 Retail shopping commercial center and apartment buildings in forest (upper left) close to entrance–exit of north–south multi-lane highway. Town edge (far right) separated from highway by cemetery areas. Infrared photo; southeastern Massachusetts. US Department of Agriculture, Soil Conservation Service photo. (A black and white version of this figure will appear in some formats. For the color version, please refer to the plate section.)

noticeably contribute to a regional excess of parking places, e.g., about 2.7 parking spaces per vehicle in the Upper Midwest USA (Davis *et al.*, 2010). These extensive car parks are essentially the opposite of ones designed for ecology and aesthetics (Rushton, 2001; Higgins and Cardillo, 2011).

Shop types both duplicate and complement those in the town center. The edge commercial center serves the nearby residential area, but also serves surrounding villages, even towns, for weekly shoppers (e.g., Askim) (Powe and Hart, 2007; Vaz and Nijkamp, 2013; Friedman, 2014). Sadly, such edge commercial centers are often built on prime food-producing soils (Spilkova and Sefrna, 2010), and also may put some town center mom-and-pop shops out of business.

Basically, three types of commercial centers appear on the edge of towns. One is a *large shopping mall* normally designed and constructed by a single developer by a major highway entrance/exit to serve a regional market, and contains one or a few large buildings (Figure 9.4). Second is a *small shopping center* with similar attributes except that it is typically anchored by a food supermarket and mainly serves the nearby residential area. Third is a *strip development* where a series of stores, restaurants, gasoline (petrol) stations, and so forth are aligned along a main road.

The large shopping mall is commonly oval in shape with a large central building(s) containing many shops, some of which are often regional or national chain stores. Restaurants are present. Other medium-large buildings with stores may be present. As a relatively new building, few microhabitats for birds, insects, and other animals exist. The usual extensive flat roof is suitable for a green roof and/or solar panels. In theory certain birds could readily nest there, but the extensive paved car park and all-night lighting probably inhibit that. A small shopping center has many of the same building characteristics.

In contrast, strip development buildings are in a row on one or both sides of a main road, and are extremely heterogeneous (e.g., Socorro, Plymouth). Since most buildings are relatively new, few construction sites are present. However, the buildings' construction quality and current condition generally varies from good to poor. Thus, usually many small holes and other microsites on the buildings are available for birds, insects, and other animals. Numerous small green roofs could cover the flat tops of these buildings. Strip developments along roads characteristically have some stores selling items too heavy to carry by walking shoppers. Thus, stores selling cars, furniture, mattresses, large appliances, house-yard equipment, farm equipment, lumber, and construction supplies and equipment are common. Restaurants are also common, and small industries may be intermixed.

Strip development often has a separately paved car park in front of each store. Pedestrians are present in a large mall car park, common around a small shopping center, and nearly absent by strip development, where cars park right in front of a store. A continuous paved strip behind the line of stores provides convenient access for service trucks. Also, behind each store is a dumpster or pile of bins or boxes with rubbish accumulation. This sequence of rubbish is a major attraction for scavenger wildlife, from diurnal gulls, crows, starlings, and house sparrows to nocturnal rats, cockroaches, raccoons, even bears (Figure 9.3) (Washburn, 2012).

A strip of woods commonly separates the strip development from nearby residential housing (Ellis *et al.*, 2006). Although the woods does little to reduce noise propagation, it greatly reduces visual disturbance and light. Streetlamps and security lights of the strip development produce the "vacuum cleaner effect," sucking night insects out of their woods habitat. The woods are also habitat for some of the scavengers feeding on rubbish.

The occasional business office center generally includes considerable vegetated area which serves as a stepping stone for wildlife movement (Serret *et al.*, 2014). But green vegetated sites are small in the large shopping mall, and basically absent in the other two edge commercial centers. Tiny lawns, shrublines, and flowerbeds decorate a mall. A few remnant trees often persist around at least the smaller car parks, and a few stores grow a few flowers in front.

Air quality of edge commercial areas next to surrounding agricultural or natural land is usually fairly good. Service trucks, shoppers' vehicles, and, in the case of strip development, road traffic, reduce air quality somewhat. Particulate matter from extensive surfaces used by vehicles is lifted by wind and moving vehicles into the air. Traffic noise from a busy road may inhibit sensitive birds for hundreds of meters outward. The extensive paved car parks, access roads, and rooftops emit considerable heat, so these edge commercial areas have hot air, perhaps only exceeded by town center air.

Lighting of the car parks is commonly from tall very bright lights, with considerable light escaping upward. This noticeably extends the town's skyglow in the direction of an edge commercial center. Enormous numbers of night insects are drawn from surrounding lands to the lights, and die. Night predators and scavengers are attracted to the feast.

Strip development especially serves as a barrier or filter for wildlife moving between habitats separated by the road. Traffic, people, and noise during shopping hours, plus light throughout the night, create the barrier.

Rainwater runs off flat rooftops to the extensive surrounding paved area, where it normally rushes off as sheet flow. This flowing water picks up heat, heavy metals, hydrocarbons, trash, and other pollutants. One or more large constructed basins/ponds/wetlands should be present downslope, but these structures are often absent. Therefore, huge amounts of polluted water rush onward to the local waterbody, leading to flooding and water pollution. With a limited stormwater drainage system and pipe network, underground runways used by rats and cockroaches are limited. Surface water in commercial centers is often noticed as a small decorative pond at the large shopping mall.

Industrial Centers

Industrial centers are small districts or manufacturing sites in town. One or a few old industries which evolved with the town may remain near the town center. More commonly though, a few to several industries are clustered at a site on the edge of town. Traditionally in most regions medium and small industries were located by streams or rivers, which provided power, cooling, and waste disposal. Today all three of these can usually be provided anywhere around a town.

Especially in competition-run market economies, individual factories, industries, or companies commonly have "boom and bust" phases. The company grows and hires employees, next goes through a short or long production period of slight or no growth (or growth equal to the national economy), and then declines and disappears (Forman, 2014). A long little-or-no-growth period provides stability to the town. All three of the phases, as well as the post-production period, may have distinctive and important environmental effects on the town. Also, in the face of decline or disappearance, the resilience of the town reveals much about the town's future and its environment.

For industrial centers, we now explore: (1) types and locations; (2) environmental conditions; (3) interactions and hazards for the surroundings; and (4) company towns and the environment.

Types and Locations

Small- to medium-size companies, for example producing textiles, shoes, food, drink, or electronics, are typical in towns. Also, highly diverse start-up manufacturing companies are common. Indeed, small craft shops and workshops of, say, one or two people, may be in homes, commercial or separate buildings, or an industrial center (Brown, 1986). Industrial centers on town edges also attract employees from surrounding villages or towns.

However, several types of industries hire numerous employees and require considerable space. Mining (e.g., Falun; see Appendix; see also book cover), paper/pulp mill (e.g., Rumford/Mexico), and metal smelter (e.g., Palmerton, Pennsylvania, USA; Sudbury, Ontario, Canada) are familiar examples in towns. A single large industry effectively creates a company town. A large mine or smelter extends far beyond the town (e.g., Maria Elena, Chile), which may quickly grow into a city (e.g., Ciudad Guayana, Venezuela) (Correa, 2016).

Deep-mine entrances in town mean large heaps of mine waste accumulate after extracting the metals or coal. Surface mining for sand or gravel normally occurs in a river floodplain, leaving deep holes. Quarrying rock in town produces considerable noise and dust, and either leaves a rock-cut hillside or a deep hole (e.g., Issaquah).

Seemingly, in the early phase of an industry the environmental impacts are small, perhaps because for a new company the equipment and processes are new and the amount produced is small (Gilbert, 1991). Environmental impacts increase perhaps faster than production, as equipment ages, products increase, byproducts accumulate, and environmental regulations are less adhered to or enforced in town. When a company closes, the severe environmental impacts decrease only slowly (Forman, 2014).

Interdependent industries (or industrial ecology, industrial symbiosis) at a site have the byproduct output of one industry used as raw material input of a second industry, whose byproduct output in turn is raw material input for the first (or a third) industry (Forman, 2014). A number of industries may be linked in this way at a site.

The best several-decade example is at Kalundborg (Denmark) (Frosh and Gallopoulos, 1989; Beatley, 2000; Cohen-Rosenthal and Musnikow, 2003), where an oil refinery and a power station are interdependent and are also linked to a plasterboard factory. Several other companies use some of the byproducts of these three. In effect, interdependent industries use less raw material, produce less waste, and provide stability. However, in a competitive economy, if one company decides to seek short-term profit, the system may fall apart. The Kalundborg system has persisted in part because the power station is effectively a public utility, which the public requires and government can subsidize when needed.

Industrial centers in towns are mainly located relative to topography, railway, river/stream (Figure 9.5), and street layout, as illustrated by four Alabama (southern USA) towns (Williams, 2006). From the air, a cluster of large buildings by a railway spur or branch and/or a major highway access leading out of town, normally indicates an industrial center. Industrial centers are located on sites which originally had little chance of flooding. Since establishment of the industries, however, regional deforestation, agricultural plowing, mining-related activities, and air and water pollutants may have increased the flood hazard. The contrasting Alabama towns with industries are:

1. *Irondale* has an industrial center with a railway bisecting it on one side of town. Forested ridges are on both sides of the industrial center, and most residential

Figure 9.5 Factory by dammed river which traditionally provides power, cooling, and waste disposal. Regulations now limit water pollution. Dover-Foxcroft, Maine, USA. R. Forman photo. (A black and white version of this figure will appear in some formats. For the color version, please refer to the plate section.)

development (for potential employees) is on the upwind northwest side of one ridge, where noise, visual effects, water pollution, and air pollution are less. A railway also cuts through town so the town center of several blocks with commercial buildings surrounds the train station. The town center has hotels, as many people arrived by train.

2. *Bessemer* is located at a spot where both mineable iron and coal are located. Two industrial centers with railway access are present on opposite edges of town. Two other clusters of somewhat large buildings within the town are surrounded by residential areas and primarily served by main roads. One or both is a commercial retail center, and small/medium industries may be intermixed with the commercial buildings. The town's street grid is based on the railway northeast/southwest route, and is thus 45° off of the usual north/south, east/west grid.
3. *Cullman* also has rail tracks cutting diagonally through town. The older half of town has a diagonal street grid, while the newer half has the north/south, east/west grid. Freight warehouses, probably with some commercial and industrial uses, line both sides of the railway near the town center with the passenger station. The town center buildings near the station are set back a bit from the railway. Three small clusters of medium-size buildings on the town edge represent commercial areas probably with some industries.

4. *Decatur*'s cluster of large buildings is also by the railway. Originally the industrial center was on the town edge, but now residential growth has almost encompassed it. Decatur is a riverside town where a prominent row of warehouses lines the riverside. Raw materials for, and products from, the industrial center can be stored in the warehouses for shipping by boat. Thus, the town center is set back from the river and less subject to river flooding.

All four towns have industrial centers organized around a railway. Forest ridges help organize Irondale, Bessemer is sandwiched between two different mines, and Decatur is arranged around a river that frequently floods. In all four cases, the industry or industrial site probably strongly influences the entire town.

Environmental Conditions

Typically, industrial areas are exceedingly heterogeneous, especially if several or many industries are present. Buildings of varied shapes and sizes, diverse features such as processing facilities, yards for moving materials, conveyor belts, smokestacks, pylons, power-source structures/materials, roads, rail lines, cranes, large specialized vehicles, trucks, cars, pipelines, ditches, storage of raw materials, storage of products, piles of byproduct wastes, rubbish dumps, and pools for waste decomposition are often present. These create high microhabitat diversity.

Diverse incoming raw materials, which also bring in numerous nonnative species, add to the complexity (Gilbert, 1991). For instance, long lists of such species have been recorded for diverse industries in Britain, including the woolen industry, high-quality rag paper, oil (bean, soybean, rapeseed) processing, imported timber products, and flour mills and breweries using grain shipments. Especially around large/heavy industries, crucifer, *Melilotus*, and *Epilobium* plants are common, apparently due to both increased heat and altered substrate. In the UK, house sparrows, European starlings, and common pigeons are characteristic of industrial areas. Occasional birds include blackbird, kestrel, crow, and magpie, as well as a rather rare songbird, black redstart (*Phoenicarus ochruros*).

Warehouses for diverse goods and storehouses with food are particularly important ecologically. Often with warm and moist conditions, microbes, insects, spiders, and rodents multiply (Gilbert, 1991). Cockroaches thrive. Furs, skins, clothes, and other organic materials are fed on, e.g., by clothes moths (Tineidae), house moths (Oecophoridae), and carpet and hide beetles (Dermestidae). Beetles, moths, and mites are common residents in food-storage containers. Cockroaches, ants, and crickets typically live elsewhere in a building, but come and feed on the food. Many of the specific species present in such warehouses are not found elsewhere in the UK.

Considerable heat is generated in industrial processes, which heats both buildings and industrial wastewater. Byproducts of the industrial processes accumulate in piles or are spread over the area (Figure 9.6). Alkaline materials, cellulose fibers, steel slag, woolen cleanings, burned-coal cinders, chunks of concrete, fuel ash, and so much more are seldom benign, and often toxic. Different species are tolerant to each of these pollutants.

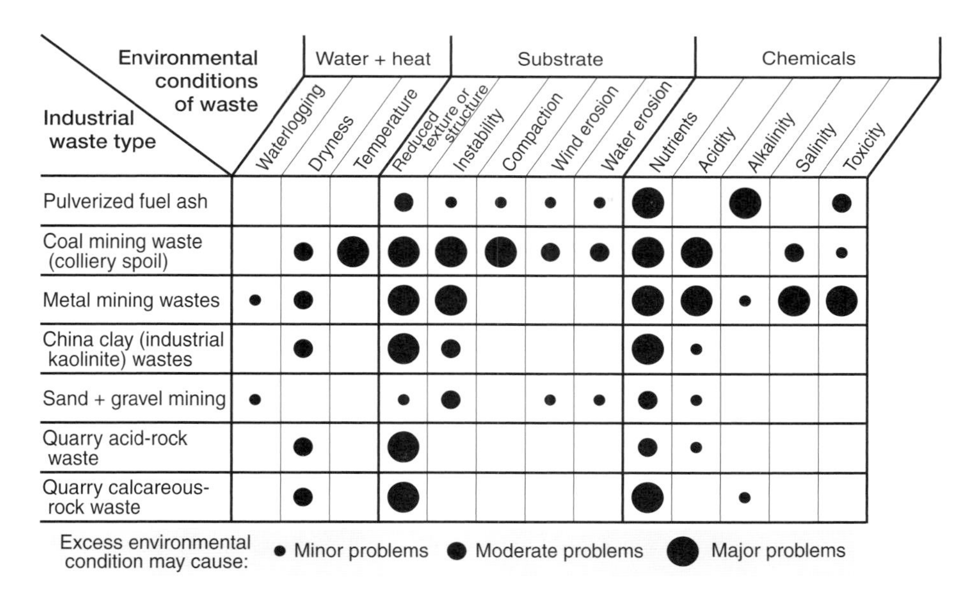

Figure 9.6 Industrial waste conditions which limit soil and vegetation development. Adapted from Wheater (1999).

For instance, calcareous plants such as some orchids, sedges, and composites grow on alkaline wastes (Gilbert, 1991).

The combination of unusual industrial structures, raw materials, processes, storage materials, and wastes creates numerous unusual microhabitats, which could support an enormous range of plant and animal species. However, the abundance of toxic-level pollutants typically spread over the industrial area severely limits the biodiversity present. Indeed, plants thriving long term on heavy metal-polluted soil are probably genetically adapted to it (Ernst, 2006; Baker *et al.*, 2010; Batty and Hallberg, 2010). As society and companies clean up their rubbish, old buildings, machinery, dumps, and handling processes, microhabitats and species probably decrease.

An overview of an industrial center focuses on its inputs and outputs (Forman, 2014). Typically, the key inputs are: raw materials (from afar); power (renewable or fossil fuel energy); transport (roads, rail); employees (from nearby); species (from afar); species (from nearby); and precipitation. Key outputs are: economic products; air pollutants (gases, particulates); heat; transport; employees; waste piles and pools; soil pollutants; stormwater pollutants; and stormwater runoff. Many of these affect surrounding areas in and beyond the town. In addition, the ecological patterns, ecological processes, and species types within the industrial area represent a key distinctive set of features (Forman, 2014). This overview of an industry's productive period differs markedly from that for the subsequent post-productive period.

After an industry closes the main inputs to the land are precipitation and species (from nearby) (Forman, 2014). Key outputs are: air pollutants; waste piles/ponds removed; stormwater pollutants; and stormwater runoff. The ecological changes on-site are even more interesting. Some species types, ecological patterns, and ecological processes continue. Semi-natural vegetation colonizes very slowly. Spontaneous town/urban vegetation colonizes slowly. Soil organic matter and structure accumulate slowly. Soil pollution decreases very slowly. Water pollution decreases slowly. Water runoff decreases. Air pollution drops rapidly. Ecological succession begins and increases. If buildings remain, start-up companies may move in (Friedman, 2014), or buildings may be converted to residential use as apartments or condominiums. Alternatively, buildings may stand derelict, gradually disintegrating.

A *brownfield* or area of contaminated soil due to former industry of course depends on the site and particular pollutants. Nevertheless, overall such sites commonly have erodible soil with poor structure (associated with compaction and waterlogging) (Wheater, 1999). Plant mineral nutrients are generally low, while the soil pH may be extremely high or low. Fuel ash soils are often salty. The chemicals in upper levels of soil are toxic to most but not all plants, so plant colonization and succession are slow.

Plants with deep roots may colonize dry areas. Limestone soil plants appear on alkaline soils. Some grasses grow on heavy metal-polluted sites, and on highly acid sites such as by coal-waste heaps. Regionally rare plant and invertebrate species are often found on such post-industrial brownfield sites. In part this relates to high tolerance of the extreme environmental conditions, and partly due to little competition by other species.

Interactions and Hazards for the Surroundings

Some town factories use resources from the surroundings as raw materials, which effectively ties town and its surrounding land together. Food, wine, weaving, tanning, leather, and furniture industries often fit this model (e.g., Chamusca) (Brown, 1986). Some companies use local materials, and export products elsewhere. Others import raw materials, and sell locally. Market days, often weekly, are a key to survival or growth for some companies selling locally. A designated well-advertised market day, when shops are open and stocked, draws large numbers of shoppers from town, but especially from surrounding villages and towns.

Town residents daily spread pollutants from an industrial center throughout the town. Employees with their hair, skin, hats, clothes, shoes, lunchboxes, and cars collect industrial chemicals at work, and daily carry them to shops, restaurants, bars/pubs, and home (Kosek, 2006). Similarly, commuting employees from villages transport pollutants yet further. Trucks and truck tires, as well as trains and freight cars/wagons, also spread the chemicals in town and beyond.

Wind and air flow also carry the industrial chemicals outward, in this case distributing them widely and fairly evenly over town, especially in downwind directions. In this way the chemicals get into "everything." Those that cover roofs and tree leaves are washed by rainwater down walls and trunks to soil, sidewalks, streets, and local waterbodies.

When entering a town, look for *lichens* on tree trunks. If these often grayish rounded blobs are abundant, this is probably a clean-air town, because most lichens are quite sensitive to air pollutants, especially sulfur dioxide, e.g., from coal burning (Gilbert, 1970; Purvis, 2010). Overall, fluffy beard-like (fruticose) lichens are most sensitive, flattened lichens (foliose) attached to the bark are intermediate, and flattened lichens in the bark (crustose) are most resistant to pollutants. One crustose species (*Lecanora conizaecoides*), unknown in Britain before 1860, became the most common species in polluted areas after the industrial revolution, and now is decreasing as air quality slowly improves (Laundon, 2003; Purvis, 2010). Lichen species composition is also sensitive to ammonia from intensive farming, NOx from traffic and high-temperature combustion processes, and heavy metal-enriched pollution (Wolseley *et al.*, 2006; Purvis, 2010). New lichen growth on twigs seems to be a good indicator of acidification and nutrient eutrophication (Wolseley and Pryor, 1999; Larsen-Vilsholm *et al.*, 2009).

Stormwater carries some chemicals from industrial centers as infiltration into the soil, where pollutants may reach groundwater and perhaps drinking-water wells. But much of the chemical material is swept as water pollutants from impervious surfaces into drainage ditches and pipe systems. These conduits in turn quickly pour the pollutants into local treasured waterbodies, such as town pond, lake, stream, river, or estuary. All of the pollutants, including heavy metals, mineral nutrients, and toxic organic substances, negatively affect some or many aquatic organisms (Batty *et al.*, 2010).

Of particular concern are groups of "emerging contaminants" whose environmental effects are little known, and which might be in the repertoire of a town's industries (Chapter 5) (Boxall, 2010). Prominent among these are: human pharmaceuticals, veterinary medicines (Capelton *et al.*, 2006; Crane *et al.*, 2008), nano materials (Boxall *et al.*, 2007), personal care products, and flame retardants. Some, when decomposed, produce persistent and/or toxic breakdown products (Boxall *et al.*, 2004; Sinclair *et al.*, 2006).

The net effect of these human, air, and water transports from the town's industrial center is to modify the chemistry of the town. More acidity degrades concrete and brick mortar, and higher levels of varied heavy metals and nutrients widely alters the plants across town. Toxic chemicals of numerous types kill countless types of microbes and animals, altering many ecological processes. So, industries commonly provide important employment, supporting the town, but watch out for the byproduct output chemicals invading life in town.

Local hazards, which are mainly visual, also affect neighborhoods (Greenberg and Schneider, 1996). For example, the town of Marcus Hook (Pennsylvania; population 2,400) lies in a northeastern US industrial region, and has neighborhoods with multiple hazards. A survey of residents found that factory-related hazards ranked high (percent of residents) among the hazards identified:

43% Odors or smoke
35% Petroleum refinery or tank farm
33% Chemical manufacturing plant
14% Abandoned factories and businesses
8% Metal or furniture manufacturing plant

Residents at greater distance, or separated by a major physical barrier, were less worried about a neighboring factory and its effects.

High-security industries with high fencing typically provide physical separation but visual access for a neighborhood (Hough, 2004). Soil berms can visually screen an industry and significantly reduce noise, whereas tree vegetation strips only provide the visual screen. Combining the tree strip with a sequence of different microhabitats, such as pond, marsh, and swamp, plus boardwalk, may provide added value.

Several modifications may convert a chemically contaminated brownfield to a rich wildlife habitat and educational resource (Hough, 2004). Protect existing wildlife. Encourage the colonization of spontaneous vegetation and ecological succession. Maintain or create diverse microhabitats, perhaps including derelict structures. Remove the worst polluted material, and minimize the spread of the remainder. Make most inorganic pollutants unavailable to microclimatic processes, e.g., by burial, and enhance the microbial and oxidative decomposition of organic pollutants. Make the place visually attractive and educationally accessible.

Company Towns and the Environment

A *company town* has a single major employer which runs a large local enterprise and dominates town patterns and activities. Factories and mining companies are typical, though some are government enterprises (Correa, 2016). A few company town examples tell the story.

Corner Brook was mainly built by a large paper/pulp company in western Newfoundland (Canada) (White, 2012). Paper manufacturing causes extensive deforestation effects, sulfur dioxide air pollution, clogged waterways, and water pollution. The company's dominance, both locally and regionally, is highlighted by the need for extensive forestland, a rail system, deep-water port, shipping, and housing for more than a thousand employees. Over its century-long history, production has vacillated in response to resource-input fluctuations, workers' strikes, wars, and competitive industrial markets. Downturns led to mayor layoffs of workers and town disruption, but also to less deforestation and air pollution. The company dominated the region, while degrading the region. Even today the company town signs are conspicuous and extensive.

Mount Ida is a large mining extraction and smelting company for lead and copper near Cloncurry, Queensland (Australia) (White, 2012). Typically, the town was composed of a few thousand workers. Ups and downs over its several decades of production were related to market and economy fluctuations, technological developments, government oversight/policy, wars, railway construction, worker strikes and labor union activities, and lead poisoning and health.

Both towns were essentially designed and built by the companies. Corner Brook was modeled on a compact "Garden City type" (Howard, 1898) English town and welcomed families, whereas Mount Ida was a workers' town with a large central plaza and spread-out houses in desert-grassland. In both towns smokestacks with smoke always towered nearby. At Corner Brook smoke contained SO_2 and particulates, and at Mount Ida smoke contained particulates especially of lead, copper, and other heavy metals. Lead is an especially bad health hazard. The Corner Brook community had somewhat better living than in many other frontier-like towns in its region, whereas Mount Ida was a rough-and-tumble outback place. Residents' dissatisfaction persisted in both towns.

Particulate pollutants mainly accumulate, and are toxic to many organisms. These company town areas end up as huge chemically contaminated brownfields.

A large concentrated supply of economically extractable mineral, energy, or terrestrial resources of international or national importance attracts a large company, which creates a company town. Maria Elena (Chile), Miami (Arizona, USA), Palmerton (Pennsylvania, USA), and Sudbury (Ontario, Canada) are other examples. Some company towns grow into cities, such as Ciudad Guayana (Venezuela) (Correa, 2016) and Johannesburg (South Africa). Others shrink or disappear when production is no longer economically viable.

Both Sudbury and Palmerton have long been company towns with mineral mining and smelting. Around Sudbury the aerial SO_2 and nickel and copper pollutants killed trees over an extensive area including tens of kilometers downwind. Most other plants and animals also disappeared and soil erosion accelerated.

The same ecological process occurred at *Palmerton*, as zinc and cadmium particles for decades poured down from the sky onto the land. The company basically molded the town, supporting and controlling most activities. That includes a large green central plaza, many civic buildings, and an attractive town center (Williams, 2006). Residences for managers are mainly located on a hill upwind of town, while workers' apartments and houses extend outward from the town center. Two clusters of large industrial buildings are at opposite ends of town, and connected by rail lines. A high ridge next to town is blanketed with pollution which killed vegetation for tens of kilometers downwind, and accumulates in the soil (Figure 9.7). Severe erosion followed vegetation loss. A small river at the base of the ridge flows through town, carrying loads of zinc, cadmium, and sediment onward downriver.

Finally, a treatise on company towns identifies conditions that produce benevolent policies toward employees, and conditions where executives have cared little about employees and communities (Green, 2010). "Benevolent" conditions are: labor shortage; need to retain skilled labor force; high profitability; remote location where a sense of community was desired; positive public relations; and a liberal/progressive national political climate. The "cared-little" conditions are: only unskilled workers needed; competitive economic margins low; commodity products that are largely under the radar screen; production at a remote location; and a conservative national political climate.

An information-technology company in Sandviken (Sweden; see Appendix) and a shoe-manufacturer in Street (Somerset, England) are examples of company towns with overall relatively benevolent policies. Large companies have become dominant in other towns, such as Adidas in Herzogenaurach (Germany), LEGO in Billund (Denmark), IKEA of Sweden in Almhult (Sweden), and Walmart in Bentonville (Arkansas, USA).

Most striking is what's missing in the preceding lists of conditions – resource overexploitation, pollution damage, and degraded health of employees and other town residents. Resource exhaustion, such as excessive deforestation for wood, typically results from an unsustainable extraction for short-term profit. Extensive pollution damage follows from a remote location, production exceeding the absorptive capacity of the environment, and the lack of, or inadequate, recycling. Degraded health of residents results from the pervasive long-term pollution in and surrounding company

Figure 9.7 Deforested and eroded mountain due to zinc and cadmium air pollutants from an adjacent smelter. During a six-decade process, trees died, decomposition slowed, and much later tree roots disintegrated and soil accelerated downslope. Small river at base received zinc/cadmium air pollutants and eroded sediment. Palmerton, Pennsylvania, USA; see Appendix. R. Forman 1975 photo.

towns. A company focus on the three key "missing" variables could transform the town from a little cared-for to a benevolent condition, creating a sustainable industrial town.

Residential Areas and the Town

Towns often contain both grid and non-grid residential areas, with grid in the center and non-grid surrounding (e.g., Cranbourne, Pompeii; see Appendix) (Williams, 2006). Some towns in the USA have: no grid (e.g., Eutaw, Alabama and Pikeville, Kentucky); a single N/S, E/W grid (e.g., Cullman, Alabama); various grids at different angles (Decatur and Selma, Alabama; Wausau/Weston); or a spider-like appearance (Welch, Mullens, and Logan, West Virginia). Grids can be analyzed by "ladder forms" between parallel streets (Williams, 2006). Variables such as how ladders fit together, number of adjoining ladders, ladder ends, and what's between rungs could be useful for environmental analyses. Still, it is hypothesized that grid patterns are about the worst environment for biodiversity.

Both older and newer residential areas of a town are composed of neighborhoods (e.g., Alice Springs, Tomar, Swakopmund). We use a physical or visual concept of *neighborhood*, as a residential area mainly of somewhat similar houses close enough

Figure 9.8 Low-density non-sprawl houses on mostly fenced plots used for chickens, goats, and other food resources. Town of Maun, Botswana. R. Forman photo. (A black and white version of this figure will appear in some formats. For the color version, please refer to the plate section.)

together to facilitate interactions among residents (Figure 9.8). Usually, many or most houses in town are built by owners, and thus appear somewhat distinctive (e.g., Sooke, St. George's), unlike housing in suburbs built together in groups by developers (e.g., Issaquah, Reykjanesbaer). The similarity in neighborhood houses implies some similarity in house age and the size of large trees across a neighborhood. Thus, along with greenspaces, neighborhoods are major components of residential areas, and house plots mainly compose neighborhoods.

Few species stay within a single house plot. Instead, species, water, and people move among house plots, helping to tie the neighborhood together, as well the residential area and town as a whole. For instance, in the city of Hamburg (Germany), swifts are abundant flying over neighborhoods of old buildings, whereas crested larks are characteristic of neighborhoods with new buildings (Sukopp, 2008). So now we briefly explore: (1) neighborhoods; (2) house plots, and (3) ecological movements around town.

Neighborhoods

Characteristic neighborhood types in towns include mobile-homes, low-income houses, attached row (terrace) houses, small-plot detached single-units, large-plot detached single-units, and low-rise apartments (Figure 9.9) (e.g., Rock Springs, Rocky Mount).

Figure 9.9 Three-level apartment with tiny vegetation spaces, characteristic of large towns and small cities. Washington Heights neighborhood, Chicago.
Source: stevegeer / E+ / Getty Images.

Large-plot detached single-units range from productive homegardens in the tropics (Chapter 7) to sprawl houses with large lawns. High-income neighborhoods, characteristic of cities and suburbs, are usually tiny or absent in towns (e.g., Astoria, Socorro, Clay Center).

Alternatively, environmental conditions can describe communities (Girling and Kellett, 2005). A green neighborhood has house clusters embedded in abundant greenspace, such as many small green patches, a green corridor network, major bisecting green corridor, green wedge projecting into town, and/or a surrounding greenbelt. "Green fabric" neighborhoods are recognized by abundant trees along streets, plus woody vegetation in front spaces/yards and in backyards (e.g., Sandviken). "Gray fabric" neighborhoods have small vegetated house plots, few street trees, and lots of paved street, sidewalk, and other spaces. Water or blue neighborhoods have one or more prominent waterbodies present.

A survey of residents in a neighborhood adjacent to strip development in a small city (College Station, Texas, USA) found that the amount of tree and shrub cover within a 450 m (1500 ft) radius of a home counteracts the negative effects of an adjoining commercial strip (Ellis *et al.*, 2006). In rural Wales, UK, low-income (poverty) neighborhoods seem to be tied together especially by community belonging and attachment to the landscape, rather than by material hardship and social exclusion (Milbourne, 2014). Low-income

(a) Spatial scales and variables measured

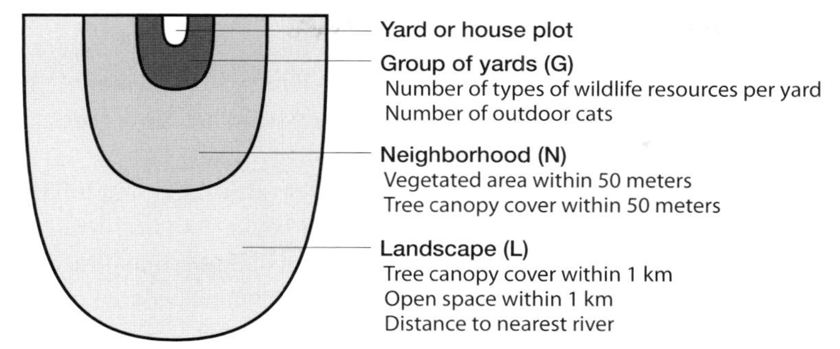

(b) Native bird species richness correlations

Yard
+ Number of wildlife resources
- Number of outdoor cats

Group of yards
+ Ratio of evergreen to deciduous trees
+ Number of yards with plants with fruits/berries
- Number of outdoor cats
+ Percent of front and back yards with trees
n.s. Number of bird feeders
n.s. Interaction of yard trees and shrubs

+ Positive correlation
- Negative correlation
n.s. Not significant

(c) Spatial scales as predictors of bird species richness

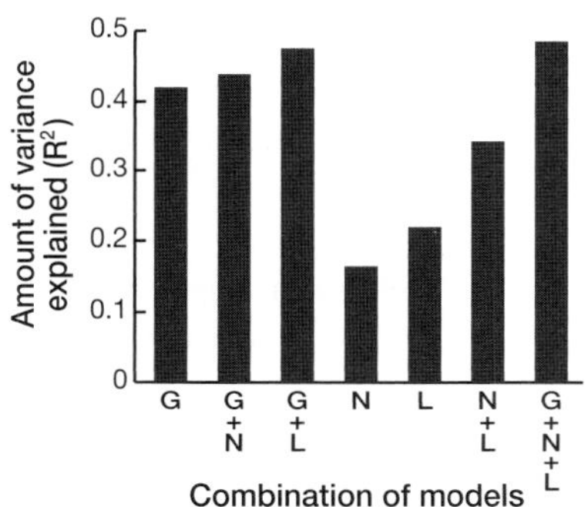

Figure 9.10 Avian diversity related to residential house-plot yard characteristics and spatial scale. Residential areas adjoining a wide forest river corridor in the fringe area of Chicago. Native bird species = 12 residents and 20 migrants during breeding season. G =group of yards; N = neighborhood; L = landscape. Migrants correlate with more trees in yards, more yard plants with fruits or berries, more wildlife resources in yards, higher ratio of evergreen to deciduous trees, fewer outdoor cats, and fewer outdoor dogs. Richness of four nonnative species correlates with more cats and more dogs. Avian richness does not correlate with number of bird feeders or insecticide use. n.s. = not significant. Adapted from Belaire *et al.* (2014).

areas of cities, such as in Phoenix (Arizona, USA), Santiago (Chile), and Rio de Janeiro (Brazil), typically have low biodiversity (Muller *et al.*, 2010; Cilliers, 2010; Lerman and Warren, 2011), though this pattern may or may not apply in towns (e.g., Swakopmund). Nevertheless, positive environmental or economic conditions are sometimes perceived negatively from a social perspective (Pejcher *et al.*, 2015).

A study of birds in suburban and exurban neighborhoods next to a major green corridor near Chicago evaluated whether avian diversity related more to the residential landscape, a neighborhood, or house-plot attributes (Belaire *et al.*, 2014). *Groups of neighboring yards* are more important for native species richness than are environmental attributes of the larger neighborhood or residential landscape scales (Figure 9.10). But native bird species richness also correlates with house-plot attributes: (1) positively, with the ratio of evergreen to deciduous trees, percent of yards with trees, and percent of yards with fruit/berry-producing plants; and (2) negatively, with the number of outdoor house cats. Migratory native birds are more abundant in house plots with these wildlife-friendly features. In contrast, nonnative birds are more frequent where more outdoor cats and more dogs are present. Thus, attributes of both house plots and groups of house plots are better predictors of avian diversity than are the broader neighborhood and residential landscape features.

Furthermore, groups of house plots (house gardens) (Smith *et al.*, 2005; Dahmus and Nelson, 2014), such as a block or a few blocks, fit better with the sizes of many vertebrate territories and home ranges. Flying insects including pollinators also move over areas larger than house plots. A block or few blocks contains a large number of microhabitat types suitable for the nests and dens of many species.

Although overall, neighborhood houses are relatively similar, dissimilar houses and house plots are normally present. Common examples are a farmhouse remaining from an earlier time (e.g., Askim), a rented house with uncut grass, untended gardens, and accumulations of junk, or successional vegetation around an abandoned house (see Figure 7.8) (Perrin and Grant, 2014) (e.g., Grande Riviere, Rock Springs). Also, in a house-plot group, many of the habitats are repeated, providing ample and diverse foods for foraging.

Housing density is a key feature of neighborhoods (Figure 9.11). Distance between houses affects both tree/shrub cover and people's social interactions (Arendt *et al.*, 1994) (e.g., Easton). Compact development produces less greenhouse gas than does low-density housing, and compact development with an extensive public transit system produces still less (Senbel *et al*, 2014).

A study of birds in 1 km^2 (250 acre) squares in the UK compared the density of different species in housing developments varying from <5 to >2,500 houses/km^2 (Tratalos *et al.*, 2007). Some species do not correlate with housing density, while others show clear preferences. The willow warbler (*Phylloscopus trochilus*) is most abundant in the lowest housing-density rural area, European robin (*Erithacus rubecula*) in intermediate housing densities (40–750/km^2), and carrion crow (*Corvus corone*) in the highest-density urban area.

Analogously in the Ottawa Region, breeding birds and also carabid beetles were measured in housing densities from <56/km^2 to >1244/km^2 (21–480/mi^2), and also in forests next to the housing developments (Gagne and Fahrig, 2011). Both bird and beetle communities are relatively similar in the several forests, but differ along the housing-density gradient. Within increasingly dense housing, bird species richness strongly declines, species composition changes, and the avian community is simplified.

A bird study in Costa Rica's Central Valley compared birds recorded before and after a 16-year period of extensive urbanization, especially by housing developments (Biamonte *et al.*, 2011). During this landscape change, 32 resident species disappeared, particularly habitat specialists. Also, 34 latitudinal migrant species are no longer detected, with their stopover habitats largely eliminated. Along with this biodiversity loss, the abundance of both resident and migratory birds has decreased. Wildlife patterns are ever changing in neighborhoods and residential areas.

Figure 9.11 Walkable compact village by road in border area of forest and pastureland. Note wooden walls and roofs, few windows, a few yurts, firewood, fences, and one electric pole on left. White Haba, Altay Xinjiang, China. (A black and white version of this figure will appear in some formats. For the color version, please refer to the plate section.)
Source: Best View Stock / Getty Images.

House Plots

In the UK, 87 percent of the 22.7 million households are estimated to have outdoor spaces/yards (domestic gardens) for plantings (Davies *et al.*, 2009). This exceeds the percentages elsewhere in Europe, such as Holland (56%), Germany (49%), and France (32%) (Gilbert, 1991). The UK house-plot spaces average 190 m^2 (2044 ft^2) in size. House-plot areas commonly contain flower gardens/beds, vegetable garden, lawn, hedges/shrub lines, rock garden/rockery, tiny pond, compost heap, walls, walkways, and driveway (Gilbert, 1991). Together this creates a rich mosaic of microhabitats. About 29 million trees are present in the UK house plots, plus 3 million (+0.5) garden ponds, 7 million households with bird feeders, and another 5 million households where birds are fed (Davies *et al.*, 2009). The cumulative effect of both house-plot habitats and supplementary food must have major and diverse effects on a nation's wildlife.

House plots are normally divided into front spaces/yards/gardens and backyards, with narrow side strips present between detached houses. *Backyard spaces* are commonly used for garden cultivation, sitting out, children's play, clothes drying, and aesthetics (Howard, 1898; Gilbert, 1991). Endless designs occur, often reflecting the differing goals

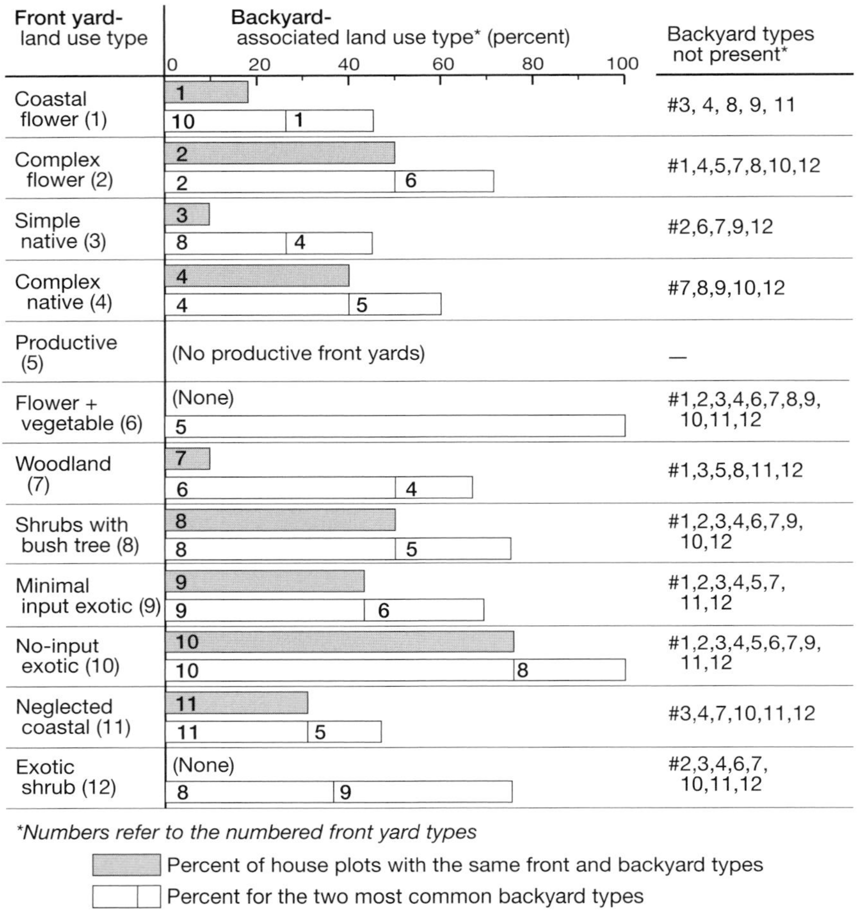

Front yard-land use type	Backyard-associated land use type* (percent) 0 20 40 60 80 100	Backyard types not present*
Coastal flower (1)	1 10 1	#3, 4, 8, 9, 11
Complex flower (2)	2 2 6	#1,4,5,7,8,10,12
Simple native (3)	3 8 4	#2,6,7,9,12
Complex native (4)	4 4 5	#7,8,9,10,12
Productive (5)	(No productive front yards)	—
Flower + vegetable (6)	(None) 5	#1,2,3,4,6,7,8,9, 10,11,12
Woodland (7)	7 6 4	#1,3,5,8,11,12
Shrubs with bush tree (8)	8 8 5	#1,2,3,4,6,7,9, 10,12
Minimal input exotic (9)	9 9 6	#1,2,3,4,5,7, 11,12
No-input exotic (10)	10 10 8	#1,2,3,4,5,6,7,9, 11,12
Neglected coastal (11)	11 11 5	#3,4,7,10,11,12
Exotic shrub (12)	(None) 8 9	#2,3,4,6,7, 10,11,12

**Numbers refer to the numbered front yard types*

Percent of house plots with the same front and backyard types

Percent for the two most common backyard types

Figure 9.12 Backyard land uses relative to front yards in house plots. 107 house plots sampled in five suburbs of Hobart, Australia. Named yard types: coastal flower = showy flowers, canopy 4–8 m high; complex flower = showy, large shrubs, canopy 4–8 m; simple native = non-showy, canopy >8 m, large shrubs; complex native = showy, canopy >8 m, deciduous trees; productive = food garden, large shrubs; flower and vegetable = food garden >450 m^2; woodland = showy evergreen trees, lawn, small and large shrub covers; shrubs with bush tree = non-showy, lawn; minimal-input exotic = non-showy, evergreen trees; non-input exotic = non-showy; neglected coastal = non-showy, herb cover, large shrubs; exotic shrubs = non-showy, lawn, small shrub cover. Adapted from Daniels and Kirkpatrick (2006).

of residents (Head and Muir, 2006). In suburban neighborhoods of Hobart, Tasmania (Australia), several types are common (Figure 9.12) (Daniels and Kirkpatrick, 2006):

1. Front space with prominent showy flowers, screening plants, and cover of small shrubs; back with productive food plants, lawn, dog, and chicken coop.
2. Front and back similar.

3. Front with showy flowers from intensive gardening; back with few productive food plants (especially in older neighborhoods).
4. Front garden somewhat showy; back with greater emphasis on showy flowers (especially in neighborhoods with high unemployment).
5. Front with showy garden; back with productive garden.

Two studies compared the front space (garden/yard) of a row house with that next door, across the street, down the block, and farther away. In a Montreal urban neighborhood, the front space of a row house mimicked that next door, and became progressively different with distance from house (Zmyslony and Gagnon, 2000). In contrast, in suburban Hobart with detached houses, no difference was found between the front garden type of a house and that of an adjacent or across the street house (Kirkpatrick *et al.*, 2009). Also, no effect of distance from house was present. The question relative to towns remains for research.

As mentioned above, native bird species richness correlates with the ratio of evergreen and deciduous trees in suburban yards, plus the percent of yards with trees or plants with fruits/berries (Figure 9.10) (Belaire *et al.*, 2014). Native bird diversity does not correlate with the presence of bird feeders, though bird abundance usually increases.

Many backyards contain cats or dogs, which may have negative effects on birds (Chapter 8) (Belaire *et al.*, 2014). Except for farmsteads, stables, and fields, backyards are normally the only places in town where farm animals, such as chickens and goats, may be found (e.g., Emporia, St. George's). Garden ponds, lawn, and walls (Chapters 6 and 7) are commonest in backyards, but may be present in front as well. Walls in front or back are often porous and discontinuous (Gilbert, 1991). Front yards along a street form a heterogeneous corridor, and backyards, with more microhabitats, typically form a more important ecological corridor in town (Rudd *et al.*, 2002).

Irrigating parts of a house plot with a hose or pipes is commonly done to increase plant growth. On a globe increasingly short of readily available freshwater, irrigating lawns and showy flowers is often wasteful. A three-year study of 134 suburban yards (median size 266 m^2) near Zaragoza (Spain) in a relatively dry climate found that the average 0.8 m^3 of irrigation water per day was 46 percent of the total household water use (Salvador *et al.*, 2011). Sixty percent of the households over-irrigated, 34 percent adequately irrigated, and 6 percent under-irrigated, relative to the standard "net irrigation requirements" there. Over-irrigation more than doubled the amount of water used annually, 1,350 mm, compared with the standard of 555 mm for the house plots sampled. If these results are representative for towns mostly covered by residences, yard irrigation is a major water use, and loss of water.

Calculating carbon, nitrogen, and phosphorus input-and-output budgets provides key insight into the most important human activities affecting flows. These element flows were calculated for 360 single-unit detached houses (households) in the suburbs of Minneapolis-St. Paul, Minnesota (USA) (Fissore *et al.*, 2011). For carbon, 85 percent of both consumption and emission is from aircraft and motor vehicle transportation plus home energy use (see Figure 5.10). Home energy use is near normally distributed across households. But one-fifth of the households produced 40 percent of vehicle emissions

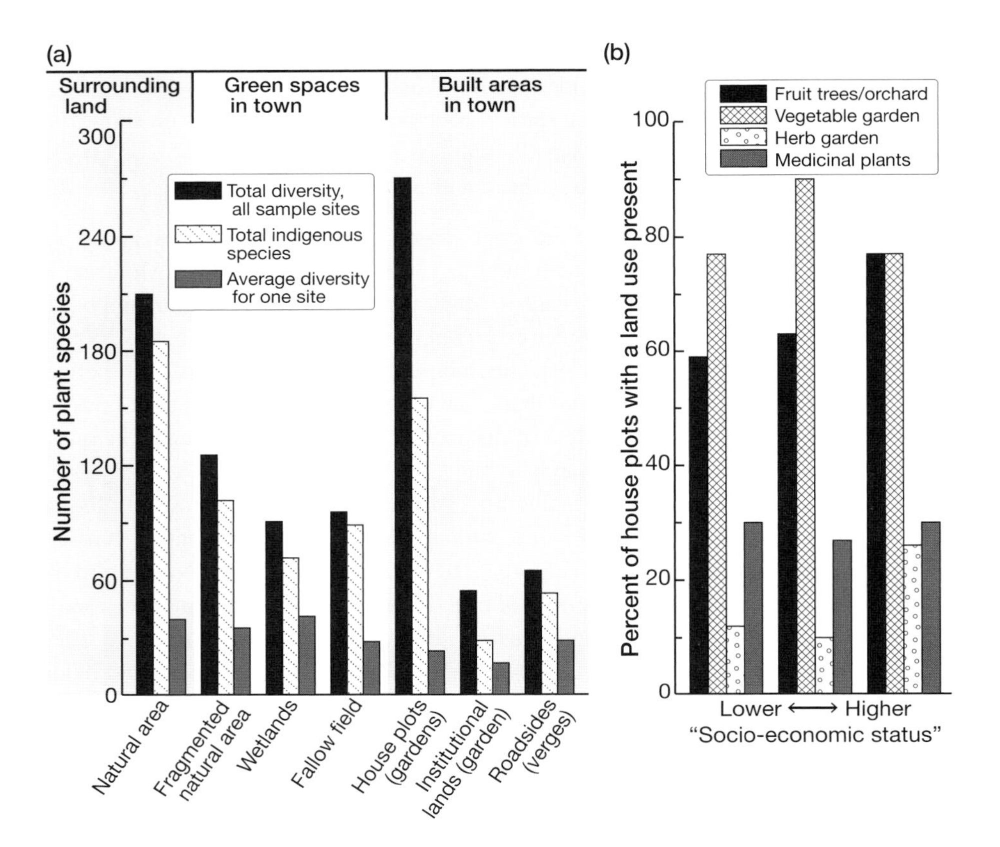

Figure 9.13 Plant diversity and productive uses of house plots. (a) Plants sampled at 100 points along five transects, including through many house plots. Total diversity = gamma diversity; one site diversity = alpha diversity. (b) Socioeconomic status determined by percent unemployment; three levels plotted have 50–56%, 21–50%, and 8–21%, respectively (also household size, number of rooms, and schooling status correlate with percent unemployment). Rural relatively low-income town or small city of Ganyesa (population about 40,000; mainly one ethnic group, Botswana), North-West Province, South Africa. Adapted from Cilliers *et al.* (2012).

and 75 percent of air-travel emissions. For nitrogen, the human diet and lawn fertilizer applications are about 65 percent of the input. Phosphorus inputs were mainly for human diet, detergent use, and pet food. Thus, a few variables contribute most of the C, N, and P fluxes. Also, for home energy use, a relatively small number of households are responsible for much of the energy consumption and emission.

House plots (house gardens) in and around the small city of Ganyesa (population 40,000; North-West Province, South Africa) are the hotspots of plant species richness for the region (Cilliers *et al.*, 2012). An individual house plot contains about the same number of plant species (alpha diversity) as the equivalent area in a natural area, or in a fragmented natural area, institutional land (garden), wetland, fallow field, or roadside (verge) (Figure 9.13). However, the total number of species in numerous house plots (gamma diversity) is more than double the number per equivalent area in every other habitat except a natural area, which is only slightly less diverse. A similar pattern exists for indigenous plant species, except that the natural area is richest. The diversity of non-native species is highest in house plots.

In addition to showy flower gardens and other uses in the Ganyesa house plots, productive vegetable gardens, fruit trees/orchard, herb gardens, and medicinal gardens are common (Molebatsi *et al.*, 2010; Cilliers *et al.*, 2012). Vegetable gardens and fruit trees/orchards are most common, medicinal gardens much less frequent, and herb gardens least common. Along a gradient of socioeconomic status, from low income to medium income, little difference exists in the relative frequency of these productive uses.

With such a species richness of plants in house plots, a high diversity of invertebrates is expected. More flower types and more flowers result in increased pollinator visits (Salisbury *et al.*, 2015). More pollinators visit native than nonnative species, though this changes as more nonnative plant species attracting pollinators flower later in the growing season. Soil animals also differ in house plots from those in nearby natural areas. For instance, in village homegardens in Sri Lanka, land snail species are a distinctive mix of native, nonnative, and nearby rainforest species (Raheem *et al.*, 2008).

Ecological Movements around Town

Business office centers or sites are characteristic of cities (Serret *et al.*, 2014) but also may be present in suburbs and large towns. A study of eight threatened butterfly species in The Netherlands focused on business sites with an average of about 10 percent low vegetation cover (e.g., lawn, fallow area, green roof) potentially suitable for butterflies (Snep *et al.*, 2011). Areas to support a local population of these species varied from 0.5 to 5 ha (1.25–12.5 acres), and dispersal distances varied from <1 to 1.5 km (0.6–0.9 mi). Considering the distance butterflies move between habitats, adding suitable habitat on the business sites might enhance about a quarter of the existing vulnerable populations of these threatened species. Furthermore, the added habitat could enhance metapopulations (Chapter 8) of many other butterfly species.

Of four groups of wildlife in The Netherlands (forest birds, small mammals, large mammals, and butterflies), butterflies are apparently least able to cross built areas (Knaapen *et al.*, 1992). Low herbaceous vegetation habitats are key sources and targets facilitating butterfly movement in towns.

All sites in town are subject to a *species rain*, that is the animals and plants that continually arrive by air or land (or water) mainly from surrounding areas. More than 3,000 species, many being invertebrates, were recorded in 15 years in a single suburban house plot in England (Owen, 1991), and more added in the next 15 years (Owen, 2012). Animals and plant seeds from outside flew, blew, or walked in during the period. Key chemicals, such as insect pheromones, PAHs (polycyclic aromatic hydrocarbons) from trees, and industrial air pollutants flow into and out of town sites. Bats and birds, cats and dogs, rats and bugs come and go at a site. Also, species range expansions, such as due to climate change, may add to a town's flora and fauna.

The amount and species composition of a species rain are strongly affected by adjacent land uses, and decrease with greater distance of species sources. Adjacent environmental hazards have a strong effect on a site (Greenberg and Schneider, 1996). In a dry climate, converting a front lawn to xeroscaping with stones and desert plants may kill adjacent street trees, which no longer receive subsurface water runoff from the irrigated lawn (William Pockman, 2016 personal communication).

Adjacency effects are also illustrated by the predation of bird nests in natural-area parks next to housing. A study of rural sites in Ohio (USA) found that the survival of nests decreased with the amount of nearby development, and strongly decreased with the number of diverse nest predators present (Rodewald et *al.*, 2011). A Colorado (USA) study found more pollinator bees and bee species in grassland if housing was nearby (Hinners *et al.*, 2012). Native desert birds in a residential area increase with proximity to natural areas, as well as with an increase in xeroscaped yards (Lerman and Warren, 2011). An inverse pattern is illustrated by housing developments degrading nearby protected natural areas.

To evaluate how important connectivity for wildlife movement is in a built area, researchers developed a data-based modeling approach for the European hedgehog (*Erinaceus europaeus*), a model species for ground-dwelling animals, in the city of Zurich (Switzerland) (Braaker *et al.*, 2014). The hedgehogs have overlapping home ranges of 10–40 ha (25–100 acres), foraging at night over 1,000–1,500 m (0.6–1.0 mi)) for a wide range of invertebrates, and roosting by day in diverse sites with dense ground-level cover. The model highlighted many connections between habitats facilitating movement, as well as "pinch points" inhibiting movement. Both habitat quality (81% importance) and habitat connectivity (19% relative importance) are keys to understanding wildlife movements in this built area. Overall, these results probably apply well to towns.

Consider six major functions of greenspace (open space) in towns (van Leeuwen, 2013):

1. Activities for production, e.g., farming or mining
2. Natural and cultural values.
3. Health, welfare, protection, and recreation benefits.
4. Public safety and hazard mitigation.
5. Infrastructure and nature corridors/networks.
6. Area for potential town expansion.

These functions highlight the town as a place for people to reside, work, and shop. Most of the greenspace functions also provide ecosystem services.

Distinct locations in town are relatively predictable sources or sinks for the movement of people. Analogously, many town sites and areas indicate animal movements, such as: (1) nighttime moths, beetles, and other flying insects between semi-natural areas and lighted town center; (2) rodents, cockroaches, and the like to and from food markets, restaurants, and dumpsters; (3) birds and other vertebrates moving away from the loud noise police/fire station area; (4) pest animals between semi-natural areas and residences; and (5) lizards and other small animals collected in yard waste (e.g., leaves, cuttings, and palm fronds), and carried from neighborhoods to leaves and brush dumping locations (e.g., Hanga Roa). Seemingly major flows and movement routes in town could be easily mapped using such an approach.

Lighting in the town center is the main source of light creating the sky dome of sky-glow (Chapter 8), visible for kilometers outside town. This central light source must attract night moths, beetles, flies, aphids, and many other flying insects to and into the town. Night birds and mammals may likewise be drawn to town.

Large old trees in town are usually rich in species of lichens, mosses, cavity-nesting vertebrates, beetles, caterpillars, and much more. But also the trees are attractants and hubs for many birds, insects, and other species moving across town. In savanna-like Napa Valley, California (USA), some 45,000 large old trees were mainly removed for modern agriculture by the 1940s (Grossinger, 2012). In 1940 about a thousand trees remained, mainly in pastures, along roads, by farmsteads, and as shade trees in towns. Perhaps only 300 stand tall today, providing unique habitat conditions and facilitating wildlife movement.

Residents can walk across a town (see book cover), unlike people in a city. Some ecological flows cross the whole town. Migrating birds, wind-blown dust, and clouds pass overhead. A stream corridor bisecting town has swimming fish, floating plankton, and silt and clay passing through, as well as an assortment of birds and mammals moving along the corridor. Somewhat similarly, a railway bisecting the town is probably followed by some flying insects and birds. Terrestrial vertebrates may move a long distance along the railway, perhaps especially at night. More noticeable are farm tractors and other equipment carrying lots of soil, small animals, seeds, microbes, and rich manure, always distributing some enroute through the town.

10 Cropland, Pastureland, and Towns

That verdant valley is too valuable for houses; it's sheep land!

Paraphrased from Jeremy Gould, conversation atop a hill in Somerset, England, 1978

Having explored ecological patterns of soil, water, air, plants, and wildlife in towns, we now broaden the perspective to landscapes around a town. This chapter introduces open cropland and pastureland, while Chapter 11 explores natural lands, i.e., forestland and aridland. Both chapters highlight how towns affect their surroundings, and inversely, how the surroundings affect towns. Chapter 12 then begins to tie towns, villages, transportation, and the land together.

Cropland and the Landscape

Evidence of agriculture appears in the Neolithic at the end of the Ice Age about 11,000 years go. Perhaps a millennium later, villages with dwellings dispersed over 1–2.5 ha (2.5–6.3 acres) had "domesticated" farming, growing grains, legumes, and fruit trees around their dwellings.

Many millennia later in medieval times, an English farm was sustained by growing similar crops with intensive labor, yet low crop yield (Figure 10.1) (Connor *et al.*, 2011). The very little soil nitrogen needed for a hectare of crop production required manure from several livestock-grazed hectares. Today modern maize (corn) production in the USA uses 17 times more nitrogen for a crop yield ten times higher.

Agricultural land surrounding a city usually includes many abandoned fields, little livestock-grazing land, some market gardening for urban use, polluted air and soil, and mostly absentee land owners (Lockeretz, 1987; Vail, 1987; Forman, 2008). In contrast, cropland around towns and villages has few abandoned fields, often considerable livestock-grazing land, no market gardening, cleaner air and soil, and few absentee landowners.

The primary goal of *farming* is to provide food for people, and secondarily to grow feed for farm animals, protect soil to sustain yields long-term, and protect the natural environment (Hart, 1992; Connor *et al.*, 2011; Foster *et al.*, 2017). The food produced is for local consumption in farmstead, village, and town, as well as for the rich and poor in urban areas. Feed is basically for farm animals in agricultural areas. Protecting the on-site farmland soil long-term generally means at least for the next generation. Protecting

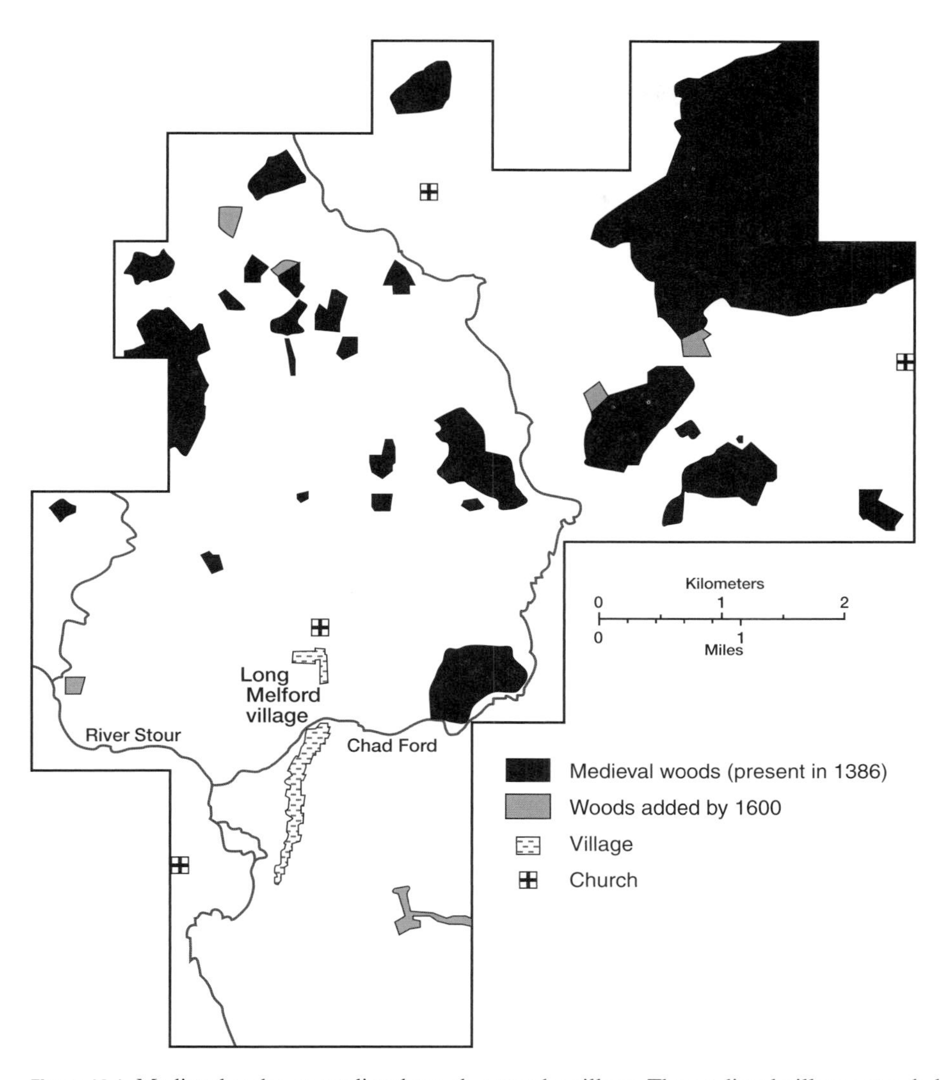

Figure 10.1 Medieval and post-medieval woods around a village. The medieval village extended 5 km (3 mi) along a north–south road (mostly northward from the 1600 village mapped). Woods mapped, including "ancient woods" and "parks," provided essential energy and many wood products for a village, so woods area limited population size. Adapted from Rackham (1976).

the natural environment is mainly local but also regional, focusing on vegetation, biodiversity, wildlife movement, surface and groundwater, diverse waterbodies, and habitat quality, size, and arrangement. Farming goals are strongly local.

Farmland of course is highly diverse depending on region, climate, soil, and distance from city. Also, agriculture continually changes, sometimes slowly, other times seemingly overnight, with technology, as well as developments in irrigation, fertilizer application, pesticide use, crop varieties, mixed cropping, and much more (Berry, 1977; Lewis, 1982; Warren *et al.*, 2008).

This section briefly highlights many important issues in farming related to: (1) the broad-scale landscape; (2) crops and biodiversity; and (3) soil and water. Many other issues, such as organic farming (mostly near urban areas) (Rundlof *et al.*, 2008; Gabriel *et al.*, 2009; Sarapakta *et al.*, 2012), agroforestry, fire use in cropland, stream quality, migrant labor effects, and greenhouse gases, could be explored. For each issue introduced, ecological dimensions are mentioned. Also, each issue introduced is tightly tied to local people, from town residents to farmers.

Cropland Heterogeneity. Cropland, while appearing monotonous at first glance, is normally highly heterogeneous. The landscape is covered with distinct different sized and shaped fields, often with different crops at different stages of growth or harvest (e.g., Chamusca, Plymouth; see Appendix). Hedgerows (or stonewalls) of different type (Thomas and White, 1980), hedgerow intersections, *field margins* (the meters-wide grassy strips where tractors turn at the end of rows; Warren *et al.*, 2008; Burchett and Burchett, 2011), *fallow flower-rich patches* (or strips) within fields (to encourage pollinators and predator insects against pests), farm roads, mud puddles, moist gullies, intermittent stream channels, isolated trees, tiny wooded patches, abandoned or fallow field, perhaps a small rock outcrop (Lindenmayer *et al.*, 2003), a wind turbine, and so forth each rewards careful observation.

Yet also the *farmstead* (farmhouse area) containing several building types (Thornbeck, 2012), diverse small yards, fences, equipment and machinery, trucks and cars, junk pile, farm dump, pond, radiating farm roads and tracks, cluster of trees, home, driveway, and perhaps a solar-panel array are all important for the local farmer to produce food. These heterogeneous objects represent different habitats and microhabitats supporting a diversity of plants and animals (Edvardsen *et al.*, 2010). Furthermore, several of the cropland features provide connectivity for movement of species in the landscape. In such a landscape in northeastern Spain, both avian diversity and rare species are correlated with the amount of cropland and of shrubland/grassland, but also with overall landscape heterogeneity/diversity (Santos *et al.*, 2008).

Agricultural Landscape and Farm Planning. Of the three rather distinct spatial levels of planning, field, farm, and broad landscape, most effort focuses on the field (Flohre *et al.*, 2011; Connor *et al.*, 2011). Yet ecological, agricultural, and economic planning for a whole farm, as well as a cropland landscape are equally valuable (Figure 10.2). Field and landscape are geometricized by numerous linear features. Both are familiar using landscape ecology (Forman, 1995; Warren *et al.*, 2008; Turner and Gardner, 2015). But only a limited ecological literature focuses on the intermediate scale, i.e., the farm (Lindenmayer *et al.*, 2003; Flohre *et al.*, 2011).

A study of more than 1,325 hay and dairy farms in northeastern Italy found that the larger the farm, the fewer species of plants, orthopterans (crickets, grasshoppers), and butterflies there are (Marini *et al.*, 2009). A study of 44 upland farms in northern England finds that bird species richness does not correlate with habitat quality or the surrounding spatial context of a farm (Dallimer *et al.*, 2009). Instead, avian diversity correlates with management activities on a farm, especially land tenure and labor inputs.

Figure 10.2 Grain elevator towers for temporary storage of locally grown grain until transported by rail to market. Town of Clay Center, Kansas, USA. See Appendix. R. Forman photo.

With a strong body of principles and extensive empirical studies, *landscape ecology* analyzes and explains patterns, processes, and changes of large areas (Forman, 1995; Turner and Gardner, 2015; Rego *et al.*, 2018). (The field replaced an earlier island-biogeography model [MacArthur and Wilson, 1967], in part because on land, diverse habitats around a patch are species sources affecting the patch, and differentially suitable for species movement in different directions, while the patch edge-to-interior ratio and arrangement of generalist edge species and specialist interior species strongly affect species richness patterns [Collinge, 2009; Forman, 2014; Laurance, 2008; Brady *et al.*, 2011].)

Mainstream principles in landscape ecology highlight the size, shape, and composition of patches; edges; corridors, networks, and connectivity; spatial arrangement; land mosaics; flows and movements relative to patterns; and changes over time. The discipline is especially useful for agricultural landscapes (Figure 10.3) (Warren *et al.*, 2008).

Landscape planning may be done by national government and local councils (UK; Vallez *et al.*, 2012), by villages (China; Rozelle *et al.*, 2005), or by non-governmental community groups (e.g., indigenous community in the tropics, and land-care groups in Australia; Scherr and McNeely 2007; Jones *et al.*, 2013). All the approaches emphasize the importance of spatial pattern and sustaining natural resources, in addition to growing food and sustaining local communities.

Natural Areas, Woods, Trees. Large trees in cropland are special microhabitats, e.g., for beetles, but also key stepping stones for bird movement across the landscape

Figure 10.3 Cropland with dispersed farmsteads and wooded corridors connected to some wooded patches. Upper New York State, USA. R. Forman photo.

(Lindenmayer *et al.*, 2003; Orlowski and Nowak, 2007; Dubois *et al.*, 2009). In Lower Silesia (southwestern Poland) 495 "champion" large trees were found in a 5,480 hectare (21 mi^2) agricultural landscape. Nearly half were along water-edge hedgerows (159) and in mid-field clumps (70). But 33 were in rural settlements and 110 along roads (avenues), with the remaining champions embellishing cemeteries, manor parks, and shelterbelts.

Non-forest woody vegetation may also be frequent in agricultural landscapes. For example, in central Bohemia (Czech Republic), over six decades (1950–2011) in an area with 12 villages, the non-forest woody vegetation decreased as follows (Demkova and Lipsky, 2013): 18 percent loss of point elements; 7 percent of patches; and 4 percent of linear elements. Ditches as linear elements extending outward from a woods in The Netherlands have a progressively decreasing density of non-grass forb ("flower") species in the first 75 m (246 ft) (Kohler *et al.*, 2008). Hoverfly (Syrphidae) pollinators decrease over the first 125 m along a ditch. Hoverflies also decrease over the first 50 m from a flower-rich patch within a crop field. Finally, few wetlands remain in croplands after filling or draining by local farmers. Thus, the ecological effects of a natural element extend well outward into cropland.

Woods (woodlots) tend to be smaller and more abundant close to a town, where they serve as stepping stones for many species entering and leaving town. A large woods provides a "rain" of species entering town (Forman, 2014). In a tropical study, more seeds (per m^2) are dropped by birds in a large woods than in adjoining and more distant farmland (Cole *et al.*, 2010). In European agriculture landscapes, the species richness of plants, birds, and five arthropod groups studied increases with the total area of semi-natural vegetation, though both plant and bird diversity decrease with more intensive fertilizer use (Billeter *et al.*, 2008). In Portugal farmland, different groups of birds (including farmland birds of conservation concern) decrease with distance from woods (Forman, 1995, 2014; Reino *et al.*, 2009). This holds whether the woods is oak-, pine-, or eucalypt-dominated, and whether the woods edge is hard or soft (Chapter 1). Many species leaving a town head in the direction of a nearby woods or other semi-natural area. Woods near a town or village may also be important for providing wood products for local workshops and home handicrafts (see Figure 11.7) (Thomas and White, 1980; Peterken, 1993; Burchett and Burchett, 2011).

Public Uses in Croplands. Community-based agricultural conservation planning or land-care planning often designates small areas in farmland as open to the public, at least local people, for various uses (Scherr and McNeely, 2007). Another approach, perhaps best illustrated in the UK, is to have a mapped and maintained walking trail network across farmland that connects villages and towns (see Figure 12.8) (Burchett and Burchett, 2011). In some areas town residents cut and collect firewood in local woods (see Figure 5.6). Hunting by local residents in both field and woods is often widespread. Fishing in local waterbodies is common. Also, to help reduce poverty in a local rural area, diverse plants, wildlife, and other resources in farmland are sometimes harvested in small amounts (Leisher *et al.*, 2013). Motorbikes and horse riders from town may regularly use a nearby trail network, and off-road vehicles (ORVs, ATVs) may have a distinct nearby degraded spot for a small network of intensively used trails (see Figure 4.11).

Near European cities apparently are a few dozen *agricultural parks*, varying in size from about 4 to 480 hectares (10–1,200 acres), primarily used for recreation and secondarily for growing food. The irrigated Parque Baix Llobregat and the Espai Rural Gallecs, both near Barcelona (Spain), and the Parco Agricolo Sud near Milan (Italy) are relatively successful examples.

Yet farmers have problems growing crops near cities (Bouraoui, 2005; Vaz *et al.*, 2013a). On 18 farms averaging 200 ha (500 acres) each on the Saclay Plateau outside Paris, farmers have: (1) difficulty in moving equipment around; (2) vagrants encamped on fields; (3) confrontations with local residents about the noise and odors of farm operations; and increasingly (4) abundant flocks of pigeons (as the city traps pigeons and lets them loose in the farmland).

Or, outside Tunis (Tunisia) on the Sijoumi Plain, with 300 farms averaging 10 ha (25 acres) each, farmers are constrained by: (1) pilfering or theft of crops; (2) "trampling" of crops by vehicles and motorbikes; (3) dumping of garbage on fields near roads; and (4) animal depredation by neighbors' cows and wandering sheep and goats. Some of these urban-related problems may occur around towns. Probably in most nations, travelers through farmland simply find a pleasant spot and spread out for lunch or rest, and then continue onward without noticeably disturbing farm operations or farmland.

Cropland fires associated with production and harvest especially occur in extensive field areas growing rice, maize (corn), wheat, tuber/root crops (excluding potato), and other cereals (Lin *et al.*, 2012). Per hectare, fire frequency is highest in cotton, tuber/root crop, oil palm, peanut, and sugarcane fields. Most cropland fires are by farmers returning nutrients to the soil. But in semi-natural vegetation, wildfires are overwhelmingly caused by local people, e.g., tossing cigarettes. Fires near towns with a fire station/company tend to be small. Irrespective, smoke from fires pollutes the air in nearby villages and towns.

Slow and Sudden Change. Towns and villages are extremely sensitive to *disasters*, such as hurricanes/cyclones, wildfires, locust plagues, and earthquakes, which often cover large areas. But, unlike urban areas, these small population centers are especially affected by tornados, lava flows, and landslides which usually damage or devastate small areas. All of these disasters are relatively sudden. Somewhat slower disasters occur, such as a conquering army or the late nineteenth-century spread of a fungus (phylloxera) decimating vineyards and towns in France (Stevenson, 1980). Generally, towns have little resistance against such disasters, but may have considerable resilience to bounce back.

Long-term change, such as town population growth or decline, also transforms local areas. Farm work evolved over time with changes in energy use and technology, i.e., from mainly human-muscle labor to farm animal to powered machine (Connor *et al.*, 2011). Both rapid and slow changes in agricultural policy, competition, and technology can alter cropland agriculture, and hence local farmers, villages, and towns (Warren *et al.*, 2008). Farmers and local residents always have chronic concerns or worries. In a 1930s Chinese village, residents constantly worried about, and tried to develop mitigations for, upcoming floods, droughts, locust plagues, the threat of new machines, and especially industrial reform that would put the town's silk factory out of business (Chapter 4) (Fei, 1939).

Population growth or decline in a village or town is conspicuous and indicative of the future. Outward *town expansion* (townification or townization, analogous to urbanization which relates to city) occurs in four ways (Chapter 3) (Forman 2008, 2014): (1) as bulges, e.g., by a commercial retail center, industrial center, more housing, or a nearby entrance/exit of a bypass highway (house-by-house concentric growth around the perimeter seems unrealistic); (2) in surrounding villages; (3) as strip/ribbon development along a main radiating road; or (4) as scattered houses or sprawl.

Highly diverse ecological changes occur along with such local human changes. For instance, in northwestern France over 55 years (1947–2003), beetles changed as tree species, microhabitat density, and broad land use changed (Dubois *et al.*, 2009). Over three decades (1970–2000) in France, weed species in field corners declined, but declined much less in field edges (Fried *et al.*, 2009). Also, in Finland, food webs of mainly invertebrates in cropland weeds changed markedly over three decades (Hyvonen and Huuseh-Veistola, 2008). In effect, most towns and villages seem timeless, yet are constantly changing along with the surrounding cropland and its ecology.

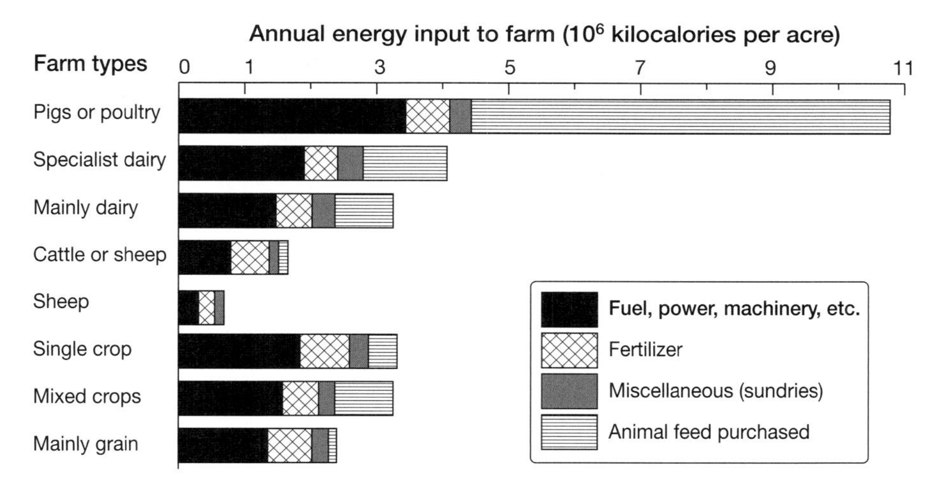

Figure 10.4 Energy input to different types of UK livestock and crop farms. Adapted from Tivy (1990).

Biodiversity, Crops, Soil, and Water

Biodiversity and Crops

Agricultural Intensification. In cropland, intensification generally involves the heavy use of fertilizers, pesticides, and fossil fuel, plus intensive plowing (tillage) of the land, to increase crop yield (Figure 10.4) (e.g., Ovcary, Clay Center) (Tivy, 1990; Connor *et al.*, 2011). In Europe, a range of "agri-environment schemes," which reduce these techniques and add others, has evolved with the basic goal of protecting and restoring nature while maintaining food production levels (Warren *et al.*, 2008; Burchett and Burchett, 2011).

Agricultural intensification in Australia involves farms with mixed crops switching to single crops (McKenzie, 2014). In Iowa (Midwestern USA) over 65 years (1937–2002), field sizes have increased and field number decreased, while field shape remained unchanged (Brown and Schulte, 2011). Meanwhile, ecological studies of agricultural intensification show a decrease in plant and avian diversity, though apparently not in the diversity of carabid beetles which are often predators on pest insects (Flohre *et al.*, 2011).

Many herbivorous insects and other species, as targets of pesticides, have doubtless genetically adapted to the repeated applications and high levels of pesticides used. Those target species that persist as pests must be more resistant to the pesticides. An extreme case of intensive agriculture is the high-technology food production under glass in large flat buildings in The Netherlands. Abundant heat, light, water, and fertilizer are used to grow food year-round. Lights are on all night, producing an eerie sight when flying over the areas before dawn.

A key alternative to intensification is *agricultural diversification* (e.g., Chamusca, Plymouth) (Connor *et al.*, 2011; Collins *et al.*, 2018), which can take many forms. "Subsistence farming" of many crops using manure, plus mainly the farmer's labor, is one approach (Ho, 2005). "Organic farming" that avoids or minimizes the use of pesticides and fertilizers is another. "Low-input farming" uses no or minimal fossil fuels, as done by the Amish farmers in Pennsylvania (USA) (Connor *et al.*, 2011). Various agro-forestry designs may be low-intensive. Overall, the presence of wide hedgerows, wide grassy field margins, scattered trees in fields, fallow/old fields (Etienne *et al.*, 1998), meadows (Peterken, 2013), and wetland spots indicate low-intensive agriculture. Small fields, which are often close to towns (e.g., Tomar, Emporia), diverse crops, and topsoil protection methods are also indicators (Fahrig *et al.*, 2015). Nevertheless, no ideal cropland farming has emerged to combine high food production and rich nature.

While farmers and farms decrease in number as large agriculture-intensified operations spread, a small but forward-thinking, educated, and sometimes technology-savvy group of young adults wanting to farm increases. Several actions may help achieve this goal, such as: (1) government programs with seed money; (2) efforts to reverse the overall rural "emptying out" process, including second jobs for farm families; (3) making suitable farming land available, rather than the worst land; (4) incentives for more families to support local shops; (5) improved-quality town school welcoming new children; and (6) arranging for some local farmers to initially encourage, teach, and aid new farmers.

Wildlife and Biodiversity in Fields. The species richness of plants, birds, and five arthropod groups all increased with the amount of semi-natural habitat scattered over an agricultural landscape (Billeter *et al.*, 2008; Gillies et *al.*, 2011; Fahrig, 2013). Each of the groups responded somewhat differently, and no single group is a good indicator of the overall pattern. The diversity of birds and carabid beetles, both related to habitat diversity, is higher (per unit area) at the whole farm scale, than in either the broader landscape or finer field scale (Flohre *et al.*, 2011). At the field scale, one may focus in on several microhabitats, i.e., the: (1) bordering hedgerow/fencerow; (2) hedgerow intersection; (3) tiny corner patch (usually missed by a tractor with equipment); (4) *field margin* (the edge grassy strip where a tractor turns at the end of rows) (Smith *et al.*, 2008; Douglas *et al.*, 2009; Burchett and Burchett, 2011); (5) central field area; (6) fallow flower-rich patch or strip within a field; and (7) wet-soil strips or patches.

Hedgerows near farmsteads are often unusual by containing food and medicinal plants, as well as a richness of birds, butterflies, and other insects (Thomas and White, 1980). Hedgerow intersections are somewhat different habitats, as well as a field's small corner patch of flower-rich vegetation. Hedgerows often contain many fruit/berry-producing plants, resulting from fruit-eating birds landing and dropping seeds. In France, weed species number declined in hedgerows over three decades (1970–2000), but changed little in field margins (Fried *et al.*, 2009). Yet in Finland, food webs associated with weeds changed in several ways over 30 years (Hyvonen and Huusela-Veistola, 2008). Earthworm density and diversity is higher in field margins than in field centers (Nieminen *et al.*, 2011).

The visual or perceptual range of an animal often helps determine the routes it takes in movement (Saunders *et al.*, 1987). Even around a farmstead, for example, rats and raccoons (*Rattus* sp., *Procyon lotor*) move relative to the arrangement of buildings (Lambert *et al.*, 2008; Beasley *et al.*, 2011).

Indeed, near farm buildings in northeastern England, habitat modification that reduces plant cover and habitat complexity decreases the survival rate and population size of rats (Figure 10.5). But altering field margins has no effect on rat populations in cropland, probably because rat home ranges are larger in fields. Rare microhabitats and clusters of microhabitats, both expected around farmsteads, indicate high species richness. More broadly, wildlife habitat and breeding areas are strongly affected by the actions of local people, irrespective of the protection status of the habitats (Daniels et *al.*, 2007).

Crop Varieties and Ecology. Three-quarters of the world's food comes from a dozen plants, and over a century their genetic variation has dropped sharply. Crop species were domesticated millennia ago and have evolved with environmental change ever since (Gepts *et al.*, 2012). Today crops grown by local farmers typically contain considerable genetic variation attuned to the region's variable environmental conditions. Such *crop genetic variation*, or *agro-biodiversity*, is especially important in subsistence farming and rural poverty alleviation (Berry, 1977; Donna, 2013; Mortimore, 2013). In addition, biodiversity of the surrounding natural species seems to be higher around fields with greater crop variation (Mortimore, 2013).

Genetically modified crops (GMCs) have been developed mostly for high yield and economic efficiency (Raybould and Gray, 1993; Thrupp, 2000). Genetic strains or cultivars are targeted to specific crop interactions, such as resistance to diseases and pests, stronger plant structure against wind, less herbicide or insecticide use, and less groundwater or surface water use. After two decades, about 90–95 percent of soybeans and maize (corn) in the USA is genetically engineered. In contrast, very little GMC exists in Europe. Europeans remember the late nineteenth-century Irish famine, when monocultures of one potato strain were wiped out by a fungus which turned potatoes to mush, and caused people to migrate.

Not surprisingly, problems with genetically modified crops have arisen. The crops are less adaptable to seasonal weather variations, and often do poorly in longer growing seasons, such as global climate change promises. In agricultural landscapes of the USA, one or more pesticide compounds now appear nearly year-round in almost all streams. Yet apparently after two decades of GMC use in the USA, crop yield has increased little and pesticide use has decreased little, compared to gains in conventional crops (Hakim, 2016).

Furthermore, ecological problems have become more prominent. GMCs may hybridize via pollinators with nearby natural plants, decreasing the plant's genetic variation, increasing the probability of invasiveness, and altering fitness. Some local farmers resist the national GMC push, saving and using diverse seed varieties of a crop to spread risk against drought, windstorm, heavy rain, and disease or pest outbreaks (Connor *et al.*, 2011), all expected to worsen with climate change.

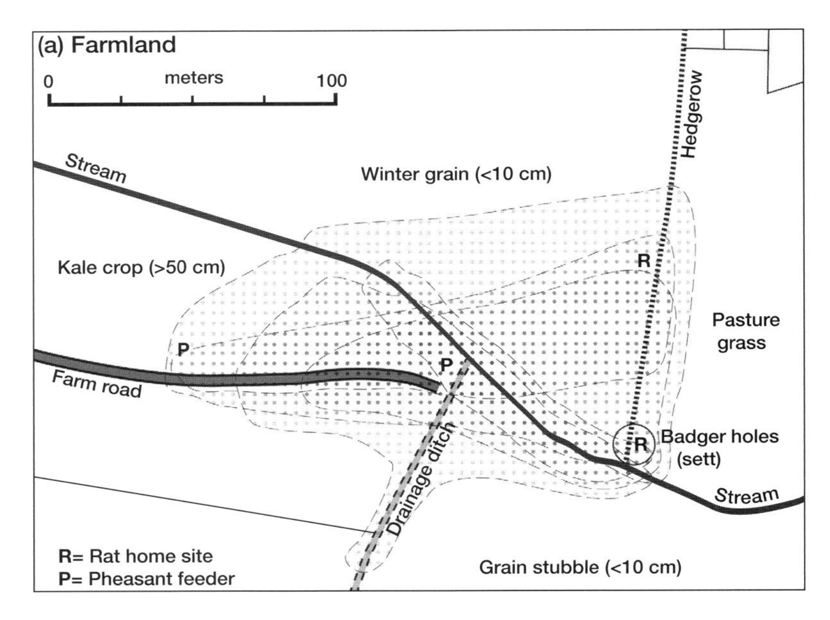

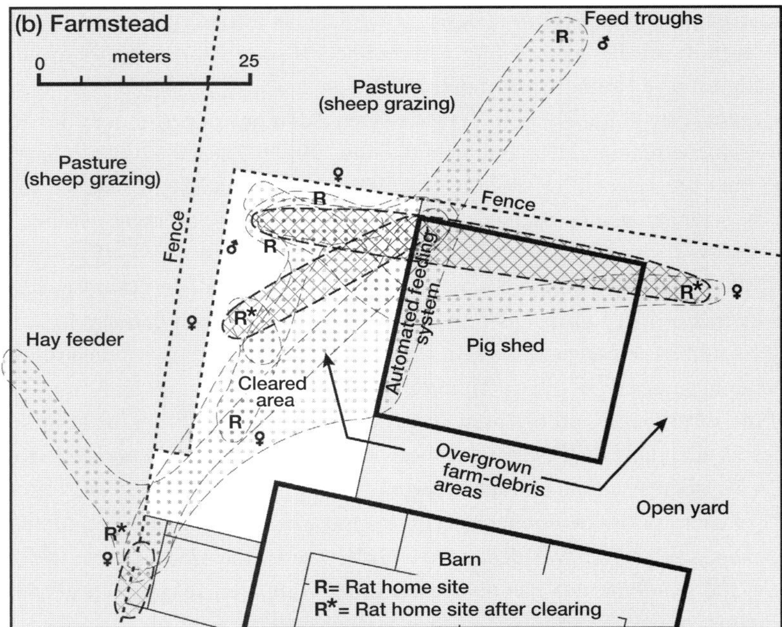

Figure 10.5 Rat home ranges in farmland and farmstead. (a) Four males and one female radio-tagged. (b) Two males and five females radio-tagged. Overgrown farm debris areas have discarded equipment, pallets, etc. with nettle (*Urtica*) and elder (*Sambucus*) probably about 1–3 m high. White area later cleared. See text. Adapted from Lambert *et al.* (2008).

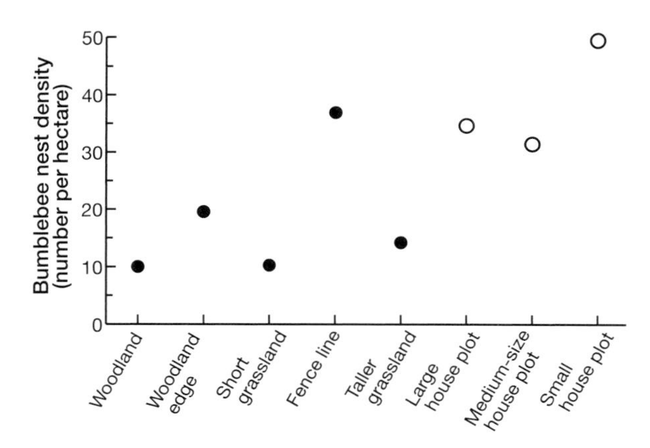

Figure 10.6 Nest sites of bumblebees (*Bombus*) in diverse countryside. 1 hectare = 2.5 acres; short grassland = <10 cm high; taller grassland = >10 cm; medium-size house plot = 100–450 m^2 (1 m = 3.3 ft); house plot = (domestic) garden in the UK. Based on the *National Bumblebee Nest Survey of 2004* (Osborne *et al.*, 2008; Peterken, 2013).

Pollination. Pollination by insects and a few flying vertebrates is essential for producing the seeds and fruits of most crops, and is therefore an important ecosystem service for society (Gemmill-Herren *et al.*, 2007). Major threats to ongoing high crop production seem to be insecticides and the loss of semi-natural habitats (hedgerows, flower-rich field margins, woodland edges) close to a crop field. Bumblebees in the UK nest abundantly in house plots/gardens (36 nests/hectare; 14/acre), especially in small house plots (Figure 10.6). Bee nests are common in linear countryside habitats (fencelines, hedgerows, woodland edges) (20–37 nests/ha) (Osborne *et al.*, 2008), and less common in non-linear countryside grassland and woodland habitats (11–15 nests/ha). In town, pollination is especially important in house plots for flower gardens, vegetable gardens, and fruit trees, as well as for town flower gardens and community vegetable gardens. But bees of many types (Hymenoptera: Apiformes) in town also move outward enhancing pollination in surrounding intensively managed crops (Samnegarel *et al.*, 2011).

In the southeastern Netherlands, pollinating hoverflies (Diptera: Syrphidae) in crop fields are more abundant within 125 m of wooded nature reserves, and within 50 m of flower-rich patches in a field (Kohler *et al.*, 2008). No such effect was recorded for bees. In the eastern USA, "complementary habitat use," in this case of a pollinator using semi-natural woody vegetation, old field (fallow) vegetation, and crops, is characteristic for bees (Mandelik *et al.*, 2012). Fifty-four wild bee species were recorded in 29 New Jersey/Pennsylvania (USA) farms in landscapes with extensive woodland (Winfree *et al.*, 2008). Three of the four primary crops grown are mainly visited by wild bees rather than nonnative honeybees (*Apis mellifera*).

In the New Jersey/Pennsylvania farms, during the early growing season most bee species use fallow flower areas within a crop field, then mainly switch mid-season to the mass-flowering crop, and in late-season move to old-field vegetation. Little use of natural forest is evident. An analogous pattern appears in oil rapeseed (*Brassica nagus*) cropland near Gottingen (Germany), where in the early season bees are especially abundant in hedgerows, mid-season in the mass-flowering crop, and late season again in hedgerows (Kovacs-Hostyanszki *et al.*, 2013). Nearby forest edges appear to be important for bee nesting throughout the season. In essence, local land patterns created by town and surrounding farmers strongly affect the natural pollination of crops, and therefore food production.

Mixed Cropping. *Mixed cropping* (multiple cropping) refers to the spatial arrangement of plants and rows within a crop field, the interspersion of different crops in neighboring fields, and the rotation of crops in a field (Tivy, 1990). Agricultural intensification reduces all three patterns. For instance, in recent years and decades, mixed-crop farms have been widely converted to single-crop farms in Australia (McKenzie, 2014), and even the number of weed species within fields in France has declined (Fried *et al.*, 2009). In contrast, for instance, herbivory on crop plants may be reduced by also planting species that: (1) provide habitat for plant pest-consuming predators, such as carabid beetles (Carabidae); (2) repel herbivores (familiar to home gardeners); or (3) attract herbivores away from the crop plants (Letourneau *et al.*, 2011).

Different crops thrive with different inputs of energy, fertilizers, and pesticides, and mixed cropping uses several approaches (Tivy, 1990). Indeed, alternating beans and grain from year to year in a field is beneficial for sustained productive yields.

Spatial patterns within a field are often varied according to the distance between plants in a row or the distance between rows (Connor *et al.*, 2011). The amount and arrangement of field margin or edge, width of grassy swales for stormwater drainage, and flower-rich patches or strips within the crop area are important variables affecting yield. The timing of plantings, various treatments, and harvesting relative to weather variation of course is important (Connor *et al.*, 2011). But also the frequency and sequence of particular crops planted within a field affects yield, as well as sustainable soil protection or enhancement (Warren *et al.*, 2008).

All of these mixed-crop activities are more effective in proximity to local farming communities. Indeed, in moist tropical agro-forestry, where both woody plants and herbaceous food crops are intermixed, and all species grow rapidly, management intensity is high and living close by is crucial (Scherr and McNeely, 2007). Mixed cropping decreases with distance from town or village.

Soil and Water

Caring for the soil is the key both to sustaining crop yields and to effects on downslope/downstream conditions. Without soil management, soil nutrients and fertility drop, critical soil organic matter shrinks, soil moisture dries up, topsoil erodes away, and/or salinity builds up (Connor *et al.*, 2011). Also, downslope sediment piles up and flooding worsens, or soil desiccates and is blown away (Figure 10.7).

Ditches, mainly covered with plants common elsewhere, are all over cropland. These ditches (Herzon and Helenius, 2008; Burchett and Burchett, 2011): (1) provide wet habitats for many aquatic and terrestrial species; (2) facilitate connectivity for movement

Figure 10.7 Wind-eroded sand accumulation around farm equipment at edge of field. Farm windmill for pumping water from groundwater to the surface. Midwestern USA. US Department of Agriculture, Soil Conservation Service photo.

across the field and farm; and (3) regulate water flow and nutrient retention, which are probably affected by the composition and structure of ditch vegetation.

Water management in cropland is gradually changing from a focus on agriculture to agriculture-and-environment (Smith and Pritchard, 2014). The linearization or geometrization of cropland, with ditches, fencerows, hedgerows, woods edges, canals, irrigation pipes/channels, and farm roads, accelerates water flows. That in turn carries away soil organic matter and nutrients, causes erosion and sedimentation, and increases flooding.

Mineral nutrient cycles of, e.g., nitrogen, phosphorus, and sulfur, show that some nutrients cycle within a field, whereas nutrient input-output budgets emphasize that most nutrient flows enter and leave a field ecosystem (Tivy, 1990). A "rain" of nitrogen from upwind soil erosion, industry, or transportation fertilizes most fields.

But each crop in each location typically requires considerable added fertilizer, either as manure or from bags, usually of nitrogen, phosphorus, and potassium (Connor *et al.*, 2011). Manure (and wet slurry) has the great advantage of containing nutrients as well as considerable organic matter, which holds the nutrients, slowly dispersing them (Tivy, 1990). In contrast, dry bagged fertilizer can be quickly washed or blown away by heavy rain or wind. Still, manure, slurry, and other farm wastes can accumulate in soil causing significant pollution effects (Tivy, 1990; Warren *et al.*, 2008).

Irrigation by canal/ditch surface flow, sprinkler (spray) pipes, or various micro-irrigation techniques is a major reason for increased crop yields worldwide, but also a major user of increasingly scarce freshwater (e.g., Pinedale, Casablanca, Tomar) (Tivy, 1990; Scherr and McNeely, 2007; Connor *et al.*, 2011). In warm regions, river floodplains with a high water table are widely irrigated for highly productive rice and other crops (Tivy, 1990), commonly in small fields (e.g., Socorro).

When and how much to irrigate depends in part on weather variations. Irrigation water is continually adjusted to: (1) avoid having crop areas with too little water (desiccation) or too much water (local flooding); (2) minimize soil erosion or salt buildup (salinization); and (3) save a limited supply of water (Tivy, 1990; Connor *et al.*, 2011).

Dryland crop production, generally in areas with less than 630 mm (25 in) of annual rainfall, especially focuses on conserving soil moisture, preventing a drop in water table, and minimizing salt buildup on the surface, all requiring careful management (Tivy, 1990; Whitford, 2002; Connor *et al.*, 2011). Attuning irrigation to the many, sometimes conflicting, goals and variables requires constant attention and maintenance, plus mainly local management (Tivy, 1990). Irrigation cannot be effectively managed from afar.

Low-impact farming, such as fine-scale mixed cropping or organic farming, and "response farming" carefully attuned to weather variations, can achieve both yield and environmental goals with local management (Warren *et al.*, 2008; Connor *et al.*, 2011). Overall, sustaining crop yield means minimizing erosion, maintaining soil organic matter and nutrients, preventing a drop in water table, preventing waterlogging and flooding, and minimizing salinization. Achieving long-term sustained yield, while minimizing environmental degradation, requires diverse micro-management by local residents. Irrigation and low-impact farming work well close to a town or village.

Finally, *local food* refers to the food produced and eaten in a town *foodshed*, i.e., the area surrounding a town which produces or could produce a significant portion of the food consumed in town (Berry, 1977; Thayer, 2003; Dallimer *et al.*, 2009). Local food is not a panacea (Desrochers and Lusk, 2016), but does greatly reduce transportation costs between crop field and food market. In addition, it strengthens the local economy and social interactions. Even a weekly farmers' market sells some locally grown products locally, and provides important social interactions between farmers and town residents that enhance life both in and surrounding a town.

Grassland, Livestock, and Grazing

Grasslands cover 25 percent of the global land surface, and virtually all grassland is grazed by livestock for human food (Neely and Hatfield, 2007; Gibson, 2009; Burchett and Burchett, 2011). The most abundant animals in order are: cattle (1.34 billion), sheep, pigs, and goats. Nearly 80 percent of the food-producing land in the USA is devoted to producing beef (i.e., cattle grazing land and cattle-feed cropland)

Figure 10.8 Texas longhorn steer, a productive breed in dry grassland plains. Transportation and industry may each affect water, soil, and atmosphere more than do cattle. Oklahoma. US Department of the Interior, Bureau of Sport Fisheries & Wildlife photo.

(Figure 10.8) (Neely and Hatfield, 2007). Other livestock require another 13 percent of the food-producing land.

As both global population and meat-eating increase, about 1.5 million hectares of forest is annually converted to livestock-grazing land. Yet worldwide, cattle production already apparently results in almost as much greenhouse gas (CO_2 and methane CH_4) effect as transportation or industry. Worse though, cattle often degrade grasslands, stream/river banks, and water quality of groundwater, rivers, and streams.

Grassland with livestock describes pastureland, rangeland, ranchland, and paddockland. Overall, grassland has less precipitation and is drier than both woodland and cropland. Some grasslands are too cold and some have unsuitable soil for efficient crop growing. During dry and/or cold periods, supplemental food in the form of hay or grain is commonly provided by farmers for the animals.

Several types of livestock farming, such as dairy cows, pigs, and chickens require intense work and care by the farmer, and therefore mainly occur close to towns and villages. Many of the principles of livestock farming are similar to those for crop farming. Thus, in the present chapter we focus on the distinctive dimensions of livestock farming.

Traditional crop–livestock systems are based on the animals moving and finding unplanted wild grass and other plants to graze (mobile pastoralism) (Neely and Hatfield, 2007). Small- to medium-size herds are typically managed by a small-landholder farmer, sometimes in unfenced land.

In contrast, *commercial crop–livestock systems* may use a *feedlot*, a small fenced area with numerous animals packed together. Huge amounts of feed grain requiring extensive animal-feed cropland (e.g., maize/corn) are provided, and massive manure piles removed. Feedlots have high inputs of feed and water use, plus high outputs of nitrogen (which pollutes surface and groundwater), manure, odors, and greenhouse gases (CO_2 and CH_4). Packing animals together facilitates disease transmission. Intensive management requires being close to a town or village, whereas the strong feedlot odor means being distant.

Traditional livestock farming requires water points (water sources) dispersed over the pastureland/rangeland and accessible to the livestock (Tivy, 1990; Neely and Hatfield, 2007). Livestock vary in the maximum period they can survive without water: e.g., cattle 0.4 days, sheep about 4 days, and camels 12 days. The distances animals (95% of a population) range from a water point also vary: sheep 3 km (1.9 mi); cattle 6 km; and camels 12 km (Saunders *et al.*, 1990; Tivy, 1990; Fensham and Fairfax, 2008). These numbers then indicate the area around a water point which is severely degraded by livestock.

A fluctuating, rather than constant, herd size is better for pastureland conditions, soil moisture, and biodiversity (Savory, 1999). Many ranches in Australia and the USA have gradually increased livestock (stock) density, while also increasing plant cover, perennial herb (forb) cover, species diversity, water quality, avian diversity, and wooded corridors for wildlife movement (Stinner *et al.*, 1997; Forge and Reid, 2006; Neely and Hatfield, 2007; Burchett and Burchett, 2011).

Rather than attempting to fence out or remove native herbivores, the trend is to have modest wildlife numbers in livestock pastureland (Figure 10.9). Indeed, many innovative approaches in livestock farming have noticeable environmental benefits, such as (Neely and Hatfield, 2007): (1) livestock in silvo-pastoral systems (trees and grass); (2) integrated crop-and-livestock systems; (3) including beekeeping; (4) harvesting wildlife (bushmeat) on pastureland; (5) mixed livestock and no-till crop farming; and (6) livestock with novel industrial technology for food and/or waste.

Managing the animals, the grassland, and the farmstead of farm buildings and home is an intensive year-round activity. For instance, chicks, lambs, and suckling pigs grow best at 29–34°C (85–93°F), so in cool climes their care is typically indoors (Tivy, 1990). Hens and calves grow optimally at 18–25°C, cows and pigs at 0–15°C, and sheep at –3 to 20°C, the best in cold climes. From birth to harvest, chickens and early lambs require 60–70 days, rabbits 60–98, pigs (pork) 115, and veal calves 131 days.

Active livestock farms may exist in the town edge, in the outside adjacent zone, or isolated in surrounding grassland. The proximity and interactions of town, grassland, and livestock are of prime interest in the following section. Thus, we explore: (1) grassland features near town and farmstead; and (2) isolated farmsteads linked to town.

Figure 10.9 Native pronghorn antelope in short-grass plain. Sevilleta National Wildlife Refuge, New Mexico, USA. R. Forman photo.

Grassland Features near Town and Farmstead

A preliminary list of 64 visible features or objects in grassland near town, village, and farmstead highlighted how heterogeneous the landscape is (Collins *et al.*, 2018). Many of the objects are most frequent near a farmstead, whether isolated, in a village, or by a town edge. Not surprisingly, most objects directly related to people and farming processes are more abundant near a farmstead. These objects are built, maintained, or regularly used by the farmer, farm workers, or family. About half of the water-related objects are most common near the farmstead, perhaps because the farmstead was located near water and every day livestock need water. Somewhat fewer objects related to livestock and wildlife, pasture and herbaceous cover, and woods and trees are mainly near the farmstead. Topography and soil objects are least related spatially to a farmstead.

Most of the grassland and livestock objects pinpointed require frequent or occasional checking and maintenance. Even the most distant property boundaries, fences, water points, and "lost" animals require trips by farmer (or farm worker). Farmers in town-edge farmsteads have essentially the same nearby objects and distant objects to be periodically checked as farmers in isolated farmsteads.

Individual objects of course are only pieces of a functioning system, actually of many systems. The following illustrate the diversity of strongly spatial systems operating on grasslands with livestock and farmstead. Some, such as a drainage basin (catchment), stream/river network, and farm road/track network, are familiar. Yet most of the systems are little studied on grassland with livestock, and represent promising research frontiers: (1) wooded patch-corridor network for wildlife; (2) grassy habitat connectivity for wildlife; (3) fenceline and hedgerow network for livestock movement and wildlife; (4) water points arrangement for livestock and wildlife; (5) habitat diversity and arrangement; (6) ecology of alternative livestock field rotations; (7) and flows among a collection of farmsteads. Each subject could be tentatively estimated or mapped, providing insight into how a grassland system with livestock, wildlife, and farmsteads works.

Isolated Farmstead Linked to Town

Features or Objects in Farmsteads. Livestock farmsteads have a barn, typically that holds some animals, some hay or feed for them, and provides some diverse work space (Thornbeck, 2012). Other structures include: a building for the temporary storage of feed grain; stacked hay in the barn or a separate building or under cover, mainly for winter or dry period use; one or more fenced corrals for farm animals; sheds for equipment, storage, and work areas; and horse stables in the barn or a separate building. Some farmsteads have a small windmill pumping water from the groundwater to tanks, perhaps in a small well-house. Various tanks or bins especially for animal feed or fertilizer are common. An array of solar energy collectors is presently uncommon, though increasing.

Specialized buildings for different livestock and farming operations are also common (Tivy, 1990; Thornbeck, 2012), such as a: vertical cylindrical silo with silage for dairy cows; long building with milking stalls for dairy cows; low building with feeding trough for hogs; low long building, perhaps with two or three levels, for chickens; well-aerated low building for sheep; and so forth. Cattle in a farmstead are mostly in fenced corrals and open buildings. From a distant road, farmers can readily identify the animals and operations of a farm from the buildings in a farmstead.

The farmstead commonly has a somewhat central courtyard with farm buildings and animals on one side and the home on the other side (Thornbeck, 2012). Thus, the courtyard serves as key space for both the farmer's work and the family's use. The family side commonly includes home, driveway, garage, car, and diverse objects associated with family activities.

Movements Among Farmsteads and Town. Some movements are between farmstead and nearby village, such as the family getting daily needs from the few shops present. In the village, family members may visit friends, the farmer may consult with village farmers, and the family joins in occasional village festivals.

Trips between town and farmstead are for more diverse purposes. The postal service delivers mail. Road maintenance vehicles and personnel keep the roads operational. Farm workers commute to the farm. The farmer goes to town for machinery parts or sales, plus numerous items required on the farm. Hay and feed may be bought or sold. Animal product preparation and transport are arranged with companies in town. Materials are needed for repairs.

School buses transport children to school. The family does weekly shopping in town, goes to government offices, visits health facilities, participates in community, religious, or school activities, visits friends, and attends the varied festivals in town. In effect, the town serves and strongly affects, not only surrounding villages, but also essentially all farmsteads dispersed over the grassland.

Ecological Patterns in Grassland

The ecological literature on grassland ecology is rich and extensive (Samson and Knopf, 1996; Gibson, 2009). Thus, simply to illustrate the kinds of insights such studies provide, especially as somewhat related to towns, villages, and farmsteads, a handful of studies are briefly mentioned. The studies are loosely grouped under (1) livestock grazing-intensity effects, and (2) microhabitats, trees, and woods. However, absorbing the main idea in each study separately may be most valuable.

Livestock Grazing-Intensity Effects

Fences or *fencelines* are normally denser, and surround smaller fields, near a town than farther away. A comparison of heavily and lightly grazed dry grassland on opposite sides of fencelines (Figure 10.10) in southern Africa found that with increased grazing: (1) plant cover decreased; (2) species composition changed; (3) species richness at a site changed little (alpha diversity); (4) alpha diversity of plant functional groups (based on annuals/perennials, grass/shrubs/etc., succulents/non-succulents) decreased; (5) compositional differences among plant communities (beta diversity) increased; and (6) differences in functional structure increased at broader scales (Gibson, 2009; Hanke *et al.*, 2014). Both livestock and some wildlife move along fences, adding dung and compacting the soil, thus producing a distinctive, often conspicuous, narrow strip of different vegetation. Fences can also block wildlife movement routes, leading to inbreeding or wildlife overabundance (Hayward and Kerley, 2009).

Overgrazing of grassland is the major cause of the extensive area of desertification on Earth (Lin *et al.*, 2012). Therefore, the effects of the intensity of livestock grazing are considered in some detail.

Lightly grazed areas in Mongolia have more pollinator species, whereas heavily grazed areas have more flower visits, more pollinators, and more ruderal plants (Yoshihara *et al.*, 2008). In Australia, plant species richness declines and nonnative species increase with heavier grazing intensity (Dorrough and Scroggie, 2008). Also in Australia, intense grazing of grass in open woodland correlates with higher soil nutrient levels and better tree health than with lighter grazing (Close *et al.*, 2008). In China, higher grazing-intensity produced vegetation patches both more fragmented and more homogeneous (Lin *et al.*, 2010). In Inner Mongolia, China, grassland changes over 25 years are mainly due to livestock grazing, especially high stocking rates, not to decreasing precipitation (Li *et al.*, 2012).

Marsh birds at natural springs and streams are reduced by livestock grazing (Richmond *et al.*, 2012). In Australian grasslands, 95 percent of the sheep are within 3 km of a water point, 6 km for cattle, and 7 km for red kangaroos (native wildlife) (Fensham and

Figure 10.10 Fenceline separating different cattle-grazing intensities in short-grass plain. Between Magdalena village and town of Socorro, New Mexico, USA. R. Forman photo.

Fairfax, 2008). The abundance of many plant species is linearly related to grazing intensity, but plant and bird species that decrease exponentially with distance from a water point are especially important far from water.

A livestock exclusion area in Argentina has lower bird density, species richness, and number of endemic species compared to a control (Garcia *et al.*, 2008). In Portugal lowland with warm nutrient-rich soil, plant diversity is high despite heavy grazing by livestock (Lomba *et al.*, 2011). In Romania, extensive grazing by sheep, goats, cattle, and horses maintains high plant species diversity (Vojta *et al.*, 2014). Species richness is highest at intermediate grazing levels, resulting in about 40 percent cover of relatively unpalatable shrubs, which presumably protect less common species (Vojta *et al.*, 2014).

Transhumance, the movement of grazing animals from winter low-elevation pastures to summer higher-elevation pastures, tends to protect both areas from overgrazing (van der Zee and Zonneveld, 2001). Finally, the combination of high agro-diversity (genetic races of crops), livestock diversity (different animals and different races), and diversity of natural species produces an especially rich agricultural ecosystem (Burchett and Burchett, 2011; Mortimer, 2013). The preceding studies emphasize the importance of grazing intensity, and hence the density of livestock, on numerous ecological dimensions of grassland.

Microhabitats, Trees, and Woods

The overall habitat diversity of a livestock farm strongly affects species diversity and movements. Thus, in a grazed grassland with scattered woods and houses in Queensland, Australia, the housing areas decrease mammal species richness and increase plant structural complexity, whereas distance from roads is relatively unimportant (Brady *et al.*, 2011). The grassland matrix apparently provides supplementary food resources and facilitates mammal movement. The habitat diversity of the grassland matrix or whole farm strongly affects overall biodiversity (van der Zee and Zonneveld, 2001; Burchett and Burchett, 2011). Maintaining, for example, both cattle (which graze by biting and pulling grass, leaving relatively tall grass) and sheep (which both bite and nibble, leaving a low grass cover), provides overall habitat diversity.

At a finer scale of microhabitats, a mosaic of microhabitats seems to be important for sustaining grassland shorebird nesting in The Netherlands. In Dutch pastures with dairy cows and sheep grazing, many plant species are most common on ditch banks within 200 m of a nature reserve (Leng *et al.*, 2009). In UK pastures, fertilization, cutting, or grazing seem to increase butterfly density, whereas bumblebees are more abundant in planted flower-rich microhabitats (Potts *et al.*, 2009). In Swedish pastures, plant species favored by livestock grazing increase with the age of a pasture (Johannson *et al.*, 2008). Rangelands lose livestock value due to habitat fragmentation, along with overgrazing and other land-use changes (Sayre *et al.*, 2013). In Colorado (USA), prairie dogs in ground colonies compete with livestock, but may carry and succumb to plague (Hartley *et al.*, 2009; Johnson *et al.*, 2011). In Australia, rock outcrops in grassland provide microhabitat heterogeneity which sustains reptile diversity (Figure 10.11) (Lindenmayer, 2003; Michael *et al.*, 2008). Microhabitat diversity is an ecological key in these grassland studies.

Similarly, scattered trees, dead or alive, and groups of trees are key microhabitats for many plants and animals in grassland (Lindenmayer and Fischer, 2006; Manning *et al*, 2009). In tropical pastureland of Costa Rica, birds drop numerous seeds in large and medium-size wooded patches, but also drop seeds under planted tree saplings and trees (Cole *et al.*, 2010). Likewise, the diversity of bird-dispersed seeds correlated with bird abundance, but not with avian diversity or forest patch size, total tree cover, or proximity to forest (Pejchar *et al.*, 2008). In similar pastureland, habitat-specialist forest birds preferentially chose routes along a riparian corridor, but used wooded stepping stones where forest was scarce (Gillies *et al.*, 2011). In contrast, forest generalist species mainly used grassland routes.

In Costa Rica, avian diversity is higher in pasture and along living fences (posts that grow) than in riparian or secondary forest (Mendoza *et al.*, 2014). Bee species richness tends to be high in pastures, while bee species composition in hedgerows is more similar to that in forest (Hannon and Sisk, 2009).

Finally, in Brazil, predators mainly kill livestock within 1,300 m of forest, but avoid areas near houses (Palmeira *et al.*, 2008). In Bhutan, livestock predation is high where horse density is high, but low with high cattle density, forest cover, and human density (Sangay and Vernes, 2008). In Sweden, grassland plant diversity is higher if the grassland

Figure 10.11 Rock outcrops, slopes, fences, scattered trees, and wooded corridors providing habitat diversity that sustains biodiversity in pastureland. An egg-laying monotreme echidna (Tachyglossidae) discovered near vehicle. Southern New South Wales, Australia. R. Forman photo, with aid of David Lindenmayer. (A black and white version of this figure will appear in some formats. For the color version, please refer to the plate section.)

is surrounded by forest rather than by cropland (Cousins and Aggemyr, 2008). Wooded patches and surrounding forest strongly affect species in grassland.

To and from Town and the Adjacent Zone

The edges of towns commonly contain one or more cemetery, dump (tip, landfill), golf course, institution, retail shopping center, industrial center, water treatment facility, sewage treatment facility, even a local airport. Strip development may extend out a main road. Farm stands selling food products from local farms are often present on a radiating main road near town (e.g., eight seasonal and two year-round farm stands in Concord; see Appendix). Farmers carry food to the farm stand, and people of the area buy and carry food from farm stand to town and village.

Farmsteads or farmhouses, formerly isolated, are often present in the edge of towns. Dogs and cats forage outward sometimes hundreds of meters from residences in the edge (Chapter 8). Invasive weeds spread outward from house plots and gardens. Town-edge farmsteads may have horse-riding stables which draw people from town and village. Grassland for grazing and periodically spreading manure may extend outward from a town-edge farmstead. Fields by these farmsteads may have specialized agriculture requiring intensive management, such as organic farming and pick-your-own (self-harvesting; food co-production) farming. Irrigation and farm animals, such as chickens and goats, are often present requiring constant attention. Roosters in town edge farmsteads awaken neighbors.

Now we look further outward and consider: (1) town to farmland movements; (2) farmland to town movements; and (3) farmland patterns in the adjacent zone. Distances are poorly known in many cases.

Town to Farmland Movements

Lots of flows and movements originate in town and move outward various distances into surrounding cropland or pastureland. Outward flows are loosely grouped into the following six categories:

Plants and Animals. Cats, outward from town edge about 200 m (650 ft) (Chapter 8). Dogs, several hundred meters. Invasive weeds from house plots. Town weeds, along roads and other radiating corridors. Town wildlife, especially in the direction of surrounding semi-natural areas. Animal-transported seeds of town plants. Pollinators.

In Air. Skyglow, from town lights at night, perhaps a few to several kilometers (Chapter 8). Also, individual lights, such as street lamps and spotlights, in the town edge, sucking insects out of habitats several hundred meters out. Town noises, commonly several hundred meters (Chapter 5). Air pollutants from industries. Wind-dispersed seeds of town plants.

Along Stream or Riverside Corridor. Stormwater runoff, especially from impervious surfaces, downstream. Stormwater pollutants downstream. Human wastewater effluent downstream. Town pollution-produced phytoplankton floating downstream. Town fish, upstream and downstream. Town wildlife in stream corridor, upstream and downstream. Town wildlife moving upslope or downslope along surrounding valley.

Along Railway, Pipeline/Powerline, Public Path. People walking, trampling, spreading trash, and disturbing wildlife, perhaps mostly within a kilometer or so of the town edge. Dog-walkers from town, often going outward a kilometer or so and disturbing more wildlife. Bicycling several kilometers along some paths.

Along Roads. Strip/ribbon development, spreading outward perhaps a kilometer on main road, and forming a filter to wildlife movement across the landscape. Commuters from town, working on farms (e.g., Casablanca). School buses, for pick up and drop off, half way to the next town. Traffic and noise. Emergency vehicle sirens, half way to the next town. Discarded cigarettes (more close to town), and fire spread (more away from town). Road maintenance activities, spreading herbicides, road salt, and dust, half way to the next town. Commuters from surrounding villages working in town.

Recreation Spots. Picnic areas, spreading trash. Natural area for nature appreciation and photography. Hunting fields or woods, reducing wildlife populations, causing infrequent accidents. Shooting range, away from town edge, usually less than half way to next town. Fishing waterbodies. Firewood cutting and wood removal. Dumping debris, especially by road, wetland, or waterbody, at town edge and further from it. Off-road vehicle (ORV, ATV) intensive-use site, away from town edge. Collecting food and medicinal plants. An agriculture-nature park, for food production and spots for recreation (e.g., near Milan and Barcelona). Typical distances from town for several activities vary widely, depending on topography and vegetation.

Farmland to Town Movements

In some regions all farmsteads (farmhouses) are in villages or towns, but perhaps in most cases single farmsteads scattered over the landscape are on the respective farm properties (Figure 10.12). Flows and movements from farmland to town are just as diverse and frequent as those in the opposite direction above, as seen in the following five groups.

Plants and Animals. Deer and other herbivores, enter from farmland. Predators, especially at night. Agricultural weeds. Pollinators. Human food subsidies such as birdfeed attract scavengers and other animals to town, particularly in a dry or cold season. Animals, some carrying seeds, continually arrive from a natural area, especially woods, near town. Linearized landscape of farm ditches, hedgerows, and fences facilitates wildlife movement into town. Insects and varied pests arrive from farm ponds and wetlands.

In Air. Wind-dispersed seeds blow in. Insect/locust plague invades town. Dust. Odors from manure, livestock, and farm operations arrive. Agricultural chemicals including NOx, fertilizer, pesticides blow in. Smoke and ash from field burning. Night moths, beetles, flies, aphids, etc. attracted to the town's skyglow, and especially individual lights in the town edge (Eisenbeis, 2006).

Along Stream and Riverside Corridor. Nitrogen in groundwater arrives from field fertilizing or livestock feedlot. Crop irrigation reduces subsurface flow to town, lowering the water table. Stream carries in sediment from field erosion, also agricultural chemicals. Farm water runoff, accelerated by linear ditches, pipes, roads, and hedgerows, reaches town, sometimes in the form of floodwater.

Farm Vehicles. Tractors and other field equipment carry soil, seeds, small animals, and manure into town (Figure 10.13). Farm equipment is transported to town for repair or replacement. Trucks carry grain to storage bins for railway or truck shipping. Trucks

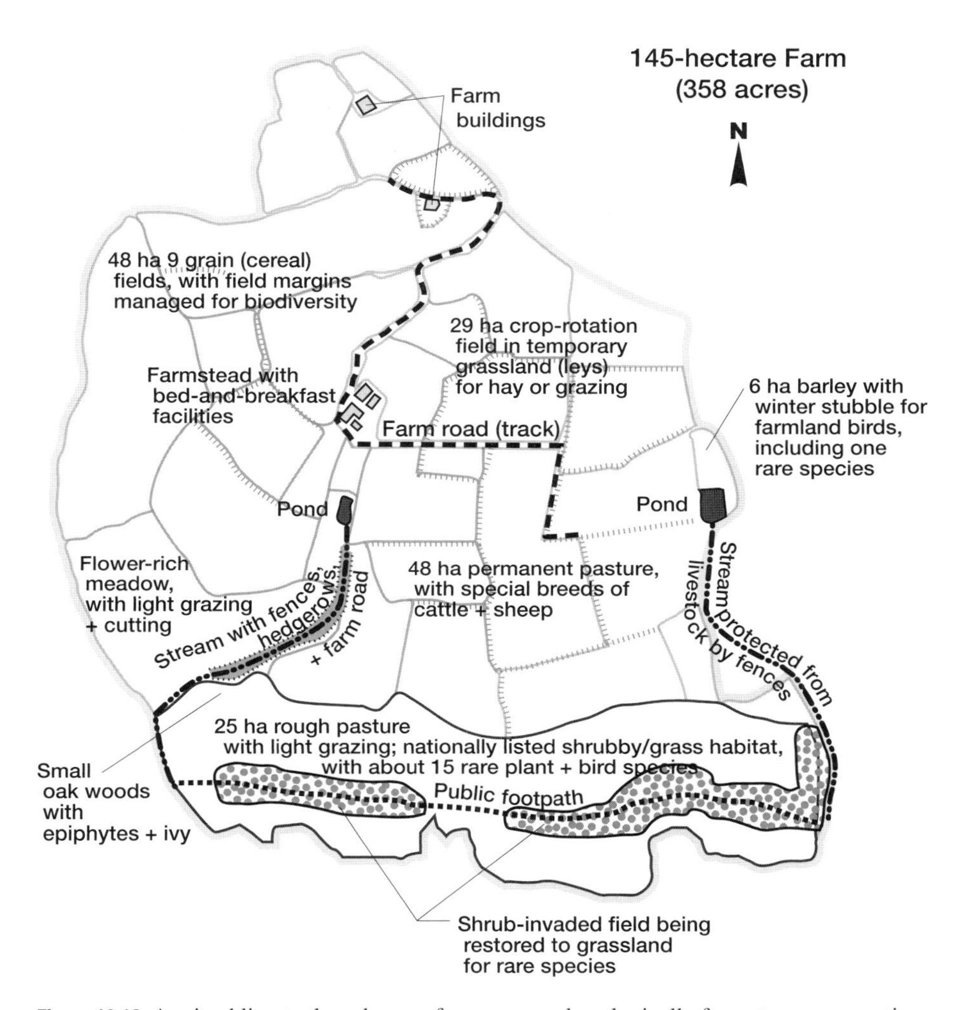

Figure 10.12 A mixed livestock-and-crops farm managed ecologically for nature conservation. Field margin = outer 5–6 m of field where tractor turns at end of rows. Southwestern England. Adapted from Burchett and Burchett (2011).

carry food products to farm stands near town. Trucks carry vegetables, fruits, flowers, and/or farm animals to a weekly farmers' market in town. Pests, diseases, and wastes are also carried.

Cars from Home. Some farmstead residents are commuters working in town. Attending secondary school in town. Weekly retail shopping is done in town, sometimes shopping more often. Occasional trips to the health center and government offices. Infrequent trips to attend special festivals and other entertainment (Dufty-Jones and Connell, 2014). Socializing that builds social capital (resource potential of social relationships) is important, and results from all the trips, especially for shopping, and perhaps time in a bar/teahouse.

Figure 10.13 Farm tractor transporting soil, stumps, and other farm material though center of town. Askim, Norway. R. Forman photo. (A black and white version of this figure will appear in some formats. For the color version, please refer to the plate section.)

Adjacent Zone and Farmland Patterns

The adjacent zone, perhaps up to a few kilometers wide, encircles the town (Chapter 1) and is noticeably different from the surrounding land beyond. This zone also differs from the exurban or periurban zone outside cities, which typically has many chemically contaminated brownfields, urban air pollution, dispersed housing developments, few active farmsteads, mainly degraded semi-natural areas, numerous abandoned fields, and land prices too high for most farming (Marzluff *et al.*, 2008; McDonnell *et al.*, 2009; Forman, 2014).

Many farmland features are present close to towns because intensive farm management for farm animals, irrigation, and orchards requires the proximity of the farmer and associated labor. The rather distinctive set of patterns is grouped into seven categories:

Water-Related. In the adjacent zone, irrigation is favored by a town's relatively high water table (which is lowered by irrigation); also, irrigation requires constant checking and care (Tivy, 1990). Rice paddies have endless problems with inputs/outputs, canals/ditches/levees, and flooding/drying out. A downstream area is likely to have town-polluted stream flow. Wetlands, long ago mainly drained or filled by farmers, are often scarce.

Fields and Crops. Small fields with a dense network of fencelines (Burchett and Burchett, 2011; Fahrig *et al.*, 2015) containing diverse crops are characteristic of the adjacent zone. Field areas, often including meadows (Peterken, 2013), are often treasured as part of the town's cultural landscape (Etienne *et al.*, 1998). Orchards (e.g., by Winters, California; Fallows and Fallows, 2018), vineyards (Madera *et al.*, 2014), and novel crops (Scherr and McNeely, 2007) requiring constant care are characteristic.

Farming and Soil. As the town probably originated on or by good farm soil, the adjacent zone often has good soil. Farming particularly attuned to detailed weather variations, such as "response farming," fits well here (Connor *et al.*, 2011). Farming on erodible soil, and with many important ditches, requires frequent management.

Woods, Hedgerows, Trees. Woods or natural areas in the adjacent zone tend to be small or tiny, elongated, and abundant, serving as stepping stones for wildlife movement. Hedgerows are often common, some wider than usual, and dominated by different tree species (Thomas and White, 1980). Large trees scattered in hedgerows, woods, and field centers may be more common than in either town or more distant cropland.

Habitats and Species. Hybrid nature–Euclid spatial patterns (Chapter 1) may be relatively frequent in the adjacent zone (Katoh *et al.*, 2009). The relatively fine scale and diverse crops produce high habitat diversity. Numerous adjacent habitats favor *complementary habitat use* by wildlife, i.e., using two or more habitats. The many small woods serving as stepping stones, and abundant hedgerow corridors, provide important connectivity for wildlife movement. Birds and other wildlife from town regularly visit this zone. In short, species richness is high.

Specialized Land Uses. All old towns have a legacy of former roads and paths radiating outward in many directions. Diverse town-related land uses of the adjacent zone include strip development, edge retail shopping center, edge industrial center, dump, cemetery, golf course, plant nursery, even a local airport. Some may be in the town edge, but affect this zone. Over time diverse farm-related activities occur in the outskirts of town (Figure 10.14). Imagine the range of ecological effects produced by a herd of livestock.

People. Walkers, dog-walkers, motorbikers, and all-terrain vehicles (ATVs, ORVs) from town move through the adjacent zone on public paths or cross-country, leaving diverse

Figure 10.14 Farmers herding goats along a main road near edge of town. North of Tolear, Madagascar. Photo courtesy and with permission of Mark Brenner.

marks on the landscape. Farm stands selling farm products are characteristic. Vehicle traffic and noise along a rather dense road network, includes raising dust on unpaved roads. Small quickly extinguished fires occur.

In effect, the outside adjacent zone differs markedly from both town and more distant outer surrounding farmland. The zone is quite diverse and intensively used as a valuable resource to sustain, rather than lose to development, for the town and its residents.

Research Frontiers. Apparently, no ecological analysis exists for this adjacent zone just beyond the town edge. Landscape ecology, supplemented by soil science, road ecology, conservation biology, and water resource ecology, provides the main conceptual framework for understanding this zone (Forman, 1995; Turner and Gardner, 2015; Rego *et al.*, 2018). Consider the following array of ecological insights as an introduction to research frontiers.

Wooded Patches. Small wooded patches and corridors in adjacent zone farmland close to a large natural woods increase the immigration rate of species to the woods, and also reduce the local extinction rate of animal species within the woods (Opdam, 1991; Forman, 1995).

A group of small patches as stepping stones, rather than a row, provides alternative routes for wildlife movement and a higher probability of reaching a target (Forman, 1995, 2014).

Wildlife move readily along, but less frequently across, straight edges, and in contrast, readily move across, but seldom along, convoluted edges (Forman, 1995). Moving across edges facilitates interactions between land uses.

The ecological width of an edge, such as farmland or forest edge, varies according to the process or type of species measured (Forman, 1995; Lindenmayer and Fischer, 2006).

A small patch of woods or other semi-natural habitat is likely to be entirely edge habitat with edge species, and contain no interior habitat.

Habitats. Small patches of many types create high habitat diversity, and thus combined high species richness (gamma diversity) (Forman, 1995).

Many habitat adjacencies in a habitat-diverse area support many multi-habitat species with complementary use of two or more habitats (Forman, 1995).

Dispersed old trees are distinctive microhabitats for several species groups including lichens, mosses, beetles, and cavity-nesting vertebrates, and serve as stepping stones for species movement across the land.

Corridors and Movement. Little-used roads are routes of predator movement, whereas heavily used roads are routes for very few species, and are barriers or filters for most wildlife (Forman *et al.*, 2003; van der Ree *et al.*, 2015).

Railways are often routes for predators, and have a series of linear microhabitat conditions compressed together, which support a somewhat distinctive flora and fauna (Forman, 2014; Borda-de-Agua *et al.*, 2018).

Species preferentially move from a source, in this case the town, across less suitable land in the direction of another habitat of similar type, especially a large habitat (Saunders *et al.*, 1987; Forman, 2014).

Wildlife movement along a stream corridor is strongly affected by corridor width and continuity (Forman, 1995). An elongated woods funnels species outward in opposite directions.

Strip or ribbon development extending out a radial main road functions as a barrier or filter inhibiting wildlife movement across the landscape (Forman, 2008, 2014).

Hedgerows and Mesh Size. Many species and ecological processes in the landscape are sensitive to the mesh size (size of enclosed fields) of the adjacent zone network. For example in northwestern France, a carabid beetle, large owl (le grand duque), and shrub are present in a less than 2 hectare (5 acre) mesh, only the owl and shrub in a 2–7 ha mesh, and only the shrub in a more than 7 ha mesh (*Les Bocages*, 1976; Forman, 1995).

Hedgerow network intersections are cooler moister microhabitats with somewhat different species than along the hedgerows, and sometimes include a few forest interior species (Forman, 1995; Burel and Baudry, 1999).

Hedgerows are often rich in animal-dispersed woody plants, which serve as sources of berries and other fruits carried by wildlife, and dropped near shrubs and trees (McDonnell and Stiles, 1983; Forman, 1995; Burel and Baudry, 1999).

Human Usage Along Routes. Dog-walkers degrade and reduce suitable wildlife habitat by flushing animals and scent-marking the area (Chapter 8). In contrast, all-terrain vehicles degrade wildlife habitat by rapid movement and noise, while also creating new disturbance microhabitats.

Moderate traffic on moderately high-speed-limit roads causes a high rate of animal roadkill, and inhibits the regular breeding by sensitive bird species (Forman *et al.*, 2002; van der Ree *et al.*, 2015). But busy multi-lane highways, with much lower roadkill rates, produce traffic noise that inhibits sensitive species for hundreds of meters outward in both directions.

Finally, which of the many characteristics of the outside adjacent zone may extend farthest outward from a town? Perhaps relatively few effects of a town would normally extend further than the surrounding towns. Examples of long-distance effects might be: hunting and firewood cutting/collecting areas; recreation and fishing spots; weeds from house plots and along railways; and sediment from town construction or industry that is washed downstream. Still, some town effects are normally limited to, even concentrated in, the first few hundred meters.

11 Forestland, Aridland, and Towns

There is something indescribably inspiriting and beautiful in the aspect of the forest skirting and occasionally jutting into the midst of new towns ... Our lives need the relief of such a background ...

Henry David Thoreau, *A Week on the Concord and Merrimack Rivers*, 1849

Towns and villages in the forest or woodland seem cozy, almost hidden. Shade, cool air, and the rustle of leaves fill the air. Walk in the surroundings and one quickly fades into the luxurious vegetation. Trees towering over us highlight the power of nature. The walk could be called "forest bathing."

The surrounding forest is a key to the economy of many towns, commonly providing clean water, wood products, and a major source for recreation. A large protected forest area next to or near a town draws urban residents and often tourists from elsewhere (see book cover) (Ode and Fry, 2006), which helps to support the shops and other businesses in town (e.g., Sooke, Issaquah; see Appendix). Cutting almost all the nearby forest causes an economic downturn in town (Fallows and Fallows, 2018).

But rural developments close to a protected area often lead to its overuse and degradation (Figure 11.1) (Gude *et al.*, 2006; Radeloff *et al.*, 2010; Wood *et al.*, 2015). More challenging still, in some regions towns and villages may be within the protected forest area (Czerny *et al.*, 2009; Prados, 2009b).

Forests are especially important for goods and services, or ecosystem services (Grebner *et al.*, 2013; Turner *et al.*, 2013; Thompson *et al.*, 2016). Goods such as logs and fuelwood (firewood) can come from almost any forest. But for services, the physical presence of a forest is, for example, of strategic or symbolic importance. Also, for services to town residents, location and access matter (Menzies, 1994).

Yet the cozy setting is not hazard-free. In dry periods, fire can sweep through the forest and damage the town (e.g., Chamusca, Vernonia) (Vince *et al.*, 2005; Theobald and Romme, 2007). Unlike the town and village, much is known about the ecology of forest (Figure 11.2) (Peterken, 1996; Barnes *et al.*, 1998; Lindenmayer and Franklin, 2002; Kimmins, 2004; Perry *et al.*, 2008a; Swanson *et al.*, 2009; Grebner *et al.*, 2013). From absorbing and storing abundant carbon (McKinley *et al.*, 2011) to supporting water flow in streams, dropping abundant leaf litter that enriches soil, and providing vertical foliage layers where different animals forage, forest is globally distinctive.

The two major sections following on town affecting forest, and vice versa, focus on the: (1) town/village and its adjacent forest zone, and (2) town/village and its outer surrounding forestland.

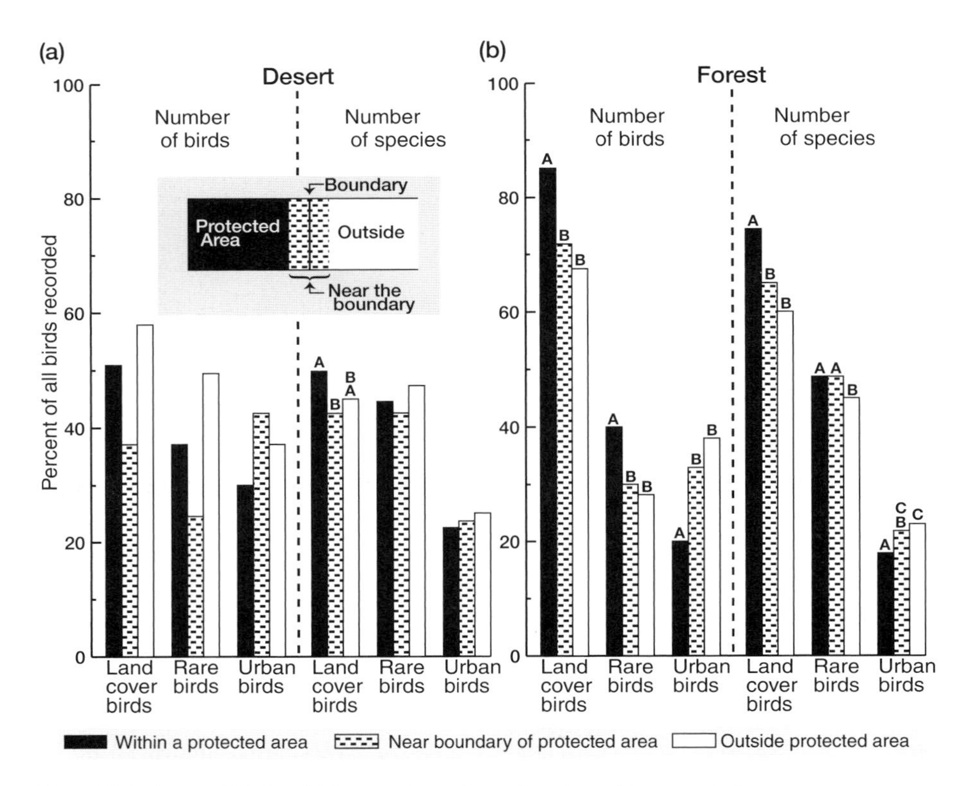

Figure 11.1 Types of birds within, near boundary of, and outside protected areas. Forest = deciduous/coniferous North Woods from northern New England to Minnesota, USA. Land cover birds = species associated with the predominant natural land-cover type sampled; rare birds = species of greatest conservation "need"; urban birds = breeding-season species (synanthropes) associated with human-modified environments. Bird data from Breeding Bird Survey based on 39.4-km roadside sampling (mainly reflecting traffic level, land use, and vegetation) entirely within 1–49% within, and entirely outside protected areas. Bars with the same letter (A, B, C) for a bird type are not significantly different. Adapted from Wood *et al.* (2015).

Town/Village and Its Adjacent Forest Zone

The *adjacent zone* is the area adjoining a town where intense environmental interactions with the town occur (e.g., Valladolid, Sooke, Sandviken; see Appendix). The width of this zone depends on the typical maximum distance numerous processes extend, such as the following. Many people will walk on a path. Many bats, birds, bees, and mid-size and large mammals move in daily foraging. Wells lower the water table. Strip development extends out main radial roads. Most nonnative plants spread. Bird feeders attract birds. Pet dogs explore. Many town noises are audible. Many direct town lights penetrate. Town industrial pollution effects are conspicuous. A cemetery or dump (tip) is located. A firebreak is burned. And seeds are dispersed by wind or

Figure 11.2 Road and vegetation attributes affecting vehicle speed, wildlife–vehicle collisions, and roadkill. Straight road promotes driver speed, collisions, and roadkill, whereas abundant shrub-layer vegetation close to road surface slows drivers. Yellowstone National Park with Grand Teton Mountains in distance; Wyoming, USA. R. Forman photo.

animals. Operationally, the adjacent zone width may typically be about 2 to a few km (1–3 mi).

Nature's boundaries on land are rarely straight. Initially, town edges may be straight, but over time they often become convoluted (De Chant *et al.*, 2010). The change may result from tree area expansion and different homeowner land uses. A convoluted boundary facilitates the movement of people and wildlife between the two land types.

We now explore the two-way effects: (1) town effects on adjacent forest/woodland zone; and inversely, (2) adjacent zone effects on town.

Town Effects on Adjacent Forest/Woodland Zone

Five ecological features illustrate the outward town effects on adjacent land: (1) water; (2) vegetation; (3) animals; (4) human structures; and (5) fire.

Water and the Adjacent Zone

Town wells, local industry wells, and homeowner wells together often slightly lower the relatively high groundwater water table under most towns. This particularly occurs where well water is widely used to sprinkle or irrigate lawns. A lowered water table in town lowers the water table a short distance into the adjacent zone.

With ample water flowing in a stream, a slightly lower water table probably has little or no effect on the stream ecosystem. But in dry periods with more town irrigation, the downstream portion of the stream in the adjacent zone is likely to have reduced flow. That produces noticeable effects, e.g., causing higher stream temperature, less dissolved oxygen, reduced fish populations, and so forth (Kalff, 2002; Lake, 2011).

Heavy rainfall on the impervious roofs, sidewalks, and streets of town sometimes produces local flooding, which may extend downstream short distances into the adjacent forest. Roads, drainage ditches, and strip development extending into the adjacent forest also may produce localized flooding. Strong turbulent winds reaching down into the open spaces and streets of a town occasionally blow down exposed trees in elongated patches adjoining a town, especially if the soil is saturated by preceding rain.

Vegetation and the Adjacent Zone

In China before 1911, villages retained forestland for community fuelwood and fodder (young branches) for livestock (Menzies, 1994). Also, forest adjacent to a village was retained to protect the soil, as well as the character of ancestral grave mounds distributed over the nearby area, especially on elevated land. The village managed the woods, and residents could collect *nontimber forest products* (NTFPs), including nuts, fruit, mushrooms, wildlife, and medicinal herbs. Sometimes a village (or a clan) would cut some of the trees to raise revenue, and tree plantations near the village were not uncommon. Wood was used for construction as well as making charcoal as fuel to burn. Often some logs were laid in rows on the ground, between which medicinal and edible fungi grew and were harvested. In some areas, fodder from oak (*Quercus*) was apparently used to cultivate silkworm caterpillars making silk. In short, the adjacent forest zone was a rich handy resource for the limited number of people in a village.

In northern Portugal the rural population is declining around a large 703 km^2 (271 mi^2) protected area (Peneda-Geres National Park) (Lourenco *et al.*, 2009). Over a decade (1991–2001), two of the villages, Melgaco and Terras de Bouro, have lost 9.5 percent and 11.5 percent respectively of their population. Meanwhile, however, the number of houses increased, by 12 percent and 27 percent respectively. This apparent paradox results from villagers moving to urban areas for jobs or education, while urban residents build weekend/holiday houses (second homes) close to both village and park.

The villagers leaving resulted in a loss of 42 ha (104 acres) of farmland and 98 ha of artificial built surface. Meanwhile, the housing increase removed 418 ha (1,032 acres) of forest, and added 418 ha of shrubby, herbaceous, and no vegetation, plus 140 ha of water bodies (presumably small ponds). In short, the village's adjacent forest zone is changing, both in human structures and in vegetation cover and species composition.

The *forest edge* adjoining a town differs ecologically from the forest interior (Chapter 1) (Lindenmayer and Fischer, 2006; Olupot, 2009), and may be widened by various town activities. Seeds of nonnative plants from residential and town plantings are carried relatively short distances by wind and animals. The density of walking trails (paths) is noticeably greater in the adjacent zone than in the surroundings beyond. Dog-walkers commonly go outward a kilometer or so. *Forest bathers* (*shinrin-yoku*), getting rejuvenated in the air containing phytoncides (essential wood oils) from trees (Li, 2010; Reeves, 2014), may commonly go one to two kilometers away from the town's residential area. Well-used trails have compacted soil with weedy plants alongside. Subtle town effects on vegetation, such as reduced pollination and reproduction of epiphytic orchids near a tropical trail (Huang *et al.*, 2009), are doubtless common though little studied.

Animals and the Adjacent Zone

Outdoor pet cats seem to explore and forage roughly 200 m from home (females less), though at night they may go farther (Chapter 8). Free-ranging pet dogs also often go farther, and dog-walkers commonly take dogs 0.5 to 1.5 km from the forest edge. Dog presence, and especially scent-marking by urine and feces, reduces wildlife populations in the woods.

Around five Panama rainforest villages, game kill sites are concentrated within 2 km of the houses (Figure 11.3) (Smith, 2008). A high deer density in and near the forest edge creates a more open ground and shrub layer. Concentrated deer browsing also changes plant species composition by targeting delicious species, and leaving unpalatable plants (such as *Senecio* ragwort and *Pteridium* bracken fern in Japan) (Takatsuki, 2009). In Rondonia, Amazonas, the species richness of birds and large mammals, as well as the number of forest interior species, is lower in rainforest within perhaps 1–2 km of rural houses (Prist *et al.*, 2012).

The presence of people (sometimes with dogs) in the adjacent forest zone may produce three wildlife responses (Hunter, 1990; Knight and Gutzwiller, 2013): (1) cause physiological stress; (2) flush the animal away from an often preferred habitat; and/or (3) cause wildlife to avoid an area. Some animals, though perhaps not most species, become *habituated* to (behaviorally used to) the frequent or continuous presence of people. Wildlife species living in the adjacent zone tend to be generalist species, whereas specialist species are mainly in the surrounding forestland further from the town (e.g., Grande Riviere).

Other town effects extend into the adjacent zone. Loud town noises of varied sorts, from ringing bells to drill hammers and police sirens, are often audible 1–2 km out of town. At night artificial lights in town may be directly visible for a few hundred meters or more, though *skyglow* (reflected and refracted light from water vapor, particulates, and aerosols in the air) extends much further (Chapter 8) (Eisenbeis and Hanel, 2009). Such light might produce the *vacuum cleaner effect*, whereby numerous insects are sucked out of their forest habitat and drawn to the artificial lights, where they mainly die. In short, forest animals in the adjacent zone are strongly impacted by the town.

Human Structures and the Adjacent Zone

The *distance-decay relationship* or equation (Chapter 7) indicates that the number of people leaving a source point, in this case a town or village, decreases exponentially with increasing distance from town into the forest. The steepness of that decrease depends in part on the relative suitability or attractiveness of the forestland.

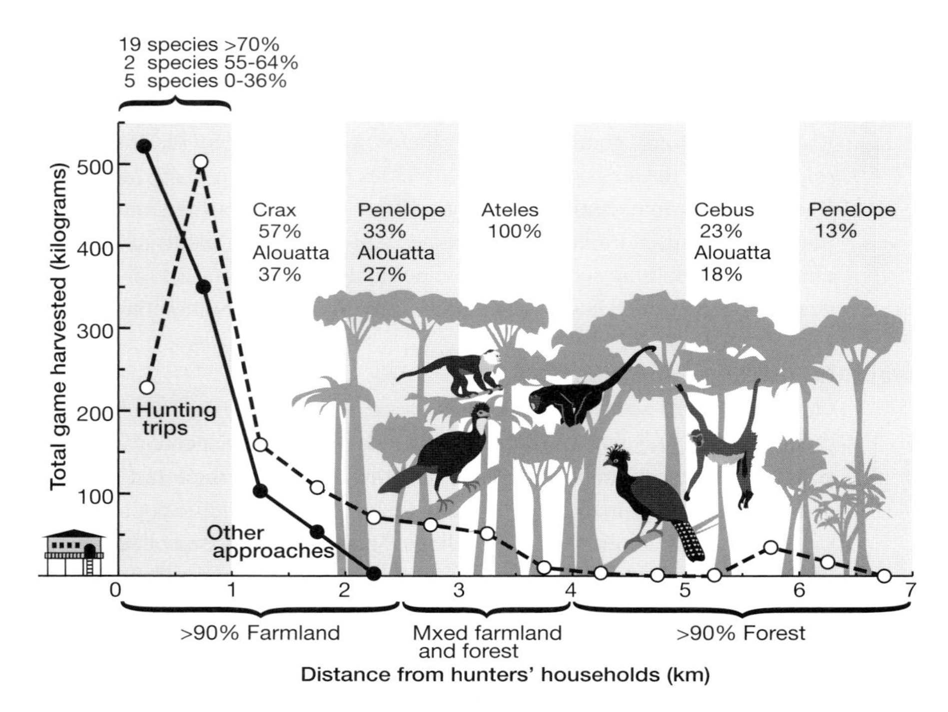

Figure 11.3 Success in harvesting rainforest animal food relative to distance from households. Based on 59 households in five clustered hamlets over 232 days. Individual or group hunting trips; other approaches are trapping, hunting in a fixed system, and opportunistic catches. Percentage of the total harvest of a species is listed by zone. In the 0–1 km zone, 19 species constituted >70% of the total game harvest; 2 species were 55% and 64%, respectively, 12 species were of intermediate importance, and 5 species were relatively unimportant. *Crax rubra* = greater curassow; *Alouatta palliata* = howler monkey; *Penelope purpureus* = crested guan; *Ateles geoffroyi* = spider monkey; *Cebus capocinus* = capuchin or white-faced monkey. Western Panama; not a protected area. 1 km = 0.62 mi; 1 kg = 2.2 lbs. Adapted from Smith (2008).

A study in southern Sweden found that four characteristics of the forest were more attractive to people leaving a settlement (i.e., the distance-decay curve was less steep) (Ode and Fry, 2006). People went farther into the forest if it: (1) is large (up to some point); (2) is mixed deciduous and coniferous (rather than only coniferous); (3) has paths present (especially near the beginning); and (4) is a protected area (for nature). Commonly many trails with walkers radiate out from a town (see Figure 12.8). All well-used trails show negative effects on soil, plants, and wildlife, though only a few key trails reach surrounding villages and towns.

Many human-made objects also enter the adjacent forest zone. Pollutants from town manufacturing and other sources blow or flow in (e.g., Tomar, Plymouth, Palmerton). Residential residents discard grass trimmings, brush, construction debris, old tires, and many other objects in the nearby forest (Matlack, 1993). Noise and artificial light were mentioned above, and may be important, especially if originating from a commercial shopping center or industrial center in the town edge, or strip development along radiating roads. Radial main roads are also strips of discarded trash or rubbish, including bottles, cans, and other debris. If a train stops in town, sparks from braking spread along tracks in the adjacent zone of the entering direction, and noise, smoke (from a steam engine), and other air pollutants are given off in the departing train direction.

Some large town features, such as a cemetery, dump, sewage treatment facility, and golf course are often located outside town in the adjacent forest zone. These of course have people, traffic, noise, and lights. Each also attracts a somewhat distinctive fauna and flora that differ from those of the adjacent zone forest.

Houses and other buildings scattered in the adjacent forest typically have people, pets, vehicles, and diverse small structures and activities. These spots support distinctive local floras and faunas, and also affect their surroundings (Hansen and Rotilla, 2001; Hersperger *et al.*, 2012).

Recent Russian dachas in the exurban area north of Moscow often have no running water or other town services, but have clean air and the opportunity to grow food in a garden (Ioffe and Nefedova, 2000; Nefedova and Pallot, 2013). Unfortunately, however, many dacha residents degrade the surrounding forest. In New England (USA) forest, nonnative plant species richness correlates more with housing characteristics than with environmental and other human features (Gavier-Pizarro *et al.*, 2010b). Such plants are especially common in exurban areas, low-density residential (usually sprawl) areas, and areas where the number of houses has noticeably changed (over six decades, 1940–2000). In brief, human structures and activities strongly alter the forest adjacent zone.

The density of roads in the adjacent zone also is higher than in the surrounding land farther out from town. A high road density (Figure 11.4) has lots of ecological implications. Wildlife movement across the area is reduced. Traffic noise is greater, and night lights more abundant. More of the forest area is open, resulting in more ditches with warm water runoff and more generalist roadside species. Unpaved roads mean more dust. More culverts are present, some of which block fish and water movement. Less natural forest habitat is present. Road density effects ramify through the adjacent zone.

Fire and the Adjacent Zone

Fires tend to be frequent in the adjacent forest zone, because of the proximity and accessibility of many fire sources related to town residents and activities (e.g., Tomar, Lake Placid). The huge 3,300-hectare (5.1 mi^2) Tijuca National Park forest, immediately on the western edge of Rio de Janeiro, is a good example. Over a decade (1991–2000), on average 75 fires per year are extinguished by firefighters (Matos *et al.*, 2002). The fire causes are: hot air balloons (24%); intentional (arson) (24%); rubbish burning (21%); and religious practices (21%). Lightning strikes are not listed as a cause, so all fires are human caused.

The small hot air balloons sent up at celebrations to float evocatively across the urban sky, plus the candles and fires of religion, are tied to cultural traditions. Although hot air balloons reach forest areas far from roads and burn more forest, more fires begin in nonnative grasses and ferns along roads slicing through the park. Because of these fires from roads, the nonnative vegetation has gradually spread into the forest. As a

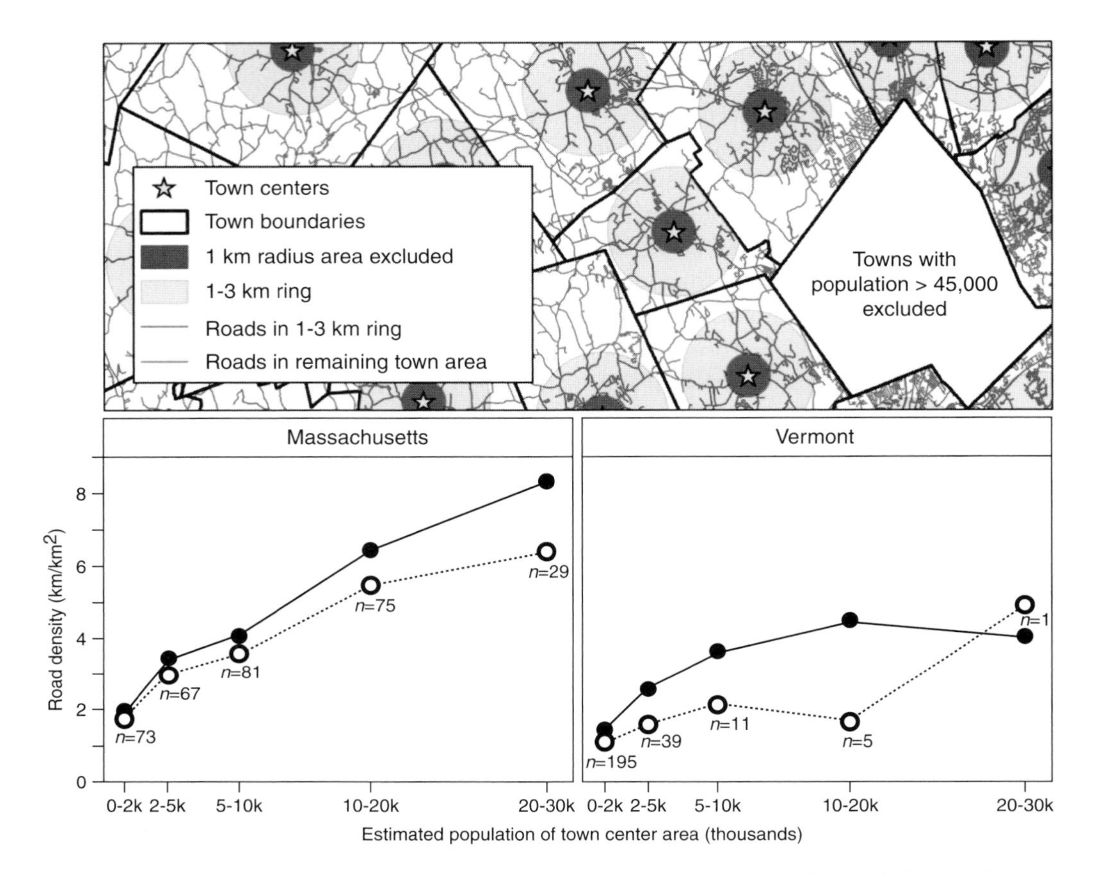

Figure 11.4 Road density in adjacent zone and outer surrounding land of towns in high and low population-density areas. In graphs, solid circles = 1–3 km ring; open circles = remaining town area. Population density: Massachusetts 336/km^2 (871/mi^2); Vermont 26/km^2 (67/mi^2). Figure courtesy of Joshua Plisinski and Jonathan Thompson, and with permission of Jonathan Thompson. (A black and white version of this figure will appear in some formats. For the color version, please refer to the plate section.)

consequence, fire managers have planted park roadsides with less flammable species, which has decreased both the nonnative species cover and the frequency of fire ignitions.

Most fires near towns originate locally, though some big fires from the surroundings may reach a town's adjacent zone. Children playing with matches, homeowners burning brush, town and commercial activities burning rubbish and other materials, and fires from industry may all spread into the adjacent forest zone. But being close by, firefighters can quickly reach and control such local fires. Consequently, most adjacent zone fires are small.

Adjacent Zone Effects on Town

Now consider the opposite direction, flows from the adjacent forest zone affecting towns. Residential areas in town, including the town edge, normally contain numerous tiny structures in house plots that attract wildlife, which may largely come from the adjacent forest zone. Bird feeders mainly attract birds during winter or the dry season. Bird houses, bat boxes, butterfly gardens, and bee hives normally attract animals within a kilometer or so. Indeed, honey yield from hives in a field within 1 km (3,280 ft) of a forest produce almost twice as much as hives more than 3 km from the forest (Sande *et al.*, 2009) (though in a residential area, numerous diverse flowers may greatly reduce this difference in honey produced). In town, dead trees and branches are often removed. Nevertheless, woodpeckers (Picidae) and other cavity-nesting birds from the nearby forest commonly forage in town. Many pollinators and seed dispersers in residential areas probably also originated in the adjacent forest zone.

Similarly, large herbivores such as deer and kangaroos tend to be most dense near and in the forest edge, from which they readily move into town. Residents commonly plant many shrubs and herbaceous plants rich in nitrogen and phosphorus that attract the herbivores.

The town of Jasper, Alberta (Canada, population 3,500) had so many large elk (*Cervus elaphus*) entering to feed that the animals essentially became dangerous residents in town (Douglas Olson, 1998 personal communication). Large males established "territories" next to a wide boulevard and in open spaces. Other animals fed on plantings around houses so people had difficulty going outdoors, even getting to their car. Some animals chased kids bicycling to school. These problems resulted from herbivores being concentrated in the adjacent zone and attracted by nutrient-rich nonnative shrubs and flowers planted throughout town. This serious elk problem was basically solved over five years, by plantings that obstructed the views of bull elk, plus the progressive removal of all nonnative shrubs. Four native shrub species present in the adjacent zone were used for plantings throughout town. Furthermore, large predators, which had also been attracted into town, decreased in number.

The net effect of an adjacent forest is to produce a *species rain* in town, whereby numerous species continually walk, fly, or are blown into town. These entering species then spread locally, such that the entire town area is continually enriched with biodiversity.

Around towns with fire-prone climates and vegetation, firebreaks of 100 to several hundred meters width are often established (Barnes *et al.*, 1998). Two landscape ecologists with fire expertise recommend a 3.2-km (2 mi) wide community protection zone around fire-hazard built areas in the USA (Theobald and Romme, 2007). These patterns relate to wildfires approaching a town or village from the surrounding forest (e.g., Chamusca, Vernonia).

Where small farm plots or a firebreak separate the adjacent forest from the town edge, forest species may be funneled into town along remnant green corridors or stepping-stone small patches. Some species movement into town may be enhanced by a double line of trees along main roads extending into towns.

Species entering town include pests that annoy people. Consider Mnisek Pod Brdy (Czech Republic, population 4,300) where the adjacent forest zone contains a large population of wild boar (*Sus scrofa*) (Z. Lipsky, 2016 personal communication). Groups of animals continually enter town damaging gardens, digging up waste places, annoying people, and threatening children. Periodically, the town designates a day when everyone can shoot pigs. Thus, over time, sudden pig reduction alternates with gradually increasing pig problems, producing an uneasy people–pigs coexistence.

Finally, scattered homes built in an adjacent forest zone pose special problems. Some houses built by residents or relatives represent town growth, though today the majority of towns worldwide are losing population. *Weekend/holiday houses* of urban people are often built in forest, especially close to an attractive town or village (see Figure 4.7). Irrespective of origin, many familiar effects emanate from such a house, including lights, noise, road, ditches, pipelines, electric powerline, chemicals from driveway or pool, damage to surrounding vegetation, and decrease in native specialist wildlife. In addition, use of the adjacent forest by town residents, who have long valued and used the forest, is truncated.

Town/Village and Its Outer Surrounding Forestland

Outer *surrounding land*, beyond the town's adjacent zone, is the area having more frequent interactions with the town than with any other town. Operationally, the outer border of a town's surrounding land may be one-half the distance to each nearby town (or small city), and the longest of those distances in other directions.

We now highlight: (1) town effects on outer surrounding land; and the reciprocal, (2) effects of surrounding land on town.

Town Effects on Outer Surrounding Land

Towns have highly diverse effects on the extensive forest or woodland surrounding the adjacent zone, conveniently recognized as: (1) resources harvested; (2) major ecological alterations; and (3) subtle often invisible effects.

Resources Harvested

Tree Cutting. Town and village residents primarily do the cutting and run the equipment in forest cutting operations. The effects of such logging reverberate widely on soil, microclimate, vegetation, wildlife, water, and streams (e.g., Rumford/Mexico, Wausau/Weston; see appendix). In western sub-Saharan Africa, even the diversity of children's diets may decrease with forest cutting.

Logging produces direct wood products, yet indirect effects happen. For instance, along the Tana River in coastal Kenya, patches totaling 25 km^2 (16 mi^2) are the only home of two endangered monkeys (Tana red colobus and Tana crested mangabey) (Marsh *et al.*, 1987; Hunter, 1990). Many of the logs are made into log canoes, essential transport for local villagers. A survey of 75 log canoes found that six tree species were cut to make the canoes. A fig species (*Ficus*), the most common tree used for a canoe, required only two weeks to make, but the relatively soft wood rotted in just over one year. Alas, this fig species was also the major food source of both monkey species.

Figure 11.5 Paper mill emitting steam, sulfur dioxide, hydrogen sulfide, and water pollution. Air and water pollutants degrade ecosystems downwind/downriver, while H_2S produces a strong smell, all effects extending for kilometers. Town of Rumford/Mexico, Maine, USA by Adroscoggin River. See Appendix. R. Forman photo. (A black and white version of this figure will appear in some formats. For the color version, please refer to the plate section.)

Finally, the solution agreed to was to make canoes out of a harder-wood ebony tree, which is widespread and little used by monkeys. An ebony canoe requires three weeks to make, but lasts three years. This solution for harvesting trees benefitted both the rare monkey species and log-canoe transportation for the villagers.

In Europe, small forest/woodland clear-cuts are typical, producing relatively fine-scale land mosaics (Peterken, 1996). A clear-cut is often followed at, say, decade intervals by thinning out smaller and misshapen trees, which are then cut for fuelwood (firewood)

Figure 11.6 Fuelwood (firewood) collectors returning to town with large pieces of wood. North of Tolear, Madagascar. Photo courtesy and with permission of Mark Brenner. (A black and white version of this figure will appear in some formats. For the color version, please refer to the plate section.)

or other wood products. Instead of this strategy of natural tree regeneration, some clear-cuts are planted to become a forest plantation (Andrieu *et al.*, 2011). Meanwhile, this overall forest mosaic (with high beta diversity) provides a rich array of resources for multiple uses, including wildlife habitat, pest resistance, walkers, hunters, and mushroom hunters.

The forest ground layer and understory vegetation changes markedly with clear-cutting. For instance, in the Canary Islands (Spain), bryophyte (moss, liverwort) communities are sharply degraded where trees are cut, whereas around rock outcrops and waterfalls, bryophyte cover changes little (Patino *et al.*, 2010). In tree-cut areas, some resistant bryophyte species survive and come successional species spread. Analogously, logging sharply alters most wildlife characteristics, including animal guilds or feeding types (Poulsen *et al.*, 2011).

Some three billion people apparently use *fuelwood* (firewood; short small-diameter pieces of wood for cooking or heating) in their daily lives (Figure 11.6) (He *et al.*, 2009). Fuelwood harvesting generally degrades forest trees, understory vegetation, wildlife habitat, and the soil (e.g., Pompeii, Rocky Mount). Distance from village into a Tiger Reserve forest (Western Gats, India) is not very important when cutting firewood to be shipped and sold for earnings (Davidar *et al.*, 2008). Yet distance is quite important when harvesting fuelwood, fodder for livestock, or green leaves for fertilizer that must be carried for household use.

In China's famous Wolong Nature Reserve, containing 10 percent of the world's giant pandas, six villages are present and largely dependent on fuelwood cutting within the reserve (He *et al.*, 2009). Both people and houses are rapidly growing in the panda reserve. Formerly, timber harvesting, poaching, and farm plots were the main problems for pandas; now firewood removal is the big threat. With the increasing loss of forest cover and changing species composition, village cutters are removing fuelwood in more remote areas, higher elevations, and steeper slopes (see Figure 8.2). This growing threat to pandas is largely driven by tourists, who come to see pandas and panda habitat. In essence, fuelwood supports the burgeoning facilities for tourists who want to see pandas, while villages and fuelwood cutters provide the fuel by reducing panda habitat.

Useful Plants. In the extensive forest surrounding a town, ecosystem services tend to be abundant, yet resources are widely distributed. This generally requires some searching to locate resources, plus a transport mechanism and efficient route to carry harvested resources. In addition to wood products, *nontimber forest products* (NTFPs) are important in many areas (Grebner *et al.*, 2013). These products include game animals, water supply, forage for some livestock, and plants for household use. Plants harvested typically provide for medicinal, food, wood, weaving, and adornment uses. Villages, as in Indonesia, are often dependent on income derived from diverse forest products (Nakagoshi *et al.*, 2014).

Wild medicinal plants are particularly familiar and important to villagers and many town residents. Thus, time is well spent gathering medicinal plants in the forest (Gavin, 2009; Turner *et al.*, 2013). Essentially all forest habitats contain food plants such as fruits, nuts, mushrooms, bulbs, rhizomes/tubers, leaves, and buds (e.g., Pompeii, Valladolid).

For instance, residents from three rainforest villages (population 300–600 each in Peru) collect useful plants in both secondary regrowth forest and primary tall mature forest (Gavin, 2009). Medicinal plants are particularly abundant in the regrowth forest. On the other hand, mature forest provides the most plants for food, wood, weaving, and adornment uses. Therefore, both accessible secondary and primary forest are important to the villages.

Hunting Game. Hunting success or yield commonly increases with greater distance from a village or town, because nearby animals have been largely eliminated (Smith, 2008). Subsistence hunting from a settlement, or several settlements with overlapping hunting zones, often alters the types (guilds) of animals present, and may cause local species extinctions (Levi *et al.*, 2011; Poulsen *et al.*, 2011).

A study of hunting from three villages found the most kills in a mosaic of forest habitats (Parry *et al.*, 2009). The harvest of game per hunting effort here was lowest in an old forest. Villagers balance time to reach a harvest or hunting site versus the expected rate of success at the site (Rist *et al.*, 2008; Dalle *et al.*, 2011).

A Gabon (West Africa) national park with overlapping populations of chimpanzees and gorillas shows a different hunting pattern. Thirteen villages are adjacent to its border, several more villages somewhat further away, and three towns (population 10,000–18,000 each) located 20, 20, and 30 km away respectively (Kuehl *et al.*, 2009). Subsistence hunting by nearby villagers has relatively little effect on the ape populations. In contrast, the populations are mainly impacted by commercial hunting, i.e., hunters from the more distant towns, who ship their "bushmeat" to urban markets. This importance of outside, rather than local, hunters on forest animal populations is emphasized by a central Africa study, where hunting pressure correlated better with proximity from multiple roads (or road density) than from the nearest road (Yackulic *et al.*, 2011).

In a Peru rainforest where hunting had eliminated all large primates and 61% of the medium-size primates (relative to a control), interestingly hunting affects the vegetation (Nunez-Iturri *et al.*, 2008). Seedlings of trees with large seeds (<2 kg, or 4.4 lbs) are dispersed by the large and medium-size primates are reduced by 46 percent. That is equivalent to losing one such tree seedling in every square meter across the forest. Concurrently, seedlings of other tree species with wind or water dispersal increased 284 percent, equivalent to eight seedlings per square meter. In effect, removing the seed-dispersing animals has changed the forest species composition, specifically by reducing the tree species which depend on the primates. Almost all towns and villages surrounded by forest have active hunters (e.g., Huntingdon/Smithfield, Valladolid).

Water Supply and Grazing. Many other resources in the surrounding forestland are harvested for towns and villages. Forest usually means the presence of ample rainfall. Thus, normally both groundwater and surface water are available. Wells, streams, and rivers provide a water supply via pipes or canals to many towns (Turner *et al.*, 2013). Yet water from the same sources is extracted for irrigation of fields, gardens, and lawns. Surface waterbodies, such as streams, rivers, lakes, and ponds support fish populations, which are readily fished for food.

Relatively open forests and woodlands in the land surrounding towns normally contain considerable grass and other herbaceous vegetation grazed by livestock in relatively low density. Consider the following uses of seven protected Tiger Reserves in India by nearby villagers (Botteron, 2001): (1) livestock grazing in the reserve (and with cropland adjacent to a reserve border); (2) livestock grazing, and bamboo harvesting every four years; (3) hunting game; (4) poaching game; (5) livestock grazing (with 87 reported tiger attacks on cattle in a year), hunting, and poaching; (6) livestock grazing, and logging of teak trees; and (7) grazing of livestock (which pass diseases to native ungulates), and poaching game. Thus, local villagers have a range of extractive activities, especially livestock grazing, within these large protected reserves containing tiger populations.

Coppicing, as an alternative way to harvest wood, cuts certain broadleaved trees down almost to ground level. Numerous sprouts grow rapidly from the cut stumps, especially in the first year (Peterken, 1993, 1996; Burchett and Burchett, 2011). The pole-like woody sprouts are then harvested in the first one to five years, and more sprouts follow. More than 75 uses of coppice wood are recognized (Figure 11.7) (Burchett and Burchett, 2011), ranging from cricket bats and firewood to charcoal, broom handles, furniture, and piano keys.

Figure 11.7 Diverse wood products from a woodland maintained with traditional coppice-and-standards management. Numerous poles cut from sprouts of large mainly underground tree "stools" provide at least a dozen household uses, while scattered tall straight trees provide lumber. Bradford Wood near Lavenham, Suffolk, UK. R. Forman photo, with aid of Oliver Rackham.

Wood is also converted to *charcoal* with a slow burn in a pit or ground-covered mound. Producing charcoal often occurs when much of the surrounding land near town has been deforested, such as by an industrial operation (e.g., Kutna Hora, Falun). Wood is heavy to transport, while charcoal, relative to the energy provided, is much easier to transport to town.

Major Ecological Alterations

The preceding harvest or removal of resources from forestland surrounding towns and villages is the beginning of the story. Consider the many major non-extractive changes in the land by nearby communities. Land patterns are altered, human features and objects are constructed on the land, human-caused large wildfires occur, and altered habitat patterns change wildlife populations.

Landscape Pattern. From an airplane window, the land around a town may appear to be *linearized* or geometricized by a history of linear features, especially roads, railways, pipelines, electric powerlines, fences, hedgerows, and canals/ditches. Hardly any species evolved in such a land of crisscrossing movement barriers and conduits. Some prominent strips, including regional pipelines and powerlines or inter-city railways, are mainly unrelated to a town. But many lead to or from the town, and others reflect logging, farming, or irrigation activities of rural residents. Geometric polygons seem to dominate the landscape. Yet at different scales, hybrid nature–Euclid patterns (Chapter 1) are commonly widespread. These hybrids suggest places where strong natural processes and human activities operate concurrently, even in conflict.

Peering beyond the obvious construction of roads and other strips raises the question of how the patterns were produced. Conceptually, five main *spatial processes* may be responsible (Forman, 1995, 2014): *dissection*; *perforation*; *fragmentation*; *shrinkage*; and *elimination*. Powerlines and roads dissect the forestland. Scattered houses and an airport perforate the land. Highways and farm fields fragment the land into pieces. Deforestation shrinks forest patches. Fire and further deforestation eliminate forest.

Other conspicuous forestland patterns result from human activities (e.g., Rocky Mount, Kutna Hora, Valladolid). Tree plantations may appear. Vertical forest structure changes as livestock clean out the shrub and ground-layer vegetation. Small farm plots perforate the forest (Dalle *et al.*, 2011). Human activities usually sharpen boundaries, changing a *soft boundary* (gradient, convolution, or strip of micro-mosaic) to an abrupt straight *hard boundary* (Forman, 1995, 2014). Also, people frequently increase or decrease habitat patchiness. Indeed, at the broad scale, usually a conspicuous mosaic of land uses appears, often reflecting the history of logging (Andrieu *et al.*, 2011). A mix of regular geometries, hybrid nature–Euclid patterns, and natural pattern seems widespread in a town's surrounding land.

Fire. Arson may be more frequent in outer surrounding forestland than in the adjacent zone near town, because the arsonist is less likely to be seen and fire is more likely to spread widely. In both Portugal and the USA, fire frequency is higher near roads and near a greater number of houses (Moreira *et al.*, 2010; Hawbaker *et al.*, 2013). However, large fires (e.g., Chamusca, Vernonia) are more likely to start away from roads. Also, fires are usually most intense where fuel buildup is high, when moving uphill, and when moving downwind.

Fire management may have three broad objectives, which are partly in conflict, specifically to protect: (1) areas with buildings; (2) forests growing wood products; and (3) rare and uncommon fire-adapted species (Forman, 2004a). Protecting the fire-adapted species, which "require" fires to survive and thrive, is often the easiest to accomplish. Simply maintain a "let-burn" policy for specific designated areas. Protecting areas with buildings is difficult in the case of arson or high wind, but otherwise generally attainable by: channeling growth into compact areas; removing isolated buildings; closing remote and spur roads to motorized vehicles; maintaining an upwind firebreak; and establishing low burn-intensity vegetation upwind of town. Most difficult is protecting forests growing wood products, which may require maintaining specific forest sizes, fire-control access roads closed to other motorized vehicles, firebreaks, and reducing fuel buildup.

Features Inserted. Lots of town-related objects or features are constructed in the surrounding forestland. Diverse types of corridors were mentioned above. Roads to town may come in different forms, from highway to dirt road which, with vehicle use in dry periods, pours dust into the forest. Roads tend to parallel rivers and cross streams, both of which produce many important ecological effects (Forman *et al.*, 2003; van der Ree *et al.*, 2015). Bridges are commonly sources of scouring, sediments, and chemical pollutants that affect fish and other aquatic organisms for a long distance downstream.

But *culverts* for water passing under roads may have a much larger ecological impact. In heavy rains, inadequate-sized culverts create local floods. Worse though, in some rural areas 20–40 percent of the culverts for streamflow block fish movement up and down stream. Such culverts commonly render many kilometers of stream length in numerous tributaries of an entire stream basin totally unavailable to the fish. Fortunately, correcting a culvert blockage is usually surprisingly simple.

Trains, especially with steam engines, may pour particles and other pollutants over the land, and create sparks that start fires. A local airport is commonly in a town's surrounding land, and loud aircraft takeoff noise extends well beyond the runway (e.g., Rock Springs, Stonington). Off-road vehicles (ORVs, ATVs) may roar long distances on trails across the land, but frequently concentrate on a particular, often sandy, area closer to town, but far enough not to disturb neighbors (see Figure 4.11) (e.g., Falun). Motorbikes and snowmobiles range widely on trails through the forestland, often with roars going uphill (e.g., Rangeley, Huntingdon/Smithfield).

Wildlife and People. In unusual cases town or village hunters target and decimate a rare species in the surrounding land. Changing landscape patterns favors, or reduces, populations of many species. For instance, in Ontario forest, lynx movement (*Lynx canadensis*, a large cat) is strongly related to human-created landscape pattern (Walpole *et al.*, 2012). Lynx mainly forage near young regrowth forest, and away from roads and logging activity. Local hunters targeting large, slowly reproducing species in low density, such as lynx, bears, and forest elephants, often significantly reduce their populations in the surrounding forestland.

Subtle, Often Invisible Effects

> There was once a town … where all life seemed to live in harmony with its surroundings …
>
> Then a strange blight crept over the area … The birds, for example – where had they gone? … Even the streams were now lifeless …
>
> … between the shingles of the roofs, a white granular powder still showed a few patches, some weeks before it had fallen like snow upon the roofs and the lawns, the fields and streams … What has already silenced the voices of spring in countless towns in America?
>
> Rachel Carson, *Silent Spring*, 1962

Pesticides did that. Subtle often invisible effects permeate both towns and surrounding forestland, as implied in the quotation (McDonnell and Pickett, 1993). The *invisibles* – microbes, gases, aerosols, sounds, vibrations, air flows, smells, underground objects, submerged objects, tops of trees, distant objects, slow flows, slow changes, and nighttime organisms and movements – are just as important as the visibles (Forman, 2014).

Trails and Recreation. Imagine walking along a rainforest trail in south China. Beautiful orchids on the trees within 10 m (33 ft) of the trail have less pollination by pollinators, and hence less fruit production, than orchids in forest away from the path (Huang *et al.*, 2009). Indeed, the arrangement of paths in a forest determines most recreational effects on the biodiversity of animals, plants, and habitats, in addition to logistic considerations such as access, slope steepness, and safety (e.g., Issaquah, Petersham, Huntingdon/Smithfield) (Ferrarini *et al.*, 2008). The rules of *path network design* are to first map the valuable and sensitive features, and then locate trails to avoid the forest interior and avoid the valuable and sensitive features.

Suppose the forest trail leads to a small picnic area. A study in southern Queensland (Australia) found that birds in the forest edge of picnic areas are relatively large and aggressive (Piper and Catterall, 2006). These birds are similar to suburban birds in the region, but unlike the numerous small insect-feeding birds of the forest interior. Also, predators disrupt more bird nests with eggs or nestlings in the picnic area forest edge than in the forest interior. Even birds in forest edges along nearby roads are more similar to forest interior birds than to birds of the picnic area.

Walking (or dog-walking, hunting, snowmobiling, and so forth) along a trail initially causes wildlife ahead to physiologically go on alert. Continued walking *flushes* an animal to move away, usually to a less suitable habitat out of sight, where it stays until well after the walker has past. Flushing distances of course are greater for noisy snowmobiles or for people with a dog which is perceived as a predator. With such trail users repeatedly passing, wildlife quickly learn to avoid the vicinity of a trail. In this way a trail network in the forest surrounding a town both reduces and subdivides suitable wildlife habitat.

People in Forest. Perhaps the greatest subtle effect of people on forestland is *road access*, which opens up a large natural area to human impacts. Closing and effectively removing a spur no-outlet road or a little-used road produces extensive ecological benefits.

People living in forest separate from a village or hamlet create subtle or indirect effects over a wide area around the house. A family forest in the USA averages 11.3 hectares (29 acres), but residents living on the property move widely through the surroundings (Figure 11.8) (Butler, 2008). Fences and fencelines often seem innocuous, but are typically both barriers to movement across for some animals, and conduits for movements (with associated effects) along a fence. Also, lots of indirect and invisible effects can be expected from community-based natural resource management from a town (Jones *et al.*, 2013).

San Juan Gbee, Oaxaca (Mexico), a village with 200 houses and 930 residents who all speak a distinctive language (zapotec), has about 5,500 ha (8.6 mi^2) of mainly dry forest with small crop fields (Hunn, 2008). Residents regularly range outward several kilometers, effectively the distance for walking out and back in a day. Some wood is cut in an adjacent communal forest, but wood is heavy for people to carry far. Residents harvest nearby firewood, stones, and adobe clay. Nontimber forest products (NTFPs), including medicinal plants, edible plants, mushrooms, and cut flowers are collected as

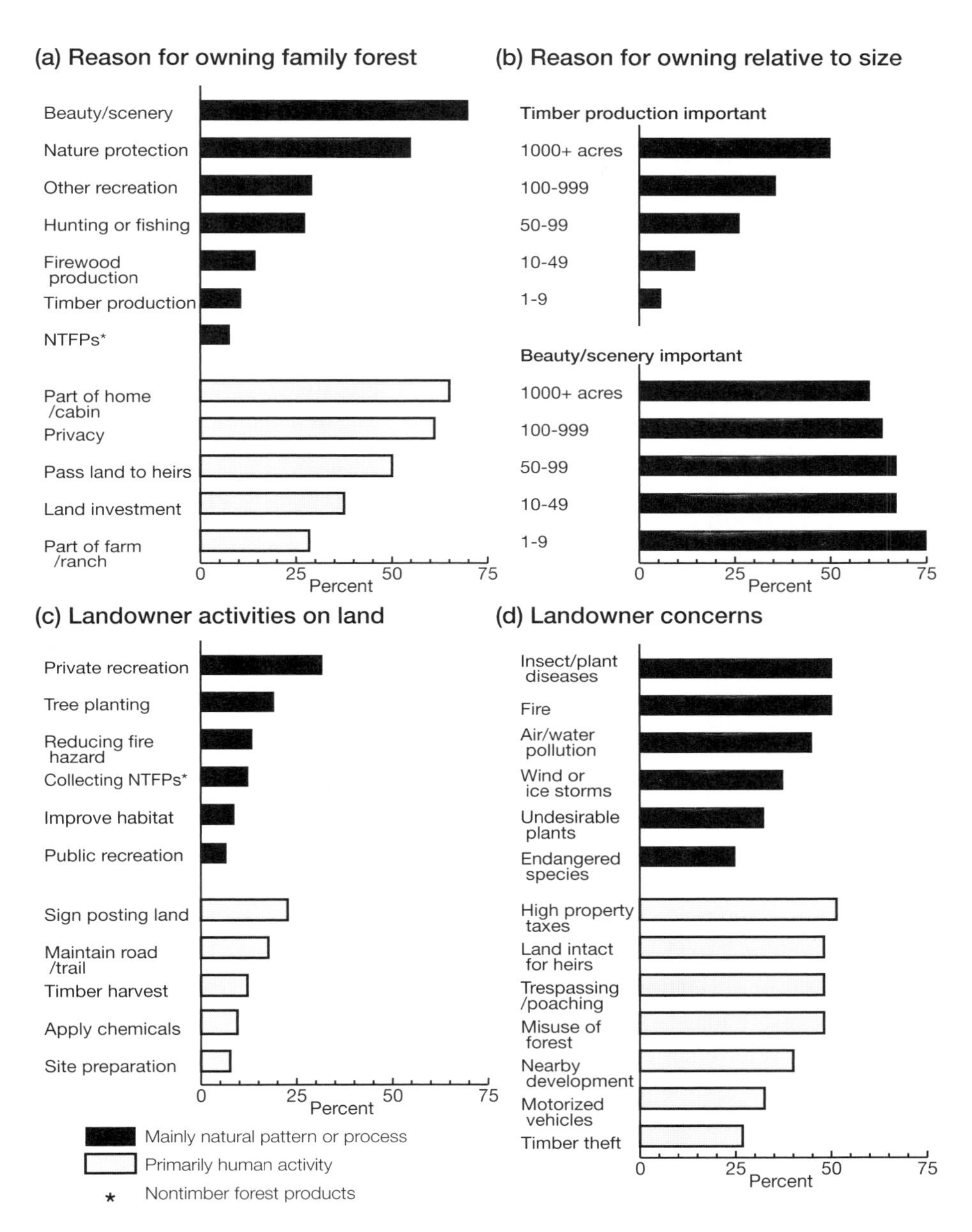

Figure 11.8 Landowner reasons for owning, activities on, and concerns for family forests in the USA. Average family forest = 11.3 ha (29 acres). Solid bar = mainly natural pattern or process; open bar = primarily human activity. * = nontimber forest products. Based on extensive surveys of landowners. Adapted from Butler (2008).

staples. Clay removal is conspicuous, whereas mushrooms collected leave no trace. San Juan Gbee is a sustainable subsistence forest village.

Air, Water, Soil. Skyglow extending outward for kilometers from a town may attract night insects, birds, and mammals to the town (Chapter 8). Irrespective, the artificial night light provides scarce light to the forest that widely affects both predators and prey (and potentially plants). The diverse forest activities originating from town and village have the cumulative effect of increasing soil erosion and reducing topsoil, as illustrated from Greek and Roman times to mid-nineteenth-century Vermont (USA) (Marsh, 1864). Groundwater variations downslope of a town are invisible, and changes in stream/river sediments and other pollutants are rarely noticed. The gradual sediment-accumulation clogging of a road culvert slowly eliminates fish habitat in upstream tributaries across a whole drainage basin (Forman *et al.*, 2003). That is an almost never noticed process.

Biodiversity and Wildlife. A well-used forest road has very few animals on it, whereas the same road rarely used by vehicles has lots of vertebrates, and is generally an important conduit for predator movement (Forman, 1995; Forman *et al.*, 2003; van der Ree *et al.*, 2015). A rare habitat may be degraded quickly or very slowly over time, but most declining rare species, almost imperceptibly, lose individuals until potentially going locally extinct. Such habitat and species loss, or gain, in biodiversity typically happens more slowly than people notice.

Town and village effects on wildlife populations and behavior in surrounding forest are often rapid, but rarely noticed, because the animals flee into escape cover (Landers *et al.*, 1979; Forman and Collinge, 1996; Knight and Gutzwiller, 2013). Human effects on wildlife occur at different scales, from a point to an entire region, such as the changing patterns of large mammals and birds across the Yucatan region (Mexico) (Urquiza-Haas *et al.*, 2009). Reducing top predator populations has trickle-down effects, permitting herbivores such as deer and kangaroos to become more abundant, which in turn often reduces vegetation cover. Also, large predators reproduce slowly so a population loss persists a long time.

Several other results of town effects on animals in the surrounding forestland are invisible or essentially so, but very important ecologically (Forman, 2014). Reproductive success changes. Behavioral patterns are altered. Movement routes shift. Diets shift. Genetic variation evolves. Adaptations appear. Animals become habituated. Specialists become generalists. Nonnative plants remain from former habitations or activities (Kuhman *et al.*, 2010). *Legacies*, that is, events or activities that happened in the past, determine current responses and patterns (e.g., Vernonia, Tomar).

Effects of Outer Surrounding Forest Land on Town

Wind and rain, regional dust and pollutants bathe a town, but do not originate in the surrounding land. However, trees in the outer surrounding forest pump water from the ground and (evapo)transpire water molecules from leaves to air. Wind then carries the atmospheric moisture to town, which is important in a dry climate. Tree leaves also give off organic chemicals which blow into town (Forman, 2014). Both inputs are generally considered to be positive.

Wildfires from the outer surrounding forest, especially with a strong wind and fire-prone vegetation, occasionally blow into town, causing extensive destruction (Theobald and Romme, 2007). Wind-driven fires may jump a firebreak in the adjacent zone, often in advance sending little burning branches into town.

Certain insect and herbivore populations sometimes build to pest proportions in the forest and readily permeate the town. Gypsy moths (*Lymantria dispar*), spruce budworms (*Choristoneura*), deer, and kangaroos are widespread examples. In town the pests continue defoliating, and sometimes killing, trees or other vegetation.

As mentioned above, artificial night lights in town produce a skyglow extending into the surrounding land that may draw in night insects. Bats may sometimes forage 10 km (6 mi) or more from their roost, and be attracted to the light. The bats have genetically "learned" that, except in cold periods, night lights in town mean lots of flying insects as good food (Struebig *et al.*, 2009).

Large predators from the outer surrounding forest come into some towns and villages, particularly when their food supply is low or they are used to finding food in town. Tigers and leopards in India, and black bears and grizzlies in North America, are familiar examples. These entering animals represent danger for residents, may result in deaths, and sometimes become pests. Towns and villages next to large forest tracts may expect to have large predator visitors.

Finally, in the dry forest of northern New Mexico (USA), two towns (Espanola, population 10,000, and Chimayo, pop. 3,200) and two nearby villages (Cordova, pop. 400 and Truchas, pop. 1,200) have the following interactions year round with the adjacent zone and surrounding land (Kosek, 2006). Analogous linkages may characterize most towns.

Town Effects on Forestland. Firewood harvested. Cattle grazed. Sheep grazed. Government land used. Pinon nuts for food collected. Metals mined, leaving piles of mine waste, with mine chemicals washed downslope and down gullies. Game hunted. Fish caught. Extensive horse riding. Fires ignited for varied purposes. Trees selectively cut. Town stormwater channeled downslope. Untreated or partially treated wastewater dispersed downslope and into gullies. Habitats degraded by diverse activities. Rare species populations reduced by hunting, collecting, and habitat degradation. Extensive off-road vehicle (ORV, ATV) use. Solid waste dumped in a few large dumps, and many tiny ones. Small orchards on optimal sites. Small farm fields in spots with good soil moisture. Irrigation ditches channeling surface water to orchards and fields. Fencing through woodland for livestock. Dogs free-ranging. Traffic noise effects. Roads as barriers to small animal movement. Roads and culverts altering surface water flows. Road-killed wildlife. Well pumping by town and homes, lowering the water table. Nonnative plants spreading into woodland. Biodiversity probably decreasing. Materials from the forest for religious and cultural purposes harvested.

Forest Effects on Town. Smoke from wildfires. Dust from wind erosion. Snow accumulations from snowstorms and blizzards. Visitors, tourists, and pilgrims arriving

(Chimayo has a revered sanctuary which attracts visitors). Pottery shoppers arriving. Hikers arriving. Loggers arriving. Firefighters arriving. Government forest trucks endlessly arriving. Campground campers arriving with needs. Environmentalists arriving. Flash floods and floodwater pouring in during heavy rains. Massive sediment accumulations deposited by floodwater. Adobe, stone, and wood carried in by residents. Bears, cougar (mountain lion), and elk periodically arriving, and deer more often. Chainsaw noise. Truck and mountain vehicle noise. Aerially sprayed insecticides for spruce budworms float and wash in. Herbicides against mistletoe enter. Ash from wildfires falls in. Radioactivity spreads into town and village, doubtless carried by both wind and residents returning from work (just upwind of the towns and villages is one of the world's largest nuclear research facilities, which has worked with radioactivity for over 70 years).

Clearly towns and their residents are totally tied to their surroundings, which provide a treasure chest of resources, problems, and activities.

Water and Livelihoods in Aridland

Every day residents of desert towns and villages see and appreciate the open landscape surrounding their town. People are culturally tied to their surroundings, perhaps more than in any other part of the globe. Over time, residents have been in many parts of the landscape and some people use the land regularly. In a dry landscape, human effects, even minor ones, are conspicuous and persist for long periods before slow ecological succession heals the desert vegetation.

Infrequent disturbances, such as extreme rainfalls, flash flooding, heavy sediment deposits (Barrow, 1991), prolonged drought conditions (Lake, 2011), extreme low temperatures, and clouds of pest insects, seem to be all too frequent. But experiencing and learning from such disturbances creates a highly adaptable population. A keen observer notes many *adaptations* that anticipate and will help the community get through future environmental changes.

Aridland refers to areas with less than 300 mm (11.8 inches) of annual precipitation, and thus may be considered approximately synonymous with *desert* (Ludwig *et al.*, 1997; Whitford, 2002; Ward *et al.*, 2016). Typically, the surface is covered by rock, sand, a biological *soil crust* (composed of mineral particles with, e.g., lichens, algae, and microbes), scattered grassy patches, and/or scattered shrubs. *Semi-arid land* or *dryland* has 300–500+ mm of precipitation, and thus may include dry grassland, shrubland, or woodland, often with grazing livestock (Chapter 10). Aridland precipitation is not only low, but erratic and unpredictable. Irrespective, *evapotranspiration* rates, of water molecules from soil and plants to the air, may be high in both arid and semi-arid lands. Most of the aridland insights ahead also apply to *desertified land* or *desertification*, which has largely lost its productivity due to human effects, especially overgrazing of semi-arid land.

Desert towns and villages occur where they can get water (Figure 11.9). *Oases* are usually in low places where groundwater may be near the surface and support luxuriant

Figure 11.9 Scattered desert oasis villages where the water table is close to the ground surface. Small irrigated rectangular fields within oases. Near Kerman, Iran. R. Forman photo, courtesy of Ed LaPorte.

plant growth. Some communities depend on water from nearby mountains or hills. Most pump deep groundwater.

Town residents tend to be familiar with the ecological processes and natural history of the desert, which are largely hidden to outsiders. For instance, holes of different sizes and locations in the ground indicate what animals are resting below (Merlin, 2003). At night, especially around scattered shrubs on small accumulations of sand, the desert often becomes alive with animals in a diverse food web.

Deciduous shrubs and subshrubs, annual grasses, forbs (e.g., herbaceous plants with conspicuous flowers), and evergreen subshrubs tell residents much about water conditions (Dregne, 1976; Whiford, 2002). Even the fine-scale surface pattern of grassy strips and patches indicates where rainwater, sediments, and plant litter flow and accumulate (Ludwig *et al.*, 1997). Virtually all long-term desert town and village residents know what to do, and importantly how to find water, when lost in the desert, an adapted talent known to few outsiders.

We now explore towns in deserts from three perspectives: (1) water in the desert; (2) flows between towns and surroundings; and (3) livelihoods in aridland.

Water in the Desert

Precipitation, either on-site or far off, provides water for a town or desert location. A light rain, e.g., less than 13 mm (half inch), mainly soaks into the soil, and then soon is largely sucked back up into the atmosphere by evapotranspiration (Whitford, 2002;

Ward *et al.*, 2016). In a heavy rain much of the water moves horizontally as surface runoff to vegetation patches, or forms puddles or shallow ponds. A very heavy rain typically causes flooding, indeed a *flash flood* of water flowing rapidly across the surface (as sheet flow). Heavy rains in nearby hills or mountains often produce flash floods across roads or into a town.

Water from rain or snow on hills and mountains is the source of streams and rivers, which in lowland desert may flow visibly over the surface in gullies and valleys. However, more often in deserts the water flows invisibly within the sandy bottoms of gullies and valleys. Some of the hill/mountain water may infiltrate into soil down to the deeper groundwater, which also flows invisibly in the ground. Most groundwater flows at some depth under the desert, though some groundwater reaches the surface as a spring or seep at the bottom of a hill or mountain.

Wells penetrating into groundwater pump water for drinking and other uses, such as mining, farming, and industry. Commonly desert well-water is rich in certain nutrients and heavy metals, sometimes at toxic levels, even with radioactivity. Nevertheless, wells provide irrigation water used for livestock, hayfields, orchards, or crops.

A desert valley may be a corridor of food-producing plots irrigated from water flowing invisibly in the valley-bottom sand. The irrigated cropland eliminates most floodplain and riparian vegetation, which typically remains as only narrow remnant strips (e.g., Socorro; see Appendix). Surface water runoff may also be partially captured in gullies with small "check dams," which form small temporary ponds used by livestock and native wildlife (Figure 11.10). Sometimes such ponds are used for temporary irrigation.

Low shrubby vegetation commonly requires little water to survive (Dregne, 1976). A vegetated area captures considerable surface runoff, much of which infiltrates into the soil. Some subsurface water then is absorbed by plant roots while some may reach the groundwater. Groundwater then may be pumped to support human activities as well as other plants (Abderrahman, 2000). Maintaining or restoring desert plants thus helps reduce water runoff flooding and erosion, but usually a greater benefit is reducing soil erosion by wind (Dregne, 1976).

Desert towns and villages have water sufficient to grow trees, cover the area with some vegetation, and often support small farm or garden plots (Figure 11.11). Occasional floods may be a problem. Since water areas are scarce, many water-dependent desert plants and animals are rare. Wherever water comes to or near the desert surface, such as at a town, that is a permanent attractant for wildlife. Consequently, in towns and villages rare species and abundant animals occupy many of the same scarce spaces.

Some towns proactively plan and monitor water uses. But most desert towns continue many wasteful practices (Dluzewska, 2009; Forman, 2014). Water shortages worldwide are worsening, and this may be especially true in water-scarce desert towns and villages (Millennium Ecosystem Assessment, 2005).

An oasis town, Ejina, Inner Mongolia (China; population 16,500), apparently owing its origin to the melting of far-off mountain glaciers due to climate warming, essentially began in 1950 (Zhang *et al.*, 2012). The population of both people and livestock grew linearly to 2004. The nearby Ejina River during this period has changed from 2 to 19

Figure 11.10 Dam creating a temporary pond for livestock within dry riverbed in desert. Trees around pond and area below the dam. Graded road (right) crosses the riverbed near a rectangular cattle-holding plot downriver from the dam, minimizing the chance of blockage or danger from a flash flood. West central Namibia. R. Forman photo.

channels, which frequently move in location during annual snowmelt periods, though apparently total water flow has changed little over the 54 years.

Microhabitat diversity in the river basin, based on different moisture regimes (determined by dendrochronology of trees dating back centuries; Woodhouse, 2004; Lara *et al.*, 2008), was greater before 1950, and has progressively decreased. This loss of habitat diversity probably results from the increase in number of small dams and reservoirs from 2 to 98 during the period. Since the 1970s, the amount of poplar (*Populus*) forest in the oasis has decreased by 50 percent, and its fragmentation into patches has increased. This vegetation change alters the oasis–desert atmospheric boundary layer, permitting hot air to penetrate the oasis. That increases evapotranspiration, which in turn reduces available groundwater for the town.

Also, strong winds and sandstorms occurred on average every three years (during 1973–1998) (Zhang *et al.*, 2012). These events alter soil and growing conditions, e.g., by exposing or burying tree-root systems. During each of the strong-wind years, or the year following, microhabitat diversity near the river was reduced.

In short, the 54-year period of desert town activities – growth in population, small dams/reservoirs, and livestock – superimposed on the presumed long-term pattern of strong winds/sandstorms, has reduced soil moisture-based habitat diversity for vegetation. Oasis forest loss and fragmentation degrades town conditions and probably lowers groundwater. Ironically, the town and its people apparently owe their existence

Figure 11.11 Desert town with no lawns, some abandoned buildings, mine and coal waste (far left), and hill providing cool air drainage. Superior, Arizona, USA. See Appendix. R. Forman photo. (A black and white version of this figure will appear in some formats. For the color version, please refer to the plate section.)

to climate warming. But when the mountain ice is gone, the river-water flow will disappear, threatening the town's very existence.

Livelihoods in Aridland

In spite of the characteristic water shortage in deserts, people have managed to thrive or survive in a variety of ways. Low-density grazing livestock may persist at the moist end of the precipitation gradient (Ludwig *et al.*, 1997). Yet livestock especially graze the often scarce grass in a desert, which leads to soil erosion.

Removing low-density cattle for 19 months from an arid Simpson Desert (central Australia) area resulted in (Frank *et al.*, 2014): a change in plant species composition in woodland; little change in grassland species composition; an increase in livestock-palatable plant cover and more flowers (forbs); no effect on reptiles; and variable

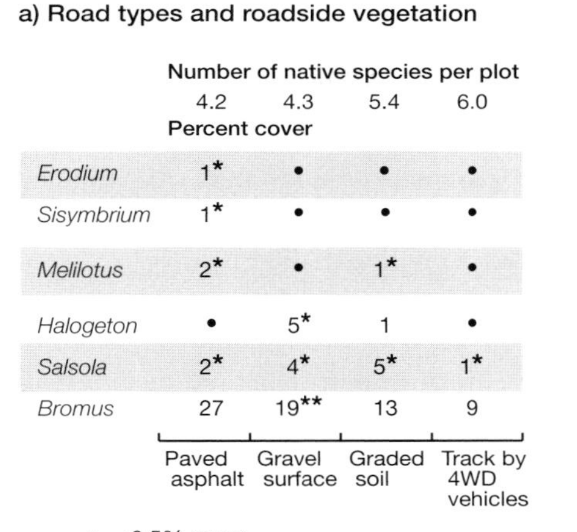

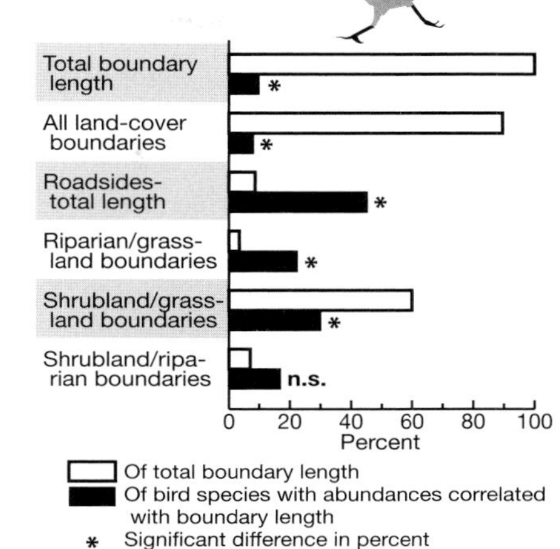

Figure 11.12 Species associated with different types of roads and boundaries in semi-arid landscapes. (a) Six grass species by different road types. Based on Gelbard and Belnap (2002). (b) Bird abundance (black bars) relative to boundary length (white bars). South Texas. Adapted from Gutzwiller and Barrow (2008).

effects on small mammals. Cattle grazing reduces grass cover which favors the spread of shrubs, though not always (Browning and Archer, 2011). Also various boundaries providing habitats are prominent in dry landscapes. In these, birds are especially abundant along roadsides and riparian/grassland boundaries, compared with other common boundary types (Figure 11.12).

Dryland farming (crop production without irrigation in semi-arid areas) sometimes works at the moist end of the desert precipitation gradient. Human-created linear features are abundant and important for the distribution of both plant and bird populations (Figure 11.12). Soil erosion is usually a problem in dryland farming. Livestock and non-irrigated crop production in deserts are normally subsistence farming, and at best support low-income residents (Millennium Ecosystem Assessment, 2005).

On the other hand, irrigated crop production, for example of hay, certain vegetables, and other somewhat salt-tolerant crops, may be quite successful economically. The water brings in mineral nutrients. But water is evaporated to the air, leaving dry nutrients as *salt buildup* (salinization) in the soil. Crop production persists until the water begins to run out, or the soil becomes too salty for the crops to grow.

Livelihoods tied to tourism and/or recreation may thrive in deserts. Tourism usually focuses on a historic site, rare water body, mountain climbing, or desert animals and plants. Hiking, biking, hunting, rock climbing, cave exploration, and extreme sports are common recreation examples (Knight and Gutzwiller, 2013). Significant constraints on tourism and recreation also exist in aridland, such as harsh climate, remoteness, water shortage, scarcity of woodland, and habitat degradation by activities (Millennium Ecosystem Assessment, 2005). Sustaining protected areas for animals with large home ranges with minimal human effects in a water-scarce land is difficult. Often biodiversity and people compete for the same scarce water.

So-called *alternative livelihoods* in deserts include: cash-crop agriculture under plastic; dryland aquaculture; and growing relatively salt-tolerant plants or aquatic organisms using somewhat salty pumped groundwater (Millennium Ecosystem Assessment, 2005). Many innovative building and development approaches highlight solar energy, wind energy, or evaporative cooling in deserts.

A range of *ecosystem services* provided by the natural ecosystems of deserts has been suggested (Costanza *et al.*, 1997; Whitford, 2002; Wall *et al.*, 2012), such as: climate regulation; water supply; erosion control and sediment retention; disturbance regime; soil formation; nutrient cycling; refugia for species; waste treatment; food production; raw materials for manufacturing; recreation; and genetic resources. Almost all, however, also apply in most other regions, where they operate at greater rates.

Finally, *desertification* or land degradation is the all too common result of living in deserts. This process causes an enduring loss of primary productivity by desert vegetation (Barrow, 1991; Whitford, 2002; Millennium Ecosystem Assessment, 2005). *Overgrazing* (by livestock densities that degrade vegetation and soil) is the major cause of desertification worldwide. Pumping groundwater to create *water points* (Chapter 10) for watering livestock has enormously expanded the area overgrazed and desertified (Whitford, 2002). Livestock grazing reduces vegetation cover, input of leaf litter, water infiltration, soil organic matter, soil moisture, root growth, and seedling establishment (Ludwig *et al.*, 1997; Whitford, 2002). Furthermore, grazing increases soil temperature, water runoff, water erosion potential, wind erosion potential, and heterogeneity of soil nutrients via dung and urine. These effects together result in a long-term loss of vegetation growth, i.e., desertification.

Other desertification processes include crop irrigation leading to salinization of soil, destabilization of sandy areas leading to sandstorms and spreading dunes, and global climate warming. Grassland and livestock experts often consider the spread of shrubs to be a desertification process. Even a four-wheel-drive vehicle cutting through the desert soil crust, or a path walked frequently for a month in the growing season, typically leaves tracks visible for decades. Some desert residents are careful to avoid such ever-present desertification processes. Yet the history of humans with livelihoods from deserts is recorded in the extensive areas of the global land surface transformed from grassland to desert-like conditions by human activity.

Flows Between Desert Town and Surroundings

Desert towns interact with the ground beneath them, for example in pumping groundwater for household use, irrigation, and industry, or mining metals from rock or sand from a floodplain. In deserts, toxic elements are often pumped to the surface. Flows downward from the surface also occur, including stormwater and its pollutants (sometimes from industrial smoke and waste), wastewater sewage, chlorine from swimming pools, and injected fracking chemicals. Since water is so scarce in deserts, groundwater pollution is serious.

A town is also tied to its surroundings by the above-ground energy, materials, and species arriving and leaving. Thus, the following two sections explore: (1) desert affecting town; and (2) town affecting desert.

Desert Affecting Town

Water, air, and species enter town, and often serve as vectors carrying other things. The town edge and the adjacent zone outside affect these inputs (see Figure 2.4).

Water. In excess, water becomes floodwater that covers low areas or much of a town. Floodwater often carries: rocks, pebbles, and sand; tree trunks and uprooted shrubs; and toxic mine waste. A surface water stream or river in normal flow condition brings in fish, plankton, leaves/twigs, sandy sediment, nutrients, and sometimes mining and other toxic chemicals. Also, the stream corridor as a whole is a major route for terrestrial animal movement, including reptiles, mammals, and birds, from desert to town.

Wind. Deserts normally have considerable wind, and strong winds occur frequently (Whitford, 2002). Wind carries into town: desert heat and cold; rain and snow; sand and dust; plus nutrients, heavy metals, and other toxic materials from desert soil and mine waste (e.g., Superior; see Appendix). Fire infrequently enters towns in moist desert areas with considerable grass or other vegetation cover. *Wind erosion*, by saltation (lifting off) of particles from desert soils, is higher on smooth surfaces with smaller particles or less moisture (Barrow, 1991; Lancaster, 2004). Blowing sand leaves sediment piles, even dunes, in town (Figure 11.13). Dust carries in chemical contaminants as well as bacteria and fungus spores.

Although the urban heat island barely applies to desert towns (Golden, 2004; Forman, 2014), plants and buildings divert and funnel air and pollutants that cool specific spaces in town and in buildings (Koch-Nielsen, 2002). Wind runs windmills pumping water and wind-energy turbines generating electricity. Woodland and, at a finer scale, shrubs protect soil from wind erosion. On a still night with no wind, cool air on a hill/mountain top close by is heavier than the warm air in town, so cool air moves downslope (*cool air drainage*; Chapter 5) (Figure 11.11) (e.g., Superior). That forces the heat and pollutants in town air upward, effectively ventilating the town (Forman, 2014).

Species. In a *mass flow* or transport process, wind and water passively carry organisms and other items into town. Thus, seeds, spores, algae, spiders, and other small organisms are blown into town (Ward *et al.*, 2016). Also, seeds enter in floodwater, and plankton in streamflow (Lake, 2011). Other animals enter by *locomotion*, using their own energy from food consumed. Bats, birds, butterflies, and diverse insects, including infrequent insect plagues, fly in. Mammals and reptiles walk or run in. Animals are attracted to the water in town, and at different times various species become abundant enough to

Figure 11.13 Transportation by two-humped camels in desert with moving sand dunes. Lobe of Gobi Desert, southern Inner Mongolia, China. R. Forman photo, courtesy of Jianguo Wu. (A black and white version of this figure will appear in some formats. For the color version, please refer to the plate section.)

be pests. Also, some animals entering desert towns are dangerous. In effect the town experiences a species rain.

Town Edge and Adjacent Zone. Conditions in these two strips may have a little or large effect on the items entering. A stream corridor slicing through town facilitates entering by many species. The adjacent zone often contains town-related features such as a dump, sewage treatment facility, golf course (e.g., Page), highway maintenance facility, bypass highway, and airport. New residents may settle on the outskirts of towns, e.g., in mobile-home neighborhoods. Informal squatter settlements form where arrivals have no money and scavenge materials for shelter near potential jobs (Figure 11.14).

All of the town edge and adjacent zone features provide habitat diversity, and have some water and other resources that attract animals from desert to town. Desert towns surrounded by a wall or other barriers have less interaction with the surroundings, whereas a convoluted town boundary facilitates movement inward and outward (see Figure 2.3) (Forman, 1995, 2014). Roads radiating outward from a town facilitate the movement inward of school buses, commuters, shoppers from villages, and certain animals along moist ditches.

Land uses within a town, such as xeroscaping, irrigated gardens, or trees, also attract or repel animals entering (Figure 11.11) (e.g., Page). For instance, birds in suburban/urban Phoenix, Arizona (USA) were studied in 40 neighborhoods located 12–12,000

Figure 11.14 Informal squatter settlement of recent arrivals in desert town edge. Government provides some widely separated public toilets and street lights, plus a straight earthen road through center with narrow perpendicular roads; residents create infrequent shops and occasional school rooms. Swakopmund, Namibia. R. Forman photo. (A black and white version of this figure will appear in some formats. For the color version, please refer to the plate section.)

m from a surrounding large desert tract (Lerman and Warren, 2011). Native birds are most abundant in neighborhoods: (1) with desert landscaping designs, such as crushed stone instead of turf grass; (2) of higher income residents; and (3) closer to large tracts of desert. Native bird species richness (mainly the surrounding Sonoran Desert species) increases with the number of desert trees, and also the number of shrubs, in a neighborhood. In effect, house plots with more desert landscaping attract more native birds and species into town.

Town Affecting Desert

Movement Along Roads. Some movements outward from town are on the several radiating main roads, and may extend shorter distances on small paved or unpaved roads projecting from town. The density of unpaved roads is often high in the area surrounding desert towns. Movement outward includes: walkers and bicyclists; commuter car traffic; school buses; and trucks carrying goods. Due to rainwater runoff, roadsides and ditches along paved desert roads are typically greener, with denser vegetation and more nonnative species than in the surroundings (Forman *et al.*, 2003; Kalwij *et al.*, 2008; Craig *et al.*, 2010). This effect decreases, and grass species change, along a gradient from paved road, graveled road, graded dirt road, and four-wheel-drive track (Figure 11.12) (Gelbard and Belnap, 2002). Road maintenance activities increase microhabitat diversity, favoring both native and nonnative species (Bugg *et al.*, 1997).

On main roads, vehicles produce traffic noise during busy periods. The noise may have significant effects on wildlife outward for a few hundred meters on both sides of a road (Chapters 6 and 12) (Forman *et al.*, 2003; Reijnen and Foppen, 2006; van der Ree *et al.*, 2015). In the Chihuahuan Desert of Texas, 46 percent of the bird species correlate, half positively and half negatively, with the total length of roadside present (Figure 11.12) (Gutzwiller and Barrow, 2008). On unpaved roads, especially during dry periods, dust covers the nearby downwind vegetation.

Wildlife roadkill is especially frequent on fast two-lane highways, such as the radiating main roads. Scavenger birds and mammals may move along roads looking for roadkill. Predators often forage along seldom used roads and tracks, but busy roads are little used as conduits for wildlife movement (Forman *et al.*, 2003).

Linearization. Infrastructure extending outward from town, such as pipelines, electric powerlines, roads, railways, and irrigation canals, linearizes or geometricizes the land (Forman, 1995; Forman *et al.*, 2003). Such a pattern provides conduits or corridors facilitating the flow of water and movement of wildlife. At the same time, the pattern disrupts flows and movements across the land, since the corridors also serve as barriers or filters.

Irrigation of Fields. Near towns, irrigation produces yet more imprints on the arid land. Examples are an elongated series of small fields along a desert floodplain, a rectangular patch with a grid of irrigation canals and ditches, or a circular imprint from central-pivot irrigation (a long pipe spraying water and rotating on wheels around a central well pump). Downwind of an irrigated area is a plume of moister air that reduces evapotranspiration and facilitates plant growth. Often downslope of irrigation, the water table is lower and surface water flows are reduced. The irrigated area itself is a major attractant for surrounding wildlife, as well as for migrating birds (e.g., Socorro).

Recreation. Recreation by town residents in the surroundings includes (Knight and Gutzwiller, 2013): hunting and fishing; hiking (e.g., Alice Springs); horse riding; shooting practice; off-road vehicle and motorbike use (e.g., Superior); plus picnicking, dog-walking, and nature walks. Fishing is targeted to a specific unusual spot(s) in a desert, and shooting practice often focuses on a particular quarry or other isolated spot. Off-road vehicle use also is usually concentrated on a specific, often sandy, small location away from town, where a network of intensely used trails has been created (see Figure 4.11). Picnicking is usually at a beautiful or appealing spot, often near water or a high place with a view. Motorbiking and nature walking usually use a larger area with a network of trails, but, being almost mutually exclusive, use different areas.

Industrial and Other Waste. Diverse wastes from town also affect the surroundings. Air pollution from a factory(s) extends outward in a downwind plume (e.g., Superior). A wastewater treatment facility and a solid-waste dump are both at discrete locations

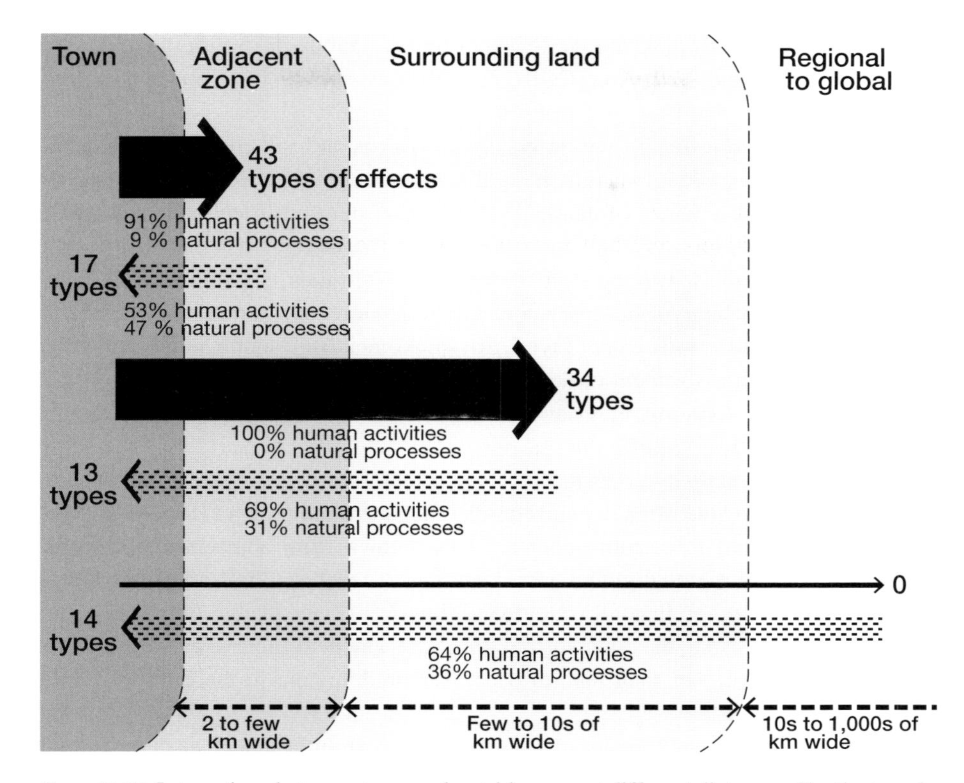

Figure 11.15 Interactions between town and outside areas at different distances. For the town's adjacent zone and outer surrounding land, approximate numbers refer to types of effects mentioned in this chapter. Percentages are for mainly human activities and mainly natural processes.

near town. A local airport used by town residents has a long flat strip usually outside town (e.g., Rock Springs, Superior), and aircraft noise extends farther in the usual upwind takeoff direction.

Wastewater-Enhanced Streams. Streams receiving treated or partially treated wastewater are increasingly important in dry climates, such as deserts and Mediterranean-climate areas (Luthy *et al.*, 2015). The addition of effluent from a wastewater treatment facility to a stream raises the level of (base) water flow. Intermittent stream sections may begin to flow perennially. The downstream benefits of adding wastewater effluent to streams may include (ironically) improved water quality, increased wildlife habitat (Halaburka *et al.*, 2013), and more recreational opportunity.

Hazards or problems are also present though, including: higher water temperature unsuitable for some fish and other aquatic organisms; altered sediments; mosquitoes; nitrites; and trace concentrations of pharmaceuticals, steroid hormones, and personal care products (Brooks *et al.*, 2006; Johnson and Sumpter, 2015). Nevertheless, benefits may outweigh shortcomings (National Research Council, 2005). Streams (in the USA) benefitting from added wastewater effluent include the South Platte (Colorado), Trinity (Texas), and Santa Ana (California) (Brooks *et al.*, 2006; Bonada and Resh, 2013).

Food Gathering. In the desert outside towns and villages, food gatherers target areas with a diversity of specific natural habitats (e.g., Alice Springs) (Walsh, 1990). Remote Aborigine villages of central Australia are unusually rich in community facilities and resources. For example, compared with a "hamlet" of less than 100 people, villages of more than 1,000 people are about 10 to 100 times more likely to have a pre-primary school, primary school, secondary school, store, accommodation for the aged, and primary health care center (Taylor, 2014). Women from such hamlets and villages typically collect an assortment of desert foods to feed village families (see Figure 7.10) (Walsh, 1990). In foraging, they have learned to search different habitats in different parts of the adjacent zone at different seasons.

It gradually emerges in this chapter that a rural town would wither without its adjacent zone, and often depends on its outer surrounding land (Figure 11.15). The number of types of flows/movement entering a town from adjacent zone, outer surrounding land, and regional/global area is similar. A third to a half of the inward flows are natural processes. In contrast, outward flows/movements are highly diverse to the adjacent zone, less so to the outer surrounding land, and negligible to the broader context. Human activities rather than natural processes predominate in the outward flows/movements from town.

The adjacent zone usually contains town essentials, such as dump, cemetery, sewage treatment facility, and fairground. It also provides the feeling of space and distinctiveness, separate from the nearby competing town, and not just a piece in sprawl. Adjacent vegetation provides the town's species rain, pollinators, and wildlife. Yet the outer surrounding land often provides the town's economic foundation of crops, livestock, or wood products. Potential disasters, such as large fires, pest outbreaks, dust/sandstorms, and desertification, and large predators come from the surrounding land. In effect, adjacent and outer surrounding land together make a town.

12 Tying Transportation, Towns, and Land Together

Optimo City, however, is not one town, it is a hundred or more towns, all very much alike, scattered across the United States ... a very familiar feature ... you have never stopped there except to buy gas ...

John Brinkerhoff Jackson, The almost perfect town. In *Landscape*, 1952

Now we explore the *land of towns and villages*, or town-centric land, as a whole. As suggested above, to a traveler from elsewhere the small population centers may seem monotonous. Yet to a local, each place is distinctive. This land of towns and villages is the massive middle between urban/suburban spots and the remote forest, desert, and tundra. All cropland, much pastureland, much forest, some desert, and little tundra constitute the land of towns and villages.

Town-centric land appears extensively or somewhat linearized by human-created corridors, as well as patches with straight boundaries (Hoskins, 1955; Rackham, 1986). Yet curvy and fine-textured boundaries of natural vegetation are also present, even predominant. A mixture of rectilinear and hybrid nature–Euclid forms (Chapter 1) is characteristic of the land of towns and villages.

In some regions walking for services predominates within towns. Yet in most regions, cars and small trucks are the prime way to get about in town (Moles *et al.*, 2006; Vigar, 2007). That is ironic because towns are normally small enough to walk across (see book cover). Walking, biking, pedi-rickshawing, and a small bus or van system are most logical for getting from home to services in town (Figure 12.1).

Based on the 55 population centers observed (see Appendix), an internal bus system is common in small cities and towns with a population more than 14,000. One exception is in a somewhat smaller hill town (Astoria) with an internal transport system. Probably all the towns and half the villages are connected to other towns and a city by a regional bus system. Passenger trains connect many medium and large communities to elsewhere (Figure 12.2). Towns with a major industrial area are connected to the outside by freight trains, and a railyard is commonly present.

Towns with a central train station are bisected by the railway (e.g., Askim, Concord, Rock Springs). A major highway bisects a few towns (e.g., Issaquah, Concord, Scappoose), and a bypass highway skirts several towns (e.g., Casablanca, Easton, Socorro). Only one highway noise wall was observed around the 55 communities. Powerline corridors cut through one or two towns (e.g., Issaquah), and presumably all towns have an electric substation.

Figure 12.1 Transportation by pedi-rickshaw, bicycle, and motorbike in built area. Abandoned building with arcade and colonizing plants on left. City of Tolear, Madagascar. Photo courtesy and with permission of Mark Brenner. (A black and white version of this figure will appear in some formats. For the color version, please refer to the plate section.)

Wildlife underpasses, though scarce, provide safe animal movement across major highways (e.g., Concord, Plymouth; Appendix). Some towns and villages are connected to an established footpath network across the land (e.g., Tetbury, Sooke). For off-road vehicles (ATVs, motorbikes), some towns have established a nearby location with a dense trail network (see Figure 4.11) (e.g., Socorro, Alice Springs, Sooke). Dryland towns and some others contain both paved and unpaved roads (e.g., Superior, Socorro, Swakopmund). Finally, the centers of some towns have streets of block or cut-stone paving, which noticeably slow traffic, are noisy, and benefit pedestrians (e.g., Hanga Roa, Falun, Kutna Hora).

However, this chapter is mainly about the routes of flows/movements across the landscape. Corridors of diverse types are conspicuous conduits for movements of wildlife and people (Forman, 1995; Bennett, 2003; Diaz Pineda and Schmitz, 2011). Streams/rivers of course carry water, fish, boats, and water-transported materials. Yet many movements are without corridors, often in curvilinear routes through a series of patches, or within an extensive matrix habitat, even crossing corridors (Jongman and Pungetti, 2004; Opdam *et al.*, 2006).

Figure 12.2 Commuter rail station with bicycles for movement within a small city. Electric train serves surrounding towns. Ghent, Belgium. R. Forman photo.

At the landscape scale, most corridors are interconnected, forming networks (Forman, 1995; Diaz Pineda and Schmitz, 2011; Jaeger, 2015). Stream networks are *dendritic* (tree-like), while road networks in flatland are usually *rectilinear* (with right-angle intersections). Mesh size, as in nets to catch small or large fish, affects the movements and species present. But many species in the enclosed network patches may be bottled up, effectively disrupted by the surrounding corridors. Conflicts between moving wildlife and busy roads abound, but solutions do also (Clevenger and Huijser, 2011). Rectilinear networks tend to benefit corridor-moving species, but imprison species of the enclosed cells.

In Europe, basically trains and cars are for people, and trucks for commerce. Roads used by walkers, horse riders, wagons, bikers, motorbikers/cyclists, buses, trucks, and increasingly cars rule in most of the world (Forman *et al.*, 2003; van der Ree *et al.*, 2015b). Roads today seemingly unravel around the globe in a "global road rush" (Laurance, 2015).

In most town-centric areas, change reigns. Most changes are visible, even conspicuous, whereas some changes result from invisible connections (Forman, 2014).

Shared sociocultural values, informal linkages between towns or villages, olfactory and audible communications, and electromagnetic/electronic connections are often invisible.

The land changes, e.g., as huge increases in farmland are needed to feed global population growth. The first cut in a remote forest is a trigger, catalyzing many more cuts by new growing communities (Laurance, 2015). Paved roads have a much greater environmental effect than do unpaved roads, the former with faster traffic, the latter with little traffic in wet places and wet periods.

The "road map" ahead explores: (1) corridors and networks; (2) towns, villages, and land; (3) flows/movements across the land; and the (4) town-centric land mosaic. Tying these together builds on foundations in landscape ecology, road ecology, agricultural ecology, natural area ecology, and urban ecology.

Corridors and Networks

This big subject is highlighted by: (1) connectivity and corridor types; (2) roads and traffic; and (3) networks and their interaction.

Connectivity and Corridor Types

Connectivity refers to the degree of linkage or connection among objects. High connectivity is extremely important for the efficient movement of animals or people between here and there. In natural land, wildlife typically take curvilinear rather than straight routes between distant points. The animal moves through more suitable habitat, avoiding less suitable places, for example to avoid predators. Ecological planners normally recognize an irregular series of stepping-stone habitats as better for wildlife movement than a straighter route through less suitable habitat, as illustrated in the concept of an ecological network (Chapter 2).

Typically a *corridor*, i.e., a strip, of suitable habitat is the optimal route connecting two points (Forman, 1995; Hess and Fischer, 2001; Bennett, 2003). Natural corridors such as a stream corridor or existing animal trail are usually curvy. In contrast, human-created corridors such as roads and hedgerows are often straight. Straight corridors are most efficient for movement if composed of suitable habitat or land use. However, long straight corridors normally contain several or many habitat types, some of which are less suitable for wildlife.

Indeed, the linearization of the land by human corridors overall accelerates flows and movements, not only of wildlife and people, but also of water and wind-transported materials. Straight boundaries between land uses such as forest and field commonly accelerate flows along the boundary. The inherent structure and texture of a landscape with natural and hybrid nature–Euclid patterns (Chapter 1) provide many connected routes for both sociocultural and biophysical processes and flows (Diaz Pineda and Schmitz, 2011). Still corridors and networks are especially important for landscape flows/movements.

Corridors have several key functions beyond being conduits for movement (Chapter 2) (Forman, 1983, 1995, 2015; Bennett, 2003). Much of our ecological understanding of movement on land comes from analyzing both the effect of, and the effect on, corridors.

The preceding attributes essentially refer to all corridors. Yet natural and human-created corridors differ noticeably. Stream and river corridors, animal trails, ridge lines, gullies, and valley bottoms are widespread natural corridors. Boundaries between rock types, soil types, and vegetation types also may function as natural corridors. *Natural corridors* are usually curvy with straight stretches, even convoluted, and variable in width with narrows.

In contrast, *human-created corridors* are typically straight or with a gentle curve, and constant in width. Some enclose an entity such as road or pipe. Common examples are roads, railways, pipelines, oil/gas seismic exploration lines (Lee and Boutin, 2006), powerlines, fences/fencing, long walls, ditches, canals, hedgerows, and major biking/walking trails. Boundaries between properties, farm fields, and forest plantations also function as human-created corridors. To glimpse the ecology of these corridors, we now consider pipelines, powerlines, and, in the following section, railways.

Pipelines, powerlines. Pipelines carry gas, oil, water supply, stormwater, human wastewater, plus communication and electric wires (Finer *et al.*, 2008; Latham and Boutin, 2015). Generally, these corridors are straight, in low density, and support economic growth. Construction and drilling machinery produce considerable noise and air pollution, with reverberating ecological effects (Bayne *et al.*, 2008; Rabanal *et al.*, 2010). To provide access for repair vehicles, trees are mainly eliminated and typically a narrow earthen track or road is present.

Earthquakes, corrosion, and defective connections or materials periodically lead to pipeline spills, and consequent toxic effects on vegetation and wildlife. The open strip provides a conduit for some wildlife movement, especially for foraging predators. The central portion and two opposite-facing edges, plus the earthen track, provide habitat heterogeneity supporting species less common in the surroundings. The open pipeline strip also serves as a barrier or filter against some wildlife moving across the landscape.

Powerline corridors, with electric transmission lines on pylons above a strip of open vegetation, have similar and additional effects to those of pipelines (Figure 12.3) (Luken *et al.*, 1992; Goosem and Marsh, 1997). Trees are removed to prevent interference with the wires, and in most stretches an earthen track or road is present for maintenance and repair vehicles. The spatial effects on vegetation and wildlife are essentially the same as for pipelines. Shrubland may predominate along powerline corridors (Dreyer and Niering, 1986; King and Byers, 2002). A study in Norway forest finds that plant species richness increases from the open corridor center to the corridor edge, with a considerable difference in species composition (Eldegard *et al.*, 2015). Bees and other pollinators tend to benefit from the connected habitats of a powerline corridor (Russell *et al.*, 2005; Wagner *et al.*, 2014).

But the tall pylons are often hit by migrating birds which die (Bevanger, 1998; Guil *et al.*, 2011; Adams, 2016). Large birds are periodically electrocuted by the wires (Lehman, 2001). A study of an eagle species in France finds that mainly juveniles and immatures are killed (Chevallier *et al.*, 2015). Terrestrial animals are reported to often

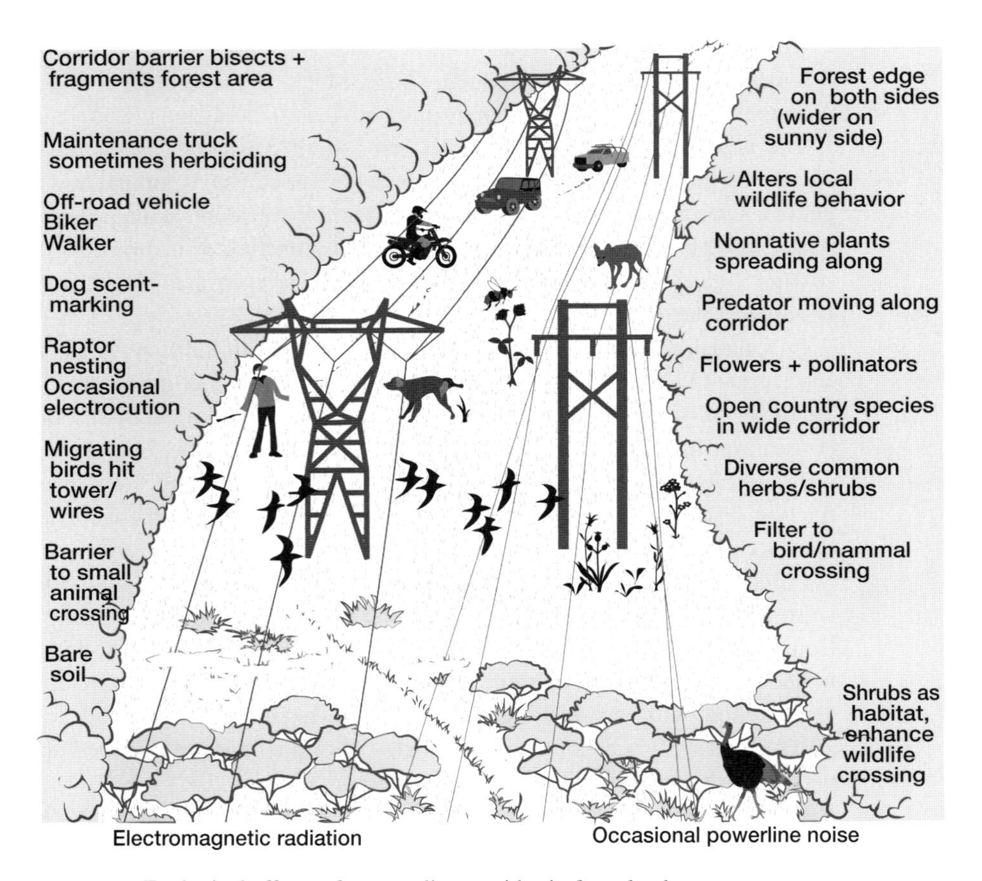

Figure 12.3 Ecological effects of a powerline corridor in forestland.

avoid powerline corridors, perhaps in part to avoid predation in the open strip through forest.

However, the corridor avoidance may more likely result from the animals seeing a wide band of ultraviolet light (invisible to us) (Carrington, 2014). In the UV light, powerlines are seen as glowing and flashing strips, due to electricity ionizing the air around the powerline cable. This reduces electric transmission efficiency, and also causes the hissing and crackling sometimes heard around powerlines.

Railways. Railway corridors are straight or with gentle curves, and differ from the preceding by having large steel objects, trains, speeding along them (Figure 12.4). Trains enroute produce loud noise, iron particles along rails, and air pollutants including heavy metals, hydrocarbons from combustion, and aerial particles (Forman, 2014). Fine materials carried in freight cars/wagons, such as heavy metals from mines or seeds from grain shipments, inevitably are spread along railways by trains. Maintenance commonly includes heavy spraying of herbicides to eliminate all plants and soil atop the rail bed. Creosote or other wood preservative used on the ties (sleepers) seeps into the rail

Figure 12.4 Long freight train on rail bed with no plants, and using its horn for road crossings. Short-grass plains east of Belen, New Mexico, USA. R. Forman photo.

bed. Annual plants often predominate on sloping sides of the rail bed, moisture-tolerant plants along drainage ditches, and shrubs and other plants on adjoining banksides.

Speeding trains, and to a lesser extent the rail bed and track, affect wildlife populations (Forman, 2014; Dorsey *et al.*, 2015; Borda-de-Agua *et al.*, 2018) by direct mortality, a habitat barrier effect, and disturbance. Disturbance includes visual disruption, noise, vibration, light, and air and soil pollutants. For example, during migration, Mongolian gazelles tend to avoid the area within 300 m (1,000 ft) of a railway (Ito *et al.*, 2005), though wildlife avoidance of railways is relatively little studied.

Apparently rather few animals, either walking or flying, move along rail beds. During deep-snow periods in boreal forest, moose and wolves may use the raised rail bed where snow is less deep. In Sweden, moose mortality on rail beds is low with fewer than 80 trains passing per day, or with fewer than 1,000 vehicles per day on roads (Seiler, 2004). Lots of mitigation approaches have been used to reduce train-wildlife mortality (Borda-de-Agua *et al.*, 2018), such as: slowing the train; wildlife crossing culverts, underpasses, and overpasses; baffles to force flying animals over a train; fencing to block movement; planting patterns of vegetation; and washing/cleaning to reduce the spread of pollutants and seeds.

Some railways pass through towns, especially for passenger trains to stop at a station, or freight trains at a local industry to leave or pick up train cars/wagons for transporting heavy goods. Freight cars/wagons, high-speed trains, and inter-city trains make loud noise (Kanda *et al.*, 2007; Guarinoni *et al.*, 2012). Yet in town, braking, starting, and acceleration create loud noise and pollution (Guarinoni *et al.*, 2012; Forman, 2014). The train in town at street level temporarily blocks vehicle and walking traffic (e.g., four crossings in Askim, and four in Concord; see Appendix). A railway to a central train station bisects the town. That may fragment town wildlife populations, though the railway may be a wildlife attractant with more suitable habitat than the adjoining built area with people. Studies of railways in urban areas find considerable habitat diversity and species richness, a pattern which might also refer to towns. Irrespective, the railway through town is probably a conduit for some wildlife movement into the center, and indeed all the way through town.

Roads and Traffic

Roads subdivide and fragment the land of species, but connect and provide access across the land for us. Roads and roadsides cover a surprising amount of land, e.g., 1–2 percent of the USA and of Sweden, and 4–5 percent of Germany. About one and a quarter billion vehicles use the world's roads, a quarter of a billion in the USA alone (Sperling and Gordon, 2009). But globally with continued population growth, major increases are expected in the number of vehicles, the total length of roads, and the amount of driving per capita. Trucks carrying an increasing portion of goods produced doubtless will also grow markedly in number. All of these increases are likely to be centered in the land of towns and villages.

The goals of transportation are safety, efficiency, and environmental sensitivity. Overall, most transportation experts sense that the first two are done well, often at the expense of the third, done poorly. *Road ecology*, the science of interactions among organisms, the physical environment, roads, and vehicles, underlies the third goal (Forman *et al.*, 2003; Davenport and Davenport, 2006; Coffin, 2007; Clevenger and Huijser, 2011; van der Ree *et al.*, 2015b).

Local low-traffic roads, some or many unpaved, are the predominant routes around towns and villages. Biodiversity around a low-traffic road is commonly higher than that of the forest interior, but lower than that of the forest edge (Figure 12.5) (Salek *et al.*, 2010). Forest edges contain generalist species of both the forest and adjoining land use. The forest roadside provides a different habitat, which combines some forest interior with some forest edge species.

Unpaved earthen roads often have limited vehicle use during wet periods, while gravel roads generally have year-round low-speed traffic use (Laurance, 2015). Water soaking into a road bed creates deep mud, even roadbed failure requiring repair. During dry periods traffic lifts dust upward that mainly settles downwind on the vegetation and ground nearby. Heavy rain on sloping unpaved roads erodes the surface, especially with vehicle use (MacDonald *et al.*, 2001; Ziegler *et al.*, 2001). Also, little-used earthen roads may be important corridors for predator movement in foraging for food.

Paving an unpaved road with asphalt/tarmac greatly reduces the plant diversity alongside (Zwaenepoel, 1997). Many of the roads on crop farms, pastureland, forestland, and desert are unpaved, with alternating dust and mud. Main roads leading into towns are typically paved, and at least some in-town roads and streets are likely to be paved, with outer roads sometimes unpaved.

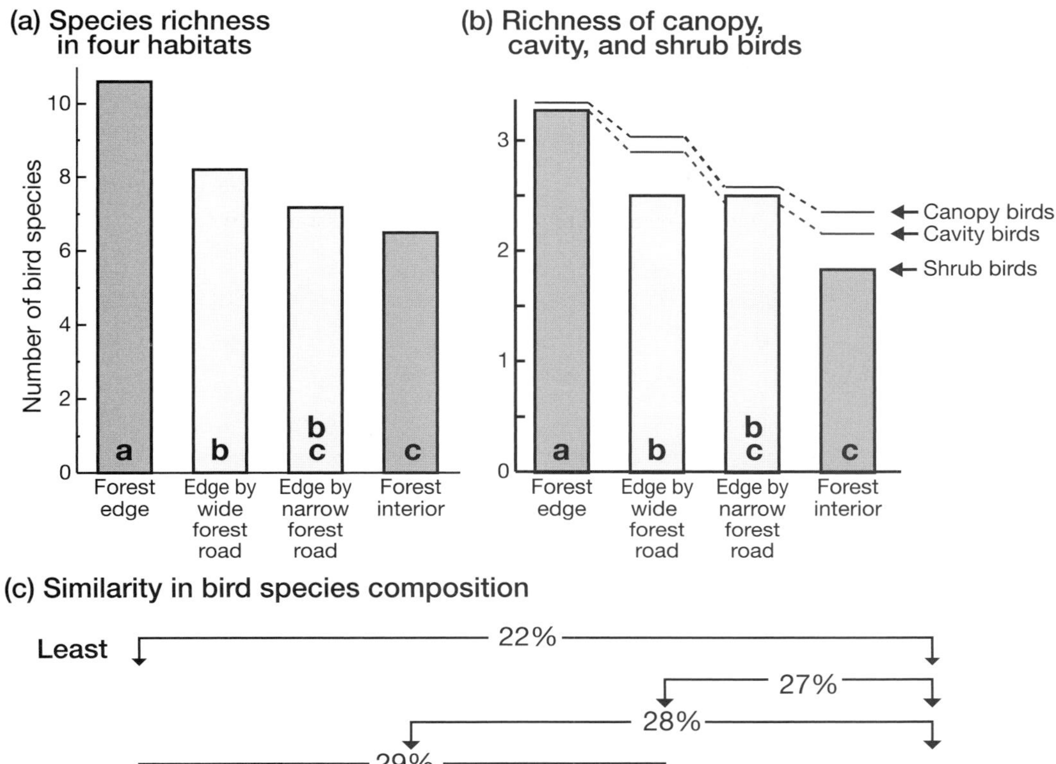

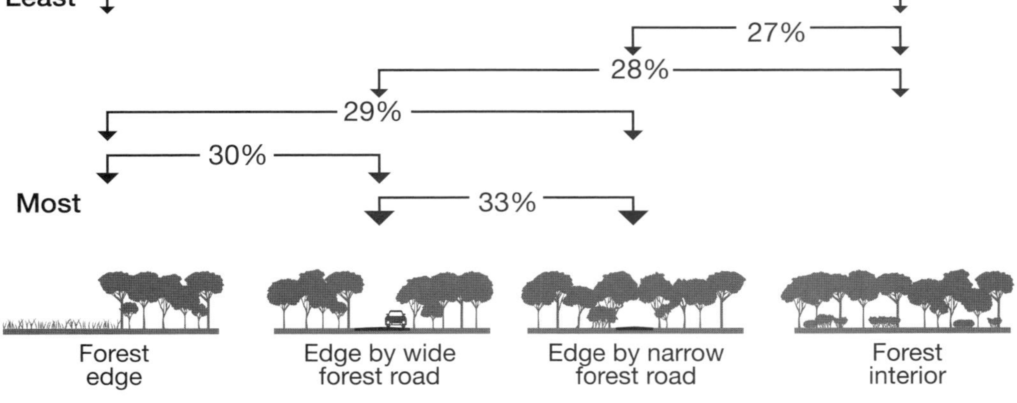

Figure 12.5 Bird species richness and composition alongside forest roads relative to forest interior and edge. Forest interior and road areas about 60 percent planted spruce and 20 percent beech (*Picea abies, Fagus sylvatica*); forest edge about 10 percent less of each, and 10 percent more of oak and pine. Wide forest roads are paved, averaging 8 m width; narrow roads unpaved 6 m wide. Eighteen cavity-nesting bird species; 16 canopy inhabitants; 9 shrub breeders. Histograms with the same small letters are not significantly different. Similarity measured by the Jaccard index. Czech Republic. Adapted from Salek *et al.* (2010).

Roadsides (verges) are notably heterogeneous with a rich sequence of microhabitats (Milton *et al.*, 2015). Combinations of different underlying rock types, soil types, slope angles, exposures, shading, and adjoining land uses or habitats create this habitat diversity (Harper-Lore and Wilson, 2000). Added to this are nitrogen, heavy metals, and hydrocarbons from vehicles, disturbance by mowing and wayward vehicles, and

Figure 12.6 Truck transport on multi-lane highway with dense-shrub median strip providing cover for wildlife crossing highway. Southern New South Wales, Australia. R. Forman photo.

maintenance by roadside scraping and/or snowplowing. A high percentage of trucks passing adds to the chemical pollution and noise, even with dense shrub cover alongside (Figure 12.6).

In intensive agricultural land, unfertilized roadsides often support uncommon habitats or rare species. Roadside ditches, with runoff water from the road surface, may appear as a narrow strip of wetland plants (Jodoin *et al.*, 2008). In arid lands, roadsides receive water runoff from road surfaces, thus producing denser taller vegetation which in turn attracts wildlife to the roadside (Lee *et al.*, 2015).

In Illinois (USA), a small mammal species (vole, *Microtus*) moved along roadsides but could not get through towns or villages (Getz *et al.*, 1978). A limited-access multi-lane highway was constructed through the countryside, and the vole then expanded its range 90 km (56 mi) westward along the highway's continuous roadside. But overall, few animal species use roads/roadsides as a conduit for movement.

Animal/vehicle collisions causing roadkill or animal mortality doubtless occur on all roads. Collisions with large animals such as deer, moose, bear, and kangaroo often damage both vehicle and people. Roadkill of large and small animals is usually highest on two-lane fast-driving roads, such as the main roads leading to and from towns. Driving at less than 70 km/hr (<45 mph) greatly lowers the roadkill rate (Seiler, 2004). Also, different animal species tend to cross at different types of locations along a road (Roger *et al.*, 2012; Gunson and Teixeira, 2015).

Examination of the windshield and front of a vehicle, which drove past a waterbody or wetland in the growing season, normally reveals an abundance and diversity of smashed insects on the vehicle. One estimate considered realistic is more than 100 individual insects killed per moving vehicle per kilometer in Germany (Gepp, 1973; Reck and van der Ree, 2015). That is a lot of animals and a pile of dead biomass.

Many wildlife species are absent or less abundant in habitats near a road. In some cases, a high roadkill rate may be the main reason. But in many cases traffic disturbance, especially traffic noise, is probably the main cause.

Traffic noise decreases the presence and density of sensitive vertebrate species (Siemers and Schaub, 2011). Along outer-suburb roads near Boston with 8,000–15,000 vehicles passing per day, the density of sensitive bird species is reduced for a few hundred meters outward on both sides (Forman *et al.*, 2002). Similar results are recorded from The Netherlands (Reijnen and Foppen, 1995, 2006). This traffic level is equivalent to that on many main roads entering and leaving towns, though of course traffic amounts vary widely.

Traffic noise from a multi-lane highway with more than 30,000 vehicles/day, such as a major highway bypassing a town, reduces sensitive birds for several hundred meters outward. Noise barriers are rare by towns, though in rural Norway some highways are bordered by long soil berms (bell-shaped in cross section) 2–5 m (6–16 ft) high to minimize traffic noise propagation. A high proportion of trucks passing significantly adds to traffic noise (Figure 12.6). Since traffic noise is low frequency, some birds have responded by singing at higher pitches (Slabbekoorn and Peet, 2003; Parris, 2015).

In warm seasons, roadside lighting in, or extending outward from, a town, or around a nearby highway entrance/exit interchange, also has widely spreading ecological effects (Chapter 8). Such lights exert the so-called *vacuum-cleaner effect*. Flying insects are drawn out of their habitats and fly to the lights, where they usually die by exhaustion from endless encircling or by predation (Eisenbeis, 2006; Eisenbeis and Hanel, 2009; Blackwell *et al.*, 2015).

Numerous mitigation approaches now operate worldwide to reduce the barrier effect of roads fragmenting habitat. Most also reduce wildlife roadkill rates and wildlife–vehicle collisions (Forman *et al.*, 2003; Iuell *et al.*, 2003; Clevenger and Huijser, 2011; van der Grift and van der Ree, 2015; Smith *et al.*, 2015; van der Ree *et al.*, 2015b). These approaches range from diverse wildlife overpasses and underpasses (e.g., Concord, Plymouth) to tunnels, culverts, fencing, plantings that funnel or divert, attractants and repellents, vegetation-covered and curved noise walls, soil berms, and quieter road surfaces. For wildlife crossing, some bridge designs providing wide horizontal visibility are effective (Figure 12.7). Fencing and walls keep animals off the road but disrupt movement patterns and wildlife home ranges (Clevenger and Huijser, 2011; Ascencao et al., 2015; van der Ree *et al.*, 2015a).

The question of how many and where to locate wildlife-crossing structures remains (Clevenger *et al.*, 2002; van der Ree *et al.*, 2015b). Cost is always important. Generally, the goal is to distribute crossings according to the sizes of wildlife home ranges. For providing options, two crossings are much better than one, three are better than two, and little is gained by more than five.

Figure 12.7 Effective wildlife underpass and passage for flash-floodwaters crossing beneath multi-lane highway. Most of the time water flows slowly through subsurface sand. San Antonio village south of Socorro, New Mexico, USA. R. Forman photo.

Also, few data exist evaluating the effects of both wildlife and people crossing on the same structure (Clevenger and Waltho, 2000; van der Ree and van der Grift, 2015). If common generalist wildlife species are the target, limited people crossing may be alright. But if uncommon specialist species are the target, human use, especially with dogs (Chapter 8), doubtless greatly reduces the effectiveness of a wildlife crossing structure.

New roads cause special problems. *Road access* into a remote area causes the most disruption to wildlife and degradation of nature (Forman *et al.*, 2003; Laurance and Balmford, 2013; Selva *et al.*, 2015). Indeed, closing such roads has the greatest ecological benefit. New roads normally bring habitat loss, spreading human land uses, and conflicts with wildlife (Wu *et al.*, 2014). A new paved road in the Amazon supports fast traffic and catalyzes a network of unpaved roads (with limited passage in wet periods) and small settlements (Fearnside, 2015; Laurance, 2015).

A *bypass highway* passing around or near towns typically degrades adjoining habitat. Traffic noise degrades the most habitat and biodiversity, and annoys people on that side of town (Reijnen and Foppen, 2006; Shindhe, 2006; Valtonen *et al.*, 2007). Commonly, an *entrance/exit intersection* close to town acts as a nucleus stimulating commercial development, including between intersection and town. In a Swiss region with many urbanized locations, development especially occurs within about 3 km (1.8 mi) of highway entrance/exit intersections (Muller *et al.*, 2010). Industrial growth is particularly prominent in that 3 km zone.

A bypass highway passing just outside a town usually protects the town from the effects of through traffic. The bypass eliminates considerable habitat close to town, and blocks wildlife and walkers on that side of town. Traffic noise and light associated with the highway and intersection further degrade surrounding habitat. Socorro, New Mexico (USA; population 8,900) built a bypass, California Street, in the edge of town to protect its central plaza area and concentrate commercial shops along the street for travelers (see Figure 1.2). Later, a major highway, Interstate-25, with two entrance/exit intersections was built just outside town.

Networks and Their Interactions

Corridors of the same type connected together form a *network*. All the corridor types just introduced form networks in the land of towns and villages. Stream networks, commonly tying a town to its surrounding land, are *dendritic* (tree-like) with acute angles, a conspicuous hierarchy, and strong directionality of flows. Almost all the other corridor types form *rectilinear* networks with a predominance of right angles indicating human origin. The ecology of most network types is little known. Edge species predominate in network corridors such as ditches, hedgerows, and roadsides (Forman, 1995; Geertsema and Spranglers, 2002). Based on modeling, the arrangement of fence networks, e.g., in pastureland, strongly affects the density and spatial patterns of key wildlife populations (Borda-de-Agua *et al.*, 2011). Also, fencing commonly separates pastures with contrasting vegetation. Livestock regularly moves along the fence leaving a path of bare soil with animal droppings.

Corridor density, the total corridor length per unit area, e.g., km/km^2, is the simplest measure of a network. Road density is widely used and quite useful near towns in rural areas. Half of Europe's land is within 1.5 km of a transportation corridor (Torres *et al.*, 2016). In Drenthe (The Netherlands) moor frogs (*Rana arvalis*), which live in ponds, are sensitive to vibration and noise, and also are often killed in crossing roads with traffic (Vos, 1997). In the highest road density portion (36 km/km^2) of the landscape, frogs are in 5 percent of the ponds, whereas in the lowest road-density area (<1 km/km^2) there is a 93 percent chance of a pond having the moor frog.

Based on genetic analysis of two amphibian species (a newt and a toad) relative to road density in northern Spain, rural paved one-to-two-lane low-traffic roads are barriers that limit crossing by toads, but not newts (Garcia-Gonzalez *et al.*, 2012). For 13 mammal species in Ontario (Canada), body size is a good predictor of the relative animal abundance around different road densities (Rytwinski and Fahrig, 2015). Road density thresholds, above which a pattern changes, have been suggested for the absence of elk (*Cervus elaphus*) (Frair *et al.*, 2008), and for various environmental effects (Jaeger, 2015).

Figure 12.8 Footpath connecting communities and other features across the landscape. Grassland with shrubland above and conifer plantations below. Skiddaw Mountain, Lake District National Park, UK. (A black and white version of this figure will appear in some formats. For the color version, please refer to the plate section.)
Source: John Finney Photography / Moment / Getty Images.

Network form is less studied but a better predictor of ecological pattern and process. Networks vary from a perfect grid (equal-sized squares) to containing both large and small enclosures, a varied proportion of right angles and acute angles, having varied shapes of enclosures, containing a hierarchy of large-to-small corridors, and so forth. A perfect grid may be ecologically the worst; essentially no species evolved with this. A network containing large enclosed natural patches can support rare interior species. Acute angles facilitate directional movement. Thus, for instance, modeling road density highlights the importance of shape and minimum size of enclosed patches for wildlife population survival (Borda-de-Agua *et al.*, 2011).

Networks and travel routes may be evaluated by distance traveled along the corridors (Figure 12.8). Other animals or people move through the network enclosures, with some delay for crossing corridors. Alternatively, one may consider time-distance or cost-distance, that is, the time or cost expended to move along a particular route (Drake and Mandrak, 2010).

Introductions to network theory provide additional insight in landscape and road ecology (Forman, 1995; Forman *et al.*, 2003; Diaz Pineda and Schmitz, 2011). Consider evaluating the road, trail, stream, and other networks between town and its nearby villages from the following network perspectives: connectivity; circuitry (proportion of loops or alternate routes); corridor-size hierarchy; branching angles; directionality; mesh or grain size; intersection type and arrangement; abundance, location, and sizes of attached nodes; and corridor surface area. The ecological and socioeconomic effects of these features of the land basically await researchers.

Enclosed patches within a network also vary in heterogeneity and habitat diversity. Enclosures are subject to the effects of movements along the corridors, such as noise, dust, fire, and people. Many wildlife species show an avoidance distance from an actively used corridor, and therefore are mainly limited to the small center of a network enclosure. Based on this reduction in animal density, the infrastructure of roads and railways in Spain affects birds over 56 percent of the land, and affects mammals over 98 percent of the land (see Figure 2.13) (Torres *et al.*, 2016).

Network *intersections* are where two corridors cross, either of the same type or of different type, and therefore are of particular importance for flows. The area immediately around the intersection of two multi-lane highways, or an entrance/exit, often has periodically mowed grassy patches surrounded by ramps with traffic. A study of 17 highway intersection areas in Finland found that butterfly and moth abundance and richness are lower than in comparable areas elsewhere in the landscape (Valtonen *et al.*, 2007). In The Netherlands where a two-lane road crossed over a multi-lane highway, a line of tree stumps placed under the bridge facilitated the movement of small mammals (voles and mice, and potentially 16 other mammal species) from one side of the smaller road to the other (Forman, 1995; van der Linden, 1997).

Intersections of different corridor types are even more interesting. Rivers and streams cross under road and railway bridges, where scouring, sediment deposition, and chemical runoff affect the water courses. Water in small gullies and intermittent channels flows under roads in culverts, which are much studied for fish and water flows (see Figure 6.14). Defective culverts block fish movement from entire upstream drainage basins. Major wildlife routes often cross unpaved and low-traffic roads, but with more traffic, wildlife–vehicle collisions and roadkill increase.

In addition to intersections, many other interactions occur between different network types. Some networks have corridors which avoid or repel one another, such as a footpath not going closely parallel to a road because of collision danger, traffic noise, and visual disturbance (Figure 12.8). Wildlife corridors avoid nearby roads for the same reasons. Some pipeline corridors avoid closely paralleling roads and railroads perhaps because of vibrations caused by vehicles and trains. Water-supply pipelines avoid the proximity of sewage wastewater pipelines due to potential leaks and contamination. On the other hand, roads, pipelines, and wildlife corridors often go along parallel to streams, even into town. Powerlines may be adjacent to railways, and roads along rivers.

A wide wildlife corridor may be relatively effective crossing a network of small low-traffic roads (Jaarsma, 2004). This interaction combines width with occasional vehicle crossing to provide both spatial and temporal options for wildlife to cross. A so-called "fused grid" for planning an urban neighborhood has through traffic on the edge, slow traffic on a network of streets with curves, and a walking network reaching all portions of the neighborhood (Kemp and Stephani, 2011).

Towns, Villages, and Land

Lots of flows and movements from a town spread outward into the adjoining land (see Figure 11.15). Examples are: noise from trucks, factory, and construction; night lights from vehicles, street lights, and ballfields; and people walking, hunting, or using an all-terrain vehicle. Of course, noise and light may continue far out from roads. Typically, villages are far enough from a town that these town flows have no or little effect on a village.

Town–Town and Town–Villages Interactions

Town–Town Interactions. In areas with considerable regional thinking and activity towns frequently interact, usually to mutual benefit. Regional approaches often change competition to collaboration between towns. In effect, the towns form a functional group for the collective good. Sometimes this includes a city which plays a predominant role. Indeed, with towns varying in population size from 2,000 to some 30,000, towns inevitably vary in importance for regional activity. The Schleswig-Holstein area of Germany has about 45 towns, a limited number of villages, and one small city, a pattern suggesting considerable inter-town cooperation.

Common interactions between towns include: commuters for jobs; weekly shopping; trucks carrying goods; entertainment; competitive games; festivals; and access to specialized services (e.g., medical specialty, musical instrument lessons, advanced computer training). These interactions enhance the local economies of the towns, as well as social interactions and cultural ties (Vaz *et al.*, 2013a).

A regional bus service generally connects towns, though most people may go in personal cars and trucks (Figure 12.9). More town-to-town traffic means more fossil fuel used, CO_2 emitted, aerial particulate matter from wear of road surface, tires, and vehicles, collisions, roadkill, and traffic noise degrading wildlife habitat.

Since many, and in some cases most, issues facing a town involve one or more other towns, regional interactions can also solve problems. Rather than the usual division of a province or state into "regions" with little power, and with most towns interacting in more than one region, a *town-centered region* approach was developed in Concord (see Appendix) (Forman *et al.*, 2004; Forman, 2008). In essence, 75 ways in which the town interacts with other towns were listed. The list revealed that the bulk of positive and negative interactions were with 16 surrounding towns. The diameter of this town-centered region was five times the town's diameter.

Maps of environmental and socioeconomic characteristics, including stream/river system, train system, regional trail systems, large natural areas, and so forth, were developed for the 17 towns as a whole. The maps were provided to all towns. Thus, all towns had the same mapped databases, could readily see how each town fit within the region, and could make decisions and plan in the local region context. With the town-centered region approach, all financial and political power remains within the town.

Town–Villages Interactions. A town and its surrounding nearby villages inherently function as a unit, of course with units changing over time. A town's *outer surrounding*

Figure 12.9 Inter-town bus being packed with goods, luggage, and people. Town north of Tolear, Madagascar. Photo courtesy and with permission of Mark Brenner. (A black and white version of this figure will appear in some formats. For the color version, please refer to the plate section.)

zone extends half the distance to each nearby town, and the longest of those distances in a direction with no nearby town. This area adjoining a town includes the villages, which normally are strongly influenced by, and tied to, the town.

Most or all of the interactions listed above for town–town interactions apply to town–village interactions. Commuting, at least weekly shopping, festivals, and so forth build economic, social, and cultural linkages between town and village. Although transport is usually by car, local buses (e.g., Tetbury, Icod de los Vinos, Falun; see Appendix) are available from some towns, even a local two-car/wagon train around Kutna Hora, Czech Republic. Villages may be interconnected by walking (e.g., Kaihsienkung) or buses (e.g., Selborne, Akumal, Grande Riviere).

The basic *gravity model* of geography is useful in considering interactions or flows between population centers. Basically the amount of interaction between two communities equals the sum of the two populations divided by the square of the distance between them. Since distance is a square function (d^2) and community size is not, distance has a greater effect on interaction than does population size. Even road quality, public transportation, and a commercial center on the edge of town are usually less important for interaction than distance. In effect, little interaction occurs between distant communities.

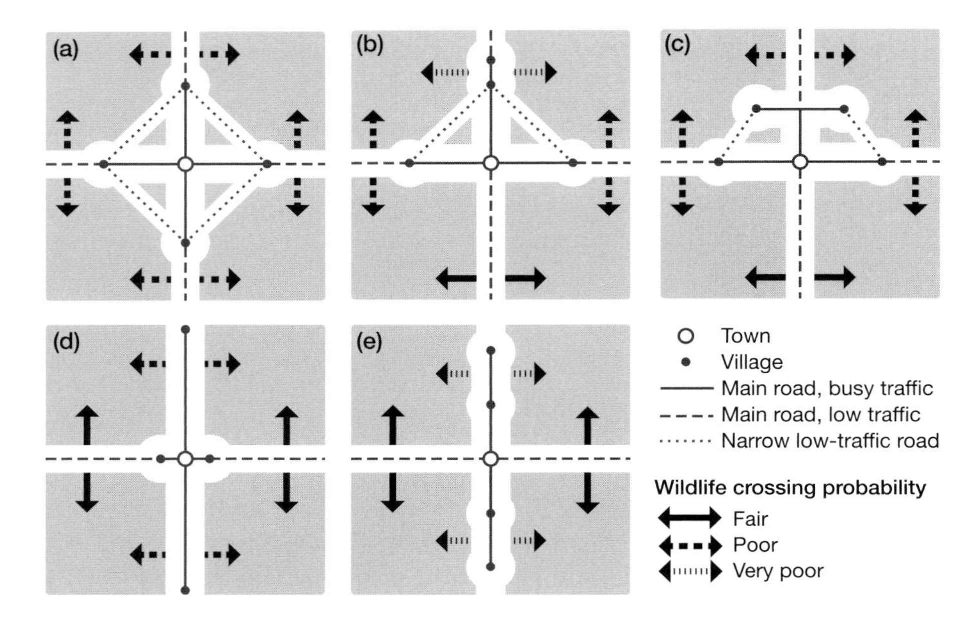

Figure 12.10 Wildlife habitat and movement relative to arrangement of town and villages. Probability of wildlife crossing a road decreases with: more traffic; wider road; shorter road length (fewer suitable locations); and less suitable target patch (e.g., surrounded by built area and busy roads).

The spatial arrangement of villages around a town is important for both people and environmental conditions. If villages are aligned along a main road, such as along a valley or coast, buses may connect them to town with public transportation (Figure 12.10*e*). In the figure, (*a*) with villages dispersed in different directions has the greatest length of roads requiring maintenance and affecting wildlife; (*f*) has the most people concentrated together; (*a*) has four, and (*b*), (*c*), and (*d*) each have two, small patches of somewhat degraded habitat; (*e*) and (*f*) have the most, and most connected, undegraded habitat. With small rural roads serving as firebreaks, in (*a*) the town is least likely to get swept by wildfire in a hot windy period. Villages are least dominated by the town in (*a*) and most dominated in (*e*).

Villages close to town are most likely to be affected by the town's light, factory pollution, and traffic (Figure 12.10*b* and *f*). Also, villages and towns close to one another are often partially connected by a ribbon of strip development (Davis and Baxter, 1977; Forman, 2014). Strip development by a town is likely to have more houses than commercial businesses. Irrespective, strip development attracts considerable slow traffic, some turning on and off the road.

Strip development also forms a barrier or filter against medium and large wildlife moving across the road. Wildlife often attempt to move across strip development under bridges and through culverts. But many animals avoid the strip development area, and would have to detour around a town to reach the other side.

Dispersed houses, such as farmsteads and weekend/holiday houses, are present in almost all town-centric landscapes (e.g., Flam, Pinedale, Easton). Farmsteads, of course, are in farmland. On house plots from, say, 2 ha (5 acres) to hundreds of hectares, residents or temporary residents value a degree of isolation and the proximity of nature.

Dispersed *weekend/holiday houses* for temporary use by urban residents are generally near a protected area, other natural resource, or an appealing village or town (see Figure 4.7) (Kline *et al.*, 2007; Prados, 2009a; Roca, 2013). Dispersed houses are often sources of nonnative plants in natural areas (Gavier-Pizarro *et al.*, 2010a, 2010b). In Europe, as illustrated near the impressive Donana National Park of southern Spain (Prados, 2009b), the best predictors (in decreasing order) of people owning weekend/holiday houses seem to be: are a residential property owner (back home); appreciate the environment and landscape; and were born in or come from a rural area.

The arrangement of dispersed houses relative to town and villages indicates much about environmental, social, and economic conditions. Of the options in Figure 12.11, clustering dispersed houses near town (*a*): most concentrates people; best supports the town's local economy; requires the least driving plus CO_2 and other pollutant production; has the least degradation effect on nature; and minimizes the town's fire hazard. In contrast, isolated houses (*e*) has the greatest effect on nature; produces the greatest fire hazard; and requires the most driving and CO_2 production.

Aggregating dispersed houses near villages (*b*) develops social interactions and supports the economy of the villages, and limits degradation of nature. Option (*f*) avoids the local communities, but clusters weekend/holiday house owners. Option (*g*) near roads limits driving and CO_2 and production of other pollutants, but tends to interrupt wildlife movement across the whole town–villages area. Overall, (*a*) seems to be ecologically best.

Note that Figure 12.11*a* may mimic low-density sprawl. However, if the area of a house plot is mainly for productive uses, such as vegetable gardening, orchard, home garden, or farm buildings, on balance it may be considered positive rather than negative. This would be especially true if patches of natural habitat and corridors for wildlife movement were well integrated with the clustered house plots (Freeman and Bell, 2011).

Arrangement of Land and Land Uses

Following the preceding focus on the arrangement of towns, villages, and isolated houses, we now explore the patterns of land and land uses. These include (1) geological landforms, (2) remote and protected areas, (3) roads in the land, and (4) energy resources in the land.

Geological Landforms

For convenience nine major land and water forms are highlighted (Forman, 2014). For soils, sand particles are coarse, silt medium-size, and clay particles are fine.

1. *Hilltop or elongated ridgetop*. Typically covered by rocks and gravelly sandy soil (e.g., Lake Placid); nutrient-poor low pH soil (excluding limestone ridges in dry

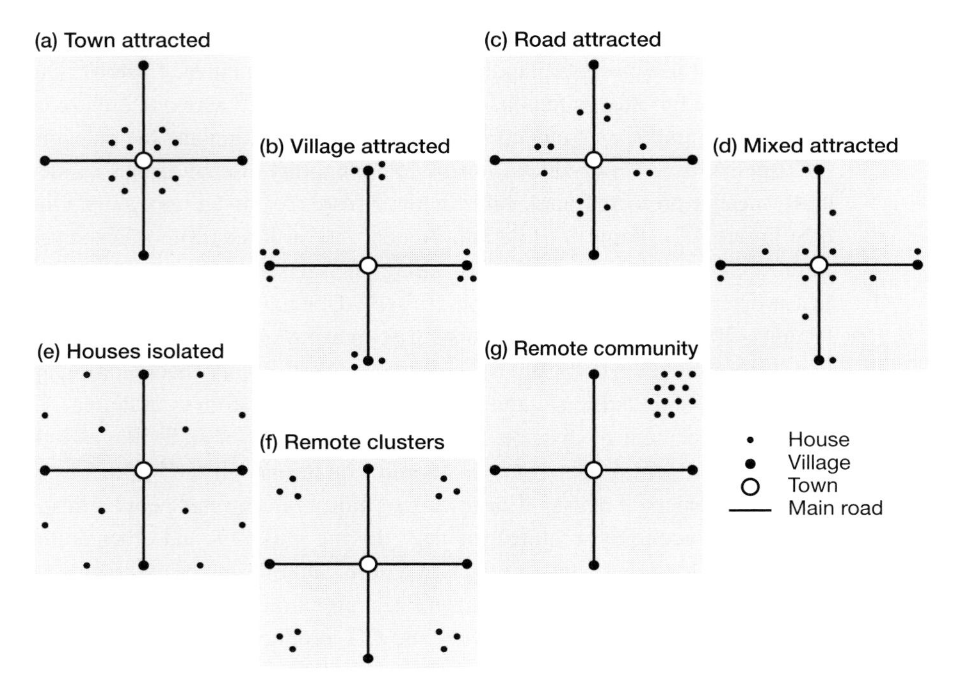

Figure 12.11 Alternative patterns of dispersed houses in town-and-villages landscape. All houses also have an access driveway/lane.

areas); uncommon plants and animals common; wildlife movement, hiking path, communication and other towers, and hikers'/tourist viewpoint along ridgetop; roads and powerlines cut across over the ridge.

2. *Slope surfaces*. Steep slope with rocks and gravel (e.g., Astoria, Icod de los Vinos, St. George's) (or sandy, silty bluff); moderate slope mostly silts and sands; eroding streams common; zones of vegetation; microclimate and vegetation differ on north- and south-facing slopes; woodland/forest and pastureland characteristic; some crops grow well.
3. *Valley bottom*. Usually sandy soils and stream present; a river deposits considerable silt (e.g., Alice Springs, Flam); crops and pastures thrive.
4. *Stream/river corridor*. Narrow or wide floodplain with silt deposits; banks usually with woody vegetation; in places streambanks degraded by livestock and human activities; streamflow disrupted by dams or bridges; floods typically frequent.
5. *A plain*. Extensive, relatively flat with silty or clayey soils (e.g., Clay Center, Casablanca, Ovcary) (sometimes sandy or organic soil); generally nutrient rich; moist plain mainly cropland, dry plain mainly pastureland.
6. *Pond/lake*. Slow-flowing freshwater from precipitation and often stream; water for livestock, or recreation uses.
7. *Wetland*. Water table at or above ground surface for prolonged period most years; water flow very slow, so some water is anaerobic (no oxygen); wooded swamp or grassy marsh common in surface depression.

8. *Coastal strip*. Very heterogeneous along its length; interacting sea and land create many habitat types; people concentrated in the few to several kilometer-wide coastal strip for residential, commercial, industrial, and recreational uses.
9. *Saltwater*. Harbor, estuary, or sea; lots of large and small boats; fishing/shellfishing; people transport; recreation water sports; repository for wastes, including stormwater, sewage treatment effluent, and industrial outputs.

A diversity of habitats and resources occurs around linkages between landforms, especially in valley bottoms and coastal strip. Locations in a plain are usually least diverse. Overall, local land uses are closely linked to topography (Danoedoro, 2001).

Protected and Remote Areas

The least usable or lowest economic value areas are ridgetops, hilltops, steep slopes, and wetlands. Therefore, these locations are most likely to be protected areas. In contrast, the highest economic value areas are the coastal strip and most valley bottoms, thus typically the least protected for nature.

Protected areas attract people for recreation and nature experiences which often leads to development close to a protected area. Classic examples are the Yellowstone/Teton national parks area (USA) and the Wolong Reserve for pandas (China) (Gude *et al.*, 2006; He *et al.*, 2009; Radeloff *et al.*, 2010). In the Yellowstone/Teton area tourists enter the parks, widely alter wildlife behavior, and degrade local sites. In the Wolong case, resident workers from communities both just outside and inside the reserve increasingly harvest wood and degrade the habitat within the reserve. In Banff National Park (Alberta, Canada) the density of residents, tourists, and vehicles in a central valley forms a filter or barrier to wildlife movement from one portion of the park to another.

Remote areas are especially important ecologically because human impacts on natural patterns, processes, and changes are minimal. Thus, constructing an *access road* into a remote area for oil/gas prospecting, recreation, or other purposes has an immediate and far-reaching effect degrading nature (Forman *et al.*, 2003; van der Ree *et al.*, 2015b). From altering behavior, movement, and population sizes of sensitive wildlife to land clearing, settlement, pollution, and hunting effects, the remote area is degraded by an access road.

Roads in the Land

The socioeconomic and environmental values of remote or roadless areas include (Selva *et al.*, 2015): wilderness experience; scientific value; ethics of protecting the Earth's diversity; sustaining interior and wide-ranging species; protecting rare habitats and combinations thereof; and maintaining the natural disturbance regimes in which species evolved. Consequently, closing and removing an access road into a remote area is a high priority.

Roads can be built across any of the landforms. Small rural low-traffic roads, for example, cause erosion on slopes, dust in dry plains, sediment and nutrient flows into streams and ponds, and altered water tables in wetlands (Forman *et al.*, 2003; van der Ree *et al.*, 2015b). Main roads with considerable traffic accentuate each of these

Figure 12.12 Winding road that slows careful drivers in forest with wetlands and rocky areas. Door County, Wisconsin, USA. Photo courtesy and with permission of Bill Claybrook, Photographer.

effects. But in addition, main roads and highways often have high levels of wildlife–vehicle collisions, road-killed animals, the barrier or filter effect fragmenting habitat, and traffic noise effects on vertebrates that degrade habitat for hundreds of meters outward on both sides. A narrow road with numerous curves slows traffic, greatly reducing effects on wildlife (Figure 12.12), and thus is particularly valuable in parks and other protected areas.

A network of roads further disrupts wildlife movement across the land and increases roadkill hazard and wildlife–vehicle collisions. With traffic noise degradation of habitat, only the isolated centers of the enclosed patches in a network may remain optimal for wildlife. A road network is especially disruptive of an ecological network (see above) for animals moving through different suitable habitats from point A to B in the land (Jongman and Pungetti, 2004; Opdam *et al.*, 2006).

Energy Resources in the Land

Oil and gas exploration and pumping create a fine-scale network of narrow unpaved roads or tracks. Individually, the little roads are of minor importance to wildlife, but combined in network form greatly degrade habitat quality as well as wildlife movement. Perhaps more important are the moving vehicles and density of drilling and pumping machinery. Noise and visual disturbance further reduce habitat quality for animals. Air pollutants from the vehicles and machinery add to the habitat degradation, and may also have downwind effects.

A group of wind-energy turbines, i.e., a *wind farm*, on farmland and ridges is visually prominent for long distances. A study of a wind farm and raptors (hawks, eagles, etc.) in Spain concluded that mortality by turbines causes an incremental decrease in a raptor's population viability (Carrete *et al.*, 2009). In a UK wind farm, turbines appear to change the distribution of only one of the winter farmland bird species present (Devereux *et al.*, 2008). Another UK wind-farm study found that 7 of 12 breeding bird species are less frequent close to wind turbines, with a 15–53 percent decrease in abundance within 500 m (1,640 ft) of a turbine (Pearce-Higgins *et al.*, 2009). Birds may avoid vehicle access tracks, and no significant avoidance of overhead electric transmission lines is present. With a rapidly growing literature, the overall ecological effects of wind turbines should soon be clearer.

Oil and gas pipelines and powerlines for electricity snake across many landscapes (e.g., Issaquah, Millstone/East Millstone). Reducing the oil spills, methane leaks, and habitat damage due to maintenance vehicles will ecologically improve the land. But reducing the barrier effect on wildlife movement by wide cut-open strips and by flashing powerlines (see above) will especially re-stitch the land together.

Although large arrays of solar-energy panels exist, most solar-energy capture is at a fine scale, appropriate to individual buildings or clusters of buildings such as a farmstead, or hamlet or village. Curiously, very few houses or other buildings in the 55 population centers observed (see Appendix) contain solar panels. As the technology evolves and solar-panel costs drop, their use in towns, villages, and isolated houses should be one of the great changes and benefits of rural living.

Flows/Movements across the Land

The preceding spatial structure of the landscape strongly affects how it works, that is, the broad-scale processes or flows and movements. As a feedback loop, flows and movements rapidly or slowly affect the spatial pattern. So here we introduce flows and movements across the landscape for (1) water and wind, and (2) animals and people.

Water and Air Flows across Land

Water overwhelmingly flows across the land as invisible groundwater and subsurface flow, or in surface channels (Winter *et al.*, 1998). Except in limestone karst areas, groundwater moving through soil particles flows slowly, often very slowly. Thus, pollutants reaching groundwater tend to persist. Nevertheless, groundwater may move long distances such as from mountains to seacoast. Subsurface water flow, above the groundwater water table and below the ground surface, generally moves faster but only to the nearest waterbody.

On the ground surface, sheet flow of surface water across a relatively bare slope is a local process mainly resulting from a heavy rainfall. Gullies and small valleys are the usual locations for intermittent-flow channels leading to the smallest perennial (first-order) streams. Intermittent and channel flows also mainly result from heavy rainfalls, sometimes combined with snowmelt or icemelt. Canals built to bring in water, and ditches to drain water away are also mostly local. However, especially on river floodplains, canal systems for growing rice or other crops may be extensive.

Otherwise, streams leading to rivers in a dendritic network carry most water across the landscape. Water reaching a stream normally will end up in the sea, unless it evaporates from the stream/river surface, is evapotranspired by emergent or streamside plants, infiltrates into a sandy soil where the water table is below the stream bottom, or is diverted by human activity (Winter *et al.*, 1998). Basically, slow groundwater and rapid stream/river flows carry water across the land. Floodwaters periodically pour enormous amounts of water across the land surface.

But the rectilinear road network in the land of towns and villages widely intersects with the stream/river dendritic system. Rivers and most streams flow under road bridges, which often constrict and briefly accelerate flow (Venturi effect). Sediments enter the stream at bridges, thus tending to smooth stream bottoms, which reduces fish habitat. Accelerated flows scour bottom sediments for a short distance. Chemicals from road surface and bridge also enter at bridges, and may degrade both water and fish downstream.

Culverts in huge numbers under roads provide passage for small streams (which have the greatest total length) and intermittent channel flows. Culverts come in many shapes and sizes, in cross-section from simple circles for pipes to arches with a natural stream bottom (Forman, 1995; Gillespie *et al.*, 2014). Culverts should function well in both high water flow and low flow periods, and for (1) water, (2) fish, and (3) movement of small and medium-size terrestrial animals. A culvert acting as a bottleneck or blockage eliminates fish access to the entire upstream drainage basin (catchment). Amphibians, reptiles, small mammals, and medium-size mammals moving efficiently through culverts avoid going over a roadbed and the hazard of roadkill.

A major rainstorm ("Irene," 2011) in Vermont, USA caused flooding that washed out numerous road culverts and many kilometers of road surface, killing six people, and isolating 13 towns and villages for multiple days. However, several culverts in the region had recently been upgraded to the arch form with a natural stream at the bottom, in order to remove blockages of fish movement up and downstream. None of the arch culverts and associated roads washed out (Gillespie *et al.*, 2014). Inserting those arch culverts saved considerable repair money and eliminated a lot of problems for many people.

A study of fish movement through various structures under roads in the Ouachita National Forest (Arkansas, USA) found that no fish passed over a concrete slab for vehicles to ford a stream (Warren and Pardew, 1998). Fish movement through culverts was about ten times less than for an open concrete box, a natural stream-bottom ford, or a stretch of natural stream. Also, the species richness of fish moving was less in culverts than in the other options. These results held during both spring high-flow periods and summer low flows.

Airflow or wind carries particles, aerosols, gases, pollen, spores, some seeds, some spiders, and other tiny organisms over the land, depositing them in specific areas. Regional wind also transports wildfire smoke, industrial air pollutants, soil erosion from dry croplands and dry pasturelands, agricultural chemicals from fields, particles from mine waste heaps, salt from the seacoast, snow, and atmospheric moisture downwind from an oasis. Snow and sand may accumulate in large piles in well-known arrangements around obstructions to airflow. Traffic and other noise, as well as artificial light at night, also extend significant distances across the land. Hurricanes, blizzards, and other windstorms produce extreme flow and transport effects. In essence, wind is a major transporter of visible and invisible materials across the landscape.

Animal and People Movement across Land

Animals mainly move across the land in foraging within a home range. Other movements may be dispersal of subadults to find a mate and establish a new home range, or seasonal cyclic migration between suitable habitat areas. Typically, the distance moved increases from foraging to dispersal to migration.

The most likely routes across the land of towns and villages for animal movement appear to be, in order (Figure 12.13):

1. A continuous high-quality habitat area.
2. The same but with small gaps, sometimes with small stepping-stone habitats.
3. A corridor of high-quality or suitable habitat, sometimes with stepping stones.
4. A curvilinear route through suitable and partially suitable habitat patches (an ecological network option; Jaarsma, 2004).

Prominent, somewhat linear geological features, i.e., stream/river, parallel to stream/river, valley, and ridge, are prime examples of animal routes. The major long, human-created linear features, highway, railway, powerline, and pipeline, are used for movement by some animals, but are apparently not major animal routes across the land.

Other linear features in the land are attractive for movement by certain animals, but avoided by others. Small low-traffic unpaved roads through forest, farmland, or desert are often important routes for predators foraging. Prey herbivores commonly avoid moving far along such roads. Similarly, predators may move along the linear edge between two land-use or habitat types, a route avoided by herbivores as too hazardous. Moose, bears, and a few other animals may move along a raised railway bed to avoid deep snow in winter, but hardly ever move along a busy road (Figure 12.14). Some migratory birds may use a major linear feature on the land as one of a number of directional cues for flying.

Attraction and especially avoidance are important in such movement (Figure 12.13). Enroute animals are doubtless attracted to high-quality habitats, particularly with cover and food. Avoiding towns, villages, busy roads, noisy places, people, hunters, and dogs is also a priority.

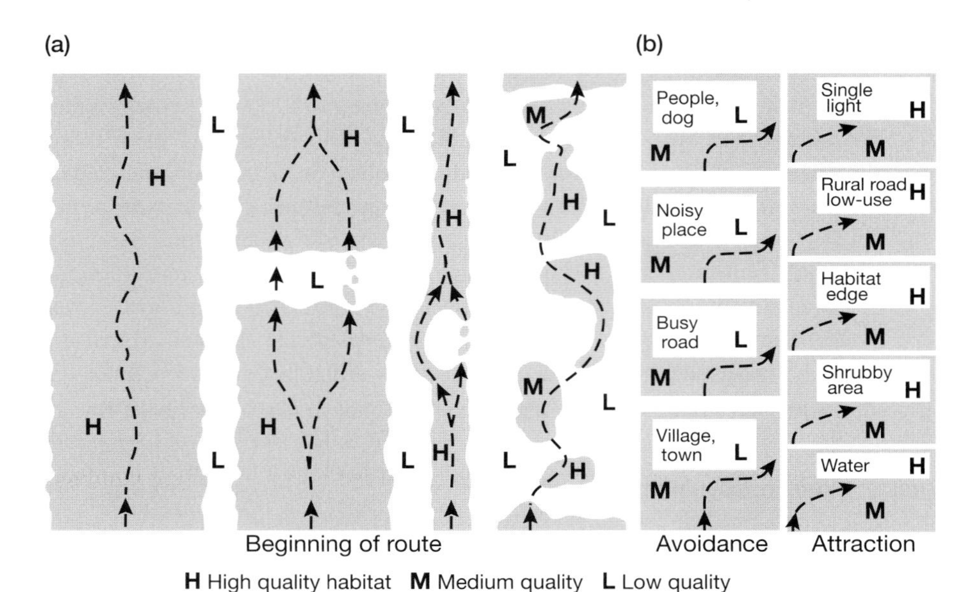

Figure 12.13 Wildlife movement routes across the land, with avoidance and attraction features. (a) Four habitat arrangements in high-to-low sequence of expected wildlife movement. (b) Features that typically modify wildlife routes.

Figure 12.14 Bull bison or buffalo (*Bison bison*) walking slowly along road with kilometer-length lines of creeping traffic in both directions. Few species move along roads or roadsides. Yellowstone National Park, Wyoming, USA. R. Forman photo. (A black and white version of this figure will appear in some formats. For the color version, please refer to the plate section.)

Artificial lights at night attract certain animals such as moths and many other insects (Chapter 8). A single light seems to attract more insects than does a cluster of lights. Bats, birds, lizards, frogs, small mammals, and other animals that feed on the night-flying insects are apparently also attracted to lights (Outen, 2002). Yet such animals feeding near the ground by street lights may be hit by moving cars. Migrating birds are commonly attracted to a bright light such as a lighthouse, airport beacon, or lighted skyscraper window.

While landscape ecology and other literature now make it easy to identify preferred animal movement routes across the landscape, major human-created corridors, especially roads, crisscross those routes (e.g., wildlife-crossing highway signs by Socorro, Concord, Lake Placid; see Appendix). Fortunately, road ecology has also mushroomed in the past 15 years, and much is known about road crossings of wildlife (Forman *et al.*, 2003; Clevenger and Huijser, 2011; van der Ree *et al.*, 2015b). Many transportation agencies now commonly address wildlife-road issues, including building underpasses for wildlife crossing near protected areas and in or near towns (e.g., Concord, Plymouth).

In urban Calgary, Alberta (Canada), forest songbirds reportedly cross diverse corridors (most to least readily): railway (30 m wide); highway (traffic noise apparently less inhibitory than traffic volume); a greater than 30 m wide gap in a forest; and river (Tremblay and St. Clair, 2009). Local rural roads, however, are closer to the home range size of many terrestrial species. Hence animals are more likely to know the locations of usable culverts, places and times with less traffic, and locations with better visibility and protective cover for safely crossing.

In central Spain, a wide range of wildlife – amphibians, lizards, snakes, small mammals, rats, hedgehogs, rabbits, wildcat, and common genet – used 17 culverts studied to cross under roads and railways (Yanes *et al.*, 1995). Different culvert designs favored different animals. Similar results were found in Alberta (Canada) (Clevenger *et al.*, 2001a).

In Alberta, wildlife–vehicle collisions especially occur near the ends of roadside fencing (Clevenger *et al.*, 2001b). In the Northern Rocky Mountain area, black bears (*Ursus americanus*) commonly cross highways at known places (Lewis *et al.*, 2011). At the landscape scale, forested areas away from development are optimal for bear crossing. At the site scale, a short distance to vegetation cover, an abundance of roadside shrubs, and a short distance to water are optimal for bears to cross roads.

In essence, wildlife movement across the landscape focuses on the connectivity of quality habitat (Figure 12.13) (Abbott *et al.*, 2012, 2015). Attraction may affect some movement, but certainly avoidance is an important determinant. Major human-created corridors, especially roads acting as filters, are almost always encountered, and are best crossed in bridges/culverts and specific, often learned, locations along the road.

Four different routes across the land are mainly used by people: (1) highway, using car, truck, bus, motorcycle; (2) local rural road, using car, truck, tractor, motorbike, bicycle, horsepower, running/jogging, walking; (3) trail/footpath, by walking (e.g., Tetbury, Kutna Hora, Falun), hiking (Alice Springs), biking (Sooke, Issaquah, Huntingdon/Smithfield); and (4) off-trail route, by nature walking, dog-walking, hunting. All four routes mainly follow raised terrain to avoid wetlands and flooded areas.

Off-trail nature walkers and hunters do not take a random route. Normally, they want to see or shoot certain wildlife, and they know in advance something about the habitat and movement patterns of the species (Dickson *et al.*, 2009; Knight and Gutzwiller, 2013). In general, wildlife are denser along forest–field and other edges between habitats or land uses. Less common specialist species often avoid such edges. A shrubby area provides protective cover. Waterbodies attract wildlife. So, the off-trail route of a nature walker or hunter targeted to species of interest is often determined by the arrangement of habitat edges, shrubby areas, and waterbodies. Hunting is also often concentrated near reserve boundaries and trails or small roads, which provide accessibility and facilitate the carrying out of game harvested (Kuehl *et al.*, 2009; Levi *et al.*, 2011; Yackulic *et al.*, 2011).

City and Town-Centric Land Mosaics

Cities, though frequently distant, of course affect the land of towns and villages. We now consider the urban relationship from three perspectives: (1) towns and megalopolis, urban region, city, periurb/suburb; (2) beyond the exurban area; (3) land-use arrangements; and the (4) prime ecological footprint.

Towns and Megalopolis, Urban Region, City, Periurb/Suburb

A *megalopolis* ties together nearby major cities with surrounding development plus remnant fragmented agricultural and semi-natural lands (Gottman, 1961; Forman, 2014). Dovetailed into this huge object are towns and villages still surrounded by undeveloped land. Though heavily affected by the surrounding cities and development, some of these population centers may retain their integrity. Protected semi-natural lands and market-gardening farmland with close by markets for food products may prevent development from overwhelming the towns and villages. Residents support their local shops which benefit with considerable support from outsiders.

Villagers mainly interact with the surrounding urban and developed areas, and not, as elsewhere, with towns. Towns also strongly interact with the surrounding population centers and do not depend on villages. In time, town and village residents probably become more and more like those in urban and periurban/suburban communities. Consequently, towns and villages in megalopolis often gradually become small suburbs engulfed in development.

The Sacramento River Valley, California (USA), about 250 km long and 100 km wide (160 x 60 mi), contains 2 medium-size cities, 13 small cities, and numerous towns and villages (Thayer, 2003). In some ways the valley functions as a unit, though porous. A central river, agricultural production, water and energy systems, the transportation network, and residents widely identifying with the valley help tie the area together in economics, social patterns, and environmentally. While rural villages are still strongly linked to a nearby town, both villages and towns are also linked to the network of small to medium-sized cities encircling them.

An *urban region*, in contrast, encompasses a city and the surrounding area strongly interacting with the city (Forman, 2008). For major cities the radius of interactions is typically about 70–100 km (45–60 mi). Much depends on distance from city, such as: inner suburbs, outer suburbs, exurban area, transportation modes, commuters, traffic, goods and services provided, air and water pollution, recreationists and tourists, inner and outer urban-region zones, and location near the urban-region boundary.

A rather rigid spatial framework exists, within which urban-region towns and villages function, characterized by: circular movement on a ring road; dividing the region with transportation barriers into pie-shaped sections; and being located between two major radii. The framework may be functionally more important, or less important, than the rural town-and-villages network.

A considerable socioeconomic literature describes the suburban area with its component towns largely surrounded by development (e.g., Issaquah, Concord, Cranbourne; see Appendix) (Stilgoe, 1988; Duany *et al.*, 2003; Caldiron, 2005; Tacoli, 2006; Dunham-Jones and Williamson, 2009). A growing literature describes the ecology of these urban-related town areas (Marzluff *et al.*, 2008; Alberti, 2008; McDonnell *et al.*, 2009; Gaston, 2010; Niemela *et al.*, 2011; Adler and Tanner, 2013; Forman, 2014).

In the *exurban or periurban zone*, usually with scattered urban-related towns, villages, dispersed developments, and dispersed houses, the city strongly affects the towns, while villages are affected by both town and city. One town sends few products to the city, though many towns ship a diversity of products to the city. Nevertheless, the city provides numerous products for every town.

Beyond the Exurban Area

This is the land of towns and villages, some in the outer urban region but most far from a city. Here a perceptive geo- or eco-detective can notice that an original settlement site was largely determined by the combination of surrounding natural resources, or in an occasional case by the crossroads of major routes (Antrop, 1988). Later, when the population grew and land uses spread outward, interactions with neighboring towns altered the patterns. Even more broadly, the distribution of the small population centers and the shapes of their influence zones strongly affect the cultural and psychological landscape around residents.

Villages may fit into this visual landscape, while towns mainly fit into the geological forms and transportation routes. A town, or indeed a village, is usefully considered as a patch in *patch-centered mosaic* models (Forman, 2014). Thus, the size, shape, heterogeneity, and composition of the patch tells much about its functioning and changes (e.g., Reykjanesbaer, Lake Placid, Rock Springs). But, to understand how a central town patch works, interactions with surrounding patches are equally or more important. What are the adjacency effects? How far are other patches of the same type? Are corridors present, and where do they lead? Are nearby stepping stones important for movement? And so forth. The patch-corridor-matrix model is the spatial or landscape ecology model to understand the pattern, processes, and changes of a town and its area, including the villages (Forman, 1995, 2014; Rego *et al.*, 2018).

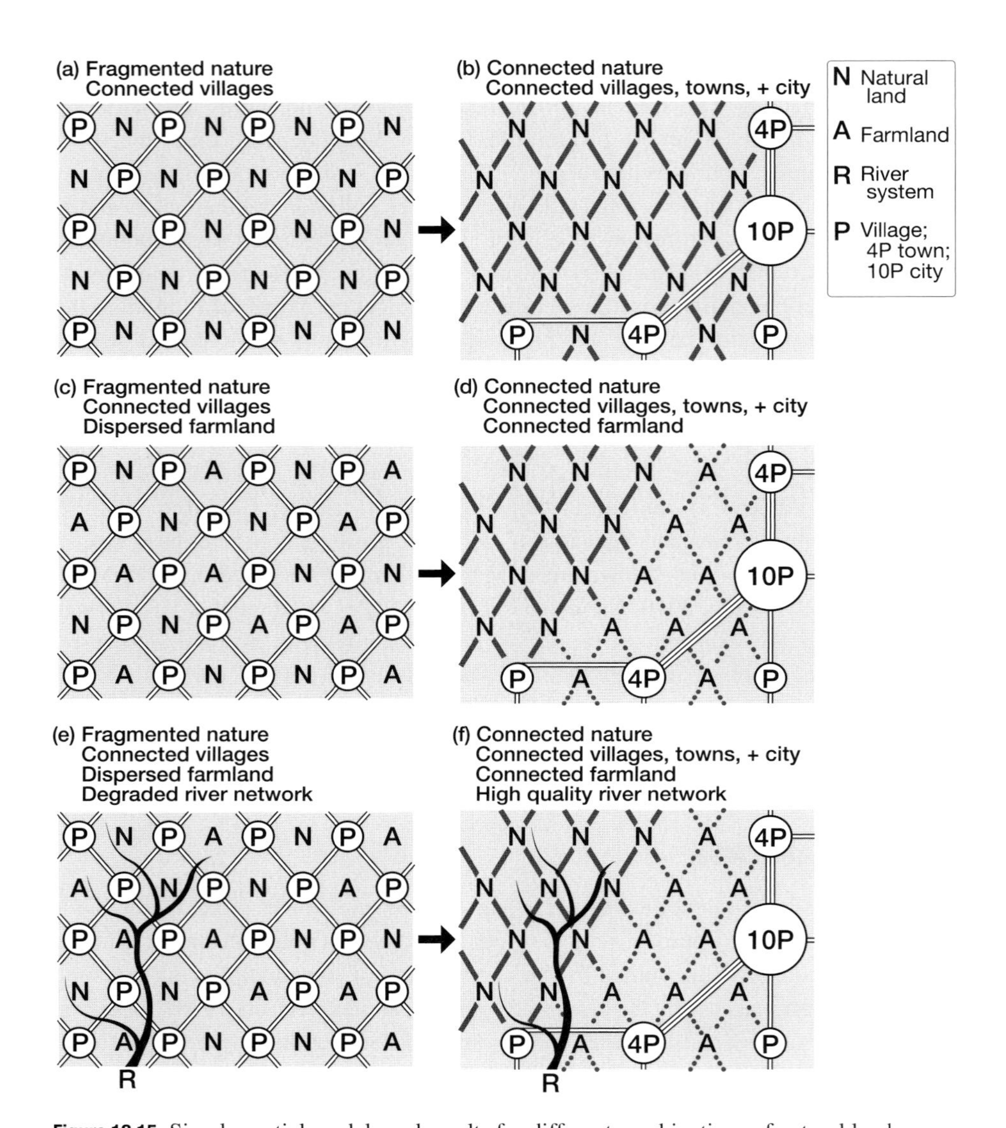

Figure 12.15 Simple spatial models and results for different combinations of natural land, farmland, river system, and population centers. In (c) all villages are next to both agricultural and natural land. In (d) agriculture is close to all the population centers. See text.

Finally, consider the population of rare wildcats (*Felis silvestris*) in Germany (Klar *et al.*, 2008). Wildcats select habitats close to forest and close to water, but away from settlements. The best multi-variable models highlight the proximity of wildcats to forest, water, and meadow, and avoidance of villages (>900 m; 2,950 ft), of dispersed houses (>200 m), and of roads (>200 m). Ecological studies such as this tie patterns, processes, and changes to a land mosaic of towns, villages, landforms, and major human-created corridors and other land uses. That produces understanding and leads to useful applications.

Nature Land-Use Arrangement Model

The mosaic of land uses (covers), e.g., in Britain, provides biodiversity conservation, carbon storage, agriculture value, and urban development potential (Moilanen *et al.*, 2011). The greatest conflicts exist between biodiversity and urban areas, while the largest carbon stocks occur where agriculture and urban pressure are low. A more detailed study of the Willamette Basin (Oregon, USA) identifies areas that sustain both high biodiversity and economic returns (Polasky *et al.*, 2008).

A simple spatial modeling approach, the *nature land-use arrangement (NLA) model* involving population centers, natural land, agricultural land, and water bodies highlights the central importance of spatially arranging land uses (Figure 12.15*a* to *f*). In (*a*), the entire area is natural land with a regular distribution of villages interconnected by roads. That produces "connected villages" and "fragmented nature." In (*b*), the population is redistributed into one city, two towns, and two villages. This creates "connected nature," providing a clean-water aquifer, habitat for large home-range wildlife, viable interior species populations, an integrated stream headwaters, and flexibility for nature's changes. Also (*b*) creates "connected villages, towns, and city," and provides city features such as symphony/opera, museums, rail hub, industrial center, large university, diverse jobs, and diverse housing.

In (*c*), agricultural land is added to the (*a*) pattern, so the land still has connected villages and fragmented nature, but now has "dispersed farmland." In (*d*), connected nature and connected villages, towns, and city are maintained, but "connected farmland" near the population centers is created.

In (*e*), water in the form of a stream network is added to pattern (c), but the streams are heavily degraded in passing by villages, roads, and farmlands. In (*f*), the connected natural land is rearranged to protect all the water except that in the connected farmland. Thus, (*a*) is the worst model, and (*f*) is the best ecological model for combining connected nature, connected farmland, connected villages, towns, and city, and clean waterbodies. Thus, the nature land-use arrangement model highlights the central importance of spatial arrangement of land uses. The approach provides alternatives to evaluate, and is easily enhanced with additional variables.

Part IV

Town Ecology Principles and Solutions

13 Toward Better Towns, Better Land

> In Java, small farmers cultivate 607 species in their home gardens. In sub-Saharan Africa, women cultivate as many as 120 different plants in the spaces left alongside the cash crops, and this is the main source of household food security ... What the world needs to feed a growing population sustainably is biodiversity intensification, not chemical intensification or genetic engineering.
>
> Vandana Shiva, *Economic Globalization Has Become a War Against Nature and the Poor*, 2000

Many years ago, I visited friends in King Arthur's Somerset (England), a productive land of sweeping hills and plains. Looking out from a former windmill atop a hill, I posed a query: "Suppose I owned a hectare in the middle of that green plain. What limitations are there on what house I could build?" My perceptive, but incredulous, host twice ignored the question posed, but after I asked a third time, he responded with vehemence: "You can't build there. That land is too valuable for houses. It's for sheep!"

Many times since, on hills with students, I've described the conversation, generating considerable thought and discussion. The verdant valley in Somerset has a rich soil, high productivity, plenty of wildlife, yet occasionally floods. Sheep supporting the wool industry have long been an important economic, cultural, and environmental force. Britain is a finite space with a dense population, and by trial-and-error learned the importance of maintaining different areas for the future. To sustain the classic farming countryside and upland wild landscapes, buildings should fit in towns and villages. The encompassing natural environment and ecological conditions are too valuable to destroy.

Having deeply explored the ecology of towns, villages, and the land of towns and villages in a dozen chapters, we now briefly look ahead. How could the communities and land be improved? What have we learned that can be put to use? We explore this challenging subject from four perspectives: (1) towns and town ecology principles; (2) ecology in planning towns and surrounding land; (3) illustrative goals/solutions for towns and land types; and (4) towns in a regional, global context.

Towns and Town Ecology Principles

To sustain natural patterns and processes, five *indispensable* land patterns have been highlighted (Figure 13.1) (Forman, 1995, 2014): (1) a few large natural patches or areas; (2) connectivity among the large patches; (3) vegetation corridors along major streams;

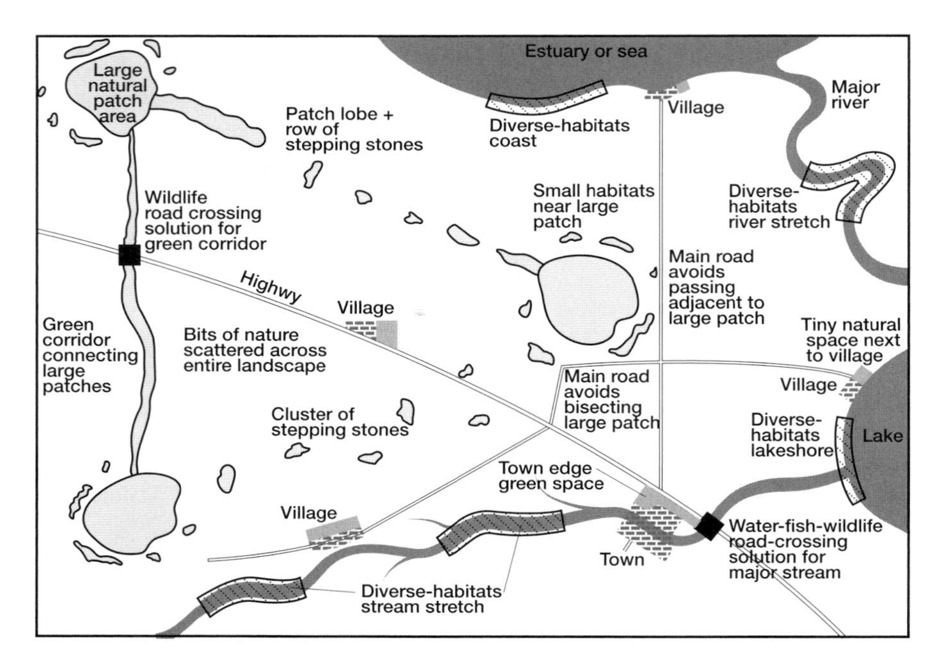

Figure 13.1 Principles illustrated for sustaining nature in a town-and-village agricultural landscape. See text.

(4) tiny natural patches/corridors spread across the matrix; and (5) small patches/corridors near the large patches. No technological or feasible alternatives exist for providing their values.

We now add four more indispensables: (6) main roads avoided through or adjacent to the large natural patches; (7) solutions provided for wildlife crossing of a main road between the large patches, and for water, fish, and wildlife crossing at a main road/stream intersection; (8) stretches of diverse natural habitat along a river; and (9) a natural patch close to, and providing a species rain to, a town or village.

The following subjects are useful in looking ahead: (1) town types; (2) common features of towns; and (3) town ecology principles.

Town Types

Twenty types of towns have been examined in this book (see Appendix):

Desert	River	Manufacturing	University/college
Cattle/sheep	Inland tourist	Company	Retirement
Farm/crop	Coastal tourist	Suburban	High income
Forest/logging	Fishing/port	Commuter	Low income
Hill/mountain	Mining	Regional shopping	Island

A few others, such as a transportation hub town, Native People's town, and exurban town, could be added. Suburban and exurban towns (much analyzed in urban literature) are little mentioned, since they relate to a city and are tightly tied to an urban area, rather than the land of towns and villages. Many of the 20 town types also refer to villages of fewer than 2,000 people, though company, suburban, regional shopping, and high-income villages would be unusual.

Common Features of Towns

Commonality of Towns

It is easy to focus on the differences among towns, as the names themselves suggest. Yet features common to most or all towns are equally interesting but little emphasized. Consider the following characteristics:

Larger than a village, and smaller than a city
Population 2,000 to 30,000
Manufacturing as part of the town economy
A town center with mainly two- to three-level commercial buildings by a central green plaza
Local waterbody and good farm soil important to most residents
Native species predominant in town
All residences close to, and readily walkable to, vegetated areas outside town
Diverse important two-way interactions with the town's adjacent zone
Outside wildlife move throughout town
Interdependence of town with surrounding villages.

Another approach to town commonality is to highlight characteristics of an archetypal town for a region. A semi-mythical town, "Optimo City," captures the idea for an extensive portion of the USA Midwest and Southwest, yet the characteristics seem to apply well in similar landscapes of other continents. It "is not one town, it is a hundred or more towns all very much alike, scattered across the United States" (Jackson, 1952, p. 2). Encapsulated it looks like the following.

Population 10,783. Overlooks a small river. Rolling countryside with grain crops and cattle. A minor spur railway for shipping. Two-lane highway passes through. Gasoline stations, motels, and occasional regional chain store near town edge. Twenty miles to the nearest town, the ballgames rival, and boasting only one paved road. Town formed by history, but not very interesting.

Prominent features all familiar. Big central courthouse. Factory making work clothes. Grain processing company. Tall grain storage elevators. A state institution. Stockyard for shipping cattle. Warehouse. Rundown slum-like neighborhood. Water tower.

North–south, east–west street grid. Topography ignored, resulting in some steep streets, and others invaded by riverside shrubs. Mostly minority workers in flimsy houses toward the river. South Main St. easygoing, loud, colorful, sometimes a bit disreputable. North Main with tree-shaded streets, river view, summer breezes, and several solid homes of leading citizens. East and west, block after block of single-level homes, no sidewalks, trees in front, little fences and hedges. Occasional church, small backyard auto-repair shop, unpaved street, and farm tractor just parked in driveway.

Courthouse square with large trees and bit of lawn. Government offices and services. Flagpoles, war memorials, fountain. Center of civic pride, even charm and dignity. Hotel and classic solid bank. Square or plaza interrupts street vistas, creates sluggish traffic. Parking scarce. A solitary traffic light.

Surrounding the town plaza are several blocks of repetitive two-level retail shop buildings. Cinema movie house. Pedestrians chatting, or walking. Cafes, pharmacies/drugstores, lawyers, doctors, dentists, insurance agents – local owners, local services. The social center.

Dust. Mud. Pigeons. Dogs. Weeds and blowing seeds. Cluster of mud-splattered pickup trucks. Countryside close. Town a part of the landscape, farmers a part of the town. Ties between country and town not yet broken.

Yet Chamber of Commerce planning never dies – tear down the old courthouse for parking; widen Main Street for more visitors, regional chain stores, restaurants, motels; squeeze out the rundown South Main houses for a riverside tourist picnic area; more store fronts, traffic, tourists; more housing developments. Break away from our agricultural past and present. Faster growth.

Driving into Optimo reveals universal patterns in towns everywhere. Rest assured that there is no architectural gem, nothing of interest in stores, no distinctive restaurant. No problem. There's another town 50 miles ahead. Such towns are in no danger of dying out.

Town Ecology Principles

Analyses of diverse town types in the preceding chapters point to a range of useful principles. Normally such principles (Forman, 2016): (1) represent an important subject; (2) are widely applicable; (3) are based on a reasonable amount of evidence; and (4) provide predictive ability. For town ecology, I find four background references with principles relative to built areas particularly useful (Beatley, 2000; Steiner, 2002; Pickett *et al.*, 2013; Vaz *et al.*, 2013a). Two other sources provide detailed lists of principles relating built areas and ecology, one at a broad scale (Forman, 2008), the other at all scales (Forman, 2016). Interestingly, of 90 urban ecology principles listed in the last reference, very few apply to the distinctive characteristics of towns.

The brief list of town ecology principles following is both preliminary and illustrative, appropriate for a subject that didn't exist before. Some principles that focus on natural land, farmland, or urban area also apply to towns, but the list here is limited to principles especially focused on towns/villages. These principles, listed in familiar groups, are useful and ready to be used for better towns, better land.

Habitats Principles

1. A green wedge protruding into town strengthens nature's "species rain" to all house plots, and provides a convenient walkable natural resource for all.

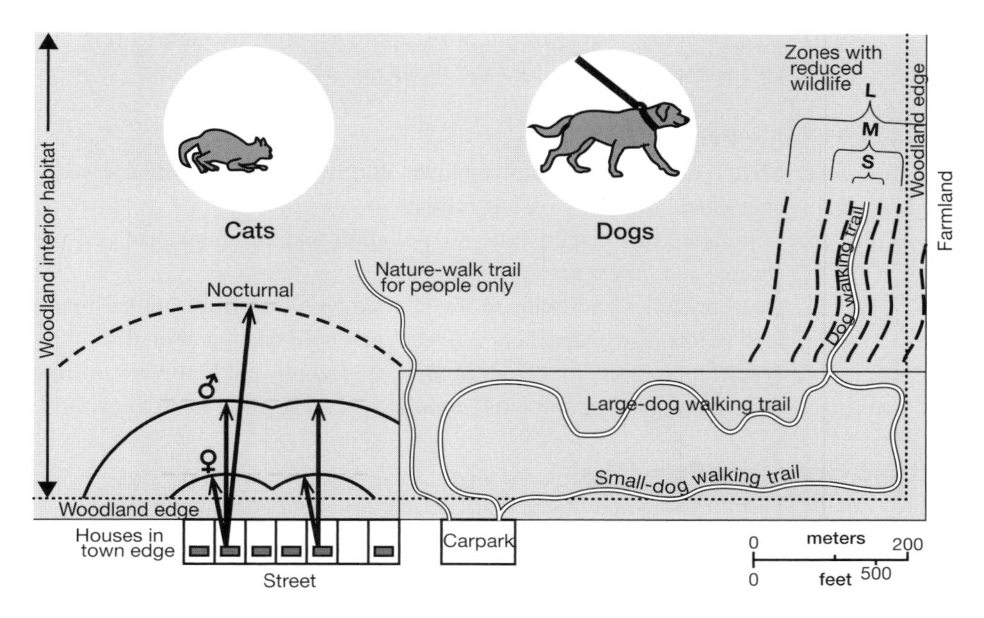

Figure 13.2 Domestic cats and dogs in a natural area. Relative foraging distances of house cats in their daytime and nocturnal home range. Rectangular entrance section with 1 km trail for large dogs proposed for intensive scent-making and most dog waste. Estimated wildlife exclusion zones: L = 200 m wide for large mammals; M = 100 m for medium-size mammals; S = 10–50 m for small mammals, birds, amphibians, reptiles. 1 m = 3.3 ft.

2. With uncommon wildlife species of a natural landscape close to town, dog-walking beyond the town markedly degrades wildlife habitat (Figure 13.2).
3. A railway in town provides considerable unusual microhabitat diversity, as well as a movement corridor for wildlife into and through town.
4. Every cluster-of-yards patch counts, as a local habitat with species dispersing outward and as a stepping stone for movement across town.
5. Rare small places and objects are distinctive microhabitats, as well as stepping stones for movement, together noticeably enriching town biodiversity.

Biodiversity Principles

6. With close by vegetation surrounding the town, a continual "species rain" enriches the entire town with farmland or natural land species.
7. Plant and animal diversity, including pests, noticeably increase with abandoned house plots, empty storefronts, and less waste removal during economic downturns.

Plants and Vegetation Principles

8. Shrubs and associated wildlife thrive in towns, including near walkways and public spaces, where human security normally is of little concern.
9. Diverse plant growth flourishes throughout towns, with a relatively high water table, heterogeneous soil conditions, and clean air.

10. Scattered large old trees are of exceptional value, by highlighting the town's heritage and supporting a flora and fauna rare elsewhere in town.

Animals/Wildlife Principles

11. Main roads radiating out from town, normally two-lane with fast traffic, are noticeable killers of road-crossing animals.
12. Pollinators thrive with little air pollution and an abundance of residential trees, shrubs, and flower gardens.
13. Dead branches and holes in trees, often resulting from limited tree maintenance, lead to lots of woodpeckers (Picidae) and hole-nesting animals.
14. A few appropriate predators in town tend to prevent herbivore overpopulation and consequent plant defoliation, mainly by scaring away herbivores.

Soil and Organisms Principles

15. Farm tractors and trucks from agricultural land continually spread seeds, soil, and wastes through town.
16. Residents' relative self-sufficiency, local food production, longer use of products, and more recycling mean less material transported into town, and less waste.

Chemicals and Organisms Principles

17. Except in limestone areas, a heterogeneous pattern of soil pH largely reflects the scattered distribution of concrete and brick-mortar objects which raise the pH.
18. Heavy metals and toxic organic chemicals from local industry and transportation disperse and accumulate in sediment and organisms of specific town areas, unless contained and appropriately treated or disposed of.

Air and Organisms Principles

19. A single street light near the edge of town produces the "vacuum cleaner effect," drawing huge numbers of night flying insects out of their habitats to die.
20. Trees on the sunny side of town-center buildings and residential houses cool the summer air, but only grow well with not too much or too little water.

Water and Organisms Principles

21. With limited groundwater pumping in town, a characteristically rather high water table supports rich plant growth, wetlands, and local streams.
22. A stream cutting through town is rarely lost into a pipe, but rather serves for industry's water power, cooling, and waste disposal, or as a park and recreational amenity.
23. Except in hills/mountains, a limited stormwater system carries some water away, but most rainwater infiltrates into the soil.
24. Flooding in the town center is a hazard, where a stream or river flows by, or where most rainwater soaks into the ground due to a limited stormwater drainage system.
25. Reducing and cleaning stormwater runoff occurs in grassy-lined swales and bioretention basins in neighborhoods, car parks, commercial sites, and other greenspaces.

Figure 13.3 Pedestrianized street for walking, biking, eating, and shopping in small city. Almost no motor vehicles. Residential and retail commercial area with car parks close by. Malmo, Sweden. (A black and white version of this figure will appear in some formats. For the color version, please refer to the plate section.)
Source: Werner Nystrand / Folio Images / Getty Images.

Residential/Commercial/Industrial Areas Principles

26. Native birds in residential areas relate more to a cluster of house plots (yard-cluster), than to either a single house plot or a neighborhood.
27. With a typical paucity of town parks and other greenspaces, wildlife habitat and movement strongly depend on residential house plots.
28. The town center functions as a distinctive habitat with heat, light, food waste, waste materials, scattered plants, and heterogeneous structures as microhabitats and hazards (Figure 13.3).
29. A retail commercial center in the town edge has bright lights at night, and puts out organic wastes, both tending to degrade nearby adjacent habitats.
30. An industrial center in the town edge usually puts out raw materials, byproducts, and other pollutants that lead to degraded adjacent habitats.
31. Homegardens, as house plots mainly covered with productive useful plants, enrich the soil, support animal diversity, and provide aesthetics.

Town-as-a-Whole Principles

32. Small infrastructure networks, with limited soil disturbance maintenance, fossil fuel use, and greenhouse gas emission, are often used as recreation footpaths around compact towns.

33. Avoiding smooth efficiency, towns retain richness, interest, and surprises in part because things break, people ignore rules, the unexpected returns.
34. Towns and villages, as prime custodians, sustain local and regional culture, especially the rich components involving nature.
35. Species and people, many representing generations in town, have adapted to disturbance regimes that provide resistance or resilience to survive future changes.

Clearly more town ecology principles could be, and will be, added ahead.

Ecology in Planning Towns and Surrounding Land

Design and physical planning effectively arrange objects to achieve an objective such as inspiration, sense of community, better transportation, rich biodiversity, or improved water systems. Context, movements, and change are normally a key to success. Thus, a particularly effective way to plan an area, such as town and country, is illustrated by a *flows-first design* process:

1. Map the flows and movements in and surrounding the area.
2. Map existing valuable places, problem places, and places likely to change anyway.
3. Then creatively arrange solutions and objects to fit the two results.

In the town and surrounding land case, major flows include groundwater and surface streams/river, occasional floodwaters, major wildlife movement corridors, predominant winds and storms, major transportation routes, and footpaths extending out from town. Planning for these greatly reduces both conflicts and investment costs. Designing around existing valuable places protects water supply, prime agricultural soils, rare habitats and biodiversity, heritage sites, and buildings. Problem places require specific solutions. Places likely to change anyway recognizes that any design/plan is for an area in flux, and that adjacencies around rapid-change areas require special attention.

For instance, in Figure 13.4, a national ecological network being developed has a new green corridor planned. Therefore, rural traffic calming that greatly reduces traffic in that area is proposed. In summary, a design/plan arrangement that fits with the above three fundamental patterns and processes means a lower annual maintenance/repair budget, and thus in effect is more sustainable.

Planned Towns

A scatter of *planned towns*, those designed and built as a whole, remains in the countryside (Steiner *et al.*, 1988; Forsyth and Crewe, 2009). The first 25 years of The Woodlands, Texas example may have been the most ecological and most successful (Morgan and King, 1987; Forsyth, 2005). A cluster of five villages was created in forested surroundings to highlight: (1) flood prevention in a flood-prone area, (2) stream water quality, (3) natural vegetation and biodiversity, (4) retail/cultural centers promoting a sense of community, (5) ready walkability/bikeability, and (6) aesthetics. Most creatively, specific soil types were protected against development to successfully absorb

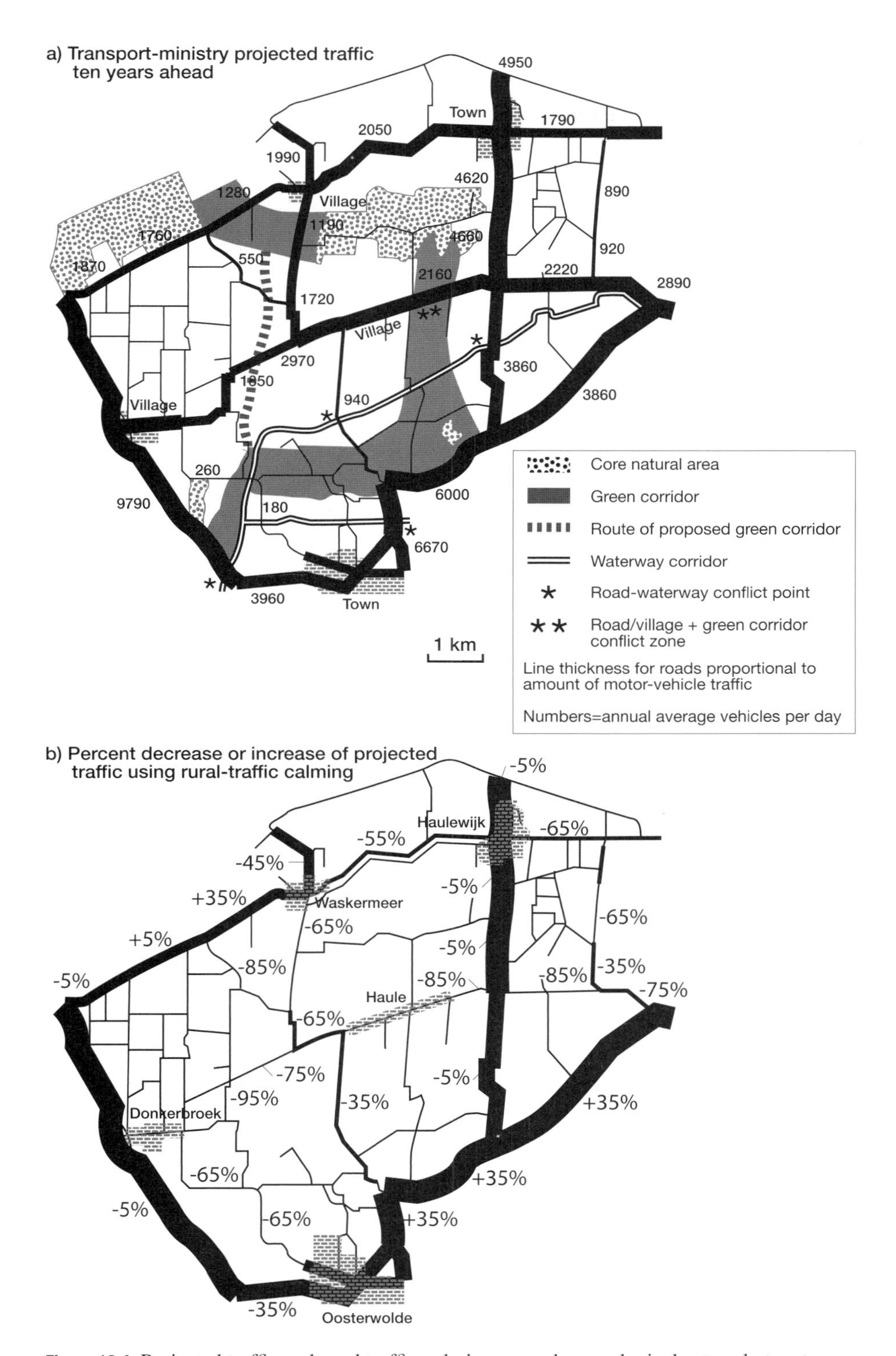

Figure 13.4 Projected traffic and rural traffic-calming around an ecological network, two towns, and three villages. Natural areas and green corridors are part of the National Ecological Network. Friesland, The Netherlands. Adapted from Jaarsma (2004).

rainwater, and prevent flooding. Natural vegetation remained in front and backyards of homes. While providing for an important rare bird from an adjacent state park was overlooked in this pre-landscape ecology project, overall The Woodlands was a remarkable ecological success.

Why not sketch out and outline a *proposed ideal town*? Here is an example from Australia (Nick Sharp, 2017: https://archive.org/details/TZYT1). The "Zillion Year Town" of about 15,000 residents. On high land to survive sea-level rise. Next to areas providing food, soft fibers, timber, and ample wilderness. Some 2,600 students from kindergarten through high school, with 200 entering per year. About 60 physicians and 300 medical staff. About 6,000 residences of varied sizes and shapes, and of intelligent design. Perhaps house lots of 1,000 m^2 (10,760 ft^2) per residence makes a town of ca. 6 km^2, with no house more than 2 km (1.2 mi) from shops, offices, and factories. Roads half the normal suburban width. People walk, cycle, skate, and scoot. Small electric vehicles or covered three-wheel rickshaws for special needs, but no cars needed. Buildings with composting toilets that take kitchen scraps. Processed human refuse used for fertilizing crops, and grey wastewater for subsurface garden irrigation. Roof capture of rain. Household water use about half of today's use. No gas. Electricity mainly from rooftop photovoltaic panels, supplemented by a wind- or solar-energy capture facility. Electric cooking.

Town semi-circular, with the diameter parallel to a nearby highway. Retail, office, garaging, and light industry functions along the diameter, with access by residents on one side, and highway vehicles on the other side. No highway traffic in town, so no pollution and almost no accidents. Employment mainly related to agriculture, secondarily knowledge workers. Products from renewable materials, especially wood and soft fibers. A town delivering a healthier and safer lifestyle, with little costly infrastructure. Another proposed future town outlined has a regular geometry, apartment living, large central courtyards, local food production, and activity areas (Mladjenovic, 1998).

Such planned towns, representative of one person's or one team's perspective, are overwhelmingly regular in geometry, urban-neighborhood like, predictable, seemingly boring, and far overplanned for a town. They lack recognition of how towns gradually and organically develop. Such plans are nearly devoid of the infinite variations, textures, surprises, unpredictability, and occasional intimidations that enrich and provide inspiration in towns. Planned towns appeal to urbanites wishing to escape the city, but basically not to town residents. Also, conspicuously absent is the importance of nature in a town. Banishing nature to the distance will not produce a real or appealing town.

Planned towns (such as new urbanist) or masterplanned communities (hundreds to thousands of hectares), which were developed at one point in time with the philosophy of one individual or team (Figure 13.5), are not included (Moudon, 1989; Duany *et al.*, 2003; Forsyth and Crewe, 2009). The environmental and ecological dimensions of such communities are generally limited or missing (or suggested by green marketing) (Lund, 2003; Forman, 2008). Furthermore, most pay little attention to patterns and flows in the surrounding land, and make little attempt to protect or restore valuable existing environmental resources.

Figure 13.5 Shaded public space with giant parade figure in city center. Planned city of Brasilia, Brazil. R. Forman photo.

Organically Evolving Towns and Plans/Designs

Paraphrasing Patrick Geddes (1915; Hyster-Rubin, 2011), an early British planner with a vegetation and ecology background: "It takes a region to make a city." Analogously, it takes a landscape of nature, productive land or water, and villages to make a town.

This book explores normal towns and villages which grew naturally over time under changing environmental, social, and economic conditions. Planning for a town and its surroundings coordinates with, and inevitably reflects the interests of, government(s) (Scarlett and McKinney, 2016; Fallows and Fallows, 2018). Getting a plan approved, and having an implemented plan that lasts, are key challenges. Three options are useful to consider. (1) A centralized government can act rapidly and boldly, but the resulting plan, with little or no public input, may or may not last. (2) Involving all stakeholders produces slow action and small change (the least common denominator), but the result is likely to last. Or (3), an agreement among the few key stakeholders, each of which brings on board secondary stakeholders, may produce a relatively rapid and bold result, which is likely to last. In any of the three scenarios, a government leader must effectively first understand the plan, second explain it, and third, the most difficult, defend the plan against critics.

Lots of books and other references explain how parts of a town should be designed or planned. An early example is Nolen's (1927) eastern USA text describing plans for many types of communities. Unfortunately, major environmental and ecological dimensions (e.g., after Earth Day 1970) are largely ignored in this and many such works, including: Reps (1969) for frontier America; Brown (1986) for market towns, social patterns, UK; Wood (1997) for New England villages, USA; Cullingworth and Nadin (1997) for planning and policy, UK; Young (1999) for social patterns, USA; Powe *et al.* (2007) for socioeconomic, market towns, UK; Kotkin (2010) for growth, economics, USA; and Zhao (2013) for land tenure, China.

Some planning references include a focus on the physical environment, such as: Smailes (1957) for physical geography, UK; Stewig (1987) for geography, Germany; Sargent *et al.* (1991) for water, life in towns, USA; Yu (1992) for Feng-Shui, China; Pezzoli (1998) for community issues, Mexico; McGregor *et al.* (2006) for peri-urban, developing nations; Tacoli (2006) for peri-urban, developing nations; Bennett *et al.* (2008) for farming, economics, China; Hu (2014) for Shan-Shui, China; Litfin (2014) for village, spiritual focus; Costanza and Kubiszewski (2014) for well-being, sustainability, global; Friedman (2014) for life in towns; Arendt (2015) for social patterns, aesthetics, USA; and Keenan and Weisz (2016) for disasters, climate change, city, USA.

Very few references for towns and their surrounding land highlight major ecological dimensions. Atypical examples are: Johnston and Don (1990) for wildlife, people, land, Australia; Schmitz *et al.* (2005) for landscape structure, Spain; Fallows and Fallows (2018) for socioeconomics, USA; and Ndubisi (2014) for conceptual planning frameworks.

Certain planning ideas have reverberated widely over time. For instance, the Letchworth (UK) town surrounded by wide *greenbelt* with farmland, town facilities, and natural areas is sometimes referred to as a Letchworth legacy, because the idea has appeared in so many plans worldwide (Howard, 1898; Parsons and Schuyler, 2000). The large towns (now small cities) of Valladolid, Yucatan (Mexico) and Boulder, Colorado (USA) effectively have greenbelts. Olmsted's *emerald necklace* in Boston, a curving greenway connecting large green patches is widely mimicked (Zaitzevsky, 1982).

The *emerald network* or ecological network of large green patches interconnected by green corridors has become widespread in ecological planning, and is particularly useful in the land of towns and villages (Forman, 1995, 2004a, 2008). For instance, green corridor networks are widely planted to connect small wooded patches in farmlands near Kutna Hora (Czech Republic; see Appendix).

Shrinking Towns

A town doesn't want to be a city, and certainly not a village. Villages don't want to turn into towns. Paradoxically, the permeating idea of continued economic growth drives towns to become cities, especially in a globe that adds a billion more people every 14 years (Forman and Wu, 2016b).

In Switzerland large municipalities use diverse innovative, incentive-based land-use policies, while small municipalities mainly rely on conventional time-tested land-use regulations (Rudolf *et al.*, 2017). In this way towns have long histories, yet adjustments seem needed to avoid becoming urbanized.

Yet *shrinkage* or *population loss* is the watchword in some regions. The vast expanse of Russia is overwhelmingly either the land of towns and villages, or natural land. Yet apparently people are dying faster than being born (MacFarquhar, 2017). In the Pskov region, maternity hospitals close. The young head to cities, while the old die. Immigrants are rarely welcome. Schools close. The general store sells vodka, but only bits of other things. Bread-but-no-butter places remain. Near cities, summerhouse dachas keep villages afloat, but not across rural Russia. Farms fold. Fields turn to shrubs and swamps. Not everywhere, but most-where, the dark forest creeps out, reinvading the land of towns and villages.

Similar patterns continue elsewhere. In central Italy, the emptying out of rural communities leads to loss of fields, loss of rare species, and rapid reforestation (Bracchetti *et al.*, 2012). In Clay County, Kansas (Midwestern USA), villages are shrinking despite the surrounding fertile farmland. Residents move to the main town in the area, Clay Center (population 4,300; see Appendix), which nevertheless has shrunk five decades in a row. Or they move to a suburb of a Kansas city, on the side toward Clay County to facilitate visits to their former home.

In Abanades, Castilla y La Mancha (Spain), an hour and a half from Madrid, the former small town has shrunk to 89 residents. A sign at the entrance says to watch out for our children. Zero children live there. But the streets are tidy and houses kept up. Every weekend and on holidays the former residents, now working or studying in the city, return to their former home. The central plaza has a new handsome playground. The mayor and village leaders want growth, more residents. But a big government facility built here some years ago was a failure and went bust. The leaders only want very slow growth, maybe one or a few families added per year.

Towns do not like big changes, big fixes. Small changes, even many, fit with a town's positive history, as well as hopes for the future.

In Donald, Victoria (Australia; population 1,700), population slowly declines as floods and droughts damage dryland agriculture (Askew *et al.*, 2014). Every 30 years farms double in size, but with half as many people working on them. Bigger, more complex machines and more skilled workers are needed. Traditional workers leave. Alas, with global markets, in good productive years prices for products drop, and in bad years prices rise. Instead of solving problems over the fence, farmers need a professional. People age, volunteers are burned out, and social linkages erode. Schools can't fill their sports teams. This Donald pattern is not everywhere. Yet much of central Australia, Northern Territory, and elsewhere are in worse shape.

Planning for a shrinkage of population is rarely done, since the assumption and desirability, sometimes necessity, for growth underlie most economics and other aspects of life (Chapter 4). Yet towns and villages losing people has become the worldwide norm. Somehow, we need to find the "Goldilocks zone" – the population, economics, and environment that is "just right."

What can be done (Dufty-Jones and Connell, 2014)?

1. Support or subsidies by a city for better water supply, wastewater treatment, park recreation, etc. in towns of the city's urban region (Forman, 2008).

Figure 13.6 Street crosswalks for children and pedestrians in residential neighborhood. Maples (*Acer*) in autumn; northeastern USA. (A black and white version of this figure will appear in some formats. For the color version, please refer to the plate section.)
Source: christopherarndt / E+ / Getty Images.

2. Encourage the elderly, with incentives such as appropriate footpaths, parks, convenient senior center, and town transportation system to return to towns and villages (Figure 13.6) (Davies, 2014).
3. Help immigrants to towns survive and thrive, with natural and economic resources attuned to their cultures (Hugo, 2014).
4. Provide freshwater security in the face of droughts and climate change.
5. Re-establish surrounding mixed farms with two or a few major products, rather than a single product.
6. Diversify socioeconomic dimensions, e.g., from agriculture or wood products, to also recreation/leisure, tourism, creative arts, and diverse festivals.
7. Optimize the use of surrounding resources in the context of both town and region (Vaz *et al.*, 2013b).
8. Target some of the global population growth (e.g., a billion added in 14 years) to promising locations in agricultural-land towns (Forman and Wu, 2016a, 2016b).

Illustrative Goals/Solutions for Towns and Land Types

"Mold the land so both nature and people thrive long-term" (Forman, 2008, 2014). This basic planning-and-action strategy is particularly appropriate for the land of towns and villages. "Mold" means do something, the *status quo* fails. "The land" means a large area. "Both nature and people" means that planning can no longer leave the leftovers to nature, but rather human land uses need to be fit around the fundamental patterns and processes of nature. "Thrive" means everyone has at least the United Nations basic human needs of housing, health, water, food, and energy. "Long-term" means several decades or over human generations. Better towns and surroundings should lie ahead.

While socioeconomic and political goals and solutions are familiar in towns, ecological ones are less common. Ecological means that organisms, especially plants and/or animals, play a major role along with the physical environment of soil, air, and/or water. So, more shops, better schools, more paved streets, a better entrance to town, and the like are not included here. Technologies and methods are little mentioned, since they are ever changing, with the new creating the obsolete.

The items below are goals, plus solutions to problems, or for achieving goals. Although plenty of items related to urban areas could be added, the illustrative items introduced are fairly distinctive of towns or villages. Most assume that the population remains similar or grows slowly over time.

Ecological Goals and Solutions in Town Areas

Goals and solutions are listed for towns that grew naturally. Each item applies to most but not all towns, and to those wishing to remain towns rather than disappear into village or city form. Items are briefly listed for the following: (1) town centers; (2) residential areas; (3) commercial and industrial areas; (4) infrastructure; (5) water systems; (6) waterbodies; (7) parks and semi-natural areas; and (8) town as a whole.

Town Centers

1. Protect rigorously a key natural feature, such as stream, rock, or vista, that highlights a town's distinctiveness (Arendt, 2004, 2015).
2. Enhance the town center environment with green roofs and green walls, which reduce flooding and heat buildup and provide resources for animals.
3. Attune the diverse groups of buildings to air flows, direct sun, heat energy patterns, and extreme weather events (Francaviglia, 1996; Girling and Kellett, 2005).
4. Create ecological buildings with natural light, energy efficiency, water efficiency, biophilia indoor plants, clean air circulation, solar panels, strategic trees, even geothermal energy.
5. Tighten the stream of food and food waste through markets, restaurants, and hotels, to minimize pest animals and disease.
6. Create pedestrianized streets enriched with plants that increase shopping and social interactions, instead of the noise and pollution of sluggish traffic and trucks.

7. Use gravel, grassy blocks, porous pavement, and the like for car parks (e.g., covered overhead with solar panels) and for sidewalks, to reduce heat buildup, downslope flooding, and polluted local waterbodies (Ferguson, 2005; Higgins and Cardillo, 2011).
8. Cool the summer air with light-colored impervious surfaces or with tree shade, green roofs, and green walls (Gartland, 2008; Forman, 2014).
9. Create the sense that plants are "everywhere" enhancing human well-being and attracting birds and other species, and especially trees with shade in warm climates.
10. Use downward-projecting street lamps and other lights, to minimize flying insect decimation and provide dark skies for migrants and star-watching.
11. Plant trees along the sunny side of buildings to cool the air, but with a dependable water supply for roots, not too much, not too little.
12. Maintain trees and other vegetation in the center of many blocks, providing habitat and stepping stones for birds (e.g., Valladolid, Mexico).

Residential Areas

1. Maintain habitat-diverse and species-rich neighborhoods, with a walkable scatter of pocket parks, historic buildings, religious/cultural structures, general store shops, and water.
2. Subdivide most large house plots, and encourage logical clusters of yards as wildlife habitats, especially connected by back-line house-plot vegetation corridors.
3. Maintain an abundance of woody plants in persistent informal squatter settlements, while constructing infrastructure for clean water, stormwater, wastewater, and light.
4. Facilitate walking, biking, and public transit vans/buses in neighborhoods and for retail shopping, thus reducing many environmental impacts of motorized traffic (Figure 13.7).
5. Require houses highly attuned to slope, soil, sun, rain, and wind, and encourage ecological houses (see town center above).
6. Reduce most house plots to, say, 10–30 per hectare (4–12/acre), sufficient for a house, car park, vegetable garden, clothes drying, children's play area, and private relaxation in greenery.
7. Create microhabitat diversity in house plots, such as gardens, green walls, solar panels over driveway, tree-shrub patches, and property-line green corridors.

Commercial and Industrial Areas

1. Shrink, disconnect, and perforate large car parks to reduce heat buildup, downslope flooding, and polluted waterbody (see town center above).
2. Distribute numerous stormwater detention and bioretention basins across commercial/industrial areas.
3. Encourage the containment and capture of industrial byproducts and pollutants, and explore local or distant markets for the materials.
4. Use strips of pollution-resistant trees, to help reduce the spread of air pollutants and shield habitat-degrading light at night.

Figure 13.7 Pickup truck converted for bus travel between villages and town. Town near Siem Reap, Cambodia. Photo courtesy and with permission of Mark Brenner. (A black and white version of this figure will appear in some formats. For the color version, please refer to the plate section.)

5. Enhance the town's habitat diversity and biodiversity by maintaining many small vegetation patches on the unusual chemical accumulations around an industrial center.
6. Reduce the number of bright lights, and direct them only downward, around town-edge commercial/industrial centers to minimize degradation of surrounding habitats by lights vacuuming out insects.
7. Establish a green corridor across a commercial/industrial center that facilitates wildlife movement from surroundings to residential areas.
8. Stop most ongoing air and water pollution, and restore areas degraded by pollutants and other human activities (e.g., Palmerton).

Infrastructure

1. Plant trees, shrubs, and flowers with genetic plasticity for a changing environment (Hunter, 2011), taking advantage of a town's heterogeneous soil, relatively high water table, and clean air.
2. Plant useful trees that provide shade, fruit, nuts, forage, vertebrate habitat, and/or positive PAHs (polycyclic aromatic hydrocarbons).

3. Convert low-usage road sections to parks for neighborhood playgrounds, meeting places, and inspiration spots.
4. Cluster different infrastructure systems (Pollalis *et al.*, 2016) to limit the strips of frequent maintenance/repair activity, and protect natural and residential areas.
5. Help walkers support local shops, and provide a public van/bus system, e.g., with figure-eight routes and cellular-phone access for off-route pickups.
6. Provide traffic noise reduction, e.g., with soil berms (common in Norway) or sunken roadway, for a busy highway slicing through or bypassing town.

Water Systems and Waterbodies

1. Survey and improve culverts, to be effective for fish and water movement at low- and high-flow times, and bridges to reduce stream degradation.
2. Channel stormwater runoff to a detention basin (pond/wetland) to slow the flow and reduce peak flooding, or to a bioretention basin to also clean the water, thus reducing pollution that reaches a waterbody (Forman, 2014; Arendt, 2015).
3. Decrease the number of stormwater pipes emptying into a stream/river, thus reducing flooding, riparian zone degradation, and poor water quality, thus effectively restoring the town's stream.
4. Use swales (grassy ditches; Rushton, 2001) and small detention/retention basins spread across a valley or slope, to eliminate both large stormwater pipes and a large downslope basin with flooding.
5. Clean polluted stream, river, and groundwater, to provide ample clean-water supply in a dry climate and in the face of climate change.
6. Build or upgrade a sewage pipe system and treatment facility, when odor, pathogenic bacteria illness, or excess phosphorus/nitrogen effects in a neighborhood or waterbody are conspicuous.

Parks and Nature

1. Maintain a semi-natural green wedge projecting into town, for movement of wildlife and walkers in and out, and for convenient inspiration.
2. Maintain a patch of nature with a soft edge, as a source of native species within a tree-lawn-bench people park with ornamentals.
3. Provide nearby nature for "forest bathing," residents' health (reducing stress, cardiovascular disease, etc.), and reconnecting with nature, benefits only partially gained in a tree-lawn-bench park (Wein, 2006; Gaston, 2010; Reeves, 2014; Williams, 2016).
4. Enhance connectivity for wildlife, with an abundance of narrow corridors and networks, and/or with sequences of stepping-stone vegetation patches (Girling and Kellett, 2005; Forman, 2014).
5. Provide microhabitat diversity and hence rich biodiversity, with an abundance of plant cover from parks and hedges to tiny spots on walls and cracks.
6. Protect the rich biodiversity of special sites, such as old trees, low-usage cemeteries, and neglected spots by industry and infrastructure.

Town as a Whole

1. Maintain affordable housing near parks and schools, to reduce transportation impacts, support temporary workers, and encourage the next generation of grown-up children to live in town.
2. Restore, and fit development around, valuable natural resources as well as distinctive heritage and cultural resources.
3. Improve conditions for health and sustainability, while minimizing greenhouse gas emission and future maintenance/repair, in all town activities and plans.
4. Avoid formal, geometric, elegant town designs characteristic of urban areas, and use a light touch with sustained small solutions (rather than a big fix), consistent with the town's natural growth (Powe *et al.*, 2015).
5. Arrange tree species, to provide microhabitats, wildlife movement, shade, cool air, clean air, aesthetics, and fruits and nuts throughout town (Nordahl, 2009; Forman, 2014).
6. Clean the air, by reducing industrial and other pollutants, with strategically located trees, and with evolving house/building-scale techniques, such as activated carbon, titanium dioxide, and air flows.
7. Support an abundance of vegetable gardens, as well as appropriate flower sources for pollinators.
8. Plan for three or more major employers, which provide diverse jobs and diverse microhabitats, but also plan the streets and neighborhoods for "seven-year-olds and 70-year-olds."
9. Respect and protect the town's setting in the landscape, as well as the town center and other core compact patterns that create a community and sense of place.
10. Link neighborhoods to a stream or river corridor with walkways and views, and connect a stream/river or railway corridor to neighborhoods with shrub/tree corridors for wildlife movement.

Ecological Goals and Solutions in Land Outside Towns

Ecologically improving the land can be expected to provide near-term enrichment and benefits, but also help sustain the towns and villages. Both a town's adjacent zone and outside surrounding land are included (Chapter 1). Almost all such areas contain some natural land, farmland, and water. If the objective were to change the predominant land use, of course quite different solutions would be needed (Curry and Owen, 2009), but the items below assume that overall the surrounding land remains the same general type. Naturally, some items will apply better around some towns than around others.

Goals and solutions in this section are presented for (1) land outside any town, and then for (2) cropland, (3) pastureland, (4) forestland, and (5) aridland.

Land Outside Any Town

1. Plan natural areas, but avoid trying to design nature, which initially resembles just plants, without the richness of nature's patterns, processes, and changes, and then develops into something else.

2. Plan for rapid or gradual change (e.g., ecological succession), which commonly erases and replaces pattern. "Change, nature's mighty law" – Robert Burns.
3. Maintain three indispensables for sustaining rich biodiversity (Forman, 1995, 2008):
   a) A few large natural patches or areas containing interior conditions and habitat diversity.
   b) Vegetation along waterbodies that protects aquatic habitats and organisms.
   c) An emerald network with corridors or stepping stones connecting the large patches and waterbodies for species movement.
4. Maintain the identity and distinctiveness of the town, so it does not coalesce with a neighboring town or city (Figure 13.8) (von Haaren and Reich, 2006).
5. Protect, on different sides of town, the nearest large natural areas, which provide a "species rain" across the town, as well as convenient recreation/inspiration places.
6. Reconnect small patches containing fragmented wildlife metapopulations (Chapter 8), thus creating larger habitats with large more sustainable populations.
7. Create multi-value edge parks in the town's adjacent zone, to support nearby neighborhoods and control outward growth (Forman, 2004a).
8. Protect a stream-system headwaters, to minimize downstream flooding, and protect the drainage basin around a water supply for sustained clean water.
9. Perforate main roads with wildlife-crossing structures for species movement across the land, and use wide natural stream-bottom culverts for suitable water and fish movement under roads (Forman *et al.*, 2003; Bissonette and Adair, 2008; Clevenger and Huijser, 2011; van der Ree *et al.*, 2015b).
10. Narrow the *road-effect zone* (area significantly impacted by road/traffic), especially along main roads/highways, to reduce habitat and built-area degradation, and recover valuable land for diverse uses (Figure 13.9) (Forman and Deblinger, 2000; Forman *et al.*, 2003; Reijnen and Foppen, 2006).
11. Create road network forms with low road density, big roads rather than many small ones, a few large semi-natural enclosures, and multiple structures for wildlife and water to effectively cross roads (Forman *et al.*, 2003; Jaarsma, 2004).
12. Close and remove low-traffic roads providing access into, and thus degrading, large natural areas and remote areas, while replacing the roads with small entrance parks and footpaths.
13. To provide human access and also protect rich nature along coastlines, limit the number of access points, and make each one wonderful so most people stay close to it (Forman, 2010).

Solutions for Cropland

1. Grow two or more major products on a farm using crop rotation, even with indigenous or novel crops, thus supporting farmer, employees, and local food consumption (Scherr and McNeely, 2007; Burchett and Burchett, 2011).
2. Use a diversity of seed and crop varieties, which are likely to withstand environmental stresses and disturbances, for sustainability (Scherr and McNeely, 2007; Burchett and Burchett, 2011; Connor *et al.*, 2011; Gepts *et al.*, 2012). As suggested at the beginning of this chapter, grow a mixture of food-producing species together.

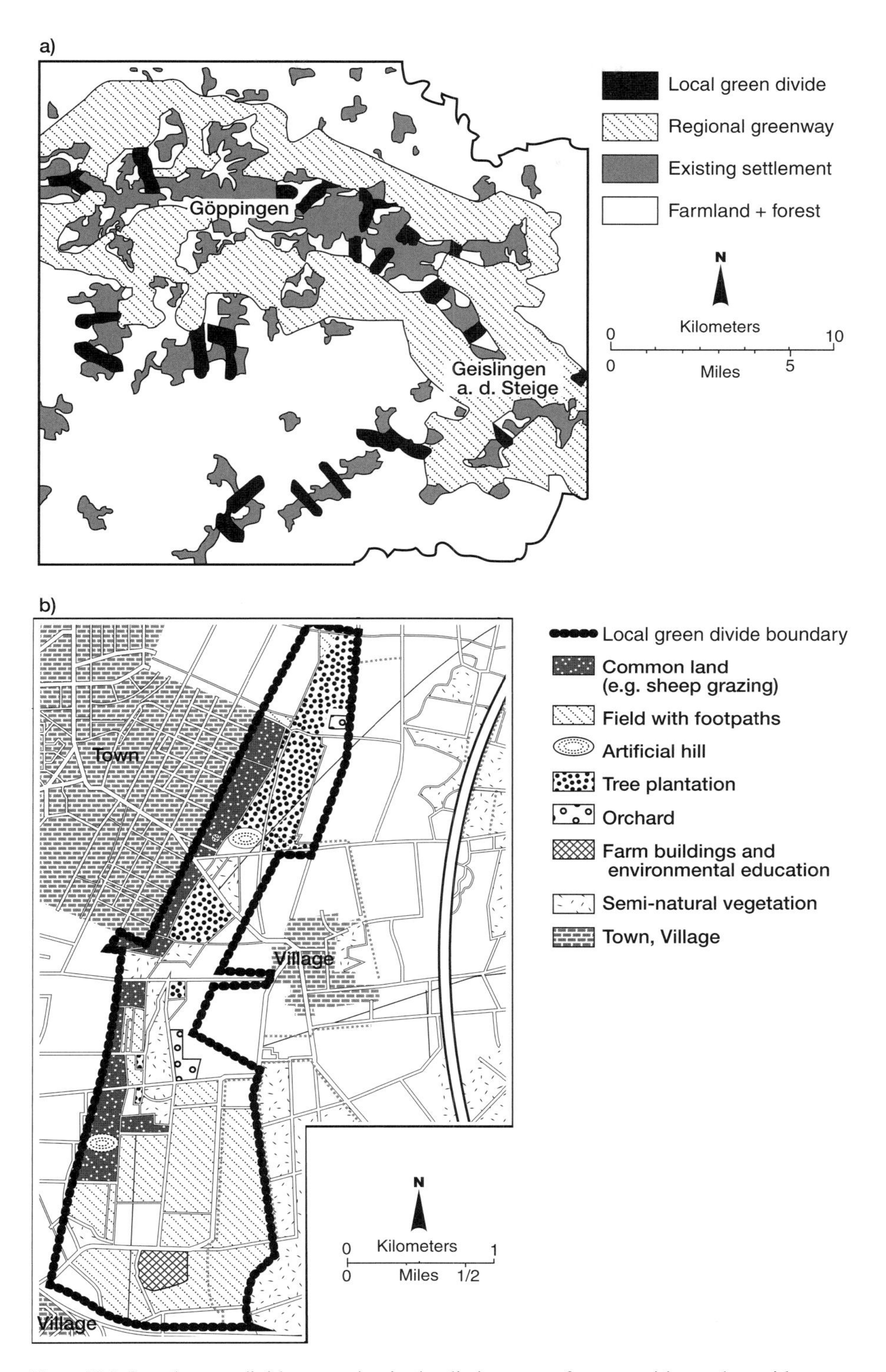

Figure 13.8 Local green divides to maintain the distinctness of communities and provide multi-value greenspace. (a) Small city, town, and villages in the Stuttgart region. (b) Local green divide in Kronsberg area of Hannover. Germany. Adapted from von Haaren and Reich (2006).

Figure 13.9 Rural road through beech (*Fagus*) tree tunnel on foggy day. Dark Hedges, Northern Ireland, UK.
Source: RelaxFoto.de / E+ / Getty Images.

3. Maintain the best agricultural soils in farming, and, as appropriate, increase soil organic matter and soil nitrogen by farming practices, minimal pesticide use, and organic fertilization (Scherr and McNeely, 2007; Warren *et al.*, 2008; Loh, 2014).
4. Minimize soil loss by wind erosion and water erosion, with plantings and soil treatment, thus also minimizing sedimentation of waterbodies.
5. Plan for lurking climate-change effects, such as more and more intense windstorms, floods, and droughts (Warren *et al.*, 2008).
6. Convert field gullies to grassy swales, and widen vegetated stream corridors, to reduce flooding, soil erosion, stream sedimentation, phosphorus, and nitrate pollution (Scherr and McNeely, 2007).
7. Minimize irrigation water use by building up soil organic matter, using buried drip irrigation, growing low water-use crops, and other techniques (Scherr and McNeely, 2007; Connor *et al.*, 2011).
8. Retain tiny semi-natural habitats across farmland, which noticeably increase habitat diversity, species richness, pollinators, and wildlife movement (Scherr and McNeely, 2007; Burchett and Burchett, 2011; Sarapatka *et al.*, 2012).
9. Use the richness of landscape ecology principles for planning and managing a whole landscape or drainage basin, including farmsteads, crop production, transportation, water, and biodiversity (Forman, 1995, 2008; Scherr and McNeely, 2007; Warren *et al.*, 2008; Turner and Gardner, 2015).

Solutions for Pastureland

1. Redesign the farmstead (ranch or paddock buildings area), to reduce dust, soil, and water pollution, and improve the health and conditions of people, livestock, other animals, and adjacent land (Tivy, 1990; Thornbeck, 2012).
2. Rotate livestock of different breeds among pastures in a grazing rotation system, to prevent overstocking (excess livestock density), overgrazing, weed encroachment, and soil erosion (Tivy, 1990; Scherr and McNeely, 2007; Gibson, 2009; Burchett and Burchett, 2011; Sarapatka *et al.*, 2012; Mortimore, 2013).
3. Retain the grassland's important carbon storage produced by dense and deep fibrous grass roots (Gibson, 2009).
4. Fence out livestock to protect ponds, wetlands, streamsides, or riversides most significant for biodiversity (and for livestock in the severest drought), and revegetate eroded degraded locations (Lindenmayer *et al.*, 2003).
5. Minimize or eliminate irrigation, which typically leads to a lower water table and soil salinization in dry pastureland (Tivy, 1990).
6. Look for signs of desertification, such as overgrazing, wind erosion, lowered water table, and salinization, and reverse the process.
7. Locate or relocate water points (holes, pump sites) far from valuable natural habitats, to minimize degradation by intense livestock trampling.
8. Protect the wide diversity and distribution of habitats in pastureland against livestock and vehicle damage, to enhance survival of native wildlife, many of which regularly use two or more habitats (Tivy, 1990).
9. Retain the richness of microhabitats, such as old trees, dead wood, shrubs, and rocks, with their distinctive biodiversity, in addition to species richness in grass pastures (Lindenmayer *et al.*, 2003; Gibson, 2009; Sarapatka *et al.*, 2012).
10. Plan for droughts, fires, floods, herds of natural herbivores, and pest outbreaks, which may become worse with climate change (Samson and Knopf, 1996; Gibson, 2009).

Solutions for Forestland

1. Recognize and sustain the several key values of forest, i.e., clean freshwater, wood products, recreation, rich biodiversity, and diverse harvested local products, when focusing on one or two (Rackham, 1976; Peterken, 1993; Lindenmayer and Franklin, 2002; Foster *et al.*, 2010, 2017; Roe *et al.*, 2013).
2. Use a tree-cutting system, such as group selection or clear-cut with retention structures, that leaves small clearings or small tree patches, to sustain biodiversity and minimize erosion and damage to remaining trees (Spurr and Barnes, 1980; Lindenmayer and Franklin, 2002).
3. Arrange forest cuts to minimize forest roads, tracks, and landings, to maintain connectivity for wildlife, and to minimize risks of blowdowns, fire, and other disturbances (Lindenmayer and Franklin, 2002).
4. Minimize, and remove when finished, logging roads, machinery tracks, and landings, and close access roads into large natural areas.

5. Eliminate houses in fire-prone forest with many fire adaptations, and spatially manage forest not only for preventing loss of timber and houses, but also to sustain fire-adapted species favored by frequent fire (Chapter 11) (Forman, 2004a).
6. Maintain the integrity of aquatic ecosystems, especially streams with dead wood, streambank shade, continuous floodplain vegetation, and fish, and expect floods to change floodplains (Lindenmayer and Franklin, 2002).
7. Protect the most valuable natural forest areas, including old growth, representative stands, rare habitats, and high habitat-diversity areas, as well as forests near town that provide a "species rain" across the town.
8. Treasure standing dead wood and logs, as well as tall trees with vertical layers, all supporting a distinctive set of animals enriching the forest (Peterken, 1996; Lindenmayer and Franklin, 2002).
9. Minimize the movement of domestic cats and dogs into the interior of natural areas, where cats may be predators on uncommon species, and dogs frighten animals and scent-mark areas degrading wildlife habitat.
10. Plan for climate change effects, such as increased fire severity, reduced water availability, decreased tree growth, altered species reproduction patterns, and changed species ranges.
11. Integrate landscape pattern, wildlife diversity, and several human dimensions in plans, as done in Rondonia, Amazonas rainforest (Figure 13.10) (Dale *et al.*, 1993, 1994a, 1994b; Dale, 2001).

In this Amazon case, settlers established homes along roads in a giant "fishbone" pattern in the forest, and ecologists collected data to model the patterns, processes, and change, with projections to 40 years. Settlers typically cut forest on a hectare or two (2.5–5 acre) plot, planted annual and perennial crops, and tended some farm animals. The farmer either remained on-site, or left after a few years when food production on the plot dropped (Figure 13.10a). Departing settlers sold their plots to a rancher who converted plots to low-quality pastureland. The "typical case" were settlers who left early because they planted annual crops, which led to more soil erosion and loss of fertility. Or these typical-case settlers depended on a few farm animals mainly for milk production, even though the plot had inadequate forage for the animals. The "best case" were settlers who primarily depended on perennial crops, which partially protected the soil and sustained the farmers long term.

Eight contrasting types of animals were studied, from top predator (jaguar, *Panthera onca*) to ants, representing major groups of the Amazon fauna. These were evaluated based on their relative (1) requirement for a large forest patch, and (2) ability to cross an open gap in the forest. Comparing these animal types with the changing landscape pattern of forest and clearings indicated that the top predator could not survive there, with either the best- or typical-case settlers present. Few of the animal groups survived with the typical-case settlers (Figure 13.10b). But with best-case settlers, all the animal groups except the top predators persisted in the forest along with the people. In essence, most rainforest biodiversity was sustained while these settlers depending on perennial crops were sustained with subsistence farming.

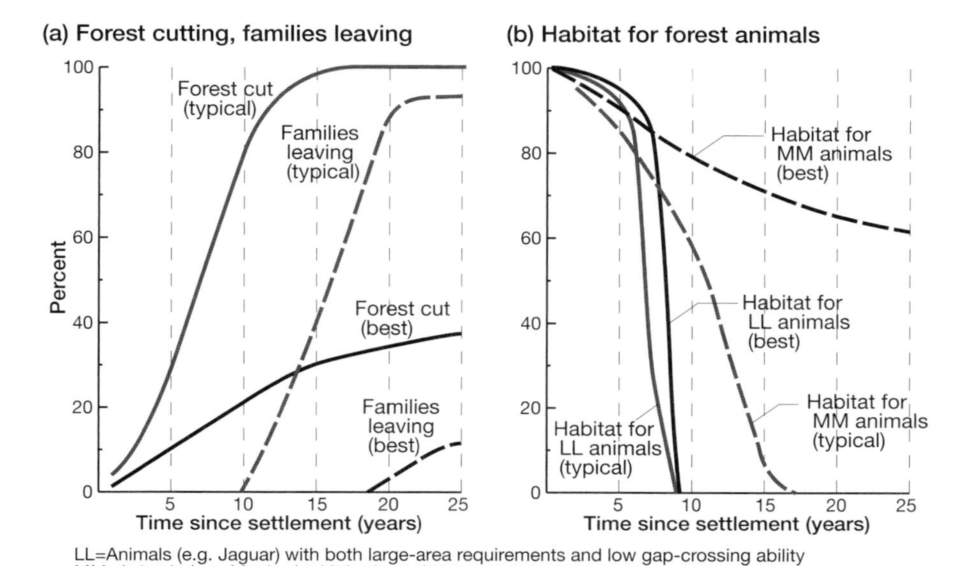

Figure 13.10 Alternative scenarios for sustaining both rainforest settlers and different rainforest animals. (a) In contrast to "typical" farmers, "best" case farmers plant many agricultural crops; grow mostly perennials rather than annuals; do minimal burning; have few or no cows; and harvest a diversity of foods from the forest. Settlers usually leave when the soil no longer produces sufficient food. (b) "LL" animals and "MM" animals plotted over time for the typical and best-case farmer conditions. All groups of smaller forest animals persisted for the entire period under both typical and best-case conditions. Ouro Preto region, Rondonia, Brazil. See text. Adapted from Dale (2001) and Dale *et al.* (1993, 1994b).

Solutions for Aridland

1. Avoid building or being in a low area downslope of mountains where scarce rain falls, producing flash floods racing dangerously across the land, sometimes as a big flood carrying almost all a year's sediment deposition in a gullies (Whitford, 2002; Ward, 2016).
2. Keep town, industrial, and mining pollutants out of groundwater, since it generally flows slowly, and often provides a town water supply and irrigation for local gardens, which may lower the water table (Sargent *et al.*, 1991; Larson *et al.*, 2013).
3. Avoid some activities around a stream or riparian vegetation downslope of a town, which may be affected by partially treated effluent from the sewage treatment facility.
4. Concentrate protection efforts on springs or oases where clean water comes to or near the surface, because distinctive vegetation, wildlife, rare species, and people are drawn to and depend on that water (Larson *et al.*, 2013; Ward, 2016).
5. Minimize damage to a biological soil crust (composed of numerous tiny species) by desert roads, all-terrain vehicles (ORV), and walkers, which may persist for decades (Ludwig *et al.*, 1997; Whitford, 2002; Ward, 2016).

6. Protect habitats rich in animals in holes by day and foraging out in a food web at night, such as around shrubs on small mounds formed by wind-blown sand deposits.
7. Prepare for more extensive desert land, as desertification by human and overgrazing activity continues, and climate warming leads to more evaporation and increased drought duration and intensity (Tivy, 1990; Ward, 2016).

Transportation a Bit Further Ahead

What may the post-car era look like? The *netway-and-pods* solution lies at the border of vision and feasibility (Forman and Sperling, 2011). This netway transportation system for inter-town, inter-city travel offers an impressive array of benefits, and all the technologies involved are relatively simple and currently operational. The system is a personal transit system (PRT), and many variants have been proposed, built as models, or tested experimentally (Sproule and Neumann, 1991; Bernasconi *et al.*, 2009; Davidheiser, 2011).

Imagine moving gloriously and quietly along in a comfortable car compartment (pod) on an elevated solid structure (netway) between trees, yet with no fossil fuel use, no greenhouse emission, and no driving or need to watch the road. Or we could move in channels just below ground level with translucent panels, solar panels, vegetable gardens, or footpaths overhead. No unpredictable drivers, no accidents. Play family games, write, do computer work, birdwatch. When ready to return to ground level, simply take manual control, and drive the pod with fully charged battery off onto local roads to work, shop, or home.

In addition to many familiar road-traffic problems in news headlines, habitat loss, road-killed wildlife, fragmented land, disrupted waterways, traffic noise effects on animals, and much more are widespread, and all addressed in the field of road ecology (Forman *et al.*, 2003; van der Ree *et al.*, 2015b). The netway-with-pods system addresses all of these in the land of towns, with links to cities.

Netways are raised or sunken structures about 7–11 m (27–35 ft) wide, somewhat wider than a football goal (Forman and Sperling, 2011). Raised netways are about 6 m (20 ft) high, equal to the tops of apple trees or the second level of windows in a building. Strong "pillars" support "spanners" (girders) about 30 m (100 ft) long, with L-shaped "wings" on each side providing the netway width. The only other essential components are two electric wires buried in the spanners that provide power for the pods moving in opposite directions. Small cranes can readily construct or replace the netway structures.

Owned or rented pods, reminiscent of a comfortable ski gondola or car interior, have a tiny electric motor at the bottom (Figure 13.11). With inductive coupling technology, the buried electric wire provides current (which can pass 30 cm [1 ft] through concrete, snow, water, etc.) to power the pod's electric motor. Automated control, analogous to some rail and aircraft systems, keeps pods moving smoothly along a netway. A small battery, charged enroute, provides energy for local off-network driving at ground level.

Renewable energy, especially solar and wind, generates the electricity running the netway pods. Small lightweight personal pods are most common. But long aerodynamic public pods with an attendant (like today's flight attendant), instead of a driver, provide "bus" service. Likewise, freight pods, which distribute weight along netways and provide flexibility for delivery to diverse locations, carry goods quietly across the land.

The netway-and-pods transportation system was primarily designed to (Forman and Sperling, 2011):

1. Recover land lost to roads and roadsides (ca. 1.5% of the USA; 4.5% of Germany) for nature and socioeconomic uses; and
2. Reconnect the rest of the land, so walkers, wildlife, livestock, streams, etc. can move and flow without blockages.

But many other significant societal benefits are also addressed with the netway-and-pods system, including:

1. No fossil fuel use.
2. Barely any unhealthful air pollution or stormwater-runoff pollution.
3. No greenhouse gas emissions.
4. No driving, no accidents, and travelers arrive rested at destinations.
5. Vegetated corridors, providing a recreational walking/biking trail network, routes for wildlife movement, and space for market gardening and community gardens near towns and cities.

I find that riding a pod on a netway is experiencing the future, gloriously (Figure 13.11).

Towns in a Regional, Global Context

A billion people are now added to the globe every 14 years. Population growth is particularly high in Asia and Africa and not expected to level off for many decades. The rural population will likely remain constant as all growth is urban. Yet the ecological footprint of humans already exceeds the total global land area.

Where on Earth is best for the next billion neighbors (Forman and Wu, 2016a, 2016b)? Not in water-stressed areas (where human needs exceed availability). Not in other drylands. Not on tundra or ice. Not in today's very high population density areas (>100 people/km^2). Not in very high biodiversity areas. Probably not in high population density areas (>50 people/km^2). And probably not in boreal forest (where making a living is difficult). However, that leaves the best (least bad) or most suitable places for another billion people in: large areas of South America; southern Canada, northern and eastern USA; south-central Africa; and scattered areas in Asia and Oceania.

Note the disconnect. Highest population growth is and will be in Asia and Africa, whereas the best areas are elsewhere. That suggests an increase in transnational migration. Such human movement creates severe problems for source areas, for town and village lands enroute, and for destinations.

Projections of population growth overwhelmingly put the new people in cities. So, where in urban regions are the best places for another billion neighbors (Forman and Wu, 2016a)? Of the 15 possible types of places, 3 are already dense (city center, ring immediately around city center, inner suburbs), 7 significantly degrade food-producing

Figure 13.11 Netway-and-pod transportation system at London airport. Automated system (no driving) with electricity from wire buried in track running a small electric motor in pod, which carries ten people with luggage. R. Forman photo. (A black and white version of this figure will appear in some formats. For the color version, please refer to the plate section.)

agricultural or natural land, both needed to support so many new people. That leaves five most suitable places, i.e., existing: outer suburbs; exurban areas; small satellite cities in natural land; the same in agricultural land; and towns and villages in agricultural land.

But this creates a second disconnect. New people go to cities, whereas the five most suitable places are outside cities. One solution could prevent this from being just more sprawl. Develop an *urban region plan*, which designates the best and worst areas for various land uses in an urban region (Forman, 2014). Then, consistent with the spatial plan, fit population growth into the five best places pinpointed above. Indeed, make an urban region plan for every city.

What can be done to improve areas while the billion newcomers are arriving (Forman and Wu, 2016a)? Why not do the following?

For Farmland

1. Protect the best soils, and enrich others.
2. Establish market-gardening areas by all cities.
3. Accelerate urban agriculture.
4. Close or relocate industries polluting valuable farmland.

5. Reduce livestock and meat production, to gain land for edible crops.
6. Reduce cropland used for fuel and animal feed.
7. Use more food species and varieties.
8. Replace annual grains with perennial grains.

For Natural Land

1. Protect the integrity of, and expand, large green areas.
2. Remove remote roads and buildings that degrade nature.
3. Connect large green areas with corridors and/or stepping stones.
4. Perforate roads/roadbeds to facilitate wildlife crossing and movement.
5. Expand vegetation along waterbodies, stream headwaters, and intermittent channels.
6. Increase bits of nature across both farmland and built areas.

For Built Communities

1. Protect or enhance the water supply with vegetation.
2. Reduce and disconnect impervious surface areas.
3. Shrink human home ranges.
4. Transform sections of local busy noisy roads to quiet neighborhood spaces.
5. Arrange town greenspaces to cool air, reduce flooding, support biodiversity, and enhance recreation.
6. Convert sprawl areas to denser compact mixed-use areas with vibrant neighborhoods.
7. Grow a tree-anchored green net along town/municipality boundaries.

With the human ecological footprint already more than filling the globe's land, unless the land is improved, the 14-year population growth scenario is a bad omen. The preceding goals and solutions were outlined for urban regions with impending population growth, yet most apply nicely in the land of towns and villages.

A *geographic region* has one or more major cities, a common macroclimate over its area, a transportation system linking all towns to the city(s), and similar cultural patterns throughout, such as town layout, architecture, language, and traditions (Forman, 2008). Towns interact with and relate to their region more than to a nation or the globe. Many towns are shrinking, though for some it is a temporary phase, and plenty of towns successfully thrive (Vaz *et al.*, 2013b).

Various global and national forces also affect towns and the land of towns and villages. Economic globalization determines or affects the markets for goods from some local industries in towns (Albrecht, 2014). For instance, the only manufacturing in the village of Kaihsienkung (China; Chapter 4; see Appendix) was a small silk factory, and the major worry of the whole village was that technology development and global production would lower the price of silk, effectively putting the local industry out of business (Fei, 1939).

Global and national chain stores often outcompete local mom-and-pop shops in a town. Globalization also strongly affects culture, eroding both local and regional cultures. Music, theatre, and the news media highlight international trends and values, while regional traditions wither.

A few percent of the global land surface is urban, and perhaps nearly 50 percent is relatively remote natural land, mainly forest, desert, and tundra (Chapter 1). That leaves

Figure 13.12 Electric car with driver's door open next to house with green roof. Village of Flam, Norway. See Appendix. R. Forman photo. (A black and white version of this figure will appear in some formats. For the color version, please refer to the plate section.)

about half the globe as the *land of towns and villages* (land of Ts & Vs). This land is about 80 percent agricultural, and the rest mainly in the edge portions of extensive natural land.

Yet several major worldwide trends are focused on the land of towns and villages. Food supply, in the face of global population growth, leaves increasing numbers of people in lives of malnutrition or starvation. In grasslands, soils are thinning, organic matter decreasing, nitrogen decreasing, and desertification marches onward. The shortage of freshwater for household and irrigation uses worsens. Natural areas shrink. Expanding road systems with more traffic fragment habitats, decreasing wildlife populations. Biodiversity is dropping. These are familiar themes experienced by residents in the land of Ts & Vs.

Are these trends largely caused by, or happening despite the care of, town and village residents? Could a trajectory of greater ecological commitment from national to local levels reverse these trends (Figure 13.12)? It seems likely, considering that all the trends are somewhat linked, and have strong ecological dimensions. Indeed, town ecology could head off global crises.

Important multi-scale issues need to be faced. Over decades, human-caused climate change has been documented in all continents, all oceans, and perhaps all nations. The greenhouse effect mainly results from accumulating CO_2 and other key gases, largely from combusting fossil fuels in transportation, industry, and other activities including cattle production. Cities are major emitters, but the multitude of towns and villages together are also emitters. Almost every household in industrial nations and many other places is a big greenhouse gas emitter. Both local and international travel are especially big emitters.

Prime Ecological Footprint

The *ecological footprint* refers to the total land and water area used or required to sustain an individual or population (Rees and Wackernagel, 1996, 2008; Adler and Tanner, 2013; Wackernagel and Rees, 2014). Area is measured as hectares (acres) per year for an individual, or km^2 (mi^2) for a population. The concept focuses on the land required for annual inputs (e.g., growing crops, paper products), plus outputs (e.g., solid-waste disposal sites, waterbodies receiving stormwater pollutants), and energy expended (e.g., in commuting, recreation). In all cases land and water area required is irrespective of location.

The per capita ecological or carbon footprint of people in industrial developed nations is often reported to be some ten times that of people in developing nations (McKinney, 2010; Nakagoshi and Mabuhay, 2014; Gurney *et al.*, 2015). Rural town and village residents tend to have slightly larger footprints than do urban residents. For instance, the average per capita ecological footprint of a UK resident is about twice that of a London resident, and a typical USA resident has more than three times the footprint of a New York City resident (footprints in Barcelona and Tokyo are less than those in London and New York). Long-distance travel, especially by aircraft, requires high energy use and expends considerable pollution, so people or populations with much long-distance travel have large footprints (Holz-Rau *et al.*, 2014). A town or village in agricultural land has added stability, by readily growing local food and having the support system of a local community (Eaton *et al.*, 2007).

Many additions to the ecological footprint concept have been proposed. The *prime footprints model*, illustrated in the following quote, seems particularly useful, both for understanding and for evaluating the sustainability of a population or place.

> Integrating urban areas with their near and distant footprint impacts, plus the transportation network linking residents to sources and sinks, would provide spatial, environmental, and economic clarity for reducing impacts and accelerating benefits. (Forman, 2014, p. 30)

A prime footprint requires identifying specific places, not just areas anywhere (Forman, 2008, 2014). It highlights the primary landscapes and sites that provide food, fiber, and other resources used, and similarly the specific major areas that receive wastes of various sorts. These dispersed areas are then linked to the population center by transport corridors, with area of transportation effects on the land calculated for the corridors. Determining the prime footprint integrates the combined network land of source areas, sink (output) areas, and transportation corridor areas.

Ecosystem services, another spatial, economic, and environmental issue, ranges in importance from local to global (Turner *et al.*, 2013). These services are provided by nature, but not adequately priced by economic markets (Figure 13.13).

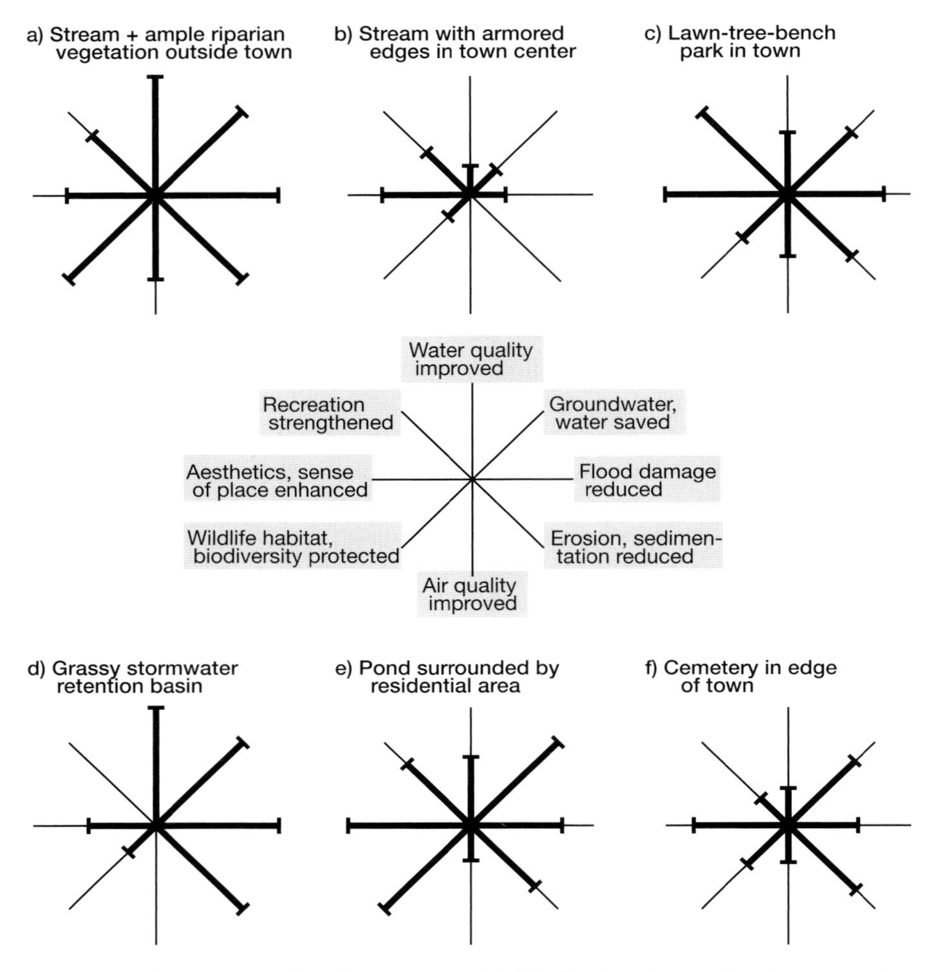

Figure 13.13 Ecosystem services for a town provided by local semi-natural and constructed water-related resources. Estimated relative per unit area amounts (distance from center point) doubtless vary somewhat for each town. Based on "flowers" of ecosystem services (nature's services) (Foley *et al.*, 2005; Larson *et al.*, 2013).

A decade study (2001–2011) of eight ecosystem services at statewide (Massachusetts, USA) and local town scales is instructive (Blumstein and Thompson, 2015). Statewide, natural ecosystems providing clean water and regulating floods changed little over the period. Habitat quality supporting biodiversity decreased, due to a gradual loss of habitat to housing development. The regulation of climate by carbon storage increased, due to accelerating land protection and continued growth of forest trees. Cultural ecosystem services in the form of outdoor recreation also increased with more protected land. "Hotspots" of ecosystem services provided are highlighted in the town-by-town analysis. Interestingly, hotspots of ecosystem services are more abundant in the rapidly changing Greater Boston region portion of the state than in the more rural areas.

Figure 13.14 Protected coastal area, with 4- to 6-m-high stone moai statues, that protects town against tsunamis. Hanga Roa, Easter Island (Isla Pascua), Chile. See Appendix. R. Forman photo.

Sustainability and Plans

Sustainability, with both environmental and socioeconomic foundations, is the other concept crossing spatial scales, but is usually considered at the local scale. How sustainable is the household, business, town, or city? The concept involves inputs and outputs, recycling, long-term planning, intergenerational equity, and more.

Sustainability for a population center would be achieved by establishing and maintaining a balance, where nature and people both thrive in the prime footprints system as a whole (Forman, 2008). Decreasing the number of input and output routes, shortening the routes, and decreasing the input-and-output amounts transported would all be useful steps toward sustainability.

Two recent developments offer particular promise (Wu, 2013, 2014). *Sustainability science* adds rigor, particularly as a place-based analysis of changing relationships between ecosystem services and human well-being. Ecology and related sciences are central, though well tied to socioeconomic dimensions. *Landscape sustainability* highlights the importance of spatial patterns and processes at the landscape and regional scales, in providing long-term ecosystem services for maintaining or improving the human condition (Figure 13.14). If these two concepts take hold, sustainability could become an even more lucid, rigorous, useful measure.

"Think globally, act locally" became buzzwords for a period. But hardly anyone really thinks globally in most decisions made. Furthermore, simply acting locally without regional context has produced the degraded and degrading land so evident around us. For years I had a bumper sticker: "Think globally, plan regionally, then act locally." I liked that, because it focused on planning general areas appropriate for different land uses, and then actively making local changes consistent with the regional features and plans.

But then a colleague pointed out that placing the globe first is an unlikely symbol for an individual to put first or actively revere. It is likely to be ignored, and if not, it leads to familiar globalization processes that homogenize economies and cultures (Jesse Keenan, 2017 personal communication). Globalization and climate change are big and long-term. Yet urbanization and loss of natural land are not only big and long-term, but also daily right in everyone's face. Moreover, development and loss of nature can be readily altered by a single person, or a local town.

The central importance of planning and action with visible results is much better highlighted as: "Plan regionally, then act locally; and remember our globe."

Towns, villages, and the land need our greatest attention. Now, with the richness of ideas and opportunities before us, what's the next step?

Appendix: Population Centers Examined

Fifty-five population centers or communities were visited in six continents, 16 nations, and 14 US states (Figure A.1). Ranging in population from 350 to 71,000, about 12 are villages, 9 small cities, and 34 towns (Figure A.2). Community selection included a range of population sizes and wide geographical distribution for each of 20 town types.

In each community, information was recorded for periods from two hours to thirty months. Population numbers are derived from Wikipedia online except for a few cases where more compelling data were available. The numbers are quite rough because they were: (1) recorded in different years from about 2002 to 2012; and (2) commonly include people in varied-size administrative units with low-density houses surrounding the "compact essentially all-built mainly residential town," as used in this book.

One to four major characteristics of a community is identified based on location, economic role, or other feature. In Figure A.2 a community is listed for each of its major characteristics.

1. Akumal, Mexico. Pop. 1,300. Town (village) type: *coastal tourist.*
2. Alice Springs, Australia. Pop. 29,000. Town type: *desert; regional shopping; inland tourist.* (Native People).
3. Askim, Norway. Pop. 15,500. Town type: *manufacturing/industrial; commuter; regional shopping.*
4. Astoria, Oregon, USA. Pop. 9,500. Town type: *fishing/port; regional shopping.*
5. Capellades, Catalunya, Spain. Pop. 5,500. Town type: *hill/mountain; commuter; river.*
6. Casablanca, Chile. Pop. 15,200. Town type: *farm-crop; cattle/sheep.*
7. Chamusca, Portugal. Pop. 3,300. Town type: *farm-crop; river.*
8. Clay Center, Kansas, USA. Pop. 3,300. Town type: *farm-crop; cattle/sheep; regional shopping; low-income.*
9. Concord, Massachusetts, USA. Pop. 12,000. Town type: *suburban; commuter; river; high-income.*
10. Cranborne, Australia. Pop. 18,600. Town type: *suburban; commuter.*
11. Easton, Maryland, USA. Pop. 16,700. Town type: *regional shopping; farm-crop.*
12. Emporia, Kansas, USA. Pop. 24,900. Town type: *farm-crop; river; regional shopping; university/college.*
13. Falun, Sweden. Pop. 37,300. Town (small city) type: *mining; manufacturing/industrial; regional shopping; university/college.*

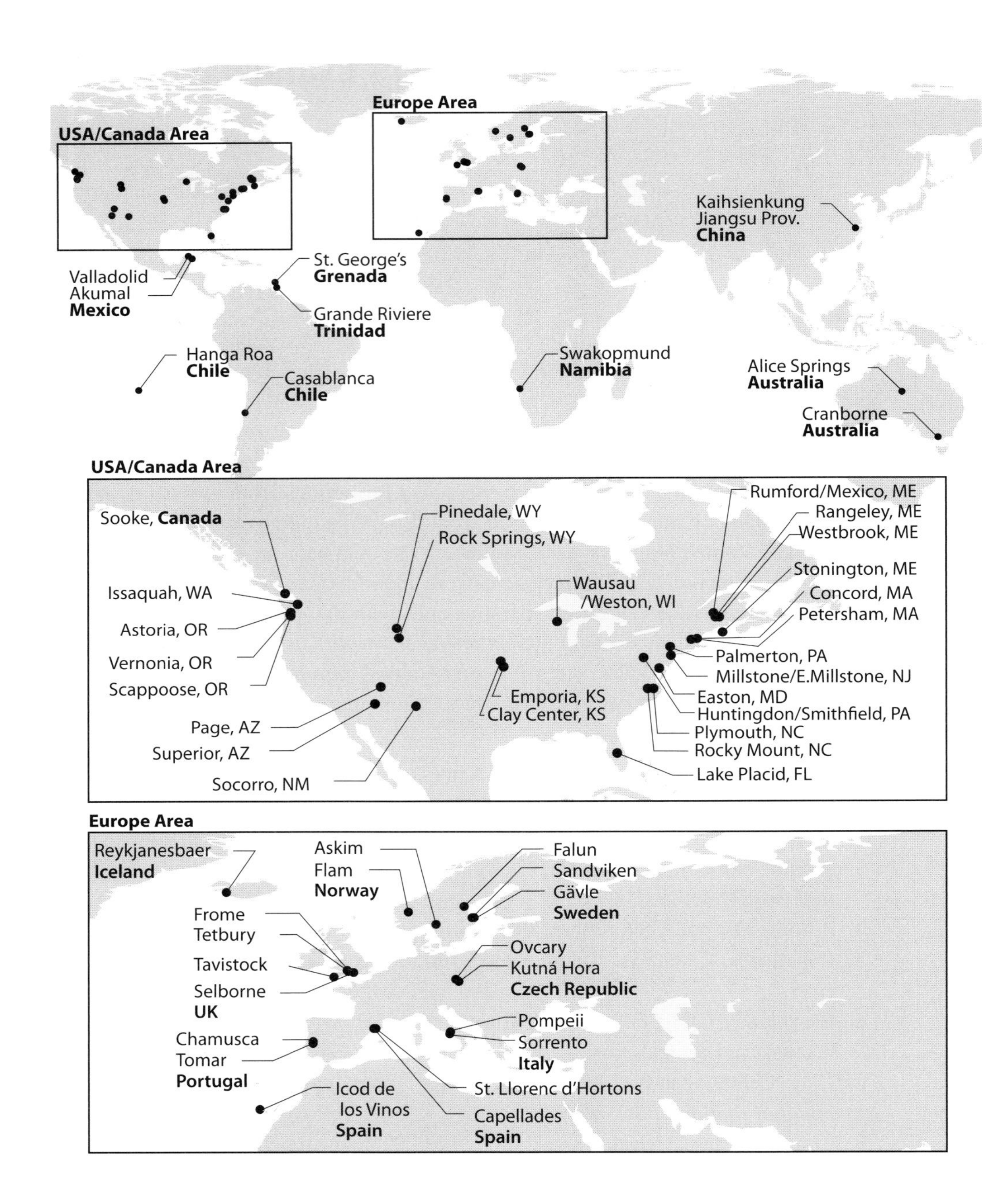

Figure A.1 Maps showing the 55 communities observed.

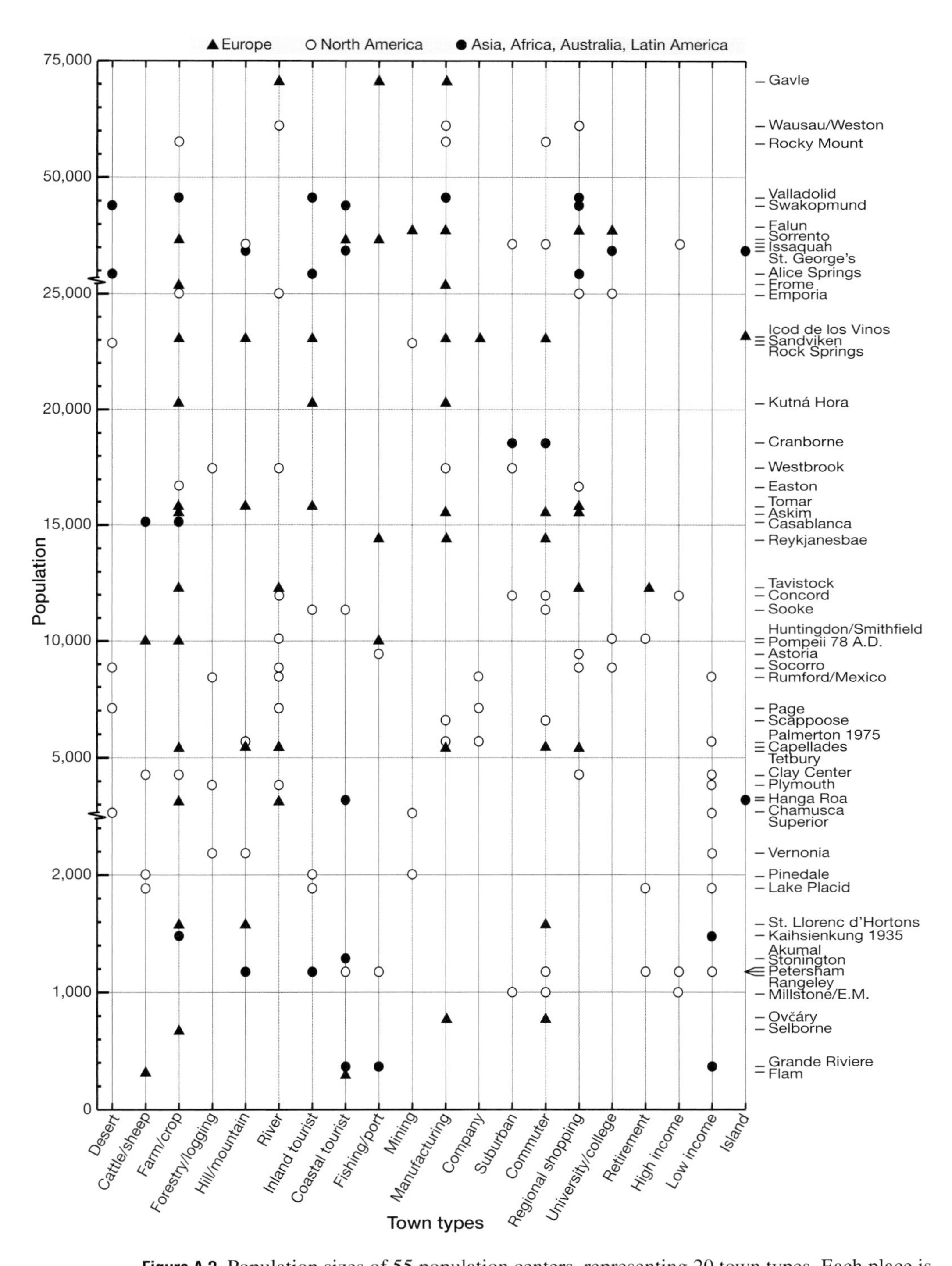

Figure A.2 Population sizes of 55 population centers, representing 20 town types. Each place is listed with its one to four major characteristics or features.

14. Flam, Norway. Pop. 350. Town (village) type: *coastal tourist; cattle/sheep.*
15. Frome, UK. Pop. 26,200. Town type: *manufacturing/industrial; farm-crop.*
16. Gavle, Sweden. Pop. 71,000. Town (small city) type: *fishing/port; river; manufacturing/industrial.*
17. Grande Riviere, Trinidad. Pop. 350. Town (village) type: *coastal tourist; fishing; low-income.*
18. Hanga Roa, Isla Pascua (Easter Island), Chile. Pop. 3,300. Town type: *coastal tourist; island.*
19. Huntingdon/Smithfield, Pennsylvania, USA. Pop. 10,100. Town type: *university/ college; retirement; river.*
20. Icod de los Vinos, Tenerife, Canary Islands, Spain. Pop. 23,100. Town type: *hill/ mountain; farm-crop; commuter; island.*
21. Issaquah, Washington, USA. Pop. 34,600. Town (small city) type: *suburban; commuter; hill/mountain; high-income.*
22. Kaihsienkung in 1935, Jiangsu Province, China. Pop. 1,500. Town (village) type: *farm-crop; low-income.* (Community no longer exists, thus not visited.)
23. Kutna Hora, Czech Republic. Pop. 20,300. Town type: *manufacturing/industrial; inland tourist; farm-crop.*
24. Lake Placid, Florida, USA. Pop. 1,900. Town type: *retirement; cattle/sheep; inland tourist; low-income.*
25. Millstone/East Millstone, New Jersey, USA. Pop. 1,000. Town (village) type: *commuter; suburban; high-income.*
26. Ovcary, Czech Republic. Pop. 800. Town (village) type: *commuter; manufacturing/ industrial.* (Visited for one hour.)
27. Page, Arizona, USA. Pop. 7,200. Town type: *desert; company* (government).
28. Palmerton, Pennsylvania (in 1975), USA. Pop. 5,600. Town type: *company; manufacturing/industrial; hill/mountain; low-income.*
29. Petersham, Massachusetts, USA. Pop. 1,200. Town (village) type: *retirement; commuter; high-income.*
30. Pinedale, Wyoming, USA. Pop. 2,000. Town type: *inland tourist; cattle/sheep; mining.*
31. Plymouth, North Carolina, USA. Pop. 3,900. Town type: *forest/logging; river; low-income.*
32. Pompeii in 78 AD, Italy. Pop. 10,000. (Colonium Cornelia Veneria Pompeiarnorum). Town type: *farm-crop; cattle/sheep; fishing/port.*
33. Rangeley, Maine, USA. Pop. 1,200. Town (village) type: *inland tourist; hill/ mountain.*
34. Reykjanesbaer, Iceland. Pop. 14,400. Town type: *fishing/port; manufacturing/ industrial; commuter.*
35. Rock Springs, Wyoming, USA. Pop. 23,000. Town type: *mining; desert.* (Transportation hub.)
36. Rocky Mount, North Carolina, USA. Pop. 57,500. Town (small city) type: *manufacturing/industrial; farm-crop; commuter.*

37. Rumford/Mexico, Maine, USA. Pop. 8,500. Town type: *forest/logging; company; river; low-income.*
38. Sandviken, Sweden. Pop. 23,000. Town type: *manufacturing/industrial; company; inland tourist.*
39. Sant Llorenc d'Hortons, Catalunya, Spain. Pop. 1,600. Town (village) type: *hill/mountain; farm-crop; commuter.*
40. Scappoose, Oregon, USA. Pop. 6,600. Town type: *commuter; manufacturing.*
41. Selborne, UK. Pop. 700. Town type: *farm-crop.*
42. Socorro, New Mexico, USA. Pop. 8,900. Town type: *desert; river; regional shopping; university/college.*
43. Sooke, British Columbia, Canada. Pop. 11,400. Town type: *coastal tourist; inland tourist; commuter.*
44. Sorrento/Sant'Agnello, Italy. Pop. 35,000. Town (small city) type: *coastal tourist; fishing/port; farm-crop.*
45. St. George's, Grenada. Pop. 34,000. Town (small city) type: *coastal tourist; fishing/port; university/college; island.*
46. Stonington, Maine, USA. Pop. 1,200. Town (village) type: *fishing/port; coastal tourist; mining; low-income.*
47. Superior, Arizona, USA. Pop. 2,800. Town type: *desert; mining; low-income.*
48. Swakopmund, Namibia. Pop. 44,700. Town (small city) type: *desert; coastal tourist; regional shopping.*
49. Tavistock, UK. Pop. 12,300. Town type: *regional shopping; river, farm-crop; retirement.*
50. Tetbury, UK. Pop. 5,500. Town type: *regional shopping; farm-crop; manufacturing/industrial.*
51. Tomar, Portugal. Pop. 15,800. Town type: *regional shopping; inland tourist; farm-crop; hill/mountain.*
52. Valladolid, Yucatan, Mexico. Pop. 45,900. Town (small city) type: *regional shopping; manufacturing/industrial; farm-crop; inland tourist.*
53. Vernonia, Oregon, USA. Pop. 2,200. Town type: *forest/logging; hill/mountain; low-income.*
54. Wausau/Weston, Wisconsin, USA. Pop. 61,500. Town (small city) type: *river; manufacturing/industrial; regional shopping.*
55. Westbrook, Maine, USA. Pop. 17,500. Town type: *suburban; manufacturing/industrial; river; forest/logging.*

References

Abbott, I. M., Berthinussen, A., Stone, E. *et al.* (2015). Bats and roads. In *Handbook of Road Ecology*, ed. R. van der Ree, D. J. Smith and C. Grilo. Chichester, UK: Wiley-Blackwell, pp. 290–299.

Abbott, I. M., Butler, F. and Harrison, S. (2012). When flyways meet highways – the relative permeability of different motorway crossing sites to functionally diverse bat species. *Landscape and Urban Planning*, 106, 293–302.

Abderrahman, W. A. (2000). Urban water management in developing arid countries. *Water Resources Development,* 16, 7–20.

Acosta-Jamett, G., Chalmers, W., Cunningham, A. *et al.* (2011). Urban domestic dog populations as a source of canine distemper virus for wild carnivores in the Coquimbo region of Chile. *Veterinary Microbiology*, 152, 247–257.

Adams, C. E. (2016). *Urban Wildlife Management*. Boca Raton, FL: CRC Press.

Adams, C. E. and Lindsey, K. J. (2011). Anthropogenic ecosystems: the influence of people on urban wildlife populations. In *Urban Ecology: Patterns, Processes, Applications*, ed. J. Niemela, J. H. Breuste, T. Elmqvist *et al.* New York: Oxford University Press, pp. 116–128.

Adams, L. W. (1994). *Urban Landscape Habitats: A Landscape Perspective.* Minneapolis, MN: University of Minnesota Press.

Adams, M., Cox, T., Moore, G. *et al.* (2006). Sustainable soundscapes: noise policy and urban experience. *Urban Studies*, 43, 2385–2398.

Adger, W. N. (2000). Social and ecological resilience: are they related? *Progress in Human Geography*, 24, 347–364.

Adler, F. R. and Tanner, C. J. (2013). *Urban Ecosystems: Ecological Principles for the Built Environment.* New York: Cambridge University Press.

Agger, P. and Brandt, J. (1988). Registration methods for studying the development of small-scale biotope structures in rural Denmark. In *Proceedings of the First International Seminar on Methodology in Landscape Ecological Research and Planning* (Volume 2), ed. J. Brandt and P. Agger. Roskilde, Denmark: Roskilde Universitetsforlag GeoRuc, pp. 61–72.

Ahern, J. (2004). Greenways in the USA: theory, trends and prospects. In *Ecological Networks and Greenways: Concept, Design, Implementation,* ed. R. H. G. Jongman and G. Pungetti. Cambridge: Cambridge University Press, pp. 34–55.

Airola, J. M. and Bucholz, K. (1984). Species structure and soil characteristics of five urban forest sites along the New Jersey Palisades. *Urban Ecology*, 8, 149–164.

Akgun, A. A., Baycan, T. and Nijkamp, P. (2013). From depreciation to appreciation of rural areas: "Beauty Idols" in Europe. In *Towns in a Rural World*, ed. T. de N. Vaz, E. van Leeuwen and P. Nijkamp. Farnham, UK: Ashgate, pp. 303–321.

Alberti, M. (2008). *Advances in Urban Ecology: Integrating Humans and Ecological Processes in Urban Ecosystems*. New York: Springer.

Alberti, M. (2016). *Cities That Think like Planets: Complexity, Resilience, and Innovation in Hybrid Ecosystems*. Seattle, WA: University of Washington Press.

Alberti, M. and Marzluff, J. (2004). Ecological resilience in urban ecosystems: linking urban patterns to human and ecological functions. *Urban Ecosystems*, 7, 241–265.

Albrecht, D. E. (2014). *Rethinking Rural: Global Community and Economic Development in the Small Town West*. Pullman, WA: Washington State University Press.

Albuquerque, U. P., Andrade, L. H. C. and Ceballero, J. (2005). Structure and floristics of home gardens in northeastern Brazil. *Journal of Arid Environments*, 62, 491–506.

Alcoforado, M. J. and Andrade, H. (2008). Global warming and the urban heat island. In *Urban Ecology: An International Perspective on the Interaction Between Humans and Nature*, ed. J. M. Marzluff, E. Shulenberger, W. Endlicher *et al.* New York: Springer, pp. 249–262.

Alexander, D. E. (2013). Resilience and disaster risk reduction: an etymological journey. *Natural Hazards and Earth Systems Science*, 13, 2707–2716.

Allan, B. F., Tallis, H. and Chaplin-Kramer, R. (2017). Can integrating wildlife and livestock enhance ecosystem services in central Kenya? *Frontiers in Ecology and the Environment*, 15, 328–335.

Allan, J. D. and Castillo, M. (2007). *Stream Ecology: Structure and Function of Running Water*. New York: Springer.

Allen, J. J., Bekoff, M. and Crabtree, R. L. (1999). An observational study of coyote (*Canis latrans*) scent-marking and territoriality in Yellowstone National Park. *Ethology*, 105, 289–302.

Anderson, B. J., Armsworth, P. R. and Eigenbrod, F. (2009). Spatial covariance between biodiversity and other ecosystem service priorities. *Journal of Applied Ecology*, 46, 888–896.

Anderson, D. C., Sartoris, J. J., Thullen, J. S. and Reusch, P. G. (2003). The effects of bird use on nutrient removal in a constructed wastewater treatment wetland. *Wetlands*, 23, 423–425.

Anderson, D. L. and Otis, R. J. (2000). Integrated wastewater management in growing urban environments. In *Managing Soils in an Urban Environment*, ed. R. B. Brown, J. H. Huddleston and J. L. Anderson. Madison, WI: American Society of Agronomy, pp. 199–250.

Anderson, M. G., Barnett, A., Clark, M. *et al.* (2016). *Resillient and Connected Landscapes for Terrestrial Conservation*. Boston, MA: The Nature Conservancy.

Andrieu, E., Ladet, S., Heintz, W. and Deconchat, M. (2011). History and spatial complexity of deforestation and logging in small private forests. *Landscape and Urban Planning*, 103, 109–117.

Anton, A., Cebrian, J., Heck, K. L. *et al.* (2011). Decoupled effects (positive to negative) of nutrient enrichment on ecosystem services. *Ecological Applications*, 21, 991–1009.

Anton, A., Maia de Souza, D., Teillard, F. and Mila i Canals, L. (2016). Addressing biodiversity and ecosystem services in life cycle assessment. In *Handbook on Biodiversity and Ecosystem Services in Impact Assessment*, ed. D. Geneletti. Cheltenham, UK: Edward Elgar, pp. 140–163.

Antos, M. J., Ehmke, G. C., Tzsros, C. L. and Weston, M. A. (2007). Unauthorized human use of an urban coastal wetland sanctuary: current and future patterns. *Landscape and Urban Planning*, 80, 173–183.

Antrop, M. (1988). Invisible connectivity in rural landscapes. In *Connectivity in Landscape Ecology*, ed. K.-F. Schreiber. Munstersche Geographische Arbeiten 29. Paderborn, Germany: F. Schoningh, pp. 57–60.

Arendt, R. (2004). *Crossroads, Hamlet, Village, Town: Design Characteristics of Traditional Neighborhoods, Old and New*. Chicago, IL: Planning Advisory Service, American Planning Association.

Arendt, R. (2015). *Rural by Design: Planning for Town and Country*. Chicago, IL: Planners Press, American Planning Association.

Arendt, R., Brabec, E. A., Yaro, R. D. *et al.* (1994). *Rural by Design: Maintaining Small Town Character*. Chicago, IL: American Planning Association.

Arifin, H. S. and Nakagoshi, N. (2011). Landscape ecology and urban biodiversity in tropical Indonesian cities. *Landscape Ecology and Engineering*, 7, 33–43.

Arifin, H. S., Kaswanto, R. L. and Nakagoshi, N. (2014). Low carbon society through *pekarangan*, traditional agroforestry practices in Java, Indonesia. In *Designing Low Carbon Societies in Landscapes*, ed. N. Nakagoshi and J. A. Mabuhay. New York: Springer, pp. 129–143.

Arnold, C. L. and Gibbons, C. J. (1996). Impervious surface coverage: the emergence of a key environmental indicator. *Journal of the American Planning Association*, 62, 243–258.

Ascencao, F., LaPoint, S. and van der Ree, R. (2015). Road traffic and verges: big problems and big opportunities for small mammals. In *Handbook of Road Ecology*, ed. R. van der Ree, D. J. Smith and C. Grilo. Chichester, UK: Wiley-Blackwell, pp. 325–333.

Ash, N. and Jenkins, M. (2007). *Biodiversity and Poverty Reduction: The Importance of Biodiversity for Ecosystem Services*. Cambridge, UK: UNEP-World Conservation Monitoring Centre.

Ashworth, G. J. and Kavaratzis, G. (2014). Cities of culture and culture in cities: the emergent uses of culture in city branding. In *Urban Planning and Design in Times of Structural and Systemic Change*, ed. T. Haas and K. Olsson. Aldershot, UK: Ashgate, pp. 72–80.

Askew, L. E., Sherval, M. and McGuirk, P. (2014). "Not just drought." Drought, rural change and more: perspectives from rural farming communities. In *Rural Change in Australia: Population, Economy, Environment*, ed. R. Dufty-Jones and J. Connell. Farnham, UK: Ashgate, pp. 235–253.

Atkinson, I. (1989). *Introduced Animals and Extinction in Conservation for the Twenty-First Century*, ed. D. Western and M. B. Pearl. New York: Oxford University Press.

Atkinson-Smith, R. and Smith, I. (2014). *Town: Small and Medium Sized Towns in Their Functional Territorial Context*. Case Study Report: Wales. Leuven, Belgium: ESPON and University of Leuven.

Austin, G. (2014). *Green Infrastructure for Landscape Planning: Integrating Human and Natural Systems*. Abingdon, UK: Routledge.

Bacon, E. N. (1974). *Design of Cities*. New York: Viking Press.

Bailey, B. R. (1982). *Main Street Northeastern Oregon: The Founding and Development of Small Towns*. Portland, OR: Oregon Historical Society.

Baker, A. J. M., Ernst, W. H. O., van der Ent, A. *et al*. (2010). Metallophytes: the unique biological resource, its ecology and conservational status in Europe, central Africa and Latin America. In *Ecology of Industrial Pollution*, ed. L. C. Batty and K. B. Hallberg. Cambridge, UK: Cambridge University Press, pp. 7–40.

Baker, J. (2006). Survival and accumulation strategies at the rural-urban interface in North-West Tanzania. In *The Earthscan Reader in Rural-Urban Linkages*, ed. C. Tacoli. London: Earthscan, pp. 41–55.

Baker, L. A., ed. (2009a). *The Water Environment of Cities*. New York: Springer.

Baker, L. A. (2009b). New concepts for managing urban pollution. In *The Water Environment of Cities*, ed. L. A. Baker. New York: Springer, pp. 69–91.

Baker, M. R., Gobush, K. and Vynne, C. (2013). Review of factors influencing stress hormones in fish and wildlife. *Journal for Nature Conservation*, 21, 309–318.

Baker, P. J., Bentley, A. J., Ansell, R. J. and Harris, S. (2005). Impact of predation by domestic cats *Felis catus* in an urban area. *Mammal Review*, 35, 302–312.

Ballard, W. B., Whitlaw, H. A., Young, S. J. *et al*. (1999). Predation and survival of white-tailed deer fawns in northcentral New Brunswick. *Journal of Wildlife Management*, 63, 574–579.

Balogh, A. L., Ryder, T. B. and Marra, P. P. (2011). Population demography of Gray Catbirds in the suburban matrix: sources, sinks and domestic cats. *Journal of Ornithology*, 152, 717–726.

Bambra, C., Robertson, S., Kasim, A. *et al*. (2014). Healthy land? An examination of the area-level association between brownfield land and morbidity and mortality in England. *Environment and Planning A*, 46, 433–454.

Banks, P. B. and Bryant, J. V. (2007). Four-legged friend or foe? Dog walking displaces native birds from natural areas. *Biology Letters*, 3, 611–613.

Banski, J. and Wesolowska, M. (2010). Transformations in housing construction in rural areas of Poland's Lublin region – influence on the spatial settlement structure and landscape aesthetics. *Landscape and Urban Planning*, 94, 116–126.

Barber, J. R., Burdett, C. L., Reed, S. E. *et al*. (2011). Anthropogenic noise exposure in protected natural areas: estimating the scale of ecological consequences. *Landscape Ecology*, 26, 1281–1295.

Barber, J. R., Crooks, K. R. and Fristrup, K. M. (2010). The costs of chronic noise exposure for terrestrial organisms. *Trends in Ecology and Evolution*, 25, 180–189.

Barkman, J. J. (1958). *Phytosociology and Ecology of Cryptogamic Epiphytes*. Assen, The Netherlands: Gorcum.

Barnard, A. (2003). Getting the facts – dog walking and visitor number surveys at Burnham Beeches and their implications for the management process. *Countryside Recreation*, 11, 16–19.

Barnes, B. V., Zak, D. R., Denton, S. and Spurr, S. H. (1998). *Forest Ecology*. New York: John Wiley.

Barnett, R. and Margetts, J. (2013). Disturbanism in the South Pacific: disturbance ecology as a basis for urban resilience in small island states. In *Resilience in Ecology and Urban Design: Linking Theory and Practice for Sustainable Cities*, ed. S. T. A. Pickett, M. L. Cadenasso and B. McGrath. New York: Springer, pp. 443–459.

Barral, M. A. (2009). Protection of beach and dune systems of the western coast of Huelva – developments in the planning and future prospect. In *Naturbanization: New Identities and Processes for Rural-Natural Areas*, ed. M. J. Prados. London: Taylor & Francis, pp. 185–204.

Barratt, D. G. (1997). Home range size, habitat utilization and movement patterns of suburban and farm cats (*Felis catus*). *Ecography*, 20, 271–280.

Barreira, A. P. (2013). Social and political determinants of the area of influence of medium-sized cities in Portugal. In *Towns in a Rural World*, ed. T. de N. Vaz, E. van Leeuwen and P. Nijkamp. Farnham, UK: Ashgate Publishing, pp. 149–172.

Barrett, G. W. and Barrett, T. L. (2001). Cemeteries as repositories of natural and cultural diversity. *Conservation Biology*, 15, 1820–1824.

Barrett, K. and Guyer, C. (2008). Differential responses of amphibians and reptiles in riparian and stream habitats to land use disturbances in western Georgia, USA. *Biological Conservation*, 141, 2290–2300.

Barrow, C. J. (1991). *Land Degradation: Development and Breakdown of Terrestrial Environments*. Cambridge, UK: Cambridge University Press.

Barten, F., Fustukian, S. and de Haan, S. (1998). The occupational health needs of workers: the need for a new international approach. In *Environmental Victories*, ed. C. Williams. London: Earthscan, pp. 142–156.

Baskin, C. C. and Baskin, J. M. (2001). *Seeds: Ecology, Biogeography, and Evolution of Dormancy and Germination*. San Diego, CA: Academic Press.

Batty, L. C. and Hallberg, K. B., eds. (2010). *Ecology of Industrial Pollution*. Cambridge, UK: Cambridge University Press.

Batty, L. C., Auladell, M. and Sadler, J. (2010). The impacts of metalliferous drainage on aquatic communities in streams and rivers. In *Ecology of Industrial Pollution*, ed. L. C. Batty and K. B. Hallberg. Cambridge, UK: Cambridge University Press.

Bauman, A. E., Russell, S. J., Furber, S. E. and Dobson, A. J. (2001). The epidemiology of dog walking: an unmet need in human and canine health. *Medical Journal of Australia*, 175, 632–634.

Bayne, E. M., Habib, L. and Boutin, S. (2008). Impacts of chronic anthropogenic noise from energy-sector activity on the abundance of songbirds in the boreal forest. *Conservation Biology*, 22, 1186–1193.

Bazzanti, M., Baldoni, S. and Seminara, M. (1996). Invertebrate fauna of a temporary pond in central Italy: composition, community parameters, and temporal succession. *Archiv fur Hydobiologie*, 137, 77–94.

Beasley, J. C., Olson, Z. H., Dharmarajan, G. *et al.* (2011). Spatio-temporal variation in the demographic attributes of a generalist mesopredator. *Landscape Ecology*, 26, 937–950.

Beatley, T. (2000). *Green Urbanism: Learning from European Cities*. Washington, DC: Island Press.

Beauchamp, V. B. and Shafroth, P. B. (2011). Floristic composition, beta diversity, and nestedness of reference sites for restoration of xeroriparian areas. *Ecological Applications*, 21, 465–476.

Beever, E. A., Hall, L. E., Varner, J. *et al.* (2017). Behavioral flexibility as a mechanism for coping with climate change. *Frontiers in Ecology and the Environment*, 15, 299–308.

Beier, P. (2006). Effects of artificial night lighting on terrestrial mammals. In *Ecological Consequences of Artificial Night Lighting*, ed. C. Rich and T. Longcore. Washington, DC: Island Press, pp. 19–42.

Belaire, J. A., Whelan, C. J. and Minor, E. S. (2014). Having our yards and sharing them too: the collective effects of yards on native bird species in an urban landscape. *Ecological Applications*, 24, 2132–2143.

Belant, J. L., Seamans, T. W., Gabrey, S. W. *et al.* (1995). Abundance of gulls and other birds at landfills in northern Ohio. *American Midland Naturalist*, 134, 30–40.

Bell, C. and Bell, B. (1969). *City Fathers: The Early History of Town Planning in Britain*. London: Barrie & Rockliff, The Cresset Press.

Bell, D. and Jayne, M. (2006). *Small Cities: Urban Experience Beyond the Metropolis*. New York: Routledge.

Benedict, M. A. and McMahon, E. T. (2006). *Green Infrastructure: Linking Landscape and Communities*. Washington, DC: Island Press.

Benitez-Lopez, A., Alkemade, R. and Verweij, P. A. (2010). The impacts of roads and other infrastructure on mammal and bird populations: a meta-analysis. *Biological Conservation*, 143, 1307–1316.

Bennett, A. (2003). *Linkages in the Landscape: The Role of Corridors and Connectivity in Wildlife Conservation*. Gland, Switzerland and Cambridge, UK: IUCN, the World Conservation Union.

Bennett, J., Wang, X. and Zhang, L., eds. (2008). *Environmental Protection in China: Land-Use Management*. Cheltenham, UK: Edward Elgar.

Benton-Short, L. and Short, J. R. (2008). *Cities and Nature*. London: Routledge.

Bergerot, B., Fontaine, B., Renard, M. *et al.* (2010). Preferences for exotic flowers do not promote urban life in butterflies. *Landscape and Urban Planning*, 96, 98–107.

Berke, P. R., Godschalk, D. R., Kaiser, E. J. and Rodriguez, D. A. (2006). *Urban Land Use Planning*. Urbana, IL: University of Illinois Press.

Berman, M. and Dunbar, I. (1983). The social behavior of free-ranging urban dogs. *Applied Animal Ethology*, 10, 5–17.

Bernasconi, C., Strager, M. P., Maskey, V. and Hasenmyer, M. (2009). Assessing public preferences for design and environmental attributes of an urban automated transportation system. *Landscape and Urban Planning*, 90, 155–167.

Bernhardt, E. S. and Palmer, M. A. (2007). Restoring streams in an urbanizing world. *Freshwater Biology*, 52, 738–751.

Berry, W. (1977). *The Unsettling of America*. San Francisco, CA: Sierra Club Books.

Bertin, R. I. (2002). Losses of native plant species from Worcester, Massachusetts. *Rhodora*, 104, 325–349.

Bertin, R. I. (2013). Changes in the native flora of Worcester County, Massachusetts. *Journal of the Torrey Botanical Society*, 140, 414–452.

Bertin, R. I. and Parise, C. M. (2014). Patterns and changes in the non-native flora of Worcester County, Massachusetts. *American Midland Naturalist*, 172, 37–60.

Bevanger, K. (1998). Biological and conservation aspects of bird mortality caused by electricity power lines: a review. *Biological Conservation*, 86, 67–76.

Bhagwat, S. and Rutte, C. (2006). Sacred groves: potential for biodiversity management. *Frontiers in Ecology and the Environment*, 4, 519–524.

Bhagwat, S. A., Kushalappa, C. G., Williams, P. H. *et al.* (2005). A landscape approach to biodiversity conservation of sacred groves in the Western Ghats of India. *Conservation Biology*, 19, 1853–1862.

Biamonte, E., Sandoval-Vargas, L., Chacon-Madrigal, E. J. *et al.* (2011). Effect of urbanization on the avifauna in a tropical metropolitan area. *Landscape Ecology*, 26, 183–194.

Biggs, J., Williams, P., Whitfield, M. *et al.* (2005). 15 years of pond assessment in Britain: results and lessons learned from the work of Pond Conservation. *Aquatic Conservation – Marine and Freshwater Ecosystems*, 15, 693–714.

Biggs, M. K. (1996). *Riparian Ecosystem Recovery in Arid Lands: Strategies and References*. Tucson, AZ: University of Arizona Press.

Biggs, T. W., Atkinson, E., Powell, R. and Ojeda-Revah, I. (2010). Land cover following rapid urbanization on the US-Mexico border: implications for conceptual models of urban watershed processes. *Landscape and Urban Planning*, 96, 78–87.

Billeter, R., Lira, J., Bailey, D. *et al.* (2008). Indicators for biodiversity in agricultural landscapes: a pan-European study. *Journal of Applied Ecology*, 45, 141–150.

Binford, M. W. and Karty, R. (2006). Riparian greenways and water resources. In *Designing Greenways: Sustainable Landscapes for Nature and People*, eds. P. C. Hellmund and D. A. Smith. Washington, DC: Island Press, pp. 110–145.

Binns, T. and Maconachie, R. (2006). Re-evaluating people–environment relationships at the rural-urban interface: how sustainable is the peri-urban zone in Kano, northern Nigeria? In *The Peri-Urban Interface: Approaches to Sustainable Natural and Human Resource Use*, ed. D. McGregor, D. Simon and D. Thompson. London: Earthscan, pp. 211–228.

Birkenholtz, T. (2013). "On the network, off the map": developing intervillage and intragender differentiation in rural water supply. *Environment and Planning D: Society and Space*, 31, 354–371.

Biro, Z., Lanszki, J., Szemethy, L. *et al.* (2005). Feeding habits of feral domestic cats (*Felis catus*), wild cats (*Felis silvestris*) and their hybrids: trophic niche overlap among cat groups in Hungary. *Journal of Zoology*, 266, 187–196.

Bissonette, J. A. and Adair, W. (2008). Restoring habitat permeability to roaded landscapes with isometrically-scaled wildlife crossings. *Biological Conservation*, 141, 482–488.

Black, M., Bewley, J. D. and Halmer, P., eds. (2006). *The Encyclopedia of Seeds: Science, Technology, and Uses*. Wallingford, UK: CABI.

Blackwell, B. E., Devault, T. E. and Seamans, T. W. (2015). Understanding and mitigating the negative effects of road lighting on ecosystems. In *Handbook of Road Ecology*, ed. R. van der Ree, D. J. Smith and C. Grilo. Chichester, UK: Wiley-Blackwell, pp. 143–150.

Blancher, P. (2013). Estimated number of birds killed by house cats (*Felis catus*) in Canada. *Avian Conservation and Ecology*, 8(2), 3.

Bland, R. L., Tully, J. and Greenwood, J. J. D. (2004). Birds breeding in British gardens: an underestimated population? *Bird Study*, 51, 96–106.

Blickley, J. L., Word, K., Krakauer, J. H. *et al.* (2012). The effect of experimental exposure to chronic noise on fecal corticosteroid metabolites in lekking male greater sage-grouse (*Centrocercus urophasianus*). *PLoS One*, 7, e50462.

Blomqvist, M. M., Vos, P., Klinkhamer, P. G. L. and ter Keurs, W. J. (2003). Declining plant species richness of grassland ditch banks – a problem of colonization or extinction? *Biological Conservation*, 109, 391–406.

Blue, R. (1996). Documentation of raptor nests on electric-utility facilities through a mail survey. In *Raptors in Human Landscapes*, eds. D. Bird, D. Varland and J. Negro. San Diego, CA: Academic Press, pp. 87–95.

Blume, H.-P. (2009). *Lehrbuch der Bodenkunde*. Berlin: Spektrum Verlag.

Blumstein, D. T. and Daniel, J. C. (2005). The loss of anti-predator behaviour following isolation on islands. *Proceedings of the Royal Society B*, 272, 1663–1668.

Blumstein, M. and Thompson, J. R. (2015). Land-use impacts on the quantity and configuration of ecosystem service provisioning in Massachusetts, USA. *Journal of Applied Ecology*, 52, 1009–1019.

Boada, M. and Capdevila, L. (2000). *Barcelona Biodiversitat urbana*. Barcelona: Ajuntament de Barcelona.

Bocek, S. (2014). Traditional fruit varieties in the Czech villages in Banat. In *Czech Villages in Romanian Banat: Landscape, Nature, and Culture*, ed. P. Madera, P. Kovar, D. Romportl *et al.* Brno, Czech Republic: Mendel University in Brno, pp. 84–97.

Bohl, C. C. (2002). *Place Making: Developing Town Centers, Main Streets, and Urban Villages*. Washington, DC: Urban Land Institute.

Boldogh, S., Dobrosi, D. and Samu, P. (2007). The effects of the illumination of buildings on house-dwelling bats and its conservation consequences. *Global Change Biology*, 21, 2467–2478.

Bonada, N. and Resh, V. H. (2013). Mediterranean-climate streams and rivers: geographically separated but ecologically comparable freshwater systems. *Hydrobiologia*, 719, 1–29.

Booth, D. B. and Bledsoe, B. P. (2009). Streams and urbanization. In *The Water Environment of Cities*, ed. L. A. Baker. New York: Springer, pp. 93–123.

Booth, D. B. and Jackson, C. R. (1997). Urbanization of aquatic systems: degradation threshold stormwater detection and the limits of mitigation. *Journal of the American Water Resources Association*, 33, 1077–1090.

Borda-de-Agua, L., Barrientos, R., Beja, P. and Pereira, H. M., eds. (2018). *Railway Ecology*. New York: Springer.

Borda-de-Agua, L., Navarro, L., Gavinhos, C. and Pereira, H. M. (2011). Spatio-temporal impacts of roads on the persistence of populations: analytic and numerical approaches. *Landscape Ecology*, 26, 253–265.

Bornstein, R. D. (1977). Urban–rural wind velocity difference. *Atmospheric Environment*, 11, 597–604.

Botelho, E. S. and Arrowood, P. C. (1996). Nesting success of western burrowing owls in natural and human-altered environments. In *Raptors in Human Landscapes*, eds. D. Bird, D. Varland and J. Negro. San Diego, CA: Academic Press, pp. 61–68.

Botteldooren, D., Coensel, B. and De Meur, T. (2004). The temporal structure of the urban soundscape. *Journal of Sound Vibrations*, 292, 105–123.

Botteron, C. A. (2001). India's "Project Tiger" reserves: the interplay between ecological knowledge and the human dimensions of policymaking for protected habitats. In *Applying Ecological Principles to Land Management*, ed. V. H. Dale and R. A. Haeuber. New York: Springer, pp. 136–162.

Bouraoui, M. (2005). Agri-urban development from a land-use planning perspective: the Saclay Plateau (France) and the Sijoumi Plain (Tunisia). In *Agropolis: The Social, Political and Environmental Dimensions of Urban Agriculture,* ed. L. J. A. Mougeot. London: Earthscan, pp. 203–227.

Bowden, J. (1982). An analysis of factors affecting catches of insects in light-traps. *Bulletin of Entomological Research*, 72, 621–629.

Bowler, D. F., Buyung-Ali, L., Knight, T. M. and Pullin, A. S. (2010). Urban greening to cool towns and cities: a systematic review of the empirical evidence. *Landscape and Urban Planning*, 97, 147–155.

Boxall, A. (2010). Impacts of emerging contaminants on the environment. In *Ecology of Industrial Pollution*, ed. L. C. Batty and K. B. Hallberg. Cambridge, UK: Cambridge University Press, pp. 101–125.

Boxall, A. B. A., Sinclair, C. J., Fenner, K. *et al.* (2004). When synthetic chemicals degrade in the environment. *Environmental Science and Technology*, 38, 369A–375A.

Boxall, A. B. A., Tiede, K. and Chaudhry, M. Q. (2007). Engineered nanomaterials in soils and water: how do they behave and could they pose a risk to human health? *Nanomedicine*, 2, 919–927.

Braaker, S., Moretti, M., Boesch, R. *et al.* (2014). Assessing habitat connectivity for ground-dwelling animals in an urban environment. *Ecological Applications*, 24, 1583–1595.

Bracchetti, L., Carotenuto, L. and Catorci, A. (2012). Land-cover changes in a remote area of central Apennines (Italy) and management directions. *Landscape and Urban Planning*, 104, 157–170.

Bradley, A. and Hall, T. (2006). The festival phenomenon: festivals, events and the promotion of small urban areas. In *Small Cities: Urban Experience Beyond the Metropolis,* ed. D. Bell and M. Jayne. New York: Routledge, pp. 77–89.

Brady, M. J., McAlpine, C. A., Possingham, H. P. *et al.* (2011). Matrix is important for mammals in landscapes with small amounts of native forest habitat. *Landscape Ecology*, 26, 617–628.

Bradybury, J. W. and Vehrencamp, S. L. (1998). *Principles of Animal Communication*. Sunderland, MA: Sinauer Associates.

Braithwaite, L. (1976). *Canals in Town*. London: Adam and Charles Black.

Brandle, J. R., Hintz, J. L. and Sturrock, J. W., eds. (1988). *Windbreak Technology*. Amsterdam: Elsevier. (Reprinted from *Agriculture, Ecosystems and Environment,* 22–23 [1988].)

Braskerud, B. C., Tonderski, K. S., Wedding, B. *et al.* (2005). Can constructed wetlands reduce the diffuse phosphorus loads to eutrophicate water in cold temperate regions? *Journal of Environmental Quality*, 34, 2145–2155.

Brauniger, C., Knapp, S., Kuhn, I. and Klotz, S. (2010). Testing taxonomic and landscape surrogates for biodiversity in an urban setting. *Landscape and Urban Planning*, 97, 283–295.

Brearley, G., McAlpine, C., Bell, S. and Bradley, A. (2012). Influence of urban edges on stress in an arboreal mammal: a case study of squirrel gliders in southeast Queensland, Australia. *Landscape Ecology*, 27, 1407–1419.

Breuste, J. H. (2010). Allotment gardens as part of urban green infrastructure: actual trends and perspectives in Central Europe. In *Urban Biodiversity and Design*, ed. N. Muller, P. Werner and J. G. Kelcey. Chichester, UK: Wiley-Blackwell, pp. 463–475.

Brickell, C., ed. (2003). *The American Horticultural Society Encyclopedia of Gardening*. New York: DK Publishing.

Briggs, M. K. (1996). *Riparian Ecosystem Recovery in Arid Lands: Strategies and References*. Tucson, AZ: University of Arizona Press.

Briggs, W. R. (2006). Physiology of plant responses to artificial lighting. In *Ecological Consequences of Artificial Night Lighting*, ed. C. Rich and T. Longcore. Washington, DC: Island Press, pp. 389–411.

Brinkmann, K., Schumacher, J., Dittrich, A. *et al.* (2012). Analysis of landscape transformation processes in and around four West African cities over the last 50 years. *Landscape and Urban Planning*, 105, 94–105.

Brittingham, M. C. and Temple, S. A. (1992). Does winter bird feeding promote dependency? *Journal of Field Ornithology*, 63, 190–194.

Brody, S. D., Zahran, S., Grover, H. and Vedlitz, A. (2008). A spatial analysis of local climate change policy in the United States: risk, stress, and opportunity. *Landscape and Urban Planning*, 87, 33–41.

Broll, G. and Keplin, B. (1995). Ecological studies on urban lawns. In *Urban Ecology as the Basis of Urban Planning*, ed. H. Sukopp, M. Numata and A. Huber. Amsterdam: SPB Academic Publishing, pp. 71–82.

Brooks, A. (1998). *Channelized Rivers: Perspectives for Environmental Management*. New York: John Wiley.

Brooks, B. W., Riley, T. M. and Taylor, R. D. (2006). Water quality of effluent-dominated ecosystems: ecotoxicological, hydrological, and management considerations. *Hydrobiologia*, 556, 365–379.

Browder, J. O. and Godfrey, B. J. (1997). *Rainforest Cities: Urbanization, Development, and Globalization of the Brazilian Amazon*. New York: Columbia University Press.

Brown, J. (1986). *The English Market Town: A Social and Economic History 1750–1914*. Marlborough, UK: The Crowood Press.

Brown, P. W. and Schulte, L. A. (2011). Agricultural landscape change (1937–2002) in three townships in Iowa, USA. *Landscape and Urban Planning*, 100, 202–212.

Browning, D. M. and Archer, S. R. (2011). Protection from livestock fails to deter shrub proliferation in a desert landscape with a history of heavy grazing. *Ecological Applications*, 21, 1629–1642.

Brumm, H. (2004). The impact of environmental noise on song amplitude in a territorial bird. *Journal of Animal Ecology*, 73, 434–440.

Buchan, U. (1992). *Wall Plants and Climbers*. North Pomfret, VT: Trafalgar Square Publishing.

Buchanan, B. W. (2006). Observed and potential effects of artificial night lighting on anuran amphibians. In *Ecological Consequences of Artificial Night Lighting*, ed. C. Rich and T. Longcore. Washington, DC: Island Press, pp. 192–220.

Budd, R., O'Green, A., Goh, K. S. *et al.* (2009). Efficacy of constructed wetlands in pesticide removal from tail water in the Central Valley, California. *Environmental Science and Technology*, 43, 2925–2930.

Bugg, R. L., Brown, C. S. and Anderson, J. H. (1997). Restoring native perennial grasses to rural roadsides in the Sacramento Valley of California: establishment and evaluation. *Restoration Ecology*, 5, 214–228.

Bunting, S. W., Kundu, N. and Mukherjee, M. (2005). Peri-urban aquaculture and poor livelihoods in Kolkata, India. In *Urban Aquaculture*, ed. B. Costa-Pierce, A. Desbonnet, P. Edwards and D. Baker. Wallingford, UK: CABI Publishing, pp. 61–76.

Burel, F. and Baudry, J. (1999). *Ecologie du paysage: Concepts, methodes et applications*. Paris: Editions TEC & DOC.

Burchett, S. and Burchett, S. (2011). *Introduction to Wildlife Conservation in Farming*. Chichester, UK: Wiley-Blackwell.

Burger, J. (1999). *Animals in Towns and Cities*. Dubuque, IA: Kendall/Hunt Publishing.

Burgi, M., Straub, A., Gimmi, U. and Salzmann, D. (2010). The recent landscape history of Limpach valley, Switzerland: considering three empirical hypotheses on driving forces of landscape change. *Landscape Ecology*, 25, 287–297.

Burley, S., Robinson, S. L. and Lundholm, J. T. (2008). Post-hurricane vegetation recovery in an urban forest. *Landscape and Urban Planning*, 85, 111–122.

Burton, M. L., Samuelson, L. J. and Mackenzie, M. D. (2009). Riparian woody plant traits across an urban-rural land use gradient and implications for watershed function with urbanization. *Landscape and Urban Planning*, 90, 42–55.

Butler, B. J. (2008). *Family Forest Owners of the United States, 2006*. General Technical Report NRS-27, Northern Research Station. Newtown Square, PA: USDA Forest Service.

Butler, D. and Davies, J. W. (2011). *Urban Drainage*. London: Spon Press.

Butler, J., Du Toit, J. and Bingham, J. (2004). Free-ranging domestic dogs (*Canis familiaris*) as predators and prey in rural Zimbabwe: threats of competition and disease to large wild carnivores. *Biological Conservation*, 115, 369–378.

Cade, T. J., Martell, M., Redig, P. *et al*. (1996). Peregrine falcons in urban North America. In *Raptors in Human Landscapes*, ed. D. M. Bird, D. E. Varland and J. J. Negro. San Diego, CA: Academic Press, pp. 3–13.

Cadenasso, M. L., Pickett, S. T. A., McGrath, B. and Marshall, V. (2013). Ecological heterogeneity in urban ecosystems: reconceptualized land cover models as a bridge to urban design. In *Resilience in Ecology and Urban Design: Linking Theory and Practice for Sustainable Cities*, ed. S. T. A. Pickett, M. L. Cadenasso and B. McGrath. New York: Springer, pp. 107–129.

Caldiron, G. (2005). *Banlieue: Vita e rivolta nelle periferie della metropolis*. Rome: Tomacelli.

Callenbach, E. (2014). Sustainable shrinkage: envisioning a smaller, stronger economy. In *Creating a Sustainable and Desirable Future: Insights from 45 Global Thought Leaders*, ed. R. Costanza and I. Kubiszewski. Hackensack, NJ: World Scientific, pp. 223–232.

Campanella, T. J. and Vale, L. J., eds. (2005). *The Resilient City: How Modern Cities Recover from Disaster*. New York: Oxford University Press.

Campbell, C. and Ogden, M. (1999). *Constructed Wetlands in the Sustainable Landscape*. New York: John Wiley.

Campos, C. B., Esteves, C. F., Ferraz, K. M. P. M. B. *et al*. (2007). Diet of free-ranging cats and dogs in a suburban and rural environment, south-eastern Brazil. *Journal of Zoology*, 273, 14–20.

Cantwell, M. D. and Forman, R. T. T. (1994). Landscape graphs: ecological modeling with graph theory to detect configurations common to diverse landscapes. *Landscape Ecology*, 8, 239–255.

Capelton, A., Conrage, C., Rumsby, P. *et al*. (2006). Prioritising veterinary medicines according to their potential indirect human exposure and toxicity profile. *Toxicology Letters*, 163, 213–223.

Carolsfeld, L., ed. (2003). *Migratory Fishes of South America: Biology, Fisheries and Conservation Status*. Victoria, BC: World Fisheries Trust.

Carpenter, D., Sinha, S. K., Brennan, K. and Slate, L. O. (2003). Urban stream restoration. *Journal of Hydraulic Engineering*, 129, 491–494.

Carrete, M., Sanchez-Zapata, J. A., Benitez, J. R. *et al*. (2009). Large scale risk-assessment of wind-farms on population viability of a globally endangered long-lived raptor. *Biological Conservation*, 142, 2954–2961.

Carrington, D. (2014). Animals see power lines as glowing, flashing bands, research reveals. *The Guardian*, March 12, 2 pp.

Carson, R. (1962). *Silent Spring*. Boston, MA: Houghton Mifflin.

Castro, P. (1990). Sacred groves and social change in Kirinyaga, Kenya. In *Social Change and Applied Anthropology*, ed. M. S. Chaiken and A. K. Fleuret. Boulder, CO: Westview Press, pp. 277–289.

Catalan, B., Sauri, D. and Serra, P. (2008). Urban sprawl in the Mediterranean? Patterns of growth and change in the Barcelona Metropolitan Region 1993–2000. *Landscape and Urban Planning*, 85, 174–184.

Catry, I., Alcazar, R. and Sutherland, W. J. (2009). Identifying the effectiveness and constraints of conservation interactions in a case study of the endangered lesser kestrel. *Biological Conservation*, 142, 2782–2791.

Catterall, C. P., Green, R. J. and Jones, D. N. (1991). Habitat use by birds across a forest-suburb interface in Brisbane: implications for corridors. In *Nature Conservation 2: The Role of Corridors*, eds. D. A. Saunders and R. J. Hobbs. Chipping Norton, NSW: Surrey Beatty, pp. 247–258.

Cheptou, P.-O., Carrue, O., Roulfed, S. and Cantarel, A. (2008). Rapid evolution of seed dispersal in an urban environment in the weed *Crepis sancta*. *Proceedings of the National Academy of Sciences (USA)*, 105, 3796–3799.

Chevallier, C., Hernandez-Matias, A., Real, J. *et al*. (2015). Retrofitting of power lines effectively reduces mortality by electrocution in large birds: an example with the endangered Bonelli's eagle. *Journal of Applied Ecology*, 52, 1465–1473.

Childs, J. E. (1986). Size dependence predation on rats (*Rattus norvegicus*) by house cats (*Felis catus*) in an urban setting. *Journal of Mammalogy*, 67, 196–199.

Chiquet, C., Dover, J. W. and Mitchell, P. (2012). Birds and the urban environment: the value of green walls. *Urban Ecosystems*, 16, 453–462.

Christaller, W. (1933). *Die Zentralen Orte in Suddeutschland.* Jena (Vienna): Gustav Fischer.

Chucas, B. and Marzluff, J. M. (2011). Coupled relationships between humans and other organisms in urban areas. In *Urban Ecology: Patterns, Processes, and Applications*, ed. J. Niemela. New York: Oxford University Press, pp. 135–147.

Churcher, P. B. and Lawton, J. H. (1989). Beware of well-fed felines: Britain's five million house cats enjoy both indoor comforts and outdoor hunting. *Natural History*, 7, 40–46.

Cilliers, S. (2010). Social aspects of urban biodiversity – an overview. In *Urban Biodiversity and Design*, ed. N. Muller, P. Werner and J. G. Kelcey. Chichester, UK: Wiley-Blackwell, pp. 81–100.

Cilliers, S., Siebert, S., Davoren, E. and Lubbe, R. (2012). Social aspects of urban ecology in developing countries, with an emphasis on urban domestic gardens. In *Applied Urban Ecology: A Global Framework*, ed. M. Richter and U. Weiland. Chichester, UK: Wiley-Blackwell, pp. 123–138.

Cinzano, P., Falchi, F. and Elvidge, C. V. (2001). The first world atlas of the artificial night sky brightness. *Monthly Notices of the Royal Astronomical Society*, 328, 689–707.

Clare, T. and Bunce, R. G. H. (2006). The potential for using trees to help define historic landscape zones: a case study in the English Lake District. *Landscape and Urban Planning*, 74, 34–45.

Clark, J. K., McChesney, R., Munroe, D. K. and Irwin, E. G. (2009). Spatial characteristics of exurban settlement pattern in the United States. *Landscape and Urban Planning*, 90, 178–188.

Clement, A. and Thomas, G. (2001). *Atlas du Paris souterrain: La doublure sombre de la Ville lumiere*. Paris: Editions Parigramme.

Clements, A. (1983). Suburban development and resultant changes in the vegetation of the bushland of the northern Sydney region. *Australian Journal of Ecology*, 8, 307–319.

Clevenger, A. P. and Ford, A. T. (2010). Wildlife crossing structures, fencing, and other highway design considerations. In *Safe Passages: Highways, Wildlife, and Habitat Connectivity,* ed. J. P. Beckmann, A. P. Clevenger, M. P. Huijser, and J. A. Hilty. Washington, DC: Island Press, pp. 17–49.

Clevenger, A. P. and Huijser, M. P. (2011). *Wildlife Crossing Structure Handbook: Design and Evaluation in North America*. Washington, DC: US Federal Highway Administration, Central Federal Lands Highway Division.

Clevenger, A. P. and Waltho, N. (2000). Factors influencing the effectiveness of wildlife underpasses in Banff National Park, Alberta, Canada. *Conservation Biology*, 14, 47–56.

Clevenger, A. P., Chruszcz, B. and Gunson, K. (2001a). Drainage culverts as habitat linkages and factors affecting passage by mammals. *Journal of Applied Ecology*, 38, 1340–1349.

Clevenger, A. P., Chruszcz, B. and Gunson, K. (2001b). Highway mitigation fencing reduces wildlife–vehicle collisions. *Wildlife Society Bulletin*, 29, 646–653.

Clevenger, A. P., Wierzchowski, J., Chruszcz, B. and Gunson, K. (2002). GIS-generated, expert-based models for identifying wildlife habitat linkages and planning mitigation passages. *Conservation Biology*, 16, 503–514.

Close, D. C., Davidson, N. J., and Watson, T. (2008). Health of remnant woodlands in fragments under distinct grazing regimes. *Biological Conservation*, 141, 2395–2402.

Coffin, A. (2007). From roadkill to road ecology: a review of the ecological effects of roads. *Journal of Transport Geography*, 15, 396–406.

Cohen-Rosenthal, E. and Musnikow, J., eds. (2003). *Eco-Industrial Strategies*. Sheffield, UK: Greenleaf Publishing.

Colburn, E. A. (2004). *Vernal Pools: Natural History and Conservation*. Blacksburg, VA: McDonald & Woodward Publishing.

Cole, R. J., Holl, K. D. and Zahawi, R. A. (2010). Seed rain under tree islands planted to restore degraded lands in a tropical agricultural landscape. *Ecological Applications*, 20, 1255–1269.

Collinge, S. K. (2009). *Ecology of Fragmented Landscapes*. Baltimore, MD: Johns Hopkins University Press.

Collins, S. L., Avolio, M. L., Gries, C. *et al.* (2018). Temporal heterogeneity increases with spatial heterogeneity in ecological communities. *Ecology*, 99, 858–865.

Coman, B. J. and Brunner, H. (1972). Food habits of the feral house cat in Victoria. *Journal of Wildlife Management*, 36, 848–853.

Compas, E. (2007). Measuring exurban change in the American West: a case study in Gallatin County, Montana, 1973–2004. *Landscape and Urban Planning*, 82, 56–65.

Connell, J. and McManus, P. (2011). *Rural Revival: Place Marketing, Tree Change and Regional Migration in Australia*. Farnham, UK: Ashgate Publishing.

Connor, D. J., Loomis, R. S. and Cassman, K. G. (2011). *Crop Ecology: Productivity and Management in Agricultural Systems*. Cambridge, UK: Cambridge University Press.

Connor, W. H. and Day, J. W., Jr. (1982). The ecology of forested wetlands in the southeastern United States. In *Wetlands: Ecology and Management*, ed. B. Gopal, R. E. Turner, R. G. Wetzel and D. F. Whigham. Jaipur, India: National Institute of Ecology and International Scientific Publications, pp. 69–87.

Cooper, J. A. (1991). Canada goose management at the Minneapolis-St. Paul International Airport. In *Wildlife Conservation in Metropolitan Environments*, ed. L. W. Adams and D. L. Leedy. Columbia, MD: National Institute for Urban Wildlife, pp. 175–183.

Cooper, P. F. and Hobson, J. A. (1989). Sewage treatment by reed bed systems: the present situation in the United Kingdom. In *Constructed Wetlands for Wastewater Treatment*, ed. D. A. Hammer. Chelsea, MI: Lewis Publishers, pp. 153–172.

Corbet, S. A., Bee, J. Dasmapatra, K. *et al.* (2001). Native or exotic? Double or single? Evaluating plants for pollinator-friendly gardens. *Annals of Botany*, 87, 219–232.

Cordell, D., Drangert, J.-O., and White, S. (2009). The story of phosphorus: global food security and food for thought. *Global Environmental Change*, 19, 292–305.

Correa, F. (2016). *Beyond the City: Resource Extraction Urbanism in South America*. Austin, TX: University of Texas Press.

Costa-Pierce, B., Desbonnet, A., Edwards, P. and Baker, D., eds. (2005). *Urban Aquaculture*. Wallingford, UK: CABI Publishing.

Costanza, R. and Kubiszewski, I., eds. (2014). *Creating a Sustainable and Desirable Future: Insights from 45 Global Thought Leaders*. London: World Scientific.

Costanza, R., d'Arge, R., de Groots, R. *et al.* (1997). The value of the world's ecosystem services and natural capital. *Nature*, 387, 253–260.

Courchamp, F., Langlais, M. and Sugihara, G. (1999). Cats protecting birds: modelling the mesopredator release effect. *Journal of Animal Ecology*, 68, 282–292.

Cousins, S. A. O. and Aggemyr, E. (2008). The influence of field shape, area and surrounding landscape on plant species richness in grazed ex-fields. *Biological Conservation*, 141, 126–135.

Cowie, R. J. and Hinsley, S. A. (1988). The provision of food and the use of bird feeders in suburban gardens. *Bird Study*, 35, 163–168.

Craig, D. J., Craig, J. E., Abella, S. R. and Vanier, C. H. (2010). Factors affecting exotic annual plant cover and richness along roadsides in the eastern Mojave Desert, USA. *Journal of Arid Environments*, 74, 702–707.

Crane, M., Boxall, A. B. A. and Barrett, K. (2008). *Veterinary Medicines in the Environment*. Boca Raton, FL: CRC Press.

Craul, P. J. (1992). *Urban Soil in Landscape Design*. New York: John Wiley.

Craul, P. J. (1999). *Urban Soils: Applications and Practices*. New York: John Wiley.

Crawford, T. W. (2007). Where does the coast sprawl the most? Trajectories of residential development and sprawl in coastal North Carolina, 1971–2000. *Landscape and Urban Planning*, 83, 294–307.

Cronk, J. K. and Fennessy, M. S. (2001). *Wetland Plants: Biology and Ecology*. Boca Raton, FL: Lewis Publishing.

Cronon, W. (1991). *Nature's Metropolis: Chicago and the Great West*. New York: W. W. Norton.

Crosetti, D. and Margoritora, F. G. (1987). Distribution and life cycles of cladocerans in temporary pools from central Italy. *Freshwater Biology*, 18, 165–175.

Crouzat, E., Mouchet, M., Turkelboom, F. *et al.* (2015). Assessing bundles of ecosystem services from regional to landscape scale: insights from the French Alps. *Journal of Applied Ecology*, 52, 1145–1155.

Cuffney, T. F., Brightbill, R. A., Maym, J. T. and Watte, I. R. (2010). Responses of benthic macroinvertebrates to environmental changes associated with urbanization in nine metropolitan areas. *Ecological Applications*, 20, 1384–1401.

Cullina, W. (2009). *Understanding Perennials: A New Look at an Old Favorite*. Boston, MA: Houghton Mifflin Harcourt.

Cullingworth, J. B. and Nadin, V. (1997). *Town and Country Planning in the UK*. London: Routledge.

Curry, N. and Owen, S. (2009). Rural planning in England: a critique of current policy. *Town Planning Review*, 80, 576–595.

Cusa, M., Jackson, D. and Mesure, A. (2015). Window collisions by migratory bird-species: urban geographical patterns and habitat associations. *Urban Ecosystems*, 18, 1427–1446.

Cushing, C. E. and Allan, J. D. (2001). *Streams: Their Ecology and Life*. San Diego, CA: Academic Press.

Cushman, S. A., Max, T., Meneses, N. *et al.* (2014). Landscape genetic connectivity in a riparian foundation tree is jointly driven by climatic gradients and river networks. *Ecological Applications*, 24, 1000–1014.

Czarna, A. and Nowinska, R. (2011). Vascular flora in cemeteries of the Roztocze region and surrounding areas (South-East Poland). *Acta Agrobotanica*, 64, 77–92.

Czerny, M., Lecka, I. and Wujek, M. (2009). The development of urbanization in the neighbourhood of Kampinoski National Park. In *Naturbanization: New Identities and Processes for Rural-Natural Areas*, ed. M. J. Prados. Leiden, The Netherlands: CRC Press/Balkema, pp. 29–43.

Dahmus, M. E. and Nelson, K. C. (2014). Yard stories: examining residents' conceptions of their yards as part of the urban ecosystem in Minnesota. *Urban Ecosystems*, 17, 173–194.

Daily, G. C., ed. (1997). *Nature's Services: Societal Dependence on Natural Ecosystems.* Washington, DC: Island Press.

Dainese, M. and Poldini, L. (2012). Plant and animal diversity in a region of the Southern Alps: the role of environmental and spatial processes. *Landscape Ecology*, 27, 417–431.

Dale, V. H. (2001). Applying ecological guidelines for land management to farming in the Brazilian Amazon. In *Applying Ecological Principles to Land Management*, ed. V. H. Dale and R. A. Haeuber. New York: Springer, pp. 213–225.

Dale, V. H., O'Neill, R. V., Pedlowski, M. A. and Southworth, F. (1993). Causes and effects of land-use change in central Rondonia, Brazil. *Photogrammetric Engineering and Remote Sensing*, 59, 997–1005.

Dale, V. H., O'Neill, R. V., Southworth, F. and Pedlowski, M. A. (1994a). Modeling effects of land management in the Brazilian Amazonian settlement of Rondonia. *Conservation Biology*, 8, 196–206.

Dale, V. H., Pearson, S. M., Offerman, H. L. and O'Neill, R. V. (1994b). Relating patterns of land-use change to faunal biodiversity in the Central Amazon. *Conservation Biology*, 8, 1027–1036.

Dalle, S. P., Pulido, M. T. and de Blois, S. (2011). Balancing shifting cultivation and forest conservation: lessons from a "sustainable landscape" in southeastern Mexico. *Ecological Applications*, 21, 1557–1572.

Dallimer, M., Acs, S., Hanley, N. *et al.* (2009). What explains property-level variation in avian diversity? An interdisciplinary approach. *Journal of Applied Ecology*, 46, 647–656.

D'Angelo, G. and van der Ree, R. (2015). Use of reflectors and auditory deterrents to prevent wildlife-vehicle collisions. In *Handbook of Road Ecology*, ed. R. van der Ree, D. J. Smith and C. Grilo. Chichester, UK: Wiley-Blackwell, pp. 213–218.

Daniels, G. D. and Kirkpatrick, J. B. (2006). Comparing the characteristics of front and back domestic gardens in Hobart, Tasmania, Australia. *Landscape and Urban Planning*, 78, 344–352.

Daniels, G. D. and Kirkpatrick, J. B. (2012). The influence of landscape context on the distribution of flightless mammals in exurban developments. *Landscape and Urban Planning*, 104, 114–123.

Daniels, T. J. and Beckoff, M. (1989). Spatial and temporal resource use by feral and abandoned dogs. *Ethology*, 81, 300–312.

Daniels, T. L., Keller, J. W., Lapping, M. B. *et al.* (2007). *The Small Town Planning Handbook.* Chicago, IL: Planners Press, American Planning Association.

Danoedoro, P. (2001). Integration of remote sensing and geographic information system in land use mapping: an Indonesia example. In *Landscape Ecology Applied in Land Evaluation, Development, and Conservation: Some Worldwide Selected Examples*, ed. D. van der Zee and I. S. Zonneveld. Enschede, The Netherlands: International Institute for Aerospace Survey and Earth Sciences, pp. 371–383.

Das, S. and Vincent, J. R. (2009). Mangroves protected villages and reduced death toll during Indian super cyclone. *Proceedings of the National Academy of Sciences (USA)*, 106, 7357–7360.

Dauer, J. T., Morensen, D. A., Luschei, E. C. *et al.* (2009). *Conyza canadensis* seed ascent in the lower atmosphere. *Agricultural and Forest Meteorology*, 149, 526–534.

Dauphine, N. and Cooper, R. J. (2011). Pick one: outdoor cats or conservation: the fight over managing an invasive predator. *Wildlife Professional*, 5, 50–57.

Davenport, J. and Davenport, J. L., eds. (2006). *The Ecology of Transportation: Managing Mobility for the Environment.* New York: Springer.

Davidar, P., Arjunan, M. and Puyravaud, J.-P. (2008). Why do local households harvest forest products? A case study from the southern Western Ghats, India. *Biological Conservation*, 141, 1876–1884.

Davidheiser, R. (2011). *The Third Generation Roadway: Metropolitan Transport for the 21st Century.* Stockton, CA: Wild Canyon Books.

Davidson, J. L., Jacobson, C., Lyth, A. *et al.* (2016). Interrogating resilience: toward a typology to improve its operationalization. *Ecology and Society*, 21, 27.

Davies, A. (2014). Urban to rural elderly migration: renewing and reinventing Australia's small rural towns. In *Rural Change in Australia: Population, Economy, Environment.* Farnham, UK: Ashgate, pp. 43–55.

Davies, W. K. D. and Baxter, T. (1997). Commercial intensification: the transformation of a highway-oriented ribbon. *Geoforum*, 28, 237–252.

Davies, Z. G., Fuller, R. A., Loram, A. *et al.* (2009). A national scale inventory of resource provision for biodiversity within domestic gardens. *Biological Conservation*, 142, 761–771.

Davis, A., Hunt, W., Traver, R. and Clar, M. (2009). Bioretention technology overview of current practice and future needs. *Journal of Environmental Engineering*, 135, 109–117.

Davis, A. P., Shokouhian, M., Sharma, H. *et al.* (2003). Water quality improvement through bioretention: lead, copper, and zinc removal. *Water Environment Research*, 75, 73–82.

Davis, A. Y., Pijanowski, B. C., Robinson, K. D. and Kidwell, P. B. (2010). Estimating parking lot footprints in the Upper Great Lakes Region of the USA. *Landscape and Urban Planning*, 96, 68–77.

Davis, C. R. and Hansen, A. J. (2011). Trajectories in land use change around US National Parks and challenges and opportunities for management. *Ecological Applications*, 21, 3299–3316.

Davis, W. K. D. and Baxter, T. (1977). Commercial intensification: the transformation of a highway-oriented ribbon. *Geoforum*, 28, 237–252.

Davlin, S. L. and VonVille, H. M. (2012). Canine rabies vaccination and domestic dog population characteristics in the developing world: a systematic review. *Vaccine*, 30, 3492–3502.

Day, J. W., Ko, J., Rybczyk, J. *et al.* (2004). The use of wetlands in the Mississippi delta for wastewater assimilation: a review. *Ocean and Coastal Management*, 47, 671–691.

De Chant, T., Gallego, A. H., Saornil, J. V. and Kelly, M. (2010). Urban influence on changes in linear forest edge structure. *Landscape and Urban Planning*, 96, 12–18.

De Groot, R. S., Wilson, M. and Boumans, R. (2002). A typology for the description, classification and valuation of ecosystem functions, goods and services. *Ecological Economics*, 41, 393–408.

de Molenaar, J. G., Sanders, M. E. and Jonkers, D. A. (2006). Road lighting and grassland birds: local influence of road lighting on a black-tailed godwit population. In *Ecological Consequences of Artificial Night Lighting*, ed. C. Rich and T. Longcore. Washington, DC: Island Press, pp. 114–136.

Decamps, H. and Decamps, O. (2001). *Mediterranean Riparian Woodlands*. Arles, France: Tour du Valat.

DeGraaf, R. M. and Witman, G. M. (1979). *Trees, Shrubs, and Vines for Attracting Birds*. Amherst, MA: University of Massachusetts Press.

DeGraaf, R. M. and Yamasaki, M. (2001). *New England Wildlife: Habitat, Natural History, and Distribution*. Hanover, NH: University Press of New England.

Del Tredici, P. (2010). *Wild Urban Plants of the Northeast: A Field Guide*. Ithaca, NY: Cornell University Press.

DeLuca, W. V., Studds, C. E., King, R. S. and Marra, P. P. (2008). Coastal urbanization and the integrity of estuarine waterbird communities: threshold responses and the importance of scale. *Biological Conservation*, 141, 2669–2678.

deMaynadier, P. G. and Hunter, M. L., Jr. (1995). The relationship between forest management and amphibian ecology: a review of the North American literature. *Environmental Review*, 3, 230–261.

deMenocal, P. B. (2001). Cultural responses to climate change during the late Holocene. *Science*, 292, 667–673.

Demkova, K. and Lipsky, Z. (2013). Changes in the extent of non-forest woody vegetation in the Novodvorsko and Zehusicko Region (Central Bohemia, Czech Republic). *AUC Geographica*, 48(1), 5–13.

Depraetere, M., Pavoine, S., Jiguet, F. *et al.* (2012). Monitoring animal diversity using acoustic indices: implementation in a temperate woodland. *Ecological Indicators*, 13, 46–54.

Desrochers, P. and Lusk, J. L. (2016). The inability and undesirability of local croplands to meet food demand. *Frontiers in Ecology and the Environment*, doi:10.1890/15.WB.016.

DeStefano, S. and Webster, C. M. (2012). Distribution and habitat of greater roadrunners in urban and suburban Arizona. In *Urban Bird Ecology and Conservation*, ed. C. A. Lepczyk and P. S. Warren. Berkeley, CA: University of California Press, pp. 155–166.

Devereux, C. L., Denny, M. J. H. and Whittingham, M. J. (2008). Minimal effects of wind turbines on the distribution of wintering farmland birds. *Journal of Applied Ecology*, 45, 1689–1694.

Dexter, R. W. (1955). The vertebrate fauna on the campus of Kent State University. *The Biologist,* 37, 84–88.

DeYoung, R. W. and Honeycutt, R. L. (2005). The molecular toolbox: genetic techniques in wildlife ecology and management. *Journal of Wildlife Management*, 69, 1362–1384.

Diaz Pineda, F. and Schmitz, M. F., coordinators. (2011). *Connectividad Ecologica Territorial: Estudios de casos de connectividad ecologica y socioecologica*. Madrid: O. A. Parques Nacionales, Ministerio de Medio Ambiente y Medio Rural y Marino.

Dickson, B., Hutton, J. and Adams, W. M., eds. (2009). *Recreational Hunting, Conservation and Rural Livelihoods: Science and Practice*. Chichester, UK: Wiley-Blackwell.

Dietz, M. and Clausen, J. (2008). A field evaluation of rain garden flow and pollutant treatment. *Water, Air and Soil Pollution*, 167, 323–328.

Dluzewska, A. (2009). The influence of the localization of tourist facilities on the dysfunction of tourism discussed on the example of southern Tunisia. In *Naturbanization: New Identities and Processes for Rural-Natural Areas*, ed. M. J. Prados. Leiden, The Netherlands: CRC Press/Balkema, pp. 125–141.

Domon, G. and Bouchard, A. (2007). The landscape history of Godmanchester (Quebec, Canada): two centuries of shifting relationships between anthropic and biophysical factors. *Landscape Ecology*, 22, 1201–1214.

Donahue, B. (2004). *The Great Meadow: Farmers and the Land in Colonial Concord*. New Haven, CT: Yale University Press.

Donna, W. (2013). Biodiversity isn't just wildlife – conserving agricultural biodiversity as a vital contribution to poverty reduction. In *Biodiversity Conservation and Poverty Alleviation: Exploring the Evidence for a Link*, ed. D. Roe, J. Elliott, C. Sandbrook and M. Walpole. Chichester, UK: Wiley-Blackwell, pp. 127–141.

Donovan, G. H., Michael, Y. L., Butry, D. T. *et al.* (2011). Urban trees and the risk of poor birth outcomes. *Health and Place*, 17, 390–393.

Donovan, G. H., Michael, Y. L. Butry, D. T. *et al.* (2012). Comment on "Green space, health inequality, and pregnancy." *Environmental International*, 33, 133.

Dorrough, J. and Scroggie, M. P. (2008). Plant responses to agricultural intensification. *Journal of Applied Ecology*, 45, 1274–1283.

Dorsey, B., Olsson, M. and Rew, L. J. (2015). Ecological effects of railways on wildlife. In *Handbook of Road Ecology*, ed. R. van der Ree, D. J. Smith and C. Grilo. Chichester, UK: John Wiley, pp. 219–227.

Douglas, D. J. T., Vickery, J. A. and Benton, T. G. (2009). Improving the value of field margins as foraging habitat for farmland birds. *Journal of Applied Ecology*, 46, 353–362.

Drake, D. A. R. and Mandrak, N. E. (2010). Least-cost transportation networks predict spatial interaction of invasion vectors. *Ecological Applications*, 20, 2286–2299.

Dregne, H. E. (1976). *Soils of Arid Regions*. New York: Elsevier.

Dreyer, G. D. and Niering, W. A. (1986). Evaluation of two herbicide techniques on electric transmission rights-of-way: development of relatively stable shrublands. *Environmental Management*, 10, 113–118.

du Pisani, P. L. (2006). Direct reclamation of potable water at Windhoek's Goreangab reclamation plant. *Desalinization*, 188, 79–88.

Duany, A., Playter-Zyberk, E. and Alminana, R. (2003). *The New Civic Art: Elements of Town Planning*. New York: Rizzoli International Publications.

Dubois, G. F., Vignon, V., Delettre, Y. R. *et al.* (2009). Factors affecting the occurrence of the endangered saproxylic beetle *Osmoderma eremita* (Scopoli, 1763) (Coleoptera: Cetoniidae) in an agricultural landscape. *Landscape and Urban Planning,* 91, 152–159.

Dufty-Jones, R. and Connell, J., eds. (2014). *Rural Change in Australia: Population, Economy, Environment*. Farnham, UK: Ashgate.

Dumyahn, S. L. and Pijanowski, B. C. (2011). Soundscape conservation. *Landscape Ecology*, 26, 1327–1344.

Dunham-Jones, E. and Williamson, J. (2009). *Retrofitting Suburbia: Urban Design Solutions for Redesigning Suburbs*. New York: John Wiley.

Duning, X. (2001). Rural landscape ecological construction in China: theory and application. In *Landscape Ecology Applied in Land Evaluation, Development and Conservation: Some Worldwide Selected Examples*, ed. D. van der Zee and I. S. Zonneveld. Enschede, The Netherlands: International Institute for Aerospace Survey and Earth Sciences (ITC), pp. 221–231.

Dunn, C. P. and Heneghan, L. (2011). Composition and diversity of urban vegetation. In *Urban Ecology: Patterns, Processes, and Applications*, eds. J. Niemela, J. H. Breuste, T. Elmqvist *et al.* New York: Oxford University Press, pp. 103–115.

Dunne, T. A. and Leopold, L. B. (1978). *Water in Environmental Planning*. San Francisco, CA: W. H. Freeman.

Dunnett, N. and Clayden, A. (2007). *Rain Gardens: Managing Water Sustainably in the Garden and Designed Landscape*. Portland, OR: Timber Press.

Dunnett, N. and Kingsbury, N. (2004). *Planting Green Roofs and Living Walls*. Portland, OR: Timber Press.

Dutton, R. A. and Bradshaw, A. D. (1982). *Land Reclamation in Cities: A Guide to Methods of Establishment of Vegetation on Urban Waste Land*. London: HMSO.

Easterling, K. (1993). *American Town Plans: A Comparative Time Line*. New York: Princeton Architectural Press.

Eaton, R. L., Hammond, G. P. and Laurie, J. (2007). Footprints on the landscape: an environmental appraisal of urban and rural living in the developed world. *Landscape and Urban Planning,* 83, 13–28.

Echols, S. (2008). Split-flow theory: stormwater design to emulate natural landscapes. *Landscape and Urban Planning*, 85, 205–214.

Eckert, M. (1987). Bargteheide und Reinfeld. In *Untersuchungen uber die Kleinstadt in Schleswig-Holstein*, ed. R. Stewig. Kiel, Germany: Im Selbstverlag des Geographischen Instituts der Universitat Kiel, pp. 321–357.

Ecological Society of America. (2001). *Water in a Changing World.* Issues in Ecology No. 9. Washington, DC: Ecological Society of America.

Edvardsen, A., Halvorsen, R., Norderhaug, A. *et al.* (2010). Habitat specificity of patches in modern agricultural landscapes. *Landscape Ecology*, 25, 1071–1083.

Eglington, S. M., Gill, J. A., Bolton, M. *et al.* (2008). Restoration of wet features for breeding waders on lowland grassland. *Journal of Applied Ecology*, 45, 305–314.

Eisenbeis, G. (2006). Artificial night lighting and insects: attraction of insects to streetlamps in a rural setting in Germany. In *Ecological Consequences of Artificial Night Lighting*, ed. C. Rich and T. Longcore. Washington, DC: Island Press, pp. 281–304.

Eisenbeis, G. and Hanel, A. (2009). Light pollution and the impact of artificial night lighting on insects. In *Ecology of Cities and Towns: A Comparative Approach*, ed. M. J. McDonnell, A. K. Hahs and J. H. Breuste. New York: Cambridge University Press, pp. 243–263.

Eisenbeis, G. and Hassel, F. (2000). Zur Anziehung nachtaktiver Insekten durch Strassenlaternen – eine Studie kommunaler Beleuchtungseinrichtungen in der Agrarlandschaft Rheinhessens [Attraction of nocturnal insects to street-lights: a study of municipal lighting systems in a rural area of Rheinhessen]. *Natur und Landschaft*, 75, 145–156.

Eldegard, K., Totland, O. and Moe, S. R. (2015). Edge effects on plant communities along power line clearings. *Journal of Applied Ecology*, 52, 871–880.

Elliott, M. A., Burdon, D. A., Hemingway, K. L. and Apitz, S. E. (2007). Estuarine, coastal and marine ecosystem restoration; confusing management and science: a revision of concepts. *Estuarine, Coastal and Shelf Science*, 74, 349–366.

Ellis, C. D., Lee, S.-W. and Kweon, B.-S. (2006). Retail land use, neighborhood satisfaction and the urban forest: an investigation into the moderating and mediating effects of trees and shrubs. *Landscape and Urban Planning*, 74, 70–78.

Enders, M. S. and Van der Wall, S. B. (2012). Black bears *Ursus americanus* are effective seed dispersers, with a little help from their friends. *Oikos*, 121, 589–596.

Erell, E., Pearlmutter, D. and Williamson, T. (2011). *Urban Microclimate: Designing the Spaces Between Buildings*. London: Earthscan.

Ernst, W. H. O. (2006). Evolution of metal tolerance in higher plants. *Forest, Snow and Landscape Research*, 80, 251–274.

Erzinclioglu, Y. Z. (1981). On the diptera associated with dog dung. *London Nature*, 60, 45–46.

Escobedo, F. J. and Nowak, D. J. (2009). Spatial heterogeneity and air pollution removal by an urban forest. *Landscape and Urban Planning*, 90, 102–110.

Ethier, K. and Fahrig, L. (2011). Positive effects of forest fragmentation, independent of forest amount, on bat abundance in eastern Ontario, Canada. *Landscape Ecology*, 26, 865–876.

Etienne, M., Aronson, J. and Floc'h, F. L. (1998). Abandoned lands and land use conflicts in southern France. In *Landscape Disturbance and Biodiversity in Mediterranean-Type Ecosystems*, ed. P. W. Rundel, G. Montenegro and F. M. Jaksic. New York: Springer, pp. 127–140.

Evans, C. V., Fanning, D. E. and Short, J. R. (2000). Human influenced soils. In *Managing Soils in an Urban Environment*, ed. R. B. Brown, J. H. Huddleston and J. I. Anderson. Madison, WI: American Society of Agronomy, pp. 33–67.

Evans, D. (2012). Binning, gifting and recovery: the conduits of disposal in household food consumption. *Environment and Planning D: Society and Space*, 30, 1123–1137.

Evans, K. L. (2010). Individual species and urbanisation. In *Urban Ecology*, ed. K. J. Gaston. New York: Cambridge University Press, pp. 53–87.

Eyre, M. D., Carr, R., McBlane, R. P. and Foster, G. N. (1992). The effects of varying site-water duration on the distribution of water beetle assemblages, adults and larvae (Coleoptera: Haliplidae, Dytiscidae, Hydrophilidae). *Archiv fur Hydrobiologie*, 124, 281–291.

Faggi, A. M., Krellenberg, K., Castro, R. *et al.* (2008). Biodiversity in the Argentinean Rolling Pampa Ecoregion: changes caused by agriculture and urbanization. In *Urban Ecology: An International Perspective on the Interaction Between Humans and Nature*, ed. J. M. Marzluff, F. Schulenberger, W. Endlicher *et al.* New York: Springer, pp. 377–389.

Faggi, A., Perepelizin, P. and Dadon, J. R. (2010). South Atlantic tourist resorts: predictors for changes induced by afforestation. In *Urban Biodiversity and Design*, ed. N. Muller, P. Werner and J. G. Kelcey. Chichester, UK: Wiley-Blackwell, pp. 363–379.

Fahrig, L. (2013). Rethinking patch size and isolation effects: the habitat amount hypothesis. *Journal of Biogeography*, 40, 1649–1663.
Fahrig, L., Girard, J., Duro, D. *et al.* (2015). Farmlands with smaller crop fields have higher within-field biodiversity. *Agriculture, Ecosystems and Environment*, 200, 219–234.
Fallows, J. and Fallows, D. (2018). *Our Towns: 100,000-Mile Journey into the Heart of America*. New York: Pantheon Books.
Fang, C.-F. and Ling, D.-L. (2005). Guidance for noise reduction provided by tree belts. *Landscape and Urban Planning*, 71, 29–34.
Farina, A. (2006). *Principles and Methods in Landscape Ecology: Towards a Science of Landscape*. New York: Springer.
Farina, A., Buscaino, G., Ceraulo, M. and Pieretti, N. (2014). The soundscape approach for the assessment and conservation of Mediterranean landscapes: principles and case studies. *Journal of Landscape Ecology*, 7, 10–22.
Farshad, A. (2001). Reconstruction of the evolution of the past agrarian landscape as a clue for assessing sustainability: a case study of Iran. In *Landscape Ecology Applied in Land Evaluation, Development and Conservation: Some Worldwide Selected Examples*, ed. D. van der Zee and I. S. Zonneveld. Enschede, The Netherlands: International Institute for Aerospace Survey and Earth Sciences (ITC), pp. 21–47.
Faulkner, W. (1957). *The Town*. New York: Random House.
Fearnside, P. M. (2015). Highway construction as a force in the destruction of the Amazon forest. In *Handbook of Road Ecology*, ed. R. van der Ree, D. J. Smith and C. Grilo. Chichester, UK: Wiley-Blackwell, pp. 414–424.
Federal Highway Administration. (1996). *Evaluation and Management of Highway Runoff Water Quality*. Publication FHWA-PD-96032. Washington, DC: US Department of Transportation.
Fei, H.-T. (1939). *Peasant Life in China*. London: Routledge & Kegan Paul.
Fensham, R. J. and Fairfax, R. J. (2008). Water-remoteness for grazing relief in Australian arid-lands. *Biological Conservation*, 141, 1447–1460.
Ferguson, B. K. (2005). *Porous Pavements*. Boca Raton, FL: Taylor & Francis.
Ferrarini, A., Rossi, G., Parolo, G. and Ferloni, M. (2008). Planning low-impact tourist paths within a Site of Community Importance through the optimization of biological and logistic criteria. *Biological Conservation*, 141, 1067–1077.
Ferreira, J. A. and Condessa, B. (2012). Defining expansion areas in small urban settlements – an application to the municipality of Tomar (Portugal). *Landscape and Urban Planning*, 107, 283–292.
Findlay, C. S. and Bourges, J. (2000). Response time of wetland biodiversity to road construction on adjacent lands. *Conservation Biology*, 14, 86–94.
Finer, M., Jenkins, C. N., Pimm, S. L. *et al.* (2008). Oil and gas projects in the western Amazon: threats to wilderness, biodiversity, and indigenous people. *PLoS One*, 3, e2932.
Fink, D. F. and Mitsch, W. J. (2004). Seasonal and storm event nutrient removal by a created wetland in an agricultural watershed. *Ecological Engineering*, 23, 313–325.
Fiorello, C. V., Noss, A. J. and Deem, S. L. (2006). Demography, hunting ecology, and pathogen exposure of domestic dogs in the Isoso of Bolivia. *Conservation Biology*, 20, 762–771.
Fissore, C., Baker, L. A., Hobbie, S. E. *et al.* (2011). Carbon, nitrogen, and phosphorus fluxes in household ecosystems in the Minneapolis-Saint Paul, Minnesota, urban region. *Ecological Applications*, 21, 619–639.
Fitzgerald, B. M. (1988). Diet of domestic cats and their impact on prey populations. In *The Domestic Cat: The Biology of Its Behavior*, ed. D. C. Turner and P. Bateson. Cambridge: Cambridge University Press, pp. 123–146.
Fletcher, N. (2007). Handbook of acoustics. In *Animal Bioacoustics*, ed. T. D. Rossing. New York: Springer, pp. 785–804.
Flink, C. A., Olka, K. and Searns, R. M. (2001). *Trails for the Twenty-First Century: Planning, Design, and Management Manual for Multi-Use Trails*. Washington, DC: Island Press.
Flohre, A., Fischer, C., Aavik, T. *et al.* (2011). Agricultural intensification and biodiversity partitioning in European landscapes comparing plants, carabids, and birds. *Ecological Applications*, 21, 1772–1781.
Florida, R. (2002). *The Rise of the Creative Class and How It's Transforming Work, Leisure, Community and Everyday Life*. New York: Basic Books.
Foley, J. A., DeFries, R., Asner, G. P. *et al.* (2005). Global consequences of land use. *Science*, 309, 570–574.
Folke, C. (2006). Resilience: the emergence of a perspective for social-ecological systems analyses. *Global Environmental Change*, 16, 253–267.

Forge, K. and Reid, N. (2006). *Wool Production and Biodiversity: A Holistic Solution for Fine Wool and Healthy Profits at "Lana."* Canberra: Australian Government, Land and Water Australia.

Forman, H. C. (1991a). *Early Nantucket and its Whale Houses*. Nantucket, MA: Mill Hill Press.

Forman, R. T. T., ed. (1979). *Pine Barrens: Ecosystem and Landscape*. New York: Academic Press.

Forman, R. T. T. (1981). Patches and structural components for a landscape ecology. *BioScience*, 31, 733–740.

Forman, R. T. T. (1983). Corridors in a landscape: their ecological structure and function. *Ekologia (Czechoslovakia)*, 2, 375–387.

Forman, R. T. T. (1991b). Landscape corridors: from theoretical foundations to public policy. In *Nature Conservation 2: The Role of Corridors*, ed. D. A. Saunders and R. J. Hobbs. Chipping Norton, NSW: Surrey Beatty, pp. 71–84.

Forman, R. T. T. (1995). *Land Mosaics: Ecology of Landscapes and Regions*. New York: Cambridge University Press.

Forman, R. T. T. (1997). *Concord's Mill Brook: Flowing through Time*. Concord, MA: Natural Resources Commission.

Forman, R. T. T. (2000). Estimate of the area affected ecologically by the road system in the United States. *Conservation Biology*, 14, 31–35.

Forman, R. T. T. (2004a). *Mosaico territorial para la region metropolitana de Barcelona*. Barcelona: Editorial Gustavo Gili.

Forman, R. T. T. (2004b). Road ecology's promise: what's around the bend? *Environment*, 46, 8–21.

Forman, R. T. T. (2008). *Urban Regions: Ecology and Planning Beyond the City*. New York: Cambridge University Press.

Forman, R. T. T. (2010). Coastal regions: spatial patterns, flows, and a people-nature solution from the lens of landscape ecology. In *La costa obliqua: Un atlante per la Puglia [The Oblique Coast: An Atlas for Puglia]*, ed. M. Mininni. Rome: Donzelli editore, pp. 249–265.

Forman, R. T. T. (2012). Infrastructure and nature: reciprocal effects and patterns for our future. In *Infrastructure and Sustainability*, ed. S. Pollalis, A. Georgoulias, S. Ramos and D. Schodek. New York: Routledge, pp. 35–49.

Forman, R. T. T. (2014). *Urban Ecology: Science of Cities*. New York: Cambridge University Press.

Forman, R. T. T. (2015). Corridores verdes ecologicos – ecological green corridors. *Paisea*, 30, 8–13.

Forman, R. T. T. (2016). Urban ecology principles: are urban ecology and natural area ecology really different? *Landscape Ecology*, 31, 1653–1662.

Forman, R. T. T. and Collinge, S. K. (1996). The "spatial solution" to conserving biodiversity in landscapes. In *Conservation of Faunal Diversity in Forested Landscapes*, ed. R. M. DeGraaf and R. I. Miller. London: Chapman & Hall, pp. 537–568.

Forman, R. T. T. and Deblinger, R. D. (2000). The ecological road-effect zone of a Massachusetts (USA) suburban highway. *Conservation Biology*, 14, 36–46.

Forman, R. T. T. and Godron, M. (1986). *Landscape Ecology*. New York: John Wiley.

Forman, R. T. T. and Moore, P. N. (1992). Theoretical foundations for understanding boundaries in landscape mosaics. In *Landscape Boundaries: Consequences for Biotic Diversity and Ecological Flows*, ed. A. J. Hansen and F. di Castri. New York: Springer, pp. 236–258.

Forman, R. T. T. and Sperling, D. (2011). The future of roads: no driving, no emissions, nature reconnected. *Solutions*, 2: 10–23. (Reprinted 2014 in *Creating a Sustainable and Desirable Future: Insights from 45 global thought leaders*, eds. R. Costanza and I. Kubiszewski. London: World Scientific, pp. 143–170.)

Forman, R. T. T. and Wu, J. (2016a). Where are the best places for the next billion people? Think globally, plan regionally. In *Handbook on Biodiversity and Ecosystem Services in Impact Assessment*, ed. D. Geneletti. Cheltenham, UK: Edward Elgar Publishing, pp. 453–473.

Forman, R. T. T. and Wu, J. (2016b). Where to put the next billion people. *Nature*, 537, 608–611.

Forman, R. T. T., Reeve, P., Beyer, H. *et al.* (2004). *Open Space and Recreation Plan 2004: Concord, Massachusetts*. Concord, MA: Natural Resources Commission.

Forman, R. T. T., Reineking, B. and Hersperger, A. M. (2002). Road traffic and nearby grassland bird patterns in a suburbanizing landscape. *Environmental Management*, 29, 782–800.

Forman, R. T. T., Sperling, D., Bissonette, J. A. *et al.* (2003). *Road Ecology: Science and Solutions*. Washington, DC: Island Press.

Forrest, A. and St. Clair, C. C. (2006). Effects of dog leash laws and habitat type on avian and small mammal communities in urban parks. *Urban Ecosystems*, 9, 51–66.

Forsyth, A. (2005). *Reforming Suburbia: The Planned Communities of Irvine, Columbia, and The Woodlands*. Berkeley, CA: University of California Press.

Forsyth, A. and Crewe, K. (2009). A typology of comprehensive designed communities since the Second World War. *Landscape Journal*, 28, 56–78.

Foster, D. R. (1992). Land-use history (1730–1990) and vegetation dynamics in central New England, USA. *Journal of Ecology*, 80, 753–772.

Foster, D. R., Donahue, B. M., Kittredge, D. B. *et al.* (2010). *Wildlands and Woodlands: A Vision for the New England Landscape*. Petersham, MA: Harvard Forest, Harvard University.

Foster, D., Lambert, K. F., Kittredge, D. *et al.* (2017). *Wildlands and Woodlands: Farmlands and Communities, Broadening the Vision for New England*. Petersham, MA: Harvard Forest, Harvard University.

Foster, D. R., Swanson, F. J., Aber, J. D. *et al.* (2003). The importance of land-use legacies to ecology and conservation. *BioScience*, 53, 77–88.

Frair, J. L., Merrill, E. H., Beyer, H. L. and Morales, J. M. (2008). Thresholds in landscape connectivity and mortality risks in response to growing road networks. *Journal of Applied Ecology*, 45, 1504–1513.

Francaviglia, R. V. (1996). *Main Street Revisited: Time, Space, and Image Building in Small-Town America*. Iowa City, IA: University of Iowa Press.

Francis, R. (2010). Wall ecology: a frontier for urban biodiversity and ecological engineering. *Progress in Physical Geography*, 35, 43–63.

Frank, A. S. K., Wardle, G. M., Dickman, C. R. and Greenville, A. C. (2014). Habitat- and rainfall-dependent biodiversity responses to cattle removal in an arid woodland-grassland environment. *Ecological Applications*, 24, 2013–2028.

Frank, K. D. (2006). Effects of artificial night lighting on moths. In *Ecological Consequences of Artificial Night Lighting*, ed. C. Rich and T. Longcore. Washington, DC: Island Press, pp. 305–344.

Franklin, J. F., Mitchell, R. J. and Palik, B. J. (2007). *Natural Disturbance and Stand Development Principles for Ecological Forestry*. USDA Forest Service General Technical Report NRS-19. Newtown Square, PA: USDA Forest Service.

Frazer, L. (2005). Paving paradise: the peril of impervious surfaces. *Environmental Health Perspective*, 113, A456–A462.

Freeman, A. S., Short, F. T., Isnain, I. *et al.* (2008). Seagrass on the edge: land-use practices threaten coastal seagrass communities in Sabah, Malaysia. *Biological Conservation*, 141, 2993–3005.

Freeman, R. C. and Bell, K. P. (2011). Conservation versus cluster subdivisions and implications for habitat connectivity. *Landscape and Urban Planning*, 101, 30–42.

Fried, G., Petit, S., Dessaint, F. and Reboud, X. (2009). Arable weed decline in northern France: crop edges as refugia for weed conservation? *Biological Conservation*, 142, 238–243.

Friedman, A. (2014). *Planning Small and Mid-Sized Towns: Designing and Retrofitting for Sustainability*. New York: Routledge.

Frosch, R. A. and Gallopoulos, N. E. (1989). Strategies for manufacturing. *Scientific American*, 261, 144–152.

Frumkin, H., Frank, L. and Jackson, R. (2004). *Urban Sprawl and Public Health: Design, Planning, and Building for Healthy Communities*. Washington, DC: Island Press.

Fuller, R. A., Irvine, K. N., Davies, Z. G. *et al.* (2012). Interactions between people and birds in urban landscapes. In *Urban Bird Ecology and Conservation*, ed. C. A. Lepczyk and P. S. Warren. Berkeley, CA: University of California Press, pp. 249–266.

Fuller, R., Warren, P., Arnsworth, P. *et al.* (2008). Garden bird feeding predicts the structure of urban bird assemblages. *Diversity and Distributions*, 14, 131–137.

Fuller, R. A., Warren, P. H. and Gaston K. J. (2007). Daytime noise predicts nocturnal singing in urban robins. *Biology Letters*, 3, 368–370.

Fynn, R. W. S., Morris, C. D. and Kirkman, K. P. (2005). Plant strategies and tradeoffs influence trends in competitive ability along gradients of soil fertility and disturbance. *Journal of Ecology,* 93, 384–394.

Gaborit, P. (2014). *European and Asian Sustainable Towns: New Towns and Their Satellite Cities in Their Metropolises*. Brussels: P. I. E. Peter Lang.

Gabriel, D., Carver, S. J., Durham, H. *et al.* (2009). The spatial aggregation of organic farming in England and its underlying environmental correlates. *Journal of Applied Ecology*, 46, 323–333.

Gagne, S. A. and Fahrig, L. (2011). Do birds and beetles show similar responses to urbanization? *Ecological Applications*, 21, 2297–2312.

Gajaseni, J. and Gajaseni, N. (1999). Ecological rationalities of the traditional home garden system in the Chao Phraya Basin, Thailand. *Agroforestry Systems*, 46, 3–23.

Garaffa, P. I., Filloy, J. and Bellocq, M. I. (2009). Bird community responses along urban–rural gradients: does the size of the urbanized area matter? *Landscape and Urban Planning*, 90, 33–41.

Garbuzov, M. and Ratnieks, F. L. W. (2013). Quantifying variation among garden plants in attractiveness to bees and other flower-visiting insects. *Functional Ecology*, doi: 10.1111/1365–2435.12178.

Garcia, C., Renison, D., Cingolani, A. M. and Fernandez-Juricic, E. (2008). Avifaunal changes as a consequence of large-scale livestock exclusion in the mountains of central Argentina. *Journal of Applied Ecology*, 45, 351–360.

Garcia-Gonzalez, C., Campo, D., Pola, I. G. and Garvic-Vazquez, E. (2012). Rural road networks as barriers to gene flow for amphibians: species-dependent mitigation by traffic calming. *Landscape and Urban Planning*, 104, 171–180.

Gartland, L. (2008). *Heat Islands: Understanding and Mitigating Heat in Urban Areas*. London: Earthscan.

Gascon, C., Lovejoy, T. E., Bierregaard, Jr., R. O. *et al.* (1999). Matrix habitat and species richness in tropical forest remnants. *Biological Conservation*, 91, 223–229.

Gaston, K. J., ed. (2010). *Urban Ecology*. New York: Cambridge University Press.

Gaston, K. J., Bennie, J., Davies, T. W. and Hopkins, J. (2013). The ecological impacts of nighttime light pollution: a mechanistic appraisal. *Biological Reviews*, 88, 912–927.

Gaston, K. J., Davies, T. W., Bennie, J. and Hopkins, J. (2012). Reducing the ecological consequences of night-time light pollution: options and developments. *Journal of Applied Ecology*, 49, 1256–1266.

Gaston, K. J., Smith, R. M., Thompson, K. and Warren, P. H. (2005). Urban domestic gardens. (II). Experimental tests of methods for increasing biodiversity. *Biodiversity and Conservation*, 14, 395–413.

Gauthreaux, Jr., S. A. and Belser, C. G. (2006). Effects of artificial night lighting on migrating birds. In *Ecological Consequences of Artificial Night Lighting,* ed. C. Rich and T. Longcore. Washington, DC: Island Press, pp. 67–93.

Gavier-Pizarro, G. I., Radeloff, V. C., Stewart, S. I. *et al.* (2010a). Rural housing is related to plant invasions in forests of southern Wisconsin, USA. *Landscape Ecology*, 25, 1505–1518.

Gavier-Pizarro, G. I., Radeloff, V. C., Stewart, S. I. *et al.* (2010b). Housing is positively associated with invasive exotic plant species richness in New England, USA. *Ecological Applications*, 20, 1913–1925.

Gavin, M. C. (2009). Conservation implications of rainforest use patterns: mature forests provide more resources but secondary forests supply more medicine. *Journal of Applied Ecology*, 46, 1275–1282.

Gaynor, K. M., Fiorella, K. J., Gregory, G. H. *et al.* (2016). War and wildlife: linking armed conflict to conservation. *Frontiers in Ecology and the Environment*, 14, 533–542.

Gearheart, R. A. (1992). Use of constructed wetlands to treat domestic wastewater, City of Arcata, California. *Water Science and Technology*, 26, 1625–1637.

Geddes, P. (1915). Cities in evolution: an introduction to the town planning movement and to the study of civics. In *Patrick Geddes: Spokesman for Man and the Environment* (1972), ed. M. Stalley. New Brunswick, NJ: Rutgers University Press, pp. 111–285.

Geertsema, W. and Sprangers, J. T. C. M. (2002). Plant distribution patterns related to species characteristics and spatial and temporal habitat heterogeneity in a network of ditch banks. *Plant Ecology*, 162, 91–108.

Gehlbach, F. R. (1996). Eastern screech owls in suburbia: a model of raptor urbanization. In *Raptors in Human Landscapes*, ed. D. Bird, D. Varland and J. Negro. San Diego, CA: Academic Press, pp. 69–74.

Geissler, A., Selenska-Pobell, S., Morris, K. *et al.* (2010). The microbial ecology of land and water contaminated with radioactive waste: towards the development of bioremediation options for the nuclear industry. In *Ecology of Industrial Pollution*, ed. L. C. Batty and K. B. Hallberg. Cambridge, UK: Cambridge University Press, pp. 226–241.

Gelbard, J. L. and Belnap, J. (2002). Roads as conduits for exotic plant invasions in a semiarid landscape. *Conservation Biology*, 17, 420–432.

Gell, P. (2010). With the benefit of hindsight: the utility of paleoecology in wetland condition assessment and identification of restoration targets. In *Ecology of Industrial Pollution*, ed. L. C. Batty and K. B. Hallberg. Cambridge, UK: Cambridge University Press, pp. 162–188.

Gemmill-Herren, B., Eardley, C., Mburu, J. *et al.* (2007). Pollinators. In *Farming with Nature: the Science and Practice of Ecoagriculture*, ed. S. J. Scherr and J. A. McNeely. Washington, DC: Island Press, pp. 166–177.

Geneletti, D., ed. (2016) *Handbook on Biodiversity and Ecosystem Services in Impact Assessment.* Cheltenham, UK: Edward Elgar Publishing.

George, S. and Crooks, K. (2006). Recreation and large mammal activity in an urban nature reserve. *Biological Conservation*, 133, 107–113.

George, W. G. (1974). Domestic cats as predators and factors in winter shortages of raptor prey. *Wilson Bulletin*, 86, 384–396.

Gepp, J. (1973). Kraftfahrzeugverkehr und fliegende Insekten [Roadway traffic and flying insects]. *Natur und Land (Austria)*, 59, 127–129.

Gepts, P., Famula, T. R., Bettinger, R. L. *et al.*, eds. (2012). *Biodiversity in Agriculture: Domestication, Evolution, and Sustainability*. Cambridge, UK: Cambridge University Press.

Germain, E., Benhamou, S. and Poulle, M.-L. (2008). Spatio-temporal sharing between the European wildcat, the domestic cat and their hybrids. *Journal of Zoology*, 276, 195–203.

Getz, L. L., Cole, F. R. and Gates, D. L. (1978). Interstate roadsides as dispersal routes for *Microtus pennsylvanicus*. *Journal of Mammalogy*, 59, 208–213.

Gibert, J., Danielopol, D. L. and Stanford, J. A., eds. (1994). *Groundwater Ecology*. San Diego, CA: Academic Press.

Gibson, C. (2014). Rural place marketing, tourism and creativity: entering the post-productivist countryside. In *Rural Change in Australia: Population, Economy, Environment*, ed. R. Dufty-Jones and J. Connell. Aldershot, UK: Ashgate Publishing, pp. 187–209.

Gibson, C., Waitt, G., Walmsley, J. and Connell, J. (2010). Cultural festivals and economic development in regional Australia. *Journal of Planning Education and Research*, 29, 280–293.

Gibson, D. J. (2009). *Grasses and Grassland Ecology*. New York: Oxford University Press.

Gilbert, O. L. (1970). A biological scale for the estimation of sulfur dioxide pollution. *New Phytologist*, 69, 629–634.

Gilbert, O. L. (1991). *The Ecology of Urban Habitats*. London: Chapman & Hall.

Gillespie, N., Unthank, A., Campbell, L. *et al.* (2014). Flood effects on road-stream crossing infrastructure: economic and ecological benefits of stream simulation designs. *Fisheries*, 39, 62–76.

Gillies, C. S., Beyer, H. L. and St. Clair, C. C. (2011). Fine-scale movement decisions of tropical forest birds in a fragmented landscape. *Ecological Applications*, 21, 944–954.

Gilroy, J. J., Anderson, G. Q. A., Grice, P. V. *et al.* (2008). Could soil degradation contribute to farmland bird declines? Links between soil penetrability and the abundance of yellow wagtails *Motacilla flava* in arable fields. *Biological Conservation*, 141, 3116–3126.

Girling, C. and Helphand, K. I. (1994). *Yard Street Park: The Design of Suburban Open Space*. New York: John Wiley.

Girling, C. and Kellett, R. (2005). *Skinny Streets and Green Neighborhoods: Design for Environment and Community*. Washington, DC: Island Press.

Gleason, H. A. (1926). The individualistic concept of the plant association. *Bulletin of the Torrey Botanical Club*, 53, 7–26.

Godde, M., Richarz, N. and Walter, B. (1995). Habitat conservation and development in the city of Dusseldorf, Germany. In *Urban Ecology as a Basis for Planning*, ed. H. Sukopp, M. Numata and A. Huber. The Hague, The Netherlands: SPB Academic Publishing, pp. 163–171.

Golden, J. S. (2004). The built environment induced urban heat island effect in rapidly urbanizing arid regions – a sustainable urban engineering complexity. *Environmental Sciences*, 1, 312–349.

Golet, F. C., Calhoun, A. J. K., DeRagon, W. R. *et al.* (1993). *Ecology of Red Maple Swamps in the Glaciated Northeast: A Community Profile*. Biological Report 12. Washington, DC: US Fish and Wildlife Service.

Gomez-Baggethun, E. and Barton, D. N. (2013). Classifying and valuing ecosystem services for urban planning. *Ecological Economics*, 86, 235–245.

Gompper, M. E. (2014). *Free-Ranging Dogs and Wildlife Conservation*. New York: Oxford University Press.

Goode, D. (2014). *Nature in Towns and Cities*. London: HarperCollins/William Collins.

Goosem, M. and Marsh, H. (1997). Fragmentation of a small-mammal community by a powerline corridor through tropical rainforest. *Wildlife Research*, 24, 613–629.

Gordon, J. C., Sampson, R. N. and Berry, J. K. (2005). The challenge of maintaining working forests at the wildlife-urban interface. In *Forests at the Wildland-Urban Interface: Conservation and Management*, ed. S. W. Vince, M. L. Duryea, E. A. Macie and L. A. Hermansen. Boca Raton, FL: CRC Press, pp. 15–23.

Gottman, J. (1961). *Megalopolis: The Urbanized Northeastern Seaboard of the United States*. New York: Twentieth Century Fund.

Grebner, D. L., Bettinger, P. and Siry, J. P. (2013). *Introduction to Forestry and Natural Resources*. San Diego, CA: Academic Press.

Green, H. (2010). *Company Towns: The Industrial Edens and Satanic Mills That Shaped the American Economy*. New York: Basic Books.

Greenberg, M. R. and Schneider, D. (1996). *Environmentally Devastated Neighborhoods: Perceptions, Policies, and Realities*. New Brunswick, NJ: Rutgers University Press.

Grek, J., Karlsson, C. and Klaesson, J. (2013). Market potential and new firm formation. In *Towns in a Rural World*, ed. T. de N. Vaz, E. van Leeuwen and P. Nijkamp. Farnham, UK: Ashgate, pp. 67–89.

Grey, G. W. and Deneke, F. J. (1992). *Urban Forestry*. Malabar, FL: Krieger Publishing.

Grisham, J. (2015). *Gray Mountain*. New York: Doubleday.

Groffman, P. M., Bain, D. J., Band, L. E. *et al.* (2003). Down by the riverside: urban riparian ecology. *Frontiers in Ecology and the Environment*, 1, 315–321.

Grossinger, R. (2012). *Napa Valley Historical Ecology Atlas: Exploring a Hidden Landscape of Transformation and Resilience*. Berkeley, CA: University of California Press.

Grote, R., Samson, R., Alonso, R. *et al.* (2016). Functional traits of urban trees: air pollution mitigation potential. *Frontiers in Ecology and the Environment*, 14, 543–550.

Guarinoni, M., Ganzleben, C., Murphy, E. and Jurkiewicz, K. (2012). *Towards a Comprehensive Noise Strategy*. Brussels, Belgium: European Union.

Gude, P. H., Hansen, A. J., Rasker, R. and Maxwell, B. (2006). Rates and drivers of rural residential development in the Greater Yellowstone. *Landscape and Urban Planning*, 77, 131–151.

Guil, E., Fernandez-Olalla, M., Moreno-Opo, R. *et al.* (2011). Minimising mortality in endangered raptors due to power lines: the importance of spatial aggregation to optimize the application of mitigation measures. *PLoS One*, 6, e28212.

Gunawardena, K. R., Wells, M. J. and Kershaw, T. (2017). Utilizing green and blue space to mitigate urban heat island intensity. *Science of the Total Environment*, 584, 1040–1055.

Gunson, K. and Teixeira, F. Z. (2015). Road-wildlife mitigation planning can be improved by identifying the patterns and processes associated with wildlife–vehicle collisions. In *Handbook of Road Ecology*, ed. R. van der Ree, D. J. Smith and C. Grilo. Chichester, UK: Wiley-Blackwell, pp. 101–109.

Gurney, K. R., Romero-Lankao, P., Seto, K. C. *et al.* (2015). Track urban emissions on a human scale. *Nature*, 525, 179–181.

Gutfreund, O. (2004). *Twentieth-Century Sprawl: Highways and the Reshaping of the American Landscape*. New York: Oxford University Press.

Gutierrez, M. (2013). *Geomorphology*. London: CRC Press, Taylor & Francis.

Gutzwiller, K. J. and Barrow, Jr., W. C. (2008). Desert bird associations with broad-scale boundary length: applications in avian conservation. *Journal of Applied Ecology*, 45, 873–882.

Gutzwiller, K. J., Kroese, E. A., Anderson, S. H. and Wilkins, C. A. (1997). Does human intrusion alter the seasonal timing of avian song during breeding periods? *Auk*, 114, 55–65.

Haase, D. (2008). Urban ecology of shrinking cities: an unrecognized opportunity. *Nature and Culture*, 3, 1–8.

Haase, J. and Lathrop, R. G. (2003). A housing-unit-level approach to characterizing residential sprawl. *Photogrammetric Engineering and Remote Sensing*, S69(9), 1021–1030.

Hakim, D. (2016). Doubts about a promised bounty: genetically modified crops have failed to lift yields and ease pesticide use. *New York Times*, October 30, CLXVI, pp. 1 and 22–23.

Halaburka, B. J., Lawrence, J. E., Bischel, H. N. *et al.* (2013). Economic and ecological costs and benefits of streamflow augmentation using recycled water in a California coastal stream. *Environmental Science and Technology*, 47, 10735–10743.

Halada, K. (2008). Concept and technology for utilizing "urban mines." *Nims Now International*, 6(5), 3–5.

Hale, J. D., Fairbrass, A. J., Matthews, T. J. *et al.* (2015). The ecological impact of city lighting scenarios: exploring gap crossing thresholds for urban bats. *Global Change Biology*, 21, 2467–2478.

Halfwerk, W., Holleman, L. J. M., Lessells, C. M. and Slabbekoorn, H. (2011). Negative impact of traffic noise on avian reproductive success. *Journal of Applied Ecology*, 48, 210–219.

Halvorsen, M. B., Casper, B. M., Woodley, C. M. *et al.* (2012). Threshold for onset of injury in Chinook salmon from exposure to impulsive pile driving sounds. *PLoS One*, 7, e38968.

Hamabata, E. (1980). Changes of herb-layer species composition with urbanization of secondary oak forests in Musashino Plain near Tokyo – studies on the conservation of suburban forest stands I. *Japanese Journal of Ecology*, 30, 347–358.

Hammer, R. B., Stewart, S. L, Winkler, R. L. *et al.* (2004). Characterizing dynamic spatial and temporal residential density patterns from 1940–1990 across the North Central United States. *Landscape and Urban Planning*, 69, 183–199.

Haninger, K., Ma, L. and Timmins, C. (2017). The value of brownfield remediation. *Journal of the Association of Environmental and Resource Economists*, 4, 197–241.

Hanke, W., Bohner, J., Dreber, N. *et al.* (2014). The impact of livestock grazing on plant diversity: an analysis across dryland ecosystems and scales in southern Africa. *Ecological Applications*, 24, 1188–1203.

Hannon, L. E. and Sisk, T. D. (2009). Hedgerows in an agri-natural landscape: potential habitat value for native bees. *Biological Conservation*, 142, 2140–2154.

Hansen, A. J. and DeFries, R. (2007). Land use change around nature reserves: implications for sustaining biodiversity. *Ecological Applications*, 17, 974–988.

Hansen, A. J. and Rotilla, J. J. (2001). Nature reserves and land use: implications of the "place" principle. In *Applying Ecological Principles to Land Management*, ed. V. H. Dale and R. A. Haeuber. New York: Springer, pp. 54–72.

Hanski, I. (1982). Distribution ecology of anthropochorous plants in villages surrounded by forest. *Annales Botanici Fennici*, 19, 1–15.

Hanson, T. (2015). *The Triumph of Seeds: How Grains, Nuts, Kernels, Pulses, and Pips Conquered the Plant Kingdom and Shaped Human History*. New York: Basic Books.

Hara, Y., Thaitakoo, D. and Takeuchi, K. (2008). Landform transformation on the urban fringe of Bangkok: the need to review land-use planning processes with consideration of the flow of fill materials to developing areas. *Landscape and Urban Planning*, 84, 74–91.

Hardoy, J. E., Mitlin, D. and Satterthwaite, D. (2001). *Environmental Problems in an Urbanizing World*. London: Earthscan.

Hardy, T. (1886). *The Mayor of Casterbridge: The Life and Death of a Man of Character.* London: Smith Elder & Co.

Harper-Lore, B. and Wilson, M. (2000). *Roadside Use of Native Plants*. Washington, DC: Island Press.

Harris, B. and McKean, C. (2014). *The Scottish Town in the Age of the Enlightenment 1740–1820*. Edinburgh: Edinburgh University Press.

Harris, R. (2015). Using Toronto to explore three suburban stereotypes, and vice versa. *Environment and Planning A*, 47, 30–49.

Hart, J. F. (1992). *The Land That Feeds Us*. New York: Norton.

Hart, T. and Powe, N. (2007). Market towns and rural employment. In *Market Towns: Roles, Challenges, and Prospects*, ed. N. Powe, T. Hart and T. Shaw. New York: Routledge, pp. 105–117.

Hartley, L. M., Detling, J. K. and Savage, L. T. (2009). Introduced plague lessens the effects of an herbivorous rodent on grassland vegetation. *Journal of Applied Ecology*, 46, 861–869.

Hartmann, W. M. (1997). *Signals, Sound and Sensation*. New York: American Institute of Physics.

Haspel, C. and Calhoon, R. E. (1991). Ecology and behavior of free-ranging cats in Brooklyn, New York. In *Wildlife Conservation in Metropolitan Environments*, ed. L. W. Adams and D. L. Leedy. Columbia, MD: National Institute for Urban Wildlife, pp. 27–30.

Hauser, S., Meixler, M. S. and Laba, M. (2015). Quantification of impacts and ecosystem services loss in New Jersey coastal wetlands due to hurricane Sandy storm surge. *Wetlands*, 35, 1137–1148.

Hawbaker, T. J., Radeloff, V. C., Stewart, S. I. *et al.* (2013). Human and biophysical influences on fire occurrence in the United States. *Ecological Applications*, 23, 565–582.

Hayward, M. W. and Kerley, G. I. H. (2009). Fencing for conservation: restriction of evolutionary potential or a riposte to threatening processes? *Biological Conservation*, 142, 1–13.

He, G., Chen, X., Beaer, S. *et al.* (2009). Spatial and temporal patterns of fuelwood collection in Wolong Nature Reserve: implications for panda conservation. *Landscape and Urban Planning*, 92, 1–9.

Head, L. and Muir, P. (2006). Suburban life and the boundaries of nature: resilience and rupture in Australian backyard gardens. *Transactions of the Institute of British Geographers*, 31, 505–524.

Heller, N. E. and Zavaleta, E. S. (2009). Biodiversity management in the face of climate change: a review of 22 years of recommendations. *Biological Conservation*, 142, 14–32.

Herrera, J. M. and Garcia, D. (2009). The role of remnant trees in seed dispersal through the matrix: being alone is not always so sad. *Biological Conservation*, 142, 149–158.

Hersperger, A. M. and Forman, R. T. T. (2003). Adjacency arrangement effects on plant diversity and composition in woodland patches. *Oikos*, 101, 279–290.

Hersperger, A. M., Langhamer, D. and Dalang, T. (2012). Inventorying human-made objects: a step towards better understanding of land use for multifunctional planning in a periurban Swiss landscape. *Landscape and Urban Planning*, 105, 307–314.

Herzon, I. and Helenius, J. (2008). Agricultural drainage ditches, their biological importance and functioning. *Biological Conservation*, 141, 1171–1183.

Hess, G. R. and Fischer, R. A. (2001). Communicating clearly about conservation corridors. *Landscape and Urban Planning*, 55, 195–208.

Hess, S. C., Banko, P. C., Goltz, D. M. *et al.* (2004). Strategies for reducing feral cat threats to endangered Hawaiian birds. *Proceedings of the Vertebrate Pest Conference*, 21, 21–26.

Hickley, P., Arlinghaus, R., Tyner, R. *et al.* (2004). Rehabilitation of urban lake fisheries for angling by managing habitat: general overview and case studies from England and Wales. *Ecohydrology and Hydrobiology*, 4, 365–378.

Higgins, A. and Cardillo, R. (2011). Parking lot garden. In *Chanticleer: A Pleasure Garden*. Philadelphia, PA: University of Pennsylvania Press.

Hill, D., Hockin, D., Price, D. *et al.* (1997). Bird disturbance: improving the quality and utility of disturbance research. *Journal of Applied Ecology*, 34, 275–288.

Hill, K. (2009). Urban design and urban water ecosystems. In *The Water Environment of Cities*, ed. L. A. Baker. New York: Springer, pp. 141–170.

Hill, K. (2015). Coastal infrastructure: a typology for the next century of adaptation to sea-level rise. *Frontiers in Ecology and the Environment*, 13, 468–476.

Hing, S., Narayan, E. J., Thompson, R. C. A. and Godfrey, S. S. (2016). The relationship between physiological stress and wildlife disease: consequences for health and conservation. *Wildlife Research*, 43, 51–60.

Hinners, S. J., Kearns, C. A. and Wessman, C. A. (2012). Roles of scale, matrix, and native habitat in supporting a diverse suburban pollinator assemblage. *Ecological Applications*, 22, 1923–1935.

Ho, P. (2005). *Developmental Dilemmas: Land Reform and Institutional Changes in China*. New York: Routledge.

Hobbs, R. J. and Askins, L. (1988). Effect of disturbance and nutrient addition on native and introduced annuals in plant communities in the Western Australian wheatbelt. *Australian Journal of Ecology*, 13, 171–179.

Hobbs, R. J., Arico, S., Aronson, J. *et al.* (2006). Novel ecosystems: theoretical and management aspects of the New Ecological World Order. *Global Ecology and Biogeography*, 15, 1–7.

Hobbs, R. J., Higgs, E. S. and Hall, C., eds. (2013). *Novel Ecosystems: Intervening in the New Ecological World Order*. New York: Wiley-Blackwell.

Hobbs, R. J., Higgs, E. and Harris, J. A. (2009). Novel ecosystems: implications for conservation and restoration. *Trends in Ecology and Evolution*, 24, 599–605.

Hodgkinson, S., Hero, J. and Warnken, J. (2007). The conservation value of suburban golf courses in a rapidly urbanizing area of Australia. *Landscape and Urban Planning*, 79, 323–337.

Hodgson, P., French, K. and Major, R. E. (2007). Avian movement across abrupt ecological edges: differential responses to housing density in an urban matrix. *Landscape and Urban Planning*, 29, 266–272.

Hoekstra, J. M., Molnar, J. M., Jennings, M. *et al.* (2010). *The Atlas of Global Conservation: Changes, Challenges and Opportunities to Make a Difference*. Berkeley, CA: University of California Press and The Nature Conservancy.

Hogan, D. M. and Welbridge, M. R. (2007). Best management practices for nutrient and sediment retention in urban stormwater runoff. *Journal of Environmental Quality*, 36, 386–395.

Hokanson, D. (1994). *Reflecting a Prairie Town: A Year in Peterson*. Iowa City, IA: University of Iowa Press.

Hole, F. D. and Campbell, J. B. (1985). *Soil Landscape Analysis*. Totowa, NJ: Rowman & Allanheld.

Holland, R. A., Eigenbrod, F., Armsworth, P. R. *et al.* (2011). Spatial covariation between freshwater and terrestrial ecosystem services. *Ecological Applications*, 21, 2034–2048.

Hollander, J., Kirkwood, N. and Gold, J. (2010). *Principles of Brownfield Regeneration*. Washington, DC: Island Press.

Holling, C. S. (1996). Surprise for science, resilience for ecosystems, and incentives for people. *Ecological Applications*, 6, 733–735.

Holling, C. S. (2001). Understanding the complexity of economic, ecological, and social systems. *Ecosystems*, 4, 390–405.

Holz-Rau, C., Scheiner, J. and Sicks, K. (2014). Travel distances in daily travel and long-distance travel: what role is played by urban form? *Environment and Planning A*, 46, 488–507.

Hong, S.-K. (2001). Factors affecting landscape changes in central Korea: cultural disturbance on the forested landscape systems. In *Landscape Ecology Applied in Land Evaluation, Development and Conservation: Some Worldwide Selected Examples*, ed. D. van der Zee and I. S. Zonneveld. Enschede, The Netherlands: International Institute for Aerospace Survey and Earth Sciences (ITC), pp. 131–147.

Hood, M. J., Clausen, J. C. and Warner, G. S. (2007). Comparison of stormwater lag times for low impact and traditional residential development. *Journal of the American Water Resources Association*, 43, 1036–1046.

Hoornweg, D., Bhada-Tata, P. and Kennedy, C. (2013). Waste production must peak this century. *Nature*, 502, 615–617.

Hopkins, G. R., Gaston, K. J., Visser, M. E. *et al.* (2018). Artificial light at night as a driver of evolution across urban-rural landscapes. *Frontiers in Ecology and the Environment*, 16: 472–479.

Horn, J. A., Mateus-Pinilia, N., Warner, R. E. and Heske, E. J. (2011). Home range, habitat use, and activity patterns of free-roaming domestic cats. *Journal of Wildlife Management*, 75, 1177–1185.

Horton, N., Brough, T. and Rochard, J. B. A. (1983). The importance of refuse tips to gulls wintering in an inland area of south-east England. *Journal of Applied Ecology*, 20, 751–765.

Horvath, G., Kriska, G., Malik, P. and Robertson, B. (2009). Polarized light pollution: a new kind of ecological photopollution. *Frontiers in Ecology and the Environment*, 7, 317–325.

Hoskins, W. G. (1955). *The Making of the English Landscape*. New York: Penguin.

Hosl, R., Strauss, P. and Glade, T. (2012). Man-made linear flow paths at catchment scale: identification, factors and consequences for the efficiency of vegetated filter strips. *Landscape and Urban Planning*, 104, 245–252.

Hough, M. (2004). *Cities and Natural Process: A Basis for Sustainability*. New York: Routledge.

Howard, E. (1898). *Tomorrow: A Peaceful Path to Real Reform*. London: Swan Sonnenschein.

Hsieh, C. and Davis, A. P. (2005). Evaluation and optimization of bioretention media for treatment of urban storm water runoff. *Journal of Environmental Engineering ASCE*, 131, 1521–1531.

Hu, J. (2014). From regional planning to site design – the application of the "Shan-shui City" concept in multi-scale landscape planning in China. In *The Ecological Design and Planning Reader*, ed. F. O. Ndubisi. Washington, DC: Island Press, pp. 470–482.

Huang, B.-Q., Sun, Y.-N., Yu, Z.-H. *et al.* (2009). Impact of proximity to a pathway on orchid pollination success in Huanglong National Park, south-west China. *Biological Conservation,* 142, 701–708.

Hudson, M.-A. R. and Bird, D. M. (2009). Recommendations for design and management of golf courses and green spaces based on surveys of breeding bird communities in Montreal. *Landscape and Urban Planning*, 92, 335–346.

Hughes, J. and Macdonald, D. W. (2013). A review of the interactions between free-roaming domestic dogs and wildlife. *Biological Conservation*, 157, 341–351.

Hugo, G. (2014). Immigrant settlement in regional Australia: patterns and processes. In *Rural Change in Australia: Population, Economy, Environment*. Farnham, UK: Ashgate Publishing, pp. 57–82.

Hunn, E. S. (2008). *A Zapotec Natural History: Trees, Herbs, and Flowers, Birds, Beasts, and Bugs in the Life of San Juan Gbee*. Tucson, AZ: University of Arizona Press.

Hunt, L. M., Arlinghaus, R., Lester, N. and Kushneriuk, R. (2011). The effects of regional angling effort, angler behavior, and harvesting efficiency on landscape patterns of overfishing. *Ecological Applications*, 21, 2555–2575.

Hunter, M. (2011). Using ecological theory to guide urban planning design: an adaptation strategy for climate change. *Landscape Journal*, 30, 2–11.

Hunter, M. L. Jr. (1990). *Wildlife, Forests, and Forestry: Principles of Managing Forests for Biological Diversity*. Englewood Cliffs, NJ: Prentice-Hall.

Hurley, S. E. and Forman, R. T. T. (2011). Stormwater ponds and biofilters for large urban sites: modeled arrangements that achieve the phosphorus reduction target for Boston's Charles River, USA. *Ecological Engineering*, 37, 850–863.

Hyster-Rubin, N. (2011). *Patrick Geddes and Town Planning: A Critical View*. New York: Routledge.

Hyvonen, T. and Huusela-Veistola, E. (2008). Arable weeds as indicators of agricultural intensity – a case study from Finland. *Biological Conservation*, 141, 2857–2864.

Ignatieva, M. (2012). Plant material for urban landscapes in the era of globalization: challenges and innovative solutions. In *Applied Urban Ecology: A Global Framework*, ed. M. Richter and U. Weiland. New York: Wiley-Blackwell, pp. 139–151.

Imhoff, M. L., Zhang, P., Wolfe, R. E. and Bounoua, L. (2010). Remote sensing of the urban heat island effect across biomes in the continental USA. *Remote Sensing of Environment*, 114, 504–513.

Ingegnoli, V. (2002). *Landscape Ecology: A Widening Foundation*. New York: Springer.

Ioffe, G. and Nefedova, T. (2000). *The Environs of Russian Cities*. Lewiston, NY: Edwin Mellon Press.

IPCC. (2007). *Climate Change 2007: Impacts, Adaptation and Vulnerability*. Intergovernmental Panel on Climate Change. New York: Cambridge University Press.

Ireland, P. H. (1989). Larval survivorship in two populations of *Ambystoma maculatum*. *Journal of Herpetology*, 23, 209–215.

Irwin, A. (2018). The dark side of light. *Nature*, 553, 268–270.

Isaacs, J. (2002). *Bush Food: Aboriginal Food and Herbal Medicine*. Marleson, SA: JB Books.

Ito, S., Nakayama, R. and Buckley, G. P. (2004). Effects of previous land-use on plant species diversity in semi-natural and plantation forests in a warm-temperate region in southeastern Kyushu, Japan. *Forest Ecology and Management*, 196, 213–225.

Ito, T., Miura, N., Lhagvasuren, B. *et al.* (2005). Preliminary evidence of a barrier effect of a railroad on the migration of Mongolian gazelles. *Conservation Biology*, 19, 945–948.
Iuell, B., Bekker, H. (G. J.), Cuperus, R. *et al.*, eds. (2003). *Habitat Fragmentation Due to Transportation Infrastructure: Wildlife and Traffic: A European Handbook for Identifying Conflicts and Designing Solutions*. COST 341. Brussels: KNNV Publishers.
Ivanov, E. D., Manakos, I., and van der Knaap, W. (2009). Conservation of coastal habitats in Mediterranean areas: a combined analytical framework for case studies. In *Naturbanization: New Identities and Processes for Rural-Natural Areas*, ed. M. J. Prados. London: Taylor & Francis, pp. 205–224.
Iverson, L. R. and Prasad, A. M. (2001). Potential changes in tree species richness and forest community types following climate change. *Ecosystems*, 4, 186–199.
Ives, C. D., Hose, G. C., Nipperess, D. A. and Taylor, M. P. (2011). Environmental and landscape factors influencing ant and plant diversity in suburban riparian corridors. *Landscape and Urban Planning*, 103, 372–382.
Jaarsma, C. F. (2004). Ecological "black spots" within the ecological network: an improved design for rural road network amelioration. In *Ecological Networks and Greenways: Concept, Design, Implementation*, ed. R. H. G. Jongman and G. Pungetti. New York: Cambridge University Press, pp. 171–187.
Jackle, J. (1982). *The American Small Town: Twentieth Century Place Images*. Hamden, CT: The Shoe String Press.
Jackson, H. B. and Fahrig, L. (2012). What size is a biologically relevant landscape? *Landscape Ecology*, 27, 929–941.
Jackson, J. B. (1952). The almost perfect town. *Landscape*, 2, 2–8.
Jackson, W. (2014). *Becoming Native to this Place*. Lexington, KY: University Press of Kentucky.
Jackson, W. B. (1951). Food habits of Baltimore, Maryland, cats in relation to rat populations. *Journal of Mammalogy*, 32, 458–461.
Jaeger, J. A. G. (2015). Improving environmental impact assessment and road planning at the landscape scale. In *Handbook of Road Ecology*, ed. R. van der Ree, D. J. Smith, and C. Grilo. Chichester, UK: Wiley-Blackwell, pp. 32–42.
Janhall, S. (2015). Review on urban vegetation and particle air pollution – deposition and dispersion. *Atmospheric Environment*, 105, 130–137.
Jarvis, P. J. (2011). Feral animals in the urban environment. In *The Routledge Handbook of Urban Ecology*, ed. I. Douglas, D. Goode, M. C. Houck and R. Wang. New York: Routledge, pp. 361–370.
Jeb, K.-P. (1987). Garding, Tonning, Friedrichstadt. In *Untersuchungen uber die Kleinstadt in Schleswig-Holstein*, ed. R. Stewig. Kiel, Germany: Im Selbstverlag des Geographischen Instituts der Universitat Kiel, pp. 55–95.
Jefferson, T. A., Hung, S. K. and Wursig, B. (2009). Protecting small cetaceans from coastal development: impact assessment and mitigation experience in Hong Kong. *Marine Policy*, 33, 305–311.
Jenny, H. (1980). *The Soil Resource: Origin and Behavior*. New York: Springer.
Jensen, M. and Thompson, H. (2004). Natural sounds: an endangered species. *George Wright Forum*, 21, 10–13.
Jensen, O. B. (2007). Culture stories: understanding cultural urban branding. *Planning Theory*, 6, 211–236.
Jim, C. Y. (2010). Old masonry walls as ruderal habitats for biodiversity conservation and enhancement in urban Hong Kong. In *Urban Biodiversity and Design*, ed. N. Muller, P. Werner and J. G. Kelcey. Chichester, UK: Wiley-Blackwell, pp. 323–347.
Jodoin, Y., Lavoie, C., Villeneuve, P. *et al.* (2008). Highways as corridors and habitats for the invasive common reed *Phragmites australis* in Quebec, Canada. *Journal of Applied Ecology*, 45, 459–466.
Johansson, L. J., Hall, K., Prentice, H. C. *et al.* (2008). Semi-natural grassland continuity, long-term land-use change and plant species richness in an agricultural landscape on Oland, Sweden. *Landscape and Urban Planning*, 84, 200–211.
Johnson, A. and Sumpter, J. P. (2015). Putting pharmaceuticals into the wider context of challenges to fish populations in rivers. *Philosophical Transactions of the Royal Society B*, 369(1656), 20130581.
Johnson, T. L., Cully, J. F., Collinge, S. K. *et al.* (2011). Spread of sylvatic plague among black-tailed prairie dogs is associated with colony spatial characteristics. *Journal of Wildlife Management*, 75, 357–368.
Johnston, P. and Don, A. (1990). *Grow Your Own Wildlife: How to Improve Your Local Environment*. Canberra, ACT: Greening Australia.
Johnstone, J. F., Allen, C. D., Franklin, J. F. *et al.* (2016). Changing disturbance regimes, ecological memory, and forest resilience. *Frontiers in Ecology and the Environment*, 14, 369–378.

Jokimaki, J. and Kaisanlahti-Jokimaki, M. L. (2003). Spatial similarity of urban bird communities: a multiscale approach. *Journal of Biogeography*, 30, 1183–1193.

Jomaa, I., Auda, Y., Abi Saleh, B. *et al.* (2008). Landscape spatial dynamics over 38 years under natural and anthropogenic pressures in Mount Lebanon. *Landscape and Urban Planning*, 87, 67–75.

Jones, B. T. B., Davis, A., Diez, L. and Diggle, R. W. (2013). Community-based natural resource management (CBNRM) and reducing poverty in Namibia. In *Biodiversity Conservation and Poverty Alleviation: Exploring the Evidence for a Link*, ed. D. Roe, J. Elliott, C. Sandbrook and M. Walpole. Chichester, UK: Wiley-Blackwell, pp. 191–205.

Jones, K. B., Slonecker, E. T., Nash, M. S. *et al.* (2010). Riparian habitat changes across the continental United States (1972–2003) and potential implications for sustaining ecosystem services. *Landscape Ecology*, 25, 1261–1275.

Jongman, R. H. G. (2004). The context and concept of ecological networks. In *Ecological Networks and Greenways: Concept, Design, Implementation*, ed. R. H. G. Jongman and G. Pungetti. Cambridge, UK: Cambridge University Press, pp. 7–33.

Jongman, R. H. G. and Pungetti, G., eds. (2004). *Ecological Networks and Greenways: Concept, Design, Implementation*. New York: Cambridge University Press.

Jordan, Y. C., Ghulam, A. and Herrmann, R. B. (2012). Floodplain ecosystem response to climate variability and land-cover and land-use change in Lower Missouri River basin. *Landscape Ecology*, 27, 843–857.

Jutila, H. M. (2003). Germination in Baltic coastal wetland meadows: similarities between vegetation and seed bank. *Plant Ecology*, 166, 275–293.

Kadlec, R. H. (1999). Constructed wetlands for treating landfill leachate. In *Constructed Wetlands for the Treatment of Landfill Leachates*, ed. G. Mulamoottil, E. A. McBean, and F. Rovers. Boca Raton, FL: Lewis Publishers, pp. 17–31.

Kadlec, R. H. and Hey, D. L. (1994). Constructed wetlands for river water quality improvement. *Water Science and Technology*, 29, 159–168.

Kadlec, R. H. and Knight, R. L. (1996). *Treatment Wetlands*. New York: Lewis Publishers.

Kaisanlahti-Jokimaki, M.-L., Jokimaki, J., Huhta, E. and Siikamaki, P. (2012). Impacts of seasonal small-scale urbanization on nest predation and bird assemblages at tourist destinations. In *Urban Bird Ecology and Conservation*, ed. C. A. Lepczyk and P. S. Warren. Berkeley, CA: University of California Press, pp. 93–110.

Kalff, J. (2002). *Limnology*. Upper Saddle River, NJ: Prentice-Hall.

Kalwij, J. M., Milton, S. J. and McGeoch, M. A. (2008). Road verges as invasion corridors? A spatial hierarchical test in an arid ecosystem. *Landscape Ecology*, 23, 439–451.

Kamm, U., Gugerli, F., Rotach, P. *et al.* (2010). Open areas in a landscape enhance pollen-mediated gene flow of a tree species: evidence from northern Switzerland. *Landscape Ecology*, 25, 903–911.

Kamra, S. (1982). *Indian Towns in Transition: A Study in Social Ecology*. New Delhi: Cosmo Publications.

Kanda, H., Tsuda, H., Ichikawa, K. and Yoshida, S. (2007). Environmental noise reduction of Tokaido Shinkansen and future prospect. In *Noise and Vibration Mitigation for Rail Transportation Systems*, ed. B. S. Werning, D. Thompson, P. E. Gautier *et al.* Berlin: Springer, pp. 1–8.

Kareiva, P., Tallis, H., Ricketts, T. H., eds. (2011). *Natural Capital: Theory and Practice of Mapping Ecosystem Services*. New York: Oxford University Press.

Kaswanto, R. L. and Nakagoshi, N. (2014). Landscape ecology-based approach for assessing pekarangan condition to preserve protected area in West Java. In *Designing Low Carbon Societies in Landscapes*, ed. N. Nakagoshi and J. A. Mabuhay. New York: Springer, pp. 289–311.

Kath, J., Maron, M. and Dunn, P. K. (2009). Interspecific competition and small bird diversity in an urbanising landscape. *Landscape and Urban Planning*, 92, 72–79.

Katoh, K., Sakai, S. and Takahashi, T. (2009). Factors maintaining species diversity in *satoyama*, a traditional agricultural landscape of Japan. *Biological Conservation*, 142, 1930–1936.

Kats, L. B. and Dill, L. M. (1998). The scent of death: chemosensory assessment of predation risk by prey animals. *EcoScience*, 5, 361–394.

Kavaratzis, M. and Ashworth, G. (2015). Hijacking culture: the disconnection between place culture and place brands. *Town Planning Review*, 86, 155–176.

Kayranli, B., Scholz, M., Mustafa, A. and Hedmark, A. (2010). Carbon storage and fluxes within freshwater wetlands: a critical review. *Wetlands*, 30, 111–124.

Kays, R. W. and DeWan, A. A. (2004). Ecological impact of inside/outside house cats around a suburban nature preserve. *Animal Conservation*, 7, 273–283.

Kazemi, F., Beecham, S. and Gibbs, J. (2011). Streetscape biodiversity and the role of bioretention swales in an Australian urban environment. *Landscape and Urban Planning*, 101, 139–148.

Kazemi, F., Beecham, S., Gibbs, J. and Clay, R. (2009). Factors affecting terrestrial invertebrate diversity in bioretention basins in an Australian urban environment. *Landscape and Urban Planning*, 92, 304–313.

Keddy, P. A. (2000). *Wetland Ecology: Principles and Conservation*. New York: Cambridge University Press.

Keenan, J. M. and Weisz, C., eds. (2016). *Blue Dunes: Climate Change by Design*. New York: Columbia University Press.

Kehlenbeck, K., Arifin, H. S. and Maass, B. (2007). Plant diversity in home gardens in a socio-economic agro-ecological context. In *Stability of Tropical Rainforest Margins*, ed. T. Tscharntke, C. Leuschner, C. Zeller *et al*. New York: Springer, pp. 295–317.

Kellert, S. R. (1996). *The Value of Life: Biological Diversity and Human Society*. Washington, DC: Island Press.

Kellert, S. R. (2005). *Building for Life: Designing and Understanding the Human-Nature Connection*. Washington, DC: Island Press.

Kellert, S. R., Heerwagen, J. H. and Mador, M. L., eds. (2008). *Biophilic Design: The Theory, Science, and Practice of Bringing Buildings to Life*. New York: John Wiley.

Kemp, R. L. and Stephani, C. J. (2011). *Cities Going Green: A Handbook of Best Practices*. Jefferson, NC: McFarland & Company.

Kendig, L., Connor, S., Byrd, C. and Heyman, J. (1980). *Performance Zoning*. Washington, DC: Planners Press, American Planning Association.

Kertson, B. N., Spencer, R. D., Marzluff, J. M. *et al*. (2011). Cougar space use and movements in the wildland-urban landscape of western Washington. *Ecological Applications*, 21, 2866–2881.

Kienast, F., Wildi, O. and Ghosh, S. (2007). *A Changing World: Challenges for Landscape Research*. New York: Springer.

Kight, C. R. and Swaddle, J. P. (2011). How and why environmental noise impacts animals: an intensive, mechanistic review. *Ecology Letters*, 14, 1052–1061.

Kight, C. R., Saha, M. S. and Swaddle, J. P. (2012). Anthropogenic noise is associated with reductions in the productivity of breeding Eastern Bluebirds (*Sialia sialis*). *Ecological Applications*, 25, 1986–1989.

Killham, K. (2010). The microbial ecology of mediating industrially contaminated land: sorting out the bugs in the system. In *Ecology of Industrial Pollution*, ed. L. C. Batty and K. B. Hallberg. New York: Cambridge University Press, pp. 242–254.

Kimmins, J. P. (2004). *Forest Ecology: A Foundation for Sustainable Forest Management and Environmental Ethics in Forestry*. Englewood Cliffs, NJ: Prentice-Hall.

King, D. I. and Byers, B. E. (2002). An evaluation of powerline rights-of-way as habitat for early-successional shrubland birds. *Wildlife Society Bulletin*, 30, 868–874.

King, D. I., Chandler, R. B., Collins, J. M. *et al*. (2009). Effects of width, edge and habitat on the abundance and nesting success of scrub-shrub birds in powerline corridors. *Biological Conservation*, 142, 2672–2680.

Kirk, R. W., Bolstad, P. V. and Manson, S. M. (2012). Spatio-temporal trend analysis of long-term development patterns (1900–2030) in a southern Appalachian county. *Landscape and Urban Planning*, 104, 47–58.

Kirkby, R., Bradbury, I. and Shen, G. (2006). The small town and urban context in China. In *The Earthscan Reader in Rural-Urban Linkages*, ed. C. Tacoli. London: Earthscan, pp. 184–197.

Kirkpatrick, J., Daniels, G. and Davison, A. (2009). An antipodean test of spatial contagion in front garden character. *Landscape and Urban Planning*, 93, 103–110.

Kitts-Morgan, S. E., Caires, K. C., Bohannon, L. A. *et al*. (2015). Free-ranging farm cats: home range size and predation on a livestock unit in northwest Georgia. *PLoS One*, 10, e0120513.

Klar, N., Fernandez, N., Kramer-Schadt, S. *et al*. (2008). Habitat selection models for European wildcat conservation. *Biological Conservation*, 141, 308–319.

Klein, R. D. (1979). Urbanization and stream quality impairment. *Water Resources Bulletin*, 15, 948–963.

Kline, J. D., Moses, A., Lettman, G. J. and Azuma, D. L. (2001). Modeling forest and range land development in rural locations, with examples from eastern Oregon. *Landscape and Urban Planning*, 80, 320–332.

Knaapen, J. P., Scheffer, M. and Harms, B. (1992). Estimating habitat isolation in landscape planning. *Landscape and Urban Planning*, 23, 1–16.

Knight, M. E., Martin, A. P., Bishop, S. *et al*. (2005). An interspecific comparison of foraging range and nest density of four bumblebee (*Bombus*) species. *Molecular Ecology*, 14, 1811–1820.

Knight, R. L. (2008). *Treatment Wetlands*. Boca Raton, FL: CRC Publishing.

Knight, R. L. and Gutzwiller, K. J., eds. (2013). *Wildlife and Recreationists: Coexistence through Management and Research*. Washington, DC: Island Press.

Knight, R. L., Walton, W. E., O'Meara, G. F. *et al.* (2003). Strategies for effective mosquito control in constructed treatment wetlands. *Ecological Engineering*, 21, 211–232.

Kobayashi, Y. and Koike, F. (2010). Separating the effects of land-use history and topography on the distribution of woody plant populations in a traditional rural landscape in Japan. *Landscape and Urban Planning*, 95, 34–43.

Kobringer, N. P. (1984). *Sources and Mitigation of Highway Runoff Pollutants – Executive Summary*. Volume 1, FHWA/RD-84/057. Milwaukee, WI: Federal Highway Administration and Rexnord EnvironEnergy Technology Center.

Koch-Nielsen, H. (2002). *Stay Cool: A Design Guide for the Built Environment in Hot Climates*. London: James & James.

Kohler, F., Verhulst, J., van Klink, R. and Kleijn, D. (2008). At what spatial scale do high-quality habitats enhance the diversity of forbs and pollinators in intensively farmed landscapes? *Journal of Applied Ecology*, 45, 753–762.

Kohn, M. H., York, E. C., Kamradt, D. A. *et al.* (1999). Estimating population size by genotyping faeces. *Proceedings of the Royal Society of London*, 266, 657–663.

Kolbe, J. J., VanMiddlesworth, P., Battles, A. C. *et al.* (2016). Determinants of spread in an urban landscape by an introduced lizard. *Landscape Ecology*, 31, 1795–1813.

Kolligs, D. (2000). Okologische Auswirkungen kunstlicher Lichtquellen auf nachtaktive Insekten, insbesondere Schmetterlinge (Lepidoptra) [Ecological effects of artificial light sources on nocturnally active insects, in particular on moths (Lepidoptera)]. *Faunistisch-Okologische Mitteilungen*, Supplement, 28, 1–136.

Konecny, M. J. (1987). Home range and activity patterns of feral house cats in the Galapagos Islands. *Oikos*, 50, 17–23.

Kontoleon, K. J. and Eumorfopoulou, E. A. (2010). The effect of the orientation and proportion of a plant-covered wall layer on the thermal performance of a building zone. *Building and Environment*, 45, 1287–1303.

Kosek, J. (2006). *Understories: The Political Life of Forests in Northern New Mexico*. Durham, NC: Duke University Press.

Kotkin, J. (2010). *The Next Hundred Million: America in 2050*. New York: Penguin Press.

Kovacs-Hostyanszki, A., Haenke, S., Batary, P. *et al.* (2013). Contrasting effects of mass-flowering crops on bee pollination of hedge plants at different spatial and temporal scales. *Ecological Applications*, 23, 1938–1946.

Kovar, P. (1995). Is plant community organization level relevant to monitoring landscape heterogeneity? Two case studies of mosaic landscapes in the suburban zones of Prague, Czech Republic. *Landscape and Urban Planning*, 32, 137–151.

Kowarik, I. and von der Lippe, M. (2011). Secondary wind dispersal enhances long-distance dispersal of an invasive species in urban road corridors. *NeoBiota*, 9, 49–70.

Krebs, C. J. (1999). *Ecological Methodology*. New York: Harper & Row.

Kretser, H. E., Sullivan, P. J. and Knuth, B. A. (2008). Housing density as an indicator of spatial patterns of reported human-wildlife interactions in northern New York. *Landscape and Urban Planning*, 84, 282–292.

Kuehl, H. S., Nzeingui, C., Yeno, S. L. D. *et al.* (2009). Discriminating between village and commercial hunting of apes. *Biological Conservation*, 142, 1500–1506.

Kuhman, T. R., Pearson, S. M. and Turner, M. G. (2010). Effects of land-use history and the contemporary landscape on non-native plant invasion at local and regional scales in the forest-dominated southern Appalachians. *Landscape Ecology*, 25, 1433–1445.

Kuhn, I. and Klotz, S. (2006). Urbanization and homogenization – comparing the floras of urban and rural areas in Germany. *Biological Conservation*, 127, 292–300.

Kumar, B. M. (2006). Carbon sequestration potential of tropical home gardens. In *Tropical Home gardens: A Time-Tested Example of Sustainable Agroforestry*, ed. B. M. Kumar and P. K. R. Nair. New York: Springer, pp. 185–204.

Kumar, B. M. (2011). Species richness and aboveground carbon stocks in the home gardens of central Kerala, India. *Agriculture, Ecosystems, and Environment*, 140, 430–440.

Kunick, W. (1990). Spontaneous woody vegetation in cities. In *Urban Ecology: Plants and Communities in Urban Environments*, ed. H. Sukopp, H. Hejny and I. Kowarik. The Hague, The Netherlands: SPB Academic Publishing, pp. 167–174.

Kuo, F. E. and Sullivan, W. C. (2001). Aggression and violence in the inner city: effects of environment via mental fatigue. *Environment and Behavior*, 33, 543–571.

Kuttler, W. (2008). The urban climate – basic and applied aspects. In *Urban Ecology: An International Perspective on the Interaction Between Humans and Nature*, ed. J. M. Marzluff, E. Schulenberger, W. Endlicher *et al.* New York: Springer, pp. 233–248.

Ladson, A. R., Walsh, J. and Fletcher, T. D. (2006). Improving stream health in urban areas by reducing runoff frequency from impervious surfaces. *Australasian Journal of Water Resources*, 10, 23–33.

Lake, P. S. (2011). *Drought and Aquatic Ecosystems: Effects and Responses.* Chichester, UK: John Wiley.

Lake, P. S., Bayly, I. A. E., and Morton, D. W. (1989). The phenology of a temporary pond in western Victoria, Australia, with special reference to invertebrate succession. *Archiv fur Hydrobiologie*, 115, 171–202.

Lambert, M. S., Quy, R. J., Smith, R. H. and Cowan, D. P. (2008). The effect of habitat management on home-range size and survival of rural Norway rat populations. *Journal of Applied Ecology*, 45, 1753–1761.

Lancaster, N. (2004). *Geomorphology of Desert Dunes*. Hoboken, NJ: Taylor & Francis.

Landers, J. L., Hamilton, R. J., Johnson, A. S. and Marchinton, R. L. (1979). Foods and habitat of black bears in southeastern North Carolina. *Journal of Wildlife Management*, 43, 143–153.

Landry, S. B. (1994). *Peterson First Guide to Urban Wildlife*. Boston, MA: Houghton Mifflin.

Landsberg, H. (1981). *Urban Climate*. New York: Academic Press.

Lara, A., Villalba, R. and Urrutia, R. (2008). A 400-year tree-ring record of the Puelo River summer-fall streamflow in the Valdivian Rainforest eco-region, Chile. *Climate Change*, 86, 351–356.

Larsen-Vilsholm, R., Wolseley, P. A., Sochting, U. and Chimonides, P. J. (2009). Biomonitoring with lichens on twigs. *Lichenologist*, 41, 189–202.

Larson, D. W., Matthes, U. and Kelly, P. E. (2000). *Cliff Ecology*. New York: Cambridge University Press.

Larson, E. K., Earl, S., Hagen, E. M. *et al.* (2013). Beyond restoration and into design: hydrologic alterations in aridland cities. In *Resilience in Ecology and Urban Design: Linking Theory and Practice for Sustainable Cities*, ed. S. T. A. Pickett, M. L. Cadenasso and B. McGrath. New York: Springer, pp. 183–210.

Latham, A. D. M. and Boutin, S. (2015). Impacts of utility and other industrial linear corridors on wildlife. In *Handbook of Road Ecology*, ed. R. van der Ree, D. J. Smith and C. Grilo. Chichester, UK: Wiley-Blackwell, pp. 228–236.

Laundon, J. R. (2003). Six lichens of the *Lecanora varia* group. *Nova Hedwigia*, 76, 83–111.

Laundre, J. (1977). The daytime behavior of domestic cats in a free-roaming population. *Animal Behavior*, 25, 990–998.

Laurance, W. F. (2008). Theory meets reality: how habitat fragmentation research has transcended island biogeographic theory. *Biological Conservation*, 141, 1731–1744.

Laurance, W. F. (2015). Bad roads, good roads. In *Handbook of Road Ecology,* ed. R. van der Ree, D. J. Smith and C. Grilo. Chichester, UK: Wiley-Blackwell, pp. 10–15.

Laurance, W. F. and Balmford, A. (2013). A global map for road building. *Nature*, 495, 308–309.

Laurance, W. F., Goosem, M. and Laurance, S. G. (2009). Impacts of roads and linear clearings on tropical forests. *Trends in Ecology and Evolution*, 24, 659–669.

Lavielle, D. and Petterson, T. (2007). Evolution of pollutant removal efficiency in storm water ponds due to changes in pond morphology. *Highway and Urban Environment*, 12, 429–439.

Law, I. B. (2003). Advanced reuse – from Windhoek to Singapore and beyond. *Water*, May, 31–36.

Law, N. L., Band, L. E. and Grove, J. M. (2004). Nitrogen input from residential lawn care practices in suburban watersheds in Baltimore County, MD. *Journal of Environmental Planning and Management*, 47, 737–755.

Le Viol, I., Mocq, J., Julliard, R. and Kerbiriou, C. (2009). The contribution of motorway stormwater retention ponds to the biodiversity of aquatic macroinvertebrates. *Biological Conservation*, 142, 3163–3171.

LeClerc, J. E. and Cristol, D. A. (2005). Are golf courses providing habitat for birds of conservation concern in Virginia? *Wildlife Society Bulletin*, 33, 463–470.

Lee, E., Croft, D. B. and Achiron-Frumkin, T. (2015). Roads in the arid lands: issues, challenges and potential solutions. In *Handbook of Road Ecology*, ed. R. van der Ree, D. J. Smith and C. Grilo. Chichester, UK: Wiley-Blackwell, pp. 382–390.

Lee, H.-S., Shepley, M. and Huang, C.-S. (2009). Evaluation of off-leash dog parks in Texas and Florida: a study of use patterns, user satisfaction, and perception. *Landscape and Urban Planning*, 92, 314–324.

Lee, J. G. and Heaney, J. P. (2003). Estimation of urban imperviousness and its impacts on storm water systems. *Journal of Water Resources Planning and Management*, 29, 419–426.

Lee, P. and Boutin, S. (2006). Persistence and developmental transition of wide seismic lines in the western Boreal Plains of Canada. *Journal of Environmental Management*, 78, 240–250.

Lee, S. L., Lee, A. N. and Cho, Y. C. (2008). Restoration planning for the Seoul metropolitan area, Korea. In *Ecology, Planning, and Management of Urban Forests*, ed. M. M. Carreiro, Y.-C. Song and J. Wu. New York: Springer, pp. 393–419.

Lehman, R. N. (2001). Raptor electrocution on powerlines: current issues and outlook. *Wildlife Society Bulletin*, 29, 804–813.

Lehner, P. N., McCluggage, C., Mitchell, D. R. and Neil, D. H. (1983). Selected parameters of the Fort Collins, Colorado, dog population, 1979–80. *Applied Animal Ethology*, 10, 19–25.

Leibovitz, J. (2006). Jumping scale: from small town politics to a "regional presence"? Re-doing economic governance in Canada's technology triangle. In *Small Cities: Urban Experience Beyond the Metropolis*, ed. D. Bell and M. Jayne. New York: Routledge, pp. 45–58.

Leinwand, I. I. F., Theobald, D. M., Mitchell, J. and Knight, R. L. (2010). Landscape dynamics at the public-private interface: a case study in Colorado. *Landscape and Urban Planning*, 97, 182–193.

Leisher, C., Sanjayan, M., Blockhus, J. *et al.* (2013). Does conserving biodiversity work to reduce poverty? A state of knowledge review. In *Biodiversity Conservation and Poverty Alleviation: Exploring the Evidence for a Link*, ed. D. Roe, J. Elliott, C. Sandbrook and M. Walpole. Chichester, UK: Wiley-Blackwell, pp. 145–159.

Leng, X., Musters, C. J. M. and de Snoo, G. R. (2009). Restoration of plant diversity on ditch banks: seed and site limitation in response to agri-environment schemes. *Biological Conservation*, 142, 1340–1349.

Lenth, B. E., Knight, R. I. and Brennan, M. E. (2008). The effects of dogs on wildlife communities. *Natural Areas Journal*, 28, 218–227.

Lepczyk, C. A., Mertig, A. G. and Liu, J. (2003). Landowners and cat predation across rural-to-urban landscapes. *Biological Conservation*, 115, 191–201.

Lepczyk, C. A., Warren, P. S., Machabee, L. *et al.* (2012). Who feeds the birds: a comparison across regions. In *Urban Bird Ecology and Conservation*, ed. C. A. Lepczyk and P. S. Warren. Berkeley, CA: University of California Press, pp. 267–284.

Lerman, S. B. and Warren, P. S. (2011). The conservation value of residential yards: linking birds and people. *Ecological Applications*, 21, 1327–1339.

Les Bocages: Histoire, Ecologie, Economie. (1976). Rennes, France: Centre National de la Recherche Scientifique, et Universite de Rennes.

Lethlean, H., Van Dongen, W. F. D., Kostoglu, K. *et al.* (2017). Joggers cause greater avian disturbance than walkers. *Landscape and Urban Planning*, 159, 42–47.

Letourneau, D. K., Armbrecht, I., Rivera, B. S. *et al.* (2011). Does plant diversity benefit agroecosystems? A synthetic review. *Ecological Applications*, 21, 9–21.

Leutscher, A. (1975). *The Ecology of Towns*. London: Franklin Watts.

Levi, T., Shepard, Jr., G. H., Ohl-Schacheren, J. *et al.* (2011). Spatial tools for modeling the sustainability of subsistence hunting in tropical forests. *Ecological Applications*, 21, 1802–1818.

Levin, S. A. (1999). *Fragile Dominion: Complexity and the Commons*. Reading, UK: Perseus.

Levin, S. A. (2015). Foreword. In *History of Landscape Ecology in the United States*, ed. G. W. Barrett, T. L. Barrett and J. Wu. New York: Springer, pp. v–viii.

Levy, J. K., Woods, J. E., Turick, S. L. *et al.* (2003). Number of unowned free-roaming cats in a college community in the southern United States and characteristics of community residents who feed them. *Journal of the American Veterinary Medical Association*, 225, 1354–1360.

Lewis, J. S., Rachlow, J. L., Horne, J. S. *et al.* (2011). Identifying habitat characteristics to predict highway crossing areas for black bears within a human-modified landscape. *Landscape and Urban Planning*, 101, 99–107.

Lewis, P. (1972). Small town in Pennsylvania. *Annals of the American Association of Geographers*, 62, 323–373.

Lewis, P. (1982). Axioms for reading the landscape. In *Material Culture Studies in America*, ed. T. J. Schlereth. Nashville, TN: American Association for State and Local History, pp. 175–182.

Li, A., Wu, J. and Huang, J. (2012). Distinguishing between human-induced and climate-driven vegetation changes: a critical application of RESTREND in Inner Mongolia. *Landscape Ecology*, 27, 969–982.

Li, Q. (2010). Effect of forest bathing trips on human immune function. *Environmental Health and Preventive Medicine*, 15, 9–17.

Liberg, O. (1980). Spacing patterns in a population of free roaming domestic cats. *Oikos*, 35, 336–349.

Liddle, M. (1997). *Recreation Ecology*. London: Chapman & Hall.

Lilly, E. L. and Wortham, C. D. (2013). High prevalence of *Toxoplasma gondii* oocyst shedding in stray and pet cats (*Felis catus*) in Virginia, United States. *Parasites & Vectors*, 6, 266–276.

Lin, H.-W., Jin, Y., Giglio, L. *et al.* (2012). Evaluating greenhouse gas emissions inventories for agricultural burning using satellite observations of active fires. *Ecological Applications*, 22, 1345–1364.

Lin, Y., Han, G. D., Zhao, M. L and Chang, S. X. (2010). Spatial vegetation patterns as early signs of desertification: a case study of a desert steppe in Inner Mongolia. *Landscape Ecology*, 25, 1519–1527.

Linda, H. G. and Peter, B. R. (2007). Air pollution and climate gradients in western Oregon and Washington indicated by epiphytic macrolichens. *Environmental Pollution*, 145, 203–218.

Lindenmayer, D. B. and Fischer, J. (2006). *Habitat Fragmentation and Landscape Change: An Ecological and Conservation Synthesis.* Washington, DC: Island Press.

Lindenmayer, D. B. and Franklin, J. F. (2002). *Conserving Forest Biodiversity: A Comprehensive Multiscaled Approach.* Washington, DC: Island Press.

Lindenmayer, D., Claridge, A., Hazell, D. *et al.* (2003). *Wildlife on Farms: How to conserve native animals.* Collingswood, Vic.: CSIRO Publishing.

Lindsay, R. (1995). *Bogs: The Ecology, Classification and Conservation of Ombrotrophic Mires.* Edinburgh: Scottish Natural Heritage.

Lipsky, Z. and Kukla, P. (2012). Mapping and typology of unused lands in the territory of the town Kutna Hora (Czech Republic). *AUC Geographica*, 47, 65–71.

Lisowska, M. (2011). Lichen recolonization in an urban-industrial area of southern Poland as a result of air quality improvement. *Environmental Monitoring and Assessment*, 179, 177–190.

Litfin, K. T. (2014). *Eco-Villages: Lessons for Sustainable Community*. Cambridge, UK: Polity Press.

Little, D. C. and Bunting, S. W. (2005). Opportunities and constraints to urban aquaculture, with a focus on South and Southeast Asia. In *Urban Aquaculture*, ed. B. Costa-Pierce, A. Desbonnet, P. Edwards and D. Baker. Wallingford, UK: CABI Publishing, pp. 25–44.

Liu, J., Daily, G. C., Ehrlich, P. R. and Luck, G.W. (2003). Effects of household dynamics on resource consumption and biodiversity. *Nature*, 42, 530–533.

Liu, J., Kang, J., Luo, T. *et al.* (2013). Spatiotemporal variability of soundscapes in a multiple functional urban area. *Landscape and Urban Planning*, 115, 1–9.

Liu, X., Li, X., Chen, Y. *et al.* (2010). A new landscape index for quantifying urban expansion using multi-temporal remotely sensed data. *Landscape Ecology*, 25, 671–682.

Lobova, T., Geiselman, C. and Mori, S. (2009). *Seed Dispersal by Bats in the Neotropics*. New York: New York Botanical Garden.

Lockeretz, W., ed. (1987). *Sustaining Agriculture near Cities*. Ankeny, IA: Soil and Water Conservation Society.

Loh, C. (2014). Fighting poverty by healing the environment. In *Creating a Sustainable and Desirable Future: Insights from 45 Global Thought Leaders*, ed. R. Costanza and I. Kubiszewski. London: World Scientific, pp. 293–298.

Lomba, A., Bunce, R. G. H., Jongman, R. H. G. *et al.* (2011). Interactions between abiotic filters, landscape structure and species traits as determinants of dairy farmland plant diversity. *Landscape and Urban Planning*, 99, 248–258.

Long, R. A., Donovan, T. M., MacKay, P. *et al.* (2011). Predicting carnivore occurrence with noninvasive surveys and occupancy modeling. *Landscape Ecology*, 26, 327–340.

Longstreth, R. (1987). *The Buildings of Main Street: A Guide to American Commercial Architecture*. Washington, DC: National Trust for Historic Preservation.

Lopez-Flores, V., MacGregor-Fors, I. and Schondube, J. E. (2009). Artificial nest predation along a neotropical urban gradient. *Landscape and Urban Planning*, 92, 90–95.

Loppi, S. and Pirintsos, S. A. (2000). Effect of dust on epiphytic lichen vegetation in the Mediterranean area (Italy and Greece). *Israel Journal of Plant Sciences*, 48, 91–95.

Loppi, S. and Pirintsos, S. A. (2003). Epiphytic lichens as sentinels for heavy metal pollution at forest ecosystems (central Italy). *Environmental Pollution*, 121, 327–332.

Losch, A. (1944). *Die Raumliche Ordnung der Wirtschaft*, 2nd edition. Jena, Germany: Gustav Fischer.

Loss, S. R. and Marra, P. P. (2017). Population impacts of free-ranging domestic cats on mainland vertebrates. *Frontiers in Ecology and the Environment*, 15, 502–509.

Loss, S. R., Niemi, G. J. and Blair, R. B. (2012). Invasions of non-native earthworms related to population declines of ground-nesting songbirds across a regional extent in northern hardwood forests of North America. *Landscape Ecology*, 27, 683–696.

Loss, S. R., Will, T. and Marra, P. P. (2013). The impact of free-ranging domestic cats on wildlife of the United States. *Nature Communications*, 4, 1–23.

Lourenco, J. M., Quental, N. and Barros, F. (2009). Naturbanization and sustainability at Peneda-Geres National Park. In *Naturbanization: New Identities and Processes for Rural-Natural Areas*, ed. M. J. Prados. Leiden, The Netherlands: CRC Press/Balkema, pp. 45–73.

Lousley, J. E. (1961). A census list of wool aliens found in Britain, 1946–1960. *Proceedings of the Botanical Society of the British Isles*, 4, 221–247.

Louv, R. (2005). *Last Child in the Woods: Saving Our Children from Nature-Deficit Disorder*. Chapel Hill, NC: Algonquin Books.

Lovejoy, T. E. (2005). Conservation with a changing climate. In *Climate Change and Biodiversity*, ed. T. E. Lovejoy and L. Hannah. New Haven, CT: Yale University Press, pp. 325–328.

Lovett, G. M., Traynor, M. M., Pouyat, R. *et al.* (2000). Atmospheric deposition to oak forests along an urban–rural gradient. *Environmental Science and Technology*, 34, 4284–4300.

Lowry, D. A. and McArthur, K. L. (1978). Domestic dogs as predators on deer. *Wildlife Society Bulletin*, 6, 38–39.

Lucarelli, A. and Berg, P. O. (2011). City branding: a state-of-the-art review of the research domain. *Journal of Place Management and Development*, 4, 9–27.

Lucas, M. and Baras, E. (2001). *Migration of Freshwater Fishes*. Malden, MA: Blackwell Science.

Ludwig, J., Tongway, D., Freudenberger, D. *et al.*, eds. (1997). *Landscape Ecology: Function and Management: Principles from Australia's Rangelands*. Collingwood, Vic.: CSIRO Publishing.

Luken, J. O., Hinton, A. C. and Baker, D. G. (1992). Response of woody plant communities in powerline corridors to frequent anthropogenic disturbance. *Ecological Applications*, 2, 356–362.

Lund, H. (2003). Testing the claims of new urbanism: local access, pedestrian travel and neighboring. *Journal of the American Planning Association*, 69, 414–429.

Lundholm, J., Macivor, J. S., MacDougall, Z. and Ranalli, M. (2010). Plant species and functional group combinations affect green roof ecosystem functions. *PLoS ONE*, 5, e9677.

Luther, D. and Baptista, L. (2010). Urban noise and the cultural evolution of bird songs. *Biological Sciences*, 277, 469–473.

Luthy, R. G., Sedlak, D. L., Plumlee, M. H. *et al.* (2015). Wastewater-effluent-dominated streams as ecosystem-management tools in a drier climate. *Frontiers in Ecology and the Environment*, 13, 477–485.

Lynch, K. (1981). *A Theory of Good City Form*. Cambridge, MA: MIT Press.

Ma, K. M., Fu, B. J. and Guo, X. D. (2001). Impact of urbanization in rural areas on plant diversity: a case study in Zunhua City. *Chinese Journal of Applied Ecology,* 12, 837–841.

MacArthur, R. H. and Wilson, E. O. (1967). *The Theory of Island Biogeography*. Princeton, NJ: Princeton University Press.

MacDonald, L. H., Sampson, R. W. and Anderson, D. M. (2001). Runoff and road erosion at the plot and road segment scales, St. John, US Virgin Islands. *Earth Surface Processes and Landforms*, 26, 251–272.

MacFarquhar, N. (2017). Russia's villages, and their culture, are "melting away." *New York Times*, July 30, 6.

MacGregor-Fors, I. and Schondube, J. E. (2011). Gray vs. green urbanization: relative importance of urban features for urban bird communities. *Basic and Applied Ecology*, 12, 372–381.

Maconachie, R. (2007). *Urban Growth and Land Degradation in Developing Cities: Change and Challenges in Kano, Nigeria*. Aldershot, UK: Ashgate Publishing.

Madera, P., Kovar, P., Romportl, D. *et al.* (2014). *Czech Villages in Romanian Banat: Landscape, Nature, and Culture*. Brno, Czech Republic: Mendel University in Brno.

Madre, F., Vergnes, A., Machon, N. and Clergeau, P. (2014). Green roofs as habitats for wild plant species in urban landscapes: first insights from a large-scale sampling. *Landscape and Urban Planning*, 122, 100–107.

Maestas, J. D., Knight, R. L. and Gilgert, W. C. (2003). Biodiversity across a rural land-use gradient. *Conservation Biology*, 24, 36–42.

Magura, T., Horvath, R. and Tothneresz, B. (2010). Effects of urbanization on ground-dwelling spiders in forest patches, in Hungary. *Landscape Ecology*, 25, 621–629.

Mahoney, D. L., Mort, M. A. and Taylor, B. E. (1990). Species richness of calanoid copepods, cladocerans, and other brachiopods in Carolina Bay temporary ponds. *American Midland Naturalist*, 123, 244–258.

Malhotra, K. C., Gokhale, Y., Chatterjee, S. and Srivastav, S. (2007). *Sacred Groves in India*. New Delhi: Aryan Books International.

Mallin, M. A., Ensign, S. H., Wheeler, T. L. and Mayes, D. B. (2002). Pollutant removal efficiency of three wet detention ponds. *Journal of Environmental Quality*, 31, 654–660.

Mandelik, Y., Winfree, R., Neeson, T. and Kremen, C. (2012). Complementary habitat use by wild bees in agro-natural landscapes. *Ecological Applications*, 22, 1535–1546.

Manning, A. D., Gibbons, P. and Lindenmayer, D. B. (2009). Scattered trees: a complementary strategy for facilitating adaptive responses to climate change in modified landscapes? *Journal of Applied Ecology*, 46, 915–919.

Manning, R. E., Ballinger, N. L., Marion, J. and Roggenbuck, J. (1996). Recreation management in natural areas: problems and practices, status and trends. *Natural Areas Journal*, 16, 142–146.

Mannle, K. and Ladle, R. J. (2012). Specific-species taboos and biodiversity conservation in northern Madagascar. In *Sacred Species and Sites: Advances in Biocultural Conservation*, ed. G. Pungetti, G. Oviedo and D. Hooke. New York: Cambridge University Press, pp. 291–304.

Marcarelli, A. M., Van Kirk, R. W. and Baxter, C. V. (2010). Predicting effects of hydrologic alteration and climate change on ecosystem metabolism in a western US river. *Ecological Applications*, 20, 2081–2088.

Marini, L., Fontana, P., Klimek, S. *et al.* (2009). Impact of farm size and topography on plant and insect diversity of managed grasslands in the Alps. *Biological Conservation*, 142, 394–403.

Marks, B. K. and Duncan, R. S. (2009). Use of forest edges by free-ranging cats and dogs in an urban forest fragment. *Southeastern Naturalist*, 8, 427–436.

Marler, P. and Slabbekoorn, H. (2004). *Nature's Music: The Science of Birdsong*. San Diego, CA: Elsevier Academic Press.

Marra, P. P. and Santella, C. (2016). *Cat Wars: The Devastating Consequences of a Cuddly Killer*. Princeton, NJ: Princeton University Press.

Marris, E. (2011). *Rambunctious Garden: Saving Nature in a Post-Wild World*. New York: Bloomsbury.

Marschall, E. A., Mather, M. E., Parrish, D. L. *et al.* (2011). Migration delays caused by anthropogenic barriers: modeling dams, temperature, and success of migrating salmon smolts. *Ecological Applications*, 21, 3014–3031.

Marsh, C. W., Johns, A. D. and Ayers, J. M. (1987). Effects of habitat disturbance on rain forest primates. In *Primate Conservation in the Tropical Rain Forest*, ed. C. W. Marsh and R. A. Mittermeier. New York: Alan Liss, pp. 83–107.

Marsh, G. P. (1864). *Man and Nature*. New York: Charles Scribner.

Marsh, W. M. (2010). *Landscape Planning: Environmental Applications*. New York: John Wiley.

Martins, V. F., Guimaraes, Jr., P. R., Haddad, C. R. B. and Semir, J. (2009). The effect of ants on the seed dispersal cycle of the typical myrmecochorous *Ricinus communis*. *Plant Ecology*, 205, 213–222.

Marzluff, J. M. (2012). Urban evolutionary ecology. In *Urban Bird Ecology and Conservation: Studies in Avian Biology*, ed. C. A. Lepczyk and P. S. Warren. Berkeley, CA: University of California Press, pp. 287–308.

Marzluff, J. M., Schulenberger, E., Endlicher, W. *et al.*, eds. (2008). *Urban Ecology: An International Perspective on the Interaction Between Humans and Nature*. New York: Springer.

Mascaro, J., Harris, J. A., Lach, L. *et al.* (2013). Origins of the novel ecosystems concept. In *Novel Ecosystems: Intervening in the New Ecological World Order*, ed. R. J. Hobbs, E. S. Higgs, and C. M. Hall. Chichester, UK: John Wiley, pp. 45–57.

Maslo, B. and Lockwood, J. L. (2014). *Coastal Conservation*. New York: Cambridge University Press.

Massey, A. (1990). Notes on the reproductive ecology of red-spotted newts (*Notophthalmus viridescens*). *Journal of Herpetology*, 24, 106–107.

Matlack, G. R. (1993). Sociological edge effects: the spatial distribution of human impact in suburban forest fragments. *Environmental Management*, 17, 829–835.

Matos, D. M. S., Santos, C. J. F. and Chevalier, D. de R. (2002). Fire and restoration of the largest urban forest of the world in Rio de Janeiro City, Brazil. *Urban Ecosystems*, 6, 151–161.

Matter, H. C. and Daniels, T. J. (2000). Dog ecology and population biology. In *Dogs: Zoonoses and Public Health*, ed. C. N. L. MacPherson, F. X. Meslin and A. I. Wandeler. London: CABI Publishing, pp. 17–62.

Mayor Farguell, X. (2008). *Connectivitat ecologica: elements teorics, determinacio i aplicacio*. Barcelona: Generalitat de Catalunya, CADS.

Mazaris, A. D., Kallimanis, A. S., Chatzigianidis, G. *et al.* (2009). Spatiotemporal analysis of an acoustic environment: interactions between landscape features and sounds. *Landscape Ecology*, 24, 817–831.

McCarthy, R. J., Levine, S. H., and Reed, J. M. (2013). Estimation of effectiveness of three methods of feral cat population control by use of a simulation model. *Journal of the American Veterinary Medical Association*, 243, 502–511.

McClure, C. J. W., Ware, H. E., Carlisle, J. *et al.* (2013). An experimental investigation into the effects of traffic noise on distributions of birds: avoiding the phantom road. *Proceedings of the Royal Society B*, 280, DOI.10.1098/rspb.2013.2290.

McCormack, G. R., Rock, M., Sandalack, B. and Uribe, F. A. (2011). Access to off-leash parks, street pattern and dog walking among adults. *Public Health*, 125, 540–546.

McDonald, R. I. (2015). *Conservation for Cities: How to Plan and Build Infrastructure*. Washington, DC: Island Press.

McDonald, R. I. (2016). Estimating watershed degradation over the last century and its impact on water-treatment costs for the world's large cities. *Proceedings of the National Academy of Sciences*, 113, 9117–9122.

McDonald, R. I., Forman, R. T. T., Kareiva, P. *et al.* (2009). Urban effects, distance, and protected areas in an urbanizing world. *Landscape and Urban Planning*, 93, 63–75.

McDonald, R. I., Kareiva, P. and Forman, R. T. T. (2008). The implications of urban growth for global protected areas and biodiversity conservation. *Biological Conservation*, 141, 1695–1703.

McDonnell, M. J. and Stiles, E. W. (1983). The structural complexity of old-field vegetation and the recruitment of bird-dispersed plant species. *Oecologia*, 56, 109–116.

McDonnell, M. J., Hahs, A. K. and Breuste, J. H., eds. (2009). *Ecology of Cities and Towns: A Comparative Approach*. New York: Cambridge University Press.

McDonnell, M. J. and Pickett, S. T. A., eds. (1993). *Components of Ecosystems: The Ecology of Subtle Human Effects and the Ecology of Populated Areas*. New York: Springer.

McGranahan, G. and Marcotullio, P. (2005). Ecosystems and human well-being: current state and trends: findings of the condition and trends working group. In *Millennium Ecosystem Assessment*, ed. R. Scholes and N. Ash. Washington, DC: Island Press, pp. 795–825.

McGregor, D., Simon, D. and Thompson, D, eds. (2006). *The Peri-Urban Interface: Approaches to Sustainable Natural and Human Resource Use*. London: Earthscan.

McHarg, I. L. and Sutton, J. (1975). Ecological planning for the Texas coastal plain: the Woodlands New Town Experiment. *Landscape Architecture*, 65, 80–90.

McKenzie, F. (2014). Trajectories of change in rural landscapes: the end of the mixed farm? In *Rural Change in Australia: Population, Economy, Environment*, ed. R. Dufty-Jones and J. Connell. Farnham, UK: Ashgate, pp. 151–167.

McKenzie, P., Cooper, A., McCann, T. and Rogers, D. (2011). The ecological impact of rural building on habitats in an agricultural landscape. *Landscape and Urban Planning*, 101, 262–268.

McKinley, D. C., Ryan, M. G., Birdsey, R. A. *et al.* (2011). A synthesis of current knowledge on forests and carbon storage in the United States. *Ecological Applications*, 21, 1902–1924.

McKinney, M. L. (2010). Urban futures. In *Urban Ecology*, ed. K. J. Gaston. New York: Cambridge University Press, pp. 287–308.

McKinney, R. A., Raposa, K. B. and Cournoyer, R. M. (2011). Wetlands as habitat in urbanizing landscapes: patterns of bird abundance and occupancy. *Landscape and Urban Planning*, 100, 144–152.

McMullen, L. E. and Lytle, D. A. (2012). Quantifying invertebrate resistance to floods: a global-scale meta-analysis. *Ecological Applications*, 22, 2164–2175.

Mehring, A. S. and Levin, L. A. (2015). Potential roles of soil fauna in improving the efficiency of rain gardens used as natural stormwater treatment systems. *Journal of Applied Ecology*, 52, 1445–1454.

Mendes, S., Colino-Rabanal, V. J. and Peris, S. J. (2011). Bird song variations along an urban gradient: the case of the European blackbird (*Turdus merula*). *Landscape and Urban Planning*, 99, 51–57.

Mendoza, S. V., Harvey, C. A., Saenz, J. C. *et al.* (2014). Consistency in bird use of tree cover across tropical agricultural landscapes. *Ecological Applications*, 24, 158–168.

Menzies, N. K. (1994). *Forest and Land Management in Imperial China*. New York: St. Martin's Press.

Merlin, P. (2003). *A Field Guide to Desert Holes*. Tucson, AZ: Arizona-Sonora Desert Museum Press.

Metosi, M. V. (2000). *The Sanitary City: Urban Infrastructure in America from Colonial Times to the Present*. Baltimore, MD: Johns Hopkins University Press.

Meyburg, B.-U., Manowsky, O. and Meyburg, C. (1996). The osprey in Germany: its adaptation to environments altered by man. In *Raptors in Human Landscapes*, ed. D. Bird, D. Valant and J. Negro. San Diego, CA: Academic Press, pp. 125–135.

Meyer, L. A. and Sullivan, S. M. P. (2013). Bright lights, big city: influences of ecological light pollution on reciprocal stream-riparian invertebrate fluxes. *Ecological Applications*, 23, 1322–1330.

Meynecke, J.-O., Lee, S. Y. and Duke, N. C. (2008). Linking spatial metrics and fish catch reveals the importance of coastal wetland connectivity to inshore fisheries in Queensland, Australia. *Biological Conservation*, 141, 981–996.

Michael, D. R., Cunningham, R. B. and Lindenmayer, D. B. (2008). A forgotten habitat? Granite inselbergs conserve reptile diversity in fragmented agricultural landscapes. *Journal of Applied Ecology*, 45, 1742–1752.

Milbourne, P. (2014). Poverty, place, and rurality: material and sociocultural disconnections. *Environment and Planning A*, 46, 566–580.

Millar, C. I., Stephenson, N. L, and Stephens, S. L. (2007). Climate change and forests of the future: managing in the face of uncertainty. *Ecological Applications*, 17, 2145–2151.

Millennium Ecosystem Assessment. (2005). *Ecosystems and Human Well-Being: A Framework for Assessment.* Washington, DC: Island Press.

Miller, J. R. and Hobbs, N. T. (2000). Recreational trails, human activity, and nest predation in lowland riparian areas. *Landscape and Urban Planning*, 50, 227–236.

Miller, S. G., Knight, R. I. and Miller, C. K. (1998). Influence of recreational trails on breeding bird communities. *Ecological Applications*, 8, 162–169.

Miller, S. G., Knight, R. I. and Miller, C. K. (2001). Wildlife responses to pedestrians and dogs. *Wildlife Society Bulletin*, 29, 124–132.

Mills, G. (2004). *The Urban Canopy Layer Heat Island.* International Association for Urban Climate, www.urban-climate.org.

Miltner, R. J., White, D. and Yoder, C. (2004). The biotic integrity of streams in urban and suburbanizing landscapes. *Landscape and Urban Planning*, 69, 87–100.

Milton, S. J., Dean, W. R. J., Sielecki, L. E. and van der Ree, R. (2015). The function and management of roadside vegetation. In *Handbook of Road Ecology*, ed. R. van der Ree, D. J. Smith and C. Grilo. Chichester, UK: Wiley-Blackwell, pp. 373–381.

Mininni, M. (2010). *La costa obliqua: Un atlante per la Puglia [The Oblique Coast: An Atlas for Puglia].* Rome: Donzelli editore.

Minnaar, C., Boyles, J. G., Minnaar, I. A. *et al.* (2015). Stacking the odds: light pollution may shift the balance in an ancient predator-prey arms race. *Journal of Applied Ecology*, 52, 522–531.

Misgav, A., Perl, N. and Avnimelech, Y. (2001). Selecting a compatible open space use for a closed landfill site. *Landscape and Urban Planning*, 55, 95–111.

Mitchell, J. H. (1984). *Ceremonial Time.* Boston, MA: Houghton Mifflin.

Mitsch, W. J. and Gosselink, J. G. (2007). *Wetlands.* New York: John Wiley.

Mitsch, W. J., Cronk, J. K., Wu, X. *et al.* (1995). Phosphorus retention in constructed freshwater riparian marshes. *Ecological Applications*, 5, 830–845.

Mitsch, W. J., Zhang, L., Anderson, C. J. *et al.* (2005). Creating riverine wetlands: ecological succession, nutrient retention, and pulsing effects. *Ecological Engineering*, 25, 510–527.

Mizon, B. (2002). *Light Pollution: Responses and Remedies.* New York: Springer.

Mladjenovic, I. (1998). "Habitat Ecology" IM-192 and IM-32TP as a small town of the future. In *Urban Ecology*, ed. J. Breuste, H. Feldmann and O. Uhlmann. Berlin: Springer, pp. 398–400.

Moffatt, S. F., McLachlan, S. M. and Kenkel, N. C. (2004). Impacts of land use on riparian forest along an urban-rural gradient in southern Manitoba. *Plant Ecology*, 174, 119–135.

Moilanen, A., Anderson, B. J., Eigenbrod, F. *et al.* (2011). Balancing alternative land uses in conservation prioritization. *Ecological Applications*, 21, 1419–1426.

Molebatsi, L. Y., Siebert, S. J., Cilliers, S. S. *et al.* (2010). The Tswara Tshimo: a homegarden system of useful plants with a particular layout and function. *African Journal of Agricultural Research*, 5(21), 2952–2963.

Moles, R., Foley, W. and O'Regan, B. (2006). The environmental impacts of private car transport on the sustainability of Irish settlements. In *The Ecology of Transportation: Managing Mobility for the Environment*, ed. J. Davenport and J. L. Davenport. New York: Springer, pp. 119–164.

Molina, M., McCarthy, J., Wall, D. *et al.* (2014). *What We Know: The Reality, Risks, and Response to Climate Change.* What We Know Initiative. Washington, DC: American Association for the Advancement of Science.

Molsher, R., Newsome, A. and Dickman, C. (1999). Feeding ecology and population dynamics of the feral cat (*Felis catus*) in relation to the availability of prey in central-eastern New South Wales. *Wildlife Research*, 26, 593–607.

Monahan, P. (2017). White lights make birds lose sleep. *Frontiers in Ecology and the Environment*, 15, 286–296.

Montagnini, F. (2006). Home gardens of Mesoamerica: biodiversity, food security, and nutrient management. In *Tropical Home gardens*, ed. B. Komar and P. Nair. Dordrecht: Springer, pp. 61–84.

Montevecchi, W. A. (2006). Influences of artificial light on marine birds. In *Ecological Consequences of Artificial Night Lighting*, ed. C. Rich and T. Longcore. Washington, DC: Island Press, pp. 94–113.

Moran, J. M. and Morgan, M. D. (1994). *Meteorology: The Atmosphere and the Science of Weather.* New York: Macmillan.

Moreira, F., Catry, F. X., Rego, F. and Bacao, F. (2010). Size-dependent pattern of wildfire ignitions in Portugal: when do ignitions turn into big fires? *Landscape Ecology*, 25, 1405–1417.

Morgan, G. T. and King, J. O. (1987). *The Woodlands: New Community Development, 1964–1983*. College Station, TX: Texas A&M University Press.
Morgan, S. A., Hansen, C. M., Ross, J. G. *et al.* (2009). Urban cat (Felis catus) movement and predation activity associated with a wetland reserve in New Zealand. *Wildlife Research*, 36, 574–580.
Morin, P. J. (2011). *Community Ecology*. Oxford: Blackwell Science.
Mortimore, M. (1989). *Adapting to Drought: Farmers, Famines, and Desertification in West Africa*. New York: Cambridge University Press.
Mortimore, M. (2013). Linking biodiversity and poverty alleviation in the drylands – the concept of "useful" biodiversity. In *Biodiversity Conservation and Poverty Alleviation*, ed. D. Roe, J. Elliott, C. Sandbrook and M. Walpole. Chichester, UK: Wiley-Blackwell, pp. 113–126.
Motzkin, G., Eberhardt, R., Hall, B. *et al.* (2002). Vegetation variation across Cape Cod, Massachusetts: environmental and historical determinants. *Journal of Biogeography*, 29, 1439–1454.
Moudon, A. V., ed. (1989). *Master-Planned Communities: Shaping Exurbs in the 1990s*. Seattle, WA: University of Washington, College of Architecture and Urban Planning.
Moustafa, M. Z., Chimney, M. J., Fontaine, T. D. *et al.* (1996). The response of a freshwater wetland to long-term "low level" nutrient loads – marsh efficiency. *Ecological Engineering*, 7, 15–33.
Muhlenbach, J. (1979). Contributions to the synanthropic (adventive) flora of the railroads in St. Louis, Missouri, USA. *Annals of the Missouri Botanical Garden*, 66, 1–108.
Muller, D. K. (2013). Second homes and outdoor recreation: a Swedish perspective on second home use and complementary spaces. In *Second Home Tourism in Europe: Lifestyle Issues and Policy Responses*, ed. Z. Roca. Farnham, UK: Ashgate Publishing, pp. 121–140.
Muller, N. (2010). Most frequently occurring vascular plants and the role of non-native species in urban areas – a comparison of selected cities in the Old and the New Worlds. In *Urban Biodiversity and Design*, ed. N. Muller, P. Werner and J. G. Kelcey. Oxford: Blackwell.
Muller, N. and Werner, P. (2010). Urban biodiversity and the case for implementing the Convention on Biodiversity in towns and cities. In *Urban Biodiversity and Design*. Chichester, UK: Wiley-Blackwell, pp. 3–33.
Muller, N., Werner, P. and Kelcey, J. G., eds. (2010). *Urban Biodiversity and Design*. Chichester, UK: Wiley-Blackwell.
Mumford, L. (1961). *The City in History*. New York: Harcourt.
Munchow, B. and Schramm, M. (1998). Permeable pavements – an appropriate method to reduce stormwater flow in urban sewer systems? In *Urban Ecology*, ed. J. Breuste, H. Feldmann and O. Uhlmann. Berlin: Springer, pp. 183–186.
Munguia, P., Osman, R. W., Hamilton, J. *et al.* (2011). Changes in habitat heterogeneity alter marine sessile benthic communities. *Ecological Applications*, 21, 925–935.
Muriuki, G., Seabrook, L., McAlpine, C. *et al.* (2011). Land cover change under unplanned human settlements: a study of the Chyulu Hills squatters, Kenya. *Landscape and Urban Planning*, 99, 154–165.
Murphy, S. M., Augustine, B. C., Ulrey, W. A. *et al.* (2017a). Consequences of severe habitat fragmentation on density, genetics, and spatial capture-recapture analysis of a small bear population. *PLoS ONE*, 12(7), e0181849.
Murphy, S. M., Ulrey, W. A., Guthrie, J. M. *et al.* (2017b). Food habits of a small Florida black bear population in an endangered ecosystem. *Ursus*, 28, 92–104.
Murray, D., ed. (1986). *Seed Dispersal*. San Diego, CA: Academic Press.
Nagendra, H., Southworth, J. and Tucker, C. (2003). Accessibility as a determinant of landscape transformation in western Honduras: linking pattern and process. *Landscape Ecology*, 18, 141–158.
Nakagoshi, N. and Mabuhay, J. A., eds. (2014). *Designing Low Carbon Societies in Landscapes*. New York: Springer.
Nakagoshi, N. and Ohta, Y. (2001). Landscape dynamics of the orange producing islands in the Seto Inland Sea, Japan. In *Landscape Ecology Applied in Land Evaluation, Development and Conservation*, ed. D. van der Zee and I. S. Zonneveld. Enschede, The Netherlands: International Institute for Aerospace Survey and Earth Sciences (ITC), pp. 95–114.
Nakagoshi, N., Suheri, H. and Amelgia, R. (2014). Community aspects of forest ecosystems in the Guning Gede Pangrango National Park UNESCO Biosphere Reserve, Indonesia. In *Designing Low Carbon Societies in Landscapes*, ed. N. Nakagoshi and J. A. Mabuhay. New York: Springer, pp. 271–287.
Nakamura, E. and Sato, K. (2011). Managing the scarcity of chemical elements. *Nature Materials*, 10, 158–161.

Nams, V. O. (2012). Shape of patch edges affects edge permeability for meadow voles. *Ecological Applications*, 22, 1827–1837.

Nassauer, J. I. (2011). Care and stewardship: from home to planet. *Landscape and Urban Planning*, 100, 321–323.

Nathan, R., Schurr, F. M., Spiegel, O. *et al.* (2008). Mechanisms of long-distance seed dispersal. *Trends in Ecology and Evolution*, 23, 638–647.

National Research Council. (2005). *The Science of Instream Flows: A Review of the Texas Instream Field Program.* Washington, DC: National Academies Press.

Natuhara, Y. (2008). Evaluation and planning of wildlife habitat in urban landscape. In *Landscape Ecological Applications in Man-Influenced Areas: Linking Man and Nature Systems*, ed. S.-K. Hong, N. Nakagoshi, B. Fu and Y. Morimoto. New York: Springer, pp. 129–147.

Ndubisi, F. O., ed. (2014). *The Ecological Design and Planning Reader*. Washington, DC: Island Press.

Neely, C. L. and Hatfield, R. (2007). Livestock systems. In *Farming with Nature: The Science and Practice of Ecoagriculture*, ed. S. J. Scherr and J. A. McNeely. Washington, DC: Island Press, pp. 121–142.

Nefedova, T. and Pallot, J. (2013). The multiplicity of second home development in the Russian Federation: a case of "seasonal suburbanization"? In *Second Home Tourism in Europe: Lifestyle Issues and Policy Responses*, ed. Z. Roca. Farnham, UK: Ashgate Publishing, pp. 91–119.

Nemeth, E. and Brumm, H. (2009). Blackbirds sing higher-pitched songs in cities: adaptation to habitat acoustics or side-effect of urbanization. *Animal Behaviour*, 78, 637–641.

Newman, J. M., Clausen, J. C. and Neafsey, J. A. (2000). Seasonal performance of a wetland constructed to process dairy milkhouse wastewater in Connecticut. *Ecological Engineering*, 14, 181–198.

Nielsen, A., Trolle, D., Sondergaard, M. *et al.* (2012). Watershed land use effects on lake water quality in Denmark. *Ecological Applications*, 22, 1187–1200.

Niemela, J., Breuste, J. H., Elmqvist, T. *et al.*, eds. (2011). *Urban Ecology: Patterns, Processes, and Applications*. New York: Oxford University Press.

Nieminen, M., Ketoia, E., Mikola, J. *et al.* (2011). Local land use effects and regional environmental limits on earthworm communities in Finnish arable landscapes. *Ecological Applications*, 21, 3162–3177.

Nightingale, B., Longcore, T. and Simenstad, C. A. (2006). Artificial night lighting and fishes. In *Ecological Consequences of Artificial Night Lighting*, ed. C. Rich and T. Longcore. Washington, DC: Island Press, pp. 257–276.

Nilsson, J. (1995). A phosphorus budget for a Swedish municipality. *Journal of Environmental Management*, 45, 243–253.

Nogales, M., Martin, A., Tershy, B. R. *et al.* (2004). A review of feral cat eradication on islands. *Conservation Biology*, 18, 310–319.

Nolen, J. (1927). *New Towns for Old: Achievements in Civic Improvement in Some American Small Towns and Neighborhoods*. Boston, MA: Marshall Jones Company. (2005 edition; Amherst, MA: University of Massachusetts Press.)

Nordahl, D. (2009). *Public Produce: The New Urban Agriculture*. Washington, DC: Island Press.

Northam, R. M. (1979). *Urban Geography*. New York: John Wiley.

Nowak, D. J. (2010). Urban biodiversity and climate change. In *Urban Biodiversity and Design*, ed. N. Muller, P. Werner and J. G. Kelcey. Chichester, UK: Wiley-Blackwell, pp. 101–117.

Nowak, D. J., Crane, D. E. and Stevens, J. C. (2006). Air pollution removal by urban trees and shrubs in the United States. *Urban Forestry and Urban Greening*, 4, 115–123.

Nowak, D. J., Hoehn, R. E., Crane, D. E. *et al.* (2008). A ground-based method of assessing urban forest structure and ecosystem services. *Arboriculture and Urban Forestry*, 34, 347–358.

Nunez-Iturri, G., Olsson, O. and Howe, H. F. (2008). Hunting reduces recruitment of primate-dispersed trees in Amazonian Peru. *Biological Conservation*, 141, 1536–1546.

Ode, A. and Fry, G. (2006). A model for quantifying and predicting urban pressure on woodland. *Landscape and Urban Planning*, 77, 17–27.

Odell, E. and Knight, R. L. (2001). Songbird and medium-sized mammal communities associated with exurban development in Pitkin County, Colorado. *Conservation Biology*, 15, 1143–1150.

Oke, T. R. (1987). *Boundary Layer Climates*. New York: Methuen.

Olupot, W. (2009). A variable edge effect on trees of Bwindi Impenetrable National Park, Uganda, and its bearing on measurement parameters. *Biological Conservation*, 142, 789–797.

Oneal, A. S. and Rotenberry, J. T. (2009). Scale-dependent habitat relations of birds in riparian corridors in an urbanizing landscape. *Landscape and Urban Planning*, 92, 264–275.

Opdam, P. (1991). Metapopulation theory and habitat fragmentation: a review of Holarctic breeding bird studies. *Landscape Ecology*, 5, 93–106.

Opdam, P. and Wascher, D. (2004). Climate change meets habitat fragmentation: linking landscape and biogeographical scale levels in research and conservation. *Biological Conservation*, 117, 285–297.

Opdam, P., Steingrover, E. and Rooij, S. (2006). Ecological networks: a spatial concept for multi-actor planning of sustainable landscapes. *Landscape and Urban Planning,* 75, 322–332.

2015 Open Space and Recreation Plan. (2015). Town of Concord, Massachusetts, USA.

Orlowski, G. and Nowak, L. (2007). The importance of marginal habitats for the conservation of old trees in agricultural landscapes. *Landscape and Urban Planning,* 79, 77–83.

Ormsby, A. (2012). Cultural and conservation values of sacred forests in Ghana. In *Sacred Species and Sites: Advances in Biocultural Conservation*, ed. G. Pungetti, G. Oviedo and D. Hooke. New York: Cambridge University Press, pp. 335–350.

Osborn, F. J. and Whittick, A. (1963). *The New Towns: The Answer to Megalopolis*. London: Leonard Hill.

Osborne, J. L., Martin, A. P., Shortall, C. R. *et al.* (2008). Quantifying and comparing bumblebee nest densities in gardens and countryside habitats. *Journal of Applied Ecology*, 45, 784–792.

Outen, A R. (2002). The ecological effects of road lighting. In *Wildlife and Roads: The Ecological Effect*, ed. B. Sherwood, D. Cutler and J. A. Burton. London: Imperial College Press, pp. 133–155.

Owen, J. (1991). *The Ecology of a Garden: The First Fifteen Years.* Cambridge, UK: Cambridge University Press.

Owen, J. (2012). *Wildlife of a Garden: A Thirty Year Study*. London: Royal Horticultural Society.

Oxley, D. J., Fenton, M. B. and Carmody, G. R. (1974). The effects of roads on populations of small mammals. *Journal of Applied Ecology*, 11, 51–59.

Ozinga, W. A., Schaminee, J. H. J., Bekker, R. M. *et al.* (2005). Predictability of plant species composition from environmental conditions is constrained by dispersal limitation. *Oikos*, 108, 555–561.

Pacione, M. (2005). *Urban Geography: A Global Perspective*. New York: Routledge.

Palmeira, F. B. L., Crawshaw, Jr., P. G., Haddad, C. M. *et al.* (2008). Cattle depredation by puma (*Puma concolor*) and jaguar (*Panthera onca*) in central-western Brazil. *Biological Conservation*, 141, 118–125.

Palmer, J. R. (1984). The naturalization of oil-milling adventive plants in the Thames estuary. *London Naturalist*, 63, 68–70.

Palomino, D. and Carrascal, L. M. (2007). Habitat associations of a raptor community in a mosaic landscape of central Spain under urban development. *Landscape and Urban Planning*, 83, 268–274.

Palta, M. M., Grimm, N. B. and Groffman, P. M. (2017). "Accidental" urban wetlands: ecosystem functions in unexpected places. *Frontiers in Ecology and the Environment*, 15, 248–256.

Panetta, F. D. and Hopkins, A. J. M. (1991). Weeds in corridors: invasion and management. In *Nature Conservation 2: The Role of Corridors*, ed. D. A. Saunders and R. J. Hobbs. Chipping Norton, NSW: Surrey Beatty, pp. 341–351.

Pardo, L. H., Fenn, M. E., Goodale, C. L. *et al.* (2011). Effects of nitrogen deposition and empirical nitrogen critical loads for ecoregions of the United States. *Ecological Applications*, 21, 3049–3082.

Park, B. J., Tsunetsugu, Y., Kasetani, T. *et al.* (2010). Physiological effects of *Shinrin-yoku*: evidence from field experiments in 24 forests across Japan. *Environmental Health and Preventive Medicine*, 15, 18–26.

Parker, J. W. (1996). Urban ecology of the Mississippi kite. In *Raptors in Human Landscapes*, ed. D. Bird, D. Varland and J. Negro. San Diego, CA: Academic Press, pp. 45–52.

Parris, K. M. (2015). Ecological impacts of road noise and options for mitigation. In *Handbook of Road Ecology*, ed. R. van der Ree, D. J. Smith and C. Grilo. Chichester, UK: Wiley-Blackwell, pp. 151–158.

Parris, K. M. and McCarthy, M. A. (2013). Predicting the effect of urban noise on the active space of avian vocal signals. *American Naturalist*, 182, 452–464.

Parris, K. M., Velik-Lord, M. and North, J. M. A. (2009). Frogs call at a higher pitch in traffic noise. *Ecology and Society*, 14(1), 1–24.

Parry, L., Barlow, J. and Peres, C. A. (2009). Allocation of hunting effort by Amazonian smallholders: implications for conserving wildlife in mixed-use landscapes. *Biological Conservation*, 142, 1777–1786.

Parsons, H., French, K. and Major, R. E. (2003). The influence of remnant bushland on the composition of suburban bird assemblages in Australia. *Landscape and Urban Planning*, 66, 43–56.

Parsons, K. C. and Schuyler, D. (2000). *From Garden City to Greencity: The Legacy of Ebenezer Howard.* Baltimore, MD: Johns Hopkins University Press.

Pataki, D. E., McCarthy, H. R., Litvak, E. and Pincetl, S. (2011). Transpiration of urban forests in the Los Angeles metropolitan area. *Ecological Applications*, 21, 661–677.

Patino, J., Hylander, K. and Gonzalez-Mancero, J. M. (2010). Effect of forest clear-cutting on subtropical bryophyte communities in waterfalls, on dripping walls, and along streams. *Ecological Applications*, 20, 1648–1663.

Paton, D., Romero, F., Cuenca, J. and Escudero, J. C. (2012). Tolerance to noise in 91 bird species from 27 urban gardens of Iberian peninsula. *Landscape and Urban Planning*, 104, 1–8.

Paton, J. A. and Sheahan, M. C. (2013). *Lophocolea brookwoodiana* (Jungermanniales: Geocalycaceae), a new species in Britain. *Journal of Bryology*, 28, 163–166.

Patten, M. A. and Bolger, D. T. (2003). Variation in top-down control in avian reproductive success across a fragmentation gradient. *Oikos*, 101, 479–488.

Paul, M. J. and Meyer, J. L. (2001). Streams in the urban landscape. *Annual Review of Ecology and Systematics*, 32, 333–365.

Paul, M. J. and Meyer, J. L. (2008). Streams in the urban landscape. In *Urban Ecology: An International Perspective on the Interaction Between Humans and Nature*, ed. J. M. Marzluff, E. Schulenberger, W. Endlicher *et al.* New York: Springer, pp. 207–231.

Pearce-Higgins, J. W., Stephen, L., Langston, R. H. W. *et al.* (2009). The distribution of breeding birds around upland wind farms. *Journal of Applied Ecology*, 46, 1323–1331.

Pearson, O. P. (1964). Carnivore-mouse predation: an example of its intensity and bioenergetics. *Journal of Mammalogy*, 45, 177–188.

Pejchar, L., Pringle, R. M., Ranganathan, J. *et al.* (2008). Birds as agents of seed dispersal in a human-dominated landscape in southern Costa Rica. *Biological Conservation*, 141, 536–544.

Pejchar, L., Reed, S. E., Bixler, P. *et al.* (2015). Consequences of residential development for biodiversity and human well-being. *Frontiers in Ecology and the Environment*, 13, 146–153.

Pelletier, F. (2006). Effects of tourist activities on ungulate behaviour in a mountain protected area. *Journal of Mountain Ecology*, 8, 15–19.

Pellissier, V., Cohen, M., Boulay, A. and Clergeau, P. (2012). Birds are also sensitive to landscape composition and configuration within the city centre. *Landscape and Urban Planning*, 104, 181–188.

Pellissier, V., Roze, F. and Clergeau, P. (2010). Constraints of urbanization on vegetation dynamics in a growing city: a chronological framework in Rennes (France). In *Urban Biodiversity and Design*, ed. N. Muller, P. Werner and J. G. Kelcey. Chichester, UK: Wiley-Blackwell, pp. 206–226.

Perkin, J. S. and Gido, K. B. (2012). Fragmentation alters stream fish community structure in dendritic ecological networks. *Ecological Applications*, 22, 2176–2187.

Perrin, L. and Grant, J. L. (2014). Perspectives on mixing housing types in the suburbs. *Town Planning Review*, 85, 363–385.

Perry, D. A., Oren, R. and Hart, S. C. (2008a). *Forest Ecosystems*. Baltimore, MD: Johns Hopkins University Press.

Perry, G. and Fisher, R. N. (2006). Night lights and reptiles: observed and potential effects. In *Ecological Consequences of Artificial Night Lighting,* ed. C. Rich and T. Longcore. Washington, DC: Island Press, pp. 169–191.

Perry, G., Buchanan, B. W., Fisher, R. N. *et al.* (2008b). Effects of artificial night lighting on amphibians and reptiles in urban environments. In *Urban Herpetology: Herpetological Conservation*, ed. J. C. Mitchell, R. E. Jung Brown and B. Bartholomew. Volume 3. Salt Lake City, UT: Society for the Study of Amphibians and Reptiles, pp. 239–256.

Perry, T. and Nawaz, R. (2008). An investigation into the extent and impacts of hard surfacing of domestic gardens in an area of Leeds, United Kingdom. *Landscape and Urban Planning*, 86, 1–13.

Pesout, P. and Hosek, M. (2014). An ecological network in the Czech Republic. *Ochrona Prirody (Nature Conservation)*, 69 Supplement, 2–9.

Peterken, G. F. (1993). *Woodland Conservation and Management*. London: Chapman & Hall.

Peterken, G. F. (1996). *Natural Woodland: Ecology and Conservation in Northern Temperate Regions*. New York: Cambridge University Press.

Peterken, G. F. (2013). *Meadows*. Totnes, UK: British Wildlife Publishing.

Peterson, J. T. (1981). Game farming and inter-ethnic relations in northeastern Luzon, Philippines. *Human Ecology*, 9, 1–22.

Petrie, W. M. F. (1891). *Illahun, Kahun and Gurob, 1889–90*. London: David Nutt.

Pezzoli, K. (1998). *Human Settlements and Planning for Ecological Sustainability*. Cambridge, MA: MIT Press.

Phillips, L. E. (1993). *Urban Trees: A Guide for Selection, Maintenance, and Master Planning*. New York: McGraw-Hill.

Pickett, S. T. A., Cadenasso, M. L. and McGrath, B., eds. (2013). *Resilience in Ecology and Urban Design: Linking Theory and Practice for Sustainable Cities*. New York: Springer.

Pickett, S. T. A., Power, M. E., Collins, S. L. *et al.* (2015). Earth stewardship: an initiative by the Ecological Society of America to foster engagement to sustain planet earth. In *Earth Stewardship: Linking Ecology and Ethics in Theory and Practice*, ed. R. Rozzi, F. S. Chapin III, J. B. Callicott *et al.* Washington, DC: Ecological Society of America, pp. 173–184.

Pijanowski, B. C., Farina, A., Gage, S. H. *et al.* (2011a). What is soundscape ecology? An introduction and overview of an emerging new science. *Landscape Ecology*, 26, 1213–1232.

Pijanowski, B. C., Villanueva-Rivera, L. J., Dumyahn, S. L. *et al.* (2011b). Soundscape ecology: the science of sound in the landscape. *BioScience*, 61, 203–216.

Pilkey, O. H. and Dixon, K. L. (1996). *The Corps and the Shore*. Washington, DC: Island Press.

Pimentel, D., Zuniga, R. and Morrison, D. (2005). Update on the environmental and economic costs associated with alien-invasive species in the United States. *Ecological Economics*, 52, 273–288.

Pineda, F. D., Schmitz, M. F., de Aranzabal, I. *et al.* (2011). *Connectividad Ecologico Territorial. Estudio de casos de connectividad ecologica y socioecological*. O. A. Parques Nacionales. Madrid: Ministerio de Medio Ambiente y Medio Rural y Marina.

Piper, S. D. and Catterall, C. P. (2006). Impacts of picnic areas on bird assemblages and nest predation activity within Australian eucalypt forests. *Landscape and Urban Planning*, 78, 251–262.

Platt, R. H. (2004). *Land Use and Society: Geography, Law, and Public Policy*. Washington, DC: Island Press.

Polasky, S., Nelson, E., Camm, J. *et al.* (2008). Where to put things? Spatial land management to sustain biodiversity and economic returns. *Biological Conservation*, 141, 1505–1524.

Pollalis, S., Zofnass, P. and Zofnass, J. (2016). *Planning Sustainable Cities: An Infrastructure Based Approach*. London: Routledge.

Porter, E. E., Bulluck, J. and Blair, R. B. (2005). Multiple spatial-scale assessment of the conservation value of golf courses for breeding birds in southwestern Ohio. *Wildlife Society Bulletin*, 33, 494–506.

Potts, S. G., Woodcock, B. A., Roberts, S. P. M. *et al.* (2009). Enhancing pollinator biodiversity in intensive grasslands. *Journal of Applied Ecology*, 46, 369–379.

Poulsen, J. R., Clark, C. J. and Bolker, B. M. (2011). Decoupling the effects of logging and hunting on an Afrotropical animal community. *Ecological Applications*, 21, 1819–1836.

Powe, N. A. and Gunn, S. (2008). Housing development in market towns: making a success of "local service centres"? *Town Planning Review*, 79, 125–148.

Powe, N. and Hart, T. (2007). Market town characteristics. In *Market Towns: Roles, Challenges and Prospects*, ed. N. Powe, T. Hart and T. Shaw. New York: Routledge, pp. 11–26.

Powe, N., Hart, T. and Shaw, T. (2007). *Market Towns: Roles, Challenges and Prospects*. New York: Routledge.

Powe, N., Pringle, R. and Hart, T. (2015). Matching the process to the challenge within small town regeneration. *Town Planning Review*, 86, 177–202.

Prados, M. J., ed. (2009a). *Naturbanization: New Identities and Processes for Rural-Natural Areas*. Boca Raton, FL: CRC Press.

Prados, M. J. (2009b). Conceptual and methodological framework of naturbanization. In *Naturbanization: New Identities and Processes for Rural-Natural Areas*, ed. M. J. Prados. Boca Raton, FL: CRC Press, pp. 11–28.

Prasad, A. M., Iverson, L. R., Peters, M. P. *et al.* (2010). Modeling the invasive emerald ash borer risk of spread using a spatially explicit cellular model. *Landscape Ecology*, 25, 353–369.

Prevett, P. T. (1991). Movement paths of koalas in the urban-rural fringes of Ballarat, Victoria: implications for management. In *Nature Conservation 2: The Role of Corridors*, ed. D. A. Saunders and R. J. Hobbs. Chipping Norton, NSW: Surrey Beatty, pp. 259–272.

Price, L. G., Johnson, P. S. and Bland, D., eds. (2007). *Water Resources of the Middle Rio Grande: San Acacia to Elephant Butte*. Socorro, NM: New Mexico Bureau of Geology and Mineral Resources.

Primack, R. B. (2004). *A Primer of Conservation Biology*. Sunderland, MA: Sinauer Associates.

Prist, P. R., Michalski, F. and Metzger, J. P. (2012). How deforestation pattern in the Amazon influences vertebrate richness and community composition. *Landscape Ecology*, 27, 799–812.

Pronello, C. (2003). The measurement of train noise: a case study in northern Italy. *Transportation Research, Part D*, 8, 113–128.

Proppe, D. S., Sturdy, C. S. and Cassady St. Clair, C. (2013). Anthropogenic noise decreases urban songbird diversity and may contribute to homogenization. *Global Climate Biology*, 19, 1075–1084.

Pulliam, H. R. and Danielson, B. J. (1991). Sources, sinks, and habitat selections: a landscape perspective on population dynamics. *American Naturalist*, 137 (Supplements), 50–66.

Pungetti, G., Oviedo, G. and Hooke, D., eds. (2012). *Sacred Species and Sites: Advances in Biocultural Conservation*. New York: Cambridge University Press.

Purvis, O. W. (2010). Lichens and industrial pollution. In *Ecology of Industrial Pollution*, ed. L. C. Batty and K. B. Hallberg. New York: Cambridge University Press, pp. 41–69.

Pysek, P. and Hejny, S. (1995). Flora and vegetation of various types of settlements in the Czech Republic: a concise comparison. In *Urban Ecology as the Basis of Urban Planning*, ed. H. Sukopp, M. Numata and A. Huber. Amsterdam: SPB Academic Publishing, pp. 151–160.

Pysek, P. and Mandak, B. (1997). Fifteen years of changes in the representation of alien species in Czech village flora. In *Plant Invasions: Studies from North America and Europe*, ed. J. H. Brock, M. Wade, P. Pysek and D. Green. Leiden, Belgium: Backhuys Publishers, pp. 183–190.

Pysek, P. and Pysek, A. (1990). Comparison of the vegetation and flora of the West Bohemian villages and towns. In *Urban Ecology: Plants and Plant Communities in Urban Environments*, ed. H. Sukopp, H. Heijny and I. Kowarik. The Hague, The Netherlands: SPB Academic Publishing, pp. 105–112.

Qian, J., Qian, L. and Zhu, H. (2012). Subjectivity, modernity, and the politics of difference in a periurban village in China: towards a progressive sense of place? *Environment and Planning D: Society and Space*, 30, 1064–1082.

Quinn, T. (1995). Using public sightings to investigate coyote use of urban habitat. *Journal of Wildlife Management*, 59, 238–245.

Rabanal, L. I., Kuehl, H. S., Mundry, R. *et al.* (2010). Oil prospecting and its impact on large rainforest mammals in Loango National Park, Gabon. *Biological Conservation*, 143, 1017–1024.

Rabeni, C. F. and Sowa, S. P. (2002). A landscape approach to managing the biota of streams. In *Integrating Landscape Ecology into Natural Resource Management*, ed. J. Liu and W. W. Taylor. New York: Cambridge University Press, pp. 114–142.

Rabin, L. A., McGowan, B., Hooper, S. L. and Owings, D. H. (2003). Anthropogenic noise and its effect on communication: an interface between comparative psychology and conservation biology. *International Journal of Comparative Psychology*, 16, 172–192.

Raciti, S. M., Groffman, P. M., Jenkins, J. C. *et al.* (2011). Nitrate production and availability in residential soils. *Ecological Applications*, 21, 2357–2366.

Rackham, O. (1976). *Trees and Woodland in the British Landscape*. London: J. M. Dent & Sons.

Rackham, O. (1986). *The History of the Countryside*. London: J. M. Dent.

Radeloff, V. C., Stewart, S. I. and Hawbaker, T. J. (2010). Housing growth in and near United States protected areas limits their conservation value. *Proceedings of the National Academy of Sciences (USA)*, 107, 940–945.

Raeymaekers, J. A. M., Raeymaekers, D., Kolzumi, I. *et al.* (2009). Guidelines for restoring connectivity around water mills: a population genetic approach to the management of riverine fish. *Journal of Applied Ecology*, 46, 562–571.

Raheem, D. C., Naggs, F., Preece, R. C. *et al.* (2008). Structure and conservation of Sri Lankan land-snail assemblages in fragmented lowland rainforest and village home gardens. *Journal of Applied Ecology*, 45, 1019–1028.

Rahman, M. L., Tarrant, S., McCollin, D. and Ollerton, J. (2012). Influence of habitat quality, landscape structure and food resources on breeding skylark (*Alauda arvensis*) territory distribution on restored landfill sites. *Landscape and Urban Planning*, 105, 281–285.

Raikow, D. F., Walters, D. M., Fritz, K. M. and Mills, M. A. (2011). The distance that contaminated aquatic subsidies extend into lake riparian zones. *Ecological Applications*, 21, 983–990.

Rajvanshi, A. (2016). Cumulative effects of dams on biodiversity. In *Handbook on Biodiversity and Ecosystem Services in Impact Assessment*, ed. D. Geneletti. Cheltenham, UK: Edward Elgar Publishers, pp. 321–341.

Rajvanshi, A., Mathur, V. B., Teleki, G. C. and Mukherjee, S. K. (2001). *Roads, Sensitive Habitats and Wildlife*. Derahdun, India: Wildlife Institute of India.

Rapoport, E. H. (1993). The process of plant colonization in small settlements and large cities. In *Humans as Components of Ecosystems: The Ecology of Subtle Human Effects and Populated Areas*, ed. M. J. McDonnell and S. T. A. Pickett. New York: Springer, pp. 190–207.

Rascio, N. and Navari-Izzo, F. (2011). Heavy metal hyperaccumulating plants: how and why do they do it? And what makes them so interesting? *Plant Science*, 180, 169–181.

Ratajczak, W. (2013). The role of small and medium-sized towns in local and regional economies. In *Towns in a Rural World*, ed. T. de N. Vaz, E. van Leeuwen and P. Nijkamp. Farnham, UK: Ashgate, pp. 25–65.

Raybould, A. F. and Gray, A. J. (1993). Genetically modified crops and hybridization with wild relatives: a UK perspective. *Journal of Applied Ecology*, 30, 199–219.
Reck, H. and van der Ree, R. (2015). Insects, snails and spiders: the role of invertebrates in road ecology. In *Handbook of Road Ecology*, ed. R. van der Ree, D. J. Smith and C. Grilo. Chichester, UK: Wiley-Blackwell, pp. 247–257.
Reddy, K. R., Kadlec, R. H., Chuimney, M. J. and Mitsch, W. J., eds. (2006). *The Everglades Nutrient Removal Project*. Special Issue of *Ecological Engineering*, 27, 265–379.
Reed, S. E. and Merenlender, A. M. (2011). Effects of management of domestic dogs and recreation on carnivores in protected areas in northern California. *Conservation Biology*, 25, 504–513.
Rees, M., Roe, J. H. and Georges, A. (2009). Life in the suburbs: behavior and survival of a freshwater turtle in response to drought and urbanization. *Biological Conservation*, 142, 3172–3181.
Rees, W. E. and Wackernagel, M. (1996). *Our Ecological Footprint: Reducing Human Impact on the Earth*. Philadelphia, PA: New Society Books.
Rees, W. E. and Wackernagel, M. (2008). Urban ecological footprints: why cities cannot be sustainable – and why they are a key to sustainability. In *Urban Ecology: An International Perspective on the Interaction Between Humans and Nature*, ed. J. M. Marzluff, E. Schulenberger, W. Endlicher *et al*. New York: Springer, pp. 537–555.
Reeves, F. (2014). *Planet Heart: How an Unhealthy Environment Leads to Heart Disease*. Vancouver, BC: Greystone Books.
Rego, F. C., Bunting, S. C., Strand, E. K. and Godinho-Ferreira, P. (2018). *Applied Landscape Ecology*. New York: John Wiley.
Reijnen, R. and Foppen, R. (1995). The effects of car traffic on breeding bird populations. IV. Influence of population size on the reduction of density of woodland breeding birds. *Journal of Applied Ecology*, 32, 481–491.
Reijnen, R. and Foppen, R. (2006). Impact of road traffic on breeding bird populations. In *The Ecology of Transportation: Managing Mobility for the Environment*, ed. J. Davenport and J. L. Davenport. New York: Springer, pp. 255–274.
Reino, L., Beja, P., Osborne, P. E. *et al*. (2009). Distance to edges, edge contrast and landscape fragmentation: interactions affecting farmland birds around forest plantations. *Biological Conservation*, 142, 824–838.
Reps, J. W. (1965). *The Making of Urban America: A History of City Planning in the United States*. Princeton, NJ: Princeton University Press.
Reps, J. W. (1969). *Town Planning in Frontier America*. Princeton, NJ: Princeton University Press.
Rich, C. and Longcore, T., eds. (2006). *Ecological Consequences of Artificial Night Lighting*. Washington, DC: Island Press.
Richardson, J. L. and Vepraskas, M. J. (2001). *Wetland Soils: Genesis, Hydrology, Landscapes, and Classification*. Boca Raton, FL: Lewis Publishing.
Richman, T., Worth, J., Dawe, P. *et al*. (1997). *Start at the Source: Residential Site Planning and Design Guidance Manual for Stormwater Quality Protection*. Los Angeles, CA: Bay Area Stormwater Management Agency Association.
Richmond, O. M. W., Tecklin, J. and Beissinger, S. R. (2012). Impact of cattle grazing on the occupancy of a cryptic threatened rail. *Ecological Applications*, 22, 1655–1664.
Riffell, S. K., Gutzwiller, K. J. and Anderson, S. H. (1996). Does repeated human intrusion cause cumulative declines in avian richness and abundance? *Ecological Applications*, 6, 492–505.
Rifkind, C. (1977). *Main Street: The Face of Urban America*. New York: Harper Colophon Books.
Riley, S. P. D., Pollinger, J. P., Sauvajot, R. M. *et al*. (2006). A southern California freeway is a physical and social barrier to gene flow in carnivores. *Molecular Ecology*, 15, 1733–1741.
Rist, J., Rowcliffe, M., Cowlishaw, G. and Milner-Gulland, E. J. (2008). Evaluating measures of hunting effort in a bushmeat system. *Biological Conservation*, 141, 2086–2099.
Riva-Murray, K., Riemann, R., Murdoch, P. *et al*. (2010). Landscape characteristics affecting streams in urbanizing regions of the Delaware River Basin (New Jersey, New York, and Pennsylvania, US). *Landscape Ecology*, 25, 1489–1503.
Roast, S., Gannicliffe, T., Ashton, D. K. *et al*. (2010). An ecological risk assessment framework for assessing risks from contaminated land in England and Wales. In *Ecology of Industrial Pollution*, ed. L. C. Batty and K. B. Hallberg. New York: Cambridge University Press, pp. 189–204.
Roberts, B. K. (1982). *Village Plans*. Aylesbury, UK: Shire Publications.
Robinson, W. H. (1996). *Urban Entomology: Insect and Mite Pests in the Human Environment*. London: Chapman & Hall.

Roca, Z., ed. (2013). *Second Home Tourism in Europe: Lifestyle Issues and Policy Responses*. Farnham, UK: Ashgate Publishing.

Rockstrom, J., Falkenmark, M., Folke, C. *et al.* (2014). *Water Resilience for Human Prosperity*. New York: Cambridge University Press.

Rodewald, A. D. and Bakermans, M. H. (2006). What is the appropriate paradigm for riparian forest conservation? *Biological Conservation*, 128, 193–200.

Rodewald, A. D., Kearns, L. J. and Shustack, D. P. (2011). Anthropogenic resource subsidies decouple predator-prey relationships. *Ecological Applications*, 21, 936–943.

Rodriguez, A., Andren, H. and Jansson, G. (2001). Habitat-mediated predation risk and decision making of small birds at forest edges. *Oikos*, 95, 383–396.

Roe, D., Elliott, J., Sandbrook, C. and Walpole, M., eds. (2013). *Biodiversity Conservation and Poverty Alleviation*. Chichester, UK: Wiley-Blackwell.

Roger, E., Bino, G. and Ramp, D. (2012). Linking habitat suitability and road mortalities across geographic ranges. *Landscape Ecology*, 27, 1167–1181.

Rohde, S., Hostmann, M., Peter, A. and Ewald, K. C. (2006). Room for rivers: an integrative search strategy for floodplain restoration. *Landscape and Urban Planning*, 78, 50–70.

Romportl, D., Havlicek, M., Chuman, T. *et al.* (2014). The development of land use and landscape structure. In *Czech Villages in Romanian Banat: Landscape, Nature, and Culture*, ed. P. Madera, P. Kovar, D. Romportl *et al.* Brno, Czech Republic: Mendel University in Brno, pp. 277–291.

Rood, S. B., Braatne, J. H. and Goater, L. A. (2010). Favorable fragmentation: river reservoirs can impede downstream expansion of riparian weeds. *Ecological Applications*, 20, 1664–1677.

Rooney, R. C., Bayley, S. E., Creed, I. F. and Wilson, M. J. (2012). The accuracy of land cover-based wetland assessments is influenced by landscape extent. *Landscape Ecology*, 27, 1321–1335.

Rosenfield, R. N., Bielefeldt, J., Affeldt, J. L. and Bechmann, D. J. (1996). Urban nesting biology of Cooper's hawks in Wisconsin. In *Raptors in Human Landscapes*, ed. D. Bird, D. Varland and J. Negro. San Diego, CA: Academic Press, pp. 41–44.

Rosi-Marshall, E. J., Kincaid, D. W., Bechtold, H. A. *et al.* (2013). Pharmaceuticals suppress algal growth and microbial respiration and alter bacterial communities in stream biofilms. *Ecological Applications*, 23, 583–593.

Ross, N. (2011). Modern tree species composition reflects ancient Maya "forest gardens" in northwest Belize. *Ecological Applications*, 21, 75–84.

Ross, T. E. (1987). A comprehensive bibliography of the Carolina Bays literature. *Journal of the Elisha Mitchell Scientific Society*, 103, 28–42.

Rousseau, D. P. L., Vanrolleghem, P. A. and Pauw, N. de M. (2004). Model-based design of horizontal subsurface flow constructed treatment wetlands: a review. *Water Reources*, 38, 1484–1493.

Rout, T. K., Ebhin Masto, R., Padhy, P. K. *et al.* (2014). Heavy metals in dusts from commercial and residential areas of Jharia coal mining town. *Environmental Earth Sciences*, 73, 347–359.

Royal Commission. (1991). *Shoreline Regeneration for the Greater Toronto Bioregion*. Toronto, ON: The Royal Commission on the Future of the Toronto Waterfront.

Rozelle, S., Brandt, L., Guo, L. and Huang, J. (2005). Land tenure in China: facts, fictions and issues. In *Developmental Dilemmas: Land Reform and Traditional Change in China*, ed. P. Ho. New York: Routledge, pp. 107–132.

Rubin, H. D. and Beck, A. M. (1982). Ecological behavior of free-ranging urban pet dogs. *Applied Animal Ecology*, 8, 161–168.

Rudd, H., Vala, J. and Schaefer, V. (2002). Importance of backyard habitat in a comprehensive biodiversity conservation strategy: a connectivity analysis of urban green spaces. *Restoration Ecology*, 10, 368–375.

Rundlof, M., Bengtsson, J. and Smith, H. G. (2008). Local and landscape effects of organic farming on butterfly species richness and abundance. *Journal of Applied Ecology*, 45, 813–820.

Rudolf, S. C., Kienast, F. and Hersperger, A. M. (2017). Planning for compact urban forms: local growth-management approaches and their evolution over time. *Journal of Environmental Planning and Management*, 61, 474–492.

Rushton, B. T. (2001). Low-impact parking lot design reduces runoff and pollutant loads. *Journal of Water Resources Planning and Management*, 127, 172–179.

Russell, E. W. (1961). *Soil Conditions and Plant Growth*. London: Longmans.

Russell, K. N., Ikerd, H. and Droege, S. (2005). The potential conservation value of unmowed powerline strips for native bees. *Biological Conservation*, 124, 133–148.

Russo, D. G. (1998). *Town Origins and Development in Early England c. 400–950 A.D.* London: Greenwood Press.

Rydell, J. (2006). Bats and their insect prey at streetlights. In *Ecological Consequences of Artificial Night Lighting*, ed. C. Rich and T. Longcore. Washington, DC: Island Press, pp. 43–60.

Rytwinski, T. and Fahrig, L. (2015). The impacts of roads and traffic on terrestrial animal populations. In *Handbook of Road Ecology*, ed. R. van der Ree, D. J. Smith and C. Grilo. Chichester, UK: Wiley-Blackwell, pp. 237–246.

Sacchi, R., Gentilli, A., Razzetti, E. and Barbieri, F. (2002). Effects of building features on density and flock distribution of feral pigeons, *Columba livia* var. *domestica*, in an urban environment. *Canadian Journal of Zoology*, 80, 48–54.

Sackett, P. D. (2014). Endangered elements: conserving the building blocks of life. In *Creating a Sustainable and Desirable Future; Insights from 45 Global Thought Leaders*, ed. R. Costanza and I. Kubiszewski. Hackensack, NJ: World Scientific, pp. 239–247.

Saeed, T. and Sun, G. (2012). A review on nitrogen and organics removal mechanisms in subsurface flow constructed wetlands: dependency on environmental parameters, operating conditions and supporting media. *Journal of Environmental Management*, 112, 429–448.

Sakamoto, K. (1988). *Remnant Forms of Ulmaceae Woods and Trees in Urban Areas*. Monograph 2. Bulletin of Revegetation Research. Kyoto, Japan: Association of Revegetation Research.

Salek, M., Svobodova, J. and Zasadil, P. (2010). Edge effect of low-traffic forest roads on bird communities in secondary production forests in central Europe. *Landscape Ecology*, 25, 1113–1124.

Salisbury, A., Armitage, J., Bostock, H. *et al.* (2015). Enhancing gardens as habitats for flower-visiting aerial insects (pollinators): should we plant native or exotic species? *Journal of Applied Ecology*, 52, 1156–1164.

Salmon, M. (2006). Protecting sea turtles from artificial night lighting at Florida's oceanic beaches. In *Ecological Consequences of Artificial Night Lighting*, ed. C. Rich and T. Longcore. Washington, DC: Island Press, pp. 141–168.

Salvador, R., Bautista-Capetillo, C. and Playan, E. (2011). Irrigation performance in private urban landscapes: a study case in Zaragoza (Spain). *Landscape and Urban Planning*, 100, 302–311.

Salvati, L., Munafo, M., Morelli, V. G. and Sabbi, A. (2012). Low-density settlements and land use changes in a Mediterranean urban region. *Landscape and Urban Planning*, 105, 43–52.

Samnegard, U., Persson, A. S. and Smith, H. G. (2011). Gardens benefit bees and enhance pollination in intensively managed farmland. *Biological Conservation*, 144, 2602–2606.

Samson, F. R. and Knopf, F. L., eds. (1996). *Prairie Conservation: Preserving North America's Most Endangered Ecosystem*. Washington, DC: Island Press.

Sande, S. O., Crewe, R. M., Raina, S. K. *et al.* (2009). Proximity to a forest leads to higher honey yield: another reason to conserve. *Biological Conservation*, 142, 2703–2709.

Sangay, T. and Vernes, K. (2008). Human-wildlife conflict in the Kingdom of Bhutan: patterns of livestock predation by large mammalian carnivores. *Biological Conservation*, 141, 1272–1282.

Sannucks, P. and Balkenhol, N. (2015). Incorporating landscape genetics into road ecology. In *Handbook of Road Ecology*, ed. R. van der Ree, D. J. Smith and C. Grilo. Chichester, UK: Wiley-Blackwell, pp. 110–118.

Sannucks, P. and Taylor, A. C. (2008). The application of genetic markers to landscape management. In *Landscape Analysis and Visualization: Spatial Models for Natural Resource Management and Planning*, ed. C. Pettit, W. Cartwright, J. Bishop *et al.* New York: Springer, pp. 211–234.

Santos, K. C., Pino, J., Roda, F. *et al.* (2008). Beyond the reserves: the role of non-protected rural areas for avifauna conservation in the area of Barcelona (NE of Spain). *Landscape and Urban Planning*, 84, 140–151.

Sarapatka, B., Niggli, U., Cizkova, S. *et al.* (2012). *Agriculture and Landscape: The Way to Mutual Harmony*. Olomouc, Czech Republic: Palacky University.

Sargent, F. O., Lusk, P., Rivera, J. A. and Varela, M. (1991). *Rural Environmental Planning for Sustainable Communities*. Washington, DC: Island Press.

Sarlov-Herlin, I. L. and Fry, G. L. A. (2000). Dispersal of woody plants in forest edges and hedgerows in a northern Sweden agricultural area: the role of site and landscape structure. *Landscape Ecology*, 15, 229–242.

Sass, C. K. and Keane, T. D. (2016). Riparian corridor change in a northeastern Kansas watershed. *Landscape Journal*, 35, 57–77.

Sathessan, S. M. (1996). Raptors associated with airports and aircraft. In *Raptors in Human Landscapes*, ed. D. Bird, D. Varland and J. Negro. San Diego, CA: Academic Press, pp. 315–323.

Satterthwaite, D. (2006). Small urban centres and large villages: the habitat for much of the world's low-income population. In *The Earthscan Reader in Rural-Urban Linkages*, ed. C. Tacoli. London: Earthscan, pp. 15–38.

Sattler, T., Obrist, M. K., Duelli, P. and Moretti, M. (2011). Urban arthropod communities: added value or just a blend of surrounding biodiversity? *Landscape and Urban Planning*, 103, 347–361.

Saunders, D. A., Arnold, G. W., Burbidge, A. A. and Hopkins, A. J. M., eds. (1987). *Nature Conservation: The Role of Remnants of Native Vegetation*. Chipping Norton, NSW: Surrey Beatty.

Saunders, D. A., Hopkins, A. J. M. and How, R. A., eds. (1990). *Australian Ecosystems: 200 Years of Utilization, Degradation and Reconstruction*. Chipping Norton, NSW: Surrey Beatty and Ecological Society of Australia.

Savory, A. (1999). *Holistic Management: A New Framework for Decision Making*. Washington, DC: Island Press.

Sayre, N. F., McAllister, R. R. J., Bestelmeyer, B. T. *et al.* (2013). Earth stewardship of rangelands: coping with ecological, economic, and political marginality. *Frontiers in Ecology and the Environment*, 11, 348–354.

Scarlett, L. and McKinney, M. (2016). Connecting people and places: the emerging role of network governance in large landscape conservation. *Frontiers in Ecology and the Environment*, 14, 116–125.

Schaafsma, J. A., Baldwin, A. H. and Streb, C. A. (2000). An evaluation of a constructed wetland to treat wastewater from a dairy farm in Maryland, USA. *Ecological Engineering*, 14, 199–206.

Schaetzl, R. and Sanderson, S. (2005). *Soils: Genesis and Geomorphology*. New York: Cambridge University Press.

Scheibe, M. A. (1999). Uber die Attraktivitat von Strassenbeleuchtungen auf Insekten aus nahegelegenen Gewassern unter Berucksichtigung unterschiedlicher UV-Emission der Lampen [On the attractiveness of roadway lighting to insects from nearby water with consideration of the different UV emission of the lamps]. *Natur und Landschaft*, 74, 144–146.

Scheifele, P. M., Browning, D. G. and Collins-Scheifele, L. M. (2003). Analysis and effectiveness of deer whistles for motor vehicles: frequencies, levels and animal threshold responses. *Acoustics Research Letters Online*, 4, 71–76.

Scherr, S. J. and McNeely, J. A., eds. (2007). *Farming with Nature: The Science and Practice of Ecoagriculture*. Washington, DC: Island Press.

Scheuerell, M. D. and Schindler, D. E. (2004). Changes in the spatial distribution of fishes in lakes along a residential development gradient. *Ecosystems*, 7, 98–106.

Schindelbeck, R. R., van Es, H. M., Abawi, G. S. *et al.* (2008). Comprehensive assessment of soil quality for landscape and urban management. *Landscape and Urban Planning*, 88, 73–80.

Schmid, J. A. (1975). Urban vegetation: a review and Chicago case study. Department of Geography Research Paper 161, University of Chicago.

Schmiedel, J. (2001). Auswirkungen kunstlicher Beleuchtung auf die Tierwelt – ein Uberblick [Effects of artificial lighting on the animal world: an overview]. *Schriftenreibe fur Landschaftspflege und Naturschutz*, 67, 19–51.

Schmitz, M. F., Pineda, F. D., Castro, H. *et al.* (2005). *Cultural Landscape and Socioeconomic Structure: Environmental Value and Demand for Tourism in a Mediterranean Territory*. Sevilla, Spain: Junta de Andalucia.

Scholz, M. and Grabowiecki, P. (2007). Review of permeable pavement systems. *Building and Environment*, 42, 3830–3836.

Schwarz, N. (2010). Urban form revisited – selecting indicators for characterising European cities. *Landscape and Urban Planning*, 96, 29–47.

Sears, A. R. and Anderson, S. H. (1991). Correlation between birds and vegetation in Cheyenne, Wyoming. In *Wildlife Conservation in Metropolitan Environments*, ed. L. W. Adams and D. L. Leedy. Columbia, MD: National Institute for Urban Wildlife, pp. 75–80.

Seaward, M. R. D. (1979). Lower plants and the urban landscape. *Urban Ecology*, 4, 217–225.

Seiler, A. (2004). Trends and spatial patterns in ungulate–vehicle collisions in Sweden. *Wildlife Biology*, 10, 301–313.

Selva, N., Switalski, A., Kreft, S. and Ibisch, P. L. (2015). Why keep areas road-free? The importance of roadless areas. In *Handbook of Road Ecology*, ed. R. van der Ree, D. J. Smith and C. Grilo. Chichester, UK: Wiley-Blackwell, pp. 16–26.

Semlitsch, R. D. (1987). Relationship of pond drying to the reproductive success of the salamander, *Ambystoma talpoideum*. *Copeia*, 1987, 61–69.

Senbel, M., Giratalla, W., Zhang, K. and Kissinger, M. (2014). Compact development without transit: life-cycle GHG emissions from four variations of residential density in Vancouver. *Environment and Planning A*, 46, 1226–1243.

Serpell, J. (1996). *In the Company of Animals: A Study of Human-Animal Relationships*. New York: Cambridge University Press.

Serrano, L. and Toja, J. (1995). Limnological description of four temporary ponds in the Donano National Park (Spain). *Archiv fur Hydrobiologie*, 113, 497–516.

Serret, H., Raymond, R., Foltete, J.-C. *et al.* (2014). Potential contributions of green spaces at business sites to the ecological network in an urban agglomeration: the case of the Ile-de-France Region, France. *Landscape and Urban Planning*, 131, 27–35.

Sewell, J. (2009). *The Shape of the Suburbs: Understanding Toronto's Sprawl*. Toronto: University of Toronto Press.

Sewell, S. R. and Catterall, C. P. (1998). Bushland modification and styles of urban development: their effects on birds in south-east Queensland. *Wildlife Research*, 25, 41–63.

Shanahan, P. (2009). Groundwater in the urban environment. In *The Water Environment of Cities*, ed. L. A. Baker. New York: Springer, pp. 29–48.

Shandas, V. and Alberti, M. (2009). Exploring the role of vegetation fragmentation on aquatic conditions: linking upland with riparian areas in Puget Sound lowland streams. *Landscape and Urban Planning*, 90, 66–75.

Shannon, G., Cordes, L. S., Hardy, A. R. *et al.* (2014). Behavioral responses associated with a human-mediated predator shelter. *PLoS ONE*, 9, e94630.

Sharma-Laden, P. and Thompson, P. (2014). Raising gross national happiness through agroforestry. In *Creating a Sustainable and Desirable Future: Insights from 45 global thought leaders*, ed. R. Costanza and I. Kubiszewski. London: World Scientific, pp. 305–308.

Sharp, T. (1946). *The Anatomy of the Village*. Harmondsworth, UK: Penguin Books.

Shashua-Bar, L., Pearlmutter, D. and Erell, E. (2009). The cooling efficiency of urban landscape strategies in a hot dry climate. *Landscape and Urban Planning*, 92, 179–186.

Shaw, R. (2014). Streetlighting in England and Wales: new technologies and uncertainty in the assemblage of streetlighting infrastructure. *Environment and Planning A*, 46, 2228–2242.

Shedlock, R. J. (1993). Interactions between ground water and wetlands: southern shore of Lake Michigan, USA. *Journal of Hydrology*, 141, 127–153.

Sheldon, F., Peterson, E. E., Boone, E. L. *et al.* (2012). Identifying the spatial scale of land use that most strongly influences overall river ecosystem health score. *Ecological Applications*, 22, 2188–2203.

Sheridan, M. J. and Nyamweru, C., eds. (2007). *African Sacred Groves: Ecological Dynamics and Social Change*. Athens, OH: Ohio University Press.

Shindhe, K. C. (2006). The national highway bypass around Hubli-Dharwad and its impact on peri-urban livelihoods. In *The Peri-Urban Interface: Approaches to Sustainable Natural and Human Resource Use*, ed. D. McGregor, D. Simon and D. Thompson. London: Earthscan, pp. 181–195.

Shueler, T. (2000). Microbes and urban watersheds: concentrations, sources, and pathways. *Watershed Protection Techniques*, 3, 1–12.

Sieghardt, M., Mursch-Radlgruber, E., Paoletti, E. *et al.* (2005). The abiotic urban environment: impact of urban growing conditions on urban vegetation. In *Urban Forests and Trees*, ed. C. C. Konijnendijk, K. Nilsson, T. B. Randrup and J. Schipperijn. New York: Springer, pp. 281–323.

Siemers, B. M. and Schaub, A. (2011). Hunting at the highway: traffic noise reduces efficiency in acoustic predators. *Proceedings of the Royal Society B*, 278, 1646–1652.

Silva Lucas, P., Gomes de Carvalho, R. and Grilo, C. (2018). Railway disturbances on wildlife types, effects, and mitigation measures. In *Railway Ecology*, ed. L. Borda-de-Agua, R. Barrientos, P. Beja and H. M. Pereira. New York: Springer.

Silva-Rodriguez, E. A. and Sieving, K. E. (2012). Domestic dogs shape the landscape-scale distribution of a threatened forest ungulate. *Biological Conservation*, 150, 103–110.

Silverman, R. (2015). Remnants of a failed utopia. *National Geographic*, December, 140–142.

Simonen, K. (2014). *Life Cycle Assessment*. New York: Routledge.

Sims, W., Evans, K., Newson, S. E. *et al.* (2007). Avian assemblage structure and domestic cat densities in urban environments. *Diversity and Distributions*, 14, 387–399.

Sinclair, C. J., Boxall, A. B. A., Parsons, S. A. and Thomas, M. (2006). Prioritization of pesticide environmental transformation products in drinking water supplies. *Environmental Science and Technology*, 40, 7283–7289.

Slabbekoorn, H. and den Boer-Visser, A. (2006). Cities change the songs of birds. *Current Biology*, 16, 2326–2331.

Slabbekoorn, H. and Peet, M. (2003). Birds sing at a higher pitch in urban noise. *Nature*, 424, 267.

Slabbekoorn, H. and Smith, T. B. (2002). Habitat-dependent song divergence in the Little Greenbul: an analysis of environmental selection pressures on acoustic signals. *Evolution*, 56, 1849–1858.

Smailes, A. E. (1957). *The Geography of Towns*. London: Hutchinson & Co.

Smallbone, L. T., Luck, G. W. and Wassens, S. (2011). Anuran species in urban landscapes: relationships with biophysical, built environment and socio-economic factors. *Landscape and Urban Planning*, 101, 43–51.

Smallwood, S. K., Nakamoto, B. J. and Geng, S. (1996). Association analysis of raptors on a farming landscape. In *Raptors in Human Landscapes*, ed. D. Bird, D. Valant and J. Negro. San Diego, CA: Academic Press, pp. 177–190.

Smith, D. A. (2008). The spatial patterns of indigenous wildlife use in western Panama: implications for conservation management. *Biological Conservation*, 141, 925–937.

Smith, D. J., van der Ree, R. and Rosell, C. (2015). Wildlife crossing structures: an effective strategy to restore or maintain wildlife connectivity across roads. In *Handbook of Road Ecology*, ed. R. van der Ree, D. J. Smith and C. Grilo. Chichester, UK: Wiley-Blackwell, pp. 172–183.

Smith, E. F. and Pritchard, B. (2014). Water reform in the 21st century: the changed status of Australian agriculture. In *Rural Change in Australia: Population, Economy, Environment*, ed. R. Dufty-Jones and J. Connell. Farnham, UK: Ashgate Publishing, pp. 169–186.

Smith, G. C. and Carlile, N. (1993). Food and feeding ecology of breeding silver gulls (*Larus novaehollandiae*) in urban Australia. *Colonial Waterbirds*, 16, 9–17.

Smith, J., Potts, S. G., Woodcock, B. A. and Eggleton, P. (2008). Can arable field margins be managed to enhance their biodiversity, conservation and functional value for soil macrofauna? *Journal of Applied Ecology*, 45, 269–278.

Smith, R. M., Gaston, K. J., Warren, P. H. and Thompson, K. (2005). Urban domestic gardens (V): relationships between landcover composition, housing and landscape. *Landscape Ecology*, 20, 235–253.

Smith, R. M., Thompson, K., Hodgson, J. G. *et al.* (2006). Urban domestic gardens (IX): composition and richness of the vascular plant flora, and implications for native biodiversity. *Biological Conservation*, 129, 312–322.

Smith, W. H. (1981). *Air Pollution and Forests: Interaction Between Air Contaminants and Forest Ecosystems*. New York: Springer.

Snep, R. P. H., WallisDeVries, M. F. and Opdam, P. (2011). Conservation where people work: a role for business districts and industrial areas in enhancing endangered butterfly populations? *Landscape and Urban Planning*, 103, 94–101.

Soloman, D., Lehmann, J., Fraser, J. A. *et al.* (2016). Indigenous African soil enrichment as a climate-smart sustainable agriculture alternative. *Frontiers in Ecology and the Environment*, 14, 71–76.

Solntsiev, V. N. (1974). O niekotorikn fundamentalnykh svoistakh gheosistemnoi struktury. In *Methody kompleksnykh issledovanii gheosistem*. Irkutsk, Soviet Union: Akademya Nauk SSSR.

Sorace, A. and Gustin, M. (2009). Distribution of generalist and specialist predators along urban gradients. *Landscape and Urban Planning*, 90, 111–118.

Sorace, A. and Visentin, M. (2007). Avian diversity on golf courses and surrounding landscapes in Italy. *Landscape and Urban Planning*, 81, 81–90.

Soule, M. E. (1991). Wildlife planning and wildlife maintenance: guidelines for conserving wildlife in an urban landscape. *Journal of the American Planning Association*, 57, 313–323.

Sow, N. A. (2001). Range inventory and evaluation for domestic livestock and wildlife: a case study in Mali around Djoumara (Kaarta). In *Landscape Ecology Applied in Land Evaluation, Development and Conservation: Some Worldwide Selected Examples*, ed. D. van der Zee and I. S. Zonneveld. Enschede, The Netherlands: International Institute for Aerospace Survey and Earth Sciences (ITC), pp. 255–272.

Spengler, J. D., Samet, J. M. and McCarthy, J. F., eds. (2001). *Indoor Air Quality Handbook*. New York: McGraw-Hill.

Sperling, D. and Gordon, D. (2009). *Two Billion Cars: Driving Toward Sustainability*. New York: Oxford University Press.

Spier, L., Van Dobben, H. and Van Dort, K. (2010). Is bark pH more important than tree species in determining the composition of nitrophytic or acidophytic lichen floras? *Environmental Pollution*, 158, 3607–3611.

Spilkova, J. and Sefrna, L. (2010). Uncoordinated new retail development and its impact on land use and soils: a pilot study on the urban fringe of Prague, Czech Republic. *Landscape and Urban Planning*, 84, 141–148.

Sproule, W. J. and Neumann, E. (1991). The Morgantown PRT: it is still running at West Virginia University. *Journal of Advances in Transportation*, 25, 269–279.

Spurr, S. H. and Barnes, B. V. (1980). *Forest Ecology*. New York: John Wiley.

Stankowich, T. (2008). Ungulate flight response to human disturbance: a review and meta-analysis. *Biological Conservation*, 141, 2159–2173.

Stanley, M. C., Beggs, J. R., Bassett, I. E. *et al.* (2015). Emerging threats in urban ecosystems: a horizon scanning exercise. *Frontiers in Ecology and the Environment,* 13, 553–560.

Stavins, R. N., ed. (2012). *Economics of the Environment: Selected Readings*. Sixth edition. New York: Norton.

Steele, M. K. and Heffernan, J. B. (2014). Morphological characteristics of urban water bodies: mechanisms of change and implications for ecosystem function. *Ecological Applications*, 24, 1070–1084.

Steiner, F. (2002). *Human Ecology: Following Nature's Lead*. Washington, DC: Island Press.

Steiner, F., Young, G. and Zube, E. (1988). Ecological planning: retrospect and prospect. In *The Ecological Planning and Design Reader* (2014), ed. F. O. Ndubisi. Washington, DC: Island Press, pp. 72–90.

Steiner, G. R. and Freeman, R. J., Jr. (1989). Configuration and substrate design considerations for constructed wetlands for wastewater treatment. In *Constructed Wetlands for Wastewater Treatment*, ed. D. A. Hammer. Chelsea, MI: Lewis Publishers, pp. 363–378.

Steiner, W. A. (1994). The influence of air pollution on moss-dwelling animals. 1. Methodology and the composition of flora and fauna. *Revue Suisse de Zoologie*, 101, 533–556.

Steinitz, C. and McDowell, S. (2001). Alternative futures for Monroe County, Pennsylvania: a case study in applying ecological principles. In *Applying Ecological Principles to Land Management*, ed. V. H. Dale and R. A. Haeuber. New York: Springer, pp. 165–193.

Stelzer, R. J., Chittka, L., Carlton, M. and Ings, T. C. (2010). Winter active bumblebees (*Bombus terrestris*) achieve high foraging rates in urban Britain. *PLoS ONE*, 5, e9559.

Sterl, P., Brandenburg, C. and Arnberger, A. (2008). Visitors' awareness and assessment of recreational disturbance of wildlife in the Donau-Auen National Park. *Journal for Nature Conservation,* 16, 135–145.

Steven, R., Pickering, C. and Castley, J. G. (2011). A review of the impacts of nature-based recreation on birds. *Journal of Environmental Management,* 92, 2287–2294.

Stevenson, I. (1980). The diffusion of disaster: the phylloxera outbreak in the department of the Herault, 1862–1880. *Journal of Historical Geography*, 6, 47–63.

Stewart, D. and Streishinsky, T. (1990). Nothing goes to waste in Arcata's teeming marshes. *Smithsonian*, 21, 174–179.

Stewart, I. D. (2011). A systematic review and scientific critique of methodology in modern urban heat island literature. *International Journal of Climatology*, 31, 200–217.

Stewig, R. (1987). *Untersuchungen uber die Kleinstadt in Schleswig-Holstein*. Kiel, Germany: Im Selbstverlag des Geographischen Instituts der Universitat Kiel.

Stilgoe, J. R. (1988). *Borderland: Origins of the American Suburb, 1820–1939*. New Haven, CT: Yale University Press.

Stinner, D., Stinner, B. and Martsolf, E. (1997). Biodiversity as an organizing principle in agroecosystem management: case studies of holistic resource management practitioners in the USA. *Agriculture, Ecosystems and Environment*, 62, 199–213.

Stone, E. L., Harris, S. and Jones, G. (2015). Impacts of artificial lighting on bats. *Mammalian Biology*, 80, 213–219.

Stone, E. L., Jones, G. and Harris, S. (2012). Conserving energy at a cost to biodiversity? Impacts of LED lighting on bats. *Global Change Biology*, 18, 2458–2465.

Storm, G. L., Andrews, R. D., Phillips, R. L. *et al.* (1976). Morphology, reproduction, dispersal and mortality of Midwestern red fox populations. *Wildlife Monographs*, 49, 5–82.

Struebig, M. J., Kingston, T., Zubaid, A. *et al.* (2009). Conservation importance of limestone karst outcrops for palaeotropical bats in a fragmented landscape. *Biological Conservation*, 142, 2089–2096.

Styers, D. M., Chappelka, A. H., Marzen, L. J. and Somers, G. L. (2010). Developing a land-cover classification to select indicators of forest ecosystem health in a rapidly urbanizing landscape. *Landscape and Urban Planning*, 94, 158–165.

Suarez-Rubin, M., Lookingbill, T. R. and Wainger, L. A. (2012). Modeling exurban development near Washington, DC, USA: comparison of a pattern-based model and a spatially-explicit econometric model. *Landscape Ecology*, 27, 1045–1061.

Sukopp, H. (2008). The city as a subject for ecological research. In *Urban Ecology: An International Perspective on the Interaction Between Humans and Nature*, ed. J. M. Marzluff, E. Schulenberger, W. Endlicher *et al.* New York: Springer, pp. 281–298.

Sundermann, A., Stoll, S. and Haase, P. (2011). River restoration success depends on the species pool of the immediate surroundings. *Ecological Applications*, 21, 1962–1971.

Sutherland, W. J., Bailey, M. J., Bainbridge, I. P. *et al.* (2008). Future novel threats and opportunities facing UK biodiversity identified by horizon scanning. *Journal of Applied Ecology*, 45, 821–833.

Suzan, G. and Ceballos, G. (2005). The role of feral mammals on wildlife infectious disease prevalence in two nature reserves within Mexico City limits. *Journal of Zoo and Wildlife Medicine*, 36, 479–484.

Swain, H. M. and Martin, P. A. (2014). Saving the Florida scrub ecosystem: translating science into conservation action. In *Conservation Catalysts: The Academy as Nature's Agent*, ed. J. N. Levitt. Cambridge, MA: Lincoln Institute of Land Policy, pp. 63–74.

Swan, L. W. (1992). The aeolian biome. *BioScience*, 42, 262–270.

Swanson, F. J. and Chapin III, F. W., (2009). Forest systems: living with long-term change. In *Principles of Ecosystem Stewardship: Resilience-Based Natural Resource Management in a Changing World*. New York: Springer, pp. 149–170.

Swetnam, R. D. (2007). Rural land use in England and Wales between 1930 and 1998: mapping trajectories of change with a high resolution spatio-temporal dataset. *Landscape and Urban Planning*, 81, 91–103.

Tacoli, C., ed. (2006). *The Earthscan Reader in Rural-Urban Linkages*. London: Earthscan.

Takatsuki, S. (2009). Effects of sika deer on vegetation in Japan: a review. *Biological Conservation*, 142, 1922–1929.

Tanner, C. C. (1996). Plants for constructed wetland treatment systems – a comparison of the growth and nutrient uptake of eight emergent species. *Ecological Engineering*, 7, 59–83.

Tarrant, S., Ollerton, J., Rahman, M. L. *et al.* (2013). Grassland restoration on landfill sites in the East Midlands, United Kingdom: an evaluation of floral resources and pollinating insects: flowers and pollinating insects on restored landfills. *Restoration Ecology,* 21, 560–568.

Tarutis, W. J., Stark, L. R. and Williams, F. M. (1999). Sizing and performance estimation of coal mine drainage wetlands. *Ecological Engineering*, 12, 353–372.

Taylor, A. R. and Knight, R. L. (2003). Wildlife responses to recreation and associated visitor perceptions. *Ecological Applications*, 13, 951–963.

Taylor, B. E., Wyngaard, G. A. and Mahoney, D. L. (1990). Hatching of *Diaptomus stagnalis* eggs from a temporary pond after a prolonged dry period. *Archiv fur Hydrobiologie*, 117, 271–278.

Taylor, J. (2014). Difference or equality? Settlement dilemmas on the Indigenous Estate. In *Rural Change in Australia: Population, Economy, Environment*, ed. R. Dufty-Jones and J. Connell. Farnham, UK: Ashgate Publishing, pp. 103–125.

Taylor, S. L., Roberts, S. C., Walsh, C. J. and Hatt, B. E. (2004). Catchment urbanization and increased benthic algal biomass in streams: linking mechanisms to management. *Freshwater Biology*, 49, 835–851.

Temby, I. D. (2004). Urban wildlife issues in Australia. In *Proceedings of the 4th International Symposium on Urban Wildlife Conservation*. Tucson, AZ, pp. 26–34.

Tennent, J. and Downs, C. T. (2008). Abundance and home ranges of feral cats in an urban conservancy where there is supplemental feeding: a case study from South Africa. *African Zoology*, 43, 218–229.

Terrasa-Soler, J. J. (2006). Landscape change and ecological corridors in Puerto Rico: towards a master plan of ecological networks. *Acta Cientifica*, 20, 57–62.

Thaitakoo, D., McGrath, B., Srithanyarat, S. and Palopakan, Y. (2013). Bangkok: the ecology and design of an aqua-city. In *Resilience in Ecology and Urban Design: Linking Theory and Practice*, ed. S. T. A. Pickett, M. L. Cadenasso and B. McGrath. New York: Springer, pp. 289–297.

Thayer, R. L., Jr. (2003). *Life Place: Bioregional Thought and Practice*. Berkeley, CA: University of California Press.

Theobald, D. M. (2001). Land-use dynamics beyond the American urban fringe. *Geographical Review*, 91, 544–564.

Theobald, D. M. (2005). Landscape patterns of exurban growth in the USA from 1980 to 2020. *Ecology and Society*, 10(1), 32. DOI.10.5751/ES-01390-100132.

Theobald, D. M. and Romme, W. H. (2007). Expansion of the US wildland-urban interface. *Landscape and Urban Planning*, 83, 340–354.

Theobald, D. M., Miller, J. R. and Hobbs, N. T. (1997). Estimating the cumulative effects of development on wildlife habitat. *Landscape and Urban Planning*, 39, 25–36.

Thiere, G., Milenkovski, S., Lindgren, P.-E. *et al.* (2009). Wetland creation in agricultural landscapes: biodiversity benefits on local and regional scales. *Biological Conservation*, 142, 964–973.

Thomas, E. and White, J. T. (1980). *Hedgerow*. New York: William Morrow and Co.

Thomas, K., Kvitek, R. G. and Bretz, C. (2003). Effects of human activity on the foraging behavior of sanderlings *Calidris alba*. *Biological Conservation*, 109, 67–71.

Thomas, R. L., Fellowes, M. D. E. and Baker, P. J. (2012). Spatio-temporal variation in predation by urban domestic cats (*Felis catus*) and the acceptability of possible management actions in the UK. *PLoS ONE*, 7, 1–13.

Thompson, J., Lambert, K. F., Foster, D. *et al.* (2014). *Changes to the Land: Four Scenarios for the Future of the Massachusetts Landscape*. Petersham, MA: Harvard Forest, Harvard University.

Thompson, J. R., Lambert, K. F., Foster, D. R. *et al.* (2016). The consequences of four land-use scenarios for forest ecosystems and the services they provide. *Ecosphere*, 7, e01469.

Thompson, J. W. and Solvig, K. (2000). *Sustainable Landscape Construction: A Guide to Green Building Outdoors*. Washington, DC: Island Press.

Thompson, K., Colsell, S., Carpenter, J. *et al.* (2005). Urban domestic gardens (VII): a preliminary survey of soil seed banks. *Seed Science Research*, 15, 133–141.

Thompson, K., Hodgson, J. G., Smith, R. M. *et al.* (2004). Urban domestic gardens (III): composition and diversity of lawn floras. *Journal of Vegetation Science*, 15, 373–378.

Thornbeck, D. (2012). *Rural Design: A New Design Discipline*. New York: Routledge.

Thrupp, L. A. (2000). Linking agricultural biodiversity and food security: the valuable role of agrobiodiversity for sustainable agriculture. *International Affairs*, 76, 283–297.

Thurlow, C. (1983). *Improving Street Climate through Urban Design*. New York: American Planning Association, Planning Advisory Service.

Tiffney, B. (2004). Vertebrate dispersal of seed plants through time. *Annual Review of Ecology, Evolution, and Systematics*, 35, 1–29.

Tilghman, N. G. (1987). Characteristics of urban woodlands affecting breeding bird diversity and abundance. *Landscape and Urban Planning*, 14, 481–495.

Tivy, J. (1990). *Agricultural Ecology*. Harlow, UK: Longman.

Todd, D. K. and Mays, L. W. (2005). *Groundwater Hydrology*. New York: John Wiley.

Toner, G. C. (1956). House cat predation on small animals. *Journal of Mammalogy*, 37(1), 119.

Tonosaki, K., Kawai, S. and Tokoro, K. (2014). Cooling potential of urban green spaces in summer. In *Designing Low Carbon Societies in Landscapes*, ed. N. Nakagoshi and J. A. Mabuhay. New York: Springer, pp. 15–34.

Torres, A., Jaeger, J. A. G. and Alonso, J. C. (2016). Assessing large-scale wildlife responses to human infrastructure development. *Proceedings of the National Academy of Science*, 113, 8472–8477.

Tratalos, J., Fuller, R. A., Evans, K. L. *et al.* (2007). Bird densities are associated with household densities. *Global Climate Change*, 13, 1685–1695.

Tremblay, M. A. and St. Clair, C. C. (2009). Factors affecting the permeability of transportation and riparian corridors to the movements of songbirds in an urban landscape. *Journal of Applied Ecology*, 46, 1314–1322.

Trenkel, V. M., Ressler, P. H., Jech, M. *et al.* (2011). Underwater acoustics for ecosystem-based management: state of the science and proposals for ecosystem indicators. *Marine Ecology Progress Series*, 442, 285–301.

Turker, M., Aydin, I. Z. and Aydin, T. (2014). Ecotourism activities for sustainability and management of forest protected areas: a case of Camili Biosphere Reserve Area, Turkey. In *Designing Low Carbon Societies in Landscapes*, ed. N. Nakagoshi and J. A. Mabuhay. New York: Springer, pp. 253–269.

Turner, B. (2010). *The Statesman's Yearbook 2010*. New York: Macmillan.

Turner, M. G. and Gardner, R. H. (2015). *Landscape Ecology in Theory and Practice: Pattern and Process*. New York: Springer.

Turner, M. G., Donato, D. C., Hansen, W. D. *et al.* (2016). Climate change and novel disturbance regimes in national park landscapes. In *Science, Conservation, and National Parks*, ed. S. R. Beissinger, D. D. Ackerly, H. Doremus and G. Machlis. Chicago, IL: University of Chicago Press, pp. 77–101.

Turner, M. G., Donato, D. C. and Romme, W. H. (2013). Consequences of spatial heterogeneity for ecosystem services in changing forest landscapes: priorities for future research. *Landscape Ecology*, 28, 1081–1097.

Turner, W. R., Brandon, K., Brooks, T. M. *et al.* (2013). The potential, realised and essential ecosystem service benefits of biodiversity conservation. In *Biodiversity Conservation and Poverty Alleviation: Exploring the Evidence for a Link*, ed. D. Roe, J. Elliott, C. Sandbrook and M. Walpole. Chichester, UK: Wiley-Blackwell, pp. 21–35.

Ulrich, R. (1984). View through a window may influence recovery from surgery. *Science*, 224, 420–421.

Ulrich, R., Zimring, C., Zhu, X. *et al.* (2008). A review of the research literature on evidence-based health care design. *Herd-Health Environments Research and Design Journal*, 1(3), 61–125.

Unwin, R. (1911). *Town Planning in Practice: An Introduction to the Art of Designing Cities and Suburbs*. London: T. Fisher Unwin.

Urban Land Institute. (2008). *Creating Great Town Centers and Urban Villages*. Washington, DC: Urban Land Institute.

Urquiza-Haas, T., Peres, C. A. and Dolman, P. M. (2009). Regional scale effects of human density and forest disturbance on large-bodied vertebrates throughout the Yucatan peninsula, Mexico. *Biological Conservation*, 142, 134–148.

US Census Bureau. (2016). *Characteristics of the US Population by Generational Status: 2013*. Washington, DC: Department of Commerce.

US Environmental Protection Agency. (1993). *Guidance Specifying Management Measures for Sources of Nonpoint Source Pollution in Coastal Waters*. Report No. 840-B-92-002, Office of Water. Washington, DC: US EPA.

US Environmental Protection Agency. (2010). *Green Infrastructure Case Studies: Municipal Policies for Managing Stormwater with Green Infrastructure*. Report No. 841-F-10-004. Washington, DC: US EPA.

US Environmental Protection Agency. (2012). *Municipal Solid Waste Generation, Recycling, and Disposal in the United States: Facts and Figures for 2012*. Report 530-F-14-001. Washington, DC: US EPA.

US Environmental Protection Agency. (2018). *Soak Up the Rain: Rain Gardens*. Washington, DC: US-EPA.

USDA. (2001). *Urban Soils*. United States Department of Agriculture, Natural Resource Conservation Service. Lincoln, NE: National Soil Survey Center.

Vail, D. (1987). Suburbanization of the countryside and the revitalization of small farms. In *Sustaining Agriculture near Cities*, ed. W. Lockeretz. Ankeny, IA: Soil and Water Conservation Society.

Valeix, M., Loveridge, A. J., Davidson, Z. *et al.* (2010). How key habitat features influence large terrestrial carnivore movements: waterholes and African lions in a semi-arid savanna of north-western Zimbabwe. *Landscape Ecology*, 25, 337–351.

Valiela, I. (2006). *Global Coastal Change*. Oxford: Blackwell.

Valitski, S. A., D'Angelo, G. J., Gallagher, G. R. *et al.* (2009). Deer responses to sounds from a vehicle-mounted sound-production system. *Journal of Wildlife Management*, 73, 1072–1076.

Vallez, D., Phelps, N. A. and Wood, A. M. (2012). Planning for growth? The implications of localism for "Science Vale," Oxfordshire UK. *Town Planning Review*, 83, 457–487.

Valtonen, A., Saarinen, K. and Jantunen, J. (2007). Intersection reservations as habitats for meadow butterflies and diurnal moths: guidelines for planning and management. *Landscape and Urban Planning*, 79, 201–209.

van Aalst, I. and van Melik, R. (2011). City festivals and urban development: does place matter? *European Urban and Regional Studies*, 19, 195–206.

van der Grift, E. A. and van der Ree, R. (2015). Guidelines for evaluating use of wildlife crossing structures. In *Handbook of Road Ecology*, ed. R. van der Ree, D. J. Smith and C. Grilo. Chichester, UK: Wiley-Blackwell, pp. 119–128.

van der Linden, P. J. H. (1997). A wall of tree-stumps as a fauna-corridor. In *Habitat fragmentation and infrastructure*, ed. K. Canters, A. Piepers and D. Hendriks-Heersma. Proceedings of the international conference. Delft, The Netherlands: Ministry of Transport, Public Works and Water Management, pp. 409–417.

van der Ree, R. and van der Grift, E. A. (2015). Recreational co-use of wildlife-crossing structures. In *Handbook of Road Ecology*, ed. R. van der Ree, D. J. Smith and C. Grilo. Chichester, UK: Wiley-Blackwell, pp. 184–189.

van der Ree, R., Gagnon, J. W. and Smith, D. J. (2015a). Fencing: a valuable tool for reducing wildlife–vehicle collisions and funnelling fauna to crossing structures. In *Handbook of Road Ecology*, ed. R. van der Ree, D. J. Smith and C. Grilo. Chichester, UK: Wiley-Blackwell, pp. 159–171.

van der Ree, R., Smith, D. J. and Grilo, C., eds. (2015b). *Handbook of Road Ecology*. Chichester, UK: Wiley-Blackwell.

Van der Ryn, S. (1978). *The Toilet Papers*. Santa Barbara, CA: Capra Press.

van der Zee, D. and Zonneveld, I. S., eds. (2001). *Landscape Ecology Applied in Land Evaluation, Development and Conservation: Some Worldwide Selected Examples*. Enschede, The Netherlands: International Institute for Aerospace Survey and Earth Sciences (ITC).

van Leeuwen, E. (2010). *Urban-Rural Interactions: Towns as Focus Points in Rural Development*. Contributions to Economics. Heidelberg, Germany: Physica Verlag.

van Leeuwen, E. (2013). Towns today and their multifunctional activities. In *Towns in a Rural World*, ed. T. de N. Vaz, E. van Leeuwen and P. Nijkamp. Farnham, UK: Ashgate Publishing, pp. 11–22.

Vaz, T. de N. and Nijkamp, P. (2013). Small towns of hope and glory. In *Towns in a Rural World*, ed. T. de N. Vaz, E. van Leeuwen and P. Nijkamp. Farnham, UK: Ashgate Publishing, pp. 3–9.

Vaz, T. de N., van Leeuwen, E. and Nijkamp, P., eds. (2013a). *Towns in a Rural World*. Farnham, UK: Ashgate Publishing.

Vaz, T. de N., van Leeuwen, E. and Nijkamp, P. (2013b). Lessons from successful towns. In *Towns in a Rural World*, ed. T. de N. Vaz, E. van Leeuwen and P. Nijkamp. Farnham, UK: Ashgate Publishing, pp. 367–371.

Verbeeck, K., Van Orshoven, J. and Hermy, M. (2011). Measuring extent, location and change of imperviousness in urban domestic gardens in collective housing projects. *Landscape and Urban Planning*, 100, 57–66.

Vermonden, K., Leuven, R. S. E. W., van der Velde, G. *et al.* (2009). Urban drainage systems: an undervalued habitat for aquatic macroinvertebrates. *Biological Conservation*, 142, 1105–1115.

Veysey, J. S., Mattfeldt, S. D. and Babbitt, K. J. (2011). Comparative influence of isolation, landscape, and wetland characteristics on egg-mass abundance of two pool-breeding amphibian species. *Landscape Ecology*, 26, 661–672.

Vigar, G. (2007). Transport and mobility in the English market town. In *Market Towns: Roles, Challenges and Prospects*, ed. N. Powe, T. Hart and T. Shaw. New York: Routledge, pp. 59–68.

Vince, S. W., Duryea, M. L., Macie, E. A. and Hermansen, L. A., eds. (2005). *Forests at the Wildland-Urban Interface: Conservation and Management*. Boca Raton, FL: CRC Press.

Violin, C. R., Cada, P., Sudduth, E. B. *et al.* (2011). Effects of urbanization and urban stream restoration on the physical and biological structure of stream ecosystems. *Ecological Applications*, 21, 1932–1949.

Vira, B. and Kontoleon, A. (2013). Dependence of the poor on biodiversity: which poor, what biodiversity? In *Biodiversity Conservation and Poverty Alleviation*, ed. D. Roe, J. Elliott, C. Sandbrook and M. Walpole. Chichester, UK: Wiley-Blackwell, pp. 52–84.

Vojta, J., Kovar, P. and Volarik, D. (2014). Patterns of grazing and plant species diversity in the pasturelands. In *Czech Villages in Romanian Banat: Landscape, Nature, and Culture*, ed. P. Madera, P. Kovar, D. Romportl *et al.* Brno, Czech Republic: Mendel University in Brno, pp. 153–163.

Vollertsen, J., Astebol, S. O., Coward, J. E. *et al.* (2007). Monitoring and modeling the performance of a wet pond for treatment of highway runoff in cold climates. *Highway and Urban Environment*, 12, 499–509.

von Haaren, C. and Reich, M. (2006). The German way to greenways and habitat networks. *Landscape and Urban Planning*, 76, 7–22.

von Stulpnagel, A., Horbert, A. and Sukopp, H. (1990). The importance of vegetation for the urban climate. In *Urban Ecology*, ed. H. Sukopp, S. Hejny and J. Kowarik. The Hague, The Netherlands: SPB Academic Publishing, pp. 175–193.

Voosen, P. (2016). Anthropocene pinned to postwar period. *Science*, 353, 852–854.

Vos, C. (1997). Effects of road density: a case study of the moor frog. In *Habitat Fragmentation and Infrastructure*, ed. K. Canters, A. Piepers and D. Hendriks-Heersma. Delft, The Netherlands: Ministry of Transport, Public Works and Water Management, pp. 93–97.

Vymazal, J. (2010). Constructed wetlands for wastewater treatment. *Water*, 2–3, 530–539.

Vymazal, J. H., Brix, P. F., Cooper, M. B. *et al.*, eds. (1998). *Constructed Wetlands for Wastewater Treatment in Europe*. Leiden, The Netherlands: Backhuys Publishers.

Vymazal, J., Greenway, M., Tonderski, K. *et al.* (2006). Constructed wetlands for wastewater treatment. In *Wetland and Natural Resource Management*, ed. J. T. A. Verhoeven. New York: Springer, pp. 69–96.

Wackernagel, M. and Rees, W. (2014). Ecological footprints for beginners: our ecological footprint: reducing human impact on the Earth (1966). In *The Ecological Design and Planning Reader*, ed. F. O. Ndubisi. Washington, DC: Island Press, pp. 501–505.

Wagner, D. L., Metzler, K. J., Leicht-Young, S. A. and Motzkin, G. (2014). Vegetation composition along a New England transmission line corridor and its implications for other trophic levels. *Forest Ecology and Management*, 327, 231–239.

Walker, B. H. and Salt, D. (2006). *Resilience Thinking: Sustaining Ecosystems and People in a Changing World*. Washington, DC: Island Press.

Wall, D. H., Bardgett, R. D., Behan-Pelletier, V. *et al.*, eds. (2012). *Soil Ecology and Ecosystem Services*. Oxford: Oxford University Press.

Walpole, A. A., Bowman, J., Murray, D. L. and Wilson, P. J. (2012). Functional connectivity of lynx at their southern range periphery in Ontario, Canada. *Landscape Ecology*, 27, 761–773.

Walsh, C. J., Fletcher, T. D. and Ladson, A. R. (2005a). Stream restoration in urban catchments through redesigning stormwater systems: looking to the catchment to save the stream. *Journal of the North American Benthological Society*, 24, 690–705.

Walsh, C. J., Roy, A. H., Feminella, J. W. *et al.* (2005b). The urban stream syndrome: current knowledge and the search for a cure. *Journal of the North American Benthological Society*, 24, 706–723.

Walsh, F. J. (1990). An ecological study of traditional Aboriginal use of "country": Martu in the Great and Little Sandy Deserts, Western Australia. *Proceedings of the Ecological Society of Australia*, 16, 23–37.

Walsh, J. (2011). *Unleashed Fury: The Political Struggle for Dog-Friendly Parks*. West Lafayette, IN: Purdue University Press.

Walters, A. W. and Post, D. M. (2011). How low can you go? Impacts of a low-flow disturbance on aquatic insect communities. *Ecological Applications*, 21, 163–174.

Wang, Y., Meng, D., Zhu, Y. and Zhang, F. (2009). Impacts of regional urbanization development on plant diversity within boundary of built-up areas of different settlement categories in Jinzhong Basin, China. *Landscape and Urban Planning*, 91, 212–218.

Ward, D. (2016). *The Biology of Deserts*. New York: Oxford University Press.

Ward, J. V., Tockner, K., Arscott, D. B. and Claret, C. (2002). Riverine landscape diversity. *Freshwater Biology*, 47, 517–539.

Warhurst, J. R., Parks, K. E., McCulloch, L. and Hudson, M. D. (2014). Front gardens to car parks: changes in garden permeability and effects on flood regulation. *Science of the Total Environment*, 485, 329–339.

Warner, R. E. (1985). Demography and movement of free-ranging domestic cats in rural Illinois. *Journal of Wildlife Management*, 49, 340–346.

Warren, J., Lawson, C. and Belcher, K. (2008). *The Agri-Environment*. New York: Cambridge University Press.

Warren, Jr., M. L. and Pardew, M. G. (1998). Road crossings as barriers to small-stream fish movement. *Transactions of the North American Fisheries Society*, 127, 617–644.

Warren, P. S., Katti, M., Ermann, M. and Brazel, A. (2006). Urban bioacoustics: it's not just noise. *Animal Behaviour*, 71, 491–502.

Washburn, B. E. (2012). Avian use of solid waste transfer stations. *Landscape and Urban Planning*, 104, 388–394.

Waterman, E., Tulp, I., Reijnen, R. *et al.* (2002). Disturbance of meadow birds by railway noise in The Netherlands. *Geluid*, 1, 2–3.

Watters Wilkes, M. (2016). *Parker's Revenge Archaeological Project: Minute Man National Historical Park, Lexington, Massachusetts*. Concord, MA: Friends of Minute Man National Historical Park.

Watts, M. T. (1975). *Reading the Landscape: An Adventure in Ecology*. New York: Macmillan.

Way, D. S. (1978). *Terrain Analysis: A Guide to Site Selection Using Aerial Photographic Interpretation*. Stroudsburg, PA: Dowden, Hutchinson & Ross.

Webster's College Dictionary. (1991). New York: Random House.

Wegner, J. and Merriam, G. (1979). Movements by birds and small mammals between a wood and adjoining farmland habitats. *Journal of Applied Ecology*, 16, 349–358.

Wein, R. W., ed. (2006). *Coyotes Still Sing in My Valley: Conserving Biodiversity in a Northern City*. Edmonton, AB: Spotted Cow Press.

Welborn, G. A., Skelly, D. K. and Werner, E. E. (1996). Mechanisms creating community structure across a freshwater habitat gradient. *Annual Review of Ecology and Systematics*, 27, 337–363.

Welty, C. (2009). The urban water budget. In *The Water Environment of Cities*, ed. L. A. Baker. New York: Springer, pp. 17–28.

Wessolek, G. (2008). Sealing of soils. In *Urban Ecology: An International Perspective on the Interactions Between Humans and Nature*, ed. J. M. Marzluff, E. Shulenberger, W. Endlicher *et al.* New York: Springer, pp. 161–179.

Westerhoff, P. and Crittendon, J. (2009). Urban infrastructure and use of mass balance models for water and salt. In *The Water Environment of Cities*, ed. L. A. Baker. New York: Springer, pp. 49–68.

Western, D., Groom, R. and Worden, J. (2009). The impact of subdivision and sedentarization of pastoral lands on wildlife in an African savanna ecosystem. *Biological Conservation*, 142, 2538–2546.

Weston, M. A., Fitzsimmons, J., Wescott, G. *et al.* (2014). Bark in the park: a review of domestic dogs in parks. *Environmental Management*, 54, 373–382.

Wezel, A. and Bender, S. (2003). Plant species diversity of home gardens of Cuba and its significance for household food supply. *Agroforestry Systems*, 57, 39–49.

Wheater, C. P. (1999). *Urban Habitats*. New York: Routledge.

White, J. S., Bayley, S. E. and Curtis, P. J. (2000). Sediment storage of phosphorus in a northern prairie wetland receiving municipal and agro-industrial wastewater. *Ecological Engineering*, 14, 127–138.

White, M. D. and Greer, K. A. (2006). The effects of watershed urbanization on the stream hydrology and riparian vegetation of Los Penasquitos Creek, California. *Landscape and Urban Planning*, 74, 125–138.

White, N. (2012). *Company Towns: Corporate Order and Community*. Toronto: University of Toronto Press.

Whitford, W. G. (2002). *Ecology of Desert Systems*. San Diego, CA: Academic Press.

Whitney, G. G. and Adams, S. D. (1980). Man as maker of new plant communities. *Journal of Applied Ecology*, 17, 431–448.

Whittaker, R. H. (1975). *Communities and Ecosystems*. New York: Macmillan.

Wickham, J. D., Wade, T. G. and Riitters, K. H. (2011). An environmental assessment of United States drinking water watersheds. *Landscape Ecology*, 26, 605–616.

Wieder, R. K. (1989). A survey of constructed wetlands for acid coal mine drainage treatment in the United States. *Wetlands*, 9, 299–315.

Wilder, T. (2013). *Our Town: A Play in Three Acts*. New York: HarperPerennial.

Williams, F. (2016). The power of parks: a yearlong exploration. *National Geographic*, January, 54–58.

Williams, J. (2006). *East 40 Degrees: An Interpretive Atlas*. Charlottesville, VA: University of Virginia Press.

Williams, K. J., Weston, M. A., Henry, S. and Maguire, G. S. (2009). Birds and beaches, dogs and leashes: dog owners' sense of obligation to leash dogs on beaches in Victoria, Australia. *Human Dimensions of Wildlife*, 14, 89–101.

Williams, P., Whitfield, M., Biggs, J. *et al.* (2003). Comparative biodiversity of rivers, streams, ditches and ponds in an agricultural landscape in Southern England. *Biological Conservation*, 115, 329–341.

Wilson, E. O. (1984). *Biophilia: The Human Bond with Other Species*. Cambridge, MA: Harvard University Press.

Winfree, R., Bartomeus, I. and Cariveau, D. P. (2011). Native pollinators in anthropogenic habitats. *Annual Review of Ecology and Systematics*, 42, 1–22.

Winfree, R., Williams, N. M., Gaines, H. *et al.* (2008). Wild bee pollinators provide the majority of crop visitation across land-use gradients in New Jersey and Pennsylvania, USA. *Journal of Applied Ecology*, 45, 793–802.

Winter, T. C., Harvey, J. W., Franke, O. L. and Alley, W. M. (1998). *Groundwater and Surface Water: A Single Resource*. US Geological Survey Circular 1139. Denver, CO: USDI Geological Survey.

Winters, C., Dodd, J. and Harrison, K. (2013). The role of universities for economic development in urban poles. In *Towns in a Rural World*, ed. T. de N. Vaz, E. van Leeuwen and P. Nijkamp. Farnham, UK: Ashgate Publishing, pp. 213–229.

Wise, S. E. and Buchanan, B. W. (2006). Influence of artificial illumination on the nocturnal behavior and physiology of salamanders. In *Ecological Consequences of Artificial Night Lighting*, ed. C. Rich and T. Longcore. Washington, DC: Island Press, pp. 221–251.

With, K. A. and Pavuk, D. M. (2011). Habitat trumps fragmentation effects on arthropods in an experimental landscape system. *Landscape Ecology*, 26, 1035–1048.

Wittig, R. (2010). Biodiversity of urban-industrial areas and its evaluation – a critical review. In *Urban Biodiversity and Design*, ed. N. Muller, P. Werner and J. G. Kelcey. Oxford: Blackwell, pp. 37–55.

Wohl, E. (2004). *Disconnected Rivers: Linking Rivers to Landscapes*. New Haven, CT: Yale University Press.

Woinarski, J. C. Z., Murphy, B. P., Legge, S. M. *et al.* (2017). How many birds are killed by cats in Australia? *Biological Conservation*, 214, 76–87.

Wolseley, P. A., James, P. W., Theobold, M. R. and Sutton, M. A. (2006). Detecting changes in epiphytic lichen communities at sites affected by atmospheric ammonia from agricultural sources. *Lichenologist*, 38, 161–176.

Wolseley, P. A. and Pryor, K. V. (1999). The potential of epiphytic twig communities on *Quercus petraea* in a Welsh woodland site (Tycanol) for evaluating environmental changes. *Lichenologist*, 31, 41–61.

Wolter, C. (2008). Towards a mechanistic understanding of urbanization's impacts on fish. In *Urban Ecology: An International Perspective on the Interaction Between Humans and Nature*, ed. J. M. Marzluff, E. Shulenberger, W. Endlicher *et al.* New York: Springer, pp. 425–436.

Wood, E. L., Pidgeon, A. M., Radeloff, V. C. *et al.* (2015). Long-term avian community response to housing development at the boundary of US protected areas: effect size increases with time. *Journal of Applied Ecology*, 52, 1227–1236.

Wood, J. S. (1997). *The New England Village*. Baltimore, MD: Johns Hopkins University Press.

Woodhouse, C. A. (2004). A paleo perspective on hydroclimatic variability in the Western United States. *Aquatic Sciences*, 66, 346–356.

Woodroffe, C. D. (2002). *Coasts: Form, Process and Evolution*. New York: Cambridge University Press.

Woods, M., McDonald, R. A. and Harris, S. (2003). Predation of wildlife by domestic cats *Felis catus* in Great Britain. *Mammal Review*, 33, 174–188.

Worster, D. (2016). *Shrinking the Earth: The Rise and Decline of American Abundance*. Oxford: Oxford University Press.

Wu, J. (2013). Landscape sustainability science: ecosystem services and human well-being in changing landscapes. *Landscape Ecology*, 28, 999–1023.

Wu, J. (2014). Urban ecology and sustainability: the state-of-the-science and future directions. *Landscape and Urban Planning*, 125, 209–221.

Wu, J. and Wu, T. (2013). Ecological resilience as a foundation for urban design and sustainability. In *Resilience in Ecology and Urban Design: Linking Theory and Practice for Sustainable Cities*, ed. S. T. A. Pickett, M. L. Cadenasso and B. McGrath. New York: Springer, pp. 211–229.

Wu, Z.-I., He, Q.-C. and Wu, Q.-J. (2014). The neglect of traditional ecological knowledge on wild elephant-related problems in Xishuangbanna, SW China. In *Designing Low Carbon Societies in Landscapes*, ed. N. Nakagoshi and J. A. Mabuhay. New York: Springer, pp. 163–176.

Wurst, S., De Deyn, G. B. and Orwin, K. (2012). Soil biodiversity and functions. In *Soil Ecology and Ecosystem Services*, ed. D. H. Wall, R. D. Bardgett, V. Behan-Pelletier *et al.* Oxford: Oxford University Press, pp. 28–44.

Wuthrow, R. (2013). *Small-Town America: Finding Community, Shaping the Future*. Princeton, NJ: Princeton University Press.

Yackulic, C. B., Strindberg, S., Maisels, F. and Blake, S. (2011). The spatial structure of hunter access determines the local abundance of forest elephants (*Loxodonta africana cyclotis*). *Ecological Applications*, 21, 1296–1307.

Yan, H. Y., Anraku, K. and Baharan, R. P. (2010). Hearing in marine fish and its application in fisheries. In *Behavior of Marine Fishes: Capture Processes and Conservation Challenges*, ed. P. He. Chichester, UK: Wiley-Blackwell, pp. 45–64.

Yanes, M., Velasco, J. M. and Suarez, F. (1995). Permeability of roads and railways to vertebrates: the importance of culverts. *Biological Conservation*, 71, 217–222.

Yang, B. and Li, M.-H. (2011). Assessing planning approaches by watershed streamflow modeling: case study of The Woodlands, Texas. *Landscape and Urban Planning*, 99, 9–22.

Yasuda, M. and Koike, F. (2006). Do golf courses provide a refuge for flora and fauna in Japanese urban landscapes? *Landscape and Urban Planning*, 75, 58–68.

Yasuda, M. and Koike, F. (2009). The contribution of the bark of isolated trees as habitat for ants in an urban landscape. *Landscape and Urban Planning*, 92, 276–281.

Yates, A. G. and Bailey, R. C. (2010). Improving the description of human activities potentially affecting rural stream ecosystems. *Landscape Ecology*, 25, 371–382.

Yoshihara, Y., Chimeddorj, B., Buuveibaatar, B. *et al.* (2008). Effects of livestock grazing on pollination on a steppe in eastern Mongolia. *Biological Conservation*, 141, 2376–2386.

Young, F. W. (1999). *Small Towns in Multilevel Society*. Lanham, MD: University Press of America.

Yu, K.-J. (1992). Experience of basin landscapes in Chinese agriculture has led to ecologically prudent engineering. In *Human Responsibility and Global Change*, ed. L. O. Hansson and B. Jungen. Goteborg, Sweden: University of Goteborg, pp. 289–299.

Yu, K., Wang, S. and Li, D. (2011). The negative approach to urban growth planning of Beijing, China. *Journal of Environmental Planning and Management*, 54, 1209–1236.

Yudelson, J. (2008). *The Green Building Revolution*. Washington, DC: Island Press.

Zaitzevsky, D. (1982). *Frederick Law Olmsted and the Boston Park System*. Cambridge, MA: Belknap Press of Harvard University Press.

Zelinski, W. (1992). *The Cultural Geography of the United States: Study Guide*. Englewood Cliffs, NJ: Prentice-Hall.

Zhang, D. Q., Gersberg, R. M. and Keat, T. S. (2009). Constructed wetlands in China. *Ecological Engineering*, 35, 1367–1378.

Zhang, Q.-B., Li, Z., Liu, P. and Xiao, S. (2012). On the vulnerability of oasis forest to changing environmental conditions: perspectives from tree rings. *Landscape Ecology*, 27, 343–353.

Zhao, Y. (2013). *China's Disappearing Countryside: Towards Sustainable Land Governance for the Poor*. Farnham, UK: Ashgate Publishing.

Zhi, W. and Ji, G. (2012). Constructed wetlands, 1991–2011: a review of research development, current trends, and future directions. *Science of the Total Environment*, 441, 19–27.

Zhu, W.-X. and Carreiro, M. M. (2004). Variations of soluble organic nitrogen and microbial nitrogen in deciduous forest soils along an urban–rural gradient. *Soil Biology and Biochemistry*, 36, 279–288.

Ziegler, A. D., Sutherland, R. A. and Giambelluca, T. W. (2001). Interstorm surface preparation and sediment detachment by vehicle traffic on unpaved mountain roads. *Earth Surface Processes and Landforms*, 26, 235–250.

Zimmerman, F. J. and Carter, M. R. (2003). Asset smoothing, consumption smoothing and the reproduction of inequality under risk and subsistence constraints. *Journal of Development Economics*, 71, 233–260.

Zmyslony, J. and Gagnon, D. (2000). Path analysis of spatial predictors of front yard vegetation in an anthropogenic environment. *Landscape Ecology*, 15, 357–371.

Zurcher, A. A., Sparks, D. W. and Bennett, V. J. (2010). Why the bat did not cross the road? *Acta Chiropterologica*, 12, 337–340.

Zwaenepoel, A. (1997). Floristic impoverishment by changing unimproved roads into metalled roads. In *Habitat Fragmentation and Infrastructure*, ed. K. Canters, A. Piepers and D. Hendriks-Heersma. Delft, The Netherlands: Ministry of Transport, Public Works and Water Management, pp. 127–137.

Index

Locators in italics refer to figures and plates. Place names followed by (A) are listed in the Appendix and are towns unless indicated otherwise. Where there is a direction to *see/see also* specific towns, the reader should refer to the Appendix for relevant towns of that category.